CONCEPTUAL
Physics
SIXTH EDITION

Written & Illustrated by

Paul G. Hewitt

City College of San Francisco

D0068929

HarperCollinsPublishers

Produced by The Compage Company

Cover design: Paul G. Hewitt
Cover photo: Don Briggs
Cover logo design and title page design: Ernie Brown
Interior design and page makeup: Christy Butterfield and Elizabeth Victor
Editorial production: Pearl C. Vapnek
Photo research: Lindsay Kefauver and Caroline Pincus
Copyediting: Judith Hibbard
Indexing: Steven M. Sorensen
Composition: GTS

Library of Congress Cataloging-in-Publication Data

Hewitt Paul G.
    Conceptual physics.

    Bibliography: p.
    Includes index.
    1. Physics. I. Title
QC23.H56 1989   530   88-29704
ISBN 0-673-39847-1

8 9 10 -RRC- 94 93 92

Printed in the United States of America

								0
								2 **He** Helium 4.003
			IIIA	**IVA**	**VA**	**VIA**	**VIIA**	
			5 **B** Boron 10.81	6 **C** Carbon 12.011	7 **N** Nitrogen 14.007	8 **O** Oxygen 15.999	9 **F** Fluorine 18.998	10 **Ne** Neon 20.17
IB	**IIB**		13 **Al** Aluminum 26.98	14 **Si** Silicon 28.09	15 **P** Phosphorus 30.974	16 **S** Sulfur 32.06	17 **Cl** Chlorine 35.453	18 **Ar** Argon 39.948
28 **Ni** Nickel 58.71	29 **Cu** Copper 63.546	30 **Zn** Zinc 65.38	31 **Ga** Gallium 69.735	32 **Ge** Germanium 72.59	33 **As** Arsenic 74.922	34 **Se** Selenium 78.96	35 **Br** Bromine 79.904	36 **Kr** Krypton 83.80
46 **Pd** Palladium 106.4	47 **Ag** Silver 107.868	48 **Cd** Cadmium 112.41	49 **In** Indium 114.82	50 **Sn** Tin 118.69	51 **Sb** Antimony 121.75	52 **Te** Tellurium 127.60	53 **I** Iodine 126.904	54 **Xe** Xenon 131.30
78 **Pt** Platinum 195.09	79 **Au** Gold 196.967	80 **Hg** Mercury 200.59	81 **Tl** Thallium 204.37	82 **Pb** Lead 207.2	83 **Bi** Bismuth 208.98	84 **Po** Polonium (209)	85 **At** Astatine (210)	86 **Rn** Radon (222)

63 **Eu** Europium 151.96	64 **Gd** Gadolinium 157.25	65 **Tb** Terbium 158.93	66 **Dy** Dysprosium 162.50	67 **Ho** Holmium 164.93	68 **Er** Erbium 167.26	69 **Tm** Thulium 168.93	70 **Yb** Ytterbium 173.04	71 **Lu** Lutetium 174.967

95 **Am** Americium (243)	96 **Cm** Curium (247)	97 **Bk** Berkelium (247)	98 **Cf** Californium (251)	99 **Es** Einsteinium (254)	100 **Fm** Fermium (257)	101 **Md** Mendelevium (258)	102 **No** Nobelium (259)	103 **Lr** Lawrencium (260)

Photo and Illustration Credits

417   *Top*: *PSSC Physics*, 2nd edition, 1965. D. C. Heath & Company with Education Development Center, Inc., Newton, MA *Bottom*: Japanese National Railways

420   Dave Vasquez

424   *Top*: National Center for Atmospheric Research/ National Science Foundation *Bottom*: Lila Lee

428   Dave Vasquez

445   Nina Leen/Life Magazine © Time, Inc.

446   © Frank Siteman/ Stock, Boston

453   Dave Vasquez

454   Dave Vasquez

457   Edwin R. Lewis, College of Engineering, University of California, Berkeley

464   Jerry Hosken

467   © Paul Conklin

474   Joan Venticinque

476   David E. Hewitt

477   *Top*: Dave Vasquez *Bottom*: Leonard Lee Rue III/ Animals Animals

482   Dave Vasquez

487   George Hall

488   Institute of Paper Chemistry

491   Ted Mahieu

492   Robert Greenler

501   *Top*: Ronald B. FitzGerald *Bottom*: Karen Tweedy-Holmes/ Animals Animals

504   Ronald B. FitzGerald

511   © Michael P. Gadomski/ Earth Scenes

512   Ronald B. FitzGerald

513   *Top*: Dave Vasquez *Bottom*: Burndy Library

515   Courtesy Education Development Center, Inc., Newton, MA

517   *Left*: © Richard Megna/ Fundamental Photographs *Bottom*: Chuck Manka

519   *Top*: Thomas Jung *Bottom*: Courtesy Education Development Center, Inc., Newton, MA

520   *Top*: Thomas Young, *A Course of Lectures on Natural Philosophy and the Mechanical Arts* (London: Taylor and Walton, 1845)/AIP Niels Bohr Library

524   *Top*: Courtesy Bausch & Lomb

529   *Top*: David E. Hewitt *Bottom*: Nancy Rodger

530   *Bottom*: Jack M. Williams, *Journal of Chemical Education*, Vol. 52, p. 210 (April 1975). Used by permission.

531   John Dennis

540   Sam Cadelinia

545   Inframetrics, Inc., Bedford, MA

554   Paul Robinson

555   *Top*: Lawrence Livermore Laboratory

560   Burndy Library/AIP Niels Bohr Library

564   Courtesy Albert Rose

565   Elisha Huggins

566   AIP Niels Bohr Library

567   *Top left*: J. Valasek, *Introduction to Theoretical and Experimental Optics* (New York: Wiley, 1949) *Top right*: H. Raether, "Elektroninterferenzen," *Handbuch der Physik*, Vol. 32 (Berlin: Springer-Verlag, 1957) *Bottom left*: M. Ohtsuki/ Laboratory of Albert V. Crewe, Enrico Fermi Institute, University of Chicago *Bottom right*: © David Scharf, 1977. All rights reserved./ Peter Arnold, Inc.

568   P. G. Merli, G. F. Missiroli, and G. Pozzi, "On the Statistical Aspect of Electron Interference Phenomena," *American Journal of Physics*, Vol. 44, No. 3, March 1976. Copyright © 1976 by the American Association of Physics Teachers.

570   Archives for History of Quantum Physics/ AIP Niels Bohr Library

577   Lawrence Berkeley Laboratory

579   AIP Niels Bohr Library

580   Margrethe Bohr Collection/ AIP Niels Bohr Library

585   Ullstein/AIP Niels Bohr Library

590   Photo of geyser: Werner Krutein

591   AIP Niels Bohr Library

592   New York Hospital

599   Courtesy of Nucleus, Inc. Photos: Pasco Scientific

600   Lawrence Berkeley Laboratory

617   Fermi Film Collection/ AIP Niels Bohr Library

618   Argonne National Laboratory

627   Lawrence Livermore Laboratory

628   Lawrence Livermore Laboratory

635   California Institute of Technology Archives

636   Mary Jew

639   William Kellogg/ AIP Niels Bohr Library

664   United Nations from UK/AEA

Back endpapers   *Verso*: Bausch & Lomb *Recto left and right*: Dave Vasquez

The following were photographed by the author: 85, 95, 97, 104, 110, 128, 176, 179, 230, 235, 260, 276, 281, 300 *left*, 308, 310, 323, 340, 345, 349, 356 *bottom*, 364 *top*, 393 *top*, 437, 498, 506, 590 (photo of boy), 661 (special effects by Dave Vasquez).

The photos on pages 128 and 235 also appear in *Conceptual Physics: A High School Physics Program* by Paul G. Hewitt (Addison-Wesley Publishing Company, Inc., 1987).

To Richard P. Feynman
for his inspiration
and for teaching us to see
the beauty in physics

Contents in Brief

Contents in Detail

PART 2   PROPERTIES OF MATTER   175

PART 3   HEAT

PART 6   LIGHT   445

PART 7   ATOMIC AND NUCLEAR PHYSICS   577

To the Student

You know you can't enjoy a game unless you know its rules—whether it's a ball game, a computer game, or simply a party game. Likewise, you can't fully appreciate your surroundings until you understand the rules of nature. Physics is about the rules of nature—so beautifully elegant that it can be neatly described mathematically. That's why many physics courses are treated as applied mathematics. But introductory physics that emphasizes computation misses something essential—*comprehension*—a gut feeling for the concepts. This book emphasizes comprehension rather than computation. We treat physics *conceptually*—in down-to-earth *English* rather than in mathematical language. You'll see the mathematical structure of physics in frequent equations, but you'll see the equations as *guides to thinking* rather than as recipes for computation.

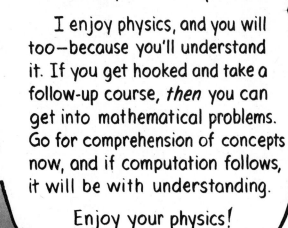

I enjoy physics, and you will too—because you'll understand it. If you get hooked and take a follow-up course, *then* you can get into mathematical problems. Go for comprehension of concepts now, and if computation follows, it will be with understanding.

Enjoy your physics!

Paul G. Hewitt

To the Instructor

Because physics is the basic science—the foundation of chemistry, biology, and all disciplines of science—it should be part of the educational mainstream for both science and nonscience students. Unfortunately, its mathematical language deters the average nonscience student. But when the ideas of physics are presented conceptually and when equations are seen to be guides to thinking rather than recipes for algebraic manipulation, our discipline is accessible to all students. And for students who will continue in the study of physics, I am convinced that the ideas of physics should be first understood conceptually before being used as a base for applied mathematics.

This book seeks to build that conceptual base. For the nonscience student, it is a base from which to view nature more perceptively—to see that surprisingly few relationships make up its rules. For the science student, it is this as well as being a springboard to a greater involvement in physics. A first-semester overview of Newtonian and modern physics for science majors will help to correct a missing essential in physics education: the practice of conceptualizing before calculating. For nonscience and science students alike, a conceptual way of looking at physics shapes analytical thinking.

New to This Edition

Although the sequence from classical mechanics to modern physics and the overall organization of this edition are much the same as in previous editions, this edition has been almost completely rewritten. The chapter on astrophysics has been omitted to make room for three new chapters, on nonlinear motion, satellite motion, and the properties of light. Part I begins with linear motion in Chapter 2 and is followed with nonlinear motion in the new Chapter 3. In this new chapter, projectile motion extends to satellite motion, but unlike the previous edition, a thorough treatment of satellite motion is deferred to its own Chapter 9, which follows the chapter on gravity. Since spacefaring activities are of general interest and already capture the imagination of our students, the brief introduction of satellite motion in Chapter 3 can build an early interest in physics. Vectors, which were relegated to an appendix in previous editions, are introduced in Chapter 3. Only simple cases of velocity vectors are treated in this early chapter, and a more general treatment is in Appendix III. As with the fifth edition, the chapter on momentum logically follows the chapter on Newton's laws, so there is no gap between Newton's third law and momentum conservation. Appendix IV of the fifth edition, "The Universal Gravitational Constant, G," is now incorporated in Chapter 8, on gravity. There are no major changes in the order of topics in Parts 2, 3, and 4. In Part 5 minor reordering of topics occurs in the chapters on magnetism and electromagnetic induction. Part 6 now begins with a new chapter on the properties of light. The chapter on color now precedes the chapter on reflection and refraction. Part 6 ends with an introduction to quantum physics, which carries into the first chapter in Part 7. Part 8 is confined to special and general relativity, with no chapter on astrophysics. I feel that this edition is a smoother and more readable treatment than the previous edition, with many new insights sprinkled throughout that I hope your students will enjoy.

Pedagogy

An important change concerns the review questions at the end of each chapter. All important ideas are framed in relatively easy-to-answer review questions and are cited by chapter sections. They are, as the name implies, a review of chapter material. Their purpose is simply to provide a structured way to review the chapter. They are not meant to challenge the student's intellect, for in the vast majority of cases, the answers can be simply looked up. The exercises, on the other hand, play a different role. These have been streamlined, with new ones added. Some are moderately simple and are designed to prompt the application of physics to everyday situations, while others are more sophisticated and call for considerable critical thinking. Some are quantitative and involve simple, straightforward calculations that will help your students capture the idea being treated without requiring algebraic skills. The challenge to your students will be in the conceptual reasoning and critical thinking that are called for in the exercises.

As in previous editions, units of measurement are not emphasized. When used, they are almost exclusively expressed in SI (exceptions include such units as calories, grams per centimeter cubed, and light years). Mathematical derivations are avoided in the main body of the text and appear in footnotes or in the appendixes.

Ancillary Materials

More than enough material is included for a one-semester course, which allows for a variety of course designs to fit your taste. These are suggested in the *Instructor's Manual*, which you'll find to be different from most instructor's manuals. It contains many lecture ideas and topics not treated in the textbook, as well as teaching tips and suggested step-by-step lectures and demonstrations.

Be sure to get the ancillary packet, which includes, among other important items, transparency masters titled "Next-Time Questions." These are like the "Figuring Physics" cartooned questions and answers that appear each month in *The Physics Teacher*. New to this edition are "Conceptual Physics Illustrations," which can help make your chalkboard presentations more interesting. The "Test Bank" booklet has been expanded and is also available on upgraded computer disks not only for Apple II and IBM PCs, but for the Macintosh as well.

Last but not least, there is finally a lab manual for *Conceptual Physics*, written by Paul Robinson. In addition to interesting laboratory experiments, it includes a range of activities similar to the home projects in *Conceptual Physics*. These guide students to experience phenomena before they quantify the same phenomena in a follow-up laboratory experiment.

Go to it! Your conceptual physics course really can be the most interesting, informative, and worthwhile science course available to your students.

Acknowledgments

For contributions to every chapter in this edition, I especially wish to thank Charlie Spiegel, research assistant, California State University, Domingues Hills, whose great help with this book is just a small part of his effort to bring mathematics, science, and language closer together. I am also indebted to former student Ronald E. Lindemann, whose many valuable ideas are sprinkled all throughout this book. For researching information on thermodynamics I thank student Clifford M. Braun (left foreground, Figure 7-39), San Francisco State University. I am grateful for the help of my friend Paul Robinson, Computech, Fresno, California. For supplying 3-D figures and helpful information, I thank my friend Marshall Ellenstein, Maine West High School, Chicago. I am especially grateful to Charlie Hibbard, Galileo High School, San Francisco, for critiquing the entire book with emphasis on improving the chapter-end review questions and exercises. I am grateful to my colleagues at City College of San Francisco (CCSF) for their suggestions and assistance: Jim Conley, Jim Court, Frank Creese, Jerry Hosken, Jim Kurck, Will Maynez (center, Figure 5-14), Dave Wall, and Norman Whitlatch. Others who have contributed to this edition include Eric Bergmark, Sellwood Middle School, Oregon; Gabe Espinda and Ron Hipschman, the Exploratorium, San Francisco; Tenny Lim (Figure 6-4), Jet Propulsion Laboratory, California; architect Wade Porter, Wyoming; Joseph Scherrer, Germantown Academy, Pennsylvania; John J. Spokas, Illinois Benedictine College; Jearl Walker, Cleveland State University; Dick Walton, Rocky Mountain College, Montana; and Dean Zollman, Kansas State University. I am most grateful to the many students, both at CCSF and at the world's most wonderful place to teach physics, the Exploratorium, for their valuable feedback.

For reviewing the entire manuscript, I thank Dack Lee, CCSF; John Hubisz, College of the Mainland, Texas; and Charlie Spiegel. For reviewing various chapters, I am indebted to Paul Doherty, the Exploratorium; Henry A. Garon, Loyola University; Chelcie Liu, CCSF; Keith Stowe, California Polytechnic State University, San Luis Obispo; and my sister Marjorie Hewitt Suchocki, Wesley Theological Seminary.

I am grateful to those whose own books initially served as principal influences and references: Theodore Ashford, *From Atoms to Stars*; Albert Baez, *The New College Physics—A Spiral Approach*; John N. Cooper and Alpheus W. Smith, *Elements of Physics*; Richard P. Feynman, *The Feynman Lectures on Physics*; Kenneth Ford, *Basic Physics*; Eric Rogers, *Physics for the Inquiring Mind*; Alexander Taffel, *Physics: Its Methods and Meanings*; UNESCO, *700 Science Experiments for Everyone*; and Harvey E. White, *Descriptive College Physics*.

Since this is the first edition written on a computer, the initial step of copying the previous edition into a computer was spread among several CCSF students; thanks to Alma Brady (second from right, Figure 5-14), Maria Burchard, Ann Chang (Figure 23-5), Winnie Fong (second from left, Figure 5-14), Angela Hirano, Chong Ho, and Pauline Ng (far right, Figure 5-14). For proofreading galleys, I thank my former students, Michael W. Moore and Ronald E. Lindemann; and for proofreading pages, Tom Hassett.

Special thanks to my friend and CCSF colleague Annette Rappleyea for helping with the test-bank questions and for writing and upgrading the computer program for the test bank. I thank my photographer-type friend Dave Vasquez (Figure 6-16) for his many photos, which add a nice touch to this edition. For supplying the cover photo of Havasu Creek in Grand Canyon, I thank Don Briggs. Thanks go to my lifelong friend Ernie Brown for designing the physics logo. I also thank my nephew and roommate Robert Baruffaldi (far left, Figure 5-14; far right, Figure 7-39) for his assistance not only in researching material, but in keeping my work and living environment orderly during the many months of writing and illustrating. For shading many of the illustrations, I am grateful to my daughter Leslie (Figure 10-1). I thank most of all Helen Yan (Figures 6-16, 13-8, and 15-11), physics graduate, University of California, Berkeley, for her graphics help and hand lettering—creating a more readable upper- and lowercase style that adorns the entire book and also the monthly column "Figuring Physics" in *The Physics Teacher*, the magazine of the American Association of Physics Teachers.

A special note of appreciation is due editor Ron Pullins, of Scott, Foresman/Little, Brown, for his very professional concern and assistance. Special thanks to Christy Butterfield for designing the book, and to Ken Burke, Pearl C. Vapnek, and Judith Hibbard, of The Compage Company, who produced the book.

San Francisco *Paul G. Hewitt*

1 About Science

First of all, science is the body of knowledge about nature that represents the collective efforts, findings, insights, and wisdom of the human race. Second, science is a human activity, with the functions of discovering the orderliness of nature and finding the causes that govern this order. Science had its beginnings before recorded history, when people first discovered regularities and relationships in nature, such as star patterns in the night sky and weather patterns—when the rainy season started or the days grew longer. From these regularities, people learned to make predictions that gave them some control over their surroundings.

Science made great headway in Greece in the seventh century BC and spread throughout the Mediterranean world. Scientific advance came to a halt in Europe when the Roman Empire fell in the fifth century AD. Barbarian hordes destroyed almost everything in their paths as they over-ran Europe and ushered in the Dark Ages. During this time the Chinese were charting the stars and the planets and Arab nations were developing mathematics. Greek science was reintroduced to Europe by Islamic influences that penetrated into Spain during the tenth, eleventh, and twelfth centuries. Universities emerged in Europe in the thirteenth century, and the introduction of gunpowder changed the social and political structure of Europe in the fourteenth century. The fifteenth century saw art and science beautifully blended by Leonardo da Vinci. Scientific thought was furthered in the sixteenth century with the advent of the printing press.

The sixteenth-century Polish astronomer Copernicus caused great controversy when he published a book proposing that the sun was stationary and that the earth revolved around the sun. These ideas conflicted with the popular view that the earth was the center of the universe. They also conflicted with Church teachings and were banned for 200 years. The Italian physicist Galileo was arrested for popularizing the Copernican theory and for his other contributions to scientific thought. Yet a century later Copernican advocates were accepted.

This kind of cycle happens age after age. In the early 1800s geologists met with violent condemnation because they differed with the Genesis account of creation. Later in the same century geology was accepted, but theories of evolution were condemned and the teaching of them forbidden. Every age has its groups of intellectual rebels who are persecuted, condemned, or suppressed at the time but who later seem harmless and often essential to the elevation of human conditions. "At every crossway on the road that leads to the future, each progressive spirit is opposed by a thousand men appointed to guard the past."*

Science and human conditions advanced dramatically after the discovery almost 4 centuries ago that nature could be analyzed and described mathematically. When the ideas of science are expressed in mathematical terms, they are unambiguous. They don't have the double meanings that so often confuse the discussion of ideas expressed in common language. When findings in nature are expressed mathematically, they are easier to verify or disprove by experiment. The methods of mathematics and experimentation led to the enormous success of science.†

The Scientific Method

The Italian physicist Galileo Galilei and the English philosopher Francis Bacon are usually considered the principal founders of the **scientific method**—a method that is extremely effective in gaining, organizing, and applying new knowledge. This method, introduced in the sixteenth century, is essentially as follows:

1. Recognize a problem.
2. Make an educated guess—a **hypothesis**.
3. Predict the consequences of the hypothesis.
4. Perform experiments to test predictions.
5. Formulate the simplest general rule that organizes the three main ingredients—hypothesis, prediction, experimental outcome—into a **theory**.

Although this method has a certain appeal, it has not always been the key to the discoveries and advances in science. Many cases of trial and error, experimentation without guessing, and just plain accidental discovery have accounted for much of the progress in science. The success of science has more to do with an attitude common to scientists than with a particular method. This attitude is one of inquiry, experimentation, and humility before the facts.

The Scientific Attitude

It is common to think of a fact as something that is unchanging and absolute. But in science, a **fact** is generally a close agreement by competent observers of a series of observations of the same phenomena. For example, where it was once a recognized fact that the earth was flat, today it is a fact that the earth is round. A scientific hypothesis, on the

---

*From Count Maurice Maeterlinck's "Our Social Duty."

†The mathematical structure of physics is evident in the equations you will encounter throughout this book. You will see that equations are simply shortcut expressions of relationships that can be expressed in words. The focus of this book is on understanding concepts—in English. Equations are compact statements that help guide thinking and are not used in this book as recipes for mathematical problem solving. A premature effort at mathematical problem solving often tends to obscure the physics, so emphasis on this kind of problem solving is best postponed until you understand the concepts—perhaps in a follow-up course. Conceptual physics puts comprehension comfortably before any computation.

other hand, is an educated guess that is only presumed to be factual until demonstrated by experiments. When a hypothesis has been tested over and over again and has not been contradicted, it may become known as a **law** or *principle*.

If a scientist believes a certain hypothesis, law, or principle is true but finds contradicting evidence, then, in the scientific spirit, the hypothesis, law, or principle is changed or abandoned with little regard for the reputation or authority of the persons advocating it. For example, the greatly respected Greek philosopher Aristotle (384–322 BC) claimed that an object falls at a speed proportional to its weight. This false idea was held to be true for more than 2000 years because of Aristotle's compelling authority. In the scientific spirit, however, a single verifiable experiment to the contrary outweighs any authority, regardless of reputation or the number of followers or advocates. In modern science, argument by appeal to authority has little value.

Scientists must accept their experimental findings even when they would like them to be different. They must strive to distinguish between what they see and what they wish to see, for scientists, like most people, have a vast capacity for fooling themselves.* People have always tended to adopt general rules, beliefs, creeds, ideas, and hypotheses without thoroughly questioning their validity and to retain them long after they have been shown to be meaningless, false, or at least questionable. The most widespread assumptions are often the least questioned. Most often, when an idea is adopted, particular attention is given to cases that seem to support it, while cases that seem to refute it are distorted, belittled, or ignored.

Scientists use the word *theory* in a way that differs from its usage in everyday speech. In everyday speech a theory is no different from a hypothesis—a supposition that has not been verified. A scientific **theory**, on the other hand, is a synthesis of a large body of information that encompasses well-tested and verified hypotheses about certain aspects of the natural world. Physicists, for example, speak of the theory of the atom, chemists speak of the molecular theory, and biologists work with the theory of cells.

The theories of science are not fixed, but undergo change. Scientific theories evolve as they go through stages of redefinition and refinement. During the last hundred years, for example, the theory of the atom has been repeatedly refined as new evidence on atomic behavior has been gathered. Similarly, chemists have refined their view of the way molecules bond together, and biologists have refined the cell theory. The refinement of theories is a strength of science, not a weakness. Many people feel that it is a sign of weakness to change their minds. Competent scientists must be experts at changing their minds. They change their minds, however, only when confronted with solid experimental evidence

Facts are revisable data about the world.

Theories interpret facts

---

*In your education it is not enough to be aware that other people may try to fool you; it is more important to be aware of your own tendency to fool yourself.

to the contrary or when a conceptually simpler hypothesis forces them to a new point of view. More important than defending beliefs is improving them. Better hypotheses are made by those who are honest in the face of fact.

Away from their profession, scientists are inherently no more honest or ethical than most other people. But in their profession they work in an arena that puts a high premium on honesty. The cardinal rule in science is that all hypotheses must be testable—they must be susceptible, at least in principle, to being proved *wrong*. In science, it is more important that there be a means of proving an idea wrong than that there be a means of proving it right. This is a major factor that distinguishes science from nonscience. At first this may seem strange, for when we wonder about most things, we concern ourselves with ways of finding out whether they are true. Scientific hypotheses are different. In fact, if you want to distinguish whether a hypothesis is scientific or not, look to see if there is a test for proving it wrong. If there is no test for its possible wrongness, then the hypothesis is not scientific.

Consider the biologist Darwin's hypothesis that life forms evolve from simpler to more complex forms. This could be proved wrong if paleontologists found that more complex forms of life appeared before their simpler counterparts. Einstein hypothesized that light is bent by gravity. This might be proved wrong if starlight that grazed the sun were undeflected from its normal path. As it turns out, less complex life forms are found to precede their more complex counterparts and starlight is found to bend as it passes close to the sun, which support the claims. If and when a hypothesis or scientific claim is confirmed, it is regarded as useful and a stepping-stone to additional knowledge.

Consider on the other hand the hypothesis that "intelligent life exists on other planets somewhere in the universe." At present, this hypothesis is not scientific. Reasonable or not, it is *speculation*. Although it can be proved correct by the verification of a single instance of intelligent life existing elsewhere in the universe, there is no way to prove the hypothesis wrong if no life is ever found. If we search the far reaches of the universe for eons and find no life, we would not prove that it doesn't exist around the next corner. A hypothesis that is capable of being proved right but not capable of being proved wrong is not a scientific hypothesis. Many such statements are quite reasonable and useful, but they lie outside the domain of science.

None of us has the time, energy, or resources to test every idea, so most of the time we take somebody's word. How do we know whose word to take? To reduce the likelihood of error, scientists accept the word only of those whose ideas, theories, and findings are testable—if not in practice, at least in principle. Speculations that cannot be tested are regarded as "unscientific." This has the long-run effect of compelling honesty—findings widely publicized among fellow scientists are generally subjected to further testing. Sooner or later, mistakes (and deception) are found out; wishful thinking is exposed. A discredited scientist does not get a second chance in the community of scientists. Honesty, so important to the progress of science, thus becomes a matter of self-interest to scientists. There is relatively little bluffing in a game where all bets are called. In fields of study where right and wrong are not so easily established, the pressure to be honest is considerably less.

Question ▶ Which of these is a scientific hypothesis?
a. Atoms are the smallest particles of matter that exist.
b. Space is permeated with an essence that is undetectable.
c. Albert Einstein is the greatest physicist of the twentieth century.

The ideas and concepts most important to our everyday life are often unscientific; their correctness or incorrectness cannot be determined in the laboratory. Interestingly enough, it seems that people honestly believe their own ideas about things to be correct, and almost everyone is acquainted with people who hold completely opposite views—so the ideas of some (or all) must be incorrect. How do you know whether or not *you* are one of those holding erroneous beliefs? There is a test. Before you can be reasonably convinced that you are right about a particular idea, you should be sure that you understand the objections and the positions of your most articulate antagonists. You should find out whether your views are supported by sound knowledge of opposing ideas or by your *misconceptions* of opposing ideas. You make this distinction by seeing whether or not you can state the objections and positions of your opposition to *their* satisfaction. Even if you can successfully do this, you cannot be absolutely certain of being right about your own ideas, but the probability of being right is considerably higher if you pass this test.

Although the notion of being familiar with counter points of view seems reasonable to most thinking people, just the opposite—shielding ourselves and others from opposing ideas—has been more widely practiced. We have been taught to discredit unpopular ideas without understanding them in proper context. With the 20/20 vision of hindsight, we can see

▶ **Answer**

Only **a** is scientific, because there is a test for falseness. The statement not only is *capable* of being proved wrong, but in fact *has* been proved wrong. Statement **b** has no test for possible wrongness and is therefore unscientific. Likewise for any principle or concept for which there is no means, procedure, or test whereby it can be shown to be wrong (if it is wrong). Some pseudoscientists and other pretenders of knowledge will not even consider a test for the possible wrongness of their statements. Statement **c** is an assertion that has no test for possible wrongness. If Einstein was not the greatest physicist, how could we know? It is important to note that because the name Einstein is generally held in high esteem, it is a favorite of pseudoscientists. So we should not be surprised that the name of Einstein, like that of Jesus and other highly respected sources, is cited often by charlatans who wish to bring respect to themselves and their points of view.

that many of the "deep truths" that were the cornerstones of whole civilizations were shallow reflections of the prevailing ignorance of the time. Many of the problems that plagued societies stemmed from this ignorance and the resulting misconceptions; much of what was held to be true simply wasn't true. Are we different today?

Question   ▶   Suppose in a disagreement between two people, A and B, you note that person A only states and restates one point of view, whereas person B clearly states both her own position and that of person A. Who is more likely to be correct?

Science, Art, and Religion

The search for order and meaning in the world around us has taken different forms: one is science, another is art, and another is religion. Although the roots of all three go back thousands of years, the traditions of science are relatively recent. More important, the domains of science, art, and religion are different, although they often overlap. Science is principally engaged with discovering and recording natural phenomena, the arts are concerned with the value of human interactions as they pertain to the senses, and religion addresses the source, purpose, and meaning of it all.

The principal values of science and the arts are comparable. In literature we find what is possible in human experience. We can learn about emotions ranging from anguish to love, even if we haven't yet experienced them. The arts do not necessarily give us those experiences, but describe them to us and suggest what may be in store for us. A knowledge of science similarly tells us what is possible in nature. Scientific knowledge helps us predict possibilities in nature even before these possibilities have been experienced. It provides us with a way of connecting things, of seeing relationships between and among them, and of making sense of the myriad natural events we find around us. Science broadens our perspective of the natural environment of which we are a part. A knowledge of both the arts and the sciences makes for a wholeness that affects the way we view the world and the decisions we make about it and ourselves. A truly educated person is knowledgeable in both.

Science and religion are different from each other. Science is both a body of knowledge and a method of probing nature's secrets; religion has

▶ **Answer**

Who knows for sure? Person B may have the cleverness of a lawyer who can state various points of view and still be incorrect. We can't be sure about the "other guy." The test for correctness or incorrectness suggested here is not a test of others, but of and for *you*. It can aid your personal development. As you attempt to articulate the ideas of your antagonists, be prepared, like scientists who are prepared to change their minds, to discover evidence counter to your own ideas—evidence that may alter your views. Intellectual growth often comes in this way.

to do not with nature, but with meaning and its implications for personal and communal life. Religious beliefs and practices normally have to do with the faith and worship of God and the creation of human community, not with the experimental practices of science. In this respect, science and religion are as different as apples and oranges and do not contradict each other. While science is concerned with the workings of cosmic processes, religion addresses itself to the purpose of the cosmos. The two complement rather than contradict each other.

When we study the nature of light later in this book, we will treat light first as a wave and then as a particle. To the person who knows a little bit about physics, waves and particles are contradictory; light can be only one or the other, and we have to choose between them. But to the enlightened physicist, waves and particles complement each other and provide a deeper understanding of light. In a similar way, it is mainly people who are either uninformed or misinformed about the deeper natures of both science and religion who feel that they must choose between believing in religion and believing in science. Unless one has a shallow understanding of either or both, there is no contradiction in being religious and being scientific in one's thinking.

Science and Technology

Science and technology are also different from each other. Whereas science has to do with discovering evidence and relationships for observable phenomena in nature and with establishing theories that organize and make sense of those phenomena, technology has to do with tools, techniques, and procedures for putting the findings of science to use.

Another difference between science and technology is their effect on human lives. Science excludes the human factor. Scientists who seek to comprehend the workings of nature cannot be influenced by their own or other people's likes or dislikes or by popular ideas about what is correct. What scientists discover may shock or anger people, as did Darwin's theory of evolution. But even an unpleasant truth is likely to be useful; besides, we have the option of refusing to believe it! But this is hardly so with technology once it is developed: we do not have the option of refusing to hear the sonic boom produced by a supersonic aircraft flying overhead, we do not have the option of refusing to breathe polluted air, and we do not have the option of living in a nonnuclear age. Unlike science, advances in technology must be measured in terms of the human factor, with the understanding that technology is our slave and not the reverse. The legitimate purpose of technology is to serve people—people in general, not just some people, and future generations, not just those who currently wish to gain advantage for themselves.

We are all familiar with the abuses of technology. Many people blame technology itself for widespread pollution, resource depletion, and even social decay in general—so much so that the promise of technology is obscured. That promise is a cleaner and healthier world. It is much wiser to combat the misuse of technology with knowledge than with ignorance. Wise applications of science and technology *can* lead to a better world.

Physics—
The Basic Science

Science is the present-day equivalent of what used to be called *natural philosophy*. Natural philosophy was the study of unanswered questions about nature. As the answers were found, they became part of what is now called *science*. The study of science today branches into the study of living things and nonliving things: the life sciences and the physical sciences. The life sciences branch into such areas as biology, zoology, and botany. The physical sciences branch into such areas as geology, astronomy, chemistry, and physics.

Physics is more than a part of the physical sciences. It is the basic science. It's about the nature of basic things such as motion, forces, energy, matter, heat, sound, light, and the insides of atoms. Chemistry is about how matter is put together, how atoms combine to form molecules, and how the molecules combine to make up the many kinds of matter around us. Biology is more complex and involves matter that is alive. So underneath biology is chemistry, and underneath chemistry is physics. The concepts of physics reach up to these more complicated sciences. That's why physics is the most basic science.

An understanding of science begins with an understanding of physics. The following chapters present physics conceptually so you can enjoy understanding it.

Question Which of the following activities involves the utmost human expression of passion, talent, and intelligence?

a. art   **b.** literature   **c.** music   **d.** science

▶ **Answer**

All of them! The human value of science, however, is the least understood by most individuals in our society. The reasons are varied, ranging from the common notion that science is incomprehensible to people of average ability to the extreme view that science is a dehumanizing force in our society. Most of the misconceptions about science probably stem from the confusion between the *abuses* of science and science itself.

Some people who view science as cold and impersonal seek psychic comfort in mysticism and other counterscientific concepts. Still more embrace a combination of science and mysticism, such as astronomy and astrology, the normal and the paranormal. In straddling science and superstition, they stand with one foot in the twentieth century and the other in the thirteenth. It is unfortunate that they do not see basic science as an enchanting human activity shared by a wide variety of people who, with present-day tools and know-how, are reaching further and finding out more about themselves and their environment than people in the past were ever able to do.

The more you know about science, the more passionate you feel toward your surroundings. There is physics in everything you see, hear, smell, taste, and touch!

In Perspective

Only a few centuries ago the most talented and most skilled artists, architects, and artisans of the world directed their genius and effort to the construction of the great cathedrals, synagogues, temples, and mosques. Some of these architectural structures took centuries to build, which means that nobody witnessed both the beginning and the end of construction. Even the architects and early builders who lived to a ripe old age never saw the finished results of their labors. Entire lifetimes were spent in the shadows of construction that must have seemed without beginning or end. This enormous focus of human energy was inspired by a vision that went beyond worldly concerns—a vision of the cosmos. To the people of that time, the structures they erected were their "spaceships of faith," firmly anchored but pointing to the cosmos.

Today the efforts of many of our most skilled scientists, engineers, artists, and artisans are directed to building the spaceships that already orbit the earth and others that will voyage beyond. The time required to build these spaceships is extremely brief compared to the time spent building the stone and marble structures of the past. Many people working on today's spaceships were alive before Charles Lindbergh made the first solo airplane flight across the Atlantic Ocean. Where will younger lives lead in a comparable time?

We seem to be at the dawn of a major change in human growth, for, as little Jenny suggests in the photo at the beginning of this chapter, we may be like the hatching chicken who has exhausted the resources of its inner-egg environment and is about to break through to a whole new range of possibilities. The earth is our cradle and has served us well. But cradles, however comfortable, are one day outgrown. So with the inspiration that in many ways is similar to the inspiration of those who built the early cathedrals, synagogues, temples, and mosques, we aim for the cosmos.

We live in an exciting time!

Summary of Terms

Fact A phenomenon about which competent observers who have made a series of observations are in agreement.

Hypothesis An educated guess; a reasonable explanation of an observation or experimental result that is not fully accepted as factual until tested over and over again by experiment.

Law A general hypothesis or statement about the relationship of natural quantities that has been tested over and over again and has not been contradicted. Also known as a *principle*.

Scientific method An orderly method for gaining, organizing, and applying new knowledge.

Theory A synthesis of a large body of information that encompasses well-tested and verified hypotheses about certain aspects of the natural world.

Suggested Reading

Cole, K. C. *Sympathetic Vibrations: Reflections of Physics as a Way of Life.* New York: Morrow, 1984.

Feynman, Richard P. *Surely You're Joking, Mr. Feynman.* New York: Norton, 1986.

Florman, S. C. *The Existential Pleasures of Engineering.* New York: St. Martin's, 1976.

Pagels, H. R. *The Cosmic Code: Quantum Physics as the Language of Nature.* New York: Simon & Schuster, 1982.

Review Questions

1. Briefly, what is science?

2. Throughout the ages, what has been the general reaction to new ideas about established "truths"?

3. What was the effect of the discovery 3 centuries ago that a mathematical structure underlies nature?

The Scientific Method

4. Outline the steps of the scientific method.

The Scientific Attitude

5. Distinguish among a scientific fact, a hypothesis, a law, and a theory.

6. In daily life people are often praised for maintaining some particular point of view, for the "courage of their convictions." A change of mind is seen as a sign of weakness. How is this different in science?

7. In daily life we see many cases of people who are caught misrepresenting things and who soon thereafter are excused and accepted by their contemporaries. How is this different in science?

8. What test can you perform to increase the chance in your own mind that you are right about a particular idea?

Science, Art, and Religion

9. Why are students of the arts encouraged to learn about science and science students encouraged to learn about the arts?

10. Why do many people believe they must choose between science and religion?

Science and Technology

11. Clearly distinguish between science and technology.

Physics—The Basic Science

12. Of physics, chemistry, and biology, which science is the least complex? The most complex? (At your school, which of these is the "easiest" and which is the "hardest" as a science course?)

Exercises

1. The following are arranged alphabetically. Arrange them in proper sequence for the scientific method: experiment, hypothesis, law, observation, plan, principle, test, theory, thinking.

2. Consider this statement, which recently appeared in a science journal: "Elevated levels of carbon dioxide in the atmosphere caused a global 'greenhouse' warming at the end of the Cretaceous period. The relatively high resulting temperatures would have interfered with the reproduction of dinosaurs, eventually bringing about their extinction." What elements of the scientific method do and don't appear here? What questions might be addressed to the author?

3. In answer to the question, "When a plant grows, where does the material come from?" Aristotle hypothesized by logic that all material came from the soil. Do you consider his hypothesis to be correct, incorrect, or partially correct? What experiments do you propose to support your choice?

4. The great philosopher and mathematician Bertrand Russell (1872–1970) wrote about ideas in the early part of his life that he rejected in the latter part of his life. Do you see this as a sign of weakness or a sign of strength in Bertrand Russell? (Do you speculate that your present ideas about the world about you will change as you learn and experience more, or do you speculate that further knowledge and experience will solidify your present understanding?)

5. Bertrand Russell wrote, "I think we must retain the belief that scientific knowledge is one of the glories of man. I will not maintain that knowledge can never do harm. I think such general propositions can almost always be refuted by well-chosen examples. What I will maintain—and maintain vigorously—is that knowledge is very much more often useful than harmful and that fear of knowledge is very much more often harmful than useful." Think of examples to support this statement.

PART I

MECHANICS

2 Linear Motion

More than 2000 years ago, the ancient Greek scientists were familiar with some of the ideas in physics that we study today. They had a very good understanding of some of the properties of light. But they were confused about motion. Probably the first to study motion seriously was Aristotle, the most outstanding philosopher-scientist in ancient Greece. Aristotle attempted to clarify motion by classification.

Aristotle on Motion

Aristotle divided motion into two main classes: *natural motion* and *violent motion*. We shall briefly consider each, not as study material, but only as a background to present-day ideas about motion.

Natural motion was thought to proceed from the "nature" of objects. In Aristotle's view, every object in the universe had a proper place, determined by this nature; any object not in its proper place would "strive" to get there. Being of the earth, an unsupported lump of clay properly fell to the ground; being of the air, an unimpeded puff of smoke properly rose; being a mixture of earth and air but predominantly earth, a feather properly fell to the ground but not as rapidly as a lump of clay. Larger objects were expected to strive harder. Hence, objects were thought to fall at speeds proportional to their weights: the heavier the object, the faster it was thought to fall.

Natural motion could be either straight up or straight down, as in the case of all things on earth, or it could be circular, as in the case of celestial objects. Unlike up-and-down motion, circular motion was seen as being without beginning or end, repeating itself without deviation. Aristotle believed that different rules applied in the heavens, and he asserted that celestial bodies were perfect spheres made of a perfect and unchanging substance, which he called *ether*. (The only celestial object with any detectable change or imperfection was the moon, which, being nearest the earth, was thought by medieval Christians to be somewhat contaminated by the corrupted earth.)

Violent motion, Aristotle's other class of motion, resulted from pushing or pulling forces. Violent motion was imposed motion. A person pushing a cart or lifting a heavy weight imposed motion, as did someone hurling a stone or winning a tug-of-war. The wind imposed motion on ships. Flood waters imposed it on boulders and tree trunks. The essential thing about violent motion was that it was externally caused and was imparted to objects; they moved not of themselves, but were pushed or pulled.

The concept of violent motion had its difficulties, for the pushes and pulls responsible for it were not always evident. For example, a bowstring moved an arrow until the arrow left the bow; after that, further explanation of the arrow's motion seemed to require some other pushing agent. It was imagined, therefore, that a parting of the air by the moving arrow resulted in a squeezing effect on the rear of the arrow as the air rushed back to prevent a vacuum from forming. The arrow was propelled through the air as a bar of soap is propelled in the bathtub when you squeeze one end of it.

To sum up, Aristotle taught that all motions resulted either from the nature of the moving object or from a sustained push or pull. Provided that an object was in its proper place, it would not move unless subjected to a force. Except for celestial objects, the normal state was one of rest.

Aristotle's statements about motion were a beginning in scientific thought, and although he did not consider them to be the final words on the subject, his followers for nearly 2000 years regarded his views as beyond question. Implicit in the thinking of ancient, medieval, and early Renaissance times was the notion that the normal state of objects was one of rest. Since it was evident to most thinkers until the sixteenth century that the earth must be in its proper place, and since a force capable of moving the earth was inconceivable, it seemed quite clear that the earth did not move.

Aristotle (384–322 BC)

Greek philosopher, scientist, and educator Aristotle was the son of a physician who personally served the king of Macedonia. At 17 he entered the Academy of Plato, where he worked and studied for 20 years until Plato's death. He then became the tutor of young Alexander the Great. Eight years later he formed his own school. Aristotle's aim was to systematize existing knowledge, just as Euclid had systematized geometry. Aristotle made critical observations, collected specimens, and gathered together, summarized, and classified almost all existing knowledge of the physical world. His systematic approach became the method from which Western science later arose. After his death, his voluminous notebooks were preserved in caves near his home and were later sold to the library at Alexandria. Scholarly activity ceased in most of Europe through the Dark Ages, and the works of Aristotle were forgotten and lost. Scholarship continued in the Byzantine and Islamic empires, and various texts were reintroduced to Europe during the eleventh and twelfth centuries and translated into Latin. The Church, the dominant political and cultural force in Western Europe, first prohibited the works of Aristotle and then accepted and incorporated them into Christian doctrine. Any attack on Aristotle was an attack on the Church itself.

Copernicus and the Moving Earth

It was in this climate that the astronomer Copernicus formulated his theory of the moving earth. Copernicus reasoned from his astronomical observations that the earth traveled around the sun. For years he worked without making his thoughts public—for two reasons. The first was that he feared persecution; a theory so completely different from common opinion would surely be taken as an attack on established order. The second reason was that he had grave doubts about it himself; he could not reconcile the idea of a moving earth with the prevailing ideas of motion. Finally, in the last days of his life, at the urging of close friends he sent his *De Revolutionibus* to the printer. The first copy of his famous exposition reached him on the day he died—May 24, 1543.

Most of us know about the reaction of the medieval Church to the idea that the earth traveled around the sun. Because Aristotle's views had become so formidably a part of Church doctrine, to contradict them was to question the Church itself. For many Church leaders, the idea of a moving earth threatened not only their authority but the very foundations of faith and civilization as well. Their fears were well founded. For better or for worse, this new idea was to overturn their conception of the cosmos.

Galileo and the Leaning Tower

It was Galileo, the foremost scientist of the sixteenth century, who gave credence to the Copernican view of a moving earth. He accomplished this by discrediting the Aristotelian ideas about motion. Although not the first to point out difficulties in Aristotle's views, Galileo was the first to provide conclusive refutation through observation and experiment.

Aristotle's falling-body hypothesis was easily demolished by Galileo. He is said to have dropped objects of various weights from the top of the Leaning Tower of Pisa and compared their falls. Contrary to Aristotle's assertion, he found that a stone twice as heavy as another did not fall twice as fast. Except for the small effect of air resistance, Galileo found that objects of various weights, when released at the same time, fell together and hit the ground at the same time. On one occasion, Galileo allegedly attracted a large crowd to witness the dropping of a light

Figure 2-1
Galileo's famous demonstration.

object and a heavy object from the top of the tower. Legend has it that many observers of this demonstration who saw the objects hit the ground together scoffed at the young Galileo and continued to hold fast to their Aristotelian teachings.

Galileo's Inclined Planes

Slope downward—
Speed increases

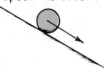

Slope upward—
Speed decreases

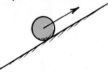

No slope—
Does speed change?

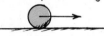

Figure 2-2
Motion of a ball on various planes.

Aristotle was an astute observer of nature, and he dealt with problems around him rather than with abstract cases that did not occur in his environment. Motion always involved a resistive medium such as air or water. He believed a vacuum to be impossible and therefore did not give serious consideration to motion in the absence of an interacting medium. That's why it was basic to Aristotle that an object require a push or pull to keep it moving. And it was this basic principle that Galileo denied when he stated that if there is no interference with a moving object, it will keep moving in a straight line forever; no push, pull, or force of any kind is necessary.

Galileo tested this hypothesis by experimenting with the motion of various objects on inclined planes. He noted that balls rolling on downward-sloping planes picked up speed, while balls rolling on upward-sloping planes lost speed (Figure 2-2). From this he reasoned that balls rolling along a horizontal plane would neither speed up nor slow down. The ball would finally come to rest not because of its "nature" but because of friction. This idea was supported by Galileo's observation of motion along smoother surfaces: when there was less friction, the motion of objects persisted for a longer time; the less the friction, the more the motion approached constant speed. He reasoned that in the absence of friction or other opposing forces, a horizontally moving object would continue moving forever.

This assertion was supported by a different experiment and another line of reasoning. Galileo placed two of his inclined planes facing each other (Figure 2-3). He observed that a ball released from a position of rest at the top of a downward-sloping plane rolled down and then up the slope of the upward-sloping plane until it almost reached its initial height. He reasoned that only friction prevented it from rising to exactly the same height, for the smoother the planes, the more nearly the ball rose to the same height. Then he reduced the angle of the upward-sloping plane. Again the ball rose to the same height, but it had to go farther. Additional reductions of the angle yielded similar results; to reach the same height, the ball had to go farther each time. He then asked, "If I have a long horizontal plane, how far must the ball go to reach the same height?" The obvious answer is "Forever—it will never reach its initial height."*

Galileo analyzed this in still another way. Because the downward motion of the ball from the first plane is the same for all cases, the speed of the ball when it begins moving up the second plane is the same for all cases. If it moves up a steep slope, it loses its speed rapidly. On a lesser slope,

---

*From Galileo's *Dialogues Concerning the Two New Sciences.*

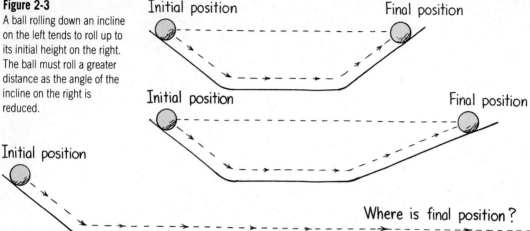

Figure 2-3
A ball rolling down an incline on the left tends to roll up to its initial height on the right. The ball must roll a greater distance as the angle of the incline on the right is reduced.

it loses its speed more slowly and rolls for a longer time. The less the upward slope, the more slowly it loses its speed. In the extreme case where there is no slope at all—that is, when the plane is horizontal—the ball should not lose any speed. In the absence of retarding forces, the tendency of the ball is to move forever without slowing down. This property of a moving object to continue moving he called **inertia**.

Galileo's concept of inertia discredited the Aristotelian theory of motion. Aristotle did not recognize the idea of inertia because he failed to imagine what motion would be like without friction. In his experience, all motion was subject to resistance, and he made this fact central to his theory of motion. Aristotle's failure to recognize friction for what it is—namely, a force like any other—impeded the progress of physics for 2000 years, until the time of Galileo. An application of Galileo's concept of inertia would show that no force was required to keep the earth in motion. The way was open for Isaac Newton to synthesize a new vision of the universe. We'll return to Newton in later chapters, and now acquaint ourselves with some of the terms that Galileo introduced to describe motion.

Question ▶ Would it be correct to say that inertia *causes* a moving object to continue in motion?

▶ **Answer**

In a strict sense, no. We don't know *why* objects exhibit this property. Nevertheless, we call the property to behave in this predictable way *inertia*. We understand many things and have labels and names for these things. There are many things we do not understand, and we have labels and names for these things also. Education consists not so much in acquiring new names and labels but in learning which we understand and which we don't.

Description of Motion

In Aristotle's description of motion, the *distance* of an object from its proper place was fundamentally important. Galileo broke with this traditional concept and realized that time was the important missing ingredient in describing motion. Galileo described motion in terms of *time rates of change*. A time *rate* of change of a quantity is a quantity divided by the time. It tells how fast something happens, or how much something changes in a certain amount of time. The rates that describe motion are *speed*, *velocity*, and *acceleration*.*

Figure 2-4
A speedometer gives readings in both miles per hour and kilometers per hour.

Speed

Things in motion travel certain distances in given times. An automobile, for example, travels so many kilometers in an hour. **Speed** is a measure of how fast something is moving. It is the rate at which distance is covered, and is always measured in terms of a unit of distance divided by a unit of time. In general,

$$\text{Speed} = \frac{\text{distance}}{\text{time}}$$

Any combination of distance and time units is legitimate for measuring speed; for motor vehicles (or long distances) the units kilometers per hour (km/h) or miles per hour (mi/h or mph) are commonly used. For shorter distances, meters per second (m/s) are often useful units. The slash symbol (/) is read as *per* and means "divided by." Throughout this book we'll primarily use meters per second (m/s). Table 2-1 shows some comparative speeds in different units.†

Table 2-1
Approximate speeds in different units

20 km/h	= 12 mi/h	= 6 m/s
40 km/h	= 25 mi/h	= 11 m/s
60 km/h	= 37 mi/h	= 17 m/s
65 km/h	= 40 mi/h	= 18 m/s
80 km/h	= 50 mi/h	= 22 m/s
100 km/h	= 62 mi/h	= 28 m/s
120 km/h	= 75 mi/h	= 33 m/s

Instantaneous Speed The speed that something has at any one instant is called *instantaneous speed*. It is the speed registered by the speedometer of a car. When we say that the speed of a car at some particular instant is 60 kilometers per hour, we are specifying its instantaneous speed, and

---

*It would be nice if this chapter helps you to master these concepts, but it will be enough for you to become familiar with them and to be able to distinguish among them. The next few chapters will sharpen your understanding of these concepts.
†Conversion is based on 1 h = 3600 s, 1 mi = 1609.344 m.

we mean that if the car continued moving as fast for an hour, it would travel 60 kilometers. If it continued at that speed for half an hour, it would cover only half the distance: 30 kilometers. If it continued for only 1 minute, it would cover only 1 kilometer.

Average Speed   A car does not always move at the same speed. On any trip the speed usually varies somewhat. We distinguish between instantaneous speed and *average speed*. The *average speed* is defined as follows:

$$\text{Average speed} = \frac{\text{total distance covered}}{\text{time interval}}$$

Average speed can be calculated rather easily. For example, if we drive a distance of 80 kilometers in a time of 1 hour, we say our average speed is 80 kilometers per hour. Likewise, if we travel 320 kilometers in 4 hours,

$$\text{Average speed} = \frac{\text{total distance covered}}{\text{time interval}} = \frac{320 \text{ km}}{4 \text{ h}} = 80 \text{ km/h}$$

We see that when a distance in kilometers (km) is divided by a time in hours (h), the answer is in kilometers per hour (km/h).

Since average speed is the whole distance covered divided by the total time of travel, it doesn't indicate the different speeds and variations that may have taken place during shorter time intervals. In practice, we experience a variety of speeds on most trips, so the average speed is often quite different from the speed at any instant, the instantaneous speed. Whether we talk about average speed or instantaneous speed, we are talking about the rates at which distance is traveled.

If we know average speed and time of travel, distance traveled is easy to find. A simple rearrangement of the definition above gives

$$\text{Total distance covered} = \text{average speed} \times \text{time}$$

If your average speed is 80 kilometers per hour on a 4-hour trip, for example, you cover a total distance of 320 kilometers.

Questions   ▶   **1.** What is the average speed of a cheetah that sprints 100 m in 4 s? How about if it sprints 50 m in 2 s?

2. If a car moves with an average speed of 60 km/h for an hour, it will travel a distance of 60 km.
   a. How far would it travel at this rate for 4 h?
   b. For 10 h?
   c. Would it be possible for the car to attain an average speed of 60 km/h and never exceed a reading of 60 km/h on the speedometer?

Velocity

Loosely speaking, we can use the words *speed* and *velocity* interchangeably. Strictly speaking, however, there is a distinction between the two. When we say that something travels at a rate of 60 kilometers per hour, we are specifying its speed. But if we say that something travels at 60 kilometers per hour to the north, we are specifying its velocity. A racecar driver is concerned with his speed—how fast he is moving. An airplane pilot is concerned with her velocity—how fast and in what direction she is moving. When we describe speed and the *direction* of motion, we are specifying **velocity**.*

We distinguish between average velocity and instantaneous velocity as we do for speed. By custom, the word *velocity* alone is assumed to mean instantaneous velocity. Likewise for the word *speed* alone. If something moves at an unchanging or constant velocity, then its average and instantaneous velocities will have the same value. The same is true for speed. When something moves at constant velocity or constant speed, then *equal distances* are covered in equal intervals of time. Constant velocity and constant speed, however, can be very different. Constant velocity means constant speed with no change in direction. A car that rounds a curve at a constant speed does not have a constant velocity—its velocity changes as its direction changes.

Figure 2-5
The car on the circular track may have a constant speed, but its velocity is changing every instant. Why?

▶ **Answers**

(Are you reading this before you have formulated a reasoned answer in your thinking? If so, do you also exercise your body by watching others do push-ups? Exercise your thinking: when you encounter the many questions as above throughout this book, think before you read the footnoted answer.)

1. In both cases the answer is 25 m/s:

$$\text{Average speed} = \frac{\text{distance covered}}{\text{time interval}} = \frac{100 \text{ m}}{4 \text{ s}} = \frac{50 \text{ m}}{2 \text{ s}} = 25 \text{ m/s}$$

2. The distance traveled is the average speed × time of travel, so
   a. Distance = 60 km/h × 4 h = 240 km
   b. Distance = 60 km/h × 10 hr = 600 km
   c. No, if the trip started from rest and ended at rest, then there are intervals with an instantaneous speed less than 60 km/h. Unless there is compensation during periods of speed greater than 60 km/h, it would not be possible to yield an average of 60 km/h. In practice, average speeds are usually appreciably less than peak instantaneous speeds.

---

*A quantity described by both magnitude (how much) and direction (which way) is called a *vector quantity*. Examples of vector quantities are velocity and acceleration, but to avoid "information overload" we won't belabor this idea yet. A quantity described only by magnitude, such as speed, is a *scalar quantity*. We will treat the vector nature of velocity in the next chapter.

Questions

1. "She moves at a constant speed in a constant direction." Say the same sentence in fewer words.

2. The speedometer of a car moving to the east reads 100 km/h. It passes another car that moves to the west at 100 km/h. Do both cars have the same speed? Do they have the same velocity?

3. During a certain period of time, the speedometer of a car reads a constant 60 km/h. Does this indicate a constant speed? A constant velocity?

Figure 2-6
We say that an object undergoes acceleration when there is a *change* in its state of motion.

Acceleration

We can change the velocity of something by changing its speed, by changing its direction, or by changing both its speed and its direction. We define the rate of change in velocity as **acceleration**:

$$\text{Acceleration} = \frac{\text{change of velocity}}{\text{time interval}}$$

We are all familiar with acceleration in an automobile. In driving, we call it "pickup" or "getaway"; we experience it when we tend to lurch toward the rear of the car. The key idea that defines acceleration is *change*. To say we are accelerating is to say we are in some way changing our state of motion. We change our velocity in a car when we step on the gas pedal, appropriately called the *accelerator*. Suppose we are driving and in 1 second we steadily increase our velocity from 30 kilometers per hour to 35 kilometers per hour, and then to 40 kilometers per hour in the next second, to 45 in the next second, and so on. We change our velocity by 5 kilometers per hour each second. This change of velocity is what we mean by acceleration.

$$\text{Acceleration} = \frac{\text{change of velocity}}{\text{time interval}} = \frac{5 \text{ km/h}}{1 \text{ s}} = 5 \text{ km/h·s}$$

In this case the acceleration is 5 kilometers per hour second (abbreviated as 5 km/h·s). Note that a unit for time enters twice: once for the unit of velocity and again for the interval of time in which the velocity is changing. Also note that acceleration is not just the total change in velocity; it is the *time rate of change*, or *change per second*, of velocity (Figure 2-6).

▶ **Answers**

1. "She moves at constant velocity."

2. Both cars have the same speed, but they have opposite velocities because they are moving in opposite directions.

3. The constant speedometer reading indicates a constant speed but not a constant velocity, because the car may not be moving along a straight-line path, in which case it is accelerating. (We'll see in Chapter 4 that whenever something accelerates, a force must be acting on it.)

Galileo Galilei (1564–1642)

Galileo was born in Pisa in the same year Shakespeare was born and Michelangelo died. He studied medicine at the University of Pisa and then changed to mathematics. He developed an early interest in the mechanics of motion and was soon at odds with his contemporaries, who held to Aristotelian ideas on falling bodies. He left Pisa to teach at the University of Padua and became an advocate of the new Copernican theory of the solar system. He was one of the first to build a telescope, and he was the first to direct it to the nighttime sky and discover mountains on the moon and the moons of Jupiter. Because he published his findings in Italian instead of in the Latin expected of so reputable a scholar, and because of the recent invention of the printing press, his ideas reached a wide readership. He soon ran afoul of the Church and was warned not to teach and hold to Copernican views. He restrained himself publicly for nearly 15 years and then defiantly published his observations and conclusions, which were counter to Church doctrine. The outcome was a trial in which he was found guilty, and he was forced to renounce his discoveries. By then an old man broken in health and spirit, he was sentenced to perpetual house arrest. Nevertheless, he completed his studies on motion and his writings were smuggled from Italy and published in Holland. Earlier he damaged his eyes looking at the sun through a telescope, which led to blindness at the age of 74. He died 4 years later.

The term *acceleration* applies to decreases as well as to increases in velocity. We say the brakes of a car, for example, produce large retarding accelerations; that is, there is a large decrease per second in the velocity of the car. We often call this *deceleration*, or *negative acceleration*. We experience deceleration when we tend to lurch toward the front of the car.

We accelerate whenever we move in a curved path, even if we are moving at constant speed because our direction and hence our velocity is changing. We experience this acceleration as we tend to lurch toward the outer part of the curve. We distinguish speed and velocity for this reason and define *acceleration* as the rate at which velocity changes, thereby encompassing changes both in speed and in direction.

Anyone who has stood in a crowded bus has experienced the difference between velocity and acceleration. Except for the effects of a bumpy road, you can stand with no extra effort inside a bus that moves at constant velocity, no matter how fast it is going. You can flip a coin and catch it exactly as if the bus were at rest. It is only when the bus accelerates—speeds up, slows down, or turns—that you experience difficulty.

In much of this book we will be concerned only with motion along a straight line. When straight-line motion is being considered, it is common to use *speed* and *velocity* interchangeably. When the direction is not changing, acceleration may be expressed as the rate at which *speed* changes.

$$\text{Acceleration (along a straight line)} = \frac{\text{change in speed}}{\text{time interval}}$$

1. A particular car can go from rest to 90 km/h in 10 s. What is its acceleration?

2. In 2.5 s, a car increases its speed from 60 km/h to 65 km/h while a bicycle goes from rest to 5 km/h. Which undergoes the greater acceleration? What is the acceleration of each vehicle?

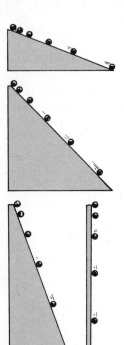

Figure 2-7
The greater the slope of the incline, the greater the acceleration of the ball. What is its acceleration if the incline is vertical?

Acceleration on Galileo's Inclined Planes

Galileo developed the concept of acceleration in his experiments on inclined planes. His main interest was falling objects, and because he lacked suitable timing devices he used inclined planes to effectively slow down accelerated motion and investigate it more carefully.

Galileo found that a ball rolling down an inclined plane will pick up the same amount of speed in successive seconds; that is, the ball will roll with uniform or constant acceleration. For example, a ball rolling down a plane inclined at a certain angle might be found to pick up a speed of 2 meters per second for each second it rolls. This gain per second is its acceleration. Its instantaneous velocity at 1-second intervals, at this acceleration, is then 0, 2, 4, 6, 8, 10, and so forth meters per second. We can see that the instantaneous speed or velocity of the ball at any given time after being released from rest is simply equal to its acceleration multiplied by the time:*

$$\text{Velocity acquired} = \text{acceleration} \times \text{time}$$

If we substitute the acceleration of the ball in this relationship, we can see that at the end of 1 second, the ball is traveling at 2 meters per second; at the end of 2 seconds, it is traveling at 4 meters per second; at the end of 10 seconds, it is traveling at 20 meters per second; and so on. The instantaneous speed or velocity at any time is simply equal to the acceleration multiplied by the number of seconds it has been accelerating.

▶ **Answers**

1. Its acceleration is 9 km/h·s. Strictly speaking, this would be its average acceleration, for there may have been some variation in its rate of picking up speed.

2. The accelerations of both the car and the bicycle are the same: 2 km/h·s.

$$\text{Acceleration}_{\text{car}} = \frac{\text{change of velocity}}{\text{time interval}} = \frac{65 \text{ km/h} - 60 \text{ km/h}}{2.5 \text{ s}} = \frac{5 \text{ km/h}}{2.5 \text{ s}} = 2 \text{ km/h·s}$$

$$\text{Acceleration}_{\text{bike}} = \frac{\text{change in velocity}}{\text{time interval}} \frac{5 \text{ km/h} - 0 \text{ km/h}}{2.5 \text{ s}} = \frac{5 \text{ km/h}}{2.5 \text{ s}} = 2 \text{ km/h·s}$$

Although the velocities involved are quite different, the rates of change of velocity are the same. Hence the accelerations are equal.

---

*Note that this relationship follows from the definition of acceleration. From $a = v/t$, simple rearrangement (multiplying both sides of the equation by t) gives $v = at$.

Galileo found greater accelerations for steeper inclines. The ball attains its maximum acceleration when the incline is tipped vertically. Then the acceleration is the same as that of a falling object (Figure 2-7). Regardless of the weight or size, Galileo discovered that when air resistance is small enough to be neglected, all material objects fall with the same constant acceleration.

Free Fall

Table 2-2
Free fall from rest

Time of fall (s)	Velocity acquired (m/s)
0	0
1	10
2	20
3	30
4	40
5	50
.	.
.	.
.	.
t	$10t$

Figure 2-8
Pretend that a falling rock is equipped with a speedometer. In each succeeding second of fall, we would find that the rock's speed increases by the same amount: 10 m/s. (Table 2-2 shows the speeds we would read at various seconds of fall.)

How Fast

Things fall because of the force of gravity. When a falling object is free of all restraints—no friction, air or otherwise—and falls under the influence of gravity alone, the object is in a state of **free fall**. (We'll consider the effects of air resistance on falling in Chapter 4). Table 2-2 shows the instantaneous speed of a freely falling object at 1-second intervals. The important thing to note in these numbers is the way the speed changes. *During each second of fall, the object gains a speed of 10 meters per second.* This gain per second is the acceleration. Free-fall acceleration is approximately equal to 10 meters per second each second, or, in short-hand notation, 10 m/s^2 (read as 10 meters per second squared). Note that the unit of time, the second, enters twice—once for the unit of speed and again for the time interval during which the speed changes.

In the case of freely falling objects, it is customary to use the letter g to represent the acceleration (because the acceleration is due to *gravity*). Although the value of g varies slightly in different parts of the world, its average value is equal to 9.8 meters per second each second, or, in shorter notation, 9.8 m/s^2. We round this off to 10 m/s^2 in our present discussion and in Table 2-2 to establish the ideas involved more clearly; multiples of 10 are more obvious than multiples of 9.8. Where accuracy is important, the value of 9.8 m/s^2 should be used.

Note in Table 2-2 that the instantaneous speed or velocity of an object falling from rest is consistent with the equation that Galileo deduced with his inclined planes:

$$\text{Velocity acquired} = \text{acceleration} \times \text{time}$$

The instantaneous velocity v of an object falling from rest* after a time t can be expressed in shorthand notation as

$$v = gt$$

The letter v symbolizes both speed and velocity. To see that this equation makes good sense, take a moment to check it with Table 2-2. Note that the instantaneous speed in meters per second is simply the acceleration $g = 10$ m/s^2 multiplied by the time t in seconds.

---

*If instead of being dropped from rest the object is thrown downward at speed v_0, the speed v after any elapsed time t is $v = v_0 + gt$. We will not be concerned with this added complication here, and will instead learn as much as we can from the most simple cases. That will be a lot!

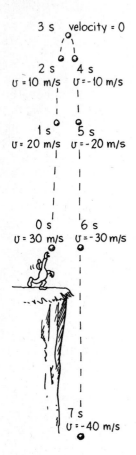

3 s   velocity = 0

2 s   4 s
$v = 10$ m/s $v = -10$ m/s

1 s   5 s
$v = 20$ m/s $v = -20$ m/s

0 s   6 s
$v = 30$ m/s $v = -30$ m/s

7 s
$v = -40$ m/s

Figure 2-9
The rate at which the velocity changes each second is the same.

So far we have been considering objects moving straight downward in the direction of gravity. How about an object thrown straight upward? Once released, it continues to move upward for a while and then comes back down. At the highest point, when it is changing its direction of motion from upward to downward, its instantaneous speed is zero. Then it starts downward just as if it had been dropped from rest at that height.

During the upward part of this motion, the object slows from its initial upward velocity to zero velocity. Its speed decreases—evidence of acceleration (or deceleration in the upward direction). How much does its speed decrease each second? It should come as no surprise that it decreases at the rate of 10 meters per second each second—the same acceleration it experiences on the way down. So interestingly enough, as Figure 2-9 shows, the instantaneous speed at points of equal elevation in the path is the same whether the object is moving upward or downward. The velocities are opposite, of course, because they are in opposite directions. During each second, the speed or the velocity changes by 10 meters per second. The acceleration is 10 meters per second squared the whole time, whether the object is moving upward or downward.

How Far

How *far* an object falls is altogether different from how *fast* it falls. With his inclined planes Galileo found that the distance a uniformly accelerating object travels is proportional to the *square of the time*. The details of this relationship are in Appendix II. We will state here only the results. The distance traveled by a uniformly accelerating object starting from rest is

$$\text{Distance traveled} = \tfrac{1}{2}(\text{acceleration} \times \text{time} \times \text{time})$$

This relationship applies to the distance something falls. We can express it for the case of a freely falling object in shorthand notation as*

$$d = \tfrac{1}{2}gt^2$$

▶ **Answer**
The speedometer readings would be 35 m/s, 60 m/s, and 1000 m/s, respectively. You can reason this from Table 2-2 or use the equation $v = gt$, where g is replaced by 10 m/s^2.

---

*d = average velocity × time = $\dfrac{\text{initial velocity} + \text{final velocity}}{2} \times \text{time} = \dfrac{0 + gt}{2} \times t$

$d = \tfrac{1}{2}gt^2$ (See Appendix II for further explanation.)

Table 2-3
Distance fallen in free fall

Time of fall (s)	Distance fallen (m)
0	0
1	5
2	20
3	45
4	80
5	125
.	.
.	.
.	.
t	$\frac{1}{2} 10\, t^2$

where d is the distance something falls when the time of fall in seconds is substituted for t and squared. If we use 10 m/s^2 for the value of g, the distance fallen for various times will be as shown in Table 2-3.

Note that an object falls a distance of only 5 meters during the first second of fall, although its speed at 1 second is 10 meters per second. This may be confusing, for we may think that the object should fall a distance of 10 meters. But for it to fall 10 meters in its first second of fall, it would have to fall at an *average* speed of 10 meters per second for the entire second. It starts its fall at 0 meters per second, and its speed is 10 meters per second only in the last instant of the 1-second interval. Its average speed during this interval is the average of its initial and final speeds, 0 and 10 meters per second. To find the average value of these or any two numbers, we simply add the two numbers and divide by 2. This equals 5 meters per second, which over a time interval of 1 second gives a distance of 5 meters. As the object continues to fall in succeeding seconds, it will fall through ever-increasing distances because its speed is continuously increasing.

Question ▶

A cat steps off a ledge and drops to the ground in $\frac{1}{2}$ second.
a. What is its speed on striking the ground?
b. What is its average speed during the $\frac{1}{2}$ second?
c. How high is the ledge from the ground?

Figure 2-10
Pretend that a falling rock is equipped with an odometer. The readings of distance fallen increase with time by $\frac{1}{2} gt^2$ and are shown in Table 2-3.

It is a common observation that all objects do not fall with equal accelerations. A leaf, a feather, or a sheet of paper may flutter to the ground slowly. The fact that air resistance is responsible for those different accelerations can be shown very nicely with a closed glass tube containing light and heavy objects—a feather and a coin, for example. In the presence of air, the feather and coin fall with quite different accelerations. But if the air in the tube is removed by a vacuum pump and

▶ **Answer**
If we round g off to 10 m/s^2, we find
a. Speed: $v = gt = 10$ m/s$^2 \times \frac{1}{2}$ s $= 5$ m/s
b. Average speed: $\bar{v} = \dfrac{\text{initial } v + \text{final } v}{2} = \dfrac{0 \text{ m/s} + 5 \text{ m/s}}{2} = 2.5$ m/s

We put a bar over the symbol to denote *average* speed—$\bar{v}$.
c. Distance: $d = \bar{v}t = 2.5$ m/s $\times \frac{1}{2}$ s $= 1.25$ m. Or equivalently,
$$d = \tfrac{1}{2} gt^2 = \tfrac{1}{2} \times 10 \text{ m/s}^2 \times (\tfrac{1}{2} \text{ s})^2 = \tfrac{1}{2} \times 10 \text{ m/s}^2 \times \tfrac{1}{4} \text{ s}^2 = 1.25 \text{ m}$$

Notice that we can find the distance by either of these equivalent relationships.

Figure 2-11
A feather and a coin fall at equal accelerations in a vacuum.

the tube is quickly inverted, the feather and coin fall with the same acceleration (Figure 2-11). Although air resistance appreciably alters the motion of things like falling feathers, the motion of heavier objects like stones and baseballs at ordinary low speeds is not appreciably affected by the air. The relationships $v = gt$ and $d = \frac{1}{2}gt^2$ can be used to a very good approximation for most objects falling in air.

How Quickly How Fast Changes

Much of the confusion that arises in analyzing the motion of falling objects comes about because it is easy to get "how fast" mixed up with "how far." When we wish to specify how fast something is falling, we are talking about speed or velocity, which is expressed as $v = gt$. When we wish to specify how far something falls, we are talking about distance, which is expressed as $d = \frac{1}{2}gt^2$. Speed or velocity (how fast) and distance (how far) are entirely different from each other.

A most confusing concept, and probably the most difficult encountered in this book, is "how quickly does how fast change"—acceleration. What makes acceleration so complex is that it is *a rate of a rate*. It is often confused with velocity, which is itself a rate (the rate of change of position). Acceleration is not velocity, nor is it even a change in velocity. Acceleration is the rate at which velocity itself changes.

Please remember that it took people nearly 2000 years from the time of Aristotle to reach a clear understanding of motion, so be patient with yourself if you find that you require a few hours to achieve as much!

Summary of Terms

Inertia The sluggishness or apparent resistance an object offers to changes in its state of motion.
Speed The distance traveled per time.
Velocity The speed of an object and specification of its direction of motion.
Acceleration The rate at which velocity changes with time; the change in velocity may be in magnitude or direction or both.
Free fall A state of fall free from air resistance and other forces except for gravity.

Review Questions

*Each chapter in this book concludes with a set of review questions and exercises. The **Review Questions** are designed to help you fix ideas and catch the essentials of the chapter material. You'll notice that answers to the questions can be found within the chapters. The **Exercises** stress thinking rather than mere recall of information and call for an understanding of the definitions, principles, and relationships of the chapter material. In many cases the intention of particular exercises is to help you to apply the ideas of physics to familiar situations. Unless you cover only a few chapters in your course, you will likely be expected to tackle only a few exercises for each chapter. The large number of exercises is to allow your instructor a wide choice of assignments.*

Aristotle on Motion

1. Contrast Aristotle's ideas of natural motion and violent motion.

2. Did Aristotle think that a force acts on the moon as it circles the earth? Did he think that a force on earth keeps a ball rolling along a smooth, level surface?

Copernicus and the Moving Earth

3. Contrast the relationship between the earth and sun as seen by Aristotle and by Copernicus.

Figure 2-12   Motion analysis.

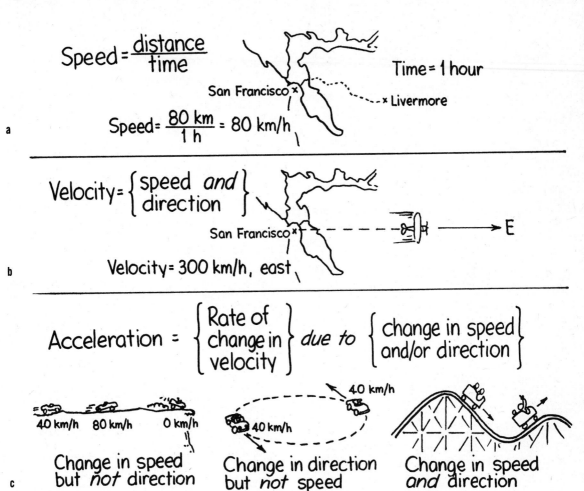

a

$$\text{Speed} = \frac{\text{distance}}{\text{time}}$$

Time = 1 hour

San Francisco ×   × Livermore

$$\text{Speed} = \frac{80 \text{ km}}{1 \text{ h}} = 80 \text{ km/h}$$

b

$$\text{Velocity} = \left\{ \begin{array}{l} \text{speed } and \\ \text{direction} \end{array} \right\}$$

San Francisco ×

→ E

Velocity = 300 km/h, east

c

$$\text{Acceleration} = \left\{ \begin{array}{l} \text{Rate of} \\ \text{change in} \\ \text{velocity} \end{array} \right\} \ due\ to\ \left\{ \begin{array}{l} \text{change in speed} \\ \text{and/or direction} \end{array} \right\}$$

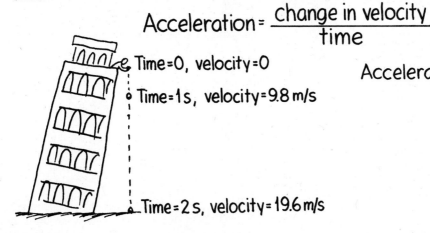

40 km/h 80 km/h 0 km/h

Change in speed
but *not* direction

40 km/h

40 km/h

Change in direction
but *not* speed

Change in speed
and direction

d

$$\text{Acceleration} = \frac{\text{change in velocity}}{\text{time}}$$

Time = 0, velocity = 0

Time = 1s, velocity = 9.8 m/s

Time = 2s, velocity = 19.6 m/s

$$\text{Acceleration} = \frac{19.6 \text{ m/s}}{2 \text{ s}}$$

$$a = 9.8 \frac{\text{m/s}}{\text{s}}$$

$$a = 9.8 \text{ m/s s}$$

$$a = 9.8 \text{ m/s}^2$$

Galileo and the Leaning Tower

4. What did Galileo discover in his legendary experiment on the Leaning Tower?

Galileo's Inclined Planes

5. What did Galileo discover in his experiments with inclined planes?

6. What does it mean to say an object has inertia? Give an example.

Description of Motion

7. What is a "time rate of change"? Give three examples.

Speed

8. What two units of measurement are necessary for describing speed?

9. Distinguish speed in general from instantaneous speed. Give an example.

10. What is the average speed of a horse that gallops a distance of 15 km in a time of 30 min?

11. How far does a horse travel if it gallops at an average speed of 25 km/h for 30 min?

Velocity

12. Distinguish between speed and velocity.

13. If a car moves with a constant velocity, does it also move with a constant speed? Explain.

14. If a car moves with a constant speed, can you say that it also moves with a constant velocity? Give an example to support your answer.

Acceleration

15. Distinguish between velocity and acceleration.

16. What is the acceleration of a car that increases its velocity from 0 to 100 km/h in 10 s?

17. What is the acceleration of a car that maintains a constant velocity of 100 km/h for 10 s? (Why do many students who correctly answer the last question get this question wrong?)

18. When does one feel the effects of velocity in a moving vehicle, when it is moving uniformly or when its motion changes?

19. Acceleration is generally defined as the time rate of change of velocity. When can it be defined as the time rate of change of speed?

Acceleration on Galileo's Inclined Planes

20. What was the relationship among velocity, acceleration, and time that Galileo discovered on his inclined planes?

21. How did Galileo tip his inclined planes to understand better the motion of falling things?

Free Fall

How Fast

22. What exactly is meant by a "freely falling" object?

23. What is the gain in speed per second for a freely falling object?

24. What is the velocity acquired by a freely falling object 5 s after being dropped from a rest position? What is it 6 s after?

25. The acceleration of free fall is about 10 m/s^2. Why does the seconds unit appear twice?

How Far

26. What relationship between distance traveled and time of travel did Galileo discover with his inclined planes?

27. What is the distance fallen for a freely falling object 5 s after being dropped from a rest position? What is it 6 s after?

How Quickly How Fast Changes

28. Consider these measurements: 10 m, 10 m/s, and 10 m/s^2. Which is a measure of distance, which of speed, and which of acceleration?

Exercises

1. A ball is rolling across the top of a billiard table and slowly rolls to a stop. How would Aristotle interpret this observation? How would Galileo interpret it?

2. Why did Galileo use inclined planes to investigate free fall?

3. Which has the greatest average speed during a game: a football, a baseball, or a golf ball?

4. What is the average speed of a jogger who jogs 2 km in 10 min?

5. One airplane travels due north at 300 km/h while another travels due south at 300 km/h. Are their speeds the same? Are their velocities the same? Explain.

6. Charlie drove his car around the block at constant velocity. True or false?

7. Cite an example of something that undergoes acceleration while moving at constant speed. Can you also give an example of something that accelerates while traveling at constant velocity? Explain.

8. Can you give an example wherein the acceleration of a body is opposite in direction to its velocity? Do so if you can.

9. If you were standing in an enclosed car moving at constant velocity, would you have to lean in some special way to compensate for the car's motion? What if the car were moving with constant acceleration? Explain.

10. On which of these hills does the ball roll down with increasing speed and decreasing acceleration? (Use this example if you wish to explain to someone the difference between speed and acceleration.)

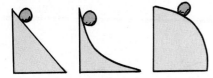

11. What is the acceleration of a car that moves at a steady velocity of 100 km/h for 100 s? Explain your answer.

12. What is the acceleration of a vehicle that changes its velocity from 100 km/h to a dead stop in 10 s?

13. One car goes from 0 to 50 km/h, and later another car goes from 0 to 60 km/h. From this information can you say which car underwent the greater acceleration? Why or why not?

14. A racing car moving at 10 m/s north increases its velocity to 16 m/s north in a time interval of 3 s. What is its average acceleration during this time interval? What is the direction of this acceleration?

15. For a freely falling object dropped from rest, what is its acceleration at the end of the 5th second of fall? The 10th second? Defend your answer.

16. Suppose that a freely falling object were somehow equipped with a speedometer. By how much would its speed reading increase with each second of fall?

17. Suppose that the freely falling object in the preceding exercise were also equipped with an odometer. Would the readings of distance fallen each second indicate equal or different falling distances for successive seconds? Explain.

18. When a ball player throws a ball straight up, by how much does the speed of the ball decrease each second while ascending? In the absence of air resistance, by how much does it increase each second while descending? How much time is required for rising compared to falling?

19. A ball is thrown straight up with an initial speed of 30 m/s. How high does it go, and how long is it in the air (neglecting air resistance)?

20. A ball is thrown with enough speed straight up so that it is in the air several seconds. (a) What is the velocity of the ball when it gets to its highest point? (b) What is its velocity 1 s before it reaches its highest point? (c) What is the change in its velocity during this 1-s interval? (d) What is its velocity 1 s after it reaches its highest point? (e) What is the change in velocity during this 1-s interval? (f) What is the change in velocity during the 2-s interval? (g) What is the acceleration of the ball during any of these time intervals and when it passes through the zero velocity point?

21. What is the instantaneous velocity of a falling object 10 s after it is released from a position of rest? What is its average velocity during this 10-s interval? How far will it go during this time?

22. A climber near the summit of a vertical cliff accidentally knocks loose a large rock. She sees it shatter at the bottom of the cliff 8 s later. What was the speed of impact? How far did the rock fall?

23. Someone standing at the edge of a cliff (as in Figure 2-9) throws a ball straight up at a certain speed and another ball straight down with the same initial speed. If air resistance is negligible, which ball will have the greater speed when it strikes the ground below?

24. If you drop an object, its acceleration toward the ground is 9.8 m/s². If you throw it down instead, would its acceleration after throwing be greater than 9.8 m/s²? Why or why not?

25. In the preceding exercise can you think of a reason why the acceleration of the object thrown downward through the air would actually be less than 9.8 m/s²?

26. If it were not for air resistance, why would it be dangerous to go outdoors on rainy days?

27. A car goes from $v = 0$ to $v = 50$ m/s in 10 s. If you wish to find the distance traveled using the equation $d = \frac{1}{2}at^2$, what value should you use for a?

28. Consider a planet where the acceleration due to gravity is 20 m/s². How fast and how far will an object at rest freely fall in 1 s? Compared to earth, how much greater and farther are the speed and distance fallen on this planet at given times?

29. Extend Tables 2-2 and 2-3 (which give values of from 0 to 5 s) to 0 to 10 s, assuming no air resistance.

30. In this chapter, we studied idealized cases of balls rolling down smooth planes and objects falling with no air resistance. Suppose a classmate complains that all this attention focused on idealized cases is valueless because idealized cases simply don't occur in the everyday world. How would you respond to this complaint? How do you suppose the author of this book would respond?

3 Nonlinear Motion

In the last chapter we discussed *linear motion*—motion along a straight line. We distinguished between two kinds of linear motion: constant velocity, such as that of a bowling ball rolling horizontally, and accelerated motion, such as vertical fall under the influence of gravity. This chapter extends these ideas to *nonlinear motion*—motion along a curved path. If you throw a ball sideways, the path it follows is a curve. In this chapter we will see that this curve is a combination of horizontal motion without acceleration and vertical motion under the acceleration of gravity. We will see that the velocity of a tossed ball at any instant is two "components" of motion combined. We will also see that these two components of motion are completely independent of each other—how the vertical part behaves doesn't depend on how the horizontal part behaves, and vice versa. To understand these ideas, we begin by studying the ideas of relative motion and the vector nature of velocity.

Motion Is Relative

In a strict sense, everything moves. Even things that appear to be at rest are moving. They move with respect to, or relative to, the sun and stars. A book that is at rest, relative to the table it lies on, is moving at about 30 kilometers per second relative to the sun. And it moves even faster relative to the center of our galaxy. When we discuss the motion of something, we describe its motion relative to something else. When we say that a space shuttle moves at 8 kilometers per second, we mean relative to the earth below. When we say an express train travels at 200 kilometers per hour, of course we mean relative to the track. Unless stated otherwise, when we discuss the speeds of things in our environment, linear or nonlinear, we mean with respect to the surface of the earth.

Consider the speed of a slow-moving airplane. If there is no wind blowing, an airplane that travels at 100 kilometers per hour relative to the air also travels at 100 kilometers per hour relative to the surface of the earth below. But if there is wind, the speed of the airplane relative to the air and its speed relative to the ground below are different. If the plane is flying in the same direction as the wind its ground speed is greater; opposite to the wind, its ground speed is less. If it flies in a crosswind direction, its speed relative to the air is likely to be different from its speed relative to the ground. These different speeds can be treated by a useful technique that involves *vectors*.

Velocity—
A Vector Quantity

Recall from Chapter 2 the difference between speed and velocity: speed is a measure of "how fast"; velocity is a measure of both how fast and "which way." If the speedometer in a car reads 100 kilometers per hour, for example, you know your speed. If there is also a compass on the dashboard indicating the car is moving due north, you know your velocity is 100 kilometers per hour north. To know your velocity is to know your speed *and* your direction.

Any quantity that requires both magnitude (the amount of, or how much) and direction (which way) for a complete description is a **vector quantity**. Thus, velocity is a vector quantity. Other examples are force and acceleration. By contrast, a quantity that can be described by magnitude only, that involves no idea of direction, is called a **scalar quantity**. Speed is a scalar quantity, as are mass and volume. So speed is a scalar quantity and velocity is a vector quantity. To avoid "information overload," we will confine ourselves in this chapter only to the vector nature of velocity.

Pictures are often much more descriptive than words. It is useful to represent a vector quantity by an arrow. Whenever the length of the arrow represents the magnitude of the quantity and the direction of the arrow represents the direction of the quantity, the arrow is called a **vector**. The vector in Figure 3-1 is scaled so that 1 centimeter represents 20 kilometers per hour; it is 3 centimeters long and points to the right, and therefore it represents a velocity of 60 kilometers per hour to the right.

Figure 3-1
This vector, scaled so that 1 cm equals 20 km/h, represents 60 km/h to the right.

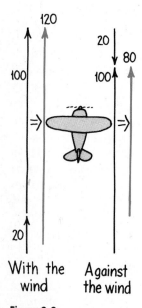

With the   Against
wind   the wind

Figure 3-2
The velocity of an airplane relative to the ground depends on its velocity relative to still air and to the wind.

Consider an airplane flying due north at 100 kilometers per hour relative to the surrounding air. We can represent this by an arrow drawn to scale—the length of the arrow corresponding to 100 kilometers per hour and its direction corresponding to the direction north. Suppose that there is a tailwind (wind from behind) that also moves due north at a velocity of 20 kilometers per hour. This example is represented with vectors in Figure 3-2 left. Here the velocity vectors are scaled so that 1 centimeter represents 20 kilometers per hour. Thus, the 100-kilometers-per-hour velocity of the airplane is shown by the 5-centimeter-long vector and the 20-kilometers-per-hour tailwind is shown by the 1-centimeter-long vector. You can see (with or without the vectors) that the resulting velocity (in color) is going to be 120 kilometers per hour. Without the tailwind, the airplane would travel 100 kilometers in 1 hour relative to the ground below. With the tailwind, it would travel 120 kilometers in 1 hour.

Suppose, instead, that the wind is a headwind (wind head-on), so the airplane flies into the wind rather than with the wind. Now the velocity vectors are in opposite directions (Figure 3-2 right). The result is 100 kilometers per hour minus 20 kilometers per hour, which equals 80 kilometers per hour (in color). Flying against a 20-kilometers-per-hour headwind, the airplane would travel only 80 kilometers relative to the ground in 1 hour.

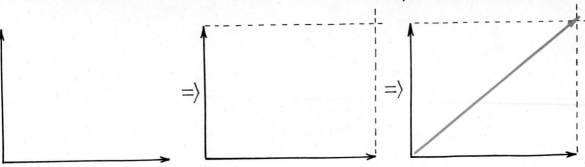

Figure 3-3
The pair of vectors at right angles to each other make two sides of a rectangle, the diagonal of which is their resultant.

Adding vectors that act along parallel directions is simple enough: if they are in the same direction, they add; if they are in opposite directions, they subtract. The sum of two or more vectors is called their **resultant**. To find the resultant of two vectors that are at angles to each other, we use the *parallelogram rule*.* Construct a parallelogram wherein the two vectors are adjacent sides; the diagonal of the parallelogram shows the resultant. This is shown in Figure 3-3, where the parallelogram is a rectangle. (In this chapter we'll only treat vectors that are either parallel or that make rectangles. A more general treatment of vectors is in Appendix III at the end of the book.)

Question ▶ Consider a motorboat that normally travels 10 km/h in still water. If the boat travels in a river that flows also at a rate of 10 km/h, what will be its velocity relative to the shore when it heads directly upstream? When it heads directly downstream?

With the vector technique we can correct for the effect of a crosswind on the velocity of an airplane. Consider a slow-moving airplane that flies north at 80 kilometers per hour and is caught in a strong crosswind of 60 kilometers per hour blowing east. Figure 3-4 shows vectors for the airplane velocity and wind velocity. The scale here is 1 centimeter to 20 kilometers per hour. The diagonal of the constructed parallelogram (a

▶ **Answer**
When the boat heads upstream its velocity is zero relative to the land (a velocity of $+10$ added to a velocity of -10 equals zero). When the boat heads directly downstream its velocity is 20 km/h in the direction of river flow (a velocity of $+10$ added to a velocity of $+10$ equals $+20$ in the same direction).

*A parallelogram is a four-sided plane figure having the opposite sides parallel and equal.

Figure 3-4
An 80-km/h airplane flying in a 60-km/h crosswind has a resultant speed of 100 km/h relative to the ground.

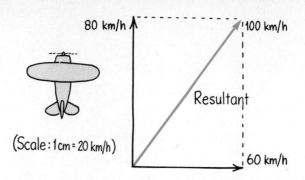

Figure 3-5
The diagonal of a square is $\sqrt{2}$ the length of one of its sides.

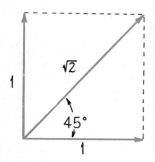

rectangle in this case) measures 5 centimeters, which represents 100 kilometers per hour. So the airplane moves at 100 kilometers per hour relative to the ground, in a direction between north and northeast.*

There is a special case of the parallelogram that often occurs. When two vectors that are equal in magnitude and at right angles to one another are to be added, the parallelogram becomes a square. Since for any square the length of a diagonal is $\sqrt{2}$, or 1.414, times one of the sides, the resultant is $\sqrt{2}$ times one of the vectors. For example, the resultant of two equal vectors of magnitude 100 acting at right angles to each other is 141.4.

We can look at the vector configuration of Figure 3-5 from another point of view. Just as a vertical and a horizontal vector combine to produce a single resultant vector at some angle between them, a single vector at any angle can be resolved into a pair of vectors that are at right angles to each other. This pair of vectors are the *components* of the single vec-

Question ▶ Again consider the motorboat that normally travels 10 km/h in still water. If the boat heads directly across the river, which also flows at a rate of 10 km/h, what will be its velocity relative to the shore?

---

*Whenever the vectors are at right angles to each other, their resultant can be found by the Pythagorean theorem, a well-known tool of geometry. It states that the square of the hypotenuse of a right-angle triangle is equal to the sum of the squares of the other two sides. Note that two right triangles are present in the parallelogram (rectangle in this case) in Figure 3-4. From either one of these triangles we get:

$$\text{Resultant}^2 = (60 \text{ km/h})^2 + (80 \text{ km/h})^2$$
$$= (3600 \text{ km/h})^2 + (6400 \text{ km/h})^2$$
$$= (10\ 000 \text{ km/h})^2$$

The square root of $(10\ 000 \text{ km/h})^2$ is 100 km/h, as expected.

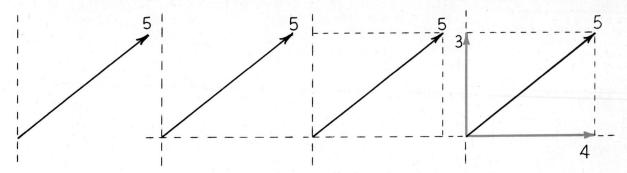

Figure 3-6
Construction of the vertical and horizontal components of a vector.

tor. This is illustrated in Figure 3-6 for a vector that is 37 degrees from the horizontal. We can resolve it into horizontal and vertical components. What would be the magnitude of each component? For this special case of 37 degrees, the horizontal is 4 and the vertical is 3. Any vector may be represented by horizontal and vertical components. Figure 3-7, as an example, shows the horizontal and vertical velocity components of a stone that is thrown into the air. Its initial path is greater than 45 degrees. Notice that the vertical component of velocity is greater than the horizontal component. Because of gravity, the vertical component of velocity will decrease as the stone rises and then increase when the stone returns to ground level. So the path of the stone is curved.

Figure 3-7
Vertical and horizontal components of the stone's velocity.

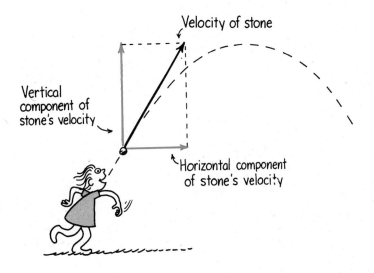

▶ **Answer**
When the boat heads cross-stream (at right angles to the river flow), its velocity is 14.14 km/h, 45° downstream (in accord with the diagram in Figure 3-5).

Projectile Motion

Figure 3-8
(a) Drop a ball, and it accelerates downward and covers a greater vertical distance each second. (b) Roll it along a level surface, and its velocity is constant because no component of gravitational force acts horizontally.

A **projectile** is any object that is projected by some means and continues in motion by its own inertia. A stone thrown into the air, a cannonball shot from a cannon, and a ball that rolls off the edge of a table are all projectiles. These projectiles follow curved paths that at first thought seem rather complicated. However, these paths will seem surprisingly simple when we look at the horizontal and vertical components of motion separately.

The horizontal component of motion for a projectile is no more complex than the horizontal motion of a bowling ball rolling freely along a level bowling alley. If the retarding effect of friction can be ignored, the bowling ball moves at constant velocity. Since there is no force acting horizontally on the ball, it rolls of its own inertia and covers equal distances in equal intervals of time. It rolls without accelerating. The horizontal part of a projectile's motion is just like the bowling ball's motion along the alley (Figure 3-8b).

b

The vertical component of motion for a projectile following a curved path is just like the motion described in Chapter 2 for a freely falling object. Like a ball dropped in mid-air, the projectile moves in the direction of earth gravity and accelerates downward (Figure 3-8a). The increase in speed in the vertical direction causes successively greater distances to be covered in each successive equal-time interval.

Interestingly enough, the horizontal component of motion for a projectile is completely independent of the vertical component of motion. Unless air drag or some other horizontal force acts, the constant horizontal velocity component is not affected by the vertical force of gravity. It is important to understand that each component acts independently of the other. Their combined effects produce the curved paths that projectiles follow.

These ideas are neatly illustrated in the simulated multiple-flash exposure in Figure 3-9, which shows equally timed successive positions for a ball rolled off a horizontal table. Investigate the figure carefully, for there's a lot of good physics there. The curved path of the ball is best analyzed by considering the horizontal and vertical components of motion separately. There are two important things to notice. The first is that the ball's horizontal component of motion doesn't change as the falling ball moves sideways. The ball travels the same horizontal distance in the equal times between each flash. That's because there is no component of gravitational force acting horizontally. Gravity acts only downward, so the only acceleration of the ball is downward. The second thing to note from

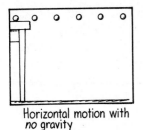

Horizontal motion with *no* gravity

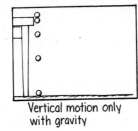

Vertical motion only with gravity

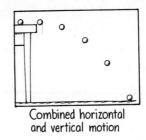

Combined horizontal and vertical motion

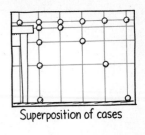

Superposition of cases

Figure 3-9
Simulated photographs of a moving ball illuminated with a strobe light.

the figure is that the vertical positions become farther apart with time. The distances traveled vertically are the same as if the ball were simply dropped. It is interesting to note that the downward motion of the ball is the same as that of free fall. Figure 3-10 is an actual strobe-light photograph of two balls released simultaneously.

The path traced by a projectile that accelerates only in the vertical direction while moving at a constant horizontal velocity is a **parabola**. When air resistance can be neglected—usually for slow-moving projectiles or ones very heavy compared to the forces of air resistance—the curved paths are parabolic.

Consider a cannonball shot at an upward angle (Figure 3-11). Pretend for a moment that there is no gravity; according to the law of inertia, the cannonball would follow the straight-line path shown by the dashed line. But there is gravity, so this doesn't happen. What really happens is that the cannonball continually falls beneath the imaginary line until it

Figure 3-10
A strobe-light photograph of two golf balls released simultaneously from a mechanism that allows one ball to drop freely while the other is projected horizontally.

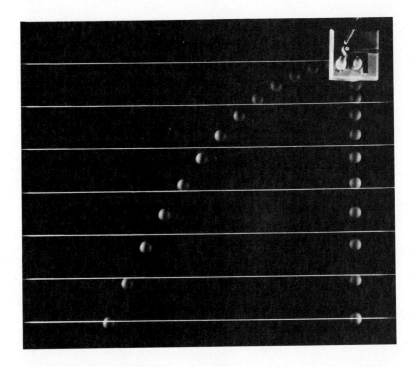

finally strikes the ground. Get this: The vertical distance it falls beneath any point on the dashed line is the same vertical distance it would fall if it were dropped from rest and had been falling for the same amount of time. This distance, as introduced in Chapter 2, is given by $d = \frac{1}{2}gt^2$, where t is the elapsed time.

We can put it another way: Shoot a projectile skyward at some angle and pretend there is no gravity. After so many seconds t, it should be at a certain point along a straight-line path. But because of gravity, it isn't. Where is it? The answer is that it's directly below this point. How far below? The answer in meters is $5t^2$ (or, more accurately, $4.9t^2$). How about that!

Question ▶

At the instant a horizontally held rifle is fired over a level range, a bullet held at the side of the rifle is released and drops to the ground. Which bullet, the one fired downrange or the one dropped from rest, strikes the ground first?

Figure 3-11
With no gravity, the projectile would follow a straight-line path (dashed line). But because of gravity, it falls beneath this line the same vertical distance it would fall if released from rest. Compare the distances fallen with those given in Table 2-3 in Chapter 2. (With $g = 9.8\ \text{m/s}^2$, these distances are more accurately 4.9 m, 19.6 m, and 44.1 m.)

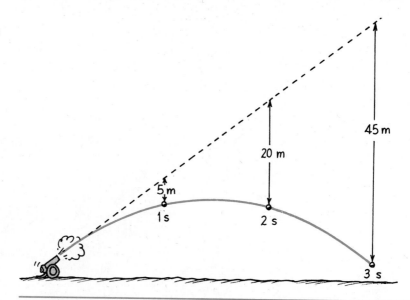

▶ **Answer**

Both bullets fall the same vertical distance with the same acceleration g due to gravity and therefore strike the ground at the same time. Can you see that this is consistent with our analysis of Figures 3-9 and 3-10? We can reason this another way by asking which bullet would strike the ground first if the rifle were pointed at an upward angle. In this case, the bullet that is simply dropped would hit the ground first. Now consider the case where the rifle is pointed downward. The fired bullet hits first. So upward, the dropped bullet hits first; downward, the fired bullet hits first. There must be some angle at which there is a dead heat—where both hit at the same time. Can you see it would be when the rifle is neither pointing upward nor downward—when it is horizontal?

Note another thing from Figure 3-11. The cannonball moves equal horizontal distances in equal time intervals. That's because no acceleration takes place horizontally. The only acceleration is vertically, in the direction of earth's gravity. The vertical distance it falls below the imaginary straight-line path during equal time intervals continually increases with time.

Questions ▶

1. Suppose the cannonball in Figure 3-11 were fired faster. How many meters below the dashed line would it be at the end of 5 s?

2. If the horizontal component of the cannonball's velocity were 20 m/s, how far downrange would the cannonball be at the end of 5 s?

In Figure 3-12 we see vectors representing both horizontal and vertical components of velocity for a projectile following a parabolic path. Notice that the horizontal component is everywhere the same, and only the vertical component changes. Note also that the actual velocity is represented by the vector that forms the diagonal of the rectangle formed by the vector components. At the top of the trajectory the vertical component vanishes to zero, so the actual velocity there is the horizontal component of velocity at all other points. Everywhere else the magnitude of velocity is greater (just as the diagonal of a rectangle is greater than either of its sides).

Figure 3-12
The velocity of a projectile at various points along its path. Note that the vertical component changes and the horizontal component is the same everywhere.

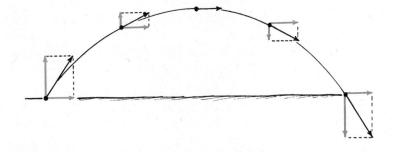

▶ **Answers**

1. Assuming $g = 10$ m/s^2, the vertical distance beneath the dashed line at the end of 5 s would be 125 m [$d = 5t^2 = 5(5)^2 = 5(25) = 125$]. (If we use $g = 9.8$ m/s^2, then this distance is 122.5 m/s). Interestingly enough, this distance doesn't depend on the angle of the cannon. If air resistance is neglected, any projectile will fall $5t^2$ meters below where it would have reached if there were no gravity.

2. In the absence of air resistance, the cannonball will travel a horizontal distance of 100 m [$d = vt = (20)(5) = 100$]. Note that since gravity acts only vertically and there is no acceleration in the horizontal direction, the cannonball travels equal horizontal distances in equal times. This distance is simply its horizontal component of velocity multiplied by the time (and not $5t^2$, which applies only to vertical motion under the acceleration of gravity).

Figure 3-13
Path for a steeper projection angle.

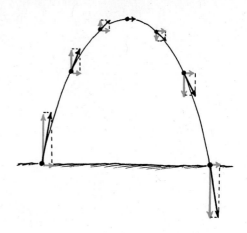

Figure 3-14
Maximum range is attained when a ball is batted at an angle of nearly 45°. (For common cases in sports in which the weight of the projectile is significant in comparison to the force of the projection, the applied force does not produce the same velocity for different projection angles, and maximum range occurs for angles less than 45°.)

Figure 3-13 shows the path traced by a projectile with the same launching speed at a steeper angle. Notice the initial velocity vector has a greater vertical component than if the projection angle is less. This greater component results in a higher trajectory. But the horizontal component is less, so the range is less.

Figure 3-15 shows the paths of several projectiles, all with the same initial speed but different projection angles. The figure neglects the effects of air resistance, so the paths are all parabolas. Notice that these projectiles reach different *altitudes*, or heights above the ground. They also have different *horizontal ranges*, or distances traveled horizontally. The remarkable thing to note from Figure 3-15 is that the same range is obtained from two different projection angles—angles that add up to 90 degrees! An object thrown into the air at an angle of 60 degrees, for example, will have the same range as if it were thrown at the same speed at an angle of 30 degrees. For the smaller angle, of course, the object remains in the air for a shorter time.

Figure 3-15
Ranges of a projectile shot at the same speed at different projection angles.

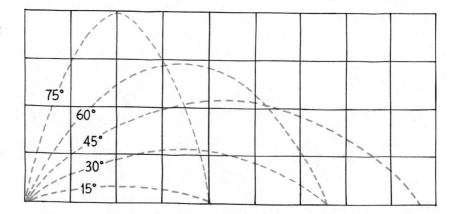

Questions ▶

1. A projectile is shot at an angle into the air. If air resistance is negligible, what is the acceleration of its vertical component of motion? Of its horizontal component of motion?

2. At what part of its trajectory does a projectile have minimum speed?

Figure 3-16
In the presence of air resistance, the trajectory of a high-speed projectile falls short of a parabolic path.

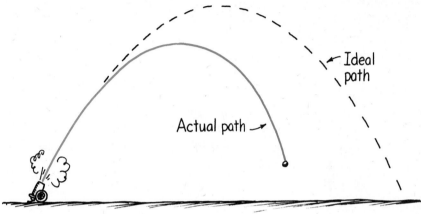

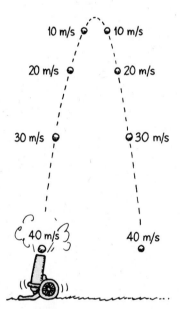

Figure 3-17
Without air resistance, speed lost while going up equals speed gained while coming down; time going up equals time coming down.

We have emphasized the special case of projectile motion without air resistance. When there is air resistance, the range of the projectile is shorter, the altitude of the projectile is less, and the path is not a true parabola (Figure 3-16).

When air resistance is small enough to be negligible, a projectile will rise to its maximum height in the same time it takes to fall from that height to the ground (Figure 3-17). This is because its deceleration by gravity while going up is the same as its acceleration by gravity while coming down. The speed it loses while going up is therefore the same as the speed it gains while coming down. So the projectile arrives at the ground with the same speed it had when it was projected from the ground.

▶ **Answers**

1. The acceleration in the vertical direction is g because the force of gravity is along the vertical direction. (We will see in the following chapter that acceleration is always in the direction of the force that acts on an object.) The acceleration is zero in the horizontal direction because no horizontal force acts on the projectile.

2. The speed of a projectile is minimum at the top of its path. If it is launched vertically, its speed at the top is zero. If it is projected at an angle, the vertical component of speed is zero at the top, leaving only the horizontal component. So the speed at the top is equal to the horizontal component of the projectile's velocity at any point. Isn't that neat?

Question ▶ The boy on the tower throws a ball 20 m downrange as shown in Figure 3-18. What is his pitching speed?

Figure 3-18
How fast is the ball thrown?

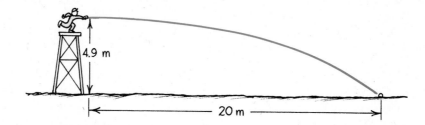

Baseball games normally take place on level ground. For the short-range projectile motion on the playing field, the earth can be considered to be flat because the flight of the baseball is not affected by the earth's curvature. For very long range projectiles, however, the curvature of the earth's surface must be taken into account. We'll now see that if an object is projected fast enough, it will fall all the way around the earth and be an earth satellite.

Fast-Moving Projectiles—Satellites

Consider the ball thrown from the cliff in Figure 3-19. If gravity did not act on the ball, the ball would follow a straight-line path shown by the dashed line. But there is gravity, so the ball falls below this straight-line path. In fact, 1 second after the ball leaves the pitcher's hand it will have fallen a vertical distance of 4.9 meters below the dashed line—whatever the pitching speed. It is important to understand this, for it is the crux of satellite motion.

Figure 3-19
Throw a stone at any speed and 1 s later it will have fallen 4.9 m below where it would have been without gravity.

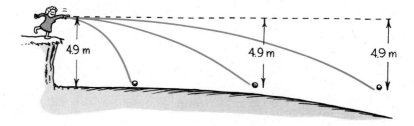

▶ **Answer**

The ball is thrown horizontally, so the pitching speed is horizontal distance divided by time. A horizontal distance of 20 m is given, but the time is not stated. However, while the ball is moving horizontally at constant speed, it is falling under gravity a vertical distance of 4.9 m, which takes 1 s. So pitching speed $v = d/t = (20 \text{ m})/(1 \text{ s})$ = 20 m/s. It is interesting to note that consideration of the equation for constant speed, $v = d/t$, guides thinking about the crucial factor in this problem—the time.

Figure 3-20
Earth's curvature—not to scale!

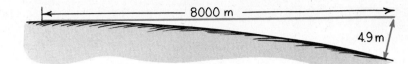

8000 m

4.9 m

An earth satellite is simply a projectile that falls *around* the earth rather than *into* it. The speed of the satellite must be great enough to ensure that its falling distance matches the earth's curvature.* A geometrical fact about the curvature of our earth is that its surface drops a vertical distance of 4.9 meters for every 8000 meters tangent to the surface (Figure 3-20).† This means that if you were floating in a calm ocean, you would be able to see only the top of a 4.9-meter mast on a ship 8 kilometers away. So if a baseball could be thrown fast enough to travel a horizontal distance of 8 kilometers during the time (1 second) it takes to fall 4.9 meters, then it would follow the curvature of the earth. A little thought will show that this speed is 8 kilometers per second. If this doesn't seem fast, convert it to kilometers per hour and you get an impressive 29 000 kilometers per hour (or 18 000 miles per hour)!

Figure 3-21
If the speed of the stone and the curvature of its trajectory are great enough, the stone may become a satellite.

At this speed, atmospheric friction would burn the baseball or even a piece of iron to a crisp. This is the fate of grains of sand and other meteorites that graze the earth's atmosphere and burn up, appearing as "falling stars." That is why satellites such as the space shuttles are launched to altitudes of 150 kilometers or so. A common misconception is that satellites orbiting at high altitudes are free from gravity. Nothing could be further from the truth. We will see later that the force of gravity on a satellite 150 kilometers above the earth's surface is nearly as great as at the surface. The high altitude is to put the satellite beyond the earth's atmosphere, not beyond the earth's gravity. We'll return to satellite motion and gravity in Chapters 8 and 9.

Figure 3-22
The space shuttle is a projectile in a constant state of free fall. Because of its tangential velocity, it falls around the earth rather than vertically into it.

---

*The conventional definition of *to fall* is "to get closer to the earth"; satellites such as the moon do not do this. In science we will find many cases where the technical definition differs from the conventional. For example, we say "the sun rises" and "the moon sets," but technically they do not.

†A tangent to a circle or to the earth's surface is a straight line that touches the circle or surface at one place, so it is parallel to the circle at the point of contact.

Isaac Newton (1642–1727)

Isaac Newton was born prematurely and barely survived on Christmas Day, 1642, the same year that Galileo died. Newton's birthplace was his mother's farmhouse in Woolsthorpe, England. His father died several months before his birth, and he grew up under the care of his mother and grandmother. As a child he showed no particular signs of brightness, and at the age of $14\frac{1}{2}$ he was taken out of school to work on his mother's farm. As a farmer he was a failure, preferring to read books he borrowed from a neighboring druggist. An uncle sensed the scholarly potential in young Isaac and prompted him to study at the University of Cambridge, which he did for 5 years, graduating without particular distinction.

A plague swept through London, and Newton retreated to his mother's farm—this time to continue his studies. At the farm, at the age of 23, he laid the foundations for the work that was to make him immortal. Seeing an apple fall to the ground led him to consider the force of gravity extending to the moon and beyond, and he formulated the law of universal gravitation (which he later proved); he invented the calculus, an indispensable mathematical tool in science; he extended Galileo's work and formulated the three fundamental laws of motion; and he formulated a theory of the nature of light and showed with prisms that white light is composed of all colors of the rainbow. It was his experiments with prisms that first made him famous.

When the plague subsided, Newton returned to Cam-

Circular Motion

Which moves faster on a merry-go-round, a horse near the outside rail or a horse near the inside rail? Which part of a phonograph record or a compact disc passes faster beneath the pickup—the part at the beginning of the disc, near the outer edge, or a later part near the center? Ask different people these questions, and you'll get different answers. That's because it's easy to get linear speed confused with rotational speed.

Linear speed is what we have been calling simply *speed*—the distance in meters or kilometers moved per unit of time. A point on the outside of a merry-go-round or record moves a greater distance in one complete rotation than a point on the inside. The linear speed is greater on the outside of a rotating object than inside and closer to the axis.

Rotational speed (often called *angular speed*) refers to the number of rotations per unit of time. All parts of a rigid merry-go-round, phono-

bridge and soon established a reputation for himself as a first-rate mathematician. His mathematics teacher resigned in his favor and Newton was appointed the Lucasian professor of mathematics. He held this post for 28 years. In 1672 he was elected to the Royal Society, where he exhibited the world's first reflector telescope. It can still be seen, preserved at the library of the Royal Society in London with the inscription: "The first reflecting telescope, invented by Sir Isaac Newton, and made with his own hands."

It wasn't until Newton was 42 that he began to write what is generally acknowledged as the greatest scientific book ever written, the *Principia Mathematica Philosophiae Naturalis*. He wrote the work in Latin and completed it in 18 months. It appeared in print in 1687 and wasn't printed in English until 1729, 2 years after his death. When asked how he was able to make so many discoveries, Newton replied that he found his solutions to problems not by sudden insight but by continually thinking very long and hard about them until he worked them out.

At the age of 46, his energies turned somewhat from science when he was elected a member of Parliament. He attended the sessions in Parliament for 2 years and never gave a speech. One day he rose and the House fell silent to hear the great man. Newton's "speech" was very brief; he simply requested that a window be closed because of a draft.

A further turn from his work in science was his appointment as warden and then as master of the mint. Newton resigned his professorship and directed his efforts toward greatly bettering the workings of the mint, to the dismay of counterfeiters who flourished at that time. He maintained his membership in the Royal Society, and was elected president and re-elected each year for the rest of his life. At the age of 62, he wrote *Opticks*, which summarized his work on light. Nine years later he wrote a second edition to his *Principia*.

Although Newton's hair turned gray at 30, it remained full, long, and wavy all his life, and unlike others in his time he did not wear a wig. He was a modest man, very sensitive to criticism, and never married. He remained healthy in body and mind into old age. At 80, he still had all his teeth, his eyesight and hearing were sharp, and his mind was alert. In his lifetime he was regarded by his countrymen as the greatest scientist who ever lived. In 1705 he was knighted by Queen Anne. Newton died at the age of 85 and was buried in Westminster Abbey along with England's kings and heroes.

Newton showed that the universe ran according to natural laws that were neither capricious nor malevolent—a knowledge that provided hope and inspiration to scientists, writers, artists, philosophers, and people of all walks of life, and that ushered in the Age of Reason. The ideas and insights of Isaac Newton truly changed the world and elevated the human condition.

graph record, or compact disc circle the axis of rotation in the same amount of time. All parts share the same rate of rotation, or number of rotations per unit of time. We express rotational rates in revolutions per minute (RPM).* A phonograph record, for example, rotates at $33\frac{1}{3}$ RPM. A ladybug sitting on the surface of the record revolves at $33\frac{1}{3}$ RPM.

---

*There is a distinction between a rotation and a revolution. When an object turns about an axis located within the object, like a skater doing a pirouette or a spinning phonograph record, the motion is a *rotation*. When an object turns about an external axis, like the earth orbiting the sun or the ladybug riding on the record surface, the rotational motion is called a *revolution*. The phonograph record *rotates* around its axis, while a ladybug sitting at its edge *revolves* around the same axis. Physics types usually describe the rate of rotation in terms of the angle turned in a unit of time. Then the symbol for rotational speed is ω (the Greek letter omega).

Figure 3-23

The entire phonograph record rotates at the same rotational speed, but ladybugs at different distances from the center travel at different linear speeds. A ladybug sitting twice as far from the center moves twice as fast.

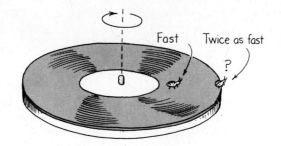

Linear speed and rotational speed are related. Have you ever ridden on a giant rotating round platform in an amusement park? The faster it turns, the faster is your linear speed. This makes sense; the greater the RPMs, the faster your speed in meters per second. More exactly, there is a direct proportion between linear speed and rotational speed. If, for example, you double the RPMs, you double your linear speed; triple the RPMs and you triple your linear speed. We say that linear speed is *directly proportional* to rotational speed.

Linear speed, unlike rotational speed, depends on the distance from the axis. At the very center of the rotating platform, you have no speed at all; you merely rotate. But as you approach the edge of the platform you find yourself moving faster and faster. Linear speed is directly proportional to distance from the axis. Move out twice as far from the rotational axis at the center and you move twice as fast. Move out three times as far and you have three times as much linear speed. If you find yourself in any rotating system whatever, your linear speed depends on how far you are from the axis of rotation. When a row of people locked arm in arm at the skating rink make a turn, the motion of "tail-end Charlie" is evidence of this greater speed.

So linear speed is directly proportional to both rotational speed and radial distance.*

Figure 3-24

The linear speed of each person equals the rotational speed of the platform multiplied by the distance from the central axis.

Question ▶

On a rotating platform like that shown in Figure 3-24, if you sit halfway between the rotating axis and the outer edge and have a rotational speed of 20 RPM and a linear speed of 2 m/s, what will be the rotational and linear speeds of your friend who sits at the outer edge?

▶ **Answer**

Since the rotating platform is rigid, all parts have the same rotational speed, so your friend also rotates at 20 RPM. Linear speed is a different story; since she is twice as far from the axis of rotation, she moves twice as fast—4 m/s.

---

*If you take a follow-up physics course, you will learn that when the proper units are used for linear speed v, rotational speed ω, and radial distance r, the direct proportion of v to ω becomes the exact equation $v = r\omega$.

Summary of Terms

Vector quantity A quantity that has both magnitude and direction. Examples are force, velocity, acceleration, and momentum.

Scalar quantity A quantity that has magnitude but not direction. Examples are mass, volume, and speed.

Vector An arrow drawn to scale used to represent a vector quantity.

Resultant The net result of a combination of two or more vectors.

Projectile Any object that is projected by some force and continues in motion by virtue of its own inertia.

Parabola The curved path followed by a projectile under the influence of gravitational attraction only.

Linear speed The distance moved per unit of time. Also called simply *speed*.

Rotational speed The number of rotations or revolutions per unit of time; often measured in rotations or revolutions per second or per minute.

Review Questions

1. Distinguish between linear motion and nonlinear motion.

Motion Is Relative

2. What is meant by saying that motion is relative?

3. If you walk at 1 m/s down the aisle of a bus that is moving at 10 m/s along the road, how fast are you moving relative to the road when you walk toward the front of the bus? Toward the rear of the bus?

Velocity—A Vector Quantity

4. What is a vector quantity? Give three examples.

5. What is a scalar quantity? Give three examples.

6. An airplane travels 200 km/h through the air. If it heads directly against a 40 km/h wind, what is its speed relative to the ground below?

7. Is a rectangle a parallelogram? Is a square a parallelogram?

8. If an airplane that travels 100 km/h through the air is caught in a 100-km/h crosswind, what will be its velocity relative to the ground?

9. If the resultant vector in Figure 3-5 has a magnitude of 141.4, what are the magnitudes of its horizontal and vertical components?

10. A rock is thrown upward at an angle. What happens to the vertical component of its velocity as it rises? As it falls?

Projectile Motion

11. What exactly is a projectile? Give three examples.

12. A rock is thrown upward at an angle. What happens to the horizontal component of its velocity as it rises? As it falls?

13. True or false: When air resistance does not affect the motion of a projectile, it covers equal horizontal distances in equal time intervals.

14. True or false: When air resistance does not affect the motion of a projectile, its horizontal and vertical components of velocity remain constant.

15. True or false: Changes in the vertical component of velocity for a projectile are proportional to changes in the horizontal component of velocity.

16. A projectile falls beneath the straight-line path it would follow if there were no gravity. How many meters does it fall below this line if it has been traveling for 1 s? For 2 s?

17. Does your answer to the last question depend on the angle at which the projectile is launched?

18. Why does the vertical component of velocity for a projectile change with time, whereas the horizontal component of velocity doesn't?

19. A projectile is launched upward at an angle of 75° from the horizontal and strikes the ground a certain distance downrange. What other angle of launch would produce the same distance?

20. A projectile is launched vertically at 100 m/s. If air resistance can be neglected, at what speed will it return to its initial level?

Fast-Moving Projectiles—Satellites

21. How can a projectile "fall around the earth"?

22. Why will a projectile that moves horizontally at 8 km/s follow a curve that matches the curvature of the earth?

23. Why is it important that the projectile in the last question be above the earth's atmosphere?

Circular Motion

24. Distinguish between linear speed and rotational speed.

25. If the rotational speed of a platform is doubled, how does the linear speed anywhere on the platform change?

26. If the rotational speed on a platform is unchanged but your distance from the center is doubled, what happens to your linear speed?

Exercises

1. Suppose you swim in a direction directly across a flowing river and end up a distance downstream that is less than the width of the river. How does your swimming speed compare to the flow rate of the river?

2. What is the ground velocity of an airplane that has an airspeed of 120 km/h when it is in a 90-km/h crosswind?

3. Why does vertically falling rain make slanted streaks on the side windows of a moving automobile? If the streaks make an angle of 45°, what does this tell you about the relative speed of the car and the falling rain?

4. Suppose you are driving along in an open car and throw a ball straight up into the air. While the ball is still in the air you step on the brakes. Where does the ball land relative to the car?

5. If you are standing in a bus that moves at constant velocity and drop a ball from your outstretched hand, you'll see its path as a vertical straight line. How will the path appear to a friend standing at the side of the road?

6. Suppose you drop an object from an airplane traveling at constant velocity, and further suppose that air resistance doesn't affect the falling object. What will be its falling path as observed by someone at rest on the ground, not directly below but off to the side where a clear view can be seen? What will be the falling path as observed by you looking downward from the airplane? Where will the object strike the ground, relative to you in the airplane? Where will it strike in the more realistic case where air resistance does affect the fall?

7. How does the downward component of the motion of a projectile compare to the motion of vertical free fall?

8. If you throw a rock off a cliff, how do the horizontal components of its motion compare for all points along its trajectory?

9. How far below an initial straight-line path will a projectile fall in 1 s? Does your answer depend on the angle of launch or on the initial speed of the projectile? Defend your answer.

10. Suppose you are on a ledge in the dark and wish to estimate the height of the ledge above the ground. So you drop a stone over the edge and hear it strike the ground in 1 s. What is the height of the ledge? How could you use the stone to estimate the height if you weren't close enough to the edge to simply drop the stone? Explain.

11. A friend claims that bullets fired by some high-powered rifles travel for many meters in a straight-line path without dropping. Another friend disputes this claim and states that all bullets from any rifle drop beneath a straight-line path a vertical distance given by $\frac{1}{2} gt^2$ and that the curved path is apparent at low velocities and less apparent at high velocities. Now it's your turn: Will all bullets drop the same vertical distance in equal times? Explain.

12. At what angle should you hold a garden hose so that the stream of water will go farthest?

13. A park ranger shoots a monkey hanging from a branch of a tree with a tranquilizing dart. The ranger aims directly at the monkey, not realizing that the dart will follow a parabolic path and thus fall below the monkey. The monkey, however, sees the dart leave the gun and lets go of the branch to avoid being hit. Will the monkey be hit anyway? Does the velocity of the 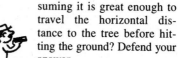 dart affect your answer, assuming it is great enough to travel the horizontal distance to the tree before hitting the ground? Defend your answer.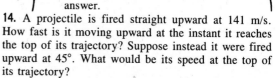

14. A projectile is fired straight upward at 141 m/s. How fast is it moving upward at the instant it reaches the top of its trajectory? Suppose instead it were fired upward at 45°. What would be its speed at the top of its trajectory?

15. Neglecting air resistance, if you throw a baseball at 20 m/s to your friend on first base, will the catching speed be greater than, equal to, or less than 20 m/s? How about if air resistance is a factor?

16. A bullet is fired horizontally with an initial velocity of 250 m/s from a tower that is 20 m high. How long will the bullet be in the air? If air resistance is neglected, how far does the bullet travel before striking the ground?

17. An airplane is flying horizontally with a speed of 1000 km/h (280 m/s) when an engine falls off. Neglecting air resistance, if it takes 30 s for the engine to hit the ground, how far horizontally does the engine travel while it falls? If the airplane somehow continues to fly as if nothing had happened, how does this distance compare to the horizontal distance the airplane travels during the same time (again assuming no air resistance)?

18. Assuming no air resistance, why does an 8-km/s horizontally fired projectile not strike the earth's surface?

19. Acceleration due to gravity on the surface of the moon is about $\frac{1}{6}$ of that on the earth's surface. The moon is about $\frac{1}{3}$ the size of the earth. Would you expect the speed of a satellite in close moon orbit to be greater than, less than, or about equal to orbital speed about the planet earth? Explain.

20. Since the moon is gravitationally attracted to the earth, why doesn't it simply crash into the earth?

21. A ladybug sits halfway between the axis and the edge of a phonograph record. What will happen to its linear speed if the RPM rate is doubled? At this doubled rate, what will happen to its linear speed if it sits at the edge?

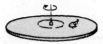

22. Does a phonograph needle ride faster or slower over the groove at the beginning or the end of the record? If fidelity increases with speed, what part of the record produces the highest fidelity?

23. Consider a bicycle that has wheels with a circumference of 2 m. What is the linear speed of the bicycle when the wheels rotate at 1 revolution per second?

24. What is the linear speed of a passenger on a Ferris wheel that has a radius of 10 m and rotates once in 30 s?

25. If you use large-diameter tires on your car, how will your speedometer reading differ?

Please let your primary goal in learning physics be acquainting yourself with the information and insights within the chapters, and *not* primarily in doing the exercises, which are intended as some mental pushups to try *after* you have studied the chapter material!

4 Newton's Laws of Motion

In 1642, several months after Galileo died, Isaac Newton was born. By the time Newton was 23, he developed his famous laws of motion, which completed the overthrow of the Aristotelian ideas that had dominated the thinking of the best minds for 2000 years. In this chapter we will consider these laws in order. The first is a restatement of the concept of inertia as proposed earlier by Galileo. The second law relates acceleration to the cause that produces it, force. The third is the well-known law of action and reaction. These three important laws first appeared in one of the most important books of all time, Newton's famous *Principia*.

Newton's First Law of Motion

Aristotle's idea that a moving object must have a force exerted on it was completely turned around by Galileo, who stated that in the *absence* of a force, a moving object will continue moving. The tendency of things to resist changes in motion was what Galileo called *inertia*. Newton refined Galileo's idea and made it his first law, appropriately called the **law of inertia**. From Newton's *Principia*:

> **Law 1: Every material object continues in its state of rest, or of uniform motion in a straight line, unless it is compelled to change that state by forces impressed upon it.**

The key word in this law is *continues*: an object *continues* to do whatever it happens to be doing unless a force is exerted upon it. If it is at rest, it *continues* in a state of rest. If it is moving, it *continues* to move without turning or changing its speed. In short, the law says that an object does not accelerate of itself; acceleration must be imposed against the tendency of an object to retain its state of motion. Things at rest tend to stay at rest—things moving tend to continue moving. This tendency of things to resist changes in motion is **inertia**.

Mass

Every material object possesses inertia; how much depends on the amount of matter in the substance of the object—the more matter, the more inertia. In speaking of how much matter something has, we use the term *mass*. The greater the mass of an object, the greater its inertia. **Mass** is a measure of the inertia of a material object.

51

Mass corresponds to our intuitive notion of **weight**. We say something has a lot of matter if it is heavy. That's because we are accustomed to measuring the quantity of matter in things by their gravitational attraction to the earth. But mass is more fundamental than weight; it is a fundamental quantity that completely escapes the notice of most people. There are times, however, when weight corresponds to our unconscious notion of inertia. For example, if you are trying to determine which of two small objects is the heavier, you might shake them back and forth in your hands or move them in some way instead of lifting them. In doing so, you are judging which of the two is more difficult to move, seeing which is the more resistant to a change in motion. You are really making a comparison of the inertias of the objects. It is easy to confuse the ideas of mass and weight, mainly because they are directly proportional to each other. If the mass of an object is doubled, its weight is also doubled; if the mass is halved, its weight is halved. But there is a distinction between the two.

Figure 4-1
Examples of inertia.

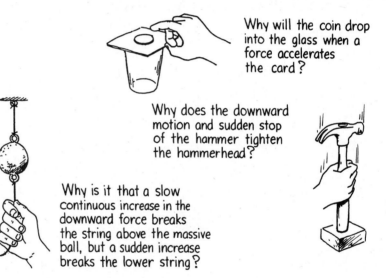

Why will the coin drop into the glass when a force accelerates the card?

Why does the downward motion and sudden stop of the hammer tighten the hammerhead?

Why is it that a slow continuous increase in the downward force breaks the string above the massive ball, but a sudden increase breaks the lower string?

We can define each as follows:

Mass: The quantity of matter in a material object. More specifically, it is the measurement of the inertia or sluggishness that an object exhibits in response to any effort made to start it, stop it, or change in any way its state of motion.

Weight: The force upon an object due to gravity.

In the United States, the quantity of matter in an object has commonly been described by the gravitational pull between it and the earth, or its weight. This has usually been expressed in *pounds*. In most of the world, however, the measure of matter is commonly expressed in a mass unit, the **kilogram**. At the surface of the earth, the mass of a 1-kilogram brick weighs 2.2 pounds. In the metric system of units, the unit of force is the

Figure 4-2
An anvil in outer space,
between the earth and
moon, for example, may be
weightless, but it is not
massless.

Figure 4-3
The astronaut in space finds
it is just as difficult to shake
the "weightless" anvil as it
would be on earth. If the
anvil is more massive than
the astronaut, which shakes
more—the anvil or the
astronaut?

newton, which is equal to a little less than a quarter pound (like the weight of a quarter-pound hamburger *after* it is cooked). A 1-kilogram brick weighs 9.8 newtons (9.8 N).* Away from the earth's surface, where the influence of gravity is less, a 1-kilogram brick weighs less. It would also weigh less on the surface of planets with less gravity than the earth. On the moon, for example, where the gravitational force on things is only $\frac{1}{6}$ as strong as on earth, 1 kilogram weighs about 1.6 newtons (or 0.36 pound). On more massive planets it would weigh more. But the mass of the brick is the same everywhere. The brick offers the same resistance to speeding up or slowing down regardless of whether the earth, moon, or anything at all is attracting it. In a space capsule located between the earth and the moon, where gravitational forces cancel each other, a brick still has mass. If placed on a scale, it wouldn't weigh anything, but its resistance to a change in motion is the same as on earth. Just as much force would have to be exerted by an astronaut in the space capsule to shake the brick back and forth as would be required to shake it back and forth while on earth. It would take the same push to start a Cadillac limousine moving on a level surface on the moon as on earth. The difficulty of lifting it against gravity (weight) is something else. Mass and weight are different from each other (Figures 4-2 and 4-3).

It is also easy to confuse mass and size—or, more specifically, **volume**. Perhaps this is because when we think of a massive object, we think of a big object—that is, a voluminous object that occupies much space. But an object can be massive—a lead storage battery, for example—without occupying much space at all. This confusion might also occur because mass and volume are often proportional to each other, at least for the same material. For example, 2 kilograms of sugar will fill a bag of twice the volume as 1 kilogram of sugar. But this does not mean that mass *is* volume. Because 2 kilograms of sugar have twice the sweetening power of 1 kilogram, we don't say that mass *is* sweetening power. Two loaves of bread may have the same mass but quite unequal volumes. You can always squeeze a loaf of bread and change its volume, but the mass doesn't change; it contains the same amount of matter. So you see that mass is neither weight nor volume.

Although Galileo introduced the idea of inertia, Newton grasped its significance. The law of inertia defines natural motion and tells us what kinds of motion are the result of applied forces. Whereas Aristotle maintained that the forward motion of an arrow through the air required a force, Newton's law of inertia instead tells us that the behavior of the arrow is natural; constant speed along a straight line (or, simply, constant

---

*So 2.2 lb equal 9.8 N, or 1 N is approximately equal to 0.2 lb—about the weight of an apple. In the metric system it is customary to specify quantities of matter in units of mass (in grams or kilograms) and rarely in units of weight (in newtons). In the United States and places that have been using the British system of units, however, quantities of matter have customarily been specified in units of weight (in pounds), and the unit of mass, the *slug*, is not well known. See Appendix I for more about systems of measurement.

velocity) requires no force. And whereas Aristotle and his followers held that the circular motions of heavenly bodies were natural and moving without applied forces, the law of inertia clearly states that in the absence of forces of some kind the planets would not move in the divine circles of ancient and medieval astronomy but would move instead in straight-line paths off into space. Newton maintained that the curved motion of the planets was evidence of some kind of force. We shall see in later chapters that his search for this force led to the law of gravitation.

Questions

1. A hockey puck sliding across the ice finally comes to rest. How would Aristotle interpret this behavior? How would Galileo and Newton interpret it? How would you interpret it?
2. Does a 2-kg iron brick have twice as much *inertia* as a 1-kg iron brick? Twice as much *mass*? Twice as much *volume*? Twice as much *weight*?
3. Would it be easier to lift a Cadillac limousine on the earth or to lift it on the moon?

Newton's Second Law of Motion

Every day we see things that do not continue in a constant state of motion: objects initially at rest later may move; moving objects may follow paths that are not straight lines; things in motion may stop. Most of the motion we observe undergoes changes and is the result of one or more applied forces. The overall net force, whether it be from a single source or a combination of sources, produces acceleration. The relationship of acceleration to force and inertia is given in Newton's second law.

▶ **Answers**

1. Aristotle would probably say that the puck slides to a stop because it seeks its proper and natural state, one of rest. Galileo and Newton would probably say that once in motion the puck would continue in motion, and that what prevents continued motion is not its nature or its proper rest state, but the friction between the puck and the ice. This friction is small compared to the friction between the puck and a wooden floor, which is why the puck slides so much farther on ice. Only you can answer the last question.
2. The answers to all parts are yes. A 2-kg iron brick has twice as many iron atoms and therefore twice the amount of matter and mass. In the same location, it also has twice the weight. And since both bricks have the same density (the same mass/volume), the 2-kg brick also has twice the volume.
3. A Cadillac limousine would be easier to lift on the moon because the gravitational force is less on the moon. When you *lift* an object, you are contending with the force of gravity (its weight) in addition to its inertia (its mass). Although its mass is the same on the earth, the moon, or anywhere, its weight is only $\frac{1}{6}$ as much on the moon, so only $\frac{1}{6}$ as much effort is required to lift it there. To move it horizontally, however, you are not pushing against gravity. When mass is the only factor, equal forces will produce equal accelerations whether the object is on the earth or the moon.

Figure 4-4
The greater the mass, the greater the force must be for a given acceleration.

Law 2: The acceleration of an object is directly proportional to the net force acting on the object, is in the direction of the net force, and is inversely proportional to the mass of the object.

In summarized form, this is

$$\text{Acceleration} \sim \frac{\text{net force}}{\text{mass}}$$

In symbol notation, this is simply

$$a \sim \frac{F}{m}$$

We shall use the wiggly line $\sim$ as a symbol meaning "is proportional to." We say that acceleration a is directly proportional to the overall net force F and inversely proportional to the mass m. By this we mean that if F increases, a increases; but if m increases, a decreases. With appropriate units of F, m, and a, the proportionality may be expressed as an exact equation:

$$a = \frac{F}{m}$$

An object is accelerated in the direction of the force acting on it. Applied in the direction of the object's motion, a force will increase the object's speed. Applied in the opposite direction, it will decrease the speed of the object. Applied at right angles, it will deflect the object. Any other direction of application will result in a combination of speed change and deflection. *The acceleration of an object is always in the direction of the net force.*

A **force**, in the simplest sense, is a push or a pull. Its source may be gravitational, electrical, magnetic, or simply muscular effort. In the second law, Newton gives a more precise idea of force by relating it to the acceleration it produces. He says in effect that *force is anything that can accelerate an object.* Furthermore, he says that the larger the force, the more acceleration it produces. For a given object, twice the force results in twice the acceleration; three times the force, three times the acceleration; and so forth. Acceleration is directly proportional to net force (Figure 4-5).

We say *net* force because oftentimes more than a single force acts on an object. Let's examine more carefully what we mean by *net*. If you and a friend pull in the same direction with equal forces on an object, the forces add to produce a net force twice as great as your single force. The combination of forces produces twice the acceleration than if you pulled alone. If, however, you each pull with equal forces but in opposite directions, the object will not accelerate. Because they are oppositely directed, the forces on the object cancel one another. One of the forces can be considered to be the negative of the other, and they add algebraically to zero. The net force is zero.

Force of hand accelerates the brick

Twice as much force produces twice as much acceleration

Twice the force on twice the mass gives the same acceleration

Figure 4-5
Acceleration is directly proportional to force.

Figure 4-6
Net force.

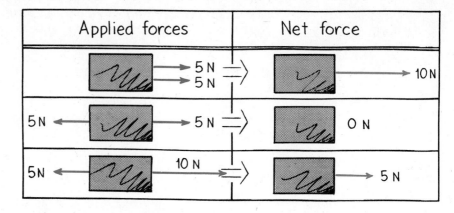

Applied forces	Net force
5 N / 5 N →	→ 10 N
5 N ← / 5 N →	0 N
5 N ← / 10 N →	→ 5 N

Suppose you pull on an object with a force of 20 newtons and your friend pulls in the opposite direction with a force of 15 newtons. Then the force and resulting acceleration is the same as if you pulled alone with a force of 5 newtons. The resulting 5 newtons is the net force. If the forces are in the same direction they are added; if in opposite directions they are subtracted (Figure 4-6). It is the net force that accelerates the object.

We see that force involves both magnitude (the amount) and direction. Force is a *vector quantity.* When two or more forces are exerted on an object, the forces combine by vector rules. When the forces are parallel, in the same or opposite directions, they simply add algebraically. But when two or more forces are exerted at angles to one another so they are neither in the same nor opposite directions, they combine via the parallelogram rule as shown in Chapter 3. To simplify our study of physics, we will not treat forces at angles in this chapter, and leave such to Appendix III.

Force and mass have opposite effects on acceleration (Figure 4-7). The more massive the object, the less its acceleration. For the same force, twice the mass results in half the acceleration; three times the mass, one-third the acceleration. Increasing the mass decreases the acceleration. For example, if we put identical Ford engines in a Cadillac and a Volkswagen, we would expect quite different accelerations even though the driving force in each car is the same. The Cadillac with its greater mass has a greater resistance to a change in velocity than does the Volkswagen. Consequently, the Cadillac requires a more powerful engine to achieve the same acceleration. For the same acceleration, a larger mass requires a correspondingly larger force. We say that acceleration is inversely proportional to mass.*

Force of hand accelerates the brick

The same force accelerates 2 bricks ½ as much

3 bricks, ⅓ as much acceleration

Figure 4-7
Acceleration is inversely proportional to mass.

---

*Mass is operationally defined as the proportionality constant between force and acceleration in Newton's second law, rearranged to read $m = F/a$. A 1-unit mass is that which requires 1 unit of force to produce 1 unit of acceleration. So 1 kg is the amount of matter that 1 N of force will accelerate 1 m/s². In British units, 1 slug is the amount of matter that 1 pound of force will accelerate 1 ft/s². We shall see later that mass is simply a form of concentrated energy.

The acceleration of an object, then, depends both on the net force exerted on the object and on the mass of the object.

Question ▶ In Chapter 2 acceleration was defined to be the time rate of change of velocity; that is, a = (change in v)/time. Are we in this chapter saying that acceleration is instead the ratio of force to mass; that is, $a = F/m$? Which is it?

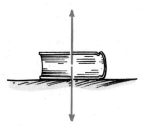

Figure 4-8
The table pushes up on the book with as much force as the downward force of gravity on the book, so the net force on the book is zero.

Figure 4-9
The sum of the upward pulls by the rings must equal her weight.

When Acceleration Is Zero—Equilibrium

When the acceleration of an object is zero, we say the object is in **mechanical equilibrium**. Whatever forces may act on it are canceled out. The net force on an object in equilibrium is zero. A book lying motionless on a table is in equilibrium because it is not accelerating. And the same book sliding at constant velocity across a smooth surface is also in equilibrium, because it also is not accelerating. In both cases the net force on the book is zero. Let's consider the case without motion first. We call this *static* equilibrium.

Consider a book lying motionless on a table. Its acceleration is zero. According to Newton's second law, the net force must also be zero. Weight acts downward, so another force must act upward on the book. The other force is the *support force* of the table (often called the *normal force*).* The table exerts an upward support force on the book that is equal in magnitude to the downward force of gravity on the book—its weight (Figure 4-8).

When you step on a bathroom scale, the downward pull of gravity and the upward support force of the floor compress a spring that is calibrated to give your weight. In effect, the scale shows the support force. Since you are not accelerating, the net force on you is zero, which means the support force and your weight are equal in magnitude. For any object that is not accelerating, either no force is acting on it or there is a combination of forces that cancel to zero. This idea that zero net force acts on things in equilibrium is often quite useful. For example, if you see that a force is exerted on something that doesn't accelerate, then you know there is another force acting—one that is equal in magnitude and opposite in direction. The net force on an object in equilibrium is always zero.

▶ **Answer**
Acceleration is the time rate of change of velocity and is *produced* by a force. How much force/mass (the cause) results in the rate change in v/time (the effect). So whereas we defined acceleration in Chapter 2, in this chapter we define the relationships that produce acceleration.

---

*This force acts at right angles to the surface. When we say "normal to" we are saying "at right angles to," which is why this force is called a normal force.

Question ▶ Suppose you stand on two bathroom scales with your weight evenly divided between the two scales. What will each scale read? How about if you stand with more of your weight on one foot than the other?

Figure 4-10
The hockey puck has practically zero acceleration after it has been struck and slides across the ice.

Consider a hockey puck that slides across the ice at constant velocity. It slides without accelerating because the net force on it is zero. Any object that moves without accelerating is said to be in *dynamic* equilibrium. A bowling ball rolling at constant velocity along a bowling alley is in dynamic equilibrium—until it interacts with the pins. When you push a crate at constant velocity across a factory floor, the crate is in dynamic equilibrium. In this case the force you apply is balanced by the force of friction between the crate and the floor. The net force is zero; hence the acceleration is zero. It's important to emphasize that zero acceleration does not mean zero velocity. Zero acceleration means that the object will maintain the velocity it happens to have, neither speeding up nor slowing down nor changing direction. Most things that are moved in our environment must be pushed or pulled to overcome friction.

Friction is a force that occurs when surfaces slide or tend to slide over one another. It depends on the kinds of material and how much they are pressed together, and it results from the mutual contact of irregularities in the surfaces. The irregularities act as obstructions to motion. Even surfaces that appear to be very smooth have irregular surfaces when viewed microscopically. The atoms cling together at many points of contact. As sliding commences, the atoms snap apart or are torn from one surface by the other.

The direction of friction force is always in a direction opposing motion. An object sliding down an incline experiences friction directed up the incline; an object that slides to the right experiences friction toward the left. Thus, if an object is to move at constant velocity, a force equal to the opposing force of friction must be applied so that the two forces exactly cancel one another. The zero net force then results in zero acceleration.

No friction exists on a crate that sits at rest on a level floor. But disturb the contact surfaces by pushing horizontally on the crate and friction

---

▶ **Answer**
The reading on both scales adds up to your weight. This is because the sum of the scale readings, which equals the support force by the floor, must counteract your weight so the net force on you will be zero. If you stand equally on each scale, each will read half your weight. If you lean more on one scale than the other, more than half your weight will be read on that scale but less on the other, so they will still add up to your weight. For example, if one scale reads two-thirds your weight, the other scale will read one-third your weight. Get it?

force is produced. How much? If the crate is still at rest, then the friction that opposes motion is just enough to cancel your push. If you push horizontally with, say, 70 newtons, the friction is 70 newtons. You push harder, say 100 newtons. The crate is on the verge of sliding—the friction between the crate and floor opposes your push with 100 newtons. If 100 newtons is the most the surfaces can muster, then when you push a bit harder the clinging gives way and the crate slides. Interestingly enough, the friction of sliding is somewhat less than the friction that builds up before sliding takes place. Physicists and engineers distinguish between static friction and sliding friction. To avoid information overload we won't pursue this distinction further, except to cite an important example—braking a car in an emergency stop. It is very important that you not jam on the brakes so as to make the tires lock in place. When tires lock, they slide, providing less friction than if they are made to roll to a stop. While the tire is rolling, its surface does not slide along the road surface, and friction is static friction—and therefore greater than sliding friction.

Figure 4-11
The crate slides to the right because of an applied force of 75 N. A friction force of 75 N opposes motion and results in a zero net force on the crate, so the crate slides at constant velocity (zero acceleration).

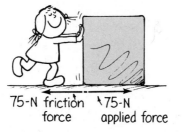

75-N friction   ⟍75-N
    force   applied force

The difference between static and sliding friction is also apparent when your car takes a corner too fast. Once the tires start to slide, the frictional force is reduced and off you go! A skilled driver keeps the tires below the threshold of breaking loose into a slide.

Question ▶ A jumbo jet cruises at a constant velocity of 1000 km/h when the thrusting force of its engines is a constant 100 000 N. What is the acceleration of the jet? What is the force of air resistance on the jet?

▶ **Answer**
The acceleration is zero because the velocity is constant. Since the acceleration is zero, it follows from Newton's second law that the net force is zero, which in turn means that the force of air resistance must just equal the thrusting force of 100 000 N and act in the opposite direction. So the air resistance on the jet is 100 000 N. (Note that we don't need to know the velocity of the jet to answer this question. We need only to know that it is constant, our clue that acceleration and therefore net force is zero.)

Figure 4-12
Friction between the tire and the ground is the same whether the tire is wide or narrow. The purpose of the greater contact area is to reduce heating and wear.

Interestingly enough, it so happens that the force of friction does not depend on speed. A car skidding at low speed has approximately the same friction at high speed. If the friction force of a crate that slides against a floor is 90 newtons at low speed, to a close approximation it is 90 newtons at a greater speed. It may be more when the crate is at rest and on the verge of sliding, but once sliding the friction force remains the same.

More interesting, friction does not depend on the area of contact. Slide the crate on its smallest surface and all you do is concentrate the same weight on a smaller area with the result that the friction is the same. So those extra wide tires you see on some cars provide no more friction than narrower tires. The wider tire simply spreads the weight of the car over more surface area to reduce heating and wear.

Friction is not restricted to solids sliding over one another. Friction occurs also in liquids and gases, collectively called *fluids*. Just as the friction between solid surfaces depends on the nature of the surfaces, fluid friction depends on the nature of the fluid: for example, it is greater in water than it is in air. But unlike the friction between solid surfaces, like the crate sliding across the floor, fluid friction does depend on speed and area of contact. This makes sense, for the amount of fluid pushed aside by a boat or airplane depends on the sizes and the shapes of the craft. A slow-moving boat or airplane encounters less friction than faster boats or airplanes. And wide boats and airplanes must push aside more fluid than narrow crafts. If the flow of fluid is relatively smooth, the friction force is approximately proportional to the speed of the object. Above a critical speed this simple proportion breaks down as fluid flow becomes erratic and friction increases dramatically.*

When Acceleration Is *g*—Free Fall

Recall from Chapter 2 that Galileo was the first to measure the acceleration of freely falling objects. Interestingly enough, although Galileo founded both the concepts of acceleration and inertia, he failed to notice the relationship of them to force. Hence he could not explain why objects of various masses fall with equal accelerations. This relationship is Newton's second law, and we'll see that it provides the explanation.

We know that a falling object accelerates toward the earth because of the gravitational force of attraction between the object and the earth. We call the force of gravity that acts on an object the *weight* of the object.†

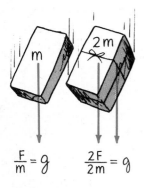

$$\frac{F}{m} = g \qquad \frac{2F}{2m} = g$$

Figure 4-13
The ratio of weight *F* to mass *m* is the same for all objects in the same locality; hence, their accelerations are the same in the absence of air resistance.

---

*Even though it may not seem so yet, most of the concepts in physics are not really complicated. But friction is different. Unlike most concepts in physics, it is a very complicated phenomenon, and the findings are empirical (gained from a wide range of experiments) and the predictions approximate (also based on experiment).

†Weight and mass are directly proportional to each other, and the constant of proportionality is *g*. We see that weight = mg, so 9.8 N = (1kg)(9.8 m/s^2).

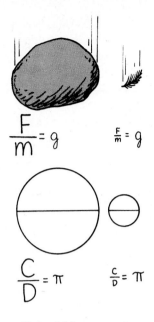

$$\frac{F}{m} = g \qquad \frac{F}{m} = g$$

$$\frac{C}{D} = \pi \qquad \frac{C}{D} = \pi$$

Figure 4-14
The ratio of weight *F* to mass *m* is the same for the large rock and the small feather; similarly, the ratio of circumference *C* to diameter *D* is the same for the large and the small circle.

When this is the only force that acts on an object—that is, when air resistance and the like are negligible—we say that the object is in a state of **free fall**. To say something is freely falling is to say it is falling free of opposing forces of any kind, usually air resistance. The only force that acts on an object in free fall is the force of gravity.

A heavier object is attracted to the earth with more force than a light object. The double brick in Figure 4-13, for example, is attracted with twice as much gravitational force as the single brick. Why then, as Aristotle supposed, doesn't the double brick fall twice as fast? The answer is that the acceleration of an object depends not only on the force—in this case, the weight—but on the mass as well. Whereas force tends to accelerate things, mass tends to resist the acceleration. So twice the force exerted on twice the inertia produces the same acceleration as half the force exerted on half the inertia. Both accelerate equally. The acceleration due to gravity is *g*.

The constant ratio of weight to mass for freely falling objects is similar to the constant ratio of circumference to diameter for circles, which is π. The ratio of weight to mass is the same for both heavy and light objects, just as the ratio of circumference to diameter is the same for both large and small circles (Figure 4-14). The acceleration of free fall is *g*. We use the symbol *g*, rather than *a*, to denote that acceleration is due to gravity alone.

So we see the acceleration of free fall is independent of weight. A boulder 100 times more massive than a pebble falls at the same acceleration as the pebble because although the force on the boulder (its weight) is 100 times greater than the force (or weight) on the pebble, its resistance to a change in motion (mass) is 100 times that of the pebble. The greater force offsets the equally greater mass.

Question ▶ In a vacuum, a coin and a feather fall equally, side by side. To explain this, would it be correct to simply say that in a vacuum equal forces of gravity act on both the coin and the feather?

▶ **Answer**
No, no, no, a thousand times no! These objects accelerate equally not because the forces of gravity on them are equal, but because the ratios of their weights to masses are equal. Although air resistance is not present in a vacuum, gravity is (you'd know this if you stuck your hand into a vacuum chamber and a Mack truck rolled over it). If you answered yes to this question, let this be an alarm to be more careful when you think physics!

Figure 4-15
When weight *mg* is greater than air resistance *R*, the falling sack accelerates. At higher speeds, *R* increases. When *R* = *mg*, acceleration reaches zero, and the sack reaches its terminal velocity.

When Acceleration Is Less Than *g*—Nonfree Fall

Objects falling in a vacuum are one thing, but what of the practical cases of objects falling in air? Although a feather and a coin will fall equally fast in a vacuum, they fall quite differently in the presence of air. How do Newton's laws apply to objects falling in air? The answer is that Newton's laws apply to *all* objects, freely falling or falling in the presence of resistive forces. The accelerations, however, are quite different for the different cases. The important thing to keep in mind is the idea of *net force*. In a vacuum or in cases where air resistance can be neglected, the net force is the weight because it is the only force acting on a falling object. In the presence of air resistance, however, the net force is the difference between the weight and force of air resistance.*

The force of air resistance acting on a falling object depends on two things. First, it depends on the size of the falling object—that is, on the amount of air the object must plow through in falling. Second, it depends on the speed of the falling object; the greater the speed, the greater the number of air molecules an object encounters per second and the greater the force of molecular impact. Air resistance depends on the size and the speed of a falling object.

A feather dropped in the air accelerates very briefly and then "floats" to the ground at constant speed. This happens because the air resistance acting against the feather quickly increases until it is equal and opposite to the weight. The net force on the feather quickly reaches zero, and no further acceleration takes place. The feather is very light and has a relatively large surface area, so it doesn't have to fall very fast for air resistance to equal its weight. The same idea applies to all objects falling in air. As a falling object gains speed, the force of air resistance finally builds up until it equals the weight of the falling object. When this happens, the net force becomes zero and the object no longer accelerates; it falls at a constant speed. When acceleration terminates, we say the object has reached its **terminal speed**. If we are concerned with direction, down for falling objects, we say the object has reached its *terminal velocity*. For a feather terminal speed is a few centimeters per second, whereas for a skydiver it is about 200 kilometers per hour. A skydiver varies

---

*In mathematical notation,

$$a = \frac{F_{net}}{m} = \frac{mg - R}{m}$$

where *mg* is the weight and *R* is the air resistance. Note that when *R* = *mg*, *a* = 0; then, with no acceleration, the object falls at constant velocity. With elementary algebra we can go another step and get

$$a = \frac{F_{net}}{m} = \frac{mg - R}{m} = g - \frac{R}{m}$$

We see that the acceleration *a* will always be less than *g* if air resistance *R* impedes falling. Only when *R* = 0 does *a* = *g*.

terminal speed by varying position. Head or feet first is a way of encountering less air resistance and attaining maximum terminal speed. Minimum terminal speed is attained by spreading oneself out like a flying squirrel.

Consider a man and woman parachuting together from the same altitude (Figure 4-16). Suppose that the man is twice as heavy as the woman and that their same-sized chutes are initially opened. The same size chute means that at equal speeds the air resistance is the same on each. Who gets to the ground first—the heavy man or the lighter woman? The answer is that the person with the greater terminal speed gets to the ground first.

Figure 4-16
The heavier parachutist must fall faster than the lighter parachutist for air resistance to cancel his greater weight.

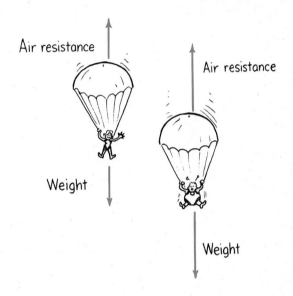

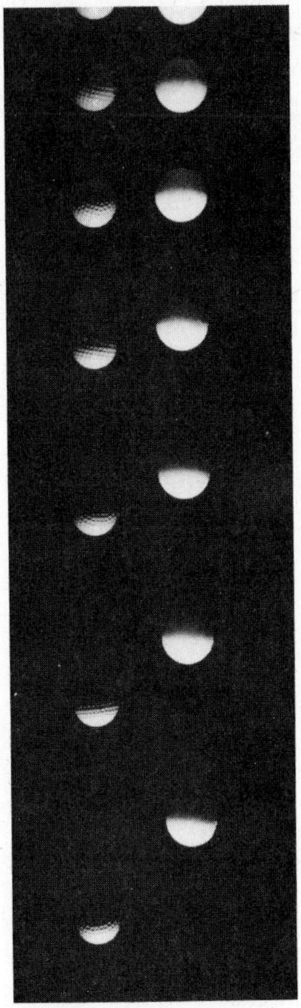

At first we might think that because the chutes are the same, the terminal speeds for each would be the same, and therefore both would reach the ground together. This doesn't happen because air resistance also depends on speed. The woman will reach her terminal speed when air resistance against her chute equals her weight. When this happens, the air resistance against the chute of the man will not yet equal his weight. Force equilibrium occurs first for her, while he continues to accelerate to greater speeds. He must fall faster than she does for air resistance to match his greater weight.* Terminal speed is greater for the heavier person, with the result that the heavier person reaches the ground first.

Consider a pair of tennis balls, one a hollow regular ball and the other filled with iron pellets. Although they are the same size, the iron-filled ball is considerably heavier than the regular ball. If you hold them above your head and drop them simultaneously, you'll see that they strike the

Figure 4-17
A stroboscopic study of a golf ball and a styrofoam ball falling in air. The air resistance is negligible for the heavier golf ball, and it accelerates essentially at g. Air resistance is not negligible for the lighter styrofoam ball, which reaches its terminal velocity sooner.

---

*Terminal speed for the man will be somewhat less than double that of the half-as-heavy woman because the retarding force of the air is not exactly proportional to the speed.

ground at the same time. But if you drop them from a greater height, say from the top of a building, you'll note the heavier ball strikes the ground first. Why? In the first case the balls did not gain much speed in their short fall. Air resistance encountered was small compared to their weights, even for the regular ball. The tiny difference in their arrival time was not noticed. But when dropped from a greater height, the greater speeds of fall are met with greater air resistance. At any given speed each ball encounters the same air resistance because each has the same size. This air resistance may be a lot compared to the weight of the lighter ball, but only a little compared to the weight of the heavier ball (like the parachutists in Figure 4-16). So even with equal air resistances the accelerations of each are different. There is a moral to be learned here. Whenever you consider the acceleration of something, use the equation of Newton's second law to guide your thinking: the acceleration is equal to the ratio of *net* force to mass. In this case, acceleration decreases as net force decreases. If air resistance builds up to equal the weight of the falling object, the net force becomes zero and acceleration terminates.

| Question | ▶ | A skydiver jumps from a high-flying helicopter. As she falls faster and faster through the air, does her acceleration increase, decrease, or remain the same? |

Newton's Third Law of Motion

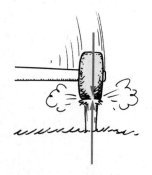

In the simplest sense, a force is a push or a pull. Looking closer, however, a force is not a thing in itself but is due to an *interaction* between one thing and another. We pull on a cart and it accelerates. A hammer hits a stake and drives it into the ground. One object interacts with another object. Which exerts the force and which receives the force? Newton's answer to this is that neither force has to be identified as "exerter" or "receiver"; he believed nature is symmetrical and concluded that both objects must be treated equally. For example, when we pull the cart, the cart in turn pulls on us, as evidenced perhaps by the tightening of the rope wrapped around our hand. And the hammer exerts a force on the stake but is itself brought to rest in the process. Such observations led Newton to his third law—the law of action and reaction.

Figure 4-18

In the interaction between the hammer and the stake, each exerts the same amount of force on the other.

▶ Answer

Acceleration decreases because the net force on her decreases. Net force is equal to her weight minus her air resistance, and since air resistance increases with increasing speed, net force and hence acceleration decrease. By Newton's second law,

$$a = \frac{F_{\text{net}}}{m} = \frac{mg - R}{m}$$

where mg is her weight and R is the air resistance she encounters. As R increases, a decreases. Note that if she falls fast enough so that $R = mg$, $a = 0$, then with no acceleration she falls at constant speed.

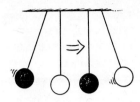

Figure 4-19
The impact force between the dark and white ball moves the white ball and stops the dark ball.

Law 3: Whenever one object exerts a force on a second object, the second object exerts an equal and opposite force on the first.

Newton's third law is often stated thus: "To every action there is always opposed an equal reaction." In any interaction there is an action and reaction pair of forces that are equal in magnitude and opposite in direction. Neither force exists without the other—forces come in *pairs*, one action and the other reaction. The action and reaction pair of forces make up the interaction between two things.

When you walk across the floor you push against the floor, and in turn the floor pushes against you. Both you and the floor push against each other simultaneously—there is an interaction between you and the floor. Likewise, the tires of a car push against the road while the road pushes back on the tires—the tires and road push against each other. In swimming you push the water backward, and the water pushes you forward—you and the water push against each other. In each case there is a pair of forces, one action and the other reaction, that make up the interaction. The reaction forces are what account for our motion in these cases. These forces depend on friction; a person or car on ice, for example, may not be able to exert the action force to produce the needed reaction force.

Figure 4-20
Action and reaction forces. Note that when action is *A exerts force on B*, the reaction is then simply *B exerts force on A*.

Action: tire pushes on road   Reaction: road pushes on tire

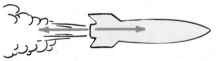

Action: rocket pushes on gas   Reaction: gas pushes on rocket

Action: man pulls on spring   Reaction: spring pulls on man

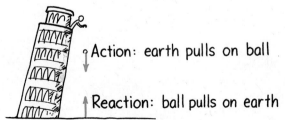

Action: earth pulls on ball

Reaction: ball pulls on earth

Figure 4-21
When an apple pulls on an orange, the orange accelerates—period! The fact that the orange pulls back on the apple affects the apple—not the orange. You can't cancel a force on the orange with a force on the apple.

Figure 4-22
The earth is pulled up by the boulder with just as much force as the boulder is pulled down by the earth.

Figure 4-23
The force exerted against the recoiling rifle is just as great as the force that drives the bullet. Why, then, does the bullet accelerate more than the rifle?

It doesn't matter which force we call *action* and which we call *reaction* as long as we remember that neither exists without the other. In our first example, we may call our push against the floor the action force and the floor pushing back the reaction force. We could just as well say the push of the floor against us is the action force, in which case our push against the floor is the reaction force.

Action and reaction forces never cancel one another because each acts on a different object. For example, when you pull on a cart, the cart pulls on you. There are two forces here, one exerted on the cart and the other exerted by the cart on you. We distinguish between forces exerted *on* something and *by* something. The force exerted *on* the cart accelerates it; the force exerted *by* the cart doesn't—the force exerted by the cart on you affects not the cart, but you. If we call the force exerted on an object the *action force*, the reaction force will be the force exerted by the object on whatever other object is involved in the interaction. Since action and reaction act on different objects, they never cancel each other.

A falling object pulls upward on the earth as much as the earth pulls downward on it. This fact may seem puzzling. The downward pull on the object is less puzzling because the acceleration of 9.8 meters per second squared is quite noticeable. The same amount of force acting upward on the huge mass of the earth, however, produces an acceleration so small it cannot be measured. But it is there. A less extreme example might make this clearer. When a rifle is fired, the force exerted on the bullet is as great as the reaction force exerted on the rifle; hence, the rifle kicks. Since the forces are equal in magnitude, why doesn't the rifle recoil with the same speed of the bullet? In analyzing changes in motion, Newton's second law reminds us that we must also consider the masses involved. Suppose we let F represent both the action and reaction force, m the mass of the bullet, and M the mass of the more massive rifle. The accelerations of the bullet and the rifle are then found by taking the ratio of force to mass. The acceleration of the bullet is given by

$$\frac{F}{m} = a$$

while the acceleration of the recoiling rifle is

$$\frac{F}{M} = a$$

We see why the change in motion of the bullet is so huge compared to the change of motion of the rifle. A given force divided by a small mass

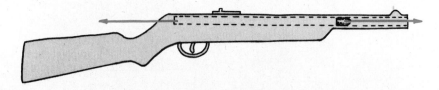

Questions

1. Does a speeding missile possess force?
2. We know that the earth pulls on the moon. Does it follow that the moon also pulls on the earth?
3. Since action and reaction are exactly equal and oppositely directed, how can there ever be a net force on an object?
4. Can you identify the action and reaction forces in the case of an object falling in a vacuum?

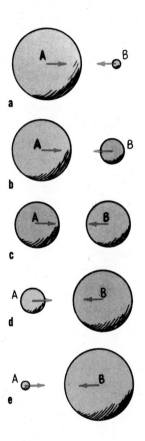

Figure 4-24
Which falls toward the other, A or B? Do the accelerations of each relate to their relative masses?

produces a large acceleration, while the same force divided by a large mass produces a small acceleration. We have used different-sized symbols to indicate the differences in relative masses and resulting accelerations. Going back to the example of the falling object, if we used similarly exaggerated symbols to represent the acceleration of the earth reacting to a falling object, the symbol m for the earth's mass would be larger than the page. This huge quantity divided into the force F, the weight of the falling object, would result in a microscopic a to represent the acceleration of the earth toward the falling object.

We can see that the earth accelerates slightly in response to a falling object in still another way by considering the exaggerated examples of two planetary bodies, a through e in Figure 4-24. The forces between

---

▶ **Answers**

1. No, a force is not something an object *has*, like mass, but is part of an interaction between one object and another. A speeding missile may possess the capability of exerting a force on another object when interaction occurs, but it does not possess force as a thing in itself. We will see in the two chapters to follow that a speeding missile possesses both momentum and kinetic energy.
2. Yes, both pulls make up an action-reaction pair of forces that make up the gravitational interaction between the earth and the moon. We can say that (1) the earth pulls on the moon, and (2) the moon likewise pulls on the earth; but it is more insightful to think of this as a single interaction—the earth and moon simultaneously pull on each other.
3. The net force on any object is a combination of forces that act *on* the object— say, the action forces. The reaction forces *by* the object act on something else. For example, when you kick a can, the can kicks back on you. But your foot provides the only force exerted on the can. The can flies through the air because it experiences only the action force; the reaction force is exerted *by* the can, not *on* the can. The reaction force decelerates your foot, which is of little interest to the can. The can experiences only what happens to it, not what it does to your foot. You can't cancel the force exerted on an object with the force it simultaneously exerts on something else.
4. To identify a pair of action-reaction forces in any situation, first identify the pair of interacting objects involved. Something is interacting with something. In this case the world is interacting (gravitationally) with the falling object. So the world pulls downward on the object (call it action), while the object pulls upward on the world (reaction).

bodies A and B are equal in magnitude and oppositely directed in each case. If acceleration of planet A is unnoticeable in *a*, then it is more noticeable in *b* where the difference between the masses is less extreme. In *c*, where both bodies have equal mass, acceleration of object A is as evident as it is for B. Continuing, we see the acceleration of A becomes even more evident in *d* and even more so in *e*. So strictly speaking, when you step off the curb, the street comes up ever so slightly to meet you.

Using Newton's third law, we can understand how a helicopter gets its lifting force. The whirling blades are shaped to force air particles down (action), and the air forces the blades up (reaction). This upward reaction force is called *lift*. When lift equals the weight of the craft, the helicopter hovers in midair. When lift is greater, the helicopter climbs upward.

This is true for birds and airplanes. A bird flies by pushing air downward. The air in turn pushes the bird upward. When the bird is soaring, the wing must be shaped so that moving air particles are deflected downward. Slightly tilted wings that deflect oncoming air downward produce the lift on an airplane. Air must be pushed downward continuously to maintain lift. This supply of air is obtained by the forward motion of the aircraft, which results from propellers that push air backward. When the propellers push air backward, the air pushes the propellers forward. For jet aircraft, the jet pushes exhaust gases backward and the exhaust gases in turn push the jet forward. We will learn in Chapter 13 that the curved surface of a wing is an airfoil, which enhances the lifting force.

We see Newton's third law at work everywhere. A fish pushes the water backward with its fins, and the water in turn pushes the fish forward. The wind pushes against the branches of a tree, and the branches in turn push back on the wind and we have whistling sounds. Forces are interactions between different things. Every contact requires at least a twoness; there is no way that an object can exert a force on nothing. Forces, whether large shoves or slight nudges, always occur in pairs, each of which is opposite to the other. Thus, we cannot touch without being touched.

Figure 4-25
You cannot touch without being touched—Newton's third law.

Summary of Newton's Three Laws

An object at rest tends to remain at rest; an object in motion tends to remain in motion at constant speed along a straight-line path (constant velocity). This tendency of objects to resist change in motion is called *inertia*. Mass is a measure of inertia. Objects will undergo changes in motion only in the presence of a net force.

When a net force is impressed upon an object, it will accelerate. The acceleration is directly proportional to the net force and inversely proportional to the mass. Symbolically, $a \sim \frac{F}{m}$. Acceleration is always in the direction of the net force. When objects fall in a vacuum, the net force is simply the weight, and the acceleration is g (the symbol g denotes that acceleration is due to gravity alone). When objects fall in air, the net force is equal to the weight minus the force of air resistance, and the acceleration is less than g. If and when the force of air resistance equals the weight of a falling object, acceleration terminates, and the object falls at constant speed.

Whenever one object exerts a force on a second object, the second object exerts an equal and opposite force on the first. This is because forces are the parts of an interaction between at least two objects. Forces occur only in *pairs*, one action and the other reaction, both of which constitute the interaction between one thing and the other. Neither force exists without the other. Since action and reaction forces act on different objects, action and reaction never cancel each other.

Questions

1. A car accelerates along a road. Strictly speaking, what is the force that moves the car?

2. A high-speed bus and an innocent bug have a head-on collision. The force of impact splatters the poor bug over the windshield. Is the corresponding force that the bug exerts against the windshield greater, less, or the same? Is the resulting deceleration of the bus greater than, less than, or the same as that of the bug?

▶ **Answers**

1. It is the road that pushes the car along. Really! Except for air resistance, only the road provides a horizontal force on the car. How does it do this? The rotating tires push back on the road (action). The road, in turn, pushes forward on the tires (reaction). How about that!

2. The magnitudes of both forces are the same, for they constitute an action-reaction force pair that makes up the interaction between the bus and the bug. The accelerations, however, are very different because the masses involved are different. The bug undergoes an enormous and lethal deceleration, while the bus undergoes a very tiny deceleration—so tiny that the very slight slowing of the bus is unnoticed by its passengers. But if the bug were more massive—another bus, for example—the slowing down would be quite evident!

Summary of Terms

Inertia The tendency of things to resist changes in motion.

Mass The quantity of matter in an object. More specifically, it is the measurement of the inertia or sluggishness that an object exhibits in response to any effort made to start it, stop it, or change in any way its state of motion.

Weight The force due to gravity on an object.

Kilogram The fundamental SI unit of mass. One kilogram (symbol kg) is the amount of mass in 1 liter (L) of water at 4°C.

Newton The SI unit of force. One newton (symbol N) is the force that will give an object of mass 1 kg an acceleration of 1 m/s^2.

Volume The quantity of space an object occupies.

Force Any influence that can cause an object to be accelerated, measured in newtons (in pounds in the British system).

Mechanical equilibrium The state of an object or system of objects for which any impressed forces cancel to zero and no acceleration occurs.

Friction The resistive forces that arise to oppose the motion or attempted motion of an object past another with which it is in contact.

Free fall Motion under the influence of gravitational pull only.

Terminal speed The speed at which the acceleration of a falling object terminates because friction balances the weight.

Review Questions

Newton's First Law of Motion

1. Is inertia the *reason* for objects to maintain their states of motion or the *name* given to this property of matter?

Mass

2. Clearly distinguish among *mass*, *weight*, and *volume*.

3. Which is fundamental, *mass* or *weight*? Does mass have weight, or does weight have mass?

4. Does a 2-kg iron brick have twice as much *inertia* as a 1-kg block of wood? Twice as much *volume*? (Why are your answers different?)

5. What kind of path would the planets follow if suddenly no force acted on them?

Newton's Second Law of Motion

6. If we say that one quantity is *proportional* to another quantity, does this mean they are *equal* to each other? Explain briefly, using mass and weight as an example.

7. A cart is pulled to the left with a force of 100 N, and to the right with a force of 30 N. What is the net force on the cart?

8. Why do we say that force is a vector quantity?

9. If the net force impressed on a sliding block is tripled, by how much does the acceleration increase?

10. If the mass of a sliding block is tripled while a constant net force is applied, by how much does the acceleration decrease?

11. If the mass of a sliding block is tripled at the same time the net force on it is tripled, how does the resulting acceleration compare to the original acceleration?

When Acceleration Is Zero—Equilibrium

12. What is the net force on something that is in mechanical equilibrium?

13. Consider a book that weighs 15 N at rest on a flat table. How many newtons of support force does the table provide? What is the net force on the book in this case?

14. Consider a woman weighing 500 N who stands with her weight evenly divided on a pair of bathroom scales. What is the reading on each scale? If she shifts her weight so one of the scales reads 300 N, what will the other scale read?

15. What is the acceleration of an object that moves at constant velocity? What is the net force on the object in this case?

16. Why is it more difficult to slide a crate from a position of rest across the floor than it is to keep it in motion once it is sliding?

17. If you push horizontally with a force of 50 N on a crate and make it slide at constant velocity, how much friction acts on the crate? If you increase your force, will the crate accelerate? Explain.

18. What effect does the speed of a sliding object have on the friction that acts on it? What effect does the area of contact have on friction?

19. What effect do speed and area have on the friction that an object experiences when moving through a fluid?

When Acceleration Is g—Free Fall

20. What is meant by *free fall*?

21. What is the net force that acts on a 10-N freely falling object?

22. Why doesn't a heavy object accelerate more than a lighter object when both are freely falling?

When Acceleration Is Less Than g—Nonfree Fall

23. What is the net force that acts on a 10-N falling object when it encounters 4 N of air resistance? 10 N of air resistance?

24. What two principal factors affect the force of air resistance on a falling object?

25. What is the acceleration of a falling object that has reached its terminal velocity?

26. Why does a heavy parachutist fall faster than a lighter parachutist who wears the same-sized parachute?

Newton's Third Law of Motion

27. Consider hitting a baseball with a bat. If we call the force on the bat against the ball the *action* force, identify the *reaction* force.

28. Why do action and reaction pairs of forces never cancel each other?

29. If the forces that act on a bullet and the recoiling gun from which it is fired are equal in magnitude, why do the bullet and gun have very different accelerations?

30. How does a helicopter get its lifting force?

Home Projects

1. Ask a friend to drive a small nail into a piece of wood placed on top of a pile of books on your head. Why doesn't this hurt you?

2. Drop a sheet of paper and a coin at the same time. Which reaches the ground first? Why? Now crumple the paper into a small, tight wad and again drop it with the coin. Explain the difference observed. Will they fall together if dropped from a second-, third-, or fourth-story window? Try it and explain your observations.

3. Drop a book and a sheet of paper and note that the book has a greater acceleration—*g*. Place the paper beneath the book and it is forced against the book as both fall, so both fall at *g*. How do the accelerations compare if you place the paper on top of the raised book and then drop both? You may be surprised, so try it and see. Then explain your observation.

4. Drop two balls of different weight from the same height, and at small speeds they practically fall together. Will they roll together down the same inclined plane?

If each is suspended from an equal length of string, making a pair of pendulums, and displaced through the same angle, will they swing back and forth in unison? Try it and see; then explain using Newton's laws.

5. The net force acting on an object and the resulting net force are always in the same direction. You can demonstrate this with a spool. If the spool is pulled horizontally to the right, in which direction will it roll?

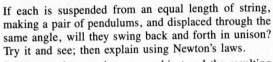

6. Hold your hand like a flat wing outside the window of a moving automobile. Then slightly tilt the front edge upward and notice the lifting effect. Can you see Newton's laws at work here?

Exercises

Please do not be intimidated by the large number of exercises in this and other meatier chapters. If your course work is to cover many chapters, your instructor will probably assign only a few exercises from each.

1. In terms of inertia, what is the disadvantage of a lightweight camera when snapping the shutter? Why is a massive tripod preferred by most photographers?

2. Your empty hand is not hurt when it bangs lightly against a wall. Why is it hurt if it does so while carrying a heavy load? Which of Newton's laws is most applicable here?

3. In tearing a paper towel or plastic bag from a roll, why is a sharp jerk more effective than a slow pull?

4. Each of the chain of bones forming your spine is separated from its neighbors by disks of elastic tissue. What happens, then, when you jump heavily on your feet from an elevated position? Can you think of a reason why you are a little taller in the morning than in the night? (*Hint:* Think about the hammerhead in Figure 4-1.)

5. Before the time of Galileo and Newton, it was thought by many learned scholars that a stone dropped from the top of a tall mast of a moving ship would fall vertically and hit the deck behind the mast by a distance equal to how far the ship had moved forward while the stone was falling. In light of your understanding of Newton's laws, what do you think about this?

6. Because the earth rotates once per 24 hours, the west wall in your room moves in a direction toward you at a linear speed that is probably more than 1000 km/h (the exact speed depends on your latitude). When you stand facing the wall you are carried along at the same speed, so you don't notice it. But when you jump upward, with your feet no longer in contact with the floor, why doesn't the high-speed wall slam into you?

7. The chimney of a stationary toy train consists of a vertical spring gun that shoots a steel ball a meter or so straight into the air—so straight that the ball always falls back into the chimney. Suppose the train moves at constant speed along the straight track. Do you think the ball will still return to the chimney if shot from the moving train? How about if the train accelerates along the straight track? How about if it moves at a constant speed on a circular track? Why are your answers different?

8. When a junked car is crushed into a compact cube, does its mass change? Its weight? Explain.

9. Gravitational force on the moon is only $\frac{1}{6}$ that of the gravitational force on the earth. What would be the weight of a 10-kg object on the moon and on the earth? What would its mass be on the moon and on the earth?

10. What is your own mass in kilograms? Your weight in newtons?

11. If a mass of 1 kg is accelerated 1 m/s^2 by a force of 1 N, what would be the acceleration of 2 kg acted on by a force of 2 N?

12. How much acceleration does a 747 jumbo jet of mass 30 000 kg experience in takeoff when the thrust for each of four engines is 30 000 N?

13. A rocket becomes progressively easier to accelerate as it travels through space. Why is this so? (*Hint:* About 90 percent of the mass of a newly launched rocket is fuel.)

14. If an object has no acceleration, can you conclude that no forces are exerted on it? Explain.

15. Can an object round a curve without any force acting on it?

16. As you stand on a floor, does the floor exert an upward force against your feet? How much force does it exert? Why are you not moved upward by this force?

17. The little girl hangs at rest from the ends of the rope as shown. How does the reading on the scale compare to her weight?

18. Harry the painter swings year after year from his bosun's chair. His weight is 500 N and the rope, unknown to him, has a breaking point of 300 N. Why doesn't the rope break when he is supported as shown to the left below? One day Harry is painting near a flagpole, and, for a change, he ties the free end of the rope to the flagpole instead of to his chair as shown to the right. Why did Harry end up taking his vacation early?

19. Consider the two forces acting on the person who stands still, namely, the downward pull of gravity and the upward support of the floor. Are these forces equal and opposite? Do they form an action-reaction pair? Why or why not?

20. Two 100-N weights are attached to a spring scale as shown. Does the scale read 0, 100, or 200 N, or give some other reading? (*Hint:* Would it read any differently if one of the ropes were tied to the wall instead of to the hanging 100-N weight?)

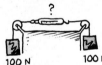

100 N   100 N

21. If we find an object that is not moving even though we know it to be acted on by a force, what inference can we draw?

22. When your car moves along the highway at constant velocity, the net force on it is zero. Why, then, do you continue running your engine?

23. A "shooting star" is usually a grain of sand from outer space that burns up and gives off light as it enters the atmosphere. What exactly causes this burning?

24. In the attempt to ascend a steep road from a standstill position in a vehicle, why is it important that the wheels not spin against the pavement?

25. Why does one experience no air resistance when soaring in a balloon?

26. What is the net force on an apple that weighs 1 N when you hold it at rest above your head? What is the net force on it when you release it?

27. Does a stick of dynamite contain force?

28. Can a dog wag its tail without the tail in turn "wagging the dog"? (Consider a dog with a relatively massive tail.)

29. When the athlete holds the barbell overhead, the reaction force is the weight of the barbell on his hand. How does this force vary when the barbell is accelerated upward? Downward?

30. Use Newton's third law to explain why when standing on a weighing scale you cannot decrease your weight by pulling upward on your boot straps.

31. Why can you exert greater force on the pedals of a bicycle if you pull up on the handlebars?

32. If the earth exerts a force of 1000 N on an orbiting communications satellite, how much force does the satellite exert on the earth?

33. Your weight is the result of a gravitational force of the earth on your body. What is the corresponding reaction force?

34. The strong man will push the two initially stationary freight cars of equal mass apart before he himself drops to the ground. Is it possible for him to give either of the cars a greater speed than the other? Why or why not?

35. Suppose two carts, one twice as massive as the other, fly apart when the compressed spring that joins them is released. How fast does the heavier cart roll compared to the lighter cart?

36. If you exert a horizontal force of 200 N to slide a crate across a factory floor at constant velocity, how much friction is exerted by the floor on the crate? Is the force of friction equal and oppositely directed to your 200-N push? Does the force of friction make up the reaction force to your push? Why not?

37. If a Mack truck and Volkswagen have a head-on collision, upon which vehicle is the impact force greater? Which vehicle undergoes the greater change in its motion? Explain your answers.

38. Two people of equal mass attempt a tug-of-war with a 12-m rope while standing on frictionless ice. When they pull on the rope, they each slide toward each other. How do their accelerations compare, and how far does each person slide before they meet?

39. Suppose in the preceding exercise that one person has twice the mass of the other. How far does each person slide before they meet?

40. A horse pulls a heavy wagon with a certain force. The wagon, in turn, pulls back with an opposite but equal force on the horse. Doesn't this mean the forces cancel each other, making acceleration impossible? Why or why not? (*Hint:* Consider the role of the ground.)

41. If the effects of friction are negligible, which will roll down a hill faster—a Cadillac or a Volkswagen? Explain.

42. How does the weight of a falling body compare to the air resistance it encounters just before it reaches terminal velocity? After?

43. Why is it that a tennis ball dropped from the top of a 50-story building will hit the ground no faster than if it were dropped from the 20th story?

44. As an object falls faster and faster through the air, where air resistance is a factor, does its acceleration increase, decrease, or remain constant?

45. If and when Galileo dropped two balls from the top of the Leaning Tower of Pisa, air resistance was not really negligible. Assuming both balls were the same size yet one much heavier than the other, which ball struck the ground first? Why?

46. If you simultaneously drop a pair of tennis balls from the top of a building, they will strike the ground at the same time. If one of the tennis balls is filled with lead pellets, will it fall faster and hit the ground first? Which of the two will encounter more air resistance? Defend your answers.

47. What is the acceleration of a rock at the top of its trajectory when thrown straight upward? (Is your answer consistent with Newton's second law?)

48. In the absence of air resistance, if a ball is thrown vertically upward with a certain initial speed, on returning to its original level it will have the same speed. When air resistance is a factor affecting the ball, will the ball fall to the ground faster, the same, or slower than if there were no air resistance? Why? (Consider the "principle of exaggeration" in formulating your answer; that is, consider the exaggerated case of a feather, not a ball, because the effect of air resistance on the feather is more pronounced and therefore easier to visualize.)

49. If a ball is thrown vertically into the air in the presence of air resistance, would you expect the time during which it rises to be longer or shorter than the time during which it falls? (Consider the "principle of exaggeration.")

50. Consider a bullet fired horizontally from a high-flying balloon. The gravitational attraction of the earth makes it turn more and more downward in its curved path. Can the path eventually become vertical? Defend your answer.

Remember, review questions provide you with a self check of whether or not you grasp the central ideas of the chapter. The exercises are extra "pushups" for you to try after you have at least a fair understanding of the chapter and can handle the review questions.

5 Momentum

In Chapter 2 we introduced Galileo's idea of inertia, and in Chapter 4 we showed how this idea was incorporated into Newton's laws of motion. We discussed inertia in terms of objects at rest and objects in motion. In this chapter we will concern ourselves only with the inertia of moving objects. When we combine the ideas of inertia and motion, we are dealing with momentum. *Momentum* refers to moving things.

Momentum

We all know that a heavy truck is harder to stop than a small car moving at the same speed. We state this fact by saying that the truck has more momentum than the car. By **momentum** we mean inertia in motion, or, more specifically, the product of the mass of an object and its velocity; that is,

$$\text{Momentum} = \text{mass} \times \text{velocity}$$

Or, in shorthand notation,

$$\text{Momentum} = mv$$

When direction is not an important factor, we can say:

$$\text{Momentum} = \text{mass} \times \text{speed}$$

which we still abbreviate mv.

Figure 5-1
Why are the engines of a supertanker normally cut off 25 km from port?

We can see from the definition that a moving object can have a large momentum if either its mass or its velocity is large, or if both its mass and its velocity are large. The truck has more momentum than the car moving at the same speed because it has a greater mass. We can see that a huge ship moving at a small velocity can have a large momentum, as can a small bullet moving at a high velocity. And, of course, a huge object moving at a high velocity, such as a massive truck rolling down a steep hill with no brakes, has a huge momentum, whereas the same truck at rest has no momentum at all.

Impulse

Figure 5-2
Impact force against a golf ball.

The momentum of an object will change if either the mass or the velocity or both the mass and the velocity change. If the momentum changes while the mass remains unchanged, as is most often the case, then the velocity changes. Acceleration occurs. And what produces an acceleration? The answer is a force. The greater the force that acts on an object, the greater will be the change in velocity and, hence, the change in momentum.

But something else is important also: time—how long the force acts. Apply a force briefly to a stalled automobile, and you produce a small change in its momentum. Apply the same force over an extended period of time, and a greater change in momentum results. A long sustained force produces more change in momentum than the same force applied briefly. So for changing the momentum of an object, both force and the time during which the force acts are important. We name the product of force and this time interval **impulse**.

Figure 5-3
A large change in momentum in a long time requires a small force.

$$\text{Impulse} = \text{force} \times \text{time interval}$$

Whenever you exert a force on something, you also exert an impulse. The resulting acceleration depends on the force; the resulting change in momentum depends on both the force and the time during which the force acts.

Figure 5-4
A large change in momentum in a short time requires a large force.

Questions ▶ **1.** Does a moving object have impulse?

2. Does a moving object have momentum?

Impulse and Momentum

The relationship of impulse to momentum comes from Newton's second law ($a = F/m$). The time interval of impulse is "buried" in the term for acceleration (change in velocity/time interval). Rearrangement of Newton's second law gives*

Force × time interval = change in (mass × velocity)

or, equivalently,

Impulse = change in momentum

We can express all terms in this relationship in shorthand notation and introduce the delta symbol Δ (Greek letter D), which stands for "change in" (or "difference in"):

$$Ft = \Delta\, mv$$

which reads, "force multiplied by the time-during-which-it-acts = change in momentum."

The impulse-momentum relationship helps us to analyze a variety of circumstances where momentum is changed. We will consider familiar examples of impulse for the cases of (1) increasing momentum, (2) decreasing momentum over a long time, and (3) decreasing momentum over a short time.

Case 1: Increasing Momentum

If you wish to increase the momentum of something as much as possible, you not only apply the greatest force you can, you also extend the time of application as much as possible. Hence the different results in pushing briefly on a stalled automobile and giving it a sustained push.

Long-range cannons have long barrels. The longer the barrel, the greater the velocity of the emerging cannonball or shell. Why? The force of exploding gunpowder in a long barrel acts on the cannonball for a longer

▶ **Answers**

1. No, impulse is not something an object has, like momentum. Impulse is what an object can *provide* when it interacts with some other object. An object cannot possess impulse just as it cannot possess force.
2. Yes, but, like velocity, in a relative sense—that is, with respect to a frame of reference, usually taken to be the earth's surface. The momentum possessed by a moving object with respect to a stationary point on earth is different from the momentum it possesses with respect to another moving object.

---

*Newton's second law can be expressed as $F = ma$, and since $a =$ (change in velocity)/ (time interval), we can say $F = m$ (change in v)/t. Simple algebraic rearrangement gives $Ft =$ change in mv, or, in delta notation, $Ft = \Delta mv$.

time. This increased impulse produces a greater momentum. Of course the force that acts on the cannonball is not steady—it is strong at first and weaker as the gases expand. In this and many other cases the forces involved in impulses vary over time. The force that acts on the golf ball in Figure 5-2, for example, increases rapidly as the ball is distorted and then diminishes as the ball comes up to speed and returns to its original shape. When we speak of impact forces in this chapter, we mean the *average* force of impact.

Case 2: Decreasing Momentum over a Long Time

Imagine you are in a car out of control, and you have a choice of slamming into either a concrete wall or a haystack. You needn't know much physics to make the better decision, but knowing some physics helps you to understand why hitting something soft is entirely different from hitting something hard. Whether you come to a stop by hitting either the wall or the haystack, your momentum will be decreased by the same impulse. The same impulse means the same product of force and time, not the same force or the same time. You have a choice. By hitting the haystack instead of the wall, you extend the time of impact—you extend the time during which your momentum is brought to zero. The longer time is compensated by a lesser force. If you extend the time of impact 100 times, you reduce the force of impact by 100. So whenever you wish the force of impact to be small, extend the time of impact.

A wrestler thrown to the floor tries to extend his time of arrival on the floor by relaxing his muscles and spreading the crash into a series of impacts as foot, knee, hip, ribs, and shoulder fold onto the floor in turn. The increased time of impact reduces the force of impact.

A person jumping from an elevated position to a floor below bends his knees upon making contact, thereby extending the time during which his momentum is being reduced by 10 to 20 times that of a stiff-legged, abrupt landing. Such knee bending reduces the forces experienced by the bones by 10 to 20 times. Of course, falling on a mat is preferable to falling on a solid floor, for this also increases the time of impact.

Figure 5-5
In both cases the impulse by the boxer's jaw reduces the momentum of the punch. (a) The boxer is moving away when the glove hits, thereby extending the time of contact. Much of the impulse therefore involves time. (b) The boxer is moving into the glove, thereby lessening the time of contact. Much of the impulse therefore involves force.

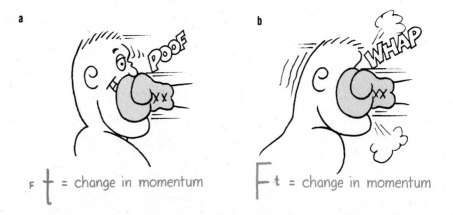

A person is better off falling on a wooden floor than a concrete floor. A wooden floor with "give" allows for a longer time of impact and therefore a lesser force of impact than a concrete floor. A safety net used by acrobats provides an obvious example of small impact force over a long time to provide the required impulse to reduce the momentum of fall.

If you're catching a fast baseball with your bare hand, you extend your hand forward so you'll have plenty of room to let your hand move backward after you make contact with the ball. You extend the time of impact and thereby reduce the force of impact. Similarly, a boxer rides or rolls with the punch to reduce the force of impact (Figure 5-5).

Questions

1. If the boxer in Figure 5-5 is able to make the duration of impact three times as long by riding with the punch, by how much will the force of impact be reduced?

2. If the boxer instead moves into the punch such as to decrease the duration of impact by half, by how much will the force of impact be increased?

3. A boxer being punched contrives to extend time for best results, whereas a karate expert delivers a force in a short time for best results. Is this a contradiction?

Case 3: Decreasing Momentum over a Short Time

When boxing, move into a punch instead of away and you're in trouble. Likewise if you catch a high-speed baseball while your hand moves toward the ball instead of away upon contact. Or when out of control in a car, drive it into a concrete wall instead of a haystack and you're really in trouble. In these cases of short impact times, the impact forces are large.

The idea of short time of contact explains how a karate expert can fracture a stack of bricks with the blow of her bare hand (Figure 5-6). She brings her arm and hand swiftly against the bricks with considerable momentum. This momentum is quickly reduced when she delivers an impulse to the bricks. The impulse is the force of her hand against the bricks multiplied by the time her hand makes contact with the bricks. By swift execution she makes the time of contact very brief and correspondingly makes the force of impact huge. If her hand is made to bounce upon impact, the force is even greater.

▶ **Answers**

1. The force of impact will be three times less than if he didn't pull back.

2. The force of impact will be two times greater than if he were hit at rest. This force is further increased because of the additional impulse produced when his momentum of approach is stopped. The increased impulse and short time of impact result in forces that account for many knockouts.

3. There is no contradiction because the best results for each are quite different. The best result for the boxer is reduced force, accomplished by maximizing time, and the best result for the karate expert is increased force, delivered in minimum time.

Figure 5-6
Cassy imparts a large impulse to the bricks in a short time and produces a considerable force.

Bouncing

You know that if a flower pot falls from a shelf onto your head, you may be in trouble. And whether you know it or not, if it bounces from your head, you're certainly in trouble. Impulses are greater when bouncing takes place. This is because the impulse required to bring something to a stop and then, in effect, "throw it back again" is greater than the impulse required merely to bring something to a stop. Suppose, for example, that you catch the falling pot with your hands. Then you provide an impulse to catch it and reduce its momentum to zero. If you were to then throw the pot upward, you would have to provide additional impulse. So it would take more impulse to catch it and throw it back up than merely to catch it. The same greater impulse is supplied by your head if the pot bounces from it.

Figure 5-7
The Pelton wheel. The curved blades cause water to bounce and make a U-turn, which produces a greater impulse to turn the wheel.

Impulse

An interesting application of the greater impulse that occurs when bouncing takes place was employed with great success in California during the gold rush days. The water wheels used in gold-mining operations were inefficient. A man named Lester A. Pelton saw that the problem had to do with their flat paddles. He designed curved-shape paddles that would cause the incident water to make a U-turn upon impact—to "bounce." In this way the impulse exerted on the water wheels was greatly increased. Pelton patented his idea and made more money from his invention, the Pelton wheel, than any of the gold miners.

Questions

1. In reference to Figure 5-6, how does the force that Cassy exerts on the bricks compare to the force exerted on her hand?

2. How will the impulse at impact differ if Cassy's hand bounces back upon striking the bricks?

Conservation of Momentum

Recall from Chapter 4 that Newton's second law tells us that if we want to accelerate an object, we must apply a force on it. We say much the same thing in this chapter when we say that to change the momentum of an object, we must apply an impulse on it. In any case, the force or impulse must be exerted on the object by something external to the object. Internal forces don't count. For example, the molecular forces within a baseball have no effect on the momentum of the baseball, just as a person sitting inside an automobile pushing against the dashboard has no effect in changing the momentum of the automobile. This is because these forces are internal forces, that is, forces that act and react within the objects themselves. An outside, or external, force acting on the baseball or automobile is required for a change in momentum. If no external force is present, then no change in momentum is possible.

When a bullet is fired from a rifle, the forces present are internal forces. The total momentum of the system comprising the bullet and rifle therefore undergoes no net change (Figure 5-8). By Newton's third law of action and reaction, the force exerted on the bullet is equal and opposite to the force exerted on the rifle. The forces acting on the bullet and rifle act for the same time, resulting in equal but oppositely directed momenta (the plural form of *momentum*). The recoiling rifle has just as

▶ **Answers**

1. In accord with Newton's third law, the forces will be equal. Only the resilience of the human hand and the training she has undergone to toughen her hand allow this feat to be performed without broken bones.

2. The impulse will be greater if her hand bounces from the bricks upon impact. If the time of impact is not correspondingly increased, a greater force is then exerted on the bricks (and her hand!).

Figure 5-8
The momentum before firing
is zero. After firing, the net
momentum is still zero,
because the momentum of
the rifle is equal and opposite to the the momentum of
the bullet.

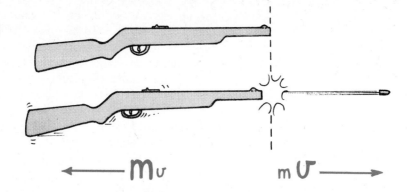

Figure 5-9
The machine gun recoils
from the bullets it fires and
climbs upward.

much momentum as the speeding bullet. Although both the bullet and rifle by themselves have gained considerable momentum, the bullet and rifle together as a system experience no net change in momentum. Before firing, the momentum was zero; after firing, the net momentum is still zero. No momentum is gained and no momentum is lost.

Two important ideas are to be learned from the rifle-and-bullet example. The first is that momentum, like velocity, is a quantity that is described by both magnitude and direction; we measure both "how much" and "which way." Momentum is a *vector quantity*. When momenta act in the same direction, they are simply added; when they act in opposite directions, they are subtracted.

The second important idea to be taken from the rifle-and-bullet example is the idea of *conservation*. The momentum before and after firing is the same. For the system of rifle and bullet, no momentum was gained; none was lost. When a physical quantity remains unchanged during a process, that quantity is said to be conserved. We say momentum is conserved.

If we extend the idea of a rifle recoiling or "kicking" from the bullet it fires, we can understand rocket propulsion. Consider a machine gun recoiling each time a bullet is fired. The momentum of recoil increases by an amount equal to the momentum of each bullet fired. If the machine gun is free to move, mounted so that it can slide along a stretched wire, for example, it will accelerate away from its target as the bullets are fired. A rocket accomplishes acceleration by the same means. It is continually "recoiling" from the ejected exhaust gases. Each molecule of exhaust gas can be thought of as a tiny bullet shot out of the rocket. If we consider the total system of rocket and exhaust gases, then, as with the rifle and bullet, the total momentum undergoes no net change. But we are not usually interested in the total momentum in such a case, and we concern ourselves with only the momentum of the rocket.

A common misconception is that a rocket is propelled by the exhaust gases pushing against the atmosphere. This is equivalent to saying that a gun kicks because the bullet pushes against the atmosphere—not true for either the recoiling gun or the recoiling rocket. In fact, a rocket works best above the atmosphere, where there is no air resistance to restrict speed. A rocket simply recoils from the gases it ejects. It will recoil best where air resistance is absent.

Questions ▶

1. A high-speed bus and an innocent bug have a head-on collision. The sudden change of momentum for the bug splatters it all over the windshield. Compared to the change in momentum experienced by the poor bug, is the change in momentum of the bus greater, less, or the same?

2. What happens to the speed of a fighter aircraft chasing another when it opens fire? What happens to the speed of the pursued aircraft when it returns the fire?

Collisions

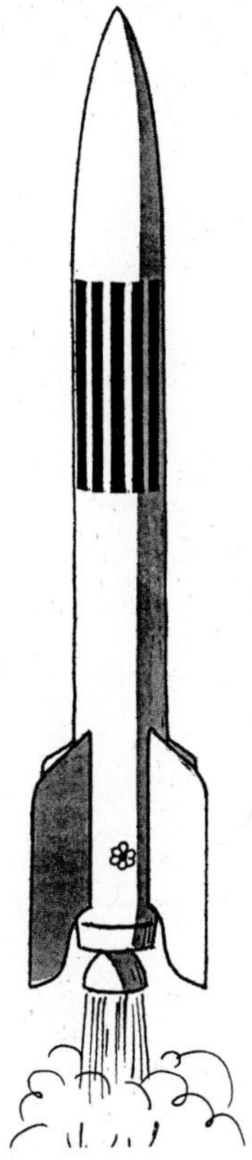

Momentum is conserved in collisions—that is, the total momentum of a system of colliding objects is unchanged before, during, and after the collision. This is because the forces that act during the collision are internal forces—forces acting and reacting within the system itself. There is only a redistribution or sharing of whatever momentum exists before the collision.

When a moving billiard ball makes a head-on collision with another billiard ball at rest, the moving ball comes to rest and the other ball moves with the speed of the colliding ball. We call this an **elastic collision**; the colliding objects rebound without lasting deformation or the generation of heat. Whatever the initial motions of colliding objects, their motions after rebound are such that they have the same total momentum (Figure 5-11).

In any collision, we can say

Total momentum before collision = total momentum after collision.

This is true even when the colliding objects become entangled during the collision. This is an **inelastic collision**, characterized by deformation or the generation of heat or both. Consider, for example, the case of a freight car moving along a track and colliding with another freight car at rest

▶ **Answers**

1. The momentum of both bug and bus change by the same amount because the amount of force and impact time, and therefore the impulse on each, is the same. Although the change of momentum for each is the same, the change of speed is another story. Because of the huge mass of the bus, the reduction of speed is very tiny—too small for the passengers to notice.

2. The momentum of the chasing aircraft is reduced by an amount equal to the momentum of the forward-directed bullets it fires. Its speed is therefore reduced. The speed of the aircraft being chased, however, is increased because its momentum is increased by an amount equal to the momentum it gives to the bullets it fires from its rear. The overall effect is to reduce the rate of closure between the two aircraft.

Figure 5-10
The rocket recoils from the "molecular bullets" it fires and climbs upward.

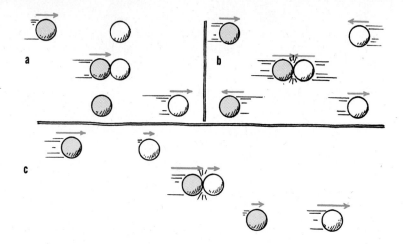

(Figure 5-12). If the freight cars are of equal mass and are coupled by
the collision, can we predict the velocity of the coupled cars after impact?

Suppose the single car is moving at 10 meters per second, and we
consider the mass of each car to be *m*. Then, from the conservation of
momentum,

$$(\text{Total } mv)_{\text{before}} = (\text{total } mv)_{\text{after}}$$

$$(m \times 10)_{\text{before}} = (2m \times ?)_{\text{after}}$$

Since twice as much mass is moving after the collision, the velocity must
be half as much as the velocity before collision, or 5 meters per second.
Both sides of the equation are then equal.

Notice the importance of direction in these cases. Momentum, like
velocity, is a vector quantity. We combine momenta as we do velocities.
Momenta in the same direction are simply added. If two objects are moving toward each other, one of the momenta is considered negative, and
we combine them by subtraction.

Figure 5-12

Inelastic collision. The
momentum of the freight car
on the left is shared with the
freight car on the right after
collision.

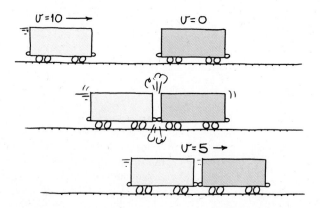

Figure 5-13
More inelastic collisions. The net momentum of the trucks before and after collision is the same.

Note the inelastic collision shown in Figure 5-13. If A and B are moving with equal momenta in opposite directions (A and B colliding head-on), then one of these is considered to be negative, and the momenta add algebraically to zero. After collision, the coupled wreck remains at the point of impact, with zero momentum.

If A and B are moving in the same direction (A catching up with B), the net momentum is simply the addition of their individual momenta.

If A, however, moves east with, say, 10 more units of momentum than B moving west (not shown in the figure), after collision the coupled wreck moves east with 10 units of momentum. The wreck will finally come to rest, of course, because of the external force of friction by the ground. The time of impact is short, however, and the impact force of the collision is so much greater than the external friction force that momentum immediately before and after the collision is practically conserved. The total momentum just before the trucks collide (10 units) is equal to the combined momentum of the crumpled trucks just after impact. When external force is absent—as it is, for example, when space vehicles dock in space—the 10 units of momentum will persist indefinitely.

Figure 5-14
An air track. Blasts of air from tiny holes provide a friction-free air cushion for the carts to glide upon.

Question ▶ Consider the air track in Figure 5-14. Suppose a gliding cart bumps into and sticks to a stationary cart that has three times the mass. Compared to the speed of the gliding cart, how fast will the coupled carts glide after collision?

More Complicated Collisions

The net momentum before and after any collision is unchanged, even when the colliding objects move at an angle to each other. Expressing the net momentum when different directions are involved can be achieved with the parallelogram rule of vector addition. We will not treat such complicated cases in great detail here but will show some simple examples to convey the idea.

Figure 5-15
Momentum is a vector quantity.

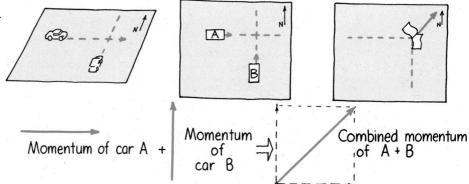

Momentum of car A + Momentum of car B ⟹ Combined momentum of A + B

In Figure 5-15 we see a collision between two cars traveling at right angles to each other. Car A has a momentum directed due east, and car B's momentum is directed due north. If their individual momenta are

Figure 5-16
After the bomb bursts, the momenta of its fragments add up (by vector addition) to the original momentum.

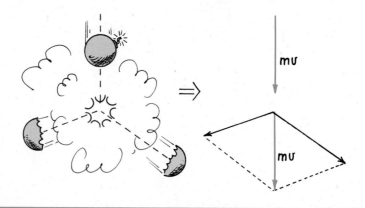

▶ **Answer**
Four times as much mass will be moving after collision, so the coupled carts will glide $\frac{1}{4}$ as fast.

equal in magnitude, then their combined momentum is in a northeast direction. This is the direction the coupled cars will travel after collision. We see that just as the diagonal of a square is not equal to the sum of two of the sides, the magnitude of the resulting momentum will not simply equal the sum of the two momenta before collision. Recall the relationship between the diagonal of a square and the length of one of its sides, Figure 3-5 in Chapter 3—the diagonal is $\sqrt{2}$ the length of the side of a square. Similarly, the magnitude of the resultant momentum will be equal to $\sqrt{2}$ of the momentum of either vehicle.

Figure 5-16 shows a falling bomb exploding into two pieces. The momenta of the bomb fragments combine by vector addition to equal the original momentum of the falling bomb. Figure 5-17b extends this idea to the microscopic realm, where the tracks of subatomic particles are revealed in a liquid hydrogen bubble chamber.

Figure 5-17

Momentum is conserved for colliding billiard balls and for colliding nuclear particles in a liquid hydrogen bubble chamber. In (a), the billiard ball A strikes billiard ball B, which was initially at rest. In (b), proton A collides with protons B, C, and D. The moving protons leave tracks of tiny bubbles.

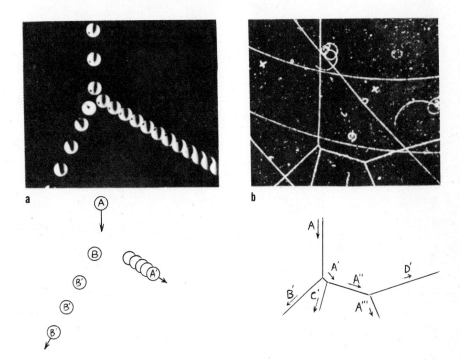

Whatever the nature of a collision or however complicated it is, the total momentum before, during, and after remains unchanged. This extremely useful concept enables us to learn much from collisions without regard to the form of the forces that interact in collisions. We will see in the next chapter that energy as well as momentum is conserved. By applying momentum and energy conservation to the collisions of subatomic particles as observed in various detection chambers, we can compute the masses of these tiny particles. We obtain this information by measuring momenta and energy before and after collisions. The forces of the collision processes, however complicated, need not be of concern.

Summary of Terms

Momentum The product of the mass of an object and its velocity.

Impulse The product of the force acting on an object and the time during which it acts.

Relationship of impulse and momentum Impulse is equal to the change in the momentum of the object that the impulse acts on. In symbol notation,

$$Ft = \Delta mv$$

Conservation of momentum When no external net force acts on an object or a system of objects, no change of momentum takes place. Hence, the momentum before an event involving only internal forces is equal to the momentum after the event:

$$mv_{(\text{before event})} = mv_{(\text{after event})}$$

Elastic collision A collision in which colliding objects rebound without lasting deformation or the generation of heat.

Inelastic collision A collision in which the colliding objects become distorted and generate heat during the collision.

Review Questions

Momentum

1. Which has a greater momentum, a heavy truck at rest or a moving skateboard?

Impulse

2. How does impulse differ from force?

3. What are the two ways that the impulse exerted on something can be increased or decreased?

Impulse and Momentum

4. What is the relationship of the impulse-momentum relationship to Newton's second law?

5. Why is it incorrect to say that impulse equals momentum?

Case 1: Increasing Momentum

6. To impart the greatest momentum to an object, should you both exert the largest force possible and extend that force for as long as possible? Explain.

7. For the same force, which cannon imparts the greatest speed to a cannonball—a long cannon or a short one? Explain.

Case 2: Decreasing Momentum over a Long Time

8. When you are in the way of a moving object and an impact force is your fate, are you better off decreasing its momentum over a short time or over a long time? Explain.

9. Why might a wine glass survive a fall onto a carpeted floor but not onto a concrete floor?

10. Why is it best to first extend your hand forward when you catch a fast-moving baseball with your bare hand?

Case 3: Decreasing Momentum over a Short Time

11. Why would it be a poor idea to have your hand against a rigid wall when you catch a fast-moving baseball with your bare hand?

12. Which has the greater momentum when they move at the same speed—a heavy truck or a skateboard? Which requires the greatest stopping force?

13. In answering the preceding question, perhaps you stated that the truck required more stopping force. Make an argument that the skateboard could require more stopping force. (Consider relative times.)

Bouncing

14. Which undergoes the greater change in momentum: (1) a moving object brought to rest, (2) the same object projected from rest to the speed it had before, or (3) the same moving object brought to rest and then projected backward to its original speed?

15. In the preceding question, in which case is the greatest impulse required?

Conservation of Momentum

16. Can you produce a net impulse to an automobile by sitting inside and pushing on the dashboard? Can the internal forces within a baseball produce an impulse on the baseball that will change its momentum?

17. Is it correct to say that if no impulse is exerted on an object, then no change in the momentum of the object will occur?

18. What does it mean to say that a quantity is *conserved*?

19. When a bullet is fired, its momentum indeed changes! Also the momentum of the recoiling rifle changes. So momentum is not conserved for the bullet, and momentum is not conserved for the rifle. In what sense do we say that momentum is conserved?

20. Would momentum be conserved for the system of rifle and bullet if momentum were not a vector quantity? Explain.

Collisions

21. Distinguish between an *elastic* collision and an *inelastic* collision. Why is momentum conserved for both types of collisions?

22. Railroad car A rolls at a certain speed against car B of the same mass, which is at rest. Compare the motions of the cars after collision if the collision is perfectly elastic.

23. In the preceding question, compare the motions of the cars before and after collision if the collision is inelastic.

More Complicated Collisions

24. Suppose a particle moving horizontally with 1 unit of momentum collides and sticks to an identical particle moving vertically with 1 unit of momentum. Why is their combined momentum not simply the arithmetic sum, 2 units?

25. In the preceding question, what is the net momentum before and after the collison?

Home Project

When you get a bit ahead in your studies, cut classes some afternoon and visit your local pool or billiards parlor and bone up on momentum conservation. Note that no matter how complicated the collision of balls, the momentum along the line of action of the cue ball before impact is the same as the combined momentum of all the balls along this direction after impact and that the components of momenta perpendicular to this line of action cancel to zero after impact, the same value as before impact in this direction. You'll see both the vector nature of momentum and its conservation more clearly when rotational skidding—English—is not imparted to the cue ball. When English is imparted by striking the cue ball off center, rotational momentum, which is also conserved, somewhat complicates analysis. But regardless of how the cue ball is struck, in the absence of external forces, both linear and rotational momentum are always conserved. Pool or billiards offers a first-rate exhibition of momentum conservation in action.

Exercises

1. To bring a supertanker to a stop, its engines are typically cut off about 25 km from port. Why is it so difficult to stop or turn a supertanker?

2. In terms of impulse and momentum, why are padded dashboards safer in automobiles?

3. In terms of impulse and momentum, why are nylon ropes, which stretch considerably under stress, favored by mountain climbers?

4. In terms of impulse and momentum, why is it important that an airplane wing be designed so that it deflects oncoming air downward?

5. It is generally much more difficult to stop a heavy truck than a skateboard when they move at the same speed. State a case where the moving skateboard could require more stopping force. (Consider relative times.)

6. If a ball is projected upward from the ground with 10 units of momentum, what is the momentum of recoil of the world? Why do we not feel this?

7. When an apple falls from a tree and strikes the ground without bouncing, what becomes of its momentum?

8. Why is a punch more forceful with a bare fist than with a boxing glove?

9. A boxer can punch a heavy bag for more than an hour without tiring, but will tire quickly when boxing with an opponent for a few minutes. Why? (*Hint:* When aimed at the bag, what supplies the impulse to stop the punches? When aimed at the opponent, what or who supplies the impulse to stop the punches that are missed?)

10. Railroad cars are loosely coupled so that there is a noticeable time delay from the time the first car is moved and last cars are moved from rest by the locomotive. Discuss the advisability of this loose coupling and slack between cars from the point of view of impulse and momentum.

11. If only an external force can change the state of motion of a body, how can the internal force of the brakes bring a car to rest?

12. A fully dressed person is at rest in the middle of a pond on perfectly frictionless ice and must get to shore. How can this be accomplished?

13. If you throw a ball horizontally while standing on roller skates, you roll backward with a momentum that matches that of the ball. Will you roll backward if you go through the motions of throwing the ball, but instead hold on to it? Explain.

14. Using examples, show that situations explained by Newton's third law can also be explained by the conservation of momentum.

15. Go back to Exercise 35 in Chapter 4 and answer it in terms of momentum conservation.

16. Your friend says that the law of momentum conservation is violated when a ball rolls down a hill and gains momentum. What do you say?

17. Why is it difficult for a fire fighter to hold a hose that ejects large amounts of water at a high speed?

18. Would you care to fire a gun that has a bullet ten times as massive as the gun? Explain.

19. The momentum of the rifle and bullet shown in Figure 5-8 is conserved. Why do we not say the *velocities* are conserved? What would happen if the mass of the bullet were equal to the mass of the rifle?

20. A railroad diesel engine weighs four times as much as a freight car. If the diesel engine coasts at 5 km per hour into a freight car that is initially at rest, how fast do the two coast after they couple together?

21. A 5-kg fish swimming 1 m/s swallows an absentminded 1-kg fish swimming toward it at 4 m/s. What is the speed of the larger fish immediately after lunch?

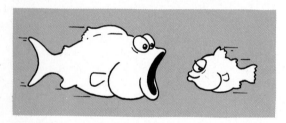

22. Before rockets in space were commonplace, a popular misconception was that a rocket needs air to push against. This of course is not true, as shown by rockets to the moon and so forth. How is a rocket propelled in a region completely devoid of an atmosphere?

23. An ice sailcraft is stalled on a frozen lake on a windless day. The skipper sets up a fan as shown. If all the wind bounces backward from the sail, will the craft be set in motion? If so, in what direction?

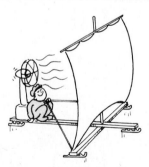

24. Will your answer to the preceding exercise be different if the air is brought to a halt by the sail without bouncing?

25. Discuss the advisability of simply removing the sail in the preceding exercises.

26. When you are traveling in your car at highway speed, the momentum of a bug is suddenly changed as it splatters onto your windshield. Compared to the change in momentum of the bug, by how much does the momentum of your car change?

27. If a Mack truck and a Volkswagen have a head-on collision, which vehicle will experience the greater force of impact? The greater impulse? The greater change in its momentum? The greater acceleration?

28. Would a head-on collision between two cars be more damaging to the occupants if the cars stuck together or if the cars rebounded upon impact?

29. Suppose there are three astronauts outside a spaceship, and two of them decide to play catch with the third man. All the astronauts weigh the same on earth and are equally strong. The first astronaut throws the second one toward the third one and the game begins. Describe the motion of the astronauts as the game proceeds. How long will the game last?

30. To throw a ball, do you exert an impulse on it? Do you exert an impulse to catch it at the same speed? About how much impulse do you exert, in comparison, if you catch it and then throw it back again? (Imagine yourself on a skateboard.)

31. In reference to Figure 5-6, how will the impulse at impact differ if Cassy's hand bounces back upon striking the bricks? In any case, how does the force she exerts on the bricks compare to the force exerted on her hand?

32. Light consists of tiny "corpuscles" called photons that possess momentum. This can be demonstrated with a radiometer, shown in the sketch. Metal vanes painted black on one side and white on the other are free to rotate about the point of a needle mounted in a vacuum. When photons are incident on the black surface, they are absorbed; when photons are incident upon the white sur-

face, they are reflected. Upon which surface is the impulse of incident light greater, and which way will the vanes rotate? (They rotate in the opposite direction in the more common radiometers where air is present in the glass chamber; your instructor may tell you why.)

33. A deuteron is a nuclear particle of unique mass made up of one proton and one neutron. Suppose it is accelerated up to a certain very high speed in a cyclotron and directed into an observation chamber, where it collides and sticks to a target particle that is initially at rest and then is observed to rebound at exactly half the speed of the incident deuteron. Why do the observers state that the target particle is itself a deuteron?

34. A billiard ball will stop short when it collides head-on with a ball at rest. The ball cannot stop short, however, if the collision is not exactly head-on—that is, if the second ball moves at an angle to the path of the first. Do you know why? (*Hint:* Consider momentum before and after the collision along the initial direction of the first ball and also in a direction perpendicular to this initial direction.)

6 Energy

Perhaps the concept most central to all of science is energy. The combination of energy and matter makes up the universe: matter is substance, and energy is the mover of substance. The idea of matter is easy to grasp. Matter is stuff that we can see, smell, and feel. It has mass and occupies space. Energy, on the other hand, is abstract. We cannot see, smell, or feel most forms of energy. Surprisingly, the idea of energy was unknown to Isaac Newton, and its existence was still being debated in the 1850s. Although energy is familiar to us, it is difficult to define, because it is not only a "thing" but both a thing and a process—as if it were both a noun and a verb. Persons, places, and things have energy, but we usually observe energy only when it is happening—only when it is being transformed. It comes to us in the form of electromagnetic waves from the sun and we feel it as heat; it is captured by plants and binds molecules of matter together; it is in the food we eat and we receive it by digestion. We begin our study of energy by observing a related concept: work.

Work

Figure 6-1
Gary does work on the barbell. If he were taller, more energy would be expended to press the barbell over his head.

In the last chapter we saw that changes in an object's motion depend on both force and how long the force acts. "How long" meant time. We called the quantity "force × time" *impulse*. But "how long" need not always mean time. It can mean distance also. When we consider the quantity *force × distance*, we are talking about an entirely different quantity—**work**.

When we lift a load against earth's gravity, work is done. The heavier the load or the higher we lift the load, the more work is done. Two things enter into every case where work is done: (1) the exertion of a force and (2) the movement of something by that force. For the simplest case, where the force is constant and the motion takes place in a straight line in the direction of the force,* we define the work done on an object by an applied force as the product of the force and the distance through which the object is moved. In shorter form:

$$\text{Work} = \text{force} \times \text{distance}$$
$$W = Fd$$

---

*More generally, work is the product of only the component of force that acts in the direction of motion and the distance moved.

93

Figure 6-2
He may expend energy when he pushes on the wall, but if it doesn't move, no work is performed on the wall.

If we lift two loads one story up, we do twice as much work as in lifting one load, because the *force* needed to lift twice the weight is twice as much. Similarly, if we lift a load two stories instead of one story, we do twice as much work because the *distance* is twice as much.

Thus the definition of work involves both a force and a distance. A weight lifter who holds a barbell weighing 1000 newtons overhead does no work on the barbell. He may get really tired doing so, but if the barbell is not moved by the force he exerts, he does no work *on* the barbell. Work may be done on the muscles by stretching and contracting, which is force times distance on a biological scale, but this work is not done on the barbell. Lifting the barbell, however, is a different story. When the weight lifter raises the barbell from the floor, work is done on the barbell.

The unit of measurement for work combines a unit of force (N) with a unit of distance (m); the unit of work is the newton-meter (N·m), also called the *joule* (J) (rhymes with *pool*). One joule of work is done when a force of 1 newton is exerted over a distance of 1 meter, as in lifting an apple over your head. For larger values we speak of kilojoules (kJ), thousands of joules, or megajoules (MJ), millions of joules. The weight-lifter in Figure 6-1 does work in kilojoules. The energy released by a kilogram of gasoline is rated in megajoules.

Power

Figure 6-3
The three main engines of a space shuttle can develop 33 000 MW of power when fuel is burned at the enormous rate of 3400 kg/s. This is like emptying an average-size swimming pool in 20 s.

The definition of work says nothing about how long it takes to do the work. The same amount of work is done when carrying a load up a flight of stairs, whether we walk up or run up. So why are we more tired after running upstairs in a few seconds than after walking upstairs in a few minutes? To understand this difference, we need to talk about a measure of how fast the work is done—**power**. Power is equal to the amount of work done per time it takes to do it:

$$\text{Power} = \frac{\text{work done}}{\text{time interval}}$$

An engine of great power can do work rapidly. An automobile engine with twice the power of another does not necessarily produce twice as much work or go twice as fast as the less powerful engine. Twice the power means it will do the same amount of work in half the time or twice the work in the same time. The main advantage of a powerful automobile engine is the acceleration it can produce. It can get the automobile up to a given speed in less time than less powerful engines.

Here's another way to look at power: a liter (L) of gasoline can do a certain amount of work, but the power produced when we burn it can be any amount, depending on how *fast* it is burned. The liter may produce 50 units of power for a half hour in an automobile or 90 000 units of power for one second in a Boeing 747.

The unit of power is the joule per second (J/s), also known as the watt (in honor of James Watt, the eighteenth-century developer of the steam engine). One watt (W) of power is expended when 1 joule of work is done in 1 second. One kilowatt (kW) equals 1000 watts. One megawatt (MW) equals 1 million watts. In the United States we customarily rate engines in units of horsepower and electricity in kilowatts, but either may be used. In the metric system of units, automobiles are rated in kilowatts. (One horsepower is about three-fourths of a kilowatt, so an engine rated at 134 horsepower is a 100-kW engine.)

Mechanical Energy

Work is done in lifting the heavy ram of a pile driver, and, as a result, the ram acquires the property of being able to do work on a piling beneath it when it falls. When work is done by an archer in drawing a bow, the bent bow has the ability of being able to do work on the arrow. When work is done to wind a spring mechanism, the spring acquires the ability to do work on various gears to run a clock, ring a bell, or sound an alarm. In each case, something has been acquired. This "something" that is given to the object enables the object to do work. This something may be a compression of atoms in the material of an object; it may be a physical separation of attracting bodies; it may be a rearrangement of electric charges in the molecules of a substance. This something that enables an object to do work is **energy**.* Like work, energy is measured in joules. It appears in many forms, which will be discussed in the following chapters. For now we focus on mechanical energy—the form of energy due to position (potential energy) or to the movement of mass (kinetic energy). Mechanical energy may be in the form of either potential energy or kinetic energy.

Potential Energy

An object may store energy because of its position. The energy that is stored and held in readiness is called **potential energy** (PE), because in the stored state it has the potential to do work. For example, a stretched or compressed spring has the potential for doing work. When a bow is drawn, energy is stored in the bow. A stretched rubber band has potential energy because of its position, for if it is part of a slingshot, it is capable of doing work.

The chemical energy in fuels is potential energy, for it is energy of position from a microscopic point of view. This energy is available when the positions of electric charges within and between molecules are altered, that is, when a chemical change takes place. Any substance that can do work through chemical action possesses potential energy. Potential energy is found in fossil fuels, electric batteries, wound clocks, and the food we eat.

Work is required to elevate objects against earth's gravity. The potential energy due to elevated positions is called gravitational potential energy.

Figure 6-4

The potential energy of Tenny's drawn bow equals the work (average force × distance) she did in drawing the arrow into position. When released, the potential energy of the drawn bow will become the kinetic energy of the arrow.

---

*Strictly speaking, that which enables an object to do work is its available energy, for not all the energy in an object can be transformed to work.

Figure 6-5

The potential energy of the 10-N ball is the same (30 J) in all three cases because the work done in elevating it 3 m is the same whether it is (*a*) lifted with 10 N of force, (*b*) pushed with 6 N of force up the 5-m incline, or (*c*) lifted with 10 N up each 1-m stair. No work is done in moving it horizontally (neglecting friction).

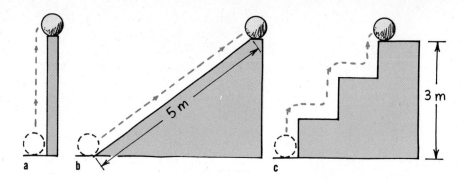

Water in an elevated reservoir, the ram of a pile driver, or simply a book perched atop a high shelf has gravitational potential energy. The amount of gravitational potential energy possessed by an elevated object is equal to the work done against gravity in lifting it. The work done equals the force required to move it upward times the vertical distance it is moved ($W = Fd$). The upward force equals the weight mg of the object. So the work done in lifting it through a height h is given by the product mgh:

$$\text{Gravitational potential energy} = \text{weight} \times \text{height}$$

$$PE = mgh$$

Note that the height h is the distance above some reference level, such as the ground or the floor of a building. The potential energy mgh is relative to that level and depends only on mg and the height h. You can see in Figure 6-5 that the potential energy of the ball at the top of the structure does not depend on the path taken to get it there. We will see that the effort we expend (the force) and the work we do are two different things.

Questions ▶

1. How much work is done on a 75-N bowling ball when you move it horizontally across a 10-m-wide room?

2. How much work is done on it when you lift it 1 m? What power is expended if you lift it this distance in 1 s?

3. What is its gravitational potential energy in the lifted position?

▶ **Answers**

1. You do no work on the ball moved horizontally, for you apply no force (except for the tiny bit to start it) in its direction of motion. It has no more PE across the room than it had initially.

2. You do 75 J of work when you lift it 1 m (Fd = 75 N·m = 75 J). Power = 75 J/1 s = 75 W.

3. It depends. With respect to its starting position its PE is 75 J; with respect to some other reference level, it would be some other value.

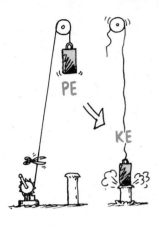

Figure 6-6
The potential energy of the elevated ram is converted to kinetic energy when released.

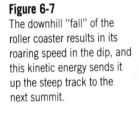

Kinetic Energy

If we push on an object, we can set it in motion. More specifically, if we do work on an object, we can change the energy of motion of that object. If an object is moving, then by virtue of that motion it is capable of doing work. We call energy of motion **kinetic energy** (KE). The kinetic energy of an object depends on mass and speed. It is equal to half the mass multiplied by the square of the speed.

$$\text{Kinetic energy} = \tfrac{1}{2} \text{ mass} \times \text{speed}^2$$

$$KE = \tfrac{1}{2} mv^2$$

When a ball is thrown, work is done on it, giving it kinetic energy. The moving ball can then hit something and push against it, doing work on what it hits. The kinetic energy of a moving object is equal to the work done in bringing it from rest to that speed or to the work it can do in being brought to rest:*

$$\text{Net force} \times \text{distance} = \text{kinetic energy}$$

$$Fd = \tfrac{1}{2} mv^2$$

Put still another way, we can say that the work done is equal to the change in kinetic energy:

$$\text{Work} = \Delta\text{KE}$$

Figure 6-7
The downhill "fall" of the roller coaster results in its roaring speed in the dip, and this kinetic energy sends it up the steep track to the next summit.

---

*This can be derived as follows: if we multiply both sides of $F = ma$ (Newton's second law) by d, we get $Fd = mad$. Recall from Chapter 2 that $d = \tfrac{1}{2} at^2$, so we can say $Fd = ma(\tfrac{1}{2} at^2) = \tfrac{1}{2} m(at)^2$; and substituting $v = at$, we get $Fd = \tfrac{1}{2}mv^2$.

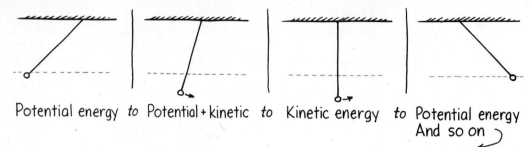

Potential energy *to* Potential+kinetic *to* Kinetic energy *to* Potential energy And so on

Figure 6-8
Energy transitions in a pendulum.

Notice that speed is squared, so if the speed of an object is doubled, its kinetic energy is quadrupled ($2^2 = 4$), and the object can do four times as much work. An object moving twice as fast as another takes four times as much work to stop. Accident investigators are well aware that an automobile going 100 kilometers per hour has four times the kinetic energy it would have at 50 kilometers per hour. This means a car going 100 kilometers per hour will skid four times as far when its brakes are locked as it would going 50 kilometers per hour. This is because speed is squared for kinetic energy.

Questions ▶

1. When the brakes of a car going 90 km/h are locked, how much farther will it skid than if the brakes lock at 30 km/h?

2. Can an object have energy?

3. Can an object have work?

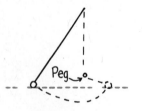

Figure 6-9
The pendulum bob will swing to its original height whether or not the peg is present.

Kinetic energy underlies other seemingly different forms of energy such as heat, sound, and light. Random molecular motion is sensed as heat, molecules vibrating in rhythmic patterns are perceived as sound, and electrons in motion make electric currents. Even light energy originates from the motion of electrons within atoms. We will find there is much in common among the various forms of energy that we will investigate.

▶ **Answers**

1. Nine times farther. The car has nine times as much energy when it travels three times as fast: $\frac{1}{2} m(3v)^2 = \frac{1}{2} m9v^2 = 9(\frac{1}{2} mv^2)$. The friction force will ordinarily be the same in either case; therefore, to do nine times the work requires nine times as much sliding distance.

2. Yes, but in a relative sense. For example, an elevated object may possess PE relative to the ground below, but none relative to a point at the same elevation. Similarly, the KE that an object has is with respect to a frame of reference, usually taken to be the earth's surface. We will see that material objects have "energy of being"—the congealed energy that makes up mass. Read on!

3. No, unlike momentum or energy, work is not something that an object *has*. Work is something that an object *does* to some other object. An object can *do* work only if it has energy.

Conservation of Energy

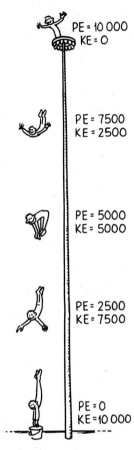

PE = 10 000	KE = 0
PE = 7500	KE = 2500
PE = 5000	KE = 5000
PE = 2500	KE = 7500
PE = 0	KE = 10 000

Figure 6-10
A circus diver at the top of a pole has a potential energy of 10,000 J. As he dives, his potential energy converts to kinetic energy. Note that at successive positions one-fourth, one-half, three-fourths, and all the way down, the total energy is constant. (Adapted from *Physics in Your World* by K. F. Kuhn and J. S. Faughn. Copyright © 1980 by W. B. Saunders Company. Reprinted by permission.)

More important than being able to state *what energy is* is understanding how it behaves—*how it transforms*. We can better understand processes or changes that occur in nature if we analyze them in terms of transformations of energy from one form to another or of transfers from one place to another.

As we draw back the stone in a slingshot, we do work in stretching the rubber band; we give the rubber band potential energy. When released, the stone has kinetic energy equal to this potential energy. It transfers this energy to its target, perhaps a wooden fence post. The slight distance the post is moved multiplied by the average force of impact doesn't quite match the kinetic energy of the stone. The energy score doesn't balance. But if we investigate further, we'll find that both the stone and fence post are a bit warmer. By how much? By the energy difference. Energy changes from one form to another. It transforms without net loss or net gain.

The study of various forms of energy and their transformations from one form into another has led to one of the greatest generalizations in physics—the law of **conservation of energy**:

Energy cannot be created or destroyed; it may be transformed from one form into another, but the total amount of energy never changes.

When we consider any system in its entirety, whether it be as simple as a swinging pendulum or as complex as an exploding galaxy, there is one quantity that doesn't change: energy. It may change form or it may simply be transferred from one place to another, but, as far as we can tell, the total energy score stays the same. This energy score takes into account the fact that the atoms that make up matter are themselves concentrated bundles of energy. When the nuclei (cores) of atoms rearrange themselves, enormous amounts of energy can be released. The sun shines because some of this energy is transformed into radiant energy. In nuclear reactors much of this energy is transformed into heat. Enormous gravitational forces in the deep, hot interior of the sun crush the cores of hydrogen atoms together to form helium atoms. This welding together of atomic cores is called *thermonuclear fusion*. This process releases radiant energy, some of which reaches the earth. Part of this energy falls on plants, and part of this in turn later becomes coal. Another part supports life in the food chain that begins with plants, and part of this energy later becomes oil. Part of the energy from the sun goes into the evaporation of water from the ocean, and part of this returns to the earth as rain that may be trapped behind a dam. By virtue of its position, the water in a dam has energy that may be used to power a generating plant below, where it will be transformed to electric energy. The energy travels through wires to homes, where it is used for lighting, heating, cooking, and operating electric gadgets. How nice that energy is transformed from one form to another!

Machines

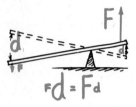

Figure 6-11
The lever.

A **machine** is a device for multiplying forces or simply changing the direction of forces. Underlying every machine is the *conservation of energy*. Consider one of the simplest machines, the *lever* (Figure 6-11). At the same time we do work on one end of the lever, the other end does work on the load. We see that the direction of force is changed, for if we push down, the load is lifted up. If the heat from friction forces is small enough to neglect, the work input will be equal to the work output.

$$\text{Work input} = \text{work output}$$

Since work equals force times distance, input force × input distance = output force × output distance.

$$(\text{Force} \times \text{distance})_{\text{input}} = (\text{force} \times \text{distance})_{\text{output}}$$

If the pivot point, or *fulcrum*, of the lever is relatively close to the load, then a small input force will produce a large output force. This is because the input force is exerted through a large distance and the load is moved over a correspondingly short distance. In this way, a lever can multiply forces. But no machine can multiply work or multiply energy. That's a conservation of energy no-no!

Figure 6-12
Applied force × applied distance = output force × output distance.

5000 N

25 cm

$$_F d = F_d$$

$$50 \times 25 = 5000 \times 0.25$$

Figure 6-13
Applied force × applied distance = output force × output distance.

A child uses the principle of the lever in jacking up the front end of an automobile. By exerting a small force through a large distance, she is able to provide a large force acting through a small distance. Consider the ideal example illustrated in Figure 6-12. Every time she pushes the jack handle down 25 centimeters, the car rises only a hundredth as far but with 100 times the force.

A block and tackle, or system of pulleys, is a simple machine that multiplies force at the expense of distance. One can exert a relatively small force through a relatively large distance and lift a heavy load through a relatively short distance. With the ideal pulley system such as that shown in Figure 6-13, the man pulls 10 meters of rope with a force of 50 newtons and lifts 500 newtons through a vertical distance of 1 meter. Neglecting the effects of friction, the energy the man expends in pulling the rope is numerically equal to the increased potential energy of the 500-newton block.

Any machine that multiplies force does so at the expense of distance. Likewise, any machine that multiplies distance, such as the construction of your forearm and elbow, does so at the expense of force. No machine or device can put out more energy than is put into it. A machine can multiply force, but not energy. No machine can create energy; it can only transfer energy from one place to another or transform it from one form to another.

Efficiency

The three previous examples were of ideal machines; 100 percent of the work input was transformed to work output. An ideal machine would operate at 100 percent efficiency. In practice, this doesn't happen and we can never expect it to happen. In any transformation some energy is dissipated to molecular kinetic energy: heat. This makes the machine warmer.

Even a lever rocks about its fulcrum and converts a small fraction of the input energy into heat. We may do 100 joules of work and get out 98 joules of work. The lever is then 98 percent efficient, and we waste only 2 joules of work input on heat. If the girl in Figure 6-12 puts in 100 joules of work and increases the potential energy of the car by 60 joules, the jack is 60 percent efficient; 40 joules of work have been used to do the work in overcoming the friction force, and these appear as heat. In a pulley system, a larger fraction of input energy goes into heat. If we do 100 joules of work, the forces of friction acting through the distances through which the pulleys turn and rub about their axles may dissipate 60 joules of energy as heat. So the work output is only 40 joules and the pulley system has an efficiency of 40 percent. The lower the efficiency of a machine, the greater the amount of energy wasted as heat.

Inefficiency exists whenever energy is transformed from one form to another. **Efficiency** can be expressed by the ratio

$$\text{Efficiency} = \frac{\text{work done}}{\text{energy used}}$$

Question ▶ Suppose a miracle car has a 100 percent efficient engine and burns fuel that has an energy content of 40 MJ/L. If the air drag and overall frictional forces on the car traveling at highway speed is 2000 N, what is the upper limit in distance per liter the car could go at this speed?

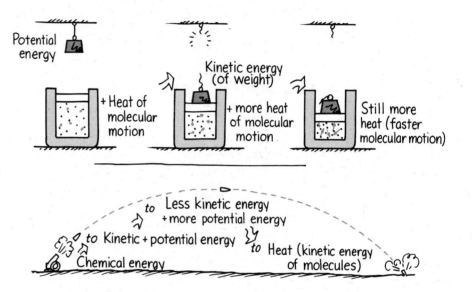

Figure 6-14
Energy transitions. The graveyard of kinetic energy is heat.

    An automobile engine is a machine that transforms chemical energy stored in fuel into mechanical energy. The bonds between the molecules in the petroleum fuel break up when the fuel burns by reacting with the oxygen in the air. Carbon atoms then bond with oxygen to form carbon monoxide, in which the bonds have less energy stored in them than they do in the original bonds. Some of the remaining energy goes into running the engine. We'd like all this energy converted into mechanical energy; that is, we'd like an engine that is 100 percent efficient. This is impossible, because some of the energy goes out in the hot exhaust gases, and nearly half is wasted in the friction of the moving engine parts. In addition to these inefficiencies, the fuel doesn't even burn completely and a certain amount of fuel energy goes unused.

▶ **Answer**

From the definition work = force × distance, simple rearrangement gives distance = work/force. If all 40 million J of energy in 1 L were used to do the work of overcoming the air drag and frictional forces, the distance would be:

$$\text{Distance} = \frac{\text{work}}{\text{force}} = \frac{40\ 000\ 000\ \text{J/L}}{2000\ \text{N}} = 20\ 000\ \text{m/L} = 20\ \text{km/L}$$

The important point here is that even with a perfect engine, there is an upper limit of fuel economy dictated by the conservation of energy.

Look at the inefficiency that accompanies transformations of energy this way: in any transformation there is a dilution of available *useful energy* to a state of wasted energy. The amount of usable energy decreases with each transformation and ultimately becomes heat. When we study thermodynamics, we'll see that the energy in the form of heat is useless for doing work unless it can be transformed to a lower temperature. Heat is the graveyard of useful energy.

Comparison of Kinetic Energy and Momentum*

Metal bullet penetrates

Rubber bullet bounces

Figure 6-15

Compared to the metal bullet with the same momentum, the rubber bullet is more effective in knocking the block over because it bounces on impact. The rubber bullet therefore undergoes the greater change in momentum and thereby imparts the greater impulse or wallop to the block. Which bullet does more damage?

Kinetic energy and momentum are both concepts concerning the motion of an object. But they are different. For example, the momentum of one object in collision with another may add or subtract. Like velocity, momentum is a vector quantity and is therefore directional and capable of being canceled entirely. But kinetic energy is a scalar quantity, like mass, and can never be canceled.† The momenta of two cars just before a head-on collision may cancel to zero, and the combined wreck after collision will have the same zero value for the momentum; but the kinetic energies add, as evidenced by the deformation and heat after collision. Or the momenta of two firecrackers approaching each other may cancel, but when they explode, there is no way their energies can cancel. Energies transform to other forms; momenta do not. The vector quantity momentum is different from the scalar quantity kinetic energy.

Another difference is the velocity dependence of the two. Whereas momentum is proportional to velocity (mv), kinetic energy is proportional to the square of velocity ($\frac{1}{2}mv^2$). An object that moves with twice the velocity of another object of the same mass has twice the momentum but four times the the kinetic energy. It can provide twice the impulse to whatever it encounters but do four times as much work.

Consider a lead bullet fired into a very large block of wood (Figure 6-15). When the bullet strikes, the bullet tends to knock the block over as the momentum of the bullet is transferred to the block. If the same bullet strikes with twice the velocity, it has twice the capability of knocking the block over; that is, if the first bullet is just barely able to knock the block over, the twice-as-fast bullet will deliver twice the impulse and be able to knock over a geometrically similar yet twice-as-massive block. But the twice-as-fast bullet has four times the kinetic energy and will penetrate four times as far. It delivers twice the wallop but four times the damage.

If our bullet is made of rubber instead of lead so that it bounces from the block instead of penetrating, it will be even more effective than a lead bullet in knocking the block over. Let's see why. If the momenta are the same for the lead and rubber bullets and they are simply brought

---

*This section may be skipped in a light treatment of mechanics.

†If you're into mathematics, you may know that when a vector quantity (velocity) is multiplied by a scalar quantity (mass), the product is also a vector (momentum). But a vector quantity squared (velocity^2) is a scalar. So KE is a scalar quantity. It can't be canceled. Although momenta can be combined in such a way to cancel to zero, there is no way to combine KEs to equal zero.

to rest, then the change in momenta and the accompanying impulses would be the same; but only the lead bullet is brought to rest. The rubber bullet bounces, which means its change in momentum and the accompanying impulse are even greater, for the bullet not only is brought to rest but also is "thrown back out again." If it bounces elastically with no loss in speed, then the change in momentum and impulse is doubled. It hits with twice as much wallop as a penetrating bullet of the same momentum.

a   b   c   d

Figure 6-16

(*a*) Dave releases the dart, which is brought to rest when it sticks into the wood block (*b*). Dave removes the metal point from the dart to expose its rubber nose and (*c*) releases it from the same height. (*d*) The dart bounces from the block, which this time is knocked over. Dave says, "The impulse is greater for bouncing because the dart is more than simply stopped; it is thrown back out again." In terms of momentum, Helen adds, "If the dart hits with positive momentum, it bounces with negative momentum. The change in momentum from positive to negative in bouncing is greater than from positive to zero in sticking. The greater change in momentum results in greater impulse."

But it does not penetrate, and, in this sense, it does little damage. This is an idealized case. In practice, such collisions are less than perfectly elastic. But now you have a better idea of why rubber bullets are used by police for knockdowns with minimum injury.

This is more safely demonstrated in Figure 6-16. When the rubber-nosed dart equipped with a sharp nail makes impact against the wooden block, the collision is inelastic. The impulse is not enough to tip the block over. But when the nail is removed to expose the rubber nose, the dart swings with the same momentum and *bounces* from the block. The nearly doubled impulse knocks the block over.

Suppose you're carrying a football and are about to be slammed into and tackled by an approaching player whose oncoming momentum is equal but opposite to your own. The combined momentum of you and the approaching player before impact cancel to zero, and, sure enough, on impact you both stop short in your tracks. The wallop or jarring exerted on each of you is the same. This is the case whether you are tackled by a slow-moving heavy player or a fast-moving light player. If the product of his mass and his velocity matches yours, you're stopped short. Stopping power is one thing, but what about damage? Every football player knows it hurts more to be stopped by a fast-moving light player than by a slow-moving heavy player. Why? Because a light player moving with the same momentum has more kinetic energy. If he has the same momentum as a heavy player but is only half as massive, he has twice the velocity. Twice the velocity doesn't give him four times the energy because his mass is half, but it gives him twice the kinetic energy of the heavier

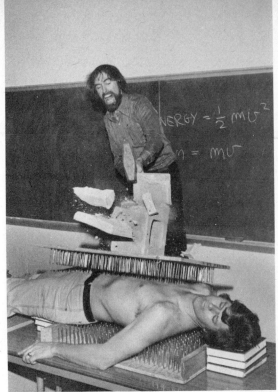

Figure 6-17
The author puts kinetic energy and momentum into the hammer, which strikes the block that rests on fellow physics instructor Paul Robinson, who is bravely sandwiched between beds of nails. Paul is not harmed. Why? Except for the flying cement fragments, every bit of the momentum of the hammer at impact is imparted to Paul, and subsequently to the table and the earth that supports him. But the momentum only provides the wallop; the energy does the damage. Most of the kinetic energy never gets to him, for it goes into smashing the block apart and into heat. What energy remains is distributed over the more than 200 nails that make contact with his body. The driving force per nail is not enough to puncture the skin.

player.* He does twice the work on you, tends to penetrate twice as far into you, and generally does twice the damage to you. Watch out for the fast-moving little guys!

Energy for Life

Every plant, animal, and living cell is a living machine and, like any other, needs an energy supply. The fuels for living organisms on this planet are various hydrocarbon compounds that react with oxygen to release energy. As with petroleum we previously discussed, the chemical bonds between hydrogen, carbon, and oxygen in food store more energy than ends up in the bonds of resulting carbon dioxide and water products after digestion. The difference sustains life.

The principal difference between the combustion of food in digestion and the combustion of fossil fuels in mechanical engines is the rate at which reactions take place. In digestion the reaction is much slower, and

---

*Note that $\frac{1}{2}(m/2)(2v)^2 = mv^2$, which is twice the value $\frac{1}{2}mv^2$ of the heavier player of mass m and speed v.

energy is released as it is required. Like the burning of fossil fuels, the reaction is self-sustaining once started as carbon combines with oxygen to form carbon dioxide. The reverse process is more difficult. Only green plants can make carbon dioxide combine with water to produce hydrocarbon compounds such as sugar. This process is photosynthesis and is energized by sunlight. Sugar is the simplest food, and all others—carbohydrates, proteins, and fats—are also synthesized compounds of carbon, hydrogen, and oxygen. How very fortunate that more energy is locked in the bonds of these hydrocarbons than in their products after combustion!

We see efficiency at work in the food chain. Larger creatures feed on smaller creatures, who in turn eat smaller creatures, and so on down the line to land plants and ocean plankton that are nourished by the sun. Moving each step up the food chain involves inefficiency. In the African bush, 10 kilograms of grass may produce 1 kilogram of gazelle. However, it will take 10 kilograms of gazelle to sustain 1 kilogram of lion. We see that each energy transformation along the food chain contributes to overall inefficiency. Interestingly enough, some of the largest creatures on the planet, the elephant and the blue whale, eat far down on the food chain. And more and more humans are considering substances such as krill or yeast as efficient sources of nourishment.

Summary of Terms

Work The product of the force and the distance through which the force moves:

$$W = Fd$$

Power The time rate of work:

$$Power = (work/time)$$

Energy The property of a system that enables it to do work.

Potential energy The stored energy that an object possesses because of its position. An elevated object has gravitational potential energy mgh.

Kinetic energy Energy of motion, described by the relationship

$$Kinetic\ energy = \tfrac{1}{2}mv^2$$

Conservation of energy Energy cannot be created or destroyed; it may be transformed from one form into another, but the total amount of energy never changes.

Machine A device that multiplies force or changes its direction. In an ideal machine, where no energy is transformed into heat,

$$Work_{input} = work_{output} \quad and \quad (Fd)_{input} = (Fd)_{output}$$

Efficiency The percent of the work put into a machine that is converted into useful work output.

Review Questions

1. When is energy most evident?

Work

2. A force sets an object in motion. When the force is multiplied by the time of its application, we call the quantity *impulse*, which changes the *momentum* of that object. What do we call the quantity *force × distance*, and what quantity does this change?

3. Cite an example where a force is exerted on an object without doing work on the object.

4. How many joules of work are done when a force of 1 N moves a book 2 m?

5. Which requires more work—lifting a 50-kg sack a vertical distance of 2 m or lifting a 25-kg sack a vertical distance of 4 m?

Power

6. If both sacks in the preceding question are lifted their respective distances in the same time, how does the power required for each compare? How about for the case where the lighter sack is moved its distance in half the time?

7. How many watts of power are expended when a force of 1 N moves a book 2 m in a time interval of 1 s?

Mechanical Energy

8. Exactly what is it that energy is capable of doing?

Potential Energy

9. A car is lifted a certain distance in a service station and therefore has potential energy with respect to the floor. If it were lifted twice as high, how much potential energy would it have?

10. Two cars are lifted to the same elevation in a service station. If one car is twice as heavy as the other, how do their potential energies compare?

11. How many joules of potential energy does a 1-kg book have when it is elevated 4 m? When it is elevated 8 m?

Kinetic Energy

12. How many joules of kinetic energy does a 1-kg book have when it is tossed across the room at a speed of 2 m/s? How much energy is imparted to the wall it accidentally encounters?

13. A moving car has kinetic energy. If it speeds up by four times, how much kinetic energy does it have in comparison? Compared to its original speed, how much work must the brakes supply to stop the four-times-as-fast car?

14. Which has the greater kinetic energy, a car traveling 30 km/h or a half-as-heavy car traveling 60 km/h?

Conservation of Energy

15. What will be the kinetic energy of an arrow shot from a bow having a potential energy of 40 J?

16. An apple hanging from a limb has potential energy. If it falls, what becomes of this energy just before it hits the ground? After it hits the ground?

17. Explain how the energy in a green leaf is the energy of sunshine.

18. Explain how the energy that operates an electric toothbrush is actually the energy of thermonuclear fusion.

Machines

19. Can a machine multiply input force? Input distance? Input energy? (If your three answers are the same, seek help, for the last question is especially important.)

20. A force of 50 N is applied to the end of a lever, which is moved a certain distance. If the other end of the lever is moved half as far, how much force can it exert?

21. If the man in Figure 6-13 pulls 1 m of rope downward with a force of 100 N, and the load rises one tenth as high (10 cm), what is the maximum load that can be lifted?

Efficiency

22. Referring to the previous question, if the load lifted is 500 N, what is the efficiency of the pulley system?

23. Referring to the previous question, how much work does the man expend? How much potential energy does the 500-N load acquire? What has become of the difference in energies?

24. Compared to work input, how much work is put out by a machine that operates at 30 percent efficiency?

25. Is a machine physically possible that has an efficiency greater than 100 percent? Discuss.

Comparison of Kinetic Energy and Momentum

26. What does it mean to say momentum is a vector quantity and energy is a scalar quantity?

27. Can momenta cancel? Can energies cancel?

28. If a moving object doubles its speed, how much more momentum does it have? Energy?

29. If a moving object doubles its speed, how much more impulse does it provide to whatever it bumps into (how much more wallop)? How much more work (how much more damage)?

Energy for Life

30. The energy we require for existence comes from the chemically stored potential energy in food, which is transformed into other forms by the process of digestion. What happens to a person whose work output is less than the energy he or she consumes? Whose work output is greater than the energy he or she consumes? Can an undernourished person perform extra work without extra food? Briefly discuss.

Home Projects

1. Fill two mixing bowls with water from the cold tap and take their temperatures. Then run an electric or hand beater in the first bowl for a few minutes. Compare the temperatures of the water in the two bowls.

2. Pour some dry sand into a tin can with a cover. Compare the temperature of the sand before and after vigorously shaking the can for a couple of minutes.

Exercises

1. Why is it easier to stop a lightly loaded truck than a heavier one that has equal speed?

2. Which requires more work to stop—a light truck or a heavy truck moving with the same momentum?

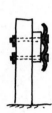

3. The design of road dividers that separate opposing traffic lanes used to consist of metal rails mounted as shown on the left. The newer design is of concrete with a cross section very deliberately shaped as shown at the right. Why is this new design more effective in slowing down a car that side-swipes it?

4. To determine the potential energy of Tenny's drawn bow (Figure 6-4), would it be an underestimate or an overestimate to multiply the force with which she holds the arrow in its drawn position by the distance she pulled it? Why do we say the work done is the *average* force × distance?

5. When a rifle with a long barrel is fired, the force of expanding gases acts on the bullet for a longer distance. What effect does this have on the velocity of the emerging bullet? (Do you see why long-range cannons have such long barrels?)

6. You and a flight attendant toss a ball back and forth in an airplane in flight. Does the KE of the ball depend on the speed of the airplane? Carefully explain.

7. Can something have energy without having momentum? Explain. Can something have momentum without having energy? Defend your answer.

8. At what point in its motion is the KE of a pendulum bob a maximum? At what point is its PE a maximum? When its KE is half its maximum value, how much PE does it have?

9. This question is typical on some driver's license exams: A car moving at 50 km/h skids 15 m with locked brakes. How far will the car skid with locked brakes at 150 km/h?

10. A physics instructor demonstrates energy conservation by releasing a heavy pendulum bob, as shown in the sketch, allowing it to swing to and fro. What would happen if in his exuberance he gave the bob a slight shove as it left his nose? Why?

11. Discuss the design of the roller coaster shown in the sketch in terms of the conservation of energy.

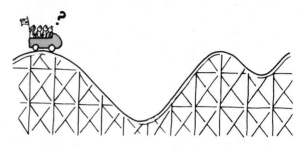

12. Suppose that you and two classmates are discussing the design of a roller coaster. One classmate says that each summit must be lower than the previous one. Your other classmate says this is nonsense, for as long as the first one is the highest, it doesn't matter what height the others are. What do you say?

13. Suppose an object is set sliding with a speed less than escape velocity on an infinite frictionless plane in contact with the surface of the earth, as shown. Describe its motion. (Will it slide forever at a constant velocity? Will it slide to a stop? In what way will its energy transfer be similar to that of a pendulum?)

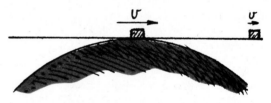

14. Does a car burn more gasoline when its lights are turned on? Does the overall consumption of gasoline depend on whether or not the engine is running? Defend your answer.

15. Matter can be recycled without loss of substance. Can energy be similarly recycled? Explain.

16. An inefficient machine is said to "waste energy." Does this mean that energy is actually lost? Explain.

17. You tell your friend that no machine can possibly put out more energy than is put into it, and your friend states that a nuclear reactor puts out more energy than is put into it. What do you say?

18. Your monthly electric bill is probably expressed in kilowatt-hours (kW·h), a unit of energy delivered by the flow of 1 kW of electricity for 1 h. How many joules of energy do you get when you buy 1 kW·h?

19. This may seem like an easy question for a physics type to answer: With what force does a rock that weighs 10 N strike the ground if dropped from a rest position 10 m high? This question does not have a straightforward numerical answer. Why?

20. Your friend is confused about ideas discussed in Chapter 4 that seem to contradict ideas discussed in this chapter. For example, in Chapter 4 we learned that the net force is zero for a car traveling along a level road at constant velocity, and in this chapter we learned that work is done in such a case. Your friend asks, "How can work be done when the net force equals zero?" What is your explanation?

21. In the absence of air resistance, a ball thrown vertically upward with a certain initial KE will return to its original level with the same KE. When air resistance is a factor affecting the ball, will it return to its original level with the same, less, or more KE? Does your answer contradict the conservation of energy? Defend your answer.

22. You're on a rooftop and you throw one ball downward to the ground below and another upward. The second ball, after rising, falls and also strikes the ground below. If air resistance can be neglected and if your downward and upward initial speeds were the same, how do the speeds of the balls compare upon striking the ground? (Use the idea of energy conservation to arrive at your answer.)

23. Scissors for cutting paper have long blades and short handles, whereas metal-cutting shears have long handles and short blades. Bolt cutters have very long handles and very short blades. Why is this so?

24. In the hydraulic machine shown, it is observed that when the small piston is pushed down 10 cm, the large piston is raised 1 cm. If the small piston is pushed down with a force of 100 N, how much force is the large piston capable of exerting?

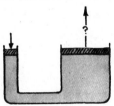

25. Discuss the fate of the physics instructor sandwiched between the beds of nails (Figure 6-17) if the block were less massive and unbreakable and the beds contained fewer nails.

26. Consider the swinging-balls apparatus. If two balls are lifted and released, momentum is conserved as two balls pop out the other side with the same speed as the released balls at impact. But momentum would also be conserved if one ball popped out at twice the speed.

Can you explain why this never happens? (And why is this exercise in Chapter 6 rather than Chapter 5?)

27. Consider the inelastic collision between the two freight cars in the last chapter (Figure 5-12). The momentum before and after the collision is the same. The KE, however, is less after the collision than before the collision. How much less, and what becomes of this energy?

28. A golf ball is hurled at and bounces from a very massive bowling ball that is initially at rest. After collision, which of the two balls has the greater momentum? (*Hint:* The golf ball rebounds with practically the same magnitude of momentum, but in the opposite direction. How great will be the momentum of the larger ball in order that the total momentum before and after collision is the same?) Which ball will have the greater kinetic energy?

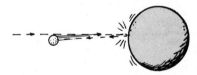

29. How many kilometers per liter will a car obtain if its engine is 25 percent efficient and it encounters an average retarding force of 1000 N? Assume that the energy content of gasoline is 40 MJ/L.

30. If an automobile had a 100 percent efficient engine, would it be warm to your touch? Would its exhaust heat the surrounding air? Would it make any noise? Would it vibrate? Would any of its fuel go unused?

31. What is the efficiency of a pulley system that will raise a 1000-N load a vertical distance of 1 m when 3000 J of effort are involved?

32. What is the efficiency of the body when a cyclist expends 1000 W of power to deliver mechanical energy to her bicycle at the rate of 100 W?

Boy, you should've seen the wild spills I used to take before I got into wide wheel bases. My first cart had a narrow wheel base, so on sharp turns the ole center of gravity didn't have to rise very high before the cart flipped over... me included. But not anymore. To stay in equilibrium, build a wide base and keep your center of gravity down to earth!

Rotational Motion

In Chapter 3 we discussed circular motion without the forces that produce it. In Chapter 4 we treated Newton's laws of motion and applied the concepts of inertia, force, and acceleration to the motion of things without any rotation they might undergo. In this chapter we extend these ideas to rotational motion—to the rotation of an object about some axis, whether inside or outside the object. We will see how rotational inertia affects rotation, how forces applied one way produce torques that tend to rotate things, and how forces applied another way tend to move things in circular paths. Then we will apply the concept of momentum to rotational motion. The concepts of this chapter are parallel to and build upon those we treated in Chapters 3 and 4—interesting stuff.

Rotational Inertia

Just as an object at rest tends to stay at rest and an object in motion tends to remain moving in a straight line, *an object rotating about an axis tends to remain rotating about the same axis unless interfered with by some external influence.* (We shall see shortly that this external influence is properly called a *torque*.) The property of an object to resist changes in its rotational state of motion is called **rotational inertia**.* Objects that

Figure 7-1
Rotational inertia depends on the distribution of mass with respect to the axis of rotation.

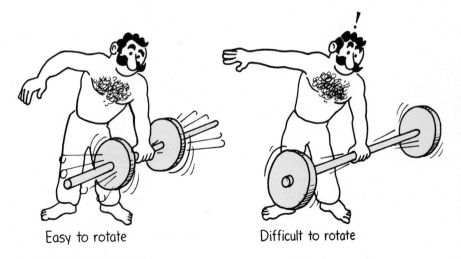

Easy to rotate   Difficult to rotate

---

*Often called *moment of inertia*.

are rotating tend to remain rotating, while nonrotating objects tend to remain nonrotating. In the absence of outside influences, a rotating top keeps rotating, while a top at rest stays at rest.

Like inertia for linear motion, the rotational inertia of an object also depends on the mass of the object. But unlike linear motion, rotational inertia depends on the distribution of the mass with respect to the axis of rotation. The greater the distance between the bulk of an object's mass and its axis of rotation, the greater the rotational inertia. For example, industrial flywheels are constructed so that most of their mass is concentrated along the rim. Once rotating, they have a greater tendency to remain rotating than they would if their mass were concentrated nearer the axis of rotation. It follows that the greater the rotational inertia of an object, the harder it is to change the rotational state of that object. If rotating, it is difficult to stop; if at rest, it is difficult to rotate. This fact is employed by a circus tightrope walker when he carries a long pole to aid balancing. Much of the mass of the pole is far from the axis of rotation, its midpoint. The pole therefore has considerable rotational inertia. If the tightrope walker starts to topple over, a tight grip on the pole rotates the pole. But the rotational inertia of the pole resists, giving the tightrope walker time to readjust his balance. The longer the pole, the better. And if our tightrope walker has no pole, he can at least increase the rotational inertia of his body by extending his arms full length.

The rotational inertia of the pole or any object depends on what axis it is rotated about.* If it rotates about an axis through its central core parallel to the length of the pole, like lead in a pencil, its mass distribution is very close to this axis and rotational inertia is very small. Rotational inertia is greater about the perpendicular midpoint axis, the axis utilized by the tightrope walker in Figure 7-2. The pole has a still greater

Figure 7-2
The tendency of the pole to resist rotation aids the acrobat.

*When the mass of an object is concentrated at or along one radial distance r from the axis of rotation, like a simple pendulum bob or a thin ring, rotational inertia I is equal to the mass m multiplied by the square of the radial distance. For this special case, $I = mr^2$. (For the general case, $I = \int r^2 dm$, which is beyond the scope of this book; it's something you may treat if you continue in your study of physics.)

Figure 7-3
Rotational inertias *I* of various objects each of mass *m*, about indicated axes and indicated centers of gravity CG.

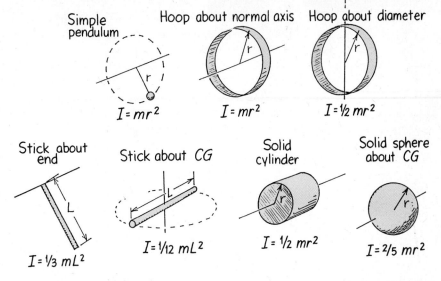

Simple pendulum
$$I = mr^2$$

Hoop about normal axis
$$I = mr^2$$

Hoop about diameter
$$I = \frac{1}{2} mr^2$$

Stick about end
$$I = \frac{1}{3} mL^2$$

Stick about CG
$$I = \frac{1}{12} mL^2$$

Solid cylinder
$$I = \frac{1}{2} mr^2$$

Solid sphere about CG
$$I = \frac{2}{5} mr^2$$

Figure 7-4
You bend your legs when you run to reduce rotational inertia.

Figure 7-5
Short legs have less rotational inertia than long legs. An animal with short legs has a quicker stride than one with long legs, just as a short pendulum swings to and fro quicker than a long one.

rotational inertia when the axis is perpendicular to one end, so it swings like a pendulum. Figure 7-3 compares rotational inertias for various shapes and axes. It is not important for you to learn these values, but you can see how they vary with the shape and axis.

A long pendulum has a greater rotational inertia than a short pendulum and therefore swings back and forth more slowly than a short one. When we walk, we allow our legs to swing with the help of gravity, that is, at a pendulum rate. Just as a long pendulum takes a long time to swing to and fro, a person with long legs tends to walk with slower strides than a person with short legs. This is also evident in animals; giraffes, horses, and ostriches run with a slower gait than dachshunds, mice, and bugs.

Figure 7-6
A solid cylinder rolls down
an incline faster than a ring,
whether or not they have the
same mass or outer diame-
ter. A ring has greater rotational inertia compared to its
mass than a cylinder does.

Ever tried running with your legs straight? It's difficult. When running, you bend your legs to reduce their rotational inertia so you can
rotate them back and forth more quickly. Bet you never looked at it that
way!

Because of rotational inertia, a solid cylinder starting from rest will
roll down an incline faster than a ring or hoop. The ring has its mass
concentrated farthest from its axis of rotation and therefore is harder to
start rolling. (That is, a ring has a greater tendency to resist a change in
its state of rotation.) Any solid cylinder will beat any ring on the same
incline. This doesn't seem plausible at first thought, but remember that
any two objects, regardless of mass, will fall together when dropped.
They will also slide together when released on an inclined plane. When
rotation is introduced, the object with the larger rotational inertia *compared to its own mass* has the greater resistance to a change in its motion.
Hence, any disk will roll faster than any ring on the same incline. Try
it and see!

Question ▶ Consider standing and balancing a hammer upright on
the tip of your finger. If the head of the hammer is
heavy and the handle long, would it be easier to balance with the end of the handle on your fingertip so
that the head is at the top, or the other way around
with the head at your fingertip and the end of the handle at the top?

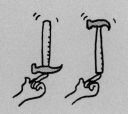

Torque

A torque (rhymes with *fork*) is the rotational counterpart of force. Force
tends to change the motion of things; torque tends to twist or change the
state of rotation of things. If you want to make a stationary object move,
apply a force. If you want to make a stationary object spin, apply a
torque. Torque differs from force just as rotational inertia and regular
inertia differ: both involve distance from the axis of rotation. In the case
of torque, this distance is called the lever arm. We define **torque** as the
product of this lever arm and the force that tends to produce rotation
about this particular distance:

$$\text{Torque} = \text{lever arm} \times \text{force}$$

▶ **Answer**

Stand the hammer with the handle at your fingertip and the head at the top. Why?
Because it will have more rotational inertia this way and be more resistant to a
rotational change. Those acrobats you see on stage who balance their friends at the
top of a long pole have an easier task when their friends are at the top of the pole.
A pole empty at the top has less rotational inertia and is more difficult to balance!

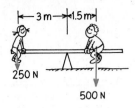

Figure 7-7
No rotation is produced
when the torques balance
each other.

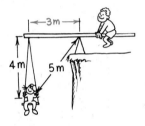

Figure 7-8
The lever arm is still 3 m.

Figure 7-9
Although the magnitudes of
the force are the same in
each case, the torques are
different.

Torques are intuitively familiar to youngsters playing on a seesaw. Kids can balance a seesaw even when their weights are unequal. Weight alone doesn't produce rotation. Torque does, and children soon learn that the distance they sit from the pivot point is every bit as important as weight (Figure 7-7). The torque produced by the boy on the right tends to produce clockwise rotation, while torque produced by the girl on the left tends to produce counterclockwise rotation. If the torques are equal, making the net torque zero, no rotation is produced. The seesaw is then in equilibrium.

Suppose that the seesaw is arranged so that the half-as-heavy girl is suspended from a 4-meter rope hanging from her end of the seesaw (Figure 7-8). She is now 5 meters from the fulcrum, and the seesaw is still balanced. We see that the lever-arm distance is still 3 meters and not 5 meters. The lever arm about any axis of rotation is the perpendicular distance from the axis to the line along which the force acts—the *line of action*, which in this case is a vertical line that extends to the earth's center. This will always be the shortest distance between the axis of rotation and the line along which the force acts.

This is why the stubborn bolt shown in Figure 7-9 is more likely to turn when the applied force is perpendicular to the handle, rather than at an oblique angle as shown in the first figure. In the first figure, the lever arm is shown by the dashed line and is less than the length of the wrench handle. In the second figure, the lever arm is equal to the length of the wrench handle. In the third figure, the lever arm is extended with a pipe to produce a greater torque.

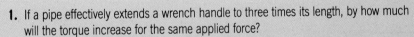

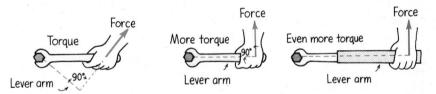

Questions ▶

1. If a pipe effectively extends a wrench handle to three times its length, by how much will the torque increase for the same applied force?

2. Consider the balanced seesaw in Figure 7-7. Suppose the girl on the left suddenly gains 50 N, such as by being handed a bag of apples. Where should she sit in order to balance, assuming the heavier boy does not move?

▶ **Answers**

1. Three times. (This method of increasing torque sometimes results in shearing off the bolt!)

2. She should sit $\frac{1}{2}$ m closer to the center. Then her lever arm is 2.5 m. This checks: 300 N × 2.5 m = 500 N × 1.5 m.

Center of Mass and Center of Gravity

Throw a baseball into the air, and it will follow a smooth parabolic trajectory. Throw a baseball bat spinning into the air, and its path is not smooth; its motion is wobbly, and it seems to wobble all over the place. But, in fact, it wobbles about a very special place, a point called the **center of mass.**

For a given object, the center of mass is the average position of all the particles of mass that constitute the object. For example, a symmetrical object such as a ball can be thought of as having all its mass concentrated at its center; by contrast, an irregularly shaped object such as a baseball bat has more of its mass toward one end. The center of mass of a baseball bat therefore is toward the thicker end. A solid cone has its center of mass exactly one-fourth of the way up from its base.

Figure 7-10
The center of gravity of the baseball and that of the bat follow parabolic trajectories.

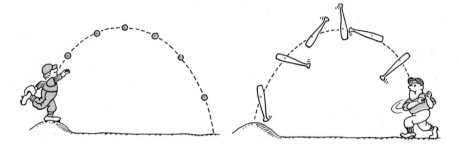

Center of gravity is a term popularly used to express center of mass. The center of gravity is simply the average position of weight distribution. Since weight and mass are proportional, center of gravity and center of mass refer to the same point of an object.* The physicist prefers to use the term *center of mass*, for an object has a center of mass whether or not it is under the influence of gravity. However, we shall use either term to express this concept and favor the term *center of gravity* when consideration of weight is involved.

Figure 7-11
The center of gravity for each object is shown by the colored dot.

 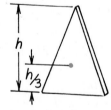 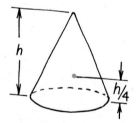

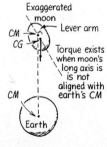
Exaggerated moon

CM
CG — Lever arm

Torque exists when moon's long axis is not aligned with earth's CM

CM

Earth

*For very large things such as the moon, the centers of mass and gravity may be in different locations. The side of the moon closest to the earth is in a region of stronger earth gravitation than the farther part, so the center of gravity is slightly closer to the earth than the center of mass. The moon is somewhat oblong, like a football, with its long axis pointed toward the earth. The centers of mass and gravity of the moon lie along this long axis. Interestingly enough, if this axis does not point directly to the earth's center of mass—that is, if the three centers do not lie on the same straight line (as shown in the sketch)—then a torque exists that tends to align them, somewhat as a compass needle experiences a torque when it is not aligned with a magnetic field. That's why the same side of the moon always faces us!

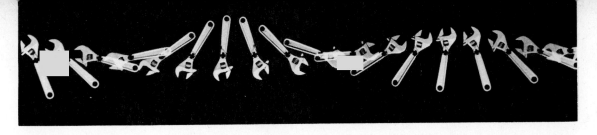

Figure 7-12
The center of mass of the spinning wrench follows a straight-line path.

The multiple-flash photograph (Figure 7-12) shows a top view of a wrench sliding across a smooth horizontal surface. Note that its center of gravity, indicated by the dark mark, follows a straight-line path, while other parts of the wrench rotate as they move across the surface. Since there is no external force acting on the wrench, its center of gravity moves equal distances in equal time intervals. The motion of the spinning wrench is the combination of the straight-line motion of its center of gravity and the rotational motion about its center of gravity.

Figure 7-13
The center of gravity of the cannonball and its fragments move along the same path before and after the explosion.

If the wrench were instead tossed into the air, no matter how it rotates its center of gravity would follow a smooth parabola. The same is true for an exploding cannonball (Figure 7-13). The internal forces that act in the explosion do not change the center of gravity of the projectile. Interestingly enough, if air resistance can be neglected, the center of gravity of the dispersed fragments as they fly though the air will be in the same place the center of gravity would be if the explosion didn't occur.

Locating the Center of Gravity

The center of gravity of a uniform object such as a meter stick is at its midpoint, for the stick acts as though its entire weight were concentrated there. Supporting that single point supports the whole stick. Balancing an object provides a simple method of locating its center of gravity. In Figure 7-14 the many small arrows represent the pull of gravity all along the meter stick. All these can be combined into a resultant force acting through the center of gravity. The entire weight of the stick may be thought of as acting at this single point. Hence, we can balance the stick by applying a single upward force in a direction that passes through this point.

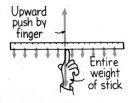

Figure 7-14
The weight of the entire stick behaves as if it were concentrated at its center.

The center of gravity of any freely suspended object lies directly beneath (or at) the point of suspension (Figure 7-15). If a vertical line is drawn through the point of suspension, the center of gravity lies somewhere along that line. To determine exactly where it lies along the line, we have only to suspend the object from some other point and draw a second vertical line through that point of suspension. The center of gravity lies where the two lines intersect.

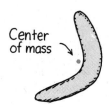

Figure 7-15
Finding the center of gravity for an irregularly shaped object.

Figure 7-16
The center of mass can be outside the mass of a body.

The center of mass of an object may be a point where no mass exists. For example, the center of mass of a ring or a hollow sphere is at the geometrical center where no matter exists. Similarly, the center of mass of a boomerang is outside the physical structure, not within the material making up the boomerang (Figure 7-16).

Figure 7-17
Dwight Stones executes a "Fosbury flop" to clear the bar while his center of gravity passes beneath the bar.

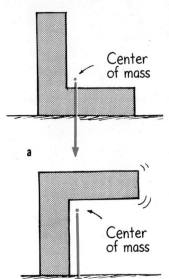

a

b

Stability

The location of the center of mass is important for stability (Figure 7-18). If we drop a line straight down from the center of mass of an object of any shape and it falls inside the base of the object, it is in stable **equilibrium**; it will balance. If it falls outside the base, it is unstable. Why doesn't the famous Leaning Tower of Pisa topple over? As we can see in Figure 7-19, a line from the center of gravity of the tower falls inside its base, so the Leaning Tower has stood for several centuries.

Figure 7-18
The center of mass of the L-shaped object is located where no mass exists. In (a), the center of mass is above a point of support, so the object is stable. In (b), it is not above a point of support, so the object is unstable and will topple over.

Figure 7-19
The center of gravity of the Leaning Tower of Pisa lies above a point of support, so the tower is in stable equilibrium.

To reduce the likelihood of tipping, it is usually advisable to design objects with a wide base and low center of gravity. The wider the base, the higher the center of gravity must be raised before the object will tip over.

Figure 7-20
The vertical distance that the center of gravity is raised in tipping determines stability. An object with a wide base and low center of gravity is more stable.

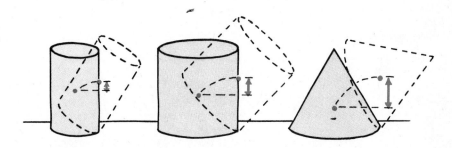

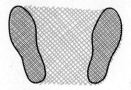

Figure 7-21
When you stand, your center of gravity is somewhere above the area bounded by your feet. Why do you keep your legs far apart when you have to stand in the aisle of a bumpy-riding bus?

Figure 7-22
You can lean over and touch your toes without falling over only if your center of gravity is above the area bounded by your feet.

When you stand erect (or lie flat), your center of gravity is within your body. Why is the center of gravity lower in an average woman than in an average man of the same height? Is your center of gravity always at the same point in your object? Is it always inside your body? What happens to it when you bend over?

If you are in reasonably good shape, you can bend over and touch your toes without bending your knees—providing you are not standing with your back against a wall. Ordinarily, when you bend over and touch your toes, you extend your lower extremity as shown in Figure 7-22 left, so that your center of gravity is above a point of support, your feet. If you attempt to do this when standing against a wall, however, you cannot counterbalance yourself, and your center of gravity soon protrudes beyond your feet as at the right in Figure 7-22. You are off-balance and you rotate.

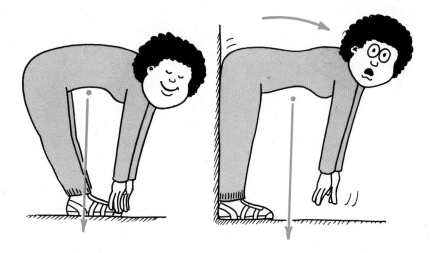

You rotate because of an unbalanced torque. This is evident in the two L-shaped objects shown in Figure 7-23. Both are unstable and will topple unless fastened to the level surface. It is easy to see that even if both shapes have the same weight, the one on the right is less stable. This is because of the greater lever arm and, hence, greater torque.

Figure 7-23
The greater torque acts on the figure in (*b*) for two reasons. What are they?

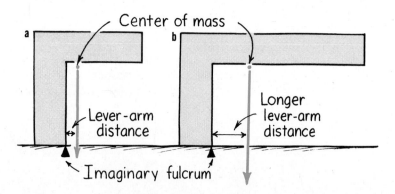

Figure 7-24
A pigeon jerks its head and neck back and forth with each step and thereby keeps its center of gravity always above the supporting foot.

Try balancing the pole end of a broom or mop on the palm of your hand. Even though rotational inertia helps, the support base is quite small and relatively far beneath the center of gravity, so it's difficult. After some practice you can do it if you learn to make slight movements of your hand to exactly respond to variations in balance. You learn to avoid under-responding or over-responding to the slightest variations in balance. Similarly, high-speed computers help massive rockets remain upright when they are launched. Variations in balance are quickly sensed. The computers regulate the firings at multiple nozzles to make corrective adjustments, in a way quite similar to the way your brain coordinates your adjustive action when balancing a long pole on the palm of your hand. Both feats are truly amazing.

Question ▶ A uniform meter stick supported at the 25-cm mark is in equilibrium when a 1-kg rock is suspended at the 0-cm end. What is the mass of the meter stick?

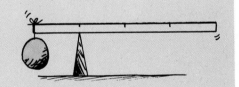

▶ **Answer**

The mass of the meter stick is the same as the mass of the rock—1 kg. Why? The system is in equilibrium, so any torques must be balanced: the torque that results from the weight of the rock is balanced by the equal but oppositely directed torque that results from the weight of the stick at its center of gravity, the 50-cm mark. The support force at the 25-cm mark is applied midway between the rock and the center of gravity of the stick, so the lever arms about the support point are equal (25 cm). This means that the weights and hence the masses of the rock and stick must also be equal. (Note that we don't have to go through the laborious task of considering the fractional parts of the stick's weight on either side of the fulcrum, for the center of gravity of the whole stick really is at one point—the 50-cm mark!) Interestingly enough, the center of gravity of the *rock and stick combination* is at the 25-cm mark—directly above the fulcrum.

Centripetal Force

Figure 7-25
The force exerted on the whirling can is toward the center.

Figure 7-26
(a) When a car goes around a curve, there must be a force pushing the car toward the center of the curve.
(b) A car skids on a curve when the centripetal force (friction of road on tires) is not great enough.

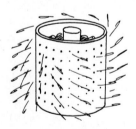

Figure 7-27
The clothes are forced into a circular path, but the water is not.

Figure 7-28
The centripetal force (adhesion of mud on the spinning tire) is not great enough to hold the mud on the tire, so it spins off in straight-line directions.

Any force that causes an object to follow a circular path is called a **centripetal force**. *Centripetal* means "center-seeking" or "toward the center." The force of a rotating carnival centrifuge on an occupant inside is a center-directed force; if it ceased to act, the occupant would no longer be held in a circular path.

If we whirl a tin can on the end of a string, we find that we must keep pulling on the string (Figure 7-25). The string transmits the centripetal force, which pulls the can into a circular path. Gravitational and electrical forces can be transmitted across empty space to produce centripetal forces. The moon, for example, is held in an almost circular orbit by gravitational force directed toward the center of the earth. The orbiting electrons in atoms experience an electrical force toward the central nuclei.

Centripetal force is not a new kind of force but is simply the name given to any force, whether string tension, gravitational, electrical, or whatever, that is directed at right angles to the path of the moving object and produces circular motion.

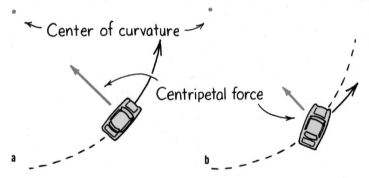

When an automobile rounds a corner, the friction between the tires and the road is the centripetal force that holds the car in a curved path (Figure 7-26). If this friction is not great enough, the car fails to make the curve and the tires slide sideways; we say the car skids.

Centripetal force plays the main role in the operation of a centrifuge. A familiar example is the spinning tub in an automatic washer (Figure 7-27). In its spin cycle, the tub rotates at high speed and produces a centripetal force on the wet clothes, which are forced into a circular path against the inner wall of the tub. The tub exerts great force on the clothes, but the holes in the tub prevent the exertion of the same force on the water in the clothes and the water escapes. Strictly speaking, the clothes are forced away from the water; the water is not forced away from the clothes. Think about that.

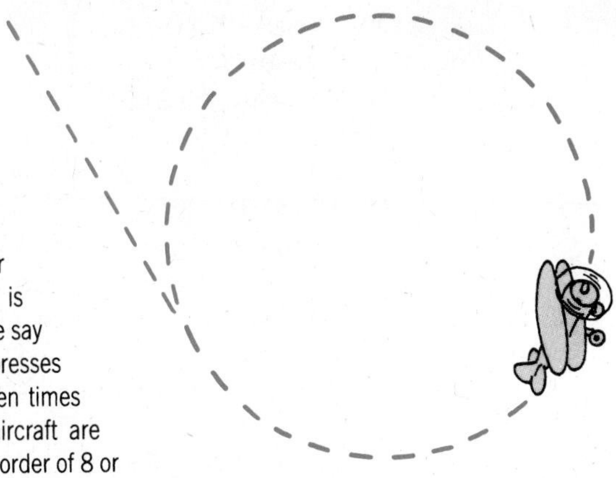

Figure 7-29
Large centripetal forces on the wings of an aircraft enable it to fly in circular loops. The acceleration away from the straight-line path the aircraft would follow in the absence of centripetal force is often several times greater than the acceleration due to gravity, g. For example, if the centripetal acceleration is 98 m/s^2 (ten times as great as 9.8 m/s^2), we say the aircraft is undergoing 10 g's. The seat presses against the pilot with an additional force ten times greater than his weight. Typical fighter aircraft are designed to withstand accelerations on the order of 8 or 9 g's. The pilot as well as the aircraft must withstand these accelerations. When an aircraft is performing an 8-g turn, for example, the force required to hold the blood in the pilot's brain must be eight times the ordinary force provided by the heart. Without it, blood may rush out of the pilot's brain and cause a blackout.

Centrifugal Force

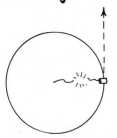

Figure 7-30
When the string breaks, the whirling can moves in a straight line, tangent to—not outward from the center of—its circular path.

Figure 7-31
The only force that is exerted on the whirling can (neglecting gravity) is directed *toward* the center of circular motion. This is a *centripetal* force. No outward force acts on the can.

In the preceding examples we have described circular motion by a center-directed force. Sometimes an outward force is attributed to circular motion. This outward force is called **centrifugal force**.* *Centrifugal* means "center-fleeing" or "away from the center." In the case of the whirling can, it is a common misconception to state that a centrifugal force pulls outward on the can. If the string holding the whirling can breaks (Figure 7-30), it is commonly stated that a centrifugal force pulls the can from its circular path. But the fact is that when the string breaks, the can goes off in a tangent straight-line path because *no* force acts on it. We illustrate this further with another example.

Suppose we are passengers in a car that suddenly stops short. We pitch forward against the dashboard. When this happens, we don't say that something forced us forward. In accord with the law of inertia, we pitched forward because of the absence of a force, which seat belts would have provided. Similarly, if we are in a car that rounds a sharp corner to the left, we tend to pitch outward to the right—not because of some outward or centrifugal force, but because there is no centripetal force holding us in circular motion (such as seat belts provide). The idea that a centrifugal force bangs us against the car door is a misconception.

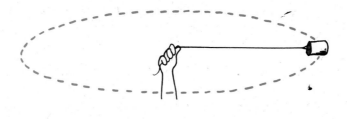

---

*Both centrifugal force and centripetal force depend on the mass m, tangential speed v, and radius of curvature r of the circularly moving object. If you take a follow-up physics course, you'll learn how the exact relationship is $F = mv^2/r$.

Likewise when we swing a tin can in a circular path. There is no force pulling the can outward, for the only force on the can is the string pulling it inward. The outward force is on the string, not on the can. Now suppose there is a ladybug inside the whirling can (Figure 7-32). The can presses against the bug's feet and provides the centripetal force that holds it in a circular path. The ladybug in turn presses against the floor of the can, but (neglecting gravity) the only force exerted on the ladybug is the force of the can on its feet. From our outside stationary frame of reference, we see there is no centrifugal force exerted on the ladybug, just as there was no centrifugal force exerted on the person who banged against the car door. The centrifugal force effect is attributed not to any real force but to inertia—the tendency of the moving object to follow a straight-line path. But try telling that to the ladybug!

Figure 7-32
The can provides the centripetal force necessary to hold the ladybug in a circular path.

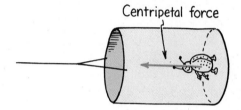

Centripetal force

Centrifugal Force in a Rotating Frame

Our frame of reference* very much influences our view of nature. When we sit on the seat of a fast-moving train, we have no speed at all relative to the train but an appreciable speed relative to the reference frame of the stationary ground outside. We have just learned that in a nonrotating reference frame, the force that holds an object in circular motion is a centripetal force. For the ladybug, the bottom of the can exerts a force on its feet. No other force was acting on the ladybug.

But nature seen from the reference frame of the rotating system is different. In the rotating frame, in addition to the force of the can on the ladybug's feet, there is a centrifugal force that is exerted on the ladybug.

Figure 7-33
From the reference frame of the ladybug inside the whirling can, it is being held to the bottom of the can by a force that is directed away from the center of circular motion. The ladybug calls this outward force a *centrifugal* force, which is as real to it as gravity.

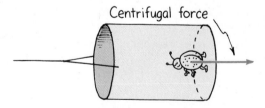

Centrifugal force

Centrifugal force in a rotating reference frame is a force in its own right, as real as the pull of gravity. However, there is a fundamental difference. Gravitational force is an interaction between one mass and another. The

---

*A nonaccelerating frame of reference is called an *inertial* frame of reference. Newton's laws are seen to hold exactly in an inertial frame.

gravity we experience is our interaction with the earth. But centrifugal force in the rotating frame has no such agent—it has no interaction counterpart. It seems like gravity, but with nothing pulling. Nothing produces it; it is a result of rotation. For this reason, physicists rank it as a fictitious force and not a real force like gravity, electromagnetism, and the nuclear forces. Nevertheless, to observers who are in a rotating system, centrifugal force seems to be, and is interpreted to be, a very real force. Just as from the surface of the earth we find gravity ever present, so within a rotating system centrifugal force is ever present.

Questions

1. A heavy iron ball is attached by a spring to the rotating platform as shown in the sketch. Two observers, one in the rotating frame and one on the ground at rest, observe its motion. Which observer sees the ball being pulled outward, stretching the spring? Which sees the spring pulling the ball in a circle?

2. The spring is observed to stretch 5 cm when the ball is midway between the rotational axis and the outer edge of the circular platform. If the spring support is moved so that the ball is directly over the edge, effectively twice the distance from the axis, will the spring be stretched more than, less than, or the same 5 cm?

Figure 7-34
The spin of the earth sets up a centrifugal force that slightly decreases our weight. Like the outside horse on the merry-go-round, we have the greatest tangential speed farthest from the earth's axis, at the equator. Centrifugal force is therefore maximum for us when we are at the equator and zero at the poles where we have no tangential speed. So, strictly speaking, if you want to lose weight, walk toward the equator!

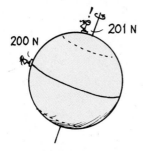

200 N   201 N

Answers

1. The observer in the reference frame of the rotating platform states that a centrifugal force pulls radially outward on the ball, which stretches the spring. The observer in the rest frame states that a centripetal force supplied by the stretched spring pulls the ball into a circle along with the rotating platform. (The rest-frame observer can in addition state that the reaction to this centripetal force is the ball pulling outward on the spring. The rotating observer, interestingly enough, can state no reaction counterpart to the centrifugal force.)

2. The ball will have twice the linear speed at twice the radial distance from the rotational axis ($v = r\omega$, Chapter 3). The greater the speed, the greater will be the centripetal/centrifugal force (which will also be twice and will stretch the spring 10 cm). (We will see in Chapter 11 that the stretch of a spring is directly proportional to the applied force.)

Simulated Gravity

Consider a colony of ladybugs living inside a bicycle tire—the old-fashioned balloon kind with plenty of room inside. If we toss the bicycle wheel through the air or drop it from an airplane high in the sky, the ladybugs will be in a weightless condition. They will float freely while the wheel is in free fall. Now spin the wheel. The ladybugs will feel themselves pressed to the outer part of the inner surface. If we spin the wheel not too fast and not too slowly but just right, we can provide the ladybugs with simulated gravity that will feel like the gravity to which they are accustomed. Gravity is simulated by centrifugal force. The "down" direction to the ladybugs will be what we would call radially outward, away from the center of the wheel.

Figure 7-35

If the spinning wheel freely falls, the ladybugs inside will experience a centrifugal force that feels like gravity when the wheel spins at the appropriate rate. To the occupants, the direction "up" is toward the center of the wheel and "down" is radially outward.

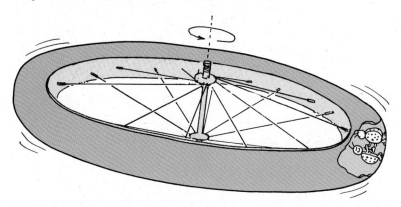

Humans today live on the outer surface of this spherical planet and are held here by gravity. The planet has been the cradle of humankind. But we will not stay in the cradle forever. We are becoming a spacefaring people. Many people in the years ahead will likely live in huge, lazily rotating habitats in space and will be held to the inner surfaces by centrifugal force. The rotating habitats will provide a simulated gravity so that the human body can function normally.

Occupants of the space shuttle are "weightless" because they lack a support force. Future space travelers need not be subject to weightlessness, however. A rotating space habitat for humans, like the rotating bicycle wheel for the ladybugs, can effectively supply a support force and neatly simulate gravity. Structures of small diameter would have to rotate at high rates to provide a simulated gravitational acceleration of 1 *g*. Sensitive and delicate organs in our middle ears sense rotation. Although

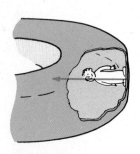

Figure 7-36

The interaction between the man and the floor as seen at rest outside the rotating system. The floor presses against the man (action) and the man presses back on the floor (reaction). The only force exerted on the man is by the floor. It is directed toward the center and is a centripetal force.

Figure 7-37

As seen from inside the rotating system, in addition to the man-floor interaction there is a centrifugal force exerted on the man at his center of mass. It seems as real as gravity. Yet, unlike gravity, it has no reaction counterpart—there is nothing out there that he can pull back on. Centrifugal force is not part of an interaction, but results from rotation. It is therefore called a fictitious force.

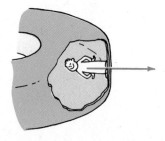

Figure 7-38
Artist's rendering of the interior of a space colony that would be occupied by a few thousand people.

there appears to be no difficulty at a single revolution per minute (RPM) or so, many people find difficulty in adjusting to rates greater than 2 or 3 RPM (although some easily adapt to 10 or so RPM). To simulate normal earth gravity at 1 RPM requires a large structure—one 2 kilometers in diameter. This is an immense structure compared to today's space shuttle vehicles.*

If the structure rotates so that inhabitants on the inside of the outer edge experience 1 *g*, then halfway to the axis they would experience 0.5 *g*. At the axis itself they would experience weightlessness at 0 *g*. The variety of fractions of *g* possible from the rim of a rotating space habitat holds promise for a most different and (at this writing) as yet unexperienced environment. We would be able to perform ballet at 0.5 *g*; diving and acrobatics at 0.2 *g* and lower-*g* states; three-dimensional soccer and new sports not yet conceived in very low *g* states.

Question ▶ If the earth were to spin faster about its axis, your weight would be less. If you were in a rotating space habitat that increased its spin rate, you'd "weigh" more. Explain why greater spin rates produce opposite effects in these cases.

▶ **Answer**

You're on the *outside* of the spinning earth, but you'd be on the *inside* of a spinning space habitat. A greater spin rate on the outside of the earth tends to throw you *off* a weighing scale and show a decrease in weight, but *against* a weighing scale inside the space habitat to show an increase in weight.

*Economics will probably dictate the size of the first inhabited structures. They are likely to be small and not rotate at all. The inhabitants will adjust to living in a weightless environment. Larger, rotating habitats will likely follow later.

Angular Momentum

Just as a mass moving in a straight line has linear momentum, a mass moving in a circular path or a mass rotating about some axis has angular momentum. **Angular momentum** is a measure of the rotational property of motion. A space colony orbiting the sun, a rock whirling at the end of a string, and the tiny electrons whirling about atomic nuclei all have angular momentum.

Angular momentum is defined as the product of rotational inertia and rotational velocity.*

Angular momentum = rotational inertia × rotational velocity

It is the counterpart of linear momentum:

Linear momentum = mass × velocity

Like linear momentum, angular momentum is a vector quantity and has direction as well as magnitude. In this book, we won't treat the vector nature of momentum except to acknowledge the remarkable action of the gyroscope. The rotating bicycle wheel in Figure 7-39 shows what happens when a torque caused by earth's gravity acts to change the direction of its angular momentum (which is along the wheel's axle). The pull of gravity that acts to topple the wheel over and change its rotational axis instead causes it to precess (move sideways) in a circular path about a vertical axis. You must do this yourself to believe it fully. You probably won't fully understand it until a later time.

For the case of an object that is small compared to the radial distance to its axis of rotation, like a tin can swinging from a long string or a

Figure 7-39
Angular momentum keeps the wheel axle horizontal when a torque supplied by earth's gravity acts on it. Instead of toppling, the torque causes the wheel to precess slowly about a vertical axis.

*Rotational velocity (often called *angular velocity*) is a vector whose magnitude is the rotational speed (symbol ω, Chapter 3). By convention, the rotational velocity vector and the angular momentum vector have the same direction and lie along the axis of rotation. If you take a followup course, you will learn that rotational velocity is measured in units of radians per second (rad/s), where 0.1 rad/s is approximately 1 RPM.

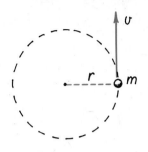

Figure 7-40
An object of concentrated mass m whirling in a circular path of radius r with a speed v has angular momentum mvr.

planet orbiting around the sun, the angular momentum can equivalently and more simply be expressed as the magnitude of its linear momentum, mv, multiplied by the radial distance, r. In shorthand notation,

$$\text{Angular momentum} = mvr$$

Just as an external net force is required to change the linear momentum of an object, an external net torque is required to change the angular momentum of an object. We restate Newton's first law of inertia for rotating systems in terms of angular momentum:

An object or system of objects will maintain its angular momentum unless acted upon by an unbalanced external torque.

We all know that it is easier to balance on a bicycle that is moving than on a bicycle at rest. Rotating wheels have angular momentum. To tip the wheels means a change in angular momentum, which requires a greater torque than tipping wheels at rest.

Conservation of Angular Momentum

Just as the linear momentum of any system is conserved if no net forces are acting on the system, angular momentum is conserved for systems in rotation. If no unbalanced external torque acts on the rotating system, the angular momentum of that system is constant. This means that the product of rotational inertia and rotational velocity at one time will be the same as at any other time.

An interesting example illustrating angular momentum conservation is shown in Figure 7-41. The man stands on a low-friction turntable with weights extended. His rotational inertia, with the help of the extended weights, is relatively large in this position. As he slowly turns, his angular momentum is the product of his rotational inertia and rotational velocity. When he pulls the weights inward, the rotational inertia of his body and the weights is considerably reduced. What is the result? His rotational speed increases! This example is best appreciated by the turning person who feels changes in rotational speed that seem to be mysterious.

Figure 7-41
Conservation of angular momentum. When the man pulls his arms and the whirling weights inward, he decreases his rotational inertia, and his rotational speed correspondingly increases.

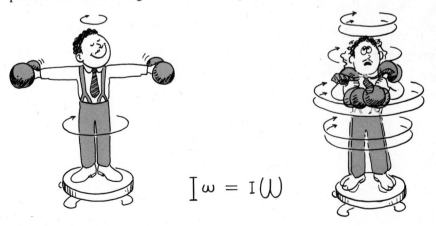

$$I\omega = I\omega$$

Figure 7-42
Rotational speed is controlled by variations in the body's rotational inertia as angular momentum is conserved during a forward somersault.

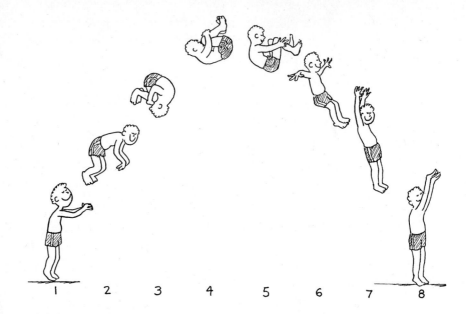

1   2   3   4   5   6   7   8

Figure 7-43
Time-lapse photo of a falling cat.

But it's straight physics! This procedure is used by a figure skater who starts to whirl with her arms and perhaps a leg extended and then draws her arms and leg in to obtain a greater rotational speed. Whenever a rotating body contracts, its rotational speed increases.

Similarly, when a gymnast is spinning freely in the absence of unbalanced torques on his body, his angular momentum does not change. However, he can change his rotational speed by simply making variations in his rotational inertia. He does this by moving some part of his body toward or away from his axis of rotation.

If a cat is held upside down and dropped, it is able to execute a twist and land upright even if it has no initial angular momentum. Zero-angular-momentum twists and turns are performed by turning one part of the body against the other. While falling, the cat rearranges its limbs and tail to change its rotational inertia. Repeated reorientations of the body configuration result in the head and tail rotating one way and the feet the other, so that the feet are downward when the cat strikes the ground. During this maneuver the total angular momentum remains zero. When it is over, the cat is at rest. This maneuver rotates the body through an angle, but it does not create continuing rotation. To do so would violate angular momentum conservation.

Humans can perform similar twists without difficulty, though not as fast as a cat. Astronauts have learned to make zero-angular-momentum rotations as they orient their bodies in preferred directions when floating freely in space. The law of momentum conservation is seen in the planetary motions and in the shape of the galaxies. This law will be a fact of everyday life to inhabitants of rotating space habitats who will head for these distant places.

It is fascinating to note that the conservation of momentum tells us that the moon is getting farther away from the earth. This is because the earth's daily rotation is slowly decreasing due to the friction of ocean waters on the ocean bottom, just as an automobile's wheels slow down when brakes are applied. This decrease in the earth's angular momentum is accompanied by an equal increase in the angular momentum of the moon in its orbital motion around the earth. This increase in the moon's angular momentum results in the moon's increasing distance from the earth and a decrease in its speed. This increase of distance amounts to one-quarter of a centimeter per rotation. Have you noticed that the moon is getting farther away from us lately? Well, it is; each time we see another full moon, it is one-quarter of a centimeter farther away!

Summary of Terms

Rotational inertia That property of an object to resist any change in its state of rotation: if at rest, the object tends to remain at rest; if rotating, it tends to remain rotating and will continue to do so unless acted upon by a net external torque.

Torque The product of force and lever-arm distance, which tends to produce rotation.

Center of mass The average position of mass or the single point associated with an object where all its mass can be considered to be concentrated.

Center of gravity The average position of weight or the single point associated with an object where the force of gravity can be considered to act.

Equilibrium The state of an object when not acted upon by a net force or net torque. An object in equilibrium may be at rest or moving at uniform velocity; that is, not accelerating.

Centripetal force A center-seeking force that causes an object to follow a circular path.

Centrifugal force An outward force that is due to rotation. In an inertial frame of reference, it is fictitious in the sense that it doesn't act on the rotating object but on whatever supplies the centripetal force; it is the reaction to centripetal force. In a rotating frame of reference, it does act on the rotating body and is fictitious in the sense that it is not an interaction with an agent or entity such as mass or charge but is a force in itself that is solely a product of rotation; it has no reaction-force counterpart.

Angular momentum A measure of an object's rotation about a particular axis; more specifically, the product of its rotational inertia and rotational velocity. For an object that is small compared to the radial distance, it is the product of mass, speed, and radial distance of rotation.

Conservation of angular momentum When no external torque acts on an object or a system of objects, no change of angular momentum takes place. Hence, the angular momentum before an event involving only internal torques is equal to the angular momentum after the event.

Suggested Reading

Brancazio, P. J. *Sport Science*. New York: Simon & Schuster, 1984.

Clarke, A. C. *Rendezvous with Rama*. New York: Harcourt Brace Jovanovich, 1973. This is the first science fiction novel to seriously consider habitation inside a spinning space facility.

NASA. *Space Settlements: A Design Study*. Washington, D.C.: U.S. Government Printing Office, 1977.

O'Neill, G. K. *The High Frontier*. New York: Morrow, 1976.

Review Questions

1. Why is it important that Chapters 3 and 4 are studied before this chapter is attempted?

Rotational Inertia

2. Inertia depends on mass; rotational inertia depends on mass and something else. What?

3. Does the rotational inertia of an object differ for different axes of rotation?

4. Why do people with long legs generally walk with a slower stride than people with short legs?

5. Which will have the greater acceleration rolling down an incline, a hoop or a solid disk?

Torque

6. Compare the effects of a force exerted on an object and a torque exerted on an object.

7. What is meant by the "lever arm" of a torque?

8. In what direction should a force be applied to produce maximum torque?

9. How do clockwise and counterclockwise torques compare when a system is balanced?

Center of Mass and Center of Gravity

10. Toss a stick into the air and it appears to wobble all over the place. Specifically, what place?

11. Where is the center of mass of a baseball? Where is its center of gravity? Where are these centers for a baseball bat?

Locating the Center of Gravity

12. If you hang at rest by your hands from a vertical rope, where is your center of gravity with respect to the rope?

13. Where is the center of mass of an empty shoe box?

Stability

14. What is the relationship between center of gravity and support base for an object in stable equilibrium?

15. How far can an object be tipped before it topples over?

16. Why doesn't the Leaning Tower of Pisa topple?

17. In terms of center of gravity, support base, and torque, why can you not stand with heels and back to a wall and then bend over to touch your toes and return to your stand-up position?

Centripetal Force

18. When you whirl a can at the end of a string in a circular path, what is the direction of the force that is exerted on the can?

19. Is it an inward force or an outward force that is exerted on the clothes during the spin cycle of an automatic washer?

Centrifugal Force

20. If the string breaks that holds a whirling can in its circular path, what kind of force causes it to move in a straight-line path—centripetal, centrifugal, or no force? What law of physics supports your answer?

21. If you are not wearing a seat belt and you slide across your seat and slam against a car door when the car rounds a curve, what kind of force is responsible—centripetal, centrifugal, or no force? Support your answer.

Centrifugal Force in a Rotating Frame

22. Identify the action and reaction forces in the interaction between the ladybug and the whirling can of Figures 7-32 and 7-33.

23. Why is centrifugal force in a rotating frame called a "fictitious force"?

Simulated Gravity

24. How can gravity be simulated in an orbiting space station?

25. Why will orbiting space stations that simulate gravity likely be large structures?

26. How will the value of g vary at different distances from the hub of a rotating space station?

Angular Momentum

27. Distinguish between linear momentum and angular momentum.

28. What is the law of inertia for angular momentum?

Conservation of Angular Momentum

29. What does it mean to say that angular momentum is conserved?

30. If a skater who is spinning pulls her arms in so as to reduce her rotational inertia to half, by how much will her angular momentum increase? By how much will her rate of spin increase? (Why are your answers different?)

Home Projects

1. Fasten a fork, spoon, and wooden match together as shown. The combination will balance nicely—on the edge of a glass, for example. This happens because the center of gravity actually "hangs" below the point of support.

2. Stand with your heels and back against a wall and try to bend over and touch your toes. You'll find you have to stand away from the wall to do so without toppling over. Compare the minimum distance of your heels from the wall with that of a friend of the opposite sex. Who can touch their toes with their heels nearer to the wall—males or females? On the average and in proportion to height, which sex has the lower center of gravity?

3. Ask a friend to stand facing a wall. With toes against the wall, ask your friend to stand on the balls of her feet without toppling backward. Your friend won't be able to do it. Now you explain why it can't be done.

4. Rest a meter stick on two fingers as shown. Slowly bring your fingers together. At what part of the stick do your fingers meet? Can you explain why this always happens, no matter where you start your fingers?

5. Swing a pail of water around rapidly in a circle at arm's length and the water will not spill. Why?

6. Place the hook of a wire coat hanger over your finger. Carefully balance a coin on the straight wire on the bottom directly under the hook. You may have to flatten the wire with a hammer or fashion a tiny platform with tape. With a surprisingly small amount of practice you can swing the hanger and balanced coin back and forth and then in a circle. Centripetal force holds the coin in place.

Exercises

1. The front wheels located far out in front of the racing vehicle help to keep the vehicle from nosing upward when it accelerates. What physics principle is illustrated here?

2. Which will roll down a hill faster—a solid cylinder or a solid sphere? A solid ball or a hollow ball? Defend your answers.

3. Why do buses and heavy trucks have large steering wheels?

4. Can an object rotate if there is no torque acting on it? Defend your answer and illustrate it with an example.

5. Is the net torque changed when a partner on a see-saw stands or hangs from her end instead of sitting? (Does the weight or the lever arm change?)

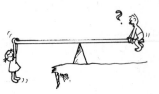

6. Which is easier for prying open a stuck cover from a can of paint—a screwdriver with a thick handle or one with a long handle? Which is easier for turning stubborn screws? Why?

7. When you pedal a bicycle, maximum torque is produced when the pedal sprocket arms are in the horizontal position, and no torque is produced when they are in the vertical position. Explain.

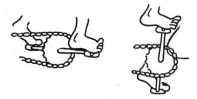

8. Why is the middle seating most comfortable in a bus traveling on a bumpy road?

9. Explain why a long pole is more beneficial to a tightrope walker if the pole droops.

10. Why is the center of mass of the solar system not at the geometrical center of the sun?

11. Why is the wobbly motion of a single star an indication that the star has a planet or system of planets?

12. Why must you bend forward when carrying a heavy load on your back?

13. Why is it easier to carry the same amount of water in two buckets, one in each hand, than in a single bucket?

14. Is it necessary that weights be placed in the middle of the pans of an equal-arm balance for accurate measurements? Why or why not?

15. Nobody at the playground wants to play with the obnoxious boy, so he fashions a seesaw as shown so he can play with himself. Explain how this is done.

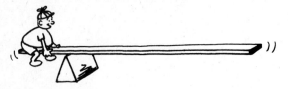

16. Using the ideas of torque and center of gravity, explain why a ball rolls down a hill.

17. Sometimes a kicked football sails through the air without rotating, and at other times it tumbles end over end as it travels. With respect to the center of mass of the ball, how is it kicked in both cases?

18. How can the three bricks be stacked so that the top brick has maximum horizontal displacement from the bottom brick? For example, stacking them like the dashed lines suggest would be unstable and the bricks would topple. (*Hint:* Start with the top brick and work down. At every interface the center of gravity of the bricks above must not extend beyond the end of the supporting brick.)

19. Why is it dangerous to roll open the top drawers of a fully loaded file cabinet that is not secured to the floor?

20. Describe the comparative stabilities of the three objects shown in Figure 7-20 in terms of work and potential energy.

21. The centers of gravity of the three trucks parked on a hill are shown by the Xs. Which truck(s) will tip over?

22. Why is less effort required in doing sit-ups when your arms are extended in front of you? Why is it more difficult when your arms are placed in back of your head?

23. The rock has a mass of 1 kg. What is the mass of the measuring stick if it is balanced by a support force at the one-quarter mark?

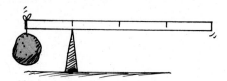

24. A long track balanced like a seesaw supports a golf ball and a more massive billiard ball with a compressed spring between the two. When the spring is released, the balls move away from each other. Does the track tip clockwise, tip counterclockwise, or remain in balance as the balls roll outward? What principles do you use for your explanation?

25. The value of g at the earth's surface is about 9.8 m/s^2. How would this value change if the earth rotated faster about its polar axis?

26. When a long-range cannonball is fired toward the equator from a northern (or southern) latitude, it lands west of its "intended" longitude. Why? (*Hint:* Consider a flea jumping from the middle to the outer edge of a moving phonograph record.)

27. If you should buy a quantity of gold in Mexico and weigh it carefully on a spring balance, would the same quantity of gold weigh more, less, or the same if weighed on the same spring balance in Alaska? Defend your answer.

28. A motorcyclist is able to ride on the vertical wall of a bowl-shaped track as shown. His weight is counteracted by the friction of the wall on the tires (vertical arrow). Is it centripetal or centrifugal force that acts on the motorcycle? On the wall?

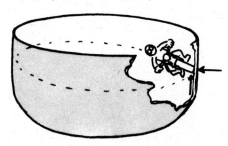

29. Your friend says that a satellite in circular orbit does not accelerate, as evidenced by its constant speed. Your other friend says it does accelerate, as evidenced by the centripetal force supplied by gravity. What do you say?

30. Does a centripetal force that acts on an object rotating in a circular path do work on the object? What is your explanation, and what evidence can you cite to support your answer?

31. A space habitat rotates at an appropriate rate to provide a comfortable acceleration of 1 *g*, earth-normal gravity, for its inhabitants. If it is rotated at twice the angular speed, by how much will the "weight" of the inhabitants increase?

32. Consider a too-small space habitat that consists of a rotating cylinder of radius 4 m. If a man standing inside is 2 m tall and his feet are at 1 *g*, what is the *g* force at the elevation of his head? (Do you see why projections call for large habitats?)

33. If the variation in *g* between one's head and feet is to be less than $\frac{1}{100}$ *g*, then compared to one's height, what should be the minimum radius of the space habitat?

34. A basketball player wishes to balance a ball on his fingertip. Will he be more successful with a spinning ball or a stationary ball? What physical principle supports your answer?

35. If you are in the center of a large, freely rotating turntable at an amusement park and you crawl toward the outer rim, does the rate of the rotation increase, decrease, or remain unchanged? What physics principle supports your answer?

36. A sizable quantity of earth is washed down the Mississippi River and deposited in the Gulf of Mexico. What effect does this tend to have on the length of a day? (*Hint:* Relate this to Figure 7-41.)

37. If the polar ice caps melted, how would this affect the length of the day?

38. Strictly speaking, as more and more skyscrapers are built on the surface of the earth, does the day tend to become longer or shorter? What physical principle supports your answer?

39. How would the length of a day be affected if all the world's inhabitants continuously walked in an easterly direction? When they stopped walking?

40. A toy train is initially at rest on a track fastened to a bicycle wheel, which is free to rotate. How does the wheel respond when the train moves clockwise?

When the train backs up? Does the angular momentum of the wheel-train system change during these maneuvers? How would the resulting motions depend on the relative masses of the wheel and train?

41. Why does a typical small helicopter with a single main rotor have a second small rotor on its tail? Describe the consequence if the small rotor fails in flight.

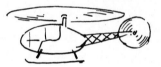

42. We believe our galaxy was formed from a huge cloud of gas and particles. The original cloud was far larger than the present size of the galaxy, was more or less spherical, and was rotating very much more slowly than the galaxy is now. In this sketch we see the original cloud and galaxy as it is now (seen edgewise). Explain how gravitation and the conservation of angular momentum contribute to the galaxy's present shape and why it rotates faster now than when it was a larger, spherical cloud.

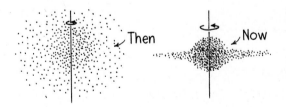

8 Gravity

For thousands of years people have looked into the nighttime sky and wondered about the stars. Having only their naked eyes, they neither saw nor dreamed that the stars are greater in number than all the grains of sand on all the deserts and beaches of the world. Our ancestors envisioned the earth at the center of the universe and considered the stars to be fixed on a great revolving crystal sphere. They distinguished the planets from the stars and placed them on inner spheres with more complicated motions. They believed that the complicated motions and the positions of the planetary spheres influenced earthly events. In the sixteenth century Copernicus discovered that the motions of the planets were much simpler if they were considered to be circling about the sun rather than the earth. The Copernican view was revolutionary, and there were great debates about whether the planets revolved around the sun or not.

The approach to these debates taken by the Danish astronomer Tycho Brahe was quite different from that of his contemporaries. Rather than carrying on with philosophical arguments, Brahe believed it would be better to measure accurately the positions of the planets in the sky and to chart their motions. This was a magnificent idea—that in formulating theories, one's efforts are best spent gathering facts and making careful measurements rather than arguing over philosophical speculations. Following this idea, Brahe built the first great observatory. Although telescopes had not yet been invented, he built huge brass protractorlike quadrants and cataloged the motions of the planets so accurately that his measurements are still used today. He recorded the planets' positions for 20 years to $\frac{1}{60}$ of a degree. Just before his death he entrusted one of his colleagues, Johannes Kepler, with editing and publishing his planetary tables. From the data, Kepler discovered some very remarkable and simple laws regarding planetary motion.

Kepler's Laws

Today we know what Brahe and Kepler did not know—that the planets of the solar system are simply falling continuously around the sun (just as earth satellites continuously fall around the earth, briefly discussed in Chapter 3). Kepler, to learn how the planets moved, performed the enormous task of transforming Brahe's earthbound observations into a path in space such as might be seen by a stationary observer outside the solar system. His conviction that the planets would circle perfectly on perfect

circles around the sun was shattered after years of effort. He found the paths to be ellipses.

Kepler also found that the planets do not go around the sun at a uniform speed but move faster when they are nearer the sun and more slowly when they are farther from the sun. They do this in such a way that an imaginary line or spoke joining the sun and the planet sweeps out equal areas of space in equal times. The triangular-shaped area swept out during a month when a planet is orbiting far from the sun (triangle ASB in Figure 8-1) is equal to the triangular area swept out during a month when the planet is orbiting closer to the sun (triangle CSD in Figure 8-1).

Ten years later Kepler discovered a third law. He had spent these years searching for a connection between the size of a planet's orbit and its period (the time required for a planet to make a complete revolution around the sun). From Brahe's data Kepler found that the square of a period is proportional to the cube of its average distance from the sun. He discovered this by noting that the fraction R^3/T^2 is the same for all planets, where R is the planet's average orbit radius and T is the planet's period in earth days. **Kepler's laws of planetary motion** are:

Law 1: **Each planet moves in an elliptical orbit with the sun at one focus.**

Law 2: **The line from the sun to any planet sweeps out equal areas of space in equal time intervals.**

Law 3: **The squares of the times of revolutions (periods) of the planets are proportional to the cubes of their average distances from the sun. ($T^2 \sim R^3$ for all planets.)**

Kepler's laws apply not only to planets but also to moons or any satellite in orbit around any body. Except for Pluto (which Kepler had no knowledge of), the elliptical orbits of the planets are very nearly circular. Only the precise measurements of Brahe showed the slight differences. Kepler had no idea why the planets traced elliptical paths about the sun and no general explanation for the mathematical relationships that he had discovered. He was familiar with Galileo's ideas about inertia and accelerated motion, but he failed to apply them to his own work. As a consequence, Kepler's thinking about forces was concerned with forces he supposed were directed in the same direction as the planets' paths to keep them moving. He never appreciated the concept of inertia. Galileo, on the other hand, never appreciated Kepler's work and held to his conviction that the planets moved in circles.* Further understanding of planetary motion required someone who could integrate the findings of these two great scientists. This task fell to Isaac Newton.

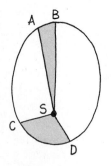

Figure 8-1
Equal areas are swept out in equal intervals of time.

---

*It is not easy to look at familiar things through the new insights of others. We tend to see only what we have learned to see or wish to see. Galileo reported that many of his colleagues were unable or refused to see the moons of Jupiter when they peered skeptically through his telescopes. Galileo's telescopes were a boon to astronomy, but more important than a new instrument to see things better was a new way of understanding what is seen. Is this true today?

Newton's Law of Universal Gravitation

From the time of Aristotle, the circular motions of heavenly bodies were regarded as natural. The ancients believed that the stars, planets, and moon moved in divine circles, free from any impressed forces. As far as the ancients were concerned, this circular motion required no explanation.

Newton recognized that a force of some kind must be acting on the planets; otherwise, their paths would be straight lines. From the law of inertia, he knew this force was not directed along the path of the planets as Kepler had speculated but was, instead, in the direction in which the planets are curved toward the sun. His analysis of Kepler's second law showed that the origin of this force was the sun. From Kepler's third law, he deduced that this force decreases as the square of the distance. A planet twice as far from the sun is pulled toward the sun with $\frac{1}{4}$ the force; three times as far, $\frac{1}{9}$; and so on. This relationship is the *inverse-square law*, which we will treat in detail shortly.

What sort of force could act from one place to another without contact? Newton was aware of such a force. When an apple falls to the ground, it is accelerated from its position on the tree. He wondered if the same force extended to the moon, holding it in an elliptical path around the earth, a path similar to a planet's path around the sun. This stroke of intuition—conceiving the same force to hold both the moon and apple—was a revolutionary break with the prevailing notion that there were two sets of natural laws, one for earthly events and another altogether for motions in the heavens.

To test this theory, Newton compared the fall of an apple with the fall of the moon. He realized that the moon falls in the sense that *it falls away from the straight line it would follow if there were no forces acting on it*. Because of its tangential speed, it "falls around" the round earth. Newton's test was to see if the moon's distance of fall per second was the same as the distance that an apple or anything at that distance would fall in 1 second. We'll go into the details of Newton's test in the next chapter and simply state here that his first calculations didn't check. Disappointed, but recognizing that brute fact must always win over a beautiful hypothesis, he placed his papers in a drawer where they remained for nearly 20 years. During this period he founded and developed the field of geometric optics for which he first became famous.

Newton's interest in mechanics was rekindled with the advent of a spectacular comet in 1680 and another 2 years later. He returned to the moon problem at the prodding of his astronomer friend, Edmond Halley, for whom the second comet was later named. He made corrections in the experimental data used in his earlier method and obtained excellent results. Only then did he publish what is one of the most far-reaching generalizations of the human mind: the **law of universal gravitation.***

Figure 8-2

The tangential velocity of the moon about the earth allows it to fall around the earth rather than directly into it. If this tangential velocity were reduced to zero, what would be the fate of the moon?

---

*This is a dramatic example of the painstaking effort and cross-checking that go into the formulation of a scientific theory. Contrast Newton's approach with the failure to "do one's homework," the hasty judgments, and the absence of cross-checking that so often characterize the pronouncements of less-than-scientific theories.

Everything pulls on everything else in a beautifully simple way that involves only mass and distance. According to Newton, every mass attracts every other mass with a force that for any two masses is directly proportional to the product of the masses involved and inversely proportional to the square of the distance separating them.

This statement can be expressed symbolically as

$$F \sim \frac{m_1 m_2}{d^2}$$

where m_1 and m_2 are the masses, and d is the distance between their centers. Thus, the greater the masses m_1 and m_2, the greater the force of attraction between them.* The greater the distance of separation d, the weaker the force of attraction, but weaker as the inverse square of the distance between their centers of mass.

The Universal Gravitational Constant, *G*

The proportionality form of the universal law of gravitation can be expressed as an exact equation when the constant of proportionality G, called the *universal gravitational constant*, is introduced. Then the equation is

$$F = G \frac{m_1 m_2}{d^2}$$

In words, the force of gravity between two objects is found by multiplying their masses, dividing by the square of the distance between their centers, and then multiplying this result by the constant G. The magnitude of G is given by the magnitude of the force between two masses of 1 kilogram each, 1 meter apart: 0.0000000000667 newton. This is an extremely weak force. G has the same extremely tiny magnitude. The units of G are such as to make the force come out in newtons. In scientific notation,†

$$G = 6.67 \times 10^{-11} \ \mathrm{N \cdot m^2/kg^2}$$

G was first measured long after the time of Newton by an English physicist, Henry Cavendish, in the eighteenth century. He accomplished this by measuring the tiny force between lead masses with an extremely sensitive torsion balance. A simpler method was later developed by Philipp von Jolly, who attached a spherical flask of mercury to one arm of a

---

*Note the different role of mass here. Thus far we have treated mass as a measure of inertia, which is called *inertial mass*. Now we see mass as a measure of gravitational force, which in this context is called *gravitational mass*. It is experimentally established that the two are equal, and, as a matter of principle, the equivalence of inertial and gravitational mass is the foundation of Einstein's general theory of relativity.

†The numerical value of G depends entirely on the units of measurements we choose for mass, distance, and time. The international system of choice is: for mass, the kilogram; for distance, the meter; and for time, the second. Scientific notation is discussed in Appendix I at the end of this book.

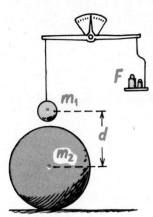

Figure 8-3
Jolly's method of measuring G. Balls m_1 and m_2 attract each other with a force F equal to the weights needed to restore balance.

sensitive balance (Figure 8-3). After the balance was put in equilibrium, a 6-ton lead sphere was rolled beneath the mercury flask. The gravitational force between the two masses was equal to the weight that had to be placed on the opposite end of the balance to restore equilibrium. All the quantities m_1, m_2, F, and d were known, from which the ratio G was calculated:

$$\frac{F}{m_1 m_2 / d^2} = 6.67 \times 10^{-11} \frac{N}{kg^2/m^2} = 6.67 \times 10^{-11} \ N \cdot m^2 / kg^2$$

The value of G tells us that the force of gravity is a very weak force. It is the weakest of the presently known four fundamental forces. (The other three are the electromagnetic force and two kinds of nuclear forces.) We sense gravitation only when masses like that of the earth are involved. The force of attraction between you and a battleship on which you stand is too weak for ordinary measurement. The force of attraction between you and the earth, however, can be measured. It is your weight.

In addition to your mass, your weight also depends on your distance from the center of the earth. At the top of a mountain your mass is no different than it is anywhere else, but your weight is slightly less than at ground level because your distance from the center of the earth is greater.

Question ▶ If there is an attractive force between all objects, why do we not feel ourselves gravitating toward massive buildings in our vicinity?

Once the value of G was known, the mass of the earth was easily calculated. The force that the earth exerts on a mass of 1 kilogram at its surface is 9.8 newtons. The distance between the 1-kilogram mass and the center of mass of the earth is the earth's radius, 6.4×10^6 meters. Therefore, from $F = G(m_1 m_2 / d^2)$, where m_1 is the mass of the earth,

$$9.8 \ N = 6.67 \times 10^{-11} \ \frac{N \cdot m^2}{kg^2} \cdot \frac{1 \ kg \times m_1}{(6.4 \times 10^6 m)^2}$$

from which the mass of the earth $m_1 = 6 \times 10^{24}$ kilograms.

▶ **Answer**
Gravity pulls us to massive buildings and everything else in the universe. Physicist Paul A. M. Dirac, winner of the 1933 Nobel Prize, put it this way: "Pick a flower on earth and you move the farthest star!" How much we are influenced by buildings or how much interaction there is between flowers is another story. The forces between us and buildings are relatively small because the masses are small compared to the mass of the earth. The forces due to the stars are small because of their great distances. These tiny forces escape our notice when they are overwhelmed by the overpowering attraction to the earth.

Figure 8-4

The inverse-square law. Paint spray travels radially away from the nozzle of the can in straight lines. Like gravity, the "strength" of the spray obeys the inverse-square law.

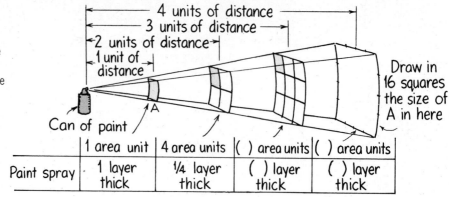

Paint spray	1 area unit	4 area units	( ) area units	( ) area units
	1 layer thick	¼ layer thick	( ) layer thick	( ) layer thick

Gravity and Distance: The Inverse-Square Law

We can better understand how gravity is diluted with distance by considering how paint from a paint gun spreads with increasing distance (Figure 8-4). Suppose we position a paint gun at the center of a sphere with a radius of 1 meter, and a burst of paint spray travels 1 meter to produce a square patch of paint that is 1 millimeter thick. How thick would the patch be if the experiment were done in a sphere with twice the radius? If the same amount of paint travels in straight lines for 2 meters, it will spread to a patch twice as tall and twice as wide. The paint would be spread over an area four times as big, and its thickness would be only $\frac{1}{4}$ millimeter. Can you see from the figure that for a sphere of radius 3 meters the thickness of the paint patch would be only $\frac{1}{9}$ millimeter? Can you see the thickness of the paint decreases as the square of the distance increases? This is known as the **inverse-square law**. The inverse-square law holds for gravity and for all phenomena wherein the effect from a localized source spreads uniformly throughout the surrounding space: the electric field about an isolated electron, light from a match, radiation from a piece of radium, and sound from a cricket.

In using Newton's equation for gravity, it is important to emphasize that the distance term d is the distance between the centers of masses of the objects that are attracted to each other. Note in Figure 8-6 that the apple that normally weighs 1 newton at the earth's surface weighs only $\frac{1}{4}$ as much when it is twice the distance from the earth's center. The greater the distance from the earth's center, the less the weight of an object. A child that weighs 300 newtons at sea level will weigh only 299 newtons atop Mt. Everest. But no matter how great the distance, the earth's gravitational force approaches, but never reaches, zero. Even if you were transported to the far reaches of the universe, the gravitational influence of home would still be with you. It may be overwhelmed by the gravitational influences of nearer and/or more massive bodies, but it is there. The gravitational influence of every material object, however small or however far, is exerted through all of space. Isn't that nice?

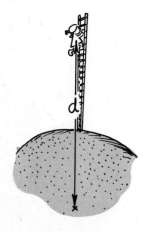

Figure 8-5

According to Newton's equation, her weight (not mass) decreases as she increases her distance from the earth's center (not surface).

Figure 8-6

If an apple weighs 1 N at the earth's surface, it weighs only $\frac{1}{4}$ N twice as far from the center of the earth. At three times the distance, it weighs only $\frac{1}{9}$ N. What would it weigh at four times the distance? Five times? Gravity versus distance is plotted in color.

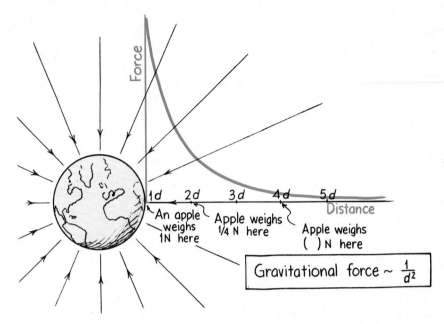

An apple weighs 1N here

Apple weighs 1/4 N here

Apple weighs ( ) N here

Gravitational force $\sim \frac{1}{d^2}$

It is interesting to note that Kepler's first and third laws are a consequence of Newton's law of universal gravitation. With it and a form of mathematics he invented called the *calculus*, Newton showed that planetary orbits are ellipses. He also showed that Kepler's third law is a consequence of the inverse-square dependence of gravitation on distance.* How nice that the single law of universal gravitation so neatly simplifies an understanding of planetary motion—and so much more.

Questions

1. By how much does the gravitational force between two objects decrease when the distance between them is doubled? Tripled? Increased tenfold?

2. Consider an apple at the top of a tree that is pulled by earth gravity with a force of 1 N. If the tree were twice as tall, would the force of gravity be only $\frac{1}{4}$ as strong? Defend your answer.

▶ **Answers**

1. It decreases to one-fourth, one-ninth, and one-hundredth.

2. No, because the twice-as-tall tree is not twice as far from the earth's center. The tree's height would have to be equal to the radius of the earth (6,370 km) before the weight of the apple reduces to $\frac{1}{4}$ N. Before its weight decreases by 1 percent, an apple or any object must be raised 32 km—nearly four times the height of Mt. Everest. So as a practical matter we disregard the effects of everyday changes in elevation.

---

*Perhaps your instructor will show that Kepler's third law results when Newton's law of gravitation is equated to centripetal force and how R^3/T^2 equals a constant that depends only on G and M, the mass of the body about which orbiting occurs. Interesting stuff!

Weight and Weightlessness

When you step on a spring balance such as a bathroom scale, you compress a spring inside. When the pointer stops, the strong electrostatic forces between the molecules inside the spring material balance the gravitational attraction between you and the earth—nothing moves as you and the scale are in static equilibrium. The pointer is calibrated to show your weight. If you stood on a bathroom scale in a moving elevator, you'd find variations in your weight. If the elevator accelerated upward, the springs inside the bathroom scale would be more compressed and your weight reading would increase. If the elevator accelerated downward, the springs inside the scale would be less compressed and your weight reading would decrease. If the elevator cable broke and the elevator fell freely, the reading on the scale would go to zero. According to the reading, you would be weightless. Would you really be weightless? We can answer this question only if we agree on what we mean by *weight*.

Figure 8-7
The sensation of weight equals the force with which you press against the supporting floor. If the floor accelerates up or down, your weight varies.

We define the weight of something as the force it exerts against the supporting floor or the weighing scales. According to this definition, you are as heavy as you feel; so in an elevator that accelerates downward, the supporting force of the floor is less and you weigh less. If the elevator is in free fall, your apparent weight is zero (Figure 8-7). Even in this weightless condition, however, there is still a gravitational force acting on you, causing your downward acceleration. But gravity now is not felt as weight because there is no support force.

Consider an astronaut in orbit. The astronaut is in a state of apparent weightlessness. He feels weightless because he is not supported by anything (Figure 8-8). There would be no compression in the springs of a bathroom scale placed beneath his feet because the bathroom scale is falling as fast as he is. Any objects that are released fall together with him and remain in his vicinity, unlike what happens on the ground. All the local effects of gravity are eliminated. The body organs respond as though gravity forces were absent, and this gives the sensation of weightlessness. The astronaut experiences the same sensation in orbit that he would feel in a falling elevator—a state of free fall.

Figure 8-8
Both are "weightless."

Figure 8-9
The half-dozen or so inhabitants in this proposed laboratory and docking facility will continually experience weightlessness in their zero-*g* environment.

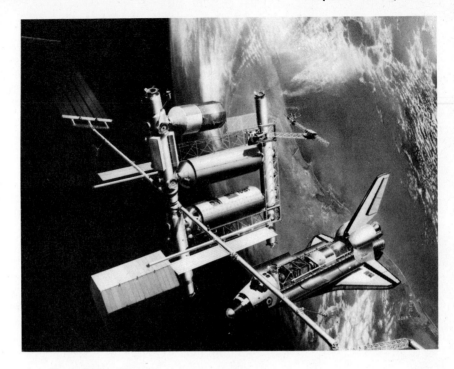

Note that the astronaut is still under the influence of gravitational force, constantly changing his direction (his velocity), thus accelerating. That's why we say he is in a state of *apparent* weightlessness. To be truly weightless, he would have to be far out in space, well away from the earth, sun, and other attracting bodies, where gravitational forces are negligible. In this truly weightless condition, any motion would be in a straight-line path rather than in the curved path of closed orbit.

The space station depicted by NASA artists in Figure 8-9 is in a state of apparent weightlessness. The shuttle, station facility, and astronauts all accelerate equally toward earth, at somewhat less than 1 *g* because of their altitude. This acceleration is not sensed at all; with respect to the station, the astronauts experience zero *g*.

Question ▶ In what sense is being infinitely far away from all celestial bodies like stepping off a table?

▶ **Answer**
In both cases you'd experience weightlessness. Far from all celestial bodies you'd be weightless in the strict sense, for you'd be away from all gravitational influences; stepping from a table, you'd momentarily experience apparent weightlessness because there would be a momentary lapse of support force.

Ocean Tides

Seafaring people have always known there was a connection between the ocean tides and the moon, but no one could offer a satisfactory theory to explain the two high tides per day. Newton showed that the ocean tides are caused by differences in the gravitational pull between the moon and the earth on opposite sides of the earth. Gravitational force between the moon and earth is strongest on the side of the earth nearest the moon, and it is weakest on the side of the earth farthest from the moon. This is simply because the gravitational force is weaker with increased distance.

Figure 8-10
Ocean tides.

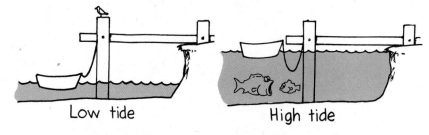

Low tide High tide

To understand why the difference in gravitational pulls by the moon on opposite sides of the earth produces tides, pretend you have a big spherical ball of Jell-O. If you exert the same force on every part of the ball, it would remain spherical as it accelerates. But if you pulled harder on one side than the other, there would be a difference in accelerations and the ball would become elongated (Figure 8-11). That is what happens to this big ball we're living on. The side closer to the moon is pulled with a greater force and has a greater acceleration toward the moon than the far side—thus the earth is somewhat football shaped. The acceleration we are talking about is actually a *centripetal* acceleration, for the earth circles the center of mass of the earth-moon system. (It is a mistake to think that the earth is at rest and the moon accelerates as it orbits about the earth, just as it would be a mistake for a moon resident to similarly think the moon is fixed and the earth accelerates toward and orbits the moon. It so happens that both the earth and moon orbit about a common point—the earth-moon center of mass.) Both the earth and the moon undergo centripetal acceleration as they circle each other about the earth-moon center of mass. The centripetal acceleration is greater for the sides of the earth and moon nearest each other than for the far sides, so both the earth and moon are slightly elongated. Elongation of the earth is mainly in its oceans, which bulge equally on opposite sides.

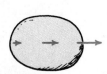

Figure 8-11
A ball of Jell-O remains spherical when all parts of it are pulled equally in the same direction. When one side is pulled more than the other, however, its shape is elongated.

On a world average, the ocean bulges are nearly 1 meter above the average surface level of the ocean. The earth spins once per day, so a fixed point on earth passes beneath both of these bulges each day. This produces two sets of ocean tides per day. Any part of the earth that passes beneath one of the bulges has a high tide. When the earth has made a quarter turn, 6 hours later, the water level at the same part of the ocean is nearly 1 meter below the average sea level. This is low tide. The water that "isn't there" is under the bulges that make up the high tides. A second high tidal bulge is experienced when the earth makes another

Figure 8-12
Two tidal bulges remain relatively fixed with respect to the moon while the earth spins daily beneath them.

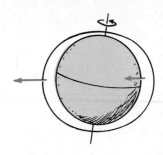

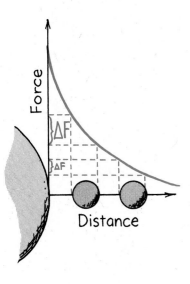

Figure 8-13
A plot of gravity versus distance. The greater the distance from the sun, the smaller the difference in gravitational pulls, ∆F, on opposite sides of a planet, and hence the smaller the tides.

Figure 8-14
When the attractions of the sun and the moon are lined up with each other, spring tides occur.

quarter turn. So we have two high tides and two low tides daily. It turns out that while the earth spins, the moon moves in its orbit and appears at the same position in our sky every 24 hours and 50 minutes, so the two-high-tide cycle is actually at 24-hour-and-50-minute intervals. That is why tides do not occur at the same time every day.

The sun also contributes to ocean tides, although it is less than half as effective as the moon in raising tides—even though its pull on the earth is 180 times greater than the pull of the moon. Why doesn't the sun cause tides 180 times greater than lunar tides? The answer has to do with a key word: *difference*. Because of the great distance of the sun, the difference in its gravitational pull on opposite sides of the earth is very small (Figure 8-13). The percentage difference in the sun's pulls across the earth is only about 0.017 percent, compared to 6.7 percent across the earth by the moon. It is only because the pull of the sun is 180 times stronger than the moon's that the sun tides are almost half as high (180 × 0.017 percent = 3 percent, nearly half of 6.7 percent).

When the sun, earth, and moon are all lined up, the tides due to the sun and the moon coincide. Then we have higher-than-average high tides and lower-than-average low tides. These are called **spring tides** (Figure 8-14). (Spring tides have nothing to do with the spring season.) You can tell when the sun, earth, and moon are aligned by the full moon or by the new moon. When the moon is full, the earth is between the sun and moon. (If all three are *exactly* in line, then you have a lunar eclipse, for the full moon is in the earth's shadow.) A new moon occurs when the moon is between the sun and earth, when the nonilluminated side of the moon faces the earth. (When this alignment is perfect, the moon blocks the sun and you have a solar eclipse.) Spring tides occur at the times of a new or full moon.

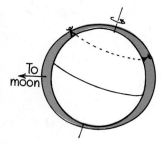

Figure 8-15
When the attractions of the sun and the moon are about 90° apart (at the time of a half moon), neap tides occur.

Figure 8-16
Inequality of the two high tides per day. Because of the earth's tilt, a person in the Northern Hemisphere may find the tide nearest the moon much lower (or higher) than the tide half a day later. Inequalities of tides vary with the positions of the moon and the sun.

All spring tides are not equally high because distances between the earth and the moon and the earth and the sun both vary; the orbital paths of the earth and the moon are not really circular, but elliptical. The moon's distance from the earth varies by about 10 percent and its effect in raising tides varies by about 30 percent. Highest spring tides occur when the moon and sun are closest to the earth.

When the moon is halfway between a new moon and a full moon, in either direction (Figure 8-15), the tides due to the sun and the moon partly cancel each other. Then, the high tides are lower than average and the low tides are not as low as average low tides. These are called **neap tides**.

Another factor that affects the tides is the tilt of the earth's axis (Figure 8-16). Even though the opposite tidal bulges are equal, the earth's tilt causes the two daily high tides experienced in most parts of the ocean to be unequal most of the time.

Our treatment of tides is quite simplified here, for tides are actually more complicated. Interfering land masses and friction with the ocean bottom, for example, complicate tidal motions. In many places the tides break up into smaller "basins of circulation," where a tidal bulge travels like a circulating wave that moves around in a small basin of water when it is properly tilted. For this reason the high tide may be hours away from an overhead moon. In mid-ocean the variation in water level—the range of the tide—is usually a meter or two. This range varies in different parts of the world; it is greatest in some Alaskan fjords and is most notable in the basin of the Bay of Fundy, between New Brunswick and Nova Scotia in southeast Canada, where tidal differences sometimes exceed 15 meters. This is largely due to the ocean floor, which funnels shoreward in a V-shape. The tide often comes in faster than a person can run. Don't dig clams near the water's edge at low tide in the Bay of Fundy!

Tides in the Earth and Atmosphere

The earth is not a rigid solid but, for the most part, is molten liquid covered by a thin solid and pliable crust. As a result, the moon-sun tidal forces produce earth tides as well as ocean tides. Twice each day the solid surface of the earth rises and falls as much as one-quarter meter! As a result, earthquakes and volcanic eruptions have a slightly higher probability of occurring when the earth is experiencing an earth spring tide—that is, near a full or new moon.

We live at the bottom of an ocean of air that also experiences tides. Because of the low mass of the atmosphere, the atmospheric tides are very small. In the upper part of the atmosphere is the ionosphere, so named because it is made up of ions—electrically charged atoms that are the result of intense cosmic ray bombardment. Tidal effects in the ionosphere produce electric currents that alter the magnetic field that surrounds the earth. These are magnetic tides. They in turn regulate the degree to which cosmic rays penetrate into the lower atmosphere. The cosmic ray penetration affects the ionic composition of our atmosphere, which in turn is evident in subtle changes in the behaviors of living things. The highs and lows of magnetic tides are greatest when the atmosphere is having its spring tides—again, near the full and new moon. During these times the atmospheric blanket about the earth changes from thinnest to thickest, all in a matter of 6 hours. Have you noticed that some of your friends seem a bit weird at the time of a full moon?

Questions ▶

1. Would there be tidal bulges on the moon if it were covered with water? How many? How often would high and low tides occur?

2. Which pulls with the greater force on the oceans of the earth, the sun or the moon?

Gravitational Fields

The earth and the moon pull on each other. This is *action at a distance*, because the earth and moon interact with each other even though they are not in contact. We can look at this in a different way: we can regard the moon as interacting with the **gravitational field** of the earth. The properties of the space surrounding any mass can be considered to be altered in such a way that another mass in this region will experience a force. This alteration of space is a gravitational field. We can think of distant rockets and space probes as interacting with gravitational fields rather than with the masses of the earth and other planets or stars that are the sources of these fields. The field concept plays an in-between role in our thinking about the forces between different masses.

A gravitational field is an example of a *force field*, for any mass in the field space experiences a force. Another force field, perhaps more familiar, is a magnetic field. Have you ever seen iron filings lined up in patterns around a magnet? (Look ahead to Figure 23-2, for example.) The pattern of the filings shows the strength and direction of the magnetic field at different points in the space around the magnet. Where the filings are closest together, the field is strongest. The direction of the filings shows the direction of the field at each point.

▶ **Answers**

1. There would be two tidal bulges for the same reason two tidal bulges occur on the earth. (The earth pulls differentially on the nearest and farthest parts of the moon. As a result, there *are* two bulges on the slightly elliptically shaped solid surface of the moon.) Since the moon takes 27.3 days to make a single revolution about its own axis (and also about the earth-moon axis), the same part of its surface faces the earth all the time. Therefore, the tidal bulges would be stationary upon the moon's surface, and no high or low tides would occur.

2. The sun, but it is so distant that it attracts all parts of the earth with almost equal strength. Hence its effectiveness in raising tides is less than that of the moon.

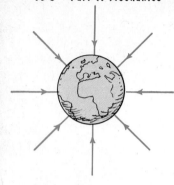

Figure 8-17
Field lines represent the gravitational field about the earth. Where the field lines are closer together, the field is stronger. Farther away, where the field lines are farther apart, the field is weaker.

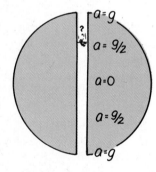

Figure 8-18
As you fall faster and faster in a hole bored completely through the earth, your acceleration decreases because the part of the earth's mass "above" you pulls in the opposite direction. At the earth's center all pulls cancel and your acceleration is zero. Momentum carries you past the center and against a growing acceleration to the opposite end of the tunnel where acceleration is again *g*.

The pattern of the earth's gravitational field can be represented by field lines (Figure 8-17). Like the iron filings around a magnet, the field lines are closer together where the gravitational field is stronger. At each point on a field line, the direction of the field at that point is along the line. Arrows show the field direction. A particle, astronaut, spaceship, or any mass in the vicinity of the earth will be accelerated in the direction of the field line at that location.

The strength of the earth's gravitational field, like the strength of its force on objects, follows the inverse-square law. It is strongest near the earth's surface and weakens with increasing distance from the earth.*

The Gravitational Field Inside a Planet†

The gravitational field of the earth exists inside the earth as well as outside. Imagine a hole drilled completely through the earth from the north pole to the south pole. Forget about impracticalities such as lava and high temperatures and consider the motion you would undergo if you fell into such a hole. If you started at the north pole end, you'd fall and gain speed all the way down to the center and then lose speed all the way "up" to the south pole. Without air drag, the one-way trip would take nearly 45 minutes. If you failed to grab the edge, you'd fall back toward the center, and return to the north pole in the same time.

Your acceleration, *a*, will be progressively less as you continue toward the center of the earth. Why? Because as you fall toward the earth's center, you are also being pulled up by the part of the earth above you. When you are at the center of the earth, the pull down is balanced by the pull up, so the net force on you is zero—$a = 0$ as you whiz with maximum speed past the center of the earth.‡ The gravitational field of the earth at its center is zero!

The composition of the earth varies, being most dense at its core and least dense at the surface. Inside a planet of uniform density, however, the field inside increases linearly, from zero at its center to *g* at the surface. We won't go into why this is so, but perhaps your instructor will provide the explanation. In any event, a plot of the gravitational field intensity inside and outside a solid planet of uniform density is shown in Figure 8-19.

---

*The strength of the gravitational field *g* at any point is measured by the force *F* exerted on a unit of mass *m* placed there. So $g = F/m$, and its units are newtons per kilogram (N/kg). The field *g* also equals the free-fall acceleration of gravity.

†This section may be skipped for a brief treatment of gravitational fields.

‡Interestingly enough, during the first few kilometers beneath the earth's surface you'd actually gain acceleration, because the density of surface material is much less than the condensed center. The mass of the earth "above" you exerts a small backward pull compared to the disproportionally greater pull when you are closer to the denser core material. This means you'd weigh slightly more for the first few kilometers beneath the earth's surface. Further in, your weight would decrease and would diminish to zero at the earth's center.

Figure 8-19
The gravitational field intensity inside a planet of uniform density is directly proportional to the radial distance from its center and is maximum at its surface. Outside, it is inversely proportional to the square of the distance from its center.

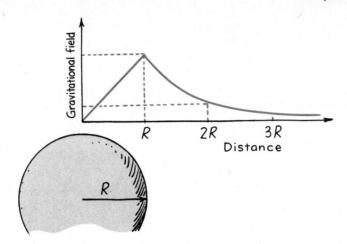

Questions ▶

1. If you stepped into a hole bored clear through the center of the earth and made no attempt to grab the edges at either end, what kind of motion would you experience?
2. Halfway to the center of the earth, would you weigh more or weigh less than you weigh at the surface of the earth?

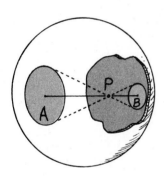

Imagine a cavern at the center of a planet. The cavern would be gravity-free because of the cancellation of gravity in every direction. Amazingly, the size of the cavern doesn't change this fact—even if it constitutes most of the volume of the planet! A hollow planet, like a huge basketball (Figure 8-20), would have no gravitational field anywhere inside it. Complete cancellation of gravity occurs everywhere inside. Consider a particle P, for example, which is twice as far from the left side of the planet as it is from the right side. If gravity depended only on distance, P would be attracted only $\frac{1}{4}$ as much to the left side as to the right side (according to the inverse-square law). But gravity also depends on mass.

Figure 8-20
The gravitational field anywhere inside a spherical shell of uniform thickness and composition is zero, because the field components from all the particles of mass in the shell cancel one another. A mass at point P, for example, is attracted just as much to the larger but farther region A as it is to the smaller but closer region B.

▶ **Answers**

1. You would oscillate back and forth, in what is called *simple harmonic motion*. Each round trip would take nearly 90 min. Interestingly enough, we will see in the next chapter that an earth satellite in close orbit about the earth also takes the same 90 min to make a complete round trip. (This is no coincidence, for if you study physics further, you'll learn that "up and down" simple harmonic motion is simply the vertical component of uniform circular motion—interesting stuff.)

2. You would weigh less, because the part of the earth's mass that pulls you "down" is counteracted by mass above you that pulls you "up." If the mass of the earth were of uniform density, halfway to the center your weight would be exactly half your surface weight. But since the earth's core is so dense (about seven times the density of surface rock), your weight would be somewhat more than half surface weight. Exactly how much depends on how the earth's density varies with depth information that is not known today.

The same solid angle subtends regions A and B, and we see that whatever the angle, A will have four times the area and therefore four times the mass as region B. Since $\frac{1}{4}$ of 4 is equal to 1, P is attracted to the farther but more massive region A with just as much force as it is to the closer but less massive region B. Cancellation occurs. More thought will show that cancellation will occur anywhere inside a planetary shell that has uniform thickness and composition. A gravitational field would exist at and beyond its outer surface and would behave as if all the mass of the planet were concentrated at its center—the center of gravity; but everywhere inside the hollow part, the gravitational field is zero. Anyone inside would feel weightless.

Gravity can be canceled inside a body or between bodies, but it cannot be shielded in the same way that electricity and magnetism can. We'll see that electricity and magnetism have repelling as well as attracting forces that enable shielding, but gravitation only attracts and therefore cannot be shielded. Eclipses provide convincing evidence for this. The moon is in the gravitational field of both the sun and the earth. During a lunar eclipse the earth is directly between the moon and the sun, and any shielding of the sun's field by the earth would result in a deviation of the moon's orbit. Even a very slight shielding effect would accumulate over a period of years and show itself in the timing of subsequent eclipses. But there have been no such discrepancies; past and future eclipses are calculated to a high degree of accuracy using only the simple law of gravitation. No shielding effect in gravitation has ever been found.

The idea of a force field has wider application in the study of electromagnetism. It also plays a central role in gravitation, as explained by Albert Einstein's general theory of relativity.

Einstein's Theory of Gravitation

A model for gravity quite unlike Newton's was presented by Einstein in his general theory of relativity. Einstein perceived a gravitational field as a geometrical warping of four-dimensional space and time; he realized that bodies put dents in space and time somewhat like a massive ball placed in the middle of a large waterbed dents the two-dimensional surface (Figure 8-21). The more massive the ball, the greater the dent or warp. If we roll a marble across the top of the bed but away from the ball, the marble will roll in a nearly straight-line path. But if we roll the marble near the ball, it will curve as it rolls across the indented surface of the waterbed. If the curve closes on itself, the marble will orbit the ball in an elliptical or circular path. From a Newtonian viewpoint, the marble curves because it is attracted to the ball. From an Einsteinian viewpoint, the marble curves not because of any force, but because the surface on which it moves curves.* In Chapter 35 we will treat Einstein's theory of gravitation in more detail.

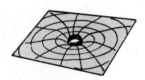

Figure 8-21
Warped space-time. Space near a star is curved in four dimensions in a way similar to the two-dimensional surface of a waterbed when a heavy ball rests on it.

---

*Don't be discouraged if you cannot visualize four-dimensional space-time. Einstein himself often told his friends, "Don't try. I can't do it either." Perhaps we are not too different from the great thinkers around Galileo who couldn't think of a moving earth!

Black Holes

Suppose you were indestructible and could go in a spaceship to the surface of a star. Your weight would depend both on your mass and the star's mass and on the distance between the star's center and your belly button. If the star were to burn out and collapse to half size with no change in its mass, your weight at the surface, determined by the inverse-square law, would be four times as much (Figure 8-22). If the star collapsed to a tenth its size, your weight at the surface would be 100 times as much. If the star kept shrinking, the gravitational field at the surface would become stronger. It would be more and more difficult for a starship to leave. The velocity required to escape, *escape velocity*, would increase. If a star such as our sun collapsed to a radius of less than 3 kilometers, the escape velocity from its surface would exceed the speed of light, and nothing—not even light—could escape! The sun would be invisible. It would be a black hole.

Figure 8-22

If a star collapses to half its radius and there is no change in its mass, gravitation at its surface would increase by 4.

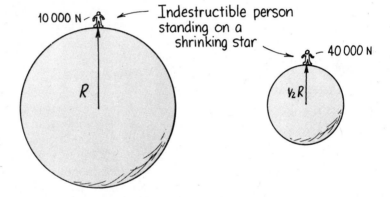

The sun, in fact, has too little mass to experience such a collapse, but when some stars with masses greater than four suns reach the end of their nuclear resources, they undergo collapse; and unless rotation is high enough, the collapse continues until the stars reach infinite densities. Gravitation near the surfaces of these shrunken stars is so enormous that light cannot escape from them. They have crushed themselves out of visible existence. The results are **black holes**, which are completely invisible.

A black hole is no more massive than the star from which it collapsed, so the gravitational field in regions at and greater than the original star's radius is no different after the star's collapse than before. But closer distances near the vicinity of a black hole are nothing less than the collapse of space itself, with a surrounding warp into which anything that passes too close—light, dust, or a spaceship—is drawn. Astronauts could enter the fringes of this warp and with a powerful spaceship still escape. After a certain distance, however, they could not, and they would disappear from the observable universe.

It has been a matter of some conjecture that material drawn into a black hole may reappear elsewhere through a "white hole." What vanishes in one place might be spewed out at another. Even if this were true,

Figure 8-23
Anything that falls into a black hole is reduced to sub-atomic particles.

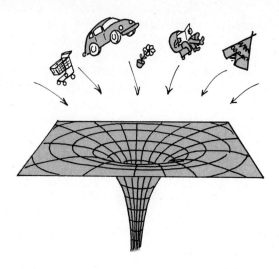

however, black holes would be no less frightening to space voyagers. Everything that falls into a black hole is reduced to elementary particles, and everything would emerge as particles. If black holes are a gate from one realm to another, they are a gate through which nothing can pass intact (Figure 8-24).

Although a black hole cannot be seen, its presence can be detected by its gravitational influence on neighboring stars. For example, it could be part of a binary star system, in which case a nearby luminous star would appear to be orbiting about an empty point. Several black holes have been tentatively identified. Information on astronomical bodies is presently being gathered faster than textbooks can report—check with your astronomy instructor for the latest update.

Figure 8-24
A black hole may be the portal to another universe.

Universal Gravitation

We all know that the earth is round. But why is the earth round? It is round because of gravitation. Everything attracts everything else, and so the earth has attracted itself together as far as it can! Any "corners" of the earth have been pulled in; as a result, every part of the surface is equidistant from the center of gravity. This makes it a sphere. Therefore, we see from the law of gravitation that the sun, the moon, and earth are spherical because they have to be (rotational effects make them slightly ellipsoidal).

If everything pulls on everything else, then the planets must pull on one another. The force that controls Jupiter, for example, is not just the force from the sun; there are also the pulls from the other planets. Their effect is small in comparison to the pull of the much more massive sun, but it still shows. When Saturn is near Jupiter, its pull disturbs the otherwise smooth ellipse traced by Jupiter. Both planets "wobble" about their expected orbits. This wobbling is called a *perturbation*. Early in the nineteenth century, unexplained perturbations were observed for the planet Uranus. Either the law of gravitation was failing at this great distance from the sun or an unknown eighth planet was perturbing Uranus. An Englishman and a Frenchman, J. C. Adams and Urbain Leverrier, each assumed Newton's law to be valid and independently calculated where an eighth planet should be. At about the same time, both sent letters to their respective observatories with instructions to search a certain area of the sky. The request by Adams was delayed by misunderstandings at Greenwich, but Leverrier's request to the director of the Berlin observatory was heeded immediately. The planet Neptune was discovered that very night!

Other perturbations of the planet Uranus led to the prediction and discovery of the ninth planet, Pluto, in 1930 at the Lowell Observatory in Arizona. Pluto takes 248 years to make a single revolution about the sun, so no one will see it in its discovered position again until the year 2178.

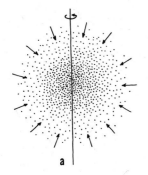

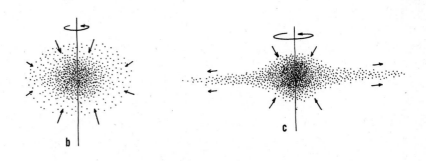

Figure 8-25
Solar system formation. A slightly rotating ball of interstellar gas (a) contracts due to mutual gravitation and (b) conserves angular momentum by speeding up. The increased momentum of individual particles and clusters of particles causes them to (c) sweep in wider paths about the rotational axis, producing an overall disk shape. The greater surface area of the disk promotes cooling and condensation of matter in swirling eddies—the birthplace of the planets.

Perturbations of double stars and the shapes of distant galaxies are evidence that the law of gravitation is true at larger distances. Over very large distances, gravitation underlies the fate of the entire universe. Current scientific speculation is that the universe originated in the explosion of a primordial fireball some 15 to 20 billion years ago. This is the **Big Bang** theory of the origin of the universe. All the matter of the universe was hurled outward from this event and continues in an outward expansion. We find ourselves in an expanding universe. This expanding may go on indefinitely, or it may eventually be overcome by the combined gravitation of all the galaxies and come to a stop. Like a stone thrown upward, whose departure from the ground comes to an end when it reaches the top of its trajectory and which then begins its descent to the place of its origin, the universe may contract and fall back into a single unity. This would be the *Big Crunch*. After that, the universe would presumably re-explode to produce a new universe. The same course of action might repeat itself and the process may well be cyclic. If this is true, we live in an oscillating universe.

We do not know whether the expansion is indefinite or cyclic because we are uncertain about whether enough mass exists to halt the expansion. The period of oscillation is estimated to be somewhat less than 100 billion years. If the universe does oscillate, who can say how many times this process has repeated? We know of no way a civilization could leave a trace of ever having existed, for all the matter in the universe is reduced to bare subatomic particles during such an event. Formation of the elements, stars, galaxies, and life again takes place. All the laws of nature, such as the law of gravitation, are rediscovered by the higher-evolving life forms. Then students (like you) of these laws read about them.

Few theories have affected science and civilization as much as Newton's theory of gravitation. The successes of Newton's ideas ushered in the so-called Age of Reason or Century of Enlightenment, for Newton had demonstrated that by observation and reason and by employing mechanical models and deducing mathematical laws, people could uncover the very workings of the physical universe. How profound that all the moons and planets and stars and galaxies have such a beautifully simple rule to govern them; namely,

$$F = G \frac{m_1 m_2}{d^2}$$

The formulation of this simple rule is one of the major reasons for the success in science that followed, for it provided hope that other phenomena of the world might also be described by equally simple laws.

This hope nurtured the thinking of many scientists, artists, writers, and philosophers of the 1700s. One of these was the English philosopher John Locke, who argued that observation and reason, as demonstrated by Newton, should be our best judge and guide in all things and that all of nature and even society should be searched to discover any "natural laws" that might exist. Using Newtonian physics as a model of reason, Locke and his followers modeled a system of government that found adherents in the thirteen British colonies across the Atlantic. These ideas culminated in the Declaration of Independence and the U.S. Constitution.

Summary of Terms

Kepler's laws of planetary motion
Law 1: Each planet moves in an elliptical orbit with the sun at one focus.
Law 2: The line from the sun to any planet sweeps out equal areas of space in equal time intervals.
Law 3: The squares of the times of revolution (or years) of the planets are proportional to the cubes of their average distances from the sun ($T^2 \sim R^3$ for all planets).

Law of universal gravitation Every mass in the universe attracts every other mass with a force that for two masses is directly proportional to the product of their masses and inversely proportional to the square of the distance separating them:

$$F = G \frac{m_1 m_2}{d^2}$$

Inverse-square law A law relating the intensity of an effect to the inverse square of the distance from the cause:

$$\text{Intensity} \sim \frac{1}{\text{distance}^2}$$

Gravity follows an inverse-square law, as do the effects of electric, magnetic, light, sound, and radiation phenomena.

Weightlessness A condition wherein apparent gravitational pull is lacking.

Spring tide A high or low tide that occurs when the sun, earth, and moon are all lined up so that the tides due to the sun and moon coincide, making the high tides higher than average and the low tides lower than average.

Neap tide A tide that occurs when the moon is midway between new and full, in either direction. Tides due to the sun and moon partly cancel, making the high tides lower than average and the low tides higher than average.

Gravitational field The space surrounding a massive body in which another mass experiences a force of attraction.

Black hole The configuration of a massive star that has undergone gravitational collapse, in which gravitation at the surface is so intense that even the star's own light cannot escape.

Big Bang The primordial explosion that is thought to have resulted in the expanding universe.

Suggested Reading

Einstein, A., and L. Infeld. *The Evolution of Physics.* New York: Simon & Schuster, 1938.

Gamow, G. *Gravity.* Science Study Series. Garden City, N.Y.: Doubleday (Anchor), 1962.

Narlikar, J. V. *The Lighter Side of Gravity.* New York: W. H. Freeman, 1982.

Valens, E. G. *The Attractive Universe: Gravity and the Shape of Space.* New York: World·Publishing, 1969. A delightful book for the lay person, heavily illustrated with photographs by Berenice Abbott.

Review Questions

1. Who gathered the data that showed that the planets travel in elliptical paths around the sun? Who discovered this fact? Who explained this fact?

Kepler's Laws

2. What relationship about the speed of planets and their distance from the sun did Kepler discover?

3. What relationship about the times of revolutions of the planets and their distance from the sun did Kepler discover?

4. In Kepler's thinking, what was the direction of force on a planet?

Newton's Law of Universal Gravitation

5. In Newton's thinking, what was the direction of force on a planet?

6. In Newton's insight, what did a falling apple have in common with the moon?

7. How can the moon "fall" without getting closer to the earth?

8. How does the force of gravity between two objects depend on their masses?

9. How does the force of gravity depend on the distance between two objects?

The Universal Gravitational Constant, G

10. How was G measured?

11. What is the magnitude of the gravitational force between the earth and your body?

12. What is the magnitude of the gravitational force between the earth and a 1-kg mass?

Gravity and Distance: The Inverse-Square Law

13. Why is the force of gravity between a pair of objects reduced to one-fourth instead of one-half when the distance between them is doubled?

14. Hanging from a tree is a certain apple that weighs 1 N. Twice as high above the ground on the same tree is another apple of the same mass. Why is its weight practically the same and not $\frac{1}{4}$ N?

15. What relationship exists between Kepler's laws and the law of gravitation?

Weight and Weightlessness

16. Would the springs inside a bathroom scale be more compressed or less compressed if you weighed yourself in an elevator that accelerated upward? Downward?

17. Would the springs inside a bathroom scale be more compressed or less compressed if you weighed yourself in an elevator that moved upward at constant velocity? Downward at constant velocity?

18. Distinguish between *weightlessness* and *apparent weightlessness.*

Ocean Tides

19. Does the moon circle the earth, does the earth circle the moon, or do both circle each other? Explain.

20. Why do both the sun and the moon exert a greater gravitational force on one side of the earth than the other?

21. Do tides depend more on the strength of gravitational pull or on the *difference* in strengths? Explain.

22. Distinguish between *spring tides* and *neap tides.*

Tides in the Earth and Atmosphere

23. How do the sizes of tides compare for the ocean and the "solid" earth?

24. What are *magnetic tides*?

25. Why are all tides greatest at the time of a full moon or new moon?

Gravitational Fields

26. What is a gravitational field, and how can its presence be detected?

The Gravitational Field Inside a Planet

27. What is the magnitude of the gravitational field at the center of the earth?

28. What would be the magnitude of the gravitational field halfway to the center of a planet with a uniform density?

29. What would be the magnitude of the gravitational field anywhere inside a hollow planet?

30. We will see later that electricity and magnetism can be shielded. Why cannot gravity be shielded?

Einstein's Theory of Gravitation

31. Newton said that the path of a planet curves because a force acts on it. Why does a planet's path curve according to Einstein?

Black Holes

32. If the earth shrank with no change in mass, what would happen to your weight at the surface?

33. What happens to the strength of the gravitational field at the surface of a star that shrinks?

34. Why is a black hole invisible?

Universal Gravitation

35. Why is the earth "round"? Why isn't it exactly spherical?

36. Distinguish between the *Big Bang* and the *Big Crunch*.

Home Projects

1. Hold up your thumb and first two fingers and make a V sign. Place a strong rubber band across your thumb and first finger. This represents the force of gravity between the sun and the earth. Place a medium-strength rubber band across your thumb and second finger to repre-  sent the force of gravity between the sun and moon. Then place a weak rubber band across your first two fingers to represent the force of gravity between the moon and earth. Note how all fingers pull on one another. Likewise for the gravitational pulls between the sun, earth, and moon.

2. Hold your hands outstretched, one twice as far from your eyes as the other, and make a casual judgment as to which hand looks bigger. Most people see them to be about the same size, while many see the nearer hand as slightly bigger. Almost nobody upon casual inspection sees the nearer hand as four times as big. But by the inverse-square law, the nearer hand should appear twice as tall and twice as wide and therefore occupy four times as much of your visual field as the farther hand. Your belief that your hands are the same size is so strong that you likely overrule this information. Now if you overlap your hands slightly and view them with one eye closed, you'll see the nearer hand as clearly bigger. This raises an interesting question: What other illusions do you have that are not so easily checked?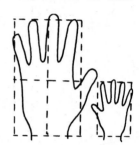

Exercises

1. Gravitational force acts on all bodies in proportion to their masses. Why, then, doesn't a heavy body fall faster than a light body?

2. Which weighs more, a sheet of aluminum foil or the same sheet crumpled into a tight wad?

3. The moon is gravitationally attracted to the sun with more than twice the force with which it is gravitationally attracted to the earth. Why, then, does the moon orbit about the earth?

4. Which planets, those closer to the sun than the earth or those farther from the sun than the earth, have a period greater than 1 earth year?

5. What are the magnitude and direction of the gravitational force that acts on a man who weighs 700 N at the surface of the earth?

6. The weight of an apple near the surface of the earth is 1 N. What is the weight of the earth in the gravitational field of the apple?

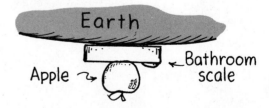

7. The earth and the moon are attracted to each other by gravitational force. Does the more massive earth attract the less massive moon with a force that is greater, smaller, or the same as the force with which the moon attracts the earth? (With an elastic band stretched between your thumb and forefinger, which is pulled more strongly by the band, your thumb or your forefinger?)

8. If the mass of the earth somehow increased, with all other factors remaining the same, would your weight also increase? (*Hint:* Let the equation for gravitational force guide your thinking.)

9. A small light source located 1 m in front of a 1-m² opening illuminates a wall behind. If the wall is 1 m behind the opening (2 m from the light source), the illuminated area covers 4 m². How many square meters will be illuminated if the wall is 3 m from the light source? 5 m? 10 m?

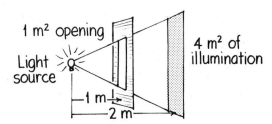

10. If you want to make a profit in buying precious material by weight at one altitude and selling it at another altitude for the same price per body weight, should you buy or sell at the higher altitude location? (Assume weighing is done on a spring scale.)

11. The planet Jupiter is more than 300 times as massive as earth, so it might seem that a body on the surface of Jupiter would weigh 300 times as much as on earth. But it so happens that a body would scarcely weigh three times as much on the surface of Jupiter as it would on the surface of the earth. Can you think of an explanation for why this is so? (*Hint:* Let the terms in the equation for gravitational force guide your thinking.)

12. From the data in the preceding exercise, estimate the size of Jupiter's diameter compared with the earth's.

13. Why do the passengers of high-altitude jet planes feel the sensation of weight while passengers in an orbiting space vehicle such as the space shuttle do not?

14. If the earth made one revolution each 90 min instead of each 24 h, would you press against the earth's surface if you were at the equator? At the poles? In the middle of the United States? Explain.

15. If you were in a car that drove off the edge of a cliff, why would you feel weightless? Would gravity still be acting on you in this state?

16. If you were in a freely falling elevator and you dropped a pencil, you'd see the pencil hovering. Is the pencil falling? Explain.

17. Explain why the following reasoning is wrong. "The sun attracts all bodies on the earth. At midnight, when the sun is directly below, it pulls on an object in the same direction as the pull of the earth on that object; at noon, when the sun is directly overhead it pulls on an object in a direction opposite to the pull of the earth. Therefore, all objects should be somewhat heavier at midnight than they are at noon." (*Hint:* Relate this to the preceding two exercises.)

18. If the mass of the earth increased, your weight would correspondingly increase. But if the mass of the sun increased, your weight would not be affected at all. Why?

19. Most people today know that the ocean tides are caused by the gravitational influence of the moon. And most people therefore think that the gravitational pull of the moon on the earth is greater than the gravitational pull of the sun on the earth. What do you think?

20. If somebody tugged on your shirt sleeve, it would likely tear. But if all parts of your shirt were tugged equally, no tearing would occur. How does this relate to tidal forces?

21. Would ocean tides exist if the gravitational pull of the moon (and sun) were somehow equal on all parts of the world? Explain.

22. Why aren't high ocean tides exactly 12 h apart?

23. With respect to spring and neap ocean tides, when are the lowest tides? That is, when is it best for digging clams?

24. Whenever the ocean tide is unusually high, will the following low tide be unusually low? Defend your answer in terms of "conservation of water." (If you slosh water in a tub so it is extra deep at one end, will the other end be extra low?)

25. The Mediterranean Sea has very little sediment churned up and suspended in its waters, mainly because of the absence of any substantial ocean tides. Why do you suppose the Mediterranean Sea has practically no tides? Similarly, are there tides in the Black Sea? Great Salt Lake? Your county reservoir? A glass of water? Explain.

26. The human body is about 80 percent water. Is it likely that the moon's gravitational pull could cause biological tides, cyclic changes among the fluid compartments of the body? (How much would the moon's pull on the near and far parts of the body differ?)

27. Since a "tidal bulge" sweeps around the earth in approximately 24 hours, does this mean that the incoming tide at the seashore comes in faster than a jet plane (which can almost beat the time zones)?

28. Exactly why do tides occur in the earth's crust and in the earth's atmosphere?

29. The value of g at the earth's surface is about 9.8 m/s^2. What is the value of g at a distance of twice the earth's radius?

30. If the earth were of uniform composition, what would the value of g be inside the earth at half its radius?

31. If the earth were of uniform composition, would your weight increase or decrease at the bottom of a deep mine shaft? Defend your answer.

32. It so happens that an actual *increase* in weight is found even in the deepest mine shafts. What does this tell us about the density of the earth's composition?

33. Which requires more fuel—a rocket going from the earth to the moon or a rocket coming from the moon to the earth? Why?

34. If you could somehow tunnel deep inside a star, would your weight increase or decrease? If, instead, you somehow stood on the surface of a shrinking star, would your weight increase or decrease? Why are your answers different?

35. If our sun shrank in size to become a black hole, show from the gravitational force equation that the earth's orbit would not be affected.

36. If the earth were hollow but still had the same mass and same radius, would your weight in your present location be more, less, or the same as it is now? Explain.

37. We say on page 152 that gravity cannot be shielded; on the same page and on page 151 we say that gravitational components cancel to zero inside a uniform shell. Is there a contradiction here? Why or why not?

38. If a massive object landed on the surface of the uniform shell of a hollow planet, would occupants inside the planet sense a gravitational attraction to it? Defend your answer.

39. Strictly speaking, you weigh a tiny bit less when you are in the lobby of a massive skyscraper. Why is this so?

40. A person falling into a black hole would probably be killed by tidal forces before ever encountering the hole itself. Explain why this is so.

9 Satellite Motion

Not too long ago, perhaps in the heyday of your grandparents, it was a fanciful and farfetched idea that humans would soon be in comfortable spaceships high above the atmosphere orbiting the planet Earth. As recently as 1969, when the first edition of this book was being written, humans first reached the moon. Today the news you see on TV from around the world is beamed to and from satellites. As you're reading this, it is quite likely that astronauts or cosmonauts are up there. The human race is right now preparing for space exploration that will take us who knows where. Our spacefaring accomplishments, amazing by present-day standards yet perhaps quaint and antiquated by tomorrow's, had their beginnings back in 1665 on a farm in Woolsthorpe, England. It was there that Issac Newton discovered the law of universal gravitation and its role in the motions of the moon, planets, and satellites.

The Falling Apple

According to legend, the sight of a falling apple triggered Newton's realization that the pull of the earth on the apple extended also to the moon. Newton had been giving a lot of thought to the fact that the moon does not follow a straight-line path, but instead circles about the earth. He understood the concept of inertia introduced earlier by Galileo, and knew that without an outside force, a moving object would continue its motion at constant speed in a straight line. He knew that if an object undergoes a change in speed or direction, then a force is responsible. Newton had the insight to see that the force that pulled the apple may be the same force that extends to the moon and pulls it into a circular path about the earth.

The Falling Moon

In the previous chapter we saw that the moon falls beneath the straight-line path it would follow if no force acted on it. Newton hypothesized that the moon is simply a projectile circling the earth under the attraction of gravity. This concept is illustrated in a drawing by Newton, shown in Figure 9-1. He compared motion of the moon to a cannonball fired from the top of a high mountain. He imagined that the mountaintop was above the earth's atmosphere, so air resistance would not impede the motion of the cannonball. If a cannonball were fired with a small horizontal speed, it would follow a parabolic path and soon hit the earth below. If it were fired faster, its path would be less curved and it would hit the earth farther away. If the cannonball were fired fast enough, Newton reasoned, the parabolic path would become a circle and the cannonball would circle indefinitely. It would be in orbit.

Figure 9-1
"The greater the velocity . . . with which [a stone] is projected, the farther it goes before it falls to the earth. We may therefore suppose the velocity to be so increased, that it would describe an arc of 1, 2, 5, 10, 100, 1000 miles before it arrived at the earth, till at last, exceeding the limits of the earth, it should pass into space without touching."—Isaac Newton, *System of the World*.

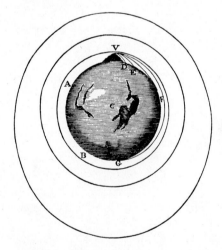

Both cannonball and moon have "sideways" velocity, or *tangential velocity*—the velocity (or component of velocity) parallel to the earth's surface—sufficient to ensure motion *around* the earth rather than *into* it. If there is no resistance to reduce its speed, the moon "falls" around and around the earth indefinitely.

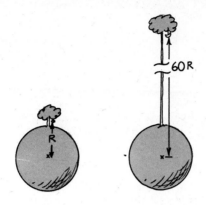

Figure 9-2
An apple falls 4.9 m during its first second of fall when it is near the earth's surface. Newton asked how far the moon would fall in the same time if it were 60 times farther from the center of the earth. His answer was (in today's units) 1.4 mm.

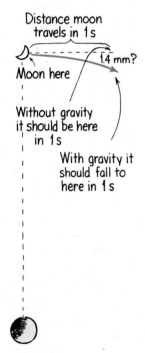

Figure 9-3
If the force that pulls apples off trees also pulls the moon into orbit, the circle of the moon's orbit should fall 1.4 mm below a point along the straight line where the moon would otherwise be one second later.

For the concept to advance from hypothesis to theory, it would have to be tested. Newton's test was to see if the moon's fall beneath its otherwise straight-line path was in correct proportion to the fall of an apple or any object at the earth's surface. He reasoned that the mass of the moon should not affect how it falls, just as mass has no effect on the acceleration of freely falling objects on earth. How far the moon falls and how far an apple at the earth's surface falls should relate only to their respective distances from the earth's center. If the distance of fall for the moon and the apple are in correct proportion, then the hypothesis that earth gravity reaches to the moon must be taken seriously.

The moon was known to be 60 times farther from the center of the earth than an apple at the earth's surface. The apple will fall nearly 5 meters in its first second of fall—or, more precisely, 4.9 meters. Newton reasoned from Kepler's laws that gravitational attraction to the earth must be "diluted" by the inverse square of distance. So if the moon is 60 times farther away, its fall toward earth should be $1/(60)^2$ the fall of an apple near the earth's surface. In one second the moon should fall $1/(60)^2$ of 4.9 meters, which is 1.4 millimeters.*

Newton confirmed his hypothesis. Every second the moon falls 1.4 millimeters beneath the tangent it would follow without gravity. It soon became clear that the earth and planets orbit the sun in the same way that the moon orbits the earth. The planets continually fall around the sun in closed paths. Why don't the planets crash into the sun? They don't because of their tangential velocities. What would happen if their tangential velocities were reduced to zero? The answer is simple enough: their motion would be straight toward the sun and they would indeed crash into it. Any objects in the solar system with insufficient tangential velocities have long ago crashed into the sun. What remains is the harmony we observe.

---

*Or, working backwards, $(.0014m)(60^2) = 4.9$ m.

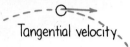

Satellite Motion

Tangential velocity

Figure 9-4
The tangential velocity of the earth about the sun allows it to fall around the sun rather than directly into it. If this tangential velocity were reduced to zero, what would be the fate of the earth?

Circular Orbits

Recall from Chapter 3 that the tangential speed a projectile needs to orbit the earth is 8 kilometers per second. That's because in the 1 second the projectile travels 8 kilometers horizontally, it falls a vertical distance of 4.9 meters—the same vertical distance the earth curves for each 8-kilometer tangent. So an 8-kilometers-per-second cannonball fired horizontally from Newton's mountain would follow the earth's curvature and coast around the earth again and again (provided the cannoneer and the cannon got out of the way). Fired slower, the cannonball would strike the earth's surface; fired faster, it would overshoot a circular orbit, as we will discuss shortly. Newton calculated the speed for circular orbit, and since such a cannon-muzzle velocity was clearly impossible, he did not foresee people launching satellites (he did not foresee multistage rockets).

Note that in circular orbit the speed of a satellite is not changed by gravity; only the direction changes. We can understand this by comparing a satellite in circular orbit with a bowling ball rolling along a bowling alley. Why doesn't the gravity that acts on the bowling ball change its speed? Because gravity is not pulling forward or backward; gravity pulls straight downward. The bowling ball has no component of gravitational force along the direction of the alley (Figure 9-5).

Likewise for a satellite in circular orbit, for it always moves perpendicularly to the earth's gravitational field—not in the direction of the field, which would increase its speed, nor against the field, which would decrease its speed. Instead, it moves at right angles to the earth's gravitational field. With no component of motion along or against the earth's field, no change in speed occurs—only change in direction. A satellite in circular orbit around the earth coasts parallel to the surface of the earth at constant speed.

Questions ▶ Consider a ball rolling along a bowling alley that completely circles the earth, and is elevated high enough so air resistance can be neglected.

1. Why would the force of gravity not change the speed of the ball?

2. If a section of the alley were cut away to leave a large gap, how fast must the ball travel to clear the gap and continue its motion as usual? At this speed, what would be the maximum gap for unchanged motion?

▶ **Answers**

1. A change in speed requires a force or component of force along the direction of travel. In this case the gravitational force on the ball is everywhere perpendicular to the alley, with no force component in the direction of motion.

2. To clear the gap without bumping into the edge of the alley, the ball must have orbital speed. Then its curved path matches that of the alley's surface. In this case the alley can be completely removed to have in effect a 360° gap, because the ball would be in earth orbit anyway!

Figure 9-5
(a) The force of gravity on the bowling ball is at 90° to its direction of motion, so it has no components of force to pull it forward or backward, and the ball rolls at constant speed. (b) The same is true even if the bowling alley is larger and remains "level" with the curvature of the earth. (c) If the ball moves at 8 km/s with no air resistance, would it need the alley?

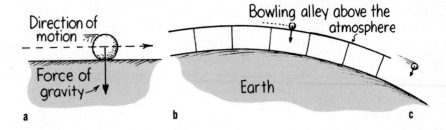

For a satellite close to the earth, the period (the time for a complete orbit about the earth) is about 90 minutes. For higher altitudes, the orbital speed is less and the period is longer. For example, communication satellites located in orbit 5.5 earth radii above the surface of the earth have a period of 24 hours. This period matches the period of daily earth rotation. For an orbit around the equator, these satellites stay above the same point on the ground. The moon is even farther away and has a period of 27.3 days. The higher the orbit of a satellite, the less its speed and the longer its period.*

Questions ▶

1. One of the beauties of physics is that there are usually different ways to explain a given phenomenon. Is the following explanation valid? Satellites remain in orbit instead of falling to the earth because they are beyond the main pull of earth's gravity.

2. Satellites in close circular orbit fall about 4.9 m during each second of orbit. Why doesn't this distance accumulate and send satellites crashing to earth?

▶ **Answers**

1. No, no, a thousand times no! If any moving object were beyond the pull of gravity, it would move in a straight line and would not curve around the earth. Satellites remain in orbit because they are being pulled by gravity, not because they are beyond it. For the altitudes of most earth satellites, the earth's gravitational field is only a few percent weaker than at the earth's surface.

2. In each second, the satellite falls about 4.9 m below the straight-line tangent it would have taken if there were no gravity. The earth's surface also curves 4.9 m beneath a straight-line 8-km tangent. The process of falling with the curvature of the earth continues from tangent line to tangent line, so the curved path of the satellite and the curve of the earth's surface "match" all the way around the earth. Satellites in fact do crash to the earth's surface from time to time, but this is principally because they encounter air resistance in the upper atmosphere that decreases their orbital speed.

---

*The speed of a satellite in circular orbit is given by $v = \sqrt{GM/d}$ and the period by $T = 2\pi\sqrt{d^3/GM}$, where G is the universal gravitational constant (see Chapter 8), M is the mass of the earth (or whatever body the satellite orbits), and d is the altitude of the satellite measured from the center of the earth or parent body.

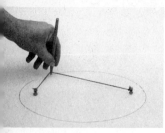

Figure 9-6
A simple method for constructing an ellipse.

Figure 9-7
The shadows cast by the ball are all ellipses, one for each lamp in the room. The point at which the ball makes contact with the table is the common focus of all three ellipses.

Elliptical Orbits

If a projectile just above the drag of the atmosphere is given a horizontal speed somewhat greater than 8 kilometers per second, it will overshoot a circular path and trace an oval-like path, an **ellipse**.

An ellipse is a specific curve; the closed path taken by a point that moves in such a way that the sum of its distances from two fixed points (called *foci*) is constant. For a satellite orbiting a planet, one focus is at the center of the planet; the other focus is empty. An ellipse can be easily constructed by using a pair of tacks, one at each focus, a loop of string, and a pencil (Figure 9-6). The closer the foci are to each other, the closer the ellipse is to a circle. When both foci are together, the ellipse is a circle. So we see that a circle is a special case of an ellipse.

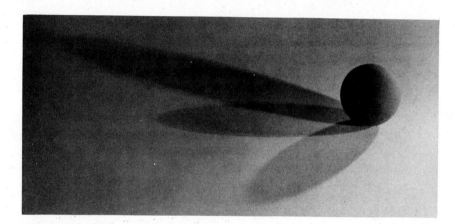

Whereas the speed of a satellite is constant in a circular orbit, speed varies in an elliptical orbit. When the initial speed is greater than 8 kilometers per second, the satellite overshoots a circular path and moves away from the earth, against the force of gravity. It therefore loses speed. Like a rock thrown into the air, it slows to a point where it no longer recedes and then begins to fall back toward the earth. The speed it loses in receding is regained as it falls back toward the earth, and it finally crosses its original path with the same speed it had initially (Figure 9-8).

Figure 9-8
Elliptical orbit. An earth satellite that has a speed somewhat greater than 8 km/s overshoots a circular orbit (a) and travels away from the earth. Gravitation slows it to a point where it no longer leaves the earth (b). It falls toward the earth gaining the speed it lost in receding (c) and overshoots as before in a repetitious cycle.

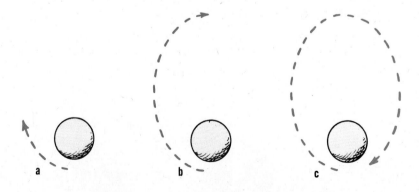

a   b   c

The procedure repeats over and over, and an ellipse is traced each cycle.

Interestingly enough, the parabolic path of a projectile such as a tossed baseball or a cannonball (Chapter 3) is actually a tiny segment of a thin ellipse that extends within and just beyond the center of the earth (Figure 9-9*a*). In Figure 9-9*b*, we see several paths of cannonballs fired from Newton's mountain. All ellipses have the center of the earth as one focus. As muzzle velocity is increased, the ellipses are less eccentric (wider); and when muzzle velocity reaches 8 kilometers per second, the ellipse rounds into a circle and does not intercept the earth's surface. The cannonball coasts in circular orbit. At greater muzzle velocities, the orbiting cannonball traces the familiar external ellipse.

Figure 9-9
(*a*) The parabolic path of the cannonball is part of an ellipse that extends within the earth. The earth's center is the far focus. (*b*) All paths of the cannonball are ellipses. For less than orbital speeds, the center of the earth is the far focus; for circular orbit, both foci are the earth's center; for greater speeds, the near focus is the earth's center.

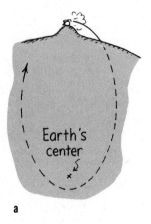

a

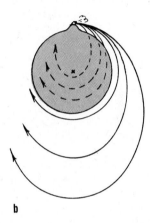

b

Question ▶ The orbital path of a satellite is shown in the sketch. In which of the marked positions A through D does the satellite have the greatest speed? Lowest speed?

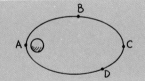

Putting a payload into earth orbit requires control over the speed and direction of the rocket that carries it above the atmosphere. A rocket initially fired vertically is intentionally tipped from the vertical course; then, once above the drag of the atmosphere, it is aimed *horizontally,* whereupon the payload is given a final thrust to 8 kilometers per second or more. This is shown in Figure 9-10, where for the sake of simplicity the payload is the entire single-stage rocket. We see that with the proper tangential velocity it falls around the earth, rather than into it, and becomes an earth satellite.

▶ **Answer**
The satellite has its greatest speed as it whips around A and has its lowest speed at position C. Beyond C it gains speed as it falls back to A to repeat its cycle.

Figure 9-10

The initial thrust of the rocket pushes it up above the atmosphere. Another thrust to a horizontal speed of at least 8 km/s is required if the rocket is to fall around rather into the earth.

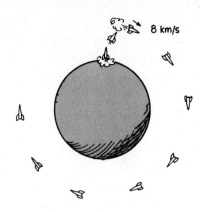

Energy Conservation and Satellite Motion

Recall from Chapter 6 that an object in motion possesses kinetic energy (KE) by virtue of its motion. An object above the earth's surface possesses potential energy (PE) by virtue of its position. Everywhere in its orbit, a satellite has both KE and PE with respect to the body it orbits. The sum of the KE and PE will be a constant all through the orbit. The simplest case occurs for a satellite in circular orbit.

Figure 9-11

The force of gravity on the satellite is always toward the center of the body it orbits. For a satellite in circular orbit, no component of force acts along the direction of motion. The speed and, thus, the KE do not change.

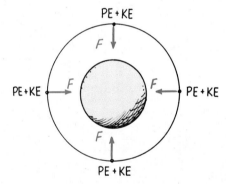

In circular orbit the distance between the body's center and the satellite does not change, which means the PE of the satellite is the same everywhere in orbit. Then, by the conservation of energy, the KE must also be constant. So a satellite in circular orbit coasts at an unchanging PE, KE, and speed (Figure 9-11).

In elliptical orbit the situation is different. Both speed and distance vary. PE is greatest when the satellite is farthest away (at the apogee) and least when the satellite is closest (at the perigee). Note that the KE will be least when the PE is most, and the KE will be most when the PE is least. At every point in the orbit, the sum of KE and PE is the same.

At all points along the orbit there is a component of gravitational force in the direction of motion of the satellite (with two exceptions, the apogee and perigee, where the force is perpendicular only). The component of

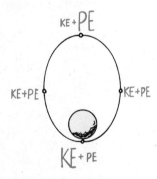

Figure 9-12

The sum of KE and PE for a satellite is a constant at all points along its orbit.

force in the direction of motion changes the speed of the satellite. Or we can say that (this component of force) × (distance moved) = ΔKE. Either way, when the satellite gains altitude and moves against this component, its speed and KE decrease. The decrease continues to the apogee. Once past the apogee, the satellite moves in the same direction as the component, and the speed and KE increase. The increase continues until the satellite whips past the perigee and repeats the cycle.

Figure 9-13
In elliptical orbit, a component of force exists along the direction of the satellite's motion. This component changes the speed and, thus, the KE. (The perpendicular component changes only the direction.)

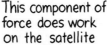

This component of force does work on the satellite

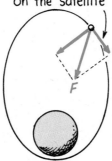

F

Questions

1. The orbital path of a satellite is shown in the sketch. In which marked positions A through D does the satellite have the greatest KE? Greatest PE? Greatest total energy?

2. Why does the force of gravity change the speed of a satellite when it is in an elliptical orbit, but not when it is in a circular orbit?

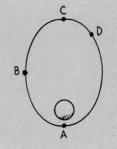

▶ **Answers**

1. KE is maximum at the perigee A; PE is maximum at the apogee C; the total energy is the same everywhere in the orbit.

2. At any point on its path, the direction of motion of a satellite is always tangent to its path. If a component of force exists along this tangent, then the acceleration of the satellite will involve a change in speed as well as direction. In circular orbit the gravitational force is always perpendicular to the direction of motion of the satellite, just as every part of the circumference of a circle is perpendicular to the radius. So there is no component of gravitational force along the tangent, and only the direction of motion changes, not the speed. But when the satellite moves in directions that are not perpendicular to the force of gravity, as in an elliptical path, there is a component of force along the direction of motion that changes the speed of the satellite. From a work-energy point of view, a component of force along the distance the satellite moves does work to change its KE.

Escape Speed

We know that a cannonball fired horizontally at 8 kilometers per second from Newton's hypothetical mountain would find itself in orbit about the earth. But what would happen if the cannonball were instead fired at the same speed *vertically*? It would do what most things that are projected upward would do. It would simply rise to some maximum height, reverse direction, and then fall back to earth. Then the old saying "What goes up must come down" would hold true, just as surely as a stone tossed skyward will be returned by gravity (unless, as we shall see, its speed is too great).

In today's spacefaring age, it is more accurate to say "What goes up *may* come down," for there is a critical speed at which a projectile is able to outrun gravity and escape the earth. This critical speed is called the **escape speed** or, if direction is involved, the *escape velocity*. From the surface of the earth, escape speed is 11.2 kilometers per second. Launch a projectile at any speed greater than that and it will leave the earth, traveling slower and slower, but never stopping due to earth gravity.* Recall the inverse-square law in the previous chapter: the ever-present force of earth gravity weakens and is less effective with increasing distance. We can understand the magnitude of this escape speed from an energy point of view.

How much work would be required to lift a payload against the force of earth gravity to a distance very very far ("infinitely far") away? We might think PE would be infinite because the distance is infinite. But gravity diminishes with distance by the inverse-square law. The force of gravity on the payload would be strong only close to the earth. Most of the work done in launching a rocket, for example, occurs near the earth's surface. It turns out that the PE of a 1-kilogram mass infinitely far away is 60 million joules (60 MJ). So to put a payload infinitely far from the earth's surface requires at least 60 million joules of energy per kilogram of load. We won't go through the calculation here, but a KE of 60 million joules for each kilogram corresponds to a speed of 11.2 kilometers per second, whatever the total mass involved. This is the escape speed from the surface of the earth.†

If we give a payload any more energy than 60 million joules per kilogram at the surface of the earth or, equivalently, any more speed than 11.2 kilometers per second, then, neglecting air resistance, the payload will escape from the earth never to return. As it continues outward, its PE increases and its KE decreases. Its speed becomes less and less, though it is never reduced to zero. The payload outruns the gravity of the earth. It escapes.

---

*Escape speed, from any planet or any body, is given by $v = \sqrt{2GM/d}$, where G is the universal gravitational constant, M is the mass of the attracting body, and d is the distance from its center. (At the surface of the body, d would simply be the radius of the body.)

†Interestingly enough, this might well be called the *maximum falling speed*. Any object, however far from earth, released from rest and allowed to fall to earth only under the influence of the earth's gravity would not exceed 11.2 km/s.

The escape speeds of various bodies in the solar system are shown in Table 9-1. Note that the escape speed from the sun is 620 kilometers per second at the surface of the sun. Even at a distance equaling that of the earth's orbit, the escape speed from the sun is 42.5 kilometers per second, considerably more than the escape speed from the earth. An object projected from the earth at a speed greater than 11.2 kilometers per second but less than 42.5 kilometers per second will escape the earth but not the sun. Rather than recede forever, it will take up an orbit around the sun.

Table 9-1

Escape speeds at the surface of bodies in the solar system

Astronomical body	Mass (earth masses)	Radius (earth radii)	Escape speed (km/s)
Sun	330 000	109	620
Sun (at a distance of the earth's orbit)	23 000		42.5
Jupiter	318	11	61.0
Saturn	95.2	9	37.0
Neptune	17.3	3.4	25.4
Uranus	14.5	3.7	22.4
Earth	1.00	1.00	11.2
Venus	0.82	0.96	10.4
Mars	0.11	0.53	5.2
Mercury	0.05	0.38	4.3
Moon	0.01	0.27	2.4

The first probe to escape the solar system, *Pioneer 10*, was launched from earth in 1972 with a speed of only 15 kilometers per second. The escape was accomplished by directing the probe into the path of oncoming Jupiter. It was whipped about by Jupiter's great gravitational field, picking up speed in the process—similar to the increase in the speed of a ball encountering an oncoming bat when it departs from the bat. Its speed of departure from Jupiter was increased enough to exceed the sun's escape speed at the distance of Jupiter. *Pioneer 10* passed the orbit of Pluto in 1984. Unless it collides with another body, it will wander indefinitely through interstellar space. Like a note in a bottle cast into the sea, *Pioneer 10* contains information about the earth that might be of interest to extraterrestrials, in hopes that it will one day wash up and be found on some distant "seashore."

It is important to point out that the escape speeds for different bodies refer to the initial speed given by a brief thrust, after which there is no force to assist motion. One could escape the earth at any sustained speed more than zero, given enough time. For example, suppose a rocket is launched to a destination such as the moon. If the rocket engines burn

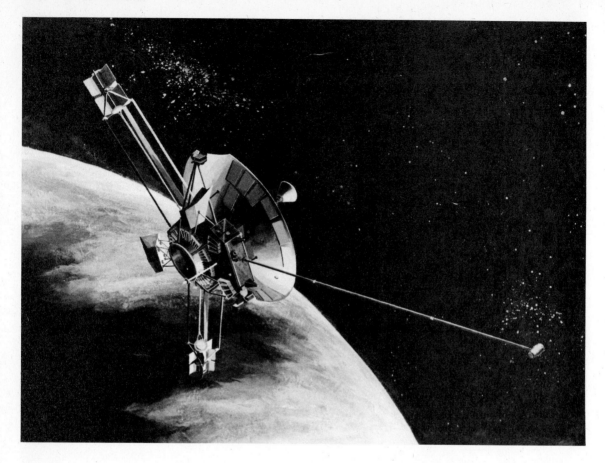

Figure 9-14

Pioneer 10, launched from earth in 1972, escaped from the solar system in 1984 and is wandering in interstellar space.

out when still close to the earth, the rocket needs a minimum speed of 11.2 kilometers per second. But if the rocket engines can be sustained for long periods of time, the rocket could go to the moon without ever attaining 11.2 kilometers per second.

It is interesting to note that the accuracy with which an unmanned rocket reaches its destination is not accomplished by staying on a pre-planned path or by getting back on that path if it strays off course. No attempt is made to return the rocket to its original path. Instead, the control center in effect asks, "Where is it now with respect to where it ought to go? What is the best way to get there from here, given its present situation?" With the aid of high-speed computers, the answers to these questions are used in finding a new path. Corrective thrusters put the rocket on this new path. This process is repeated over and over again all the way to the goal.*

---

*Is there a lesson to be learned here? Suppose you find that you are off course. You may, like the rocket, find it more fruitful to take a course that leads to your goal as best plotted from your present position and circumstances, rather than try to get back on the course you plotted from a previous position and under, perhaps, different circumstances.

Summary of Terms

Satellite A projectile or small celestial body that orbits a larger celestial body.

Ellipse The closed oval-like curve wherein the sum of the distances from any point on the curve to both foci is a constant. When the foci are together at one point, the ellipse is a circle. The farther apart the foci, the more eccentric the ellipse.

Escape speed The speed that a projectile, space probe, or similar object must reach to escape the gravitational influence of the earth or celestial body to which it is attracted.

Review Questions

The Falling Apple

1. According to Newton, what does a falling apple have in common with the moon?

The Falling Moon

2. What exactly is *tangential velocity*?

3. How far does the moon fall beneath a straight-line tangent to its motion in 1 s?

4. Why don't the planets fall into the sun?

Satellite Motion

Circular Orbits

5. Why doesn't the force of gravity change the speed of a satellite in circular orbit?

6. How much time is taken for a complete revolution of a satellite in close orbit about the earth?

7. For orbits of greater altitude, is the period greater or less?

Elliptical Orbits

8. Why does the force of gravity change the speed of a satellite in an elliptical orbit?

9. At what part of an elliptical orbit does a satellite have the greatest speed? The least speed?

Energy Conservation and Satellite Motion

10. Why is the kinetic energy a constant for a satellite in circular orbit?

11. Why is kinetic energy a variable for a satellite in an elliptical orbit?

12. With respect to the apogee and perigee of an elliptical orbit, where is the gravitational potential greatest? Least?

13. Is the sum of kinetic and potential energies a constant for satellites in circular orbits, elliptical orbits, or both?

Escape Speed

14. What is the minimum speed for orbiting the earth in close orbit? The maximum speed? What happens above this speed?

15. How was *Pioneer 10* able to escape the solar system at a speed less than escape speed?

Exercises

1. If a cannonball is fired from a tall mountain, gravity changes its speed all along its trajectory. But if it is fired fast enough to go into circular orbit, gravity does not change its speed at all. Explain.

2. The astronaut on page 161 appears to be floating free of gravity. Is he? Explain.

3. Since the moon is gravitationally attracted to the earth, why doesn't it simply crash into the earth?

4. Does the speed of a falling object depend on its mass? Does the speed of a satellite in orbit depend on its mass? Defend your answers.

5. If you have ever watched the launching of an earth satellite, you may have noticed that the rocket departs from a vertical course and continues its climb at an angle. Why?

6. Why are satellites normally sent into orbit by firing them in an easterly direction (the direction in which the earth spins)?

7. In the sketch on the left, a ball gains KE when rolling down a hill because work is done by the component of weight (*F*) that acts in the direction of motion. Sketch in the similar component of gravitational force that does work to change the KE of the satellite on the right.

8. Why is work done by the force of gravitation on a satellite when it is in an elliptical orbit, but not when it is in a circular orbit?

9. What is the shape of the orbit when the velocity of the satellite is everywhere perpendicular to the force of gravity?

10. If a pair of satellites at different altitudes but moving in the same direction are seen from the earth to pass overhead, which of the two overtakes the other?

11. If a flight mechanic drops a wrench from a high-flying jumbo jet, it crashes to earth. If an astronaut on the orbiting space shuttle drops a wrench, does it crash to earth also? Defend your answer.

12. The orbiting space shuttle travels at 8 km/s with respect to the earth. Suppose it projects a capsule rearward at 8 km/s with respect to the shuttle. Describe the path of the capsule with respect to the earth.

13. Would the speed of a satellite in close circular orbit about Jupiter be greater than, equal to, or less than 8 km/s?

14. A rocket coasts in an elliptical orbit around the earth. To attain the greatest amount of KE for escape using the least amount of fuel, should it fire its engines at the apogee or the perigee? (*Hint:* Let the formula $Fd = \Delta KE$ be your guide to thinking. Suppose the thrust F is brief and the same in either case. Then consider the distance d the rocket would travel during this brief burst at the apogee and at the perigee.)

15. The orbital velocity of the earth about the sun is 30 km/s. If the earth were suddenly stopped in its tracks, it would simply fall radially into the sun. Devise a plan whereby a rocket loaded with radioactive wastes could be fired into the sun for permanent disposal. How fast and in what direction with respect to the earth's orbit should the rocket be fired?

16. Escape speed from the surface of the earth is 11.2 km/s, but a space vehicle could escape from the earth at half this speed or less. Explain.

17. What is the maximum possible speed of impact upon the surface of the earth for a faraway body initially at rest that falls to the earth by virtue of the earth's gravity only?

18. If Pluto were somehow stopped short in its orbit, it would fall into rather than around the sun. How fast would it be moving when it hit the sun?

19. If the earth shrank in size, all other factors remaining the same, would escape velocity from its surface be greater, less, or the same as it presently is? Explain.

20. At which of the indicated positions does the satellite in elliptical orbit experience the greatest gravitational force? The greatest speed? The greatest velocity? The greatest momentum? The greatest kinetic energy? The greatest gravitational potential energy? The greatest total energy? The greatest angular momentum? The greatest acceleration?

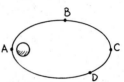

21. Many schools and hospitals could have been constructed with the funds used in the last two decades to probe and explore the solar system. How many of these schools and hospitals do you speculate would in fact have been constructed if the various space missions were not funded?

PART 2

PROPERTIES OF MATTER

10 Atomic Nature of Matter

Sometime when you have a few quiet moments, take a fantasy trip. The only ticket required is a fertile imagination. Imagine that you fall off your chair in slow motion, and while falling to the floor, you also slowly shrink in size. What would such a trip be like? What would you see? As you topple off the chair and approach the floor, you brace yourself for impact against the smooth, solid surface. And as you get nearer and nearer to it, becoming smaller and smaller all the while, you note that the floor is not as smooth as you supposed it to be, for great cracks appear. These are the microscopic irregularities found in all apparently smooth surfaces. In falling into one of these cracks, which look like canyons as you continue to shrink in size, you again brace yourself for impact against the canyon floor only to find that the bottom of the canyon is itself a myriad of cracks and crevices. Falling into one of these crevices and becoming still smaller, you note the solid walls have given way to nebulous surfaces that throb and pucker. If you could see greater detail, you would notice that the throbbing surfaces consist of hazy blobs, mostly spherical, some egg-shaped, some larger than others, and all oozing into each other, making up long chains of complicated structures. Falling still farther, you again brace yourself for impact as you approach one of these cloudy spheroids closer and closer, smaller and smaller, and—wow!—you find you have penetrated into a new universe. You fall into a sea of emptiness, occupied by occasional specks that whirl past at unbelievably high speeds. You are in an atom, as empty of matter as the solar system. You have found that the solid floor you have fallen to is, except for specks of matter here and there, empty space. If you continue falling, you might fall many kilometers before making a direct hit with a sub-atomic speck.

All matter, however solid it may appear, is made up of tiny building blocks, which themselves are mostly empty space. These are atoms—atoms that combine to form molecules that in turn combine to form the compounds and substances of matter. In this chapter we will study the atomic nature of matter. We will follow this up in succeeding chapters by investigating the properties of matter in the solid, liquid, gaseous, and plasma states.

Atoms

All manner of things—shoes, and ships, and sealing wax; cabbages and kings—anything we can think of is composed of atoms. One might think that an incredible number of different kinds of atoms exist to account for the rich variety of substances we find around us. But the number is surprisingly small. The great variety of substances results not from any great variety of atoms, but from the many ways a few types of atoms can be combined—just as in a color print three colors can be combined to form almost every conceivable color. To date (1989) we know of 109 distinct **atoms**. These are the chemical *elements*. Only 90 elements are found naturally; the others are made in the laboratory with high-energy atomic accelerators and nuclear reactors. These heaviest elements are too unstable (radioactive) to occur naturally in appreciable amounts.

Hydrogen was apparently the original element and still makes up over 90 percent of the atoms in the known universe. Elements heavier than hydrogen are manufactured in the deep interiors of stars, where enormous temperatures and pressures cause the fusion of hydrogen atoms into more complex elements. With the exception of some of the hydrogen and trace amounts of other light elements, all the elements that occur in nature are remnants of stars that exploded long before the solar system came into being.

These star remnants are the building blocks of all matter. And all matter, however complex, living or nonliving, is some combination of these elements. From a pantry having about 100 bins, each containing a different element, we have all the materials needed to make up any substance occurring in the universe. About a dozen elements compose most of the things we see every day; the majority of elements are not found in great abundance, and some are exceedingly rare.* Living things, for example, are composed primarily of four elements: carbon (C), hydrogen (H), oxygen (O), and nitrogen (N). The letters in the parentheses represent the chemical symbols for these elements.

Atoms are ageless. Atoms in your body have existed since the beginning of time, cycling and recycling among innumerable forms, both nonliving and living. When you breathe, for example, only part of the atoms that you inhale are exhaled in your next breath. The remaining atoms are taken into your body to become part of you, and they later leave your body by various means. You don't "own" the atoms that make up your body; you borrow them. We all share from the same atom pool as atoms forever migrate around, within, and throughout us. So some of the atoms in the nose you scratch today could have been part of your neighbor's ear yesterday! Not only are we all made of the same kinds of atoms, we are also made of the same atoms—atoms that cycle from person to person as we breathe, sweat, and vaporize.

---

*Most common substances are formed out of combinations of two or more of these most common elements: hydrogen (H), carbon (C), nitrogen (N), oxygen (O), sodium (Na), magnesium (Mg), aluminum (Al), silicon (Si), phosphorus (P), sulfur (S), chlorine (Cl), potassium (K), calcium (Ca), and iron (Fe).

Atoms are small, so small that there are about as many atoms of air in your lungs at any moment as there are breaths of air in the atmosphere of the whole world.* Since every breath of air you exhale eventually becomes uniformly mixed in the atmosphere (in about 6 years), every person in the world breathes an average of one of your exhaled atoms in a single breath—for each breath you exhale! Considering the many thousands of breaths people exhale, there are many atoms in your lungs at any moment that were once in the lungs of every person who ever lived. We are literally breathing each other.

It is difficult to imagine how small atoms are. Atoms are so small that they have no visible appearance. We could stack microscope on top of microscope and never "see" an atom. This is because light travels in waves, and atoms are smaller than the wavelengths of visible light. The size of a particle visible under the highest magnification must be larger than the wavelength of light. We can better understand this by considering the analogy of water waves. The wavelength of water waves is simply the distance between the crests of successive waves. Consider the ship in Figure 10-2. The ship is much larger than the crest-to-crest distance, or wavelength, of the waves incident on it. Information about the ship is easily revealed by its influence on the passing waves. Consider Figure 10-3. Waves are incident on the blades of grass. The grass shoots

Figure 10-2
Information about the boat is revealed by passing waves because the distance between wave crests is small compared to the size of the boat.

---

*There are about 10^{22} atoms in a liter of air at atmospheric pressure and a total of about 10^{22} liters of air in the atmosphere.

Figure 10-3
Information about the grass shoots is not revealed by passing waves because the distance between wave crests is large compared to the size of the blades of grass.

are much smaller than the incident waves, which pass by as if the grass were not there. Only when the blades of grass are thicker—that is, wider than the distance between wave crests—will the waves carry information regarding details of the grass. In the same way, waves of visible light are too coarse compared to the size of an atom to reveal details of the size and shape of atoms. Atoms are incredibly small.

The first somewhat direct evidence for the existence of atoms was unknowingly discovered in 1827 by a Scottish botanist, Robert Brown, while studying the spores of pollen under a microscope. He noticed that the spores were in a constant state of agitation, always moving, always jumping about. At first he thought that the spores were some sort of moving life forms. But later he found that inanimate dust particles and grains of soot also showed this motion. This perpetual jiggling of particles—called **Brownian motion**—results from bombardment by neighboring particles and atoms too small to be seen.

Question   Are there really atoms that were once a part of Albert Einstein in the brain matter of all your classmates?

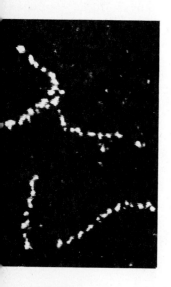

Figure 10-4
The strings of dots are chains of thorium atoms as revealed by a scanning electron microscope.

More direct evidence is available today. A photograph of individual atoms is shown in Figure 10-4. The photograph was not made with visible light but with an electron beam. An electron beam, such as that which sprays the picture on your television screen, is a stream of particles. But particles have wave properties,* and a high-energy electron beam can have an associated wavelength more than 1000 times smaller than the wavelength of visible light. (In an electron microscope, the beam is focused not by mirrors and lenses but by magnetic fields.) The historic photograph in Figure 10-4 was taken with a very powerful yet extremely thin (50 billionths of a centimeter) electron beam in a scanning electron microscope developed in 1970 by Albert Crewe at the University of Chicago's Enrico Fermi Institute. It is the first high-resolution photograph of individual atoms.

Recently, IBM researchers have developed an electron microscope small enough to be held in your hand, the scanning tunneling microscope. An

▶ **Answer**
Yes, and from Charlie Chaplin, too, although the configurations of these atoms with respect to others are now quite different! The next time you have one of those days when you feel like you'll never amount to anything, take comfort in the thought that the atoms that now compose you will live forever in the bodies of all the people on earth who are yet to be.

*More about the wave properties of particles in Chapters 28 and 31.

image of graphite taken with this remarkable instrument in 1985 is shown in Figure 10-5. The gray "hilltop" areas indicate the location of individual carbon atoms in the graphite layer.

Figure 10-5
An image of graphite obtained with a scanning tunneling microscope. The "bumps" indicate the location of individual carbon atoms.

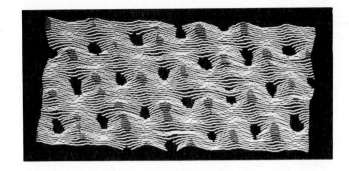

Molecules

Atoms combine to form **molecules** (Figure 10-6). Two atoms of hydrogen (H_2) combine with a single atom of oxygen (O) to produce a water molecule (H_2O). A molecule may be as simple as the two-atom combination of oxygen (O_2) or nitrogen (N_2) or as complex as the double helix of deoxyribonucleic acid (DNA), which consists of millions of atoms—the basic building block of life.

Figure 10-6
Models of simple molecules. The atoms in a molecule are not just mixed together but are joined in a well-defined way.

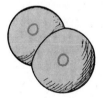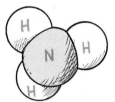

A molecule can be divided into atoms with chemical properties of their own. You can do a simple experiment to see this for yourself. Place two wires that are connected to the terminals of an ordinary battery into a glass of salted water (Figure 10-7). Position the wires on opposite sides of the glass so they don't touch, and you will see bubbles of gas forming on the wires. If you collect these gases, you will find that they have entirely different chemical properties from water vapor. One of the gases is hydrogen, and the other is oxygen. If the gases are mixed and ignited with a match, a quick fire or explosion results. The explosion is the violent combination of hydrogen and oxygen to form water again. The amount of energy that is suddenly released is the same amount of energy that was gradually drawn from the battery to separate the gases.

The energy released in common chemical burning comes from the combining of carbon and oxygen atoms to form carbon monoxide (CO) or carbon dioxide (CO_2). Carbon atoms attract oxygen much more than oxygen attracts oxygen or carbon attracts carbon. Oxygen and carbon

Figure 10-7
When electric current passes through salted water, bubbles of oxygen form at the left wire and bubbles of hydrogen form at the right wire.

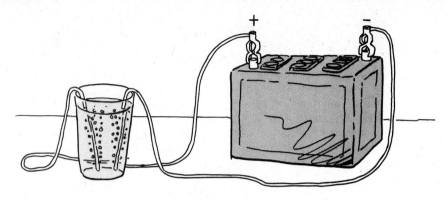

snap together and produce energy just as a slamming screen door produces sound. We get heat from the combination of oxygen and carbon, which is ordinarily in the form of molecular motion of the hot gas, but in certain circumstances the heat is so great that it generates light. This is how we get flames.

If the burning is slow enough, such as the burning of wood in a fireplace, the carbon will attach itself to two oxygen atoms and form carbon dioxide (CO_2). If the burning takes place very rapidly, as in an automobile engine, toxic carbon monoxide (CO) is also formed. (Wouldn't the air be healthier if cars burned gasoline more slowly?) These and many other arrangements release large amounts of energy and produce flames and explosions, depending on the reactions. Any process in which atoms rearrange to form different molecules is called a **chemical reaction**.

Molecules are too small to be seen with optical microscopes, but exceedingly small quantities of them are evident to our sense of smell. Our olfactory organs clearly discern noxious gases such as sulfur dioxide, ammonia, and ether. The smell of perfume is the result of molecules that rapidly evaporate (change from the liquid to the gaseous state; see Chapter 16) and bumble around haphazardly in the air until some of them accidentally work their way into our noses. Certainly they are not drawn to noses. They are just a few of the billions of jostling molecules that, in their aimless wanderings, happen to bumble up into the nose. You can get an idea of the speed of molecules in the air when you are in your bedroom and smell food almost immediately after the oven door in the kitchen has been opened.

More direct evidence of molecules is shown in the electron microscope photograph of virus molecules composed of thousands of atoms (Figure 10-8). These giant molecules are still too small to be seen with visible light. So we see again that an atom or a molecule that is invisible to light may be visible to a shorter-wavelength electron beam.

We often say that seeing is believing. But we don't always have to see something to believe in its existence. We don't have to look inside a

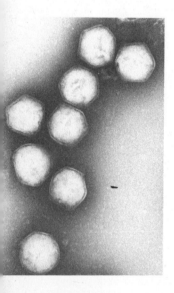

Figure 10-8
An electron microscope photo of virus molecules.

closed cardboard box that is too heavy to lift to know that something is in it. We could measure its mass without benefit of light. Atoms and molecules are too small to be seen with light, but this doesn't prevent us from measuring their masses with great precision.

Molecular and Atomic Masses

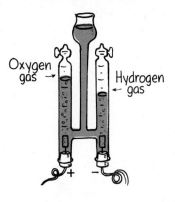

Oxygen gas →   ← Hydrogen gas

Figure 10-9
In the electrolysis of water, twice the volume of hydrogen gas as oxygen gas is collected.

The earliest determination of molecular and atomic masses was made with the measurements of gases in the early 1800s. As an example of how this is done, consider the hydrogen and oxygen gases that result when water is decomposed by electricity. Ordinarily, measurements show the mass of water before decomposition is equal to the sum of the masses of the two gases. Water is composed of two parts of hydrogen to one part of oxygen, hence the chemical formula H_2O. When the masses of these two gases are compared, we find for every 2 grams of hydrogen formed, 16 grams of oxygen are formed. Since there are twice as many hydrogen molecules as oxygen, it follows that the oxygen molecule is 16 times as heavy as the hydrogen molecule. The key assumption in this analysis is that twice as many hydrogen molecules as oxygen molecules are formed. Our observation is that twice the volume of hydrogen gas was formed. Our assumption, then, is that equal volumes of gas (at the same temperature and pressure) contain the same number of molecules.

This was first proposed in 1811 by the Italian physicist Amedeo Avogadro. **Avogadro's principle** states that equal volumes of any gas at the same temperature and pressure contain equal numbers of molecules, regardless of the size of the molecules. This statement makes sense only when it is realized that in a gas the molecules are far apart. For steam (water vapor) the distance between molecules is about twelve times the molecular diameter, so the molecules occupy about 1700 times the volume of the same mass of liquid ($12^3 = 1728$).

The theory of relative masses of molecules follows from Avogadro's hypothesis. We simply weigh equal volumes of two gases and compare the masses. The ratio of the masses of gas is identical to the ratio of the masses of the individual molecules making up the gases. For example, if gas A is twice as heavy as an equal volume of gas B, then molecule A has twice the mass of molecule B. If the chemical formula for the molecule is known, the relative masses of individual atoms can be determined.

To compare the masses of elements, one atom of hydrogen (the simplest element) was chosen to be *exactly* one **atomic mass unit** (abbreviated **amu**).* The mass of each other kind of atom was expressed relative to it. Thus the mass of a carbon atom is 12 amu; oxygen is 16 amu; uranium, one of the heaviest, is 238 amu.

---

*Subsequently, for reasons that do not concern us now (including greater precision), the carbon atom was chosen as *exactly* 12.000, and the others are *almost* integers. Hydrogen, for example, is 1.008, oxygen is 15.995, and uranium is 238.051.

Elements, Compounds, and Mixtures

Certain solids such as gold, liquids such as mercury, and gases such as neon are composed of a single element. These substances are called *elements*, for they are composed of one kind of atom. Certain other solids such as crystals of common table salt, liquids such as water, and gases such as methane are made up of elements that are chemically combined. These are called **compounds**. A compound may or may not be made of molecules. Water and carbon dioxide are compounds made of molecules. On the other hand, table salt (NaCl) is made of different kinds of atoms arranged in a regular pattern. Each chlorine atom is surrounded by six sodium atoms (Figure 10-10). In turn, every sodium atom is surrounded by six chlorine atoms. As a whole, there is one sodium atom for each chlorine atom, but there are no separate sodium-chlorine groups that can be labeled as molecules.

Figure 10-10
Table salt (NaCl) is a compound that is not made of molecules. The sodium and chlorine atoms are arranged in a repeating pattern where each atom is surrounded by six atoms of the other kind.

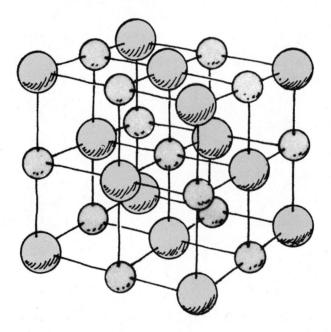

Compounds have properties different from those of their elements. Sodium is a yellowish metal that reacts violently with water and chlorine is a poisonous greenish gas, yet the compound of these two elements is the harmless white crystal (NaCl) you sprinkle on your potatoes. At ordinary temperatures water (H_2O) is a liquid, but hydrogen and oxygen are both gases. Compounds are formed only when the elements react chemically and bond with each other.

Substances mixed together without combining chemically are called **mixtures**. Hydrogen and oxygen gases form a mixture until ignited, whereupon they form the compound water. Sand combined with salt is a mixture. The most common mixture is nitrogen and oxygen together with a small percentage of carbon dioxide and traces of rare gases such as neon. It is the air we breathe.

Question ▶ A mixture of hydrogen and oxygen gases when ignited will form the compound water, H_2O. How many grams of oxygen should be mixed with 1 gram (g) of hydrogen to get pure water?

Atomic Structure

Nearly all the mass of an atom is concentrated in the *nucleus*, which occupies only a few quadrillionths the volume of an atom. The nucleus therefore is extremely dense. If bare atomic nuclei could be packed against each other into a lump 1 centimeter in diameter (about the size of a large pea), the lump would weigh 133 000 000 tons! Huge electrical forces of repulsion prevent such close packing of atomic nuclei. Each nucleus is electrically charged and repels other nuclei. Only under special circumstances are the nuclei of two or more atoms squashed into contact, and when this happens, a violent reaction takes place. This reaction is what happens in a hydrogen bomb; it will be discussed in Chapter 33.

The principal building block of the nucleus is the *nucleon*, which (as we will see in Chapter 32) is in turn composed of fundamental particles called *quarks*. When the nucleon is in its electrically neutral state, it is a *neutron*; when it is in its electrically charged state, it is a *proton*. All protons are identical; they are copies of one another. Likewise with neutrons; each neutron is like every other neutron. The lighter nuclei have roughly equal numbers of protons and neutrons; more massive nuclei have more neutrons than protons. Protons have a positive electric charge that repels other positive charges but attracts negative charges. So like kinds of electrical charges repel one another, and unlike charges attract one another. It is the positive protons in the nucleus that attract a surrounding cloud of negatively charged particles called *electrons* to make an atom.

Electrons make up the flow of electricity in electrical circuits. They are exceedingly light, almost 2000 times lighter than nucleons, and contribute very little mass to the atom. An electron in one atom is identical to any electron in or out of any other atom. Electrons are negative charges and repel other electrons, and they are attracted to the positive nucleus.

The number of protons in the nucleus is electrically balanced by an equal number of electrons whirling in the surrounding cloud. The atom itself is electrically neutral, so it normally doesn't attract or repel other atoms. But when atoms are close together, the negative electrons of one atom may at times be closer to the positive nucleus of another atom, which results in a net attraction between the atoms. That is why some atoms combine to form molecules.

▶ **Answer**

8 g. Each H_2O molecule has two atoms of H for each atom of O. The relative masses of H and O are 1 and 16. Water has 2 g of H for every 16 g of O. When two or more elements combine to form a compound, they always do so in the same proportion by mass. This is called the **law of definite proportions**, and for water it's always 8 g O to 1 g H.

Like the solar system, the atom is mostly empty space. The nucleus and surrounding electrons occupy only a tiny fraction of the atomic volume. If it weren't for the electrical forces of repulsion between the electrons of neighboring atoms, solid matter would be much more dense than it is. We and the solid floor are mostly empty space, because the atoms making up these and all materials are themselves mostly empty space. These same electrical repulsions prevent us from falling through the solid floor. Electrical forces keep atoms from caving in on each other under pressure. Atoms too close will repel (if they don't combine to form molecules), but when atoms are several atomic diameters apart, the electrical forces they exert on each other are negligible.

The main characteristic that distinguishes atoms from one another is the number of protons in the nucleus (equivalently, the number of electrons about the nucleus). Hydrogen is the simplest element and consists of a single electron orbiting a single proton. Helium has two electrons orbiting a pair of protons and neutrons. Lithium has three electrons, and the list continues with each successive element having an additional proton in the nucleus and an additional electron in the cloud of orbiting electrons (Figure 10-11). Equal numbers of protons and electrons make the atom electrically neutral.

Figure 10-11
The classical model of the atom consists of a tiny nucleus surrounded by electrons that orbit within spherical shells. As the size and charge of nuclei increase, electrons are pulled closer, and the shells become smaller.

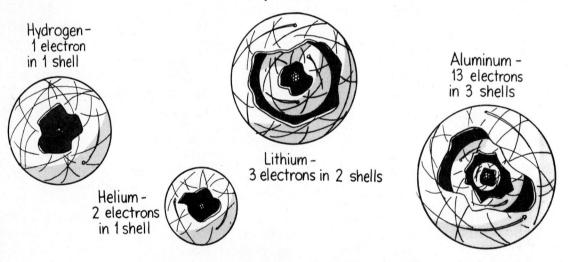

Hydrogen –
1 electron
in 1 shell

Helium –
2 electrons
in 1 shell

Lithium –
3 electrons in 2 shells

Aluminum –
13 electrons
in 3 shells

Atoms are classified by their **atomic number**, which is the same as the number of protons in the nucleus. Hydrogen has atomic number 1, helium 2, and so on in sequence to naturally occurring uranium with atomic number 92. The numbers continue through the artificially produced transuranic elements, at this writing up to 109. The arrangement of elements by their atomic numbers makes up the **periodic table**.* See the front inside cover of this book.

*The enormous human effort and ingenuity that went into finding regularity and system in the chemical elements make a fascinating atomic detective story that can be found in the books suggested at the end of the chapter.

We find that electrons revolve about the nucleus in orbits or in waves, sweeping out in concentric shells at various distances from the nucleus.* In the innermost shell there are at most two electrons; in the second shell, eight electrons; in the third shell, eighteen. In atoms having more than eighty-six electrons, there are seven shells altogether. These shells are indicated in the periodic table by rows. Note that the uppermost row consists of only two elements, hydrogen and helium. The electrons in helium complete the innermost shell. The second row contains eight elements, beginning with lithium and ending with neon, whose electrons complete the second shell, and so on. The order in which successive shells (and subshells) are filled becomes more complicated for the heavier elements.

In the periodic table the elements are arranged vertically by similarity in electronic configuration, which is then reflected in similarity in the physical and chemical properties of the elements and their compounds. It is the configuration of electrons in the outer shell of the atom that quite literally gives life and color to the world. The configuration of the outer electrons dictates whether and how atoms bond to become molecules, the melting and freezing temperatures, and the electrical conductivity, as well as the taste, texture, appearance, and color of substances.

Antimatter

Whereas matter is composed of atoms with positively charged nuclei and negatively charged electrons, **antimatter** is composed of atoms with negative nuclei and positive electrons, or *positrons* (Figure 10-12).

Positrons were first discovered in 1932, in cosmic rays bombarding the earth's atmosphere. Today antiparticles of all types are regularly produced in laboratory experiments involving nuclear physics. A positron has the same mass as an electron and the same magnitude of charge but the opposite sign. Antiprotons have the same mass as protons but are negatively charged. Antiparticles have the same mass as particles and differ only in their electromagnetic properties. All subatomic particles have been found to have corresponding antiparticles. There are even antiquarks.

Gravitational force does not distinguish between matter and antimatter. Also, there is no way to tell whether something is made of matter or antimatter by the light it emits. Only by contact could we tell, for when matter and antimatter meet, they mutually annihilate each other, and all the material is converted into radiant energy. This process, more so than any other known, results in the maximum energy output per gram of substance—$E = mc^2$ with a 100 percent mass conversion. (Nuclear fission and fusion, in contrast, convert less than 1 percent of the matter involved.)

There cannot be both matter and antimatter in our immediate environment, at least not for appreciable amounts or for appreciable times, for something made of antimatter would be completely transformed to energy

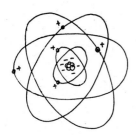

Figure 10-12
An atom of antimatter has a negatively charged nucleus surrounded by positrons.

*The wave nature of electrons is discussed in Chapter 31.

as soon as it touched matter. And it would transform an equal amount of normal matter into energy in the process. If the moon were made of antimatter, for example, a flash of energy would result as soon as one of our spaceships touched it. Both the spaceship and an equal amount of the antimatter moon would disappear in a burst of energy. We know the moon is not antimatter, but what of other galaxies? It may be that the elemental building blocks in other parts of the universe consist of anti-matter just as our part consists of matter. Or maybe not—we don't know.

States of Matter

Matter exists in four states: *solid*, *liquid*, *gaseous*, and *plasma*. In all states the atoms are perpetually moving. In the solid state the atoms and molecules vibrate about fixed positions. If the rate of molecular vibration is increased sufficiently, molecules will shake apart and wander through-out the material, vibrating in nonfixed positions. The shape of the mate-rial is no longer fixed but takes the shape of its container. This is the liquid state. If more energy is put into the material and the molecules vibrate at even greater rates, they may break away from one another and assume the gaseous state. H_2O is a common example of this changing of states. When solid, it is ice. If we heat the ice, the increased molecular motion jiggles the molecules out of their fixed positions, and we have water. If we heat the water, we can reach a stage where continued molec-ular vibration results in a separation between water molecules, and we have steam. Continued heating causes the molecules to separate into atoms; if we heat the steam to temperatures exceeding 2000°C, the atoms them-selves will be shaken apart, making a gas of free electrons and bare nuclei called **plasma**.

Although the plasma state is less common to our everyday experience, it may be considered the normal state of matter in the universe; the sun and other stars as well as much of the intergalactic matter are in the plasma state. All substances can be transformed from any state to another.

The following three chapters treat these states of matter in turn. We will continue the story of the atom in Chapter 31.

Summary of Terms

Atom   The smallest particle of an element that has all of the element's chemical properties.

Brownian motion   The haphazard movement of tiny particles suspended in a gas or liquid resulting from bombardment by the fast-moving molecules of the gas or liquid.

Molecule   The smallest particle of any substance that has all the substance's chemical properties. Atoms combine to form molecules.

Chemical reaction   The process of rearrangement of atoms from one molecule to another.

Avogadro's principle   Equal volumes of all gases at the same temperature and pressure contain the same number of molecules.

Atomic mass unit (amu)   The standard unit of atomic mass, which is equal to one-twelfth the mass of the common atom of carbon, arbitrarily given the value of exactly 12.

Compound   A chemical substance made of atoms of two or more different elements combined in a fixed proportion.

Mixture   A portion of matter consisting of two or more substances mixed together without combining chemically.

Atomic number The number of protons in the atomic nucleus; also the number of electrons surrounding the nucleus. The atomic number of an atom signifies its numerical position in the periodic table of the elements.

Periodic table A chart that lists elements by their atomic number and by their electron arrangements, so that elements with similar chemical properties are in the same column. (See inside front cover.)

Antimatter Matter composed of atoms with negative nuclei and positive electrons.

Plasma The fourth state of matter, in addition to the solid, liquid, and gaseous states, which consists of bare atomic nuclei and free electrons.

Suggested Reading

Asimov, I. *A Short History of Chemistry.* Chaps. 5 and 8. Science Studies Series. Garden City, N.Y.: Doubleday (Anchor), 1965.

Feynman, R. P., R. B. Leighton, and M. Sands. *The Feynman Lectures on Physics.* Vol. I, Chap. 1. Reading, Mass.: Addison-Wesley, 1963.

Review Questions

Atoms

1. What is the most abundant element in the known universe?

2. How many elements are known today?

3. How does the age of most atoms compare with the age of the solar system?

4. From where do the chemical elements originate?

5. How does the approximate number of atoms in the air in your lungs compare to the number of breaths of air in the atmosphere of the whole world?

6. Why cannot individual atoms be seen with visible light?

7. What causes dust particles and tiny grains of soot to move with Brownian motion?

8. Why can atoms be seen via electron beams but not via beams of light?

Molecules

9. Distinguish between an atom and a molecule.

10. How many elements compose a water molecule? How many individual atoms compose a water molecule?

11. Compared to the energy it takes to separate oxygen and hydrogen from water, how much energy is given off when they recombine? (What general physics principle is illustrated here?)

12. Is energy given off or taken in when carbon and oxygen atoms combine?

13. When the oven is opened in the kitchen, how do you know that molecules from the hot apple pie are moving?

Molecular and Atomic Masses

14. When water is decomposed, why is the volume of hydrogen gas collected twice as much as oxygen gas?

15. When water is decomposed, for every 2 g of H formed, 16 g of O are formed. Since H has twice the volume, does this mean that if a certain volume of O has a mass of 16 g, the same volume of H would have a mass of 1 g?

16. If the same volumes of gas at the same temperature and pressure have equal numbers of molecules, does this mean that an O molecule is sixteen times more massive than a molecule of H?

17. Oxygen and hydrogen molecules, O_2 and H_2, are composed of double atoms. Does it follow that a single oxygen atom has sixteen times the mass of a single hydrogen atom?

18. What does Avogadro's principle have to do with the last two questions?

19. If a volume of gas is three times as heavy as another gas of equal volume, how do the molecular masses compare? If the molecules of each gas are composed of equal numbers of atoms, how do the atomic masses compare?

20. A meter is the standard of length, a kilogram the standard of mass, and a second the standard of time. Similarly, what is the standard atomic mass unit?

Elements, Compounds, and Mixtures

21. What is a compound? Give three examples.

22. What is a mixture? Give three examples.

Atomic Structure

23. Where is most of the mass of an atom concentrated?

24. Distinguish between a proton and a neutron. What are both called?

25. What is the rule for attractive and repelling forces for positive and negative electric charges?

26. How does the mass of an electron compare to the mass of a nucleon?

27. How do the number of electrons in an electrically neutral atom compare to the number of protons in the nucleus?

28. Why do we say an atom is mostly "empty space"?

29. If atoms are mostly empty space, why don't materials made of atoms simply fall through each other?

30. What distinguishes one kind of atom from another?

31. What is meant by the *atomic number* of an atom? What is the atomic number of hydrogen? Iron? Gold? Uranium?

32. What is the *periodic table*?

33. What are *atomic shells*?

Antimatter

34. How do matter and antimatter differ?

35. What happens when a particle of matter and a particle of antimatter meet?

36. If a gram of antimatter meets a kilogram of matter, how much matter survives?

States of Matter

37. What is the role of temperature in the four states of matter?

38. What are the relative amounts of energy in H_2O when it is ice, water, and vapor?

39. Which of the four states of matter is most common in the universe?

40. How does a plasma differ from a gas?

Home Project

A candle will burn only if oxygen is present. Will a candle burn twice as long in an inverted liter jar as it will in an inverted half-liter jar? Try it and see.

Exercises

1. A cat strolls across your backyard. An hour later a dog with his nose to the ground follows the trail of the cat. Explain this occurrence from a molecular point of view.

2. If all the molecules of a body remained intact, would the body have any odor?

3. The average speed of a perfume vapor molecule at room temperature is about 650 m/s. The speed of the scent, however, is much less. Why?

4. Which are older, the atoms in the body of an elderly person or those in the body of a baby?

5. From where do the atoms that make up the body of a newborn baby originate?

6. Why is Brownian motion apparent only for microscopic particles?

7. Why are atoms visible with electron microscopes yet invisible with even ideal optical microscopes?

8. If the elements were arranged in increasing order by atomic mass rather than by atomic number, would the sequence be the same? (To answer this, check the periodic table inside the front cover.)

9. In what way do the number of protons in the atomic nucleus determine the chemical properties of an element?

10. Which of the following are pure elements: H_2, H_2O, He, Na, NaCl, H_2SO_4, U?

11. If two protons and two neutrons are removed from the nucleus of an O atom, what nucleus remains?

12. You could swallow a capsule of germanium without ill effects. But if a proton were added to each of the germanium nuclei, you would not want to swallow the capsule. Why? (Consult the periodic table of the elements.)

13. What element results if one of the neutrons in a nitrogen nucleus is converted by radioactive decay into a proton?

14. Which of the following elements would you predict to have properties most like those of silicon (Si): aluminum (Al), phosphorus (P), or germanium (Ge)? (Consult the periodic table of the elements.)

15. Which would be the more valued result: that of taking protons from or that of adding protons to the nuclei of gold atoms? Explain.

16. How many grams of O are there in 18 g of water?

17. How many grams of H are there in 16 g of methane (CH_4) gas?

18. Gas A is composed of diatomic molecules (two atoms to a molecule) of a pure element. Gas B is composed of monatomic molecules (one atom to a "molecule") of another pure element. Gas A has three times the mass of an equal volume of gas B at the same temperature and pressure. How do the masses of elements A and B compare?

19. When 50 cubic centimeters (cm^3) of alcohol is mixed with 50 cm^3 of water, the volume of the mixture is only 98 cm^3. Can you offer an explanation for this?

20. What is the principal variable that determines whether atoms form a solid or liquid or gaseous or plasma phase?

21. A teaspoon of an organic oil dropped on the surface of a quiet pond spreads out to cover almost an acre. The oil film has a thickness equal to the size of a molecule. If you know the volume of the oil and measure the film area, how can you calculate the thickness or the size of the molecule?

22. From an atomic recycling point of view, discuss the practice of cremation as opposed to confining one's remains in a sealed tomb.

23. In what sense can you truthfully say that you are a part of every person around you?

24. What are the chances that at least one of the atoms exhaled by your very first breath will be in your next breath?

25. There are approximately 10^{23} H_2O molecules in a thimbleful of water and 10^{46} H_2O molecules in the earth's oceans. Suppose that Columbus had thrown a thimbleful of water into the ocean and that the water molecules have by now mixed uniformly with all the water molecules in the oceans. Can you show that if you dip a sample thimbleful of water from anywhere in the ocean, you'll have at least one of the molecules that was in Columbus's thimble? (*Hint:* The ratio of the number of molecules in a thimble to the number of molecules in the ocean will equal the ratio of the number of molecules in question to the number of molecules the thimble will hold.)

26. There are approximately 10^{22} molecules in a single medium-sized breath of air and approximately 10^{44} molecules in the atmosphere of the whole world. Now the number 10^{22} squared is equal to 10^{44}. So how many breaths of air are there in the world's atmosphere? How does this number compare with the number of molecules in a single breath? If all the molecules from Julius Caesar's last dying breath are now thoroughly mixed in the atmosphere, how many of these on the average do we inhale with each single breath?

27. Assume that the present world population of about 4×10^9 people is about one-thirtieth the number of people who ever lived on earth. Then about how many more molecules are there in a single breath compared to the total number of people who have ever lived? Can you show that the ratio of molecules in a single breath to the total number of people who ever lived is about the same as the ratio of people who ever lived to one person?

28. Somebody told your friend that if an antimatter alien ever set foot upon the earth, the whole world would explode into pure energy. Your friend looks to you for verification or refutation of this claim. What do you say?

11 | Solids

Perhaps the classification of solids began when the earliest humans distinguished between rocks for shelter and rocks for tools. Since then the list of solids has grown without pause. Yet only within the last 80 years has the actual structure of solids been confirmed.

Prior to this it was thought that the content of a solid was what determined its characteristics—what made diamonds hard, lead soft, iron magnetic, and copper electrically conducting. It was believed that to change a substance, one merely varied the contents. We now know that the characteristics and properties of solids are determined by structure—the particular order and bonding of atoms in the material.

This change in emphasis from study of the content to study of the structure of solid materials has changed the role of investigators from that of finders and assemblers of materials to that of designers and creators of materials with new and unusual properties and uses. In the laboratories of today, people daily extend the long list of synthetic materials.

A striking example of the structure of solid matter is shown on the opposite page. The picture is a micrograph produced in 1958 by Erwin Müller. Dr. Müller used an extremely fine platinum needle, ending in a hemisphere 40-millionths of a centimeter in diameter. He enclosed it in a tube of rarefied helium and applied to it a very high positive voltage (25 000 volts). Under such electrical pressure the helium atoms "settling" on the atoms of the needle tip were ionized and torn off as positive ions, streaming away from the tip in a direction almost perpendicular to the surface at every point. They then struck a fluorescent screen, producing the picture of the needle tip, here enlarged approximately 750 000 times. Clearly, the platinum is crystalline, with the atoms arranged like oranges in a grocer's display. Although the photograph is not of the atoms themselves, it shows the positions of the atoms and reveals the microarchitecture of the solids that make up our world.

Crystal Structure

It was realized, near the turn of the nineteenth century, that many solids are composed of crystals. As the atomic theory of matter was being established, crystals themselves were thought to be made up of particular arrangements of atoms. This was confirmed in 1912 with the use of X rays. Atoms in a crystal are very close together; most are between 30

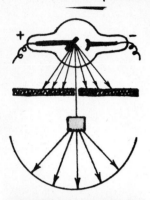

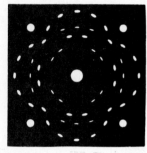

Figure 11-1

X-ray determination of crystal structure. The photograph of salt is a product of X-ray diffraction. Rays from the X-ray tube are blocked by a lead screen except for a narrow beam that hits the crystal of sodium chloride (common table salt). The radiation that penetrates the crystal to the photographic film makes the pattern shown. The white spot in the center is the main unscattered beam of X rays. The size and arrangement of the other spots result from the latticework structure of sodium and chlorine atoms in the crystal. Every crystalline structure has its own unique X-ray diffraction picture. A crystal of sodium chloride always produces this same design. (Adapted from *Matter*, Life Science Library. Copyright © 1965 by Time, Life Books. Reprinted by permission.)

and 100 nanometers apart,* about the same as the wavelength of X rays. The German physicist Max von Laue discovered that a beam of X rays directed upon a crystal is diffracted into a pattern characteristic of the crystal, called a *diffraction pattern* (Figure 11-1). The diffraction patterns made by X rays on photographic film show crystals to be neat mosaics of atoms sited on regular lattices, like three-dimensional chessboards or a child's jungle gym. Metals such as iron, copper, and gold have relatively simple crystal structures. Tin and cobalt are only a little more complex. And all are made up of interlocking crystals, each almost perfect, each with the same regular lattice but at a different inclination to the crystal nearby. These metal crystals can be seen when a metal surface is cleaned (etched) with acid (Figure 11-2). You can see them

Figure 11-2

A deeply etched surface of single-crystal antimony.

*For your information, 1 nanometer (nm) $= 10^{-9}$ meter (a billionth of a meter).

Figure 11-3
Sodium chloride crystal: the large spheres represent chlorine atoms.

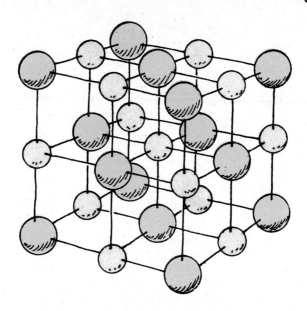

on the surface of galvanized iron that has been exposed to the weather or on brass doorknobs that have been etched by the perspiration of hands.

Von Laue's photographs of the X-ray diffraction patterns fascinated a team of English scientists, William Henry Bragg and his son William Lawrence Bragg. Through their own experimentation, they developed a mathematical formula that showed just how X rays should scatter from the various regularly spaced atomic layers in a crystal. With this formula and an analysis of the pattern of spots in a diffraction pattern, they could determine the atomic layer thicknesses and therefore the distances between the atoms in a crystal.

A solid can be amorphous or crystalline. In the *amorphous* state, the atoms and molecules exist in a random arrangement, vibrating about fixed positions. In the more common crystalline state, the atoms are in a three-dimensional orderly arrangement—a crystalline latticework of atoms (Figure 11-3). Each atom vibrates about its own fixed position and is unable to move through the lattice. Atoms are tied together in this lattice by electrical bonding forces.

We will not go into **atomic bonding** except to say there are four principal types in solids: ionic, covalent, metallic, and the weakest, van der Waals'. More can be found about them in almost any chemistry text.

Question ▶ What kind of force binds atoms together to form molecules and underlies atomic bonding in solids?

▶ **Answer**
Electrical force (more about this in Chapter 21).

Density

Figure 11-4
When the volume of the bread is reduced, its density increases.

The masses of atoms and the spacing between atoms determine the **density** of materials. We think of density as the "lightness" or "heaviness" of materials of the same size. It is a measure of the compactness of matter, of how much mass is squeezed into a given space; it is the amount of matter per unit volume:*

$$\text{Density} = \frac{\text{mass}}{\text{volume}}$$

The densities of a few materials are given in Table 11-1. Mass is measured in grams or kilograms and volume in cubic centimeters (cm^3) or cubic meters (m^3). A gram of material is defined as equal to the mass of 1 cubic centimeter of water at a temperature of 4°C.† Therefore, water has a density of 1 gram per cubic centimeter. Gold, density 19.3 grams per cubic centimeter, is therefore 19.3 times as massive as an equal volume of water. Osmium, a hard, bluish-white metallic element, is the densest natural substance on earth. Although the individual osmium atom is less massive than individual atoms of gold, mercury, lead, and uranium, the close spacing of osmium atoms in the crystalline form gives it the greatest density. More osmium atoms fit into a cubic centimeter than other more massive and more widely spaced atoms.

Questions

1. What is the density of 100 kg of water?
2. Which has the greater density—10 kg of lead or 100 kg of aluminum?
3. Which has the greater density—a single uranium atom or the whole world?

▶ **Answers**

1. The density of any amount of water is 1000 kg/m^3 (or 1 g/cm^3), which means that the mass of water that would exactly fill a cubic-meter tank would be 1000 kg. Can you see that the volume of 100 kg of water must be $\frac{1}{10}$ m^3? And that 100 kg/0.1 m^3 = 1000 kg/m^3?

2. Density is a *ratio* of weight or mass per volume, and this ratio is greater for any amount of lead than for any amount of aluminum—see Table 11-1.

3. A single uranium atom has a greater density. Only if the world were composed entirely of uranium atoms and not more closely packed in the compressed core would the densities of each be the same.

---

*Density can also be expressed by the amount of weight a body has, compared to its volume:

$$\text{Density} = \frac{\text{weight}}{\text{volume}}$$

Weight density is common to U.S. Customary units (formerly British or English), in which 1 cubic foot (ft^3) of fresh water (almost 7.5 gallons) weighs 62.4 pounds (lb). Thus, fresh water has a weight density of 62.4 lb/ft^3.

†A cubic meter is a sizable volume and contains a million cubic centimeters, so there are a million grams of water in a cubic meter (or, equivalently, a thousand kilograms of water in a cubic meter). Hence, 1 g/cm^3 = 1000 kg/m^3.

Table 11-1
Densities of some
substances

Material	Grams per cubic centimeter	Kilograms per cubic meter
Solids		
Osmium	22.5	22 480
Platinum	21.5	21 450
Gold	19.3	19 320
Uranium	19.1	19 050
Lead	11.3	11 344
Silver	10.5	10 500
Copper	8.9	8 920
Brass	8.6	8 560
Iron	7.8	7 800
Tin	7.3	7 280
Aluminum	2.7	2 702
Ice	0.92	917
Liquids		
Mercury	13.6	13 600
Glycerin	1.26	1 260
Seawater	1.03	1 025
Water at 4°C	1.00	1 000
Benzene	0.90	899
Ethyl alcohol	0.81	806

Elasticity

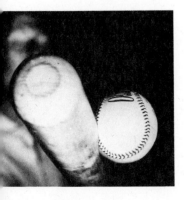

Figure 11-5
A baseball is elastic.

When an object is subjected to external forces, it usually undergoes changes in size or shape or both. The changes depend on the arrangement and bonding of the atoms in the material.

A weight hanging on a spring stretches it. Additional weight stretches the spring still more. If the weights are removed, the spring returns to its original length. We say the spring is *elastic*. When batters hit baseballs, they temporarily change their shape. When archers shoot arrows, they first bend the bow, which springs back to its original form when the arrow is released. The spring, the baseball, and the bow are examples of elastic objects. **Elasticity** is an object's property of changing shape when a deforming force acts on it and of returning to its original shape when the deforming force is removed. Not all materials return to their original shape when a deforming force is applied and then removed. Materials that do not resume their original shape after being deformed are said to be *inelastic*. Clay, putty, and dough are inelastic materials. Lead is also inelastic, since it's easy to distort it permanently.

When a weight is hung on a spring, a force (gravity) acts on it. The stretch (or compression) will be directly proportional to the applied force (Figure 11-6). This relationship, noted in the mid-seventeenth century by the British physicist Robert Hooke, a contemporary of Isaac Newton, is

Figure 11-6

The stretch of the spring is directly proportional to the applied force. If the weight is doubled, the spring stretches twice as much.

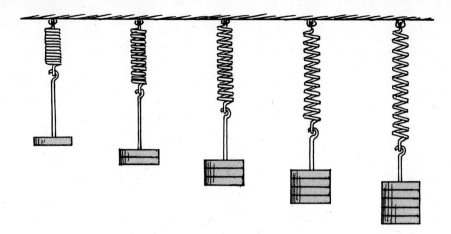

called **Hooke's law**. The amount of stretch or compression (change in length), Δx, is directly proportional to the applied force F. In shorthand notation,

$$F \sim \Delta x$$

If an elastic material is stretched or compressed beyond a certain amount, it will not return to its original state and will remain distorted. The distance beyond which permanent distortion occurs is called the *elastic limit*. Hooke's law holds only as long as the force does not stretch or compress the material beyond its elastic limit.

Questions ▶

1. A 20-kg load is hung from the end of a spring. The spring then stretches a distance of 10 cm. If, instead, a 40-kg load is hung from the same spring, how much will the spring stretch? How about if a 60-kg load were hung from the same spring? (Assume that none of these loads stretches the spring beyond its elastic limit.)

2. If a force of 10 N stretches a certain spring 4 cm, how much stretch will occur for an applied force of 15 N?

▶ **Answers**

1. A 40-kg load has twice the weight of a 20-kg load. In accord with Hooke's law, $F \sim \Delta x$, two times the applied force will result in two times the stretch, so the spring should stretch 20 cm. The weight of the 60-kg load will make the spring stretch three times as far, 30 cm. (When the elastic limit is exceeded, then the amount of stretch cannot be predicted with the information given.)

2. The spring will stretch 6 cm. By ratio and proportion, 10 N/4 cm = 15 N/6 cm, which is read 10 newtons is to 4 centimeters as 15 newtons is to 6 centimeters. If you're taking a lab, you'll learn that the ratio of force to stretch is called the spring constant k (in this case $k = 2.5$ N/cm), and Hooke's law is expressed as the equation $F = k\Delta x$.

Tension and Compression

When something is pulled on (stretched), it is said to be in *tension*. When it is pushed in on (squashed up), it is in *compression*. Compression causes things to get shorter and fatter, whereas tension causes them to get longer and thinner. This is not obvious for rigid materials because the shortening or lengthening is very small.

Steel is an excellent elastic material, for it can be stretched and compressed relatively easily. Because of its strength and elastic properties, it is used to make not only springs but construction girders as well. Vertical girders of steel used in the construction of tall buildings undergo only slight compression. A typical 25-meter-long vertical girder (column) used in high-rise construction is compressed about a millimeter when it carries a 10-ton load. Most deformation occurs when girders are used horizontally, where the tendency is to sag under heavy loads.

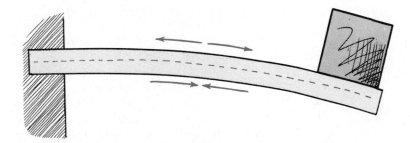

Figure 11-7
The top part of the beam is stretched and the bottom part is compressed. What happens in the middle portion, between top and bottom?

When a horizontal beam is supported at one or both ends, it is under both a tension and a compression from its weight and the load it supports. Consider the beam supported at one end (known as a *cantilever beam*) in Figure 11-7. It sags because of its own weight and because of the load it carries at its end. A little thought will show that the top part of the beam tends to be stretched. Molecules tend to be pulled apart. The top part is slightly longer because of its deformation. The top part is under tension. More thought shows that the bottom part of the beam is under compression. Molecules there are squeezed. The bottom part is slightly shorter because of the way it is bent. So the top part is under tension and the bottom part is under compression. Can you see that somewhere in between the top and bottom there will be a region where nothing happens, where there is neither tension nor compression? This is the *neutral layer.*

The horizontal beam shown in Figure 11-8 (known as a *simple beam*) is supported at both ends, and carries a load in the middle. This time

Figure 11-8
The top part of the beam is compressed and the bottom part is stretched. Where is the neutral layer (the part that is not under stress due to tension or compression)?

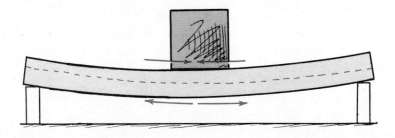

Figure 11-9

The upper half of the branch is under tension because of its weight, while the lower half is under compression. In what location is the wood neither stretched nor compressed?

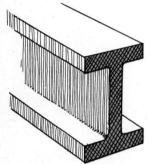

Figure 11-10

An I-beam is like a solid bar with some of the steel scooped from its middle where it is needed least. The beam is therefore lighter for nearly the same strength.

there is compression in the top of the beam and tension in the bottom part. Again, there is a neutral layer along the middle portion of the thickness of the beam all along its length.

We can see why the cross-section of steel girders has the form of the letter I (Figure 11-10). Most of the material in these I-beams is concentrated in the top and bottom flanges; the piece joining the flanges, called the *web*, is of thinner construction. Thus, when the beam is used horizontally in construction, the stress is predominantly in the top and bottom flanges—not in the central portion. One flange is squeezed while the other is stretched. Between the top and bottom flanges is a stress-free region that acts principally to connect the top and bottom flanges together. This is the neutral layer, where comparatively little material is needed. The flanges carry virtually all stresses in the beam. An I-beam is nearly as strong as a solid rectangular bar of the same overall dimensions, and its weight is considerably less. A large rectangular steel beam on a certain span might collapse under its own weight, whereas an I-beam of the same depth could carry much added load.

Question   If you had to make a hole horizontally through one of the tree branches shown in Figure 11-9 in a location that would weaken it the least, would you bore it through the top, the middle, or the bottom?

▶ **Answer**

It would be best to drill the hole in the middle, through the neutral layer. Wood fibers in the top part of the branch are being stretched, and if you drilled the hole there, that part of the branch may pull apart. Fibers in the lower part are being compressed, and if you drilled the hole there, that part of the branch might crush under compression. In between, in the neutral layer, the hole will hardly affect the strength of the branch because fibers there are being neither stretched nor compressed.

Figure 11-11
The horizontal slabs of stone of the roof cannot be too long, for stone easily breaks under tension. That is why many vertical columns are needed to support the roof.

Arches

Stones break more easily under tension than compression. The roofs of stone structures built by the Egyptians during the time of the pyramids were constructed with many horizontal stone slabs. Because of the weakness of these slabs, many vertical columns had to be erected to hold the roofs up. Likewise for the temples in ancient Greece. Then came arches—and fewer vertical columns.

Look at the top of the windows in old brick buildings. Chances are the tops are arched. Likewise for the shape of old stone bridges. When a load is placed on a structure that is properly arched (Figure 11-12), the compression strengthens rather than weakens the structure. The stones are pushed together more firmly and are held together by compressive forces. They do not even need the cement that a nonarched structure needs.

Figure 11-12
The weight of the load above the windows produces not tension but compression in the arched bricks. Bricks, like stone, break easily under tension, but are very strong under compression.

Figure 11-13
The many arches in this Roman aqueduct contribute to its strength.

Twirl an arch through a complete circle and you have a dome. The weight of the dome, like that of an arch, produces compression. Modern domes, such as the Astrodome in Houston, cover vast areas without the interruption of columns. There are shallow domes (the Jefferson Monument) and tall ones (the United States Capitol). And before these there were the famous Eskimo igloos.

Figure 11-14
The weight of the dome produces compression, not tension, so no support columns are needed in the middle.

Question ▶ Why is it easier for a chicken inside an eggshell to poke its way out than it is for a chicken on the outside to poke its way in?

▶ **Answer**
To poke its way into a shell, a chicken on the outside must contend with compression, which greatly resists shell breakage. Only the weaker shell tension must be overcome when poking from the inside. To see that the compression of the shell is strong, try squashing an egg along its long axis between your hands. Surprised? Try it along its short axis. Surprised? (Do this over a sink with some protection such as gloves for possible shell splinters.)

Scaling*

Figure 11-15

As the linear dimensions of an object change by some factor, the area changes as the square of this factor, and the volume (and hence the weight) changes as the cube of this factor. We see that when the linear dimensions are doubled (factor = 2), the area grows by $2^2 = 4$, and the volume grows by $2^3 = 8$.

Did you ever notice how strong an ant is for its size? An ant can carry the weight of several ants on its back, whereas a strong elephant couldn't even carry one elephant on its back. How strong would an ant be if it were scaled up to the size of an elephant? Would this "super ant" be several times stronger than an elephant? Surprisingly, the answer is no. Such an ant would not be able to lift its own weight off the ground. Its legs would be too thin for its greater weight and would likely break.

There is a reason for the thin legs of the ant and the thick legs of an elephant. The proportions of things in nature are in accord with their size. As the size of a thing increases, it grows heavier much faster than it grows stronger. You can support a toothpick horizontally at its ends and notice no sag. But support a tree of the same kind of wood horizontally at its ends and you'll see a noticeable sag. Compared to the toothpick, the tree is much heavier compared to its strength. **Scaling** is the study of how the volume and shape (size) of any object affect the relationship of its weight, strength, and surface.

Strength comes from the *area* of the cross-section (the "square" of the linear size), whereas weight depends on *volume* (the "cube" of the linear size). To understand this square-cube relationship, consider the

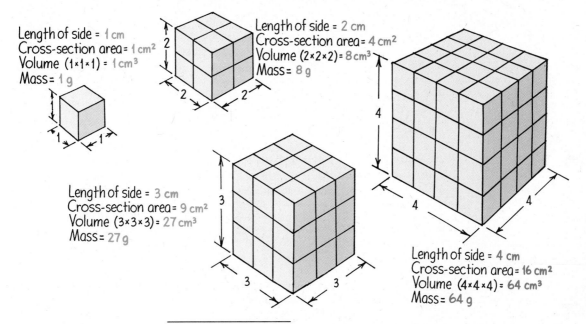

Length of side = 1 cm
Cross-section area = 1 cm²
Volume (1×1×1) = 1 cm³
Mass = 1 g

Length of side = 2 cm
Cross-section area = 4 cm²
Volume (2×2×2) = 8 cm³
Mass = 8 g

Length of side = 3 cm
Cross-section area = 9 cm²
Volume (3×3×3) = 27 cm³
Mass = 27 g

Length of side = 4 cm
Cross-section area = 16 cm²
Volume (4×4×4) = 64 cm³
Mass = 64 g

---

*Material in this section is based on two delightful and informative essays: "On Being the Right Size" from *Possible Worlds* by J. B. S. Haldane. (Copyright 1928 by Harper & Row, Publishers, Inc. Renewed 1956 by J. B. S. Haldane. By permission of Harper & Row, Publishers, Inc., and Chatto and Windus Ltd.) "On Magnitude" from *On Growth and Form* by Sir D'Arcy Wentworth Thompson. (New York: Cambridge University Press. Used with permission.)

simplest case of a solid cube of matter, 1 centimeter on a side. A 1-cubic centimeter cube has a cross section of 1 square centimeter. That is, if we sliced through the cube parallel to one of its faces, the sliced area would be 1 square centimeter. Compare this to a cube that has double the linear dimensions, a cube 2 centimeters on each side. Its cross-section area will be 2 × 2 (or 4) square centimeters and its volume will be 2 × 2 × 2 (or 8) cubic centimeters. For the same density it would have 4 times the cross section and be 8 times heavier. Careful investigation of Figure 11-15 shows that for increases of linear dimensions the cross-section area (as well as the total area) grows as the square of the increase, whereas volume and weight grow as the cube of the increase.

The volume (and thus its weight) multiplies much more than the corresponding increase of cross-sectional area. Although the figure demonstrates the simple example of a cube, the principle applies to an object of any shape. Consider a football player who can do many pushups. Suppose he could somehow be scaled up to twice his size—that is, twice as tall and twice as broad, with bones twice as thick and every linear dimension increased by a factor of two. Would he be twice as strong and be able to lift himself with twice the ease? The answer is no. Although his twice-as-thick arms would have four times the cross-sectional area and be four times as strong, he would be eight times as heavy. For comparable effort he would be able to lift only half his weight. Compared to his greater weight, he would be weaker than before.

We find in nature that large animals have disproportionately thick legs compared to those of small animals. This is because of the relationship between volume and area; the fact that volume (and weight) grow as the cube of the increase, while strength (and area) grow as the square of the increase. So we see there is a reason for the thin legs of a daddy longlegs, deer, or antelope and for the thick legs of a rhino, hippo, or elephant.

So the great strengths attributed to King Kong and other fictional giants cannot be taken seriously. The fact that the consequences of scaling are conveniently omitted is one of the differences between science and science fiction.

Questions ▶

1. Consider a 1 cubic centimeter cube scaled up to a cube 10 centimeters long on each edge.
   a. What would be the volume of the scaled-up cube?
   b. What would be its cross-sectional surface area?
   c. What would be its total surface area?
2. If you were somehow scaled up in proportion to twice your size, would you be stronger or weaker?

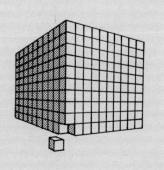

Important also is the comparison of total surface area to volume. A study of Figure 11-16 shows that as the linear size of an object increases, the volume grows faster than the total surface area. (Volume grows as the cube of the increase, and both cross-sectional area and total surface area grow as the square of the increase.) So as an object grows, its surface area and volume grow at different rates, with the result that the surface area to volume ratio *decreases*. In other words, both the surface area and the volume of a growing object increase, but the surface area grows more slowly than the volume. Relatively few people really understand this concept. The following examples may be helpful.

Figure 11-16

As the size of an object increases, there is a smaller increase in surface area than in volume; as a result, the ratio of surface area to volume decreases.

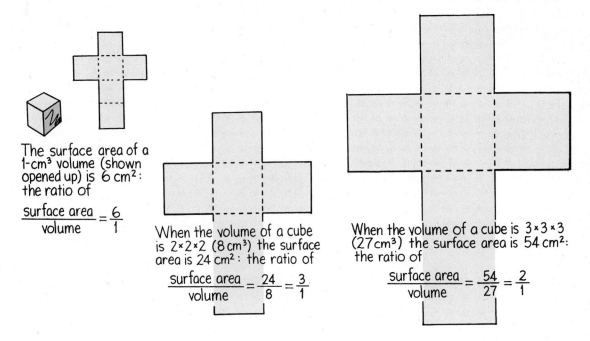

The surface area of a 1-cm³ volume (shown opened up) is 6 cm²: the ratio of

$$\frac{\text{surface area}}{\text{volume}} = \frac{6}{1}$$

When the volume of a cube is 2×2×2 (8 cm³) the surface area is 24 cm²: the ratio of

$$\frac{\text{surface area}}{\text{volume}} = \frac{24}{8} = \frac{3}{1}$$

When the volume of a cube is 3×3×3 (27 cm³) the surface area is 54 cm²: the ratio of

$$\frac{\text{surface area}}{\text{volume}} = \frac{54}{27} = \frac{2}{1}$$

▶ **Answers**

1. **a.** The volume of the scaled-up cube would be (length of side)^3 = (10 cm)^3, or 1000 cm^3.
   b. Its cross-sectional surface area would be (length of side)^2 = (10 cm)^2, or 100 cm^2.
   c. Its total surface area = 6 sides × (length of side)^2 = 600 cm^2.

2. Your scaled-up self would be four times stronger because the cross-sectional area of your twice-as-thick bones and muscles increase by four. You could lift a load four times as heavy. But your weight would be eight times as much as before, so you would not be stronger compared to your greater weight. Having four times the strength carrying eight times the weight gives you a strength-to-weight ratio of only half its former value. So if you can barely lift your own weight now, when scaled up you could only lift half your new weight. Your strength would increase, but your strength-to-weight ratio would decrease. Stay as you are!

The large ears of elephants are nature's way of making up for the small ratio of surface area to volume for these large creatures. For the large ears are not for better hearing, but for cooling. The heat that a creature radiates is proportional to its surface area. If an elephant didn't have large ears, it would not have enough surface area to cool its huge mass. The large ears of the African elephant greatly increase overall surface area, which facilitates cooling in hot climates.

At the biological level, living cells must contend with the fact that the growth of volume is faster than the growth of surface area. Cells obtain nourishment by diffusion through their surfaces. As cells grow their surface area increases, but not fast enough to keep up with volume. For example, if the surface area increases by four, the corresponding volume increases by eight. Eight times as much mass must be sustained by only four times as much nourishment. This puts a limit on the growth of a living cell. So cells divide, and there is life as we know it. That's nice.

Not so nice is the fate of large creatures when they fall. The statement "The bigger they are, the harder they fall" holds true and is a consequence of the small ratio of surface area per weight. Air resistance to movement through the air is proportional to the surface of the moving object. If you fall out of a tree, even in the presence of air resistance, your rate of fall is very nearly 1 g. You don't have enough surface area compared to your weight—unless you wear a parachute. Small creatures, on the other hand, need no parachute. They have plenty of surface area compared to their small weights. An insect can fall from the top of a tree to the ground below without harm. The surface area per weight ratio is in the insect's favor, for the insect in effect *is* its own parachute.

The fact that small things have relatively great surface areas compared to their volume, mass, or weight is evident in the kitchen. A good cook knows that more skin results from peeling 5 kilograms of small potatoes than from 5 kilograms of large potatoes. Smaller objects have more surface area per kilogram. Crushed ice will cool a drink much faster than

Figure 11-17
The long tail of the monkey not only helps maintain balance but also effectively radiates heat.

Figure 11-18
The African elephant has less surface area compared to its weight than other animals. It compensates for this with its large ears, which significantly increase its radiating surface area and promote cooling.

a single ice cube of the same mass, because crushed ice presents more surface area to the beverage. Steel wool rusts away at the sink while steel sink fixtures don't. Rusting is a surface phenomenon. Iron rusts when exposed to air, but it rusts much faster and is soon eaten away if it is in the form of small strands or filings.

Chunks of coal burn, while coal dust explodes when ignited. Thin french fries cook faster in oil than fat fries. Flat hamburgers cook faster than meatballs of the same mass. Large raindrops fall faster than small raindrops, and large fish swim faster than small fish. These are all consequences of the fact that differences in volume and differences in area are not in the same proportion to each other.

It is interesting to note that the rate of heartbeat in a mammal is proportional to the size of the mammal. The heart of a tiny shrew beats about twenty times faster than the heart of an elephant. In general, small mammals live fast and die young; larger animals live at a leisurely pace and live longer. Don't feel bad about a pet hamster who doesn't live as long as a dog. All warm-blooded animals have about the same lifespan— not in terms of years, but in the average number of heartbeats (about 800 million). Humans are the exception: we live two to three times longer than other mammals of our size.

Summary of Terms

Atomic bonding The linking together of atoms to form solids. The different kinds of bonding are ionic, covalent, and metallic bonding, and van der Waals' forces.

Density The mass of a substance per unit volume:

$$\text{Density} = \frac{\text{mass}}{\text{volume}}$$

Or we may prefer weight/volume, in which case it is called *weight density*.

$$\text{Weight density} = \frac{\text{weight}}{\text{volume}}$$

Elasticity The property of a material wherein it changes shape when a deforming force acts on it and returns to its original shape when the force is removed.

Hooke's law The amount of stretch or compression of an elastic material is directly proportional to the applied force.

Scaling The study of how volume and shape (size) affect the relationship of weight, strength, and surface.

Suggested Reading

For more about the relationships among the size, area, and volume of objects, read these essays: "On Being the Right Size" by J. B. S. Haldane and "On Magnitude" by Sir D'Arcy Wentworth Thompson. Both are in J. R. Newman, ed., *The World of Mathematics*, Vol. II. New York: Simon & Schuster, 1956.

Morrison, P. *Powers of Ten*. New York: W. H. Freeman, 1982.

Review Questions

1. Is it *content* or *structure* that determines the characteristics of a solid?

Crystal Structure

2. How does the arrangement of atoms in a crystalline substance differ from that in a noncrystalline substance?

3. What evidence can you cite for the microscopic crystal nature of certain solids? For macroscopic crystal nature?

Density

4. What happens to the volume of a loaf of bread that is squeezed? The mass? The density? (Why are all three answers different?)

5. What happens to the density of a chocolate bar when you cut it in half?

6. Uranium is the heaviest atom. Why is uranium metal not the most dense material?

7. Which has the greater density, a heavy bar of pure gold or a pure gold ring? Defend your answer.

8. Distinguish between mass density and weight density.

Elasticity

9. What is the evidence for the claim that a baseball is elastic?

10. Cite a substance that is inelastic. What is the evidence for your answer?

11. What is Hooke's law, and how does it apply to things that can be squeezed or stretched?

12. What is meant by the elastic limit for a particular object?

13. If the weight of a 1-kg mass stretches a spring by 2 cm, how much will the spring stretch when it supports a load of 3 kg? (Assume the spring has not reached its elastic limit.)

Tension and Compression

14. Distinguish between *tension* and *compression*.

15. What is the neutral layer in a beam that supports a load?

16. Why are the cross sections of metal beams in the shape of the letter I instead of solid rectangles?

Arches

17. Why were so many vertical columns needed to support the roofs of stone buildings in ancient Egypt and Greece?

18. Is it *tension* or *compression* that strengthens an arch that supports a load?

19. Why are no vertical columns needed to support the middle of the Houston Astrodome?

Scaling

20. What is the square-cube relationship in scaling?

21. Distinguish among *linear dimension, area,* and *volume.*

22. If the linear dimensions of an object are doubled, by how much does the overall area increase? By how much does the volume increase?

23. As the volume of an object is increased, its surface area also increases. Does the *ratio* of square meters to cubic meters decrease? Defend your answer.

24. Which will rust faster, a piece of iron or the same amount of iron in steel wool? Why?

25. Which has more skin, an elephant or a mouse? Which has more skin *per body weight,* an elephant or a mouse?

Home Projects

1. If you live in a region where snow falls, collect some snowflakes on black cloth and examine them with a magnifying glass. You'll see that all the many shapes are hexagonal crystalline structures, among the most beautiful sights in nature.

2. Simulate atomic close packing with a couple of dozen or so pennies. Arrange them in a square pattern so each penny inside the perimeter makes contact with four others. Then arrange them hexagonally, so contact with six others occurs. Compare the areas occupied by the same number of pennies close packed both ways.

3. Are you slightly longer while lying down than you are tall when standing up? Make measurements and see.

4. If you nail four sticks together to form a rectangle, they can be deformed into a parallelogram without too much effort. But if you nail three sticks together to form a triangle, the shape cannot be changed without actually breaking the sticks or the nails. The triangle is the strongest of all the geometrical figures, which is why you see triangular shapes in bridges. Try it and see, and then look at the triangles used in strengthening structures of many kinds.

Exercises

1. Silicon is the chief ingredient of both glass and semiconductor devices, yet the physical properties of glass are different from those of semiconductor devices. Explain.

2. What evidence can you cite to support the claim that crystals are composed of atoms that are arranged in specific patterns?

3. Is iron necessarily heavier than cork? Explain.

4. A certain spring stretches 3 additional cm when a load of 15 N is suspended from it. How much will the spring stretch if 45 N is suspended from it (and it doesn't reach its elastic limit)?

5. A spring stretches 4 additional cm when a load of 10 N is suspended from it. How much will the spring stretch if an identical spring also supports the load as shown in *a* and *b*? (Neglect the weights of the springs.)

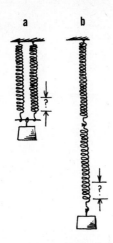

6. If a certain spring stretches 4 cm when a load of 10 N is suspended from it, how much will the spring stretch if it is cut in half and 10 N is suspended from it?

7. Why does a suspended spring stretch more at the top than at the bottom?

8. A thick rope is stronger than a thin rope of the same material. How much stronger is a twice-as-long rope?

9. Tension and compression occur in a partially supported horizontal beam when it sags due to gravity or supports a load. Make a simple sketch to show a means of supporting the beam so that tension occurs at the top part and compression at the bottom. Sketch another case where compression is at the top and tension occurs at the bottom.

10. Is a horizontal I-beam stiffer when the web is horizontal or when the web is vertical? Explain.

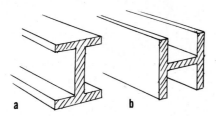

11. The sketches are of top views of a dam to hold back a lake. Which of the two designs is better? Why?

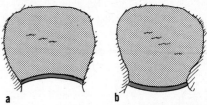

12. Consider a very large wooden barrel, such as the kind used in wineries. Should the "flat" ends be concave or convex? Why?

13. Why do you suppose that girders are so often arranged to form triangles in the construction of bridges and other structures? (Compare the stability of three sticks nailed together to form a triangle with four sticks nailed to form a rectangle, or any number of sticks to form similar geometric figures. Try it and see!)

14. Consider two bridges that are exact replicas of each other except that every dimension in the larger is exactly twice that of the other—that is, twice as long, structural elements twice as thick, etc. Which bridge is stronger?

15. A candymaker making taffy apples decides to use 100 kg of large apples rather than 100 kg of small apples. Will the candymaker need to make more or less taffy to cover the apples?

16. Will a person twice as heavy as another use more or less than twice as much suntan lotion at the beach?

17. If the linear dimensions of a storage tank are reduced to half their former values, by how much does the overall surface area of the tank decrease? By how much does its volume decrease?

18. Why is it easier to start a fire with kindling rather than with large sticks and logs of the same kind of wood?

19. Why does a chunk of coal burn when ignited, whereas coal dust explodes?

20. Why is there less heat loss in a dwelling if it is dome-shaped?

21. Why is heating more efficient in large apartment buildings than in single-family dwellings?

22. Why do thin french fries cook faster than fat fries?

23. Why do hamburgers cook faster than meatballs of the same weight?

24. If you use a batch of cake batter for cupcakes instead of cake and bake them for the time suggested for baking a cake, what will be the result?

25. Why are mittens warmer than gloves on a cold day?

26. Which parts of the body are most susceptible to frostbite? Why?

27. Why is the heartbeat of large creatures generally slower than the heartbeat of smaller creatures?

28. Nourishment is obtained from food through the inner surface area of the intestines. Why is it that a small organism, such as a worm, has a simple and relatively straight intestinal tract, while a large organism, such as a human being, has a complex and many-folded intestinal tract?

29. The human lungs have a volume of only about 4 liters, yet an internal surface area of nearly 100 m². Why is this important and how is this accomplished?

30. Why are the living cells in a whale about the same size as those in a mouse?

31. Which fall faster, large or small raindrops?

32. Why can a large fish generally swim faster than a small fish?

33. Do the effects of scaling help or hinder a small swimmer on a racing team?

34. Why doesn't a hummingbird soar like an eagle and an eagle flap its wings like a hummingbird?

35. Consider eight little spheres of mercury, each with a diameter of 1 mm. When they coalesce to form a single sphere, how big will it be? How does its surface area compare to the total surface area of the previous eight little spheres?

36. Can you relate the idea of scaling to the governance of small versus large groups of citizens? Explain.

12 Liquids

Unlike the molecules in a solid, molecules in the liquid state are not confined to fixed positions. Molecules can flow. They freely move from position to position by sliding over one another, and the liquid takes the shape of its container. Molecules of a liquid are close together and greatly resist compressive forces.

A liquid contained in a vessel exerts forces against the walls of the vessel. To discuss the interaction between the liquid and the walls, it is convenient to introduce the concept of **pressure**. Pressure is defined as force per area on which it acts:*

$$\text{Pressure} = \frac{\text{force}}{\text{area}}$$

As an illustration of the distinction between pressure and force, consider the two blocks in Figure 12-1. The blocks are identical, but one stands on its end and the other on its side. Both blocks are of equal weight and therefore exert the same force on the surface, but the upright block exerts a greater pressure against the surface. This is because the force is distributed over a smaller area, which increases the pressure. If the block were tipped up so that it made contact with a single corner, the pressure would be greater still.

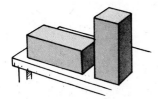

Figure 12-1
Although the weight of both blocks is the same, the upright block exerts greater pressure against the table.

Pressure in a Liquid

When you swim under water, you can feel the water pressure acting against your eardrums. The deeper you swim, the greater the pressure. What causes this pressure? It is simply the weight of the water above you pushing against you. If you swim twice as deep, there is twice the weight of water above, and you therefore feel twice the water pressure. Liquid pressure depends on depth. If you were submerged in a liquid more dense than water, the pressure would be proportionally greater. Liquid pressure

---

*Pressure may be measured in any unit of force divided by any unit of area. The standard international (SI) unit of pressure, the newton per square meter, is called the *pascal* (Pa) after the seventeenth-century theologian and scientist, Blaise Pascal. A pressure of 1 Pa is very small and approximately equals the pressure exerted by a dollar bill resting flat on a table. Science types more often use kilopascals (1 kPa = 1000 Pa).

Figure 12-2
The dependence of liquid pressure on depth is not a problem for the giraffe because of its large heart and intricate system of valves and elastic, absorbent blood vessels in the brain. Without these structures, it would faint when suddenly raising its head and would be subject to brain hemorrhaging when lowering it.

depends also on the density of the liquid. The pressure in a liquid is equal to the product of weight density and depth:*

Liquid pressure = weight density × depth

Simply put, the pressure that a liquid exerts against the sides and bottom of a container depends only on the density and depth of the liquid. At twice the depth, the liquid pressure against the bottom is twice as great; at three times the depth, the pressure is threefold; and so on. Or, if the

---

*This is derived from the definitions of pressure and density. Consider an area at the bottom of a vessel of liquid. The weight of the column of liquid directly above this area produces pressure. From the definition Weight density = weight/volume, we can express this weight of liquid as Weight = weight density × volume, where the volume of the column is simply the area multiplied by the depth. Then we get

$$\text{Pressure} = \frac{\text{force}}{\text{area}} = \frac{\text{weight}}{\text{area}} = \frac{\text{weight density} \times \text{volume}}{\text{area}}$$

$$= \frac{\text{weight density} \times \text{area} \times \text{depth}}{\text{area}} = \text{weight density} \times \text{depth}$$

liquid is two or three times as dense, liquid pressure is correspondingly two or three times as great for any given depth. Liquids are practically incompressible—that is, their volume can hardly be changed at all by pressure. So, except for changes in temperature, the density of a liquid is normally the same at all depths.*

It is important to note that the pressure does not depend on the amount of liquid. For example, you feel the same pressure whether you dunk your head a meter under water in a small pool or a meter under water in the middle of a large lake. The same is true for a fish. Refer to the connecting vases in Figure 12-3. If we hold a goldfish by its tail and dunk its head a couple of centimeters under the surface, the water pressure against the fish's head will be the same in any of the vases. If we release the fish and it swims a few centimeters deeper, the pressure on the fish will increase with depth but be the same no matter what vase the fish is in. If the fish swims to the bottom, the pressure will be greater, but it makes no difference what vase the fish is in. All vases are filled to equal depths, so the water pressure is the same at the bottom of each vase, regardless of its shape or volume. If the water pressure at the bottom of a vase were greater than the water pressure at the bottom of a neighboring, narrower vase of smaller volume, the greater pressure would force water sideways and then up the narrower vase to a higher level until the pressures at the bottom were equalized. But this doesn't happen. Pressure is depth dependent and not volume dependent, so we see there is a reason why water seeks its own level.

Figure 12-3
Liquid pressure is the same for any given depth below the surface, regardless of the shape of the containing vessel.

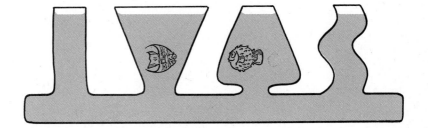

The fact that water seeks its own level can be demonstrated by filling a garden hose with water and bringing the two ends together. The levels will be equal. They will also be equal when the ends are far apart. This fact was not fully appreciated by some of the Romans who built elaborate aqueducts with tall arches and roundabout routes to ensure that water

---

*The density of fresh water is 1000 kg/m^3. Since the weight (mg) of 1000 kg is 1000 × 9.8 = 9800 N, the weight density of water is 9800 newtons per cubic meter (9800 N/m^3). The water pressure beneath the surface of a lake is simply equal to this density multiplied by the depth in meters. For example, water pressure is 9800 N/m^2 at a 1-m depth and 98 000 N/m^2 at a depth of 10 m. In SI units pressure is measured in pascals, so this would be 9800 Pa and 98 000 Pa, respectively; or, in kilopascals, 9.8 kPa and 98 kPa, respectively.

Figure 12-4
The average water pressure acting against the dam depends on the average depth of the water and not on the volume of water held back. The large shallow lake exerts only one-half the pressure that the small deep pond exerts.

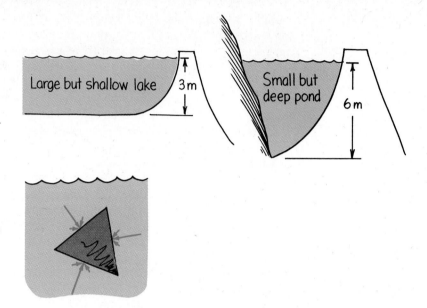

Figure 12-5
The forces in a liquid that produce pressure against a surface add up to a net force that is perpendicular to the surface.

Figure 12-6
Water pressure pushes perpendicularly to the sides of its container and increases with increasing depth.

would always flow slightly downward from the reservoir to the city. If pipes were laid in the ground and followed the natural contour of the land, in some places the water would have to flow uphill, and the Romans were skeptical of this. Although pipes were used to cross some valleys in parts of the Roman Empire, convincing experimentation was not yet the mode, so with plentiful slave labor the elaborate aqueducts were built.

Another interesting fact about liquid pressure is that it is exerted equally in all directions. For example, if we are submerged in water, no matter which way we tilt our heads we feel the same amount of water pressure on our ears. Because a liquid can flow, the pressure isn't only downward. We know pressure acts sideways when we see water spurting sideways from a leak in the side of an upright can. We know pressure also acts upward when we try to push a beach ball beneath the surface of the water. The bottom of a boat is certainly pushed upward by water pressure. Pressure in a liquid at any point is exerted in equal amounts in all directions.

When liquid presses against a surface, there is a net force directed perpendicular to the surface (Figure 12-5). If there is a hole in the surface, the liquid spurts at right angles to the surface before curving downward due to gravity (Figure 12-6). At greater depths the pressure is greater and the speed of the escaping liquid is greater.*

---

*The speed of liquid out of the hole is $\sqrt{2gh}$, where h is the depth below the free surface. Interestingly enough, this is the same speed the water or anything else would have if freely falling the same distance h.

Buoyancy

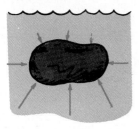

Figure 12-7
The greater pressure against the bottom of a submerged object produces an upward buoyant force.

Anyone who has ever lifted a submerged object out of water is familiar with buoyancy, the apparent loss of weight of objects submerged in a liquid. For example, lifting a large boulder off the bottom of a riverbed is a relatively easy task as long as the boulder is below the surface. When it is lifted above the surface, however, the force required to lift it is considerably more. This is because when the boulder is submerged, the water exerts an upward force on it that is exactly opposite in direction to gravity. This upward force is called the **buoyant force** and is a consequence of pressure increasing with depth. Figure 12-7 shows why the buoyant force acts upward. Pressure is exerted everywhere against the object in a direction perpendicular to its surface. The arrows represent the magnitude and directions of points of pressure. Forces that produce pressures against the sides are at equal depths and cancel one another. Pressure is greatest against the bottom of the boulder because the bottom of the boulder is at a greater depth. Since this upward pressure is greater than the downward pressure acting against the top, the pressures do not cancel, and there is a net pressure upward. This net pressure distributed over the bottom of the boulder produces a net upward force, which is the buoyant force.

Figure 12-8
When a stone is submerged, it displaces a volume of water equal to the volume of the stone.

If the weight of the submerged object is greater than the buoyant force, the object will sink. If the weight is equal to the buoyant force on the submerged object, it will remain at any level, like a fish. If the buoyant force is greater than the weight of the completely submerged object, it will rise to the surface and float.

Understanding buoyancy requires understanding the meaning cf the expression "volume of water displaced." If a stone is placed in a container brimful of water, some water will overflow (Figure 12-8). Water is *displaced* by the stone. A little thought will tell us that the *volume—* that is, the amount of space taken up or the number of cubic centimeters—*of water displaced* is equal to the *volume of the stone*. Place any object into a container partly filled with water, and the level of the surface rises (Figure 12-9). By how much? Exactly the same as if a volume of water were poured in that was equal to the volume of the submerged object. This is a good method for determining the volume of irregularly shaped objects: *A completely submerged object always displaces a volume of liquid equal to its own volume.*

Water displaced

Figure 12-9
The raised level equals that which would occur if water equal to the stone's volume were poured in.

Archimedes' Principle

The relationship between buoyancy and displaced liquid was first discovered in the third century BC by the Greek philosopher Archimedes. It is stated as follows:

> An immersed object is buoyed up by a force equal to the weight of the fluid it displaces.

This relationship is called **Archimedes' principle**. It is true of liquids and gases, which are both fluids. If an immersed object displaces 1 kilogram of fluid, the buoyant force acting on it is equal to the weight of 1 kilogram.* By *immersed*, we mean either *completely* or *partially submerged*. If we immerse a sealed 1-liter container halfway into the water,

Figure 12-10
A liter of water occupies a volume of 1000 cm³, has a mass of 1 kg, and weighs 9.8 N. Its density may therefore be expressed as 1kg/L and its weight density as 9.8 N/L. (Seawater is slightly denser, 10 N/L).

it will displace a half-liter of water and be buoyed up by the weight of a half-liter of water. If we immerse it completely (submerge it), it will be buoyed up by the weight of a full liter or 1 kilogram of water. Unless the container is compressed, the buoyant force will equal the weight of 1 kilogram at *any* depth, as long as it is completely submerged. This is because at any depth it can displace no greater volume of water than its own volume. And the weight of this volume of water (not the weight of the submerged object!) is equal to the buoyant force.

Figure 12-11
A 3-kg block weighs more in air than in water. When submerged, its loss in weight is the buoyant force, which is equal to the weight of water displaced.

---

*A kilogram is not a unit of force but a unit of mass. So, strictly speaking, the buoyant force is not 1 kg, but the *weight* of 1 kg, which is 9.8 N. We could as well say that the buoyant force is 1 *kilogram weight*, not simply 1 kg.

If a 25-kilogram object displaces 20 kilograms of fluid upon immersion, its apparent weight will be equal to the weight of 5 kilograms. Note that in Figure 12-11 the 3-kilogram block has an apparent weight equal to the weight of 1 kilogram when submerged. The apparent weight of a submerged object is its weight in air minus the buoyant force.

Questions ▶

1. Does Archimedes' principle tell us that if an immersed object displaces 10 N of fluid, the buoyant force on the object is 10 N?
2. A 1-liter (L) container completely filled with lead has a mass of 11.3 kg and is submerged in water. What is the buoyant force acting on it?
3. A boulder is thrown into a deep lake. As it sinks deeper and deeper into the water, does the buoyant force increase? Decrease?
4. Since buoyant force is the net force that a fluid exerts on an object and we learned in Chapter 4 that net forces produce accelerations, why doesn't a submerged object accelerate?

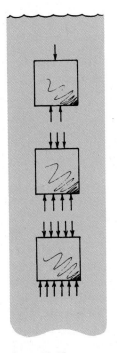

Figure 12-12
The difference in the upward and downward force acting on the submerged block is the same at any depth.

Perhaps your instructor will summarize Archimedes' principle by way of a numerical example to show that the difference between the upward-acting and the downward-acting forces due to the similar pressure differences on a submerged cube is numerically identical to the weight of fluid displaced. It makes no difference how deep the cube is placed, for although the pressures are greater with increasing depths, the *difference* in pressure up against the bottom of the cube and down against the top of the cube is the same at any depth. This is shown in Figure 12-12, not with numerical values, but with their equivalent-force vectors. The cube-shaped block makes this easier to see, but any shape produces the same result. We find that whatever the shape of the submerged object, the buoyant force is equal to the weight of fluid displaced.

▶ **Answers**

1. Yes. Looking at it another way, the immersed object pushes 10 N of fluid aside. The displaced fluid reacts by pushing back on the immersed object with 10 N.
2. The buoyant force is equal to the weight of 1 kg (9.8 N) because the volume of water displaced is 1 L, which has a mass of 1 kg and a weight of 9.8 N. The 11.3 kg of the lead is irrelevant; 1 L of anything submerged will displace 1 L of water and be buoyed upward with a force of 1 kg weight, or 9.8 N.
3. Buoyant force does not change as the boulder sinks because the boulder displaces the same volume of water at any depth. Since water is practically incompressible, its density is the same at all depths; hence, the weight of water displaced, or the buoyant force, is the same at all depths.
4. It does accelerate if the buoyant force is not balanced by other forces that act on it—the force of gravity and fluid resistance. The net force on a submerged object is the result of the net force the fluid exerts (buoyant force), the weight of the object, and, if moving, the force of fluid friction.

Density Effects on Submerged Objects

It is important to emphasize that the buoyant force that acts on a submerged object depends on the volume of the object. Small objects displace small amounts of water and are acted on by small buoyant forces. Large objects displace large amounts of water and are acted on by larger buoyant forces. It is the *volume* of the submerged object—not its *weight*—that determines buoyant force. Buoyant force is the weight of the volume of fluid displaced. (A misunderstanding of this idea is at the root of a lot of confusion that you or your friends may have about buoyancy!)

Weight does play a role, for whether an object will sink or float in a liquid depends on the relative magnitudes of buoyant force and weight. Can you see that if the buoyant force is exactly equal to the weight of the object, then the weight of the object must be the same as the weight of the water displaced? What would its density be in this case? Since the weights are the same and the volumes of object and displaced water are the same, their densities must be the same. The density of the object equals the density of water. This is true for a fish, which is "at one" with the water—it doesn't sink and it doesn't float. If the fish were somehow bloated up, it would be less dense than water and it would float to the top. If the fish swallowed a stone and became more dense than water, it would sink to the bottom. From this we see three simple rules.

1. If an object is denser than the fluid in which it is immersed, it will sink.
2. If an object is less dense than the fluid in which it is immersed, it will float.
3. If an object has a density equal to the density of the fluid in which it is immersed, it will neither sink nor float.

From these rules, what can you say about people who, try as they may, cannot float?* They're simply too dense! To float more easily, you must reduce your density. The formula Weight density = weight/volume tells you that you must either reduce your weight or increase your volume. The purpose of a life jacket is to increase volume while correspondingly adding very little to your weight.

The weight, not the volume, of a submarine is varied to achieve the desired density. Water is taken into and out of its ballast tanks. A fish regulates its density by expanding and contracting an air sac that changes its volume. The fish can move upward by increasing its volume (which decreases its density) and downward by contracting its volume (which increases its density). The overall density of a crocodile increases when it swallows stones. From 4 to 5 kilograms of stones have been found lodged in the front part of the stomach in large crocodiles. Because of this increased density, the crocodile swims lower in the water, thus exposing less of itself to its prey (Figure 12-13).

---

*Interestingly enough, the people who can't float are, nine times out of ten, males. Most males are more muscular and slightly denser than females.

Figure 12-13
(*left*) A crocodile coming toward you in the water. (*right*) A stoned crocodile coming toward you in the water.

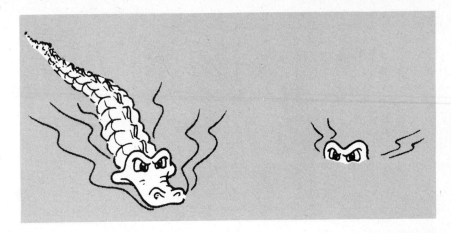

Question ▶ If a fish makes itself more dense, it will sink; if it makes itself less dense, it will rise. In terms of buoyant force, why is this so?

Flotation

Iron is much denser than water and therefore sinks, but an iron ship floats. Why is this so? Consider a solid 1-ton block of iron. Iron is nearly eight times as dense as water, so when it is submerged it will displace only $\frac{1}{8}$ ton of water, which is hardly enough to keep it from sinking. Suppose we reshape the same iron block into a bowl, as shown in Figure 12-14. It still weighs 1 ton. When we place it in the water, it settles into the water, displacing a greater volume of water than before. The deeper

Figure 12-14
An iron block sinks, while the same block shaped like a bowl floats.

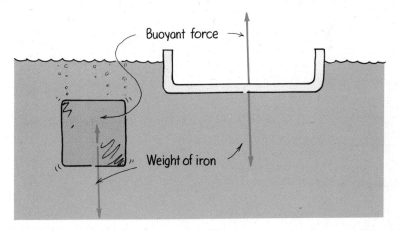

Buoyant force

Weight of iron

▶ **Answer**
When the fish increases its density by decreasing its volume, it displaces less water, so the buoyant force decreases. When the fish decreases its density by expanding, it displaces a greater volume of water, and the buoyant force increases.

Figure 12-15
The weight of a floating object equals the weight of the water displaced by the submerged part.

it is immersed, the more water it displaces and the greater the buoyant force acting on it. When the buoyant force equals 1 ton, it will sink no further.

When the iron boat displaces a weight of water equal to its own weight, it floats. This is sometimes called the **principle of flotation**, which states:

A floating object displaces a weight of fluid equal to its own weight.

Every ship, every submarine, and every dirigible must be designed to displace a weight of fluid equal to its own weight. Thus, a 10 000-ton ship must be built wide enough to displace 10 000 tons of water before it sinks too deep in the water. The same holds true in air. A dirigible or balloon that weighs 100 tons displaces at least 100 tons of air. If it displaces more, it rises; if it displaces less, it falls. If it displaces exactly its weight, it hovers at constant altitude.

Since the buoyant force upon an object equals the weight of the fluid it displaces, denser fluids will exert a greater buoyant force upon an object than less dense fluids of the *same volume*. A ship floats higher in salt water than in fresh water because salt water is slightly more dense than fresh water. Less volume of water is displaced, but not less weight. In the same way, a solid chunk of iron will sink in water but float in mercury.

It is important to stress *per volume* here, for the buoyant force on a floating object equals the weight of the object no matter what the density of the fluid.

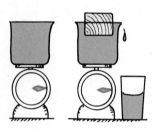

Figure 12-16
A floating object displaces a weight of fluid equal to its own weight.

Questions ▶

Fill in the blanks for these statements:

1. The volume of a submerged body is equal to the _____ of the fluid displaced.
2. The weight of a floating body is equal to the _____ of the fluid displaced.

▶ **Answers**
1. Volume.
2. Weight

Figure 12-17
The same ship empty and loaded. How does the weight of its load compare to the weight of extra water displaced?

Pascal's Principle

One of the most important facts about fluid pressure is that a change in pressure at one part of the fluid will be transmitted undiminished to other parts. For example, if the pressure of city water is increased at the pumping station by 10 units of pressure, the pressure everywhere in the pipes of the connected system will be increased by 10 units of pressure (providing the water is at rest). This rule is called **Pascal's principle**:

> **Changes in pressure at any point in an enclosed fluid at rest are transmitted undiminished to all points in the fluid and act in all directions.**

Pascal's principle was discovered by Blaise Pascal (1623–1662), a French mathematician, physicist, and theologian. The SI unit of pressure, the pascal (1 Pa = 1 N/m^2), is named after him.

Fill a U-tube with water and place pistons at each end, as shown in Figure 12-18. Pressure exerted against the left piston will be transmitted throughout the liquid and against the bottom of the right piston. (The pistons are simply "plugs" that can slide freely but snugly inside the tube.) The pressure the left piston exerts against the water will be exactly equal to the pressure the water exerts against the right piston. This is nothing to write home about. But suppose you make the tube on the right side wider and use a piston of larger area; then the result is impressive. In Figure 12-19 the piston on the left has an area of 1 square centimeter, and the piston on the right has an area fifty times as great, 50 square centimeters. Suppose there is a 1-newton load on the left piston. Then an additional pressure of 1 newton per square centimeter (1 N/cm^2) is transmitted throughout the liquid and up against the larger piston. Here is where the difference between force and pressure comes in. The additional pressure of 1 newton per square centimeter is exerted against every square centimeter of the larger piston. Since there are 50 square centimeters, the total extra force exerted on the larger piston is 50 newtons.

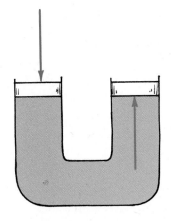

Figure 12-18
The force exerted on the left piston increases the pressure in the liquid and is transmitted to the right piston.

Figure 12-19

A 1-N load on the left piston will support 50 N on the right piston. Similarly, as shown here, a 10-kg load on the left piston will support 500 kg on the right piston.

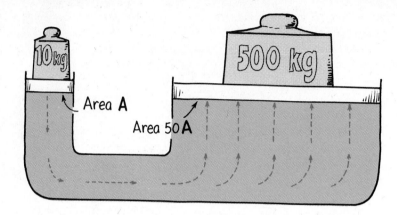

Thus, the larger piston will support a 50-newton load. This is fifty times the load on the smaller piston!

This is something to write home about, for we can multiply forces with such a device. One newton input produces 50 newtons output. By further increasing the area of the larger piston (or reducing the area of the smaller piston), we can multiply forces to any amount. Pascal's principle underlies the operation of the hydraulic press.

The hydraulic press does not violate energy conservation, because the increase in force is compensated for by a decrease in distance moved. When the small piston in the last example is moved downward 10 centimeters, the large piston will be raised only one-fiftieth of this, or 0.2 centimeters. The input force multiplied by the distance it moves is equal to the output force multiplied by the distance it moves; this is very much like the case of a mechanical lever.

Pascal's principle applies to all fluids, gases as well as liquids. A typical application of Pascal's principle for gases and liquids is the automobile lift seen in many service stations (Figure 12-20). Compressed air exerts pressure on the oil in an underground reservoir. The oil in turn transmits the pressure to a cylinder, which lifts the automobile. The relatively low pressure that exerts the lifting force against the piston is about the same as the air pressure in the tires of the automobile. This is because a low pressure exerted over a relatively large area produces a considerable force.

Question As the automobile in Figure 12-20 is being lifted, how does the change in oil level in the reservoir compare to the distance the automobile moves?

▶ **Answer**

The car moves up a greater distance than the oil level drops, since the area of the piston is smaller than the surface area of the oil in the reservoir.

Figure 12-20
Pascal's principle in a gas station.

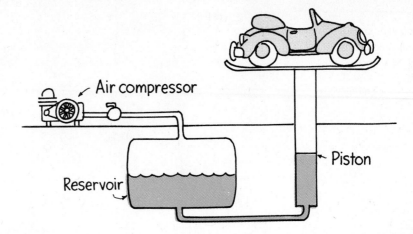

Surface Tension

Suppose we suspend a piece of clean wire from a sensitive spring, lower the wire into water, and then raise it. As we attempt to free the wire from the water surface, we see from the stretched spring that the water surface exerts an appreciable force on the wire. The water surface has a greater resistance to being stretched than the spring. The surface of water tends to contract. We note this tendency when a fine-haired paintbrush has been wet. When the brush is under water, the hairs are fluffed pretty much as they are when the brush is dry, but when the brush is lifted out, the surface film of water contracts and pulls the hairs together (Figure 12-22). We call this contractive force of the surface of liquids **surface tension**.

Surface tension accounts for the spherical shape of liquid drops. Raindrops, drops of oil, and falling drops of molten metal are all spherical because their surfaces tend to contract and force each drop into the shape having the least surface. This is a sphere, for a sphere is the geometrical figure that has the least surface for a given volume. For this reason the

Figure 12-21
When the bent wire is lowered into the water and then raised, the spring will stretch because of surface tension.

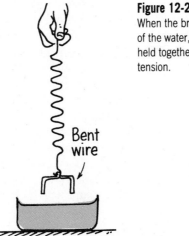

Figure 12-22
When the brush is taken out of the water, the hairs are held together by surface tension.

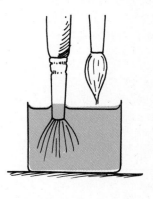

Figure 12-23

Small blobs of water are drawn by surface tension into spherelike shapes. Can you apply the ideas of Figure 11-15 in the last chapter to an explanation of why larger drops are more flattened by gravity, while the tinier drops are more spherical?

Figure 12-24

A molecule at the surface is pulled only sideways and downward by neighboring molecules. A molecule beneath the surface is pulled equally in all directions.

Figure 12-25

Tom Noddy blows smoke into an interior soap bubble to reveal its hexagonal prism shape, caused by the surface tension of surrounding bubbles.

mist and dewdrops on spider webs or on the downy leaves of plants are tiny spheres (Figure 12-23).

Surface tension results from the contraction of liquid surfaces, which in turn is caused by molecular attractions. Beneath the surface, each molecule is attracted in every direction by neighboring molecules, with the result that there is no tendency to be pulled in any preferred direction. A molecule on the surface of a liquid, however, is pulled only by neighbors to each side and downward from below; there is no pull upward (Figure 12-24). These molecular attractions thus tend to pull the molecule from the surface into the liquid. This tendency to pull surface molecules into the liquid causes the surface to become as small as possible. The surface behaves as if it were tightened into an elastic film. This is evident when dry steel needles or razor blades seem to float on water; actually, they are supported by the surface tension of the water.

Surface tension of water is greater than that of other common liquids. For example, clean water has a stronger surface tension than soapy water. We can see this when a little soap film on the surface of water is effectively pulled out over the entire surface. This minimizes the surface area of the clean water. The same thing happens for oil or grease floating on water. Oil has less surface tension than water and is drawn out into a film covering the whole surface, except when the water is hot. The surface tension of hot water decreases because the faster-moving molecules are not held as cohesively. This allows the grease or oil in hot soups to float in little bubbles on the surface of the soup. But when the soup cools and the surface tension of water increases, the grease or oil is dragged out over the surface of the soup. The soup becomes "greasy." Hot soup tastes different from cold soup primarily because the surface tension of water changes with temperature.

Capillarity

If the end of a thoroughly clean glass tube with a small inside diameter is dipped into water, the water will wet the inside of the tube and rise in it. In a tube with a bore of about $\frac{1}{2}$ millimeter in diameter, for example, the water will rise slightly higher than 5 centimeters. With a still smaller bore, the water will rise much higher (Figure 12-26). This rise or fall of a liquid in a fine hollow tube or in a narrow space is **capillarity**.

Figure 12-26
Capillary tubes.

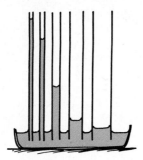

Water molecules are attracted to glass more than to each other. The attraction between unlike substances is called *adhesion*. The attraction between like substances is called *cohesion*. When a glass tube is dipped into water, the adhesion between the glass and the water causes a thin film of water to be drawn up over the surfaces of the tube (Figure 12-27a). Surface tension causes this film to contract (Figure 12-27b). The film on the inner surface continues to contract, raising water with it until the adhesive force is balanced by the weight of the water lifted (Figure 12-27c). In a small tube, the weight of the water in the tube is small and the water is lifted higher than if the tube were large.

If a paintbrush is dipped into water, the water will rise up into the narrow spaces between the bristles by capillary action. Hang your hair in the bathtub, and water will seep up to your scalp in the same way. This is how oil soaks upward in a lamp wick and water into a bath towel when one end hangs in water. Dip one end of a lump of sugar in coffee, and the entire lump is quickly wet. The capillary action in soils is important in bringing water to the roots of plants. We see capillary action at work in many phenomena of nature. How nice.

Figure 12-27
Hypothetical stages of capillary action, as seen in a cross-sectional view of a capillary tube.

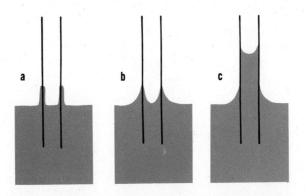

a   b   c

Summary of Terms

Pressure The ratio of force to the area over which that force is distributed:

$$\text{Pressure} = \frac{\text{force}}{\text{area}}$$

$$\text{Liquid pressure} = \text{weight density} \times \text{depth}$$

Buoyant force The net force that a fluid exerts on an immersed object.

Archimedes' principle An immersed object is buoyed up by a force equal to the weight of the fluid it displaces.

Principle of flotation A floating object displaces a weight of fluid equal to its own weight.

Pascal's principle The pressure applied to a fluid confined in a container is transmitted undiminished throughout the fluid and acts in all directions.

Surface tension The tendency of the surface of a liquid to contract in area and thus behave like a stretched rubber membrane.

Capillarity The rise or fall of a liquid in a fine, hollow tube or in a narrow space.

Suggested Reading

Rogers, E. *Physics for the Inquiring Mind*. Princeton, N.J.: Princeton University Press, 1960. Chapter 6 of this oldie-but-goodie textbook treats surface tension in interesting detail.

Review Questions

1. Distinguish between *force* and *pressure*.

Pressure in a Liquid

2. What is the relationship between liquid pressure and the depth of a liquid? Between liquid pressure and density?

3. If you swim twice as deep in water, how much more water pressure is exerted on your ears? If you swim in salt water, will the pressure at the same depth be greater than in fresh water? Why or why not?

4. How does water pressure one meter below the surface of a small pond compare to water pressure one meter below the surface of a huge lake?

5. If you punch a hole in a container filled with water, in what direction does the water initially flow outward from the container?

Buoyancy

6. Why does buoyant force act upward for an object submerged in water?

7. How does the volume of a submerged object compare to the volume of water displaced?

Archimedes' Principle

8. How does the buoyant force that acts on a fish compare to the weight of the fish?

9. If a 1-L container is immersed halfway into water, what is the volume of water displaced? What is the buoyant force on the container?

10. How does the buoyant force on a submerged object compare to the weight of water displaced?

11. What is the mass of 1 L of water? What is its weight in newtons?

12. Does the buoyant force on a submerged object depend on the weight of the object itself or on the weight of the fluid displaced by the object? Does it depend on the weight of the object itself or on its volume? Defend your answers.

Density Effects on Submerged Objects

13. If the buoyant force on a submerged object is equal to the weight of the object, how do the densities of the object and water compare?

14. If the buoyant force on a submerged object is greater than the weight of the object, how do the densities of the object and water compare?

15. If the buoyant force on a submerged object is less than the weight of the object, how do the densities of the object and water compare?

16. How is the density of a submarine controlled? How is the density of a fish controlled?

Flotation

17. How much water is displaced by a 100-ton ship? What is the buoyant force that acts on a 100-ton ship?

18. Does the buoyant force on a floating object depend on the weight of the object itself or on the weight of the fluid displaced by the object? Or are these both the same for the special case of floating? Defend your answers.

Pascal's Principle

19. What will happen to the pressure in all parts of a confined fluid if you increase the pressure in one part?

20. If the pressure in a hydraulic press is increased by an additional 10 N/cm², how much extra load will the output piston support if its cross-sectional area is 50 cm²?

Surface Tension

21. Why does the surface of a blob of water contract?

22. What geometrical shape has the least surface area?

23. Why is cold soup greasy?

Capillarity

24. Distinguish between *adhesive* and *cohesive* forces.

25. What determines how high water will climb in a capillary tube?

Home Projects

1. Try to float an egg in water. Then dissolve salt in the water until the egg floats. How does the density of an egg compare to that of tap water? To salt water?

2. Punch a couple of holes in the bottom of a water-filled container, and water will spurt out because of water pressure. Now drop the container, and as it freely falls note that the water no longer spurts out! If your friends don't understand this, could you figure it out and then explain it to them?

3. Soap greatly weakens the cohesive forces between water molecules. You can see this by putting some oil in a bottle of water and shaking it so that the oil and water mix. Notice that the oil and water quickly separate as soon as you stop shaking the bottle. Now add some soap to the mixture. Shake the bottle again and you will see that the soap makes a fine film around each little oil bead and that a longer time is required for the oil to gather after you stop shaking the bottle.

This is how soap works in cleaning. It breaks the surface tension around each particle of dirt so that the water can reach the particles and surround them. The dirt is carried away in rinsing. Soap is a good cleaner only in the presence of water.

Exercises

1. Why are persons confined to bed less likely to develop bedsores on their bodies if they use a waterbed rather than an ordinary mattress?

2. You know that a sharp knife cuts better than a dull knife. Do you know why this is so?

3. If water faucets upstairs and downstairs are turned fully on, will more water per second come out of the upstairs or downstairs faucet?

4. Which do you suppose exerts more pressure on the ground—an elephant or a lady standing on spike heels? (Which will be more likely to make dents in a linoleum floor?) Can you approximate a rough calculation for each?

5. Physics professor Joe Pizzo smashes glass bottles in front of his class and spreads the broken pieces across his lecture table. He then walks barefoot across the broken glass! What physical concepts is he demonstrating, and why is he careful that the broken pieces are small and numerous?

6. Why does your body get more rest when you're lying than when you're sitting? And why is blood pressure measured in the upper arm, at the elevation of your heart? Is blood pressure in your legs greater?

7. Which teapot holds the most liquid?

8. The sketch shows the reservoir that supplies water to a farm. It is made of wood and reinforced with metal hoops. (*a*) Why is it elevated? (*b*) Why are the hoops closer together near the bottom part of the tank?

9. A block of aluminum with a volume of 10 cm³ is placed in a beaker of water filled to the brim. Water overflows. The same is done in another beaker with a 10-cm³ block of lead. Does the lead displace more, less, or the same amount of water?

10. A block of aluminum with a mass of 1 kg is placed in a beaker of water filled to the brim. Water overflows. The same is done in another beaker with a 1-kg block of lead. Does the lead displace more, less, or the same amount of water?

11. A block of aluminum with a weight of 10 N is placed in a beaker of water filled to the brim. Water over-flows. The same is done in another beaker with a 10-N block of lead. Does the lead displace more, less, or the same amount of water? (Why are your answers to this exercise and Exercise 10 different from your answer to Exercise 9?)

12. There is a legend of a Dutch boy who bravely held back the whole Atlantic Ocean by plugging a hole in a dike with his finger. Is this possible and reasonable?

13. If you've wondered about the flushing of toilets on the upper floors of city skyscrapers, how do you suppose the plumbing is designed so that there is not an enormous impact of sewage arriving at the basement level? (Check your speculations with someone who is into architecture.)

14. Why does water "seek its own level"?

15. Suppose you wish to lay a level foundation for a home on hilly and bushy terrain. How can you use a garden hose filled with water to determine equal elevations for distant points?

16. When you are bathing on a stony beach, why do the stones hurt your feet less when you get in deep water?

17. If liquid pressure were the same at all depths, would there be a buoyant force on an object submerged in the liquid? Explain.

18. How much force is needed to push a nearly weight-less but rigid 1-L carton beneath a surface of water?

19. How does the fraction of the submerged part of a floating object compare with the ratio of its density and density of the liquid in which it floats?

20. The Himalayan Mountains are slightly less dense than the mantle material upon which they "float." Do you suppose that, like floating icebergs, they are deeper than they are high?

21. Consider the plug in your bathtub the next time you take a bath. When the tub is full, is there a buoyant force on the plug when it is in the drain? When it's pulled out and lying submerged on the tub bottom beside the drain? Defend your answers.

22. Why is it inaccurate to say that heavy objects sink and that light objects float? Give exaggerated examples to support your answer.

23. A piece of iron placed on a block of wood makes it float lower in the water. If the iron were instead suspended beneath the wood, would it float as low, lower, or higher? Defend your answer.

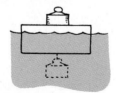

24. Compared to an empty ship, would a ship loaded with a cargo of styrofoam sink deeper into water or rise in water? Defend your answer.

25. If a submarine starts to sink, will it continue to sink to the bottom if no changes are made? Explain.

26. A barge filled with scrap iron is in a canal lock. If the iron is thrown overboard, does the water level at the side of the lock rise, fall, or remain unchanged? Explain.

27. Would the water level in a canal lock go up or down if a battleship in the lock sank?

28. A balloon is weighted so that it is barely able to float in water. If it is pushed beneath the sur-face, will it come back to the surface, stay at the depth to which it is pushed, or sink? Explain. (*Hint:* What change in density, if any, does the balloon undergo?)

29. The density of a rock doesn't change when it is submerged in water, but your density changes when you are submerged. Why is this so?

30. When floating, why do you tend to float lower in the water when you exhale?

31. In answering the question of why objects float higher in salt water than fresh water, your friend replies that the reason is that salt water is denser than fresh water. (Does your friend often answer questions by reciting only factual statements that relate to the answers but don't provide any concrete reasons?) How would you answer the same question?

32. Suppose you wear two life preservers that are iden-tical in size, first a light one filled with styrofoam and then a very heavy one filled with lead pellets. If you wear these life preservers in the water, upon which will the buoyant force be greater? Upon which will the buoyant force be ineffective? Why are your answers different?

33. The weight of the human brain is about 15 N. The buoyant force supplied by fluid around the brain is about 14.5 N. Does this mean that there must be at least 14.5 N of fluid surrounding the brain? Defend your answer.

34. The relative densities of water, ice, and alcohol are 1.0, 0.9, and 0.8, respectively. Do ice cubes float higher or lower in a mixed alcoholic drink? What can you say about a cocktail in which the ice cubes lie submerged at the bottom of the glass?

35. When an ice cube in a glass of water melts, does the water level in the glass rise, fall, or remain unchanged?

36. Suppose the ice cube in the preceding question has many air bubbles. When it melts, what happens to the water level in the glass? What if the ice cube contains many grains of heavy sand?

37. A half-filled bucket of water is on a spring scale. Will the reading of the scale increase or remain the same if a fish is placed in the bucket? (Will your answer be different if the bucket is initially filled to the brim?)

38. The weight of the container of water is equal to the weight of the stand and suspended solid iron ball as shown in *a*. When the suspended ball is lowered into the water, the balance is upset (*b*). Will the additional weight that must be put on the right side to restore balance be greater than, equal to, or less than the weight of the ball?

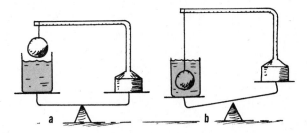

39. Would a fish float to the surface, sink, or stay at the same depth if the gravitational field of the earth increased?

40. What differences would we experience when swimming in water at the $\frac{1}{2}$-g level in an orbiting space habitat? Would we float in the water as we do on earth?

41. If you release a Ping-Pong ball beneath the surface of water, it will be buoyed to the surface. Would it do the same if the vessel of water were located at the hub of an orbiting structure in space? Explain.

42. So you're on a run of bad luck and you fall quietly into a small quiet pool as hungry crocodiles lurking at the bottom are appreciating Pascal's principle. What does Pascal's principle have to do with their delight?

43. In the hydraulic arrangement shown, the larger piston has an area which is fifty times that of the smaller piston. The strongman hopes to exert enough force on the large piston to raise the 10 kg that rest on the small piston. Do you think he will be successful? Defend your answer.

44. In the hydraulic arrangement shown in Figure 12-19, the multiplication of force is equal to the ratio of areas of the large and small pistons. Some people are surprised to learn that the area of the liquid surface in the reservoir of the arrangement shown in Figure 12-20 is immaterial. What is your explanation to clear up this confusion?

45. The piston cross-section area is 400 cm² for the hydraulic lift shown in Figure 12-20. The mass of the car and the lift itself is 2000 kg. How much pressure must be applied to the surface of the fluid in the reservoir to lift the car?

46. How much pressure do you experience when you balance a 5-kg ball on the the tip of your finger, say of area 1 cm²? How does this compare to your answer to the last question?

47. We say that the shape of a liquid is that of its container. But with no container and no gravity, what is the natural shape of a blob of water?

48. Why will hot water leak more readily than cold water through small leaks in a car radiator?

49. Would it be correct to say that the reason water rises in small, hollow tubes is because of capillarity? Explain.

50. On the surface of a pond it is common to see water striders, insects that can "walk" on the surface of water without sinking. What physics concept explains this act? Explain.

13 Gases and Plasmas

Gases as well as liquids flow; hence, both are called *fluids*. The primary difference between a gas and a liquid is the distance between molecules. In a gas, the molecules are far apart and free from the cohesive forces that dominate their motions when in the liquid and solid states. Their motions are less restricted. When the molecules in a gas collide with one another and with the walls of a container, they rebound without loss of kinetic energy. A gas expands indefinitely and fills all space available to it. Only when the quantity of gas is very large, such as the earth's atmosphere or a star, do the gravitational forces limit the size or determine the shape of the mass of gas.

The Atmosphere

Because molecules in the air are energized by sunlight and are continually in motion, and because of gravity, the earth has an atmosphere. If gas molecules were not constantly moving, our "atmosphere" would be just so much more matter on the ground. It would lie on the surface of earth just the way dormant popcorn lies at the bottom of a popcorn machine. But add heat to the popcorn and the atmospheric gases, and both will bumble their way up to higher altitudes. Pieces of popcorn in a popper attain speeds of a few kilometers per hour and are able to occupy altitudes up to a meter or two; molecules in the air move at speeds of about 1600 kilometers per hour and bumble up to many kilometers in altitude. If there were no gravity, both the popcorn and the atmospheric gases would fly into outer space. Fortunately there is an energizing sun, and there is gravity, and we have an atmosphere.

The exact height of the atmosphere has no real meaning, for the air gets thinner and thinner the higher one goes and eventually thins to emptiness into interplanetary space. Even in the vacuous regions of interplanetary space, however, there is a gas density of about 1 molecule per cubic centimeter. This is primarily hydrogen, the most plentiful element in the universe. About 50 percent of the atmosphere is below an altitude of 5.6 kilometers, 75 percent is below 11 kilometers, 90 percent is below 17.7 kilometers, and 99 percent is below about 30 kilometers (Figure 13-1). A detailed description of the atmosphere can be found in any encyclopedia.

231

Figure 13-1
The atmosphere. Air is more compressed at sea level than at higher altitudes. Like feathers in a huge pile, what's at the bottom is more squashed than what's nearer the top.

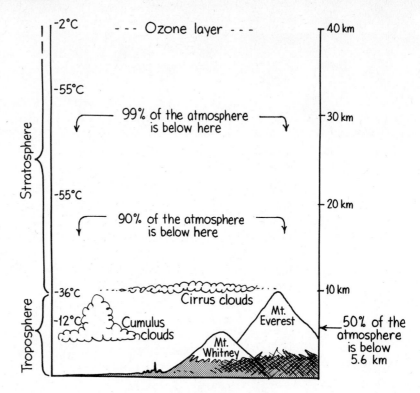

Atmospheric Pressure

Table 13-1
Densities of various gases

Gas	Density (kg/m³)*
Dry air	
0°C	1.29
10°C	1.25
20°C	1.21
30°C	1.16
Helium	0.178
Hydrogen	0.090
Oxygen	1.43

*At sea-level atmospheric pressure and at 0°C (unless otherwise specified).

We live at the bottom of an ocean of air. The atmosphere, much like the water in a lake, exerts a pressure. One of the most celebrated experiments demonstrating the pressure of the atmosphere was conducted in 1654 by Otto von Gueicke, burgomaster of Magdeburg and inventor of the vacuum pump. Von Gueicke placed together two copper hemispheres about $\frac{1}{2}$ meter in diameter to form a sphere, as shown in Figure 13-2. He set a gasket made of a ring of leather soaked in oil and wax between them to make an airtight joint. When he evacuated the sphere with his vacuum pump, two teams of eight horses each were unable to pull the hemispheres apart.

When the air pressure inside a cylinder like that shown in Figure 13-3 is reduced, there is an upward force on the piston. This force is large enough to lift a heavy weight. If the inside diameter of the cylinder is 10 centimeters or greater, a person can be lifted by this force.

What do the experiments of Figures 13-2 and 13-3 demonstrate? Do they show that air exerts pressure or that there is a "force of suction"? If we say there is a force of suction, then we assume that a vacuum can exert a force. But what is a vacuum? It is an absence of matter; it is a condition of nothingness. How can nothing exert a force? The hemispheres are not sucked together, nor is the piston holding the weight sucked upward. The hemispheres and the piston are being pushed against by the pressure of the atmosphere.

Figure 13-2
The famous "Magdeburg hemispheres" experiment of 1654, demonstrating atmosphere pressure. Two teams of horses couldn't pull the evacuated hemisphere apart. Were the hemispheres sucked together or pushed together? By what?

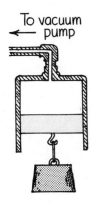

Figure 13-3
Is the piston pulled up or pushed up?

Figure 13-4
The mass of air that would occupy a bamboo pole that extends to the "top" of the atmosphere is about 1 kg. This air weighs 10 N.

Just as water pressure is caused by the weight of water, **atmospheric pressure** is caused by the weight of air. We have adapted so completely to the invisible air that we sometimes forget it has weight. Perhaps a fish similarly "forgets" about the weight of water. The reason we don't feel this weight crushing against our bodies is that the pressure inside our bodies equals that of the surrounding air. There is no net force for us to sense.

At sea level, 1 cubic meter of air has a mass of about $1\frac{1}{4}$ kilograms. So the air in your kid sister's small bedroom weighs about as much as she does! The density of air decreases with altitude. At 10 kilometers, for example, 1 cubic meter of air has a mass of about 0.4 kilograms. To compensate for this, airplanes are pressurized; the additional air needed to fully pressurize a 747 jumbo jet, for example, is more than 1000 kilograms. Air is heavy if you have enough of it.

Consider the mass of air in an upright 30-kilometer-tall bamboo pole that has an inside cross-sectional area of 1 square centimeter. If the density of air inside the pole matches the density of air outside, the enclosed mass of air would be about 1 kilogram. The weight of this much air is about 10 newtons. So air pressure at the bottom of the bamboo pole would be about 10 newtons per square centimeter (10 N/cm^2). Of course, the same is true without the bamboo pole. There are 10 000 square centimeters in 1 square meter, so a column of air 1 square meter in cross section that extends up through the atmosphere has a mass of about 10 000 kilograms. The weight of this air is about 100 000 newtons (10^5 N), which produces a pressure of 100 000 newtons per square meter—or, equivalently, 100 000 pascals, or 100 kilopascals. To be more exact, the average atmospheric pressure at sea level is 101.3 kilopascals (101.3 kPa).*

---

*The pascal or kilopascal is the SI unit of measurement. In British units, the average atmospheric pressure at sea level is 14.7 lb/in². This amount of pressure is often called 1 atmosphere.

Questions ▶

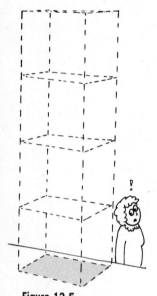

Figure 13-5
The weight of air that bears down on a 1 m² surface at sea level is about 100 000 newtons. So atmospheric pressure is about 10^5 N/m², or about 100 kPa.

760 mm

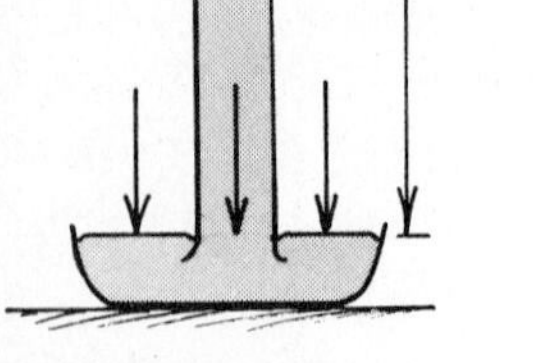

Figure 13-6
A simple mercury barometer.

The pressure of the atmosphere is not uniform. Besides altitude variations, there are variations in atmospheric pressure at any one locality due to moving air currents and storms. Measurement of changing air pressure is important to meteorologists in predicting weather.

Barometers

Instruments used for measuring the pressure of the atmosphere are called **barometers**. A simple mercury barometer is illustrated in Figure 13-6. A glass tube, longer than 76 centimeters and closed at one end, is filled with mercury and tipped upside down in a dish of mercury. The mercury in the tube runs out of the submerged open bottom until the level falls to about 76 centimeters. The empty space trapped above, except for some mercury vapor, is a pure vacuum. The vertical height of the mercury column remains constant even when the tube is tilted, unless the top of the tube is less than 76 centimeters above the level in the dish—in which case the mercury completely fills the tube.

Why does mercury behave this way? The explanation is similar to the reason a simple seesaw will balance when the weights of people at its two ends are equal. The barometer "balances" when the weight of liquid in the tube exerts the same pressure as the atmosphere outside. Whatever the width of the tube, a 76-centimeter column of mercury weighs the same as the air that would fill a supertall 30-kilometer tube of the same width. If the atmospheric pressure increases, then it pushes the mercury column higher than 76 centimeters. The mercury is literally pushed up into the tube of a barometer by atmospheric pressure.

Could water be used to make a barometer? The answer is yes, but the glass tube would have to be much longer—13.6 times as long, to be exact. You may recognize this number as the density of mercury compared to that of water. A volume of water 13.6 times that of mercury is needed to provide the same weight as the mercury in the tube (or in the imaginary tube of air outside). So the height of the tube would have to be at least 13.6 times taller than the mercury column. A water barometer would have to be 13.6 × 0.76 meter, or 10.3 meters high—too tall to be practical.

▶ **Answers**

1. The mass of air is 1000 kg. The volume of air is 200 m² × 4 m = 800 m³; each cubic meter of air has a mass of about $1\frac{1}{4}$ kg, so 800 m³ × $1\frac{1}{4}$ kg/m³ = 1000 kg.
2. Atmospheric pressure is exerted on both sides of a window, so no net force is exerted on the window. If for some reason the pressure is reduced or increased on one side only, then watch out!

Figure 13-7
Strictly speaking, they do not suck the soda up the straw. They instead reduce pressure in the straw and allow the weight of the atmosphere to press the liquid up into the straws.

What happens in a barometer is similar to what happens during the process of drinking through a straw. By sucking, you reduce the air pressure in the straw that is placed in a drink. Atmospheric pressure on the drink pushes liquid up into the reduced-pressure region. Strictly speaking, the liquid is not sucked up; it is pushed up by the pressure of the atmosphere. If the atmosphere is prevented from pushing on the surface of the drink, as in the party trick bottle with the straw through the airtight cork stopper, one can suck and suck and get no drink.

If you understand these ideas, you can understand why there is a 10.3-meter limit on the height water can be lifted with vacuum pumps. The old fashioned farm-type pump, Figure 13-8, operates by producing a partial vacuum in a pipe that extends down into the water below. The atmospheric pressure exerted on the surface of the water simply pushes the water up into the region of reduced pressure inside the pipe. Can you see that even with a perfect vacuum, the maximum height to which water can be lifted is 10.3 meters?

Question ▶ What is the maximum height to which water could be drunk through a straw?

Figure 13-8
The atmosphere pushes water from below up into a pipe that is evacuated of air by the pumping action.

▶ **Answer**
However strong your lungs may be, or whatever device you use to make a vacuum in the straw, at sea level the water could not be pushed up by the atmosphere higher than 10.3 m.

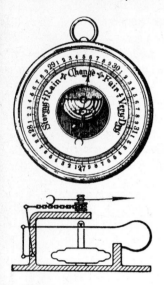

Figure 13-9
The aneroid barometer.

A small portable instrument that measures atmospheric pressure is the aneroid barometer (Figure 13-9). It uses a small metal box that is partially exhausted of air and has a slightly flexible lid that bends in or out with changes in atmospheric pressure. Motion of the lid is indicated on a scale by a mechanical spring-and-lever system. Since the atmospheric pressure decreases with increasing altitude, a barometer can be used to determine the elevation. An aneroid barometer calibrated for altitude is called an altimeter (altitude meter). Some of these instruments are sensitive enough to indicate a change in elevation less than a meter.*

The pressure (or vacuum) inside a television picture tube is about 1 ten-thousandth pascal (10^{-4} Pa). At an altitude of about 500 kilometers, artificial satellite territory, gas pressure is about 1 ten-thousandth of this (10^{-8} Pa). This is a pretty good vacuum by earthbound standards. Still greater vacuums exist in the wakes of satellites orbiting at this distance, and they can reach 10^{-13} pascals. This is a hard vacuum.

Vacuums on the earth are produced by pumps, which work by virtue of a gas tending to fill its container. If a space with less pressure is provided, gas will flow from the region of higher pressure to the one of lower pressure. A vacuum pump simply provides a region of lower pressure into which the normally fast-moving gas molecules randomly fill. The air pressure is repeatedly lowered by piston and valve action. Careful investigation of Figure 13-10 shows how this is accomplished. The best vacuums attainable with mechanical pumps are about a pascal. Better vacuums, down to 10^{-8} pascals, are attainable with vapor diffusion pumps. Greater vacuums are very difficult to attain. Technologists requiring hard vacuums are looking more to the prospects of orbiting laboratories in space.

Figure 13-10
A mechanical vacuum pump. When the piston is lifted, the intake valve opens and air moves in to fill the empty space. When the piston is moved downward, the outlet valve opens and the air is pushed out. What changes would you make to convert this pump into an air compressor?

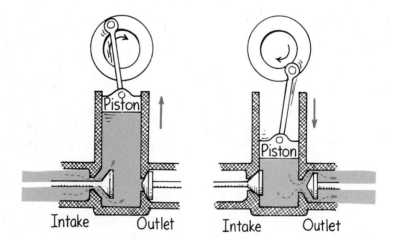

---

*Evidence of a noticeable pressure difference over a 1-m or less difference in elevation is any small helium-filled balloon that rises in air. The atmosphere really does push harder against the lower bottom than against the higher top.

Boyle's Law

The air pressure inside the inflated tires of an automobile is considerably greater than the atmospheric pressure outside. The density of air inside is also more than that of the air outside. To understand the relation between pressure and density, think of the molecules of air inside the tire. The molecules behave like tiny Ping-Pong balls, perpetually moving helter-skelter and banging against the inner walls. Their impacts produce a jittery force that appears to our coarse senses as a steady push. This pushing force averaged over a unit of area provides the pressure of the enclosed air.

Suppose there are twice as many molecules in the same volume (Figure 13-11). Then the air density is doubled. If the molecules move at the same average speed—or, equivalently, if they have the same temperature—then, to a close approximation, there are twice the number of collisions (the pressure is doubled). So pressure is proportional to density.

Figure 13-11
When the density of gas in the tire is increased, pressure is increased.

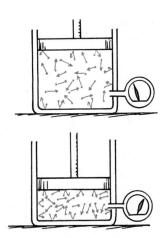

Figure 13-12
When the volume of gas is decreased, density and therefore pressure are increased.

We could instead double the density by simply compressing the air to half its volume. Consider the cylinder with the movable piston in Figure 13-12. If the piston is pushed downward so that the volume is half the original volume, the density of molecules will be doubled, and the pressure will correspondingly be doubled. Decrease the volume to a third its original value, and the pressure will be increased by three, and so forth.

Notice in these examples that the product of pressure and volume is the same. For example, a doubled pressure multiplied by a halved volume gives the same value as a tripled pressure multiplied by a one-third volume. In general, we can say that the product of pressure and volume for a given mass of gas is a constant as long as the temperature does not change. "Pressure × volume" for a quantity of gas at one time is equal to any "different pressure × different volume" at any other time. In shorthand notation,

$$P_1V_1 = P_2V_2$$

where P_1 and V_1 represent the original pressure and volume, respectively, and P_2 and V_2 the second, or final, pressure and volume. This relationship is called **Boyle's law**, after Robert Law, the seventeenth-century physicist who is credited with its discovery.* Note the similarity of this relationship to the conservation of mechanical energy, where the product of force and distance is constant: $F_1d_1 = F_2d_2$.

---

*Humor aside, Boyle's law is named after Robert Boyle.

Boyle's law applies to ideal gases, which are gases in which the disturbing effects of intermolecular forces and molecular volume are negligible. Air and other gases under normal pressures approach ideal gas conditions.*

1. A piston in an airtight pump is withdrawn so that the volume of the air chamber is increased three times. What is the change in pressure?

2. A scuba diver 10.3 m deep breathes compressed air. If she holds her breath while ascending, by how much does the volume of her lungs tend to increase?

Buoyancy of Air

A crab lives at the bottom of its ocean of water and looks upward at jellyfish and other lighter-than-water objects floating above it. Similarly, we live at the bottom of our ocean of air and look upward to balloons and other lighter-than-air objects floating above us. A balloon floats in air and a jellyfish "floats" in water for the same reason: each is buoyed upward by a displaced weight of fluid equal to its own weight. In one case the displaced fluid is air, and in the other case it is water. In water, immersed objects are buoyed upward because the pressure acting up against the bottom of the object exceeds the pressure acting down against the top. Likewise, air pressure acting up against an object immersed in air is greater than the pressure above pushing down. The buoyancy in both cases is numerically equal to the weight of fluid displaced. We can state **Archimedes' principle for air** as

> An object surrounded by air is buoyed up by a force equal to the weight of the air displaced.

We know that a cubic meter of air at ordinary atmospheric pressure and room temperature has a mass of about 1.2 kilograms, so its weight is about 12 newtons. Therefore, any 1-cubic-meter object in air is buoyed

▶ **Answers**

1. The pressure in the piston chamber is reduced to one-third. This is the principle that underlies a mechanical vacuum pump.

2. Atmospheric pressure can support a column of water 10.3 m high, so the pressure in water due to the weight of the water alone equals atmospheric pressure at a depth of 10.3 m. Taking the pressure of the atmosphere at the water's surface into account, the total pressure at this depth is twice atmospheric pressure. Unfortunately for the scuba diver, her lungs tend to inflate to twice their normal size if she holds her breath while rising to the surface. A first lesson in scuba diving is not to hold your breath when ascending. To do so can be fatal.

---

*A general law that takes temperature changes into account is $P_1V_1/T_1 = P_2V_2/T_2$, where T_1 and T_2 represent the initial and final *absolute* temperatures, measured in SI units called kelvins (Chapter 17).

up with a force of 12 newtons. If the mass of the 1-cubic-meter object is greater than 1.2 kilograms (so that its weight is greater than 12 newtons), it falls to the ground when released. If this size object has a mass less than 1.2 kilograms, it rises in the air. Any object that has a mass less than the mass of an equal volume of air will rise in air. Another way to say this is that any object less dense than air will rise in air. Gas-filled balloons that rise in air are less dense than air.

Figure 13-13
All objects are buoyed up by a force equal to the weight of air they displace. Why, then, don't all objects float like this balloon?

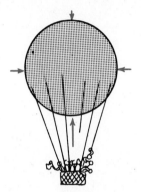

Gas is used in balloons simply because its presence prevents the atmosphere from collapsing the balloon. Helium is usually used because its mass is small enough that the combined weight of helium, balloon, and whatever the cargo happens to be is less than the weight of air it displaces.* Helium is used in a balloon for the same reason cork is used in a swimmer's life preserver. The cork possesses no strange tendency to be drawn toward the surface of water, and helium possesses no strange tendency to rise. Both are buoyed upward like anything else. They are simply light enough and large enough for the buoyant force that is exerted on them to be greater than their weights.

Unlike water, there is no sharp surface at the "top" of the atmosphere. Furthermore, unlike water, the atmosphere becomes less dense with altitude. Whereas cork will float to the surface of water, a released helium-filled balloon does not rise to any atmospheric surface. Will a lighter-than-air balloon rise indefinitely? How high will a balloon rise? We can state the answer in several ways. A balloon will rise only so long as it displaces a weight of air greater than its own weight. Since air becomes less dense with altitude, a lesser weight of air is displaced per given volume as the balloon rises. At some altitude the weight of displaced air may equal the total weight of the balloon and upward acceleration of the balloon will cease. We can also say that when the buoyant force on the balloon equals its weight, the balloon will cease rising. Equivalently, when the density of the balloon equals the density of the surrounding air, the balloon will cease rising. Helium-filled toy balloons usually break when released in the air because as the balloon rises to regions of less

*Hydrogen is the least dense gas but is highly flammable, so it is seldom used.

pressure, the helium in the balloon expands, increasing the volume and stretching the rubber until it ruptures.

Large dirigible airships are designed so that when loaded they will slowly rise in air; that is, their total weight is a little less than the weight of air displaced. When in motion, the ship may be raised or lowered by means of horizontal rudders or "elevators."

Question   Two balloons that have the same weight and volume contain equal amounts of helium. One is rigid and the other is free to expand as the pressure outside decreases. When released, which balloon will rise higher?

Thus far we have treated pressure only as it applies to stationary fluids. Motion produces an additional influence.

Bernoulli's Principle

Mark this statement true or false: Atmospheric pressure increases in a gale, tornado, or hurricane. If you answered true, sorry; the statement is false. High-speed winds may blow the roof off your house, but the pressure within the winds is actually less than for still air of the same density inside the house. As strange as it may first seem, when the speed of a fluid increases, the internal pressure decreases proportionally. This is true whether the fluid is a gas or liquid.

Daniel Bernoulli, a Swiss scientist of the eighteenth century, studied the relationship of fluid speed and pressure. When a fluid flows through a narrow constriction, its speed increases. This is easily noticed by the increased speed of a brook when it flows through the narrow parts. The fluid must speed up in the constricted region if the flow is to be continuous.

Bernoulli wondered how the fluid got the energy for this extra speed. He reasoned that it is acquired at the expense of a lowered internal pressure. His discovery, now called **Bernoulli's principle**, states:

The pressure in a fluid decreases as the speed of the fluid increases.

Bernoulli's principle is a consequence of the conservation of energy. When a fluid flows, it has kinetic energy because of its motion. It also has gravitational potential energy, or stored energy due to the earth's gravitational field. If the fluid picks up speed, or accelerates, it has more kinetic energy than before. Let's suppose that the fluid does not move up or down as it travels through the constricted region. Then its gravi-

Figure 13-14
The pressure in the spout reduces when the plug is removed.

▶ **Answer**
The balloon that is free to expand will displace more air as it rises than the balloon that is restrained. Hence, the balloon that is free to expand will experience more buoyant force and rise higher.

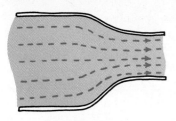

Figure 13-15
Water speeds up when it flows into the narrower pipe. The constricted streamlines indicate increased speed and decreased internal pressure.

tational potential energy does not change. How, then, does the accelerating fluid in the constricted region gain kinetic energy?

The answer is that the surrounding fluid does work on the part that goes through the constricted region. The forces that produce pressure push the accelerating fluid from behind. They do work on the accelerating fluid. The accelerating fluid has to do work on the fluid ahead of it. It turns out that when the fluid is accelerating, more work is done on it than it does on the fluid ahead. In this way, its kinetic energy increases. All through the fluid, some parts are gaining energy while others are losing energy. The net energy of the entire fluid is unchanged.

Bernoulli's principle holds for steady flow. In steady flow, the paths taken by each little region of fluid do not change as time passes. The motion of a fluid in steady flow can be represented with streamlines, which are indicated by dashed lines in Figure 13-14 and later figures. Streamlines are the smooth paths of the neighboring regions of fluid. The lines are closer together in the narrower regions, where the flow speed is greater and the pressure within the fluid is less.

If the flow speed is too great, the flow may become turbulent and follow changing, curling paths known as eddies. Then Bernoulli's principle will not hold.

Applications of Bernoulli's Principle

Hold a sheet of paper in front of your mouth, as shown in Figure 13-16. When you blow across the top surface, the paper rises. This is because the moving air pushes against the top of the paper with less pressure than the air that pushes against the lower surface, which is at rest.

We began our discussion of Bernoulli's principle by stating that atmospheric pressure decreases in a strong wind, tornado, or hurricane. As it turns out, an unvented building with airtight closed windows is in more danger of losing its roof than a well-vented building. This is because the air pressure inside may be appreciably greater than the reduced atmospheric pressure outside, and the roof is more likely to be pushed off by the relatively compressed air in the building than blown off by the wind. When the wind is blowing over a peaked roof as shown in Figure 13-17, the effect is even more pronounced. The crowding of the streamlines shows this. The difference in outside and inside pressure need not really be very much. A small pressure difference over a large area can be formidable. So if you're ever caught in an unvented building in a tornado or hurricane, consider opening the windows a bit so that the pressures inside and outside are more nearly equal.*

If we think of the blown-off roof as an airplane wing, we can better understand the lifting force that supports a heavy airliner. In both cases a greater pressure below pushes the roof and wing into a region of lesser

Figure 13-16
The paper rises when air is blown across its top surface.

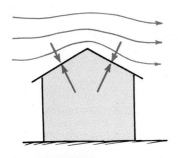

Figure 13-17
Air pressure above the roof is less than air pressure beneath the roof.

---

*A word of caution: In a building that has venting adequate to tolerate the sudden pressure drop without the need to open windows, opening the wrong window may actually increase damage.

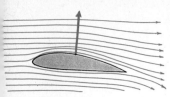

Figure 13-18
The arrow represents the net upward force (lift) that results from less air pressure above the wing than below the wing.

Figure 13-19
Where is air pressure greater—on the top or bottom surface of the hang glider?

pressure above. A cambered (arched) roof is more apt to be blown off than a flat roof. Similarly, a wing with more curvature on the top surface has greater lift than a wing with flat surfaces (like the wings of those balsa-wood gliders you used to play with). Whether it has flat or curved wings, an airplane will fly by virtue of air impact against the lower surface of wings that are tilted back slightly to deflect oncoming air downward (as discussed briefly in Chapter 4). The airfoil of a curved wing, however, adds considerably to lift and results in a greater difference in pressure between the lower and upper wing surfaces. This net upward pressure multiplied by the surface area of the wing gives the net lifting force. The lift is greater when there is a large wing area and when the plane is traveling fast. Gliders have a very large wing area so that they do not have to be going very fast for sufficient lift. At the other extreme, fighter planes designed for high speed have very small wing areas. Consequently, they must take off and land at relatively high speeds.

We all know that a baseball pitcher can throw a ball in such a way that it will curve off to one side of its trajectory. This is accomplished by imparting a large spin to the ball. Similarly, a tennis player can hit a ball that will curve. A thin layer of air is dragged around the spinning ball by friction, which is enhanced by the baseball's threads or the tennis ball's fuzz. The moving layer produces a crowding of streamlines on one side. Note in Figure 13-20b that the streamlines are more crowded at B than at A for the direction of spin shown. Air pressure is greater at A, and the ball curves as shown.

Figure 13-20
(a) The streamlines are the same on either side of a nonspinning baseball. (b) A spinning ball produces a crowding of streamlines and curves as shown by the thick arrow.

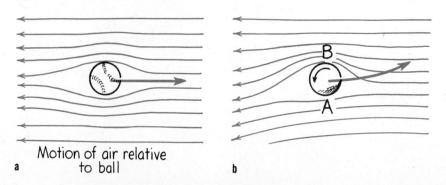

Motion of air relative to ball

Figure 13-21
Jacques Cousteau's *Alcyone* is a wind-powered ship that employs a Turbosail™ system instead of sails. Like a spinning baseball, the vertical cylinders deflect the airstream. In this design the cylinders do not spin, but instead suck in air through perforated lateral vents on one or the other side by means of an internal fan. This action simulates spinning cylinders and deflects the airstream. The airstream reacts by exerting a net force on the cylinders. The component of this net force in the forward direction propels the ship.

The net force produced by unequal pressures is not restricted to spinning balls. Bernoulli's principle has been applied with varying degrees of success to sailboats without sails since the 1920s. In the place of masts, these ships have large motor-driven vertical cylinders that rotate about their vertical axes. Like the case of the spinning baseball, a net force is produced that moves the ship, in most cases, more efficiently than canvas sails. A more recent design is Jacques Cousteau's ship, the *Alcyone* (Figure 13-21). Instead of rotating cylinders, it has fixed cylinders with special venting and equipped with an internal fan to produce the unequal pressures. Conventional diesel engines power the ship when wind speed is insufficient. With a good wind, *Alcyone* can save over 50 percent in fuel, and with winds over 25 knots (12.5 meters per second), the diesel engines can be shut off altogether for a cruising speed of 9 knots.

Bernoulli's principle accounts for the fact that passing ships run the risk of a sideways collision. Water flowing between the ships travels faster than water flowing past the outer sides. The streamlines are more compressed between the ships than outside. Water pressure acting against the hulls is therefore reduced between the ships. Unless the ships are steered to compensate for this, the greater pressure against the outer sides of the ships then forces them together.

Figure 13-22
Try this in your sink. Loosely moor a pair of toy boats side by side. Then direct a stream of water between them. The boats will draw together and collide. Why?

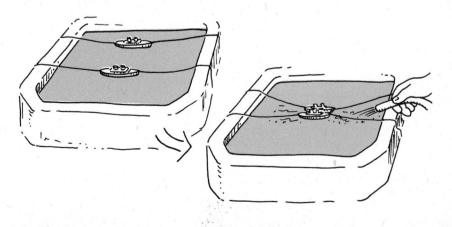

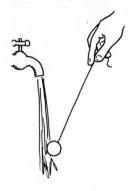

Figure 13-23
Pressure is greater in the stationary fluid (air) than in the moving fluid (water stream). The atmosphere pushes the ball into the region of reduced pressure.

You can demonstrate Bernoulli's principle quite interestingly in a stream of running water (Figure 13-23). Tape a table-tennis ball to a string and allow the ball to swing into the stream. You'll see that it will remain in the stream even when tugged slightly to the side, as shown. This is because the pressure of the stationary air next to the ball is greater than the pressure of the moving water. So we see that the ball is not sucked into the water, but is instead pushed into the region of reduced pressure by the atmosphere.

A similar thing happens to a bathroom shower curtain when the shower water is turned on full blast. Air near the water stream flows into the lower-pressure stream and is swept downward with the falling water. Air pressure inside the curtain is thus reduced, and the atmospheric pressure outside pushes the curtain inward, providing an escape route for the downward-swept air. The next time you are taking a shower and the curtain swings in against your legs, think of Daniel Bernoulli!

Figure 13-24
The curved shape of an umbrella can be disadvantageous on a windy day.

Rats to you too, Daniel Bernoulli!

Question ▶ Will the mercury level in a barometer be affected on a windy day?

Plasma

In addition to solids, liquids, and gases, there is a fourth state of matter, the least common state in our everyday environment—**plasma** (not to be confused with the clear, formless liquid part of blood, also called plasma).

A plasma looks and behaves like a high-temperature gas but with an important difference: it conducts electricity. The atoms and molecules that make it up are *ionized*, stripped of one or more electrons as a result of the more violent collisions at high temperatures. Recall that a neutral atom has as many positive protons inside the nucleus as it has negative electrons outside the nucleus. When one or more of these electrons is stripped from the atom, the atom has more positive charge than negative

▶ **Answer**
Atmospheric pressure is reduced on a windy day, so the barometer level will fall from its normal position.

charge and is called a *positive ion* (under some conditions, it may have extra electrons, in which case it is called a *negative ion*). In an ideal plasma, all atoms are completely stripped of electrons, and the mixture is composed entirely of free electrons and bare nuclei. Although the electrons and ions are themselves electrically charged, the plasma as a whole is electrically neutral because there are still equal numbers of positive and negative charges. The free electrons and ions make up an electrically conducting medium. The configuration of electrons and ions can be shaped, molded, and moved by electric and magnetic fields.

The major difference between a neutral gas and a plasma is that the particles in a plasma are ionized and can exert electromagnetic forces on one another.

Plasma in the Everyday World

If you happen to be reading this by the light emitted by a fluorescent lamp, you don't have to look far to see plasma in action. Within the glowing tube of the lamp is a plasma consisting of low-pressure mercury vapor. It is ionized by voltage across electrodes at the ends of the tube and conducts an electric current, which causes the plasma to radiate, which in turn causes the phosphor coating on the inner surface of the tube to glow.

The neon gas in an advertising sign similarly becomes a plasma when it is ionized by a high voltage and made to conduct electricity. The different colors seen in these signs correspond to different kinds of atoms glowing in the plasma phase. Vapor lamps used in street lighting also emit the light of glowing plasmas (Figure 13-25). The bluish-green white light from some of these lamps is emitted from mercury atoms in the plasma phase, and the yellow light from others is from a plasma made up of sodium atoms.

The light from a laser is emitted from a glowing plasma. Strictly speaking, a gas laser is a plasma laser once the high voltage is applied. We will treat lasers in detail in Chapter 29.

The aurora borealis (or northern lights) is emitted from glowing plasma in the upper atmosphere. Layers of low-temperature plasma encircle the whole earth. Occasionally, showerings of electrons from outer space and from radiation belts enter the "magnetic windows" at the earth's poles, crash into the layers of plasma, and produce light.

These layers, which extend upward from about 55 kilometers, make up the ionosphere and act as mirrors to low-frequency radio waves. Higher-frequency radio and TV waves pass through the ionosphere. This is why you can pick up radio stations from long distances on your lower-frequency AM radios, but you have to be in the "line of sight" of broadcasting or relay antennas to pick up higher-frequency FM and TV signals. Ever notice that at nighttime you can pick up very distant stations on your AM set? This is because the plasma layers settle closer together in the absence of the energizing sunlight and act as better radio reflectors.

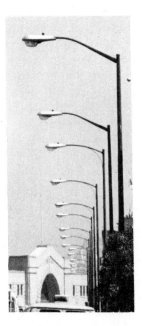

Figure 13-25
Streets are illuminated at night by the glowing plasma of vapor lamps.

Plasma Power

A higher-temperature plasma is the exhaust of a rocket engine. It is a weakly ionized plasma, but when small amounts of potassium salts or cesium metal are added, it becomes a very good conductor, and when it is directed into a magnet, electricity is generated! This is MHD power, the **magnetohydrodynamic** interaction between a plasma and a magnetic field. (We will treat the mechanics of how electricity is generated in this way in Chapter 24.) Low-pollution MHD power is now in the developmental stage and is in operation at a few places in the world already. We can expect to see more plasma power with MHD.

A more promising achievement will be plasma power of a different kind—the controlled fusion of atomic nuclei. We will treat the physics of fusion in Chapter 33. Even more important than the physics of thermonuclear fusion is the social impact of its controlled use. We have already seen the early effects of uncontrolled thermonuclear fusion, the hydrogen bomb. Like all technology, fusion can be applied for humanity's benefit as well as for its destruction. The benefits of controlled fusion may well be more far-reaching than the harnessing of electrical power in the last century. Fusion plants should ultimately not only make electrical energy abundant but provide the energy and means to recycle and even synthesize elements as well. The control of fusion may provide the setting for a new age. For the first time in our evolution, we could build a civilization having a base of abundance. This civilization should be very different from those that have had to deal with the scarcity of two important fundamentals—energy and material.

We have come a long way with our mastery of the first three states of matter. Our mastery of the fourth state should bring us ever so much farther.

Summary of Terms

Atmospheric pressure The pressure, exerted against bodies immersed in the atmosphere, that results from the weight and motion of molecules of atmospheric gases. At sea level, atmospheric pressure is about 101 kP.

Barometer Any device that measures atmospheric pressure.

Boyle's law The product of pressure and volume is a constant for a given mass of confined gas regardless of changes in either pressure or volume individually, so long as temperature remains unchanged:

$$P_1V_1 = P_2V_2$$

Archimedes' principle for air An object surrounded by air is buoyed up with a force equal to the weight of displaced air.

Bernoulli's principle The pressure in a fluid decreases with an increase in fluid velocity.

Plasma Hot matter beyond the gaseous state composed of electrically changed particles. Most of the matter in the universe is in the plasma state.

Review Questions

1. Distinguish among a *gas*, a *liquid*, and a *fluid*.

The Atmosphere

2. What is the energy source for the motion of gases in the atmosphere? What prevents atmospheric gases from flying off into space?

3. How high would you have to go in the atmosphere for half of the air to be below you?

Atmospheric Pressure

4. What is the cause of atmospheric pressure?

5. What is the mass of a cubic meter of air at room temperature (20°C)?

6. What is the approximate mass of a column of air 1 cm² in area that extends from sea level to the upper atmosphere? What is the weight of this amount of air?

7. What is the SI unit of atmospheric pressure?

8. What is the pressure at the bottom of the column of air discussed in Question 6?

Barometers

9. How does the downward pressure of the 76-cm column of mercury in a barometer compare to the pressure due to the weight of the atmosphere?

10. How does the weight of mercury in a barometer compare to the weight of an equal cross-section of air from sea level to the top of the atmosphere?

11. Why is a water barometer 13.6 times taller than a mercury barometer?

12. When you drink liquid through a straw, is it more accurate to say the liquid is pushed up the straw rather than sucked up the straw? What exactly does the pushing? Defend your answer.

13. Why will a vacuum pump not operate for a well that is more than 10.3 m deep?

14. Why is it that an aneroid barometer is able to measure altitude as well as atmospheric pressure?

15. How does a mechanical pump produce a vacuum?

Boyle's Law

16. By how much does the density of air increase when it is compressed to half its volume?

17. What happens to the air pressure inside a balloon when it is squeezed to half its volume?

18. What is an ideal gas?

Buoyancy of Air

19. How much buoyant force acts on a floating balloon that weighs 1 N? What happens if the buoyant force decreases? If it increases?

20. Does the air exert buoyant force on all objects in air or only on very light objects such as balloons that are in the air? Why is buoyancy effective in floating only things with a very low density?

21. What usually happens to a toy helium-filled balloon that rises high into the atmosphere?

Bernoulli's Principle

22. What happens to the internal pressure in a fluid flowing in a pipe when its speed increases?

23. What are streamlines? Is pressure greater or less in regions where streamlines are crowded?

Applications of Bernoulli's Principle

24. Is Bernoulli's principle the sole explanation for the flight of airplanes?

25. Why does a spinning ball curve in its flight?

26. Does atmospheric pressure increase or decrease on a windy day?

27. Does Bernoulli's principle hold for liquids? Defend your answer.

Plasma

28. How does a plasma differ from a gas?

29. Is a plasma electrically charged or electrically neutral? Explain.

Plasma in the Everyday World

30. Cite at least three examples of plasma in your daily environment.

31. What does plasma have to do with the aurora borealis (northern lights)?

32. Why is AM radio reception better at nighttime?

Plasma Power

33. What is MHD power?

34. What two fundamental products are speculated to result from fusion power plants?

Home Projects

1. You can find the weight of a car by measuring the area of contact that its four tires make with the road and then multiplying this area by the air pressure in the tires. The area can be closely approximated by tracing the edges of tire contact on sheets of paper marked with 1-cm² squares beneath each tire. If you actually did this, your calculation would be an exercise in converting units of measurement, especially if area is in square centimeters and air pressure in pounds per square inch.

2. Try this in the bathtub or when you're washing dishes. Lower a drinking glass, mouth downward, over a small floating object. What do you observe? How deep will the glass have to be pushed in order to compress the enclosed air to half its volume? (You won't be able to do this in your bathtub unless it's 10.3 m deep!)

3. You ordinarily pour water from a full glass to an empty glass by simply placing the full glass above the empty glass and tipping. Have you ever poured air from one glass to the other? The procedure is similar. Lower two glasses in water, their mouths downward. Let one fill with water by tilting its mouth upward. Then hold the water-filled glass mouth downward above the air-filled glass. Slowly tilt the lower glass and let the air escape, filling the upper glass. You will be pouring air from one glass to another!

4. Raise a filled glass of water above the waterline, but with its mouth beneath the surface. Why does the water not run out? How tall would a glass have to be before water began to run out? (You won't be able to do this indoors unless you have a 10.3-m ceiling.)

5. Place a card over the open top of a glass filled to the brim with water, and invert it. Why does the card stay intact? Try it sideways.

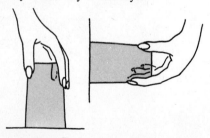

6. Invert a water-filled pop bottle or small-necked jar. Notice that the water doesn't simply fall out, but gurgles out of the container—a result of the weight of the many kilometers of air above pushing down and forcing air up into the bottle and forcing the water out. How would an inverted, water-filled bottle empty if this were done on the moon?

7. Pour about a half cup of water in a 5-or-so-liter metal can with a screw top. Place the can *open* on a stove and heat until the water boils and steam comes out of the opening. Quickly remove the can and screw the cap on tightly. Allow the can to stand and observe the results. The effect can be hastened by cooling the can with a dousing of cold water. Explain your observations.

8. Heat a small amount of water to boiling in an aluminum soda can and invert it quickly into a dish of water. Surprisingly dramatic!

9. Make a small hole near the bottom of an open tin can. Fill it with water, which proceeds to spurt from the hole. Cover the top of the can firmly with the palm of your hand and the flow stops. Explain.

10. Lower a narrow glass tube or drinking straw in water and place your finger over the top of the tube. Lift the tube from the water and then lift your finger from the top of the tube. What happens? (You'll do this often if you enroll in a chemistry lab.)

11. Fold the ends of a filing card down so that you make a little bridge. Stand it on the table and blow through the arch as shown. No matter how hard you blow, you will not succeed in blowing the card off the table (unless you blow against the side of it). Try this with your nonphysics friends. Then explain it to them!

12. Push a pin through a small card sheet and place it in the hole of a thread spool. Try to blow the card from the spool by blowing through the hole. Try it in all directions.

13. Check an encyclopedia to see how Bernoulli's principle applies to an automobile carburetor. In this and similar applications, it is called the *Venturi effect*.

14. Hold a spoon in a stream of water as shown and feel the effect of the differences in pressure.

Exercises

1. It is said that a gas fills all the space available to it. Why then doesn't the atmosphere go off into space?

2. Why is there no atmosphere on the moon?

3. Count the tires on a large tractor trailer that is unloading food at your local supermarket, and you may be surprised to count 18 or so tires. Why so many tires? (*Hint:* See Home Project 1.)

4. What is the purpose of the ridges that prevent the funnel from fitting tightly in the mouth of a bottle?

5. How would the density of air in a deep mine compare to the air density at the earth's surface?

6. Why do bubbles of gas rising in a liquid become larger as they approach the surface?

7. Why do your ears pop when you ascend to higher altitudes?

8. Two teams of eight horses each were unable to pull the Magdeburg hemispheres apart (Figure 13-2). Why? Suppose two teams of nine horses each could pull them apart. Then would one team of nine horses succeed if the other team were replaced with a strong tree? Defend your answer.

9. Before boarding an airplane, you buy a roll of camera film or any item packaged in an airtight foil package, and while in flight you notice that it is puffed up. Explain why this happens.

10. Why do you suppose that airplane windows are small compared to bus windows?

11. A half cup or so of water is poured into a 5-L can, which is put over a source of heat until most of the water has boiled away. Then the top of the can is screwed on tightly and the can is removed from the source of heat and allowed to cool. What happens to the can and why?

12. Will breaking a TV picture tube cause it to implode or to explode? Explain.

13. We can understand how pressure in water depends on depth by considering a stack of bricks. The pressure at the bottom of the first brick corresponds to the weight of the entire stack. Halfway up the stack, the pressure is half because the weight of the bricks above is half. To explain atmospheric pressure, consider compressible bricks, like foam rubber. Why is this so?

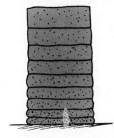

14. The "pump" in a vacuum cleaner is merely a high-speed fan. Would a vacuum cleaner pick up dust from a rug on the moon? Explain.

15. Suppose the pump shown in Figure 13-8 worked with a perfect vacuum. From how deep a well could water be pumped?

16. If a liquid only half as dense as mercury were used in a barometer, how high would its level be on a day of normal atmospheric pressure?

17. Why does the size of the cross-sectional area of a barometer not affect the height of the enclosed fluid?

18. From how deep a vessel could mercury be drawn with a siphon?

19. Would it be slightly more difficult to draw soda through a straw at sea level or on top of a very high mountain? Explain.

20. The pressure exerted against the ground by an elephant's weight distributed evenly over its four feet is less than 1 atmosphere. Why, then, would you be crushed beneath the foot of an elephant, while you're unharmed by the pressure of the atmosphere?

21. Your friend says that the buoyant force of the atmosphere on an elephant is significantly greater than the buoyant force of the atmosphere on a small helium-filled balloon. What do you say?

22. Why is it so difficult to breathe when snorkeling at a depth of 1 m, and practically impossible at a 2-m depth? Why can't a diver simply breathe through a hose that extends to the surface?

23. A block of wood and a block of iron on weighing scales each weigh 1 ton. Which has the greater mass?

24. Why does the weight of an object in air differ from its weight in a vacuum? Cite an example where this would be an important consideration.

25. Estimate the buoyant force that acts on you. (To do this, you can estimate your volume by knowing your weight and by assuming that your weight density is a bit less than that of water.)

26. A little girl sits in a car at a traffic light holding a helium-filled balloon. The windows are up and the car is relatively airtight. When the light turns green and the car accelerates forward, her head pitches backward but the balloon pitches forward. Explain why.

27. Would a bottle of helium gas weigh more or less than an identical bottle filled with air at the same pressure? Than an identical bottle with the air pumped out?

28. A steel tank filled with helium gas will not begin to float, but balloons that contain the same helium easily float. Why?

29. A balloon filled with air falls to the ground, but a balloon filled with helium rises. Why is this so?

30. The gas pressure inside an inflated rubber balloon is always greater than the air pressure outside. Why is this so?

31. Two identical balloons filled with air to the same volumes are suspended on the ends of a stick that is horizontally balanced. One of the balloons is then punctured. Is the balance of the stick upset? If so, which way does it tip?

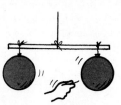

32. Two balloons that have the same weight and volume are filled with equal amounts of helium. One is rigid and the other is free to expand as the pressure outside decreases. When released, which will rise higher? Explain.

33. Imagine a huge space colony that consists of a rotating air-filled cylinder. How would the density of air at "ground level" compare to the air densities "above"?

34. Would a helium-filled balloon "rise" in the atmosphere of a rotating space habitat? Defend your answer.

35. Cite some differences in the sport of ballooning in the atmosphere of a cylindrical or spherical space colony. Do the same for hang gliding.

36. Nitrogen and oxygen in their liquid states have densities only 0.8 and 0.9 that of water. Atmospheric pressure is due primarily to the weight of nitrogen and oxygen gas in the air. If the atmosphere liquefied, would its depth be greater or less than 10.3 m?

37. Estimate the *force* that the atmosphere exerts on you, and then compare this force with the weight of something familiar in your environment. To do this, estimate your surface area from the area of your clothes, and multiply by 10^5 N/m^2. Why is your answer to this exercise so different from your answer to Exercise 25?

38. The force of the atmosphere at sea level against the outside of a 10-m^2 store window is about a million N. Why does this not shatter the window? Why might the window shatter in a strong wind?

39. Why does the fire in a fireplace burn more briskly on a windy day?

40. In a department store, an airstream from a hose connected to the exhaust of a vacuum cleaner blows upward at an angle and supports a beach ball in midair. Does the air blow under or over the ball to provide support?

41. What provides the lift to keep a Frisbee in flight?

42. Why is it that when passing an oncoming truck on the highway, your car tends to lurch toward the truck?

43. Why is it that the canvas roof of a convertible automobile bulges upward when the car is traveling at high speeds?

44. Why is it that the windows of older trains sometimes break when a high-speed train passes by on the next track?

45. A steady wind blows over the waves of an ocean. Why does the wind increase the humps and troughs of the waves?

46. In answering the question of why a flag flaps in the wind, your friend replies that it flaps because of Bernoulli's principle. (Not really a convincing explanation, is it?) How would you answer the same question?

47. Wharves are made with pilings that permit the free passage of water. Why would a solid-walled wharf be disadvantageous to ships attempting to pull alongside?

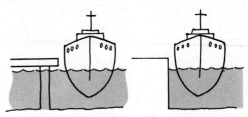

48. It is often said that fast-moving air has lower pressure than air at rest or slower-moving air. Can you make an argument that the opposite is the case; that fast-moving air is the *result*, not the *cause of*, lower pressure?

49. Under what conditions are the mercury atoms in a fluorescent lamp in a gaseous state? In a plasma state?

50. Why can you pick up faraway radio stations better at nighttime on your AM radio?

PART 3

HEAT

14 Temperature, Heat, and Expansion

All matter is composed of continually jiggling atoms or molecules. Whether the atoms and molecules combine to form solids, liquids, gases, or plasmas depends on the rate of molecular vibrations. By virtue of this vibratory motion, the molecules or atoms in matter possess kinetic energy. The average kinetic energy of the individual particles is directly related to a property you can sense: how hot something is. Whenever something becomes warmer, the kinetic energy of its particles increases. Strike a solid penny with a hammer and it becomes warm because the hammer's blow causes the molecules in the metal to jostle faster. Put a flame to a liquid and it too becomes warmer. Rapidly compress air in a tire pump and the air becomes warmer. When a solid, liquid, or gas gets warmer, its atoms or molecules move faster. The atoms or molecules have more kinetic energy.*

Temperature

Figure 14-1
Can we trust our sense of hot and cold? Will both fingers feel the same temperature when they are put in the warm water?

The quantity that tells how warm or cold an object is with respect to some standard is called **temperature**. We express the temperature of matter by a number that corresponds to the degree of hotness on some chosen scale.

Nearly all materials expand when their temperature is raised and contract when it is lowered. A thermometer is a common instrument that measures temperature by means of the expansion and contraction of a liquid, usually mercury or colored alcohol.

On the scale commonly used in laboratories, the number 0 is assigned to the temperature at which water freezes and the number 100 to the temperature at which water boils (at standard atmospheric pressure). The space between is divided into 100 equal parts called *degrees*; hence, a thermometer so calibrated has been called a *centigrade thermometer* (from *centi*, "hundredth," and *gradus*, "degree"). It is now called a *Celsius thermometer* in honor of the man who first suggested the scale, the Swed-

---

*The kinetic energy of atoms or molecules in matter is often called *thermal energy*. This term is also used to include the potential energy that is due to the forces between molecules. Because thermal energy may mean different things to different people, we avoid using the term.

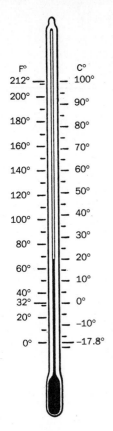

Figure 14-2
Fahrenheit and Celsius
scales on a thermometer.

ish astronomer Anders Celsius (1701–1744).

In the United States, the number 32 is assigned to the temperature at which water freezes, and the number 212 is assigned to the temperature at which water boils. Such a scale makes up a Fahrenheit thermometer, named after its illustrious originator, the German physicist G. D. Thermometer (1686–1736). The Fahrenheit scale will become obsolete if and when the United States goes metric.*

Still another temperature scale, favored by scientists, is the Kelvin scale, named after the British physicist Lord Kelvin (1824–1907). This scale is calibrated not in terms of the freezing and boiling points of water, but in terms of energy itself. The number 0 is assigned to the lowest possible temperature—**absolute zero**, at which a substance has absolutely no kinetic energy to give up. Absolute zero corresponds to $-273°C$ on the Celsius scale. Units on the Kelvin scale are the same size as degrees on the Celsius scale, so the temperature of melting ice is $+273$ kelvins. There are no negative numbers on the Kelvin scale. We won't treat this scale further until we return to it when we study thermodynamics in Chapter 17.

Arithmetic formulas are used for converting from one temperature scale to the other and are popular in classroom exams. Such arithmetic exercises are not really physics, and the probability of your having the occasion to do this task elsewhere is small, so we will not be concerned with it here. Besides, this conversion can be very closely approximated by simply reading the corresponding temperature from the side-by-side scales in Figure 14-2.

Temperature is a measure of the motion of the molecules or atoms within a substance; more specifically, it is a measure of the average kinetic energy of the molecules or atoms in a substance. Temperature is not a measure of the total kinetic energy of the particles within a substance. For example, there is twice as much molecular kinetic energy in 2 liters of boiling water as in 1 liter of boiling water—but the temperatures of both amounts of water are the same because the average kinetic energy per molecule in each is the same.

Interestingly enough, a thermometer registers its own temperature. When a thermometer is in thermal contact with something the temperature of which we wish to know, energy will flow between the two until their

---

*The conversion to Celsius will put the United States in step with the rest of the world, where the Celsius scale is the standard. Americans are slow to convert, perhaps because the Fahrenheit scale seems much better suited to everyday use. For example, its degrees are smaller ($1°F = \frac{5}{9}°C$), which gives greater accuracy when reporting the weather in whole-number temperature readings. Then, too, people somehow attribute a special significance to numbers increasing by an extra digit, so that when the temperature of a hot day is reported to reach 100°F, the idea of heat is conveyed more dramatically than by saying it is 37.7°C. Like so much of the British system of measure, the Fahrenheit scale is geared to human beings. (And, for the record, the Fahrenheit scale is named after Gabriel Daniel Fahrenheit.)

temperatures are equal and thermal equilibrium is established. If we then know the temperature of the thermometer, we also know the temperature of the something. A thermometer should be small enough that it doesn't appreciably alter the temperature of the something being measured. If you are measuring the room air temperature, then your thermometer is small enough. But if you are measuring a drop of water, its temperature after thermal contact may be quite different from its initial temperature.

Heat

Figure 14-3
There is more molecular kinetic energy in the container filled with warm water than in the small cupful of higher-temperature water.

If you touch a hot stove, energy enters your hand because the stove is warmer than your hand. When you touch a piece of ice, on the other hand, energy passes out of your hand and into the colder ice. The direction of energy transfer is always from a warmer thing to a neighboring cooler thing. The transfer of energy from one thing to another because of a temperature difference between the things is called **heat**.

It is important to point out that matter does not *contain* heat. Matter contains molecular kinetic energy, *not heat*. Heat flows and is the energy that is being transferred. Once heat has been transferred to an object or substance, it ceases to be heat. It becomes *internal energy*.

Internal energy is the grand total of all energies inside a substance. This includes potential energy due to the forces between molecules or atoms and kinetic energy due to movements of atoms and molecules. So a substance does not contain heat—it contains internal energy.

When a substance absorbs or gives off heat, internal energy in the substance changes. Thus, as a substance absorbs heat, this energy may or may not make the molecules or atoms jostle faster. In some cases, as when ice is melting, a substance absorbs heat without an increase in molecular kinetic energy. The substance undergoes a change of state, which we will cover in detail in Chapter 16.

For things in thermal contact, heat will flow from the substance at a higher temperature into a substance at a lower temperature, but it will not necessarily flow from a substance with more internal energy into a substance with less internal energy. There is more internal energy in a bowl of warm water than there is in a red-hot thumbtack; if the tack is immersed in the water, heat will not flow from the warm water to the tack. Instead, heat will flow from the hot tack to the relatively cooler

Figure 14-4
Just as water in the pipes seeks a common level (where the pressures at the bottom are the same), the thermometer and its immediate surroundings reach a common temperature (where the average molecular KE for both is the same).

water. Heat never flows of itself from a low-temperature substance into a higher-temperature substance.

How much heat flows depends not only on the temperature difference between substances but on the amount of material as well. For example, a barrelful of hot water will transfer more heat to a cooler substance than a cupful of water at the same temperature. There is more internal energy in the larger amount of water.

Figure 14-5
Although the same quantity of heat is added to both containers, the temperature increases more in the container with the smaller amount of water.

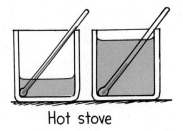

Hot stove

Questions ▶

1. Suppose you apply a flame to 1 L of water and its temperature rises by 2°C. If you apply the same flame to 2 L of water, by how much will its temperature rise?

2. If a fast marble hits a random scatter of slow marbles, does the fast marble usually speed up or slow down? Which lose(s) kinetic energy and which gain(s) kinetic energy, the initially fast-moving marble or the initially slow ones? How do these questions relate to the direction of heat flow?

Quantity of Heat

So heat is the internal energy that transfers from one thing to another by virtue of a temperature difference. The quantity of heat involved in such a transfer is measured by some change that accompanies the process. For example, in determining the energy value in food, the amount of internal energy that is released as heat is measured by burning. Fuels are rated on how much energy a certain amount of the fuel will produce when burned.

▶ **Answers**

1. Its temperature will rise by only 1°C, because there are twice as many molecules in 2 L of water and each molecule receives only half as much energy on the average. So the average kinetic energy, and thus the temperature, increases by half as much.

2. A fast-moving marble slows when it hits slower-moving marbles. It gives up some of its kinetic energy to the slower ones. Likewise with the flow of heat. Molecules with more kinetic energy that are in contact with less energetic molecules give up some of their kinetic energy to the less energetic ones. The direction of energy transfer is from hot to cold. For both the marbles and the molecules, however, the total energy before and after contact is the same.

Figure 14-6
To the weight watcher, the peanut contains 10 Calories; to the physicist, it releases 10 000 calories (or 41 870 joules) of energy when burned or digested.

Figure 14-7
The filling of hot apple pie may be too hot to eat, whereas the crust is not.

The unit of heat is defined as the energy necessary to produce some standard, agreed-on change. The most commonly used unit for heat is the calorie.* The calorie is defined as the amount of heat required to change the temperature of 1 gram of water by 1 Celsius degree.

The kilocalorie is 1000 calories (the heat required to change 1 kilogram of water by 1°C). The heat unit used in rating foods is actually a kilocalorie. To distinguish this unit from the smaller calorie, the food unit is sometimes called a *Calorie* (written with a capital *C*). It is important to remember that the calorie and the Calorie are units of energy. These names are historical carry-overs from the early idea that heat was an invisible fluid called *caloric*. This view persisted almost to the nineteenth century. We now know that heat is a form of energy, and we are presently in a transition period to the International System (SI) of units, in which the quantity of heat is measured in joules, the SI unit for all forms of energy. (The relationship between calories and joules is that 1 calorie = 4.187 joules.)

Specific Heat

Have you ever noticed that some foods remain hotter much longer than others? Boiled onions and squash on a hot dish, for example, are often too hot to eat when mashed potatoes—if not too moist—may be eaten comfortably. The filling of hot apple pie can burn your tongue while the crust will not, even when the pie has just been taken out of the oven. The aluminum covering on a frozen dinner can be peeled off with your bare fingers as soon as it is removed from the oven. A piece of toast may be comfortably eaten a few seconds after coming from the hot toaster, whereas we must wait several minutes before eating soup from a stove as hot as the toaster.

Different substances have different capacities for storing internal energy. If we heat a soup pot of water on a stove, we might find that it requires 15 minutes to raise it from room temperature to its boiling temperature. But if we put an equal mass of iron on the same flame, we would find that it would rise through the same temperature range in only about 2 minutes. For silver, the time would be less than a minute. We find that different materials require different quantities of heat to raise the temperature of a given mass of the material by a specified number of degrees. Different materials absorb energy in different ways. The energy may increase the to-and-fro vibrational motion of molecules, which raises the temperature. Or it may increase the amount of internal vibration or rotation within the molecules and go into potential energy, which does not raise the temperature. Generally there is a combination of both.

A gram of water requires 1 calorie of energy to raise the temperature 1 Celsius degree. It takes only about one-eighth as much energy to raise

*The SI unit of heat is the SI unit of energy, the joule. Another common unit of heat is the British thermal unit (Btu). The Btu is defined as the amount of heat required to change the temperature of 1 lb of water by 1 Fahrenheit degree.

the temperature of a gram of iron by the same amount. Water absorbs more heat than iron for the same change in temperature. We say water has a higher **specific heat** (sometimes called *specific heat capacity*).

> **The specific heat of any substance is defined as the quantity of heat required to change the temperature of a unit mass of the substance by 1 degree.**

Water has a much higher capacity for storing energy than all but a few uncommon materials. A relatively small amount of water absorbs a great deal of heat for a correspondingly small temperature rise. Because of this, water is a very useful cooling agent and is used in the cooling system of automobiles and other engines. If a liquid of lower specific heat were used in cooling systems, its temperature would rise higher for a comparable absorption of heat.

Question ▶	Which has a higher specific heat, water or sand?

Figure 14-8
Water has a high specific heat and is transparent, so it takes more energy to heat up than land. Solar energy incident upon land is concentrated at the surface, but upon water it extends and dilutes beneath the surface.

Water also takes a long time to cool, a fact that explains why hot-water bottles used to be employed on cold winter nights. (Electric blankets have, for the most part, taken their place.) This tendency on the part of water to resist changes in temperature improves the climate in many places. The next time you are looking at a world globe, notice the high latitude of Europe. If water did not have a high specific heat, the countries of Europe would be as cold as the northeastern regions of Canada, for both Europe and Canada get about the same amount of sunlight per square kilometer. The Atlantic current known as the Gulf Stream carries warm water northeast from the Caribbean. It holds much of its internal energy long enough to reach the North Atlantic off the coast of Europe, where it then cools. The energy released, 1 calorie per degree for each gram of water that cools, is carried by the westerly winds over the European continent. A similar effect occurs in the United States. The winds in the latitudes of North America are westerly. On the West Coast, air moves from the Pacific Ocean to the land. Because of water's high specific heat, an ocean does not vary much in temperature from summer to winter. The water is warmer than the air in the winter and cooler than the air in the summer. In winter the water warms the air that moves over and warms the coastal regions of North America. In summer, the water cools the air and the coastal regions are cooled. On the East Coast, air

▶ **Answer**

Water has the higher specific heat. The temperature of water increases less than the temperature of sand in the same sunlight. Sand's low specific heat, as evidenced by how quickly the surface warms in the morning sun and how quickly it cools at night, affects local climates.

moves from the land to the Atlantic Ocean. Land, with a lower specific heat, gets hot in the summer but cools rapidly in the winter. As a result of water's high specific heat and the wind directions, the West Coast city of San Francisco is warmer in winter and cooler in summer than the East Coast city of Washington, D.C., which is at about the same latitude.

Islands and peninsulas that are more or less surrounded by water do not have the same extremes of temperatures that are observed in the interior of a continent. The high summer and low winter temperatures common in Manitoba and the Dakotas, for example, are largely due to the absence of large bodies of water. Europeans, islanders, and people living near ocean air currents should be glad that water has such a high specific heat. San Franciscans are!

Expansion

Figure 14-9
One end of the bridge is fixed, while the end shown rides on rockers to allow for thermal expansion.

When the temperature of a substance is increased, its molecules or atoms jiggle faster and tend to move farther apart, on the average. The result is an expansion of the substance. With few exceptions, all forms of matter—solids, liquids, gases, and plasmas—generally expand when they are heated and contract when they are cooled.

In many cases the changes in the size of substances are not very noticeable, but careful observation will usually detect them. Telephone wires are longer and sag more on a hot summer day than they do on a cold winter day. Metal lids on glass fruit jars can often be loosened by heating them under hot water. If one part of a piece of glass is heated or cooled more rapidly than adjacent parts, the expansion or contraction that results may break the glass. This is especially true with thick glass. Pyrex glass is specially formulated to expand very little with increasing temperature.

The expansion of substances must be allowed for in the construction of structures and devices of all kinds. A dentist uses filling material that has the same rate of expansion as teeth. The aluminum pistons of some automobile engines are just smaller enough in diameter than the steel cylinders to allow for the much greater expansion rate of aluminum. A civil engineer uses reinforcing steel of the same expansion rate as concrete. Long steel bridges commonly have one end fixed while the other rests on rockers (Figure 14-9). The roadway itself is segmented with tongue-and-groove–type gaps called *expansion joints* (Figure 14-10). Similarly,

Figure 14-10
This gap is called an expansion joint; it allows the bridge to expand and contract.

concrete roadways and sidewalks are intersected by gaps, sometimes filled with tar, so that the concrete can expand freely in summer and contract in winter.

Questions

1. The Concorde supersonic airplane is 20 cm longer when in flight. Offer an explanation.
2. How would a thermometer be different if glass expanded more with increasing temperature than mercury?

Different substances expand at different rates. When two strips of different metals, say one of brass and the other of iron, are welded or riveted together, the greater expansion of one metal results in the bending shown in Figure 14-11. Such a compound thin bar is called a *bimetallic strip*. When the strip is heated, one side of the double strip becomes longer than the other, causing the strip to bend into a curve. On the other hand, when the strip is cooled, it tends to bend in the opposite direction, because the metal that expands more also shrinks more. The movement of the strip may be used to turn a pointer, regulate a valve, or close a switch.

Figure 14-11
A bimetallic strip. Brass expands (or contracts) more when heated (or cooled) than iron does, so the strip bends as shown.

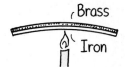

A practical application of this is the thermostat (Figure 14-12). The back-and-forth bending of the bimetallic coil opens and closes an electric circuit. When the room becomes too cold, the coil bends toward the brass side, and in so doing activates an electrical switch that turns on the heat. When the room becomes too warm, the coil bends toward the iron side, which activates an electrical contact that turns off the heating unit. Refrigerators are equipped with special thermostats to prevent them from becoming either too warm or too cold. Bimetallic strips are used in oven thermometers, electric toasters, automatic chokes on carburetors, and various other devices.

Liquids expand appreciably with increases in temperature. In most cases the expansion of liquids is greater than the expansion of solids. The gasoline overflowing a car's tank on a hot day is evidence for this. If the

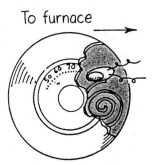

Figure 14-12
A thermostat. When the coil expands, the mercury rolls away from the electrical contacts and breaks the circuit. When the coil contracts, the mercury rolls against the contacts and completes the electrical circuit.

▶ **Answers**

1. At cruising speed (faster than the speed of sound), air friction against the Concorde raises its temperature dramatically, resulting in this significant thermal expansion.
2. The scale would have to be upside down. Do you see why?

Figure 14-13
Place a dented Ping-Pong ball in boiling water, and you'll remove the dent. Why?

tank and contents expanded at the same rate, they would expand together and no overflow would occur. Similarly, if the expansion of the glass of a thermometer were as great as the expansion of the mercury, the mercury would not rise with increasing temperature. The reason the mercury in a thermometer rises with increasing temperature is because the expansion of liquid mercury is greater than the expansion of glass.

Expansion of Water

Increase the temperature of any common liquid and it will expand. But not water at temperatures near the freezing point: ice-cold water does just the opposite! Water at the temperature of melting ice, 0°C (or 32°F), *contracts* when the temperature is increased. This is most unusual. As the water is heated and its temperature rises, it continues to *contract* until it reaches a temperature of 4°C. With further increase in temperature, the water then begins to expand; the expansion continues all the way to the boiling point, 100°C. The result of this odd behavior is shown graphically in Figure 14-14.

A given amount of water has its smallest volume—and thus its greatest density—at 4°C. The same amount of water has its largest volume—and smallest density—in its solid form, ice. (Remember, ice floats in water, evidence that it is less dense than water.) The volume of ice at 0°C is not shown in Figure 14-14. (If it were plotted to the same exaggerated scale, the graph would extend far beyond the top of the page.) After water has turned to ice, further cooling causes it to contract.

Figure 14-14
The expansion of water with increasing temperature.

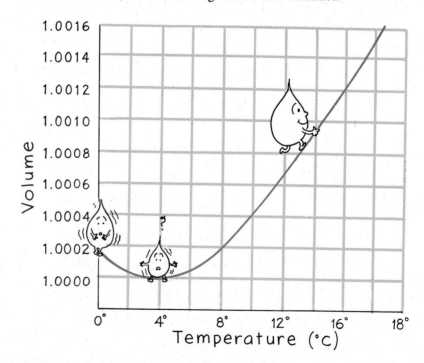

Ice has a crystalline structure. The crystals of most solids are arranged in such a way that the solid state occupies a smaller volume than the liquid state. Ice, however, has open-structured crystals (Figure 14-15). This structure results from the angular shape of the water molecules and the fact that the forces binding water molecules together are strongest at certain angles. Water molecules in this open structure occupy a greater volume than they do in the liquid state. Consequently, ice is less dense than water.

Figure 14-15
Water molecules in their crystal form have an open-structured hexagonal arrangement that results in the expansion of water upon freezing. Ice therefore is less dense than water.

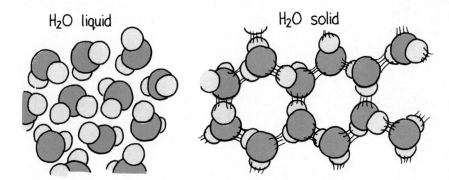

The reason for the dip in the curve of Figure 14-14 is that two types of volume changes are taking place. The open-structured crystals that make up solid ice are also present, to a much smaller extent, in ice-cold water—a "microscopic slush." At about 10°C all the ice crystals have collapsed. The left-hand graph in Figure 14-16 indicates how the volume of cold water changes because of the collapsing of the microscopic ice crystals. At the same time crystals are collapsing as the temperature increases, increased molecular motion results in expansion. This effect is shown in the center graph in Figure 14-16. Whether ice crystals are in the water or not, increased vibrational motion of the molecules increases the volume of the water. When we combine the effects of contraction and expansion, the curve looks like the right-hand graph in Figure 14-16 (or Figure 14-14).

Figure 14-16
The collapsing of ice crystals plus increased molecular motion with increasing temperature produce the overall effect of water being most dense at 4°C.

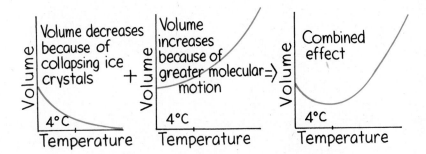

What was the precise temperature at the bottom of Lake Michigan on New Year's Eve in 1901?

This behavior of water is of great importance in nature. Suppose that the greatest density of water were at its freezing point, as is true of most liquids. Then the coldest water would settle to the bottom, and ponds would freeze from the bottom up. Pond organisms would then be destroyed in winter months. Fortunately, this does not happen. The densest water, which settles at the bottom of a pond, is 4 degrees above the freezing temperature. Water at the freezing point, 0°C, is less dense and "floats," so ice forms at the surface while the pond remains liquid below the ice. Let's examine this in more detail. Most of the cooling in a pond takes place at its surface when the surface air is colder than the water. As the surface water is cooled, it becomes more dense and sinks to the bottom. Water will "float" at the surface for further cooling only if it is equally as dense as or less dense than the water below.

Figure 14-17
As water is cooled, it sinks until the entire pond is 4°C. Only then can surface cooling to the freezing point take place without further sinking.

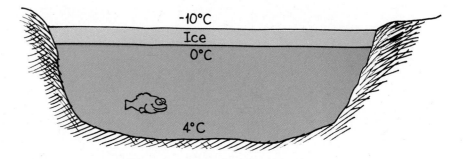

Consider a pond that is initially at, say, 10°C. It cannot possibly be cooled to 0°C without first being cooled to 4°C. And water at 4°C cannot remain at the surface for further cooling unless all the water below has at least an equal density—that is, unless all the water below is at 4°C. If the water below the surface is any temperature other than 4°C, any surface water at 4°C will be denser and will sink before it can be further cooled. So before any ice can form, all the water in a pond must be cooled to 4°C. Only when this condition is met can the surface water be cooled 3°, 2°, 1°, and 0°C without sinking. Then ice can form.

▶ **Answer**
The temperature at the bottom of any body of water that has 4°C water in it is 4°C at the bottom, for the same reason that rocks are at the bottom. Both 4°C water and rocks are more dense than water at any other temperature. Water is a poor heat conductor, so if the body of water is deep and in a region of long winters and short summers, the water at the bottom is likely a constant 4°C year round.

So we see that the water at the surface is first to freeze. Continued cooling of the pond results in the freezing of the water next to the ice, so a pond freezes from the surface downward. In a cold winter the ice will be thicker than in a milder winter. Very deep bodies of water are not ice-covered even in the coldest of winters. This is because all the water in a lake must be cooled to 4°C before lower temperatures can be reached and because the winter is not long enough for all the water to be cooled to 4°C. If only some of the water is 4°C, it lies on the bottom. Because of water's high specific heat and poor ability to conduct heat, the bottom of deep lakes in cold regions is a constant 4°C the year round. Fish should be glad that this is so.

Summary of Terms

Temperature A measure of the average kinetic energy per molecule in a substance, measured in degrees Celsius or Fahrenheit or in kelvins.

Absolute zero The lowest possible temperature that a substance may have—the temperature at which molecules of a substance have their minimum kinetic energy.

Heat The energy that flows from a substance of higher temperature to a substance of lower temperature, commonly measured in calories or joules.

Internal energy The total of all molecular energies, kinetic plus potential energy, internal to a substance.

Specific heat The quantity of heat per unit mass required to raise the temperature of a substance by 1 Celsius degree.

Review Questions

1. Why does a penny become warmer when it is struck by a hammer?

2. What happens to the temperature of air when it is rapidly compressed?

Temperature

3. What are the temperatures for freezing water on the Celsius and Fahrenheit scales? For boiling water?

4. What are the temperatures for freezing water and boiling water on the Kelvin temperature scale?

5. Why are there no negative numbers on the Kelvin scale?

6. What is meant by the statement that a thermometer measures its own temperature?

Heat

7. When you touch a cold surface, does cold travel from the surface to your hand or does energy travel from your hand to the cold surface? Explain.

8. Distinguish between temperature and heat.

9. Distinguish between heat and internal energy.

10. What determines the direction of heat flow?

Quantity of Heat

11. How is the energy value of foods determined?

12. Distinguish between a calorie and a Calorie.

13. Distinguish between a calorie and a joule.

Specific Heat

14. Which warms up faster when heat is applied, iron or silver?

15. Name three ways in which a material can absorb energy.

16. Does a substance that heats up quickly have a high or a low specific heat?

17. How does the specific heat of water compare to the specific heats of other common materials?

18. Northeastern Canada and much of Europe receive about the same amount of sunlight per unit area. Why then is Europe generally warmer in the winter?

19. By energy conservation: If ocean water cools, then does something else warm? If so, what?

20. Why is the temperature fairly constant for land masses surrounded by large bodies of water?

Expansion

21. Why will hot water poured into a drinking glass be more likely to break the glass if the glass is thick?

22. How can a bimetallic strip be used to regulate temperature?

23. Which generally expands more for increases in temperature, solids or liquids?

Expansion of Water

24. When the temperature of ice-cold water is increased slightly, does it undergo a net expansion or net contraction?

25. What is the reason for ice being less dense than water?

26. Does "microscopic slush" in water tend to make it more dense or less dense?

27. What happens to the amount of "microscopic slush" in cold water when its temperature is increased?

28. What happens to the amount of molecular motion in water when its temperature is increased?

29. At what temperature do the effects of contraction and expansion produce the smallest volume for water?

30. Why does ice form at the surface of a body of water instead of at the bottom?

Exercises

1. Why can't you establish whether you are running a high temperature by touching your own forehead?

2. Which has the greater amount of internal energy, an iceberg or a cup of hot coffee? Explain.

3. Would a common mercury thermometer be feasible if glass and mercury expanded at the same rates for changes in temperature? Explain.

4. Does it make sense to talk about the temperature of a vacuum?

5. What is *temperature* a measurement of?

6. If you drop a hot rock into a pail of water, the temperature of the rock and the water will change until both are equal. The rock will cool and the water will warm. Does this hold true if the hot rock is dropped into the Atlantic Ocean? Explain.

7. Would you expect the temperature of water at the bottom of Niagara Falls to be slightly higher than the temperature at the top of the falls? Why?

8. Why does the pressure of gas enclosed in a rigid container increase as the temperature increases?

9. Bermuda is close to North Carolina, but unlike North Carolina it has a tropical climate year round. Why is this so?

10. If the winds at the latitude of San Francisco and Washington, D.C., were from the east rather than from the west, why might San Francisco be able to grow only cherry trees and Washington, D.C., only palm trees?

11. San Francisco is warmer in winter than Washington, D.C. Why isn't it warmer in summer?

12. In addition to the to-and-fro vibrations of a molecule that are associated with temperature, some molecules can absorb large amounts of energy in the form of internal vibrations and rotations of the molecule itself. Would you expect materials composed of such molecules to have a high or a low specific heat? Explain.

13. The desert sand is very hot in the day and very cool at night. What does this tell you about its specific heat?

14. Would you or the gas company gain by having gas warmed before it passed through your gas meter?

15. A metal ball is just able to pass through a metal ring. When the ball is heated, however, it will not pass through the ring. What would happen if the ring, rather than the ball, were heated? Does the size of the hole increase, stay the same, or decrease?

16. After a machinist very quickly slips a hot, snugly fitting iron ring over a very cold brass cylinder, there is no way that the two can be separated intact. Can you explain why this is so?

17. Suppose you cut a small gap in a metal ring. If you heat the ring, will the gap become wider or narrower?

18. When a mercury thermometer is warmed, the mercury level momentarily goes down before it rises. Can you give an explanation for this?

19. One of the reasons the first light bulbs were expensive was that the electrical lead wires into the bulb were made of platinum, which expands at about the same rate as glass when heated. Why is it important that the metal leads and the glass have the same coefficient of expansion?

20. If you measure a plot of land with a steel tape on a hot day, will your measurements of the plot be larger or smaller than they actually are?

21. What was the precise temperature at the bottom of Lake Superior at 12:01 AM on October 31, 1894?

22. Suppose that water is used in a thermometer instead of mercury. If the temperature is at 4°C and then changes, why can't the thermometer indicate whether the temperature is rising or falling?

23. How does the combined volume of the billions and billions of hexagonal open spaces in the structures of ice crystals in a piece of ice compare to the portion of ice that floats above the water line?

24. How would the shape of the curve in Figure 14-14 differ if density were plotted against temperature instead of volume? Make a rough sketch.

25. Why is it important to protect water pipes so they don't freeze?

26. If cooling occurred at the bottom of a pond instead of at the surface, would a lake freeze from the bottom up? Explain.

27. If water had a lower specific heat, would ponds be more likely to freeze or less likely to freeze?

28. Suppose a bar 1 m long expands $\frac{1}{2}$ cm when heated. By how much will a bar 100 m long of the same material expand when similarly heated?

29. Steel expands 1 part in 100 000 (10^{-5}) for each Celsius degree increase in temperature. Suppose the 1.3-km main span of the Golden Gate Bridge had no expansion joints. How much longer would it be for an increase in temperature of 10°C?

30. Consider a 40 000-km steel pipe that forms a ring to fit snugly all around the circumference of the world. Suppose people along its length breathe on it so as to raise its temperature 1 Celsius degree. The pipe gets longer. It also is no longer snug. How high does it stand above ground level? (To simplify, consider only the expansion of its radial distance from the center of the earth, and apply the geometry formula that relates circumference C and radius r, $C = 2\pi r$. The result is surprising!)

Temperature and heat are quite different concepts. **Temperature** relates to the average KE of molecules — this oven and pan have a high temperature because their molecules have been energized. **Heat** is the energy that flows because of a temperature difference — my hands get more heat in a high-temperature oven than in a cooler one. But how much heat I get depends also on **conductivity**. Air is a poor conductor, so not much heat is conducted to me when I briefly stick my hands in the hot oven. Wood is a poor conductor, so I can pick up this hot pan by its wooden handle with bare hands without harm. I've seen people walk barefoot on red-hot wooden coals without harm for the same reason. Wood, whether red-hot or cold, is a poor conductor.

15 Heat Transfer

Heat transfers from warmer to cooler things. If several things of different temperatures are in contact, those that are warm become cooler and those that are cool become warmer. They tend to reach a common temperature. This equalizing of temperature occurs in three ways: *conduction*, *convection*, and *radiation*.

Conduction

Hold one end of an iron nail in a flame. It will quickly become too hot to hold. The heat enters the metal nail at the end kept in the flame and is transmitted along its whole length. The transmission of heat in this manner is called **conduction**. The fire causes the molecules at the heated end of the nail to move more rapidly. Because of this increased motion, these molecules and free electrons collide with their neighbors, and so on. This process continues until the increased motion has been transmitted to all the molecules and the entire body has become hot. Heat conduction occurs by electron and molecular collisions.

How well an object conducts heat depends on the electrical bonding of the molecular structure. Solids whose molecules have a "loose" outer electron conduct heat (and electricity) well. Metals have the "loosest" outer electrons and are the best conductors of heat and electricity for this reason. Silver is the best, copper is next, and, among the common metals, aluminum and then iron are next in order. Wool, wood, straw, paper, cork, and Styrofoam are poor conductors of heat. The outer electrons in the molecules of these materials are firmly attached. Poor conductors are called *insulators*.

Liquids and gases, in general, are poor conductors. Air is a very poor conductor. Porous substances, which have many small air spaces, are poor conductors and good insulators. The good insulating properties of such things as wool, fur, and feathers are largely due to the air spaces they contain. Be glad that air is a poor conductor; if it weren't, you'd feel quite chilly on a 20°C (68°F) day!

Snow is a poor conductor (good insulator), about the same as dry wood, and hence is popularly said to keep the earth warm. Its flakes are formed of crystals, which collect into feathery masses, imprisoning air and thereby interfering with the escape of heat from the earth's surface. Eskimo winter dwellings are shielded from the cold by their snow covering. Animals in the forest find shelter from the cold in snowbanks and

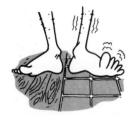

Figure 15-1
The tile floor feels colder than the wooden floor, even though both floor materials are the same temperature. This is because tile is a better conductor than wood, and heat is more readily conducted from the foot that makes contact with the tile.

269

Figure 15-2

Snow patterns on the roof of a house reveal the conduction, or lack of conduction, of heat through the roof.

in holes in the snow. The snow doesn't provide them with heat; it simply prevents the heat they generate from escaping.

Heat is transmitted from a higher to a lower temperature. We often hear people say they wish to keep the cold out of their homes. A better way to put this is to say that they want to prevent the heat from escaping. There is no "cold" that flows into a warm home. If the home becomes colder, it is because heat flows out. Homes are insulated with rock wool or spun glass to prevent heat from escaping rather than to prevent cold from entering. Interestingly enough, insulation of whatever kind does not actually prevent heat from getting through it; it simply slows the rate at which heat penetrates. Even a well-insulated warm home in winter will gradually cool. Insulation delays the transfer of heat.

Question ▶ In desert regions that are hot in the daytime and cold at nighttime, the walls of houses are usually made of mud. Why is it important that the mud walls be thick?

▶ **Answer**

A wall of appropriate thickness provides a proper amount of time for heat flow from outer to inner wall surfaces to produce maximum interior temperature during the sleeping hours, when it is cold outside, and minimum interior temperature during midday, when it is hot outside.

Convection

Liquids and gases transmit heat mainly by **convection**, which is heat transfer by the actual motion of the fluid—currents. Convection may occur in all fluids, whether liquids or gases. Whether we heat water in a pan or warm air in a room, the process is the same (Figure 15-3). If the fluid is heated from below, its molecules increase in speed; the fluid becomes less dense and is pushed up by the denser, cooler fluid that takes its place at the bottom. In this way, convection currents keep the fluid stirred up as it heats. Convection currents occur in the atmosphere, and to understand this we must first understand why warm air rises—then why it cools.

Figure 15-3

(a) Convection currents in air. (b) Convection currents in liquid.

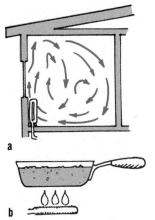

a

b

Figure 15-4

A heater at the tip of the submerged J-tube produces convection currents, which are revealed as shadows as light is deflected in water of different temperatures.

Why Warm Air Rises

We all know that warm air rises. From our study of buoyancy we understand why this is so. Warm air expands, becomes less dense than the surrounding air, and is buoyed upward like a balloon. The buoyancy is in an upward direction because the air pressure below a region of warmed air is greater than the air pressure above. And the warmed air rises because the buoyant force is greater than its weight.

We can understand the rising of warm air from a different point of view—by considering the motion of individual molecules. Consider a fairly large region of identical gas molecules. Because of gravity we would find more molecules near the bottom of our region than near the top; the gas would be slightly denser toward the ground. Suppose the region is of uniform temperature; then each molecule, on the average, has the same kinetic energy and the same average velocity. Each molecule, therefore, has the same tendency to migrate throughout the region. Suppose now that we introduce a faster-moving molecule—a "hot" one. Until it gives up its excess energy to slower-moving molecules, it will migrate farther and more rapidly than any of its neighbors. If our sample molecule is placed in the middle of our region, it will bump into and rebound from

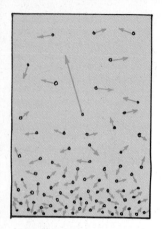

Figure 15-5
A fast-moving molecule tends to migrate toward the region of least obstruction—upward. Warm air therefore rises.

molecules in all directions. It will rebound, however, from a greater number of molecules whenever it happens to be moving downward rather than upward. This is because the density of molecules is greater below; there is more opposition to a downward migration than to an upward migration where the air is less dense. Furthermore, when our "hot" molecule moves in an upward direction, it travels farther before making a collision than when it travels downward. We say it has a longer "mean-free path" when moving upward. So our faster-moving molecule will tend to bumble upward in its random jostling.

We have simplified the idea of rising warm air by considering the behavior of a single molecule. A single fast-moving molecule would, of course, soon share its excess energy and momentum with its less energetic neighbors and would not rise very high.* However, if we start with a large cluster of energetic molecules, many of these will rise to appreciable heights before their energy and momentum dissipate.

Why Expanding Air Cools

Warm air rises, so we might expect the atmosphere to be warmer with increasing altitude; we might expect the mountaintops to be warm and green and the valleys below to be cold and snow-covered. But this is not the case; the atmosphere is cooler with increasing altitude. As strange as it may seem, a primary reason for the cooling at higher altitudes has very much to do with the fact that warm air rises. Warm air is pushed from a region of greater atmospheric pressure at the ground to a region of lesser pressure above. And while it is rising through regions of lesser pressure, it expands. Expansion of air results in cooling. The opposite process of compressing air warms it. If you've ever pumped air into a bicycle tire with a hand pump, you probably noticed that the pump soon became hot. Your pumping action served to increase the speed of the molecules of air, just as a paddle slamming against Ping-Pong balls increases the speed of the balls. The act of compressing air warms it. The converse is true when air expands.

Let's look at the compression and expansion of air in more detail. When molecules collide with others that are approaching at greater speeds, their rebound speeds are increased; the converse is that when molecules collide with others that are *receding*, their rebound speeds are lessened. This is easy to see: although a Ping-Pong ball picks up speed when struck

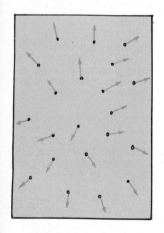

Figure 15-6
A molecule in a region of expanding air collides primarily with receding molecules, not approaching ones. Its velocity of rebound therefore lessens with each collision and results in a cooling of expanding air.

*Interestingly enough, a single helium atom will rise in the atmosphere for exactly the reasons stated in the previous paragraph, because even with no excess kinetic energy its average speed will always be considerably greater than the speeds of neighboring heavier molecules of nitrogen and oxygen. Can you see that in a mixture of gaseous particles of various masses at the same temperature, the particles of least mass have the greatest speeds? (How else would they have the same temperature, that is, the same kinetic energy?) And can you see that lightweight helium with its greater speed will migrate more than its heavier neighbors, bumble its way to the top of the atmosphere, and escape into outer space? Can you see why helium, which in fact is the seventh most common gas in the earth's atmosphere, doesn't normally exist in the lower atmosphere?

by an approaching paddle, it slows down when it rebounds from a receding paddle. In a region of air that is expanding, molecules will collide, on the average, with more molecules that are receding than are approaching. (This is just the opposite of what happens when air is compressed.) Thus, in expanding air, the average speed of the molecules decreases and the air cools. Where does the energy go in this case? It actually goes into the work done on the surrounding air as the expanding air pushes outward. In this way internal energy is diluted. So we see that the energy per volume and the temperature is less.*

A common misconception about temperature and molecular motion is that heat is produced by the number of collisions molecules make with one another—that the more frequently molecules collide, the higher the temperature of the gas will be. This is not true. Although individual molecules may gain or lose speed in approaching and receding collisions, the total number of molecules bouncing off one another have the same total energy and momentum before and after a collision. Temperature is a measure of kinetic energy, not of collision rates. The temperature of gas would be no different if all the molecules were able to move without colliding with one another. Put another way, when a gas is heated, the molecules collide more often. We can say they collide more often because they're heated. But we can *not* say that they are heated because they collide more often. ("Cause produces effect" is a one-way street!)

Thus, compression heats air; expansion cools it. But once compressed or expanded, it soon comes to a temperature equilibrium with its surroundings. Put your hand on a tank of compressed air and you will find that it has the same temperature as its surroundings. Put your hand in the path of the same air as it escapes and expands from a nozzle and you will find a considerably lower temperature. A more dramatic example occurs with steam that expands through the nozzle of a pressure cooker (Figure 15-7). The cooling effect of both expansion and rapid mixing with cooler air allows you to hold your hand comfortably in the jet of condensed vapor. (*Caution:* If you try this, be sure to place your hand high above the nozzle at first and then lower it to a comfortable distance—if you put your hand at the nozzle where no steam appears, watch out! Steam is invisible, and that's where it is before it has sufficiently expanded and cooled. The cloud of "steam" you see is actually condensed water vapor—much cooler.)

Convection currents stirring the atmosphere result in winds. Some parts of the earth's surface absorb heat from the sun more readily than others, and as a result the air near the surface is heated unevenly and convection currents form. This is evident at the seashore. In the daytime the shore warms more easily than the water; air over the shore is pushed up (we

Figure 15-7
The hot steam expands from the pressure cooker and is cool to Millie's touch.

---

*When a gas expands without heat input, the process is called *adiabatic expansion*. A gas that expands adiabatically does external work, and its internal energy and hence its temperature decrease. Conversely, temperature increases with adiabatic compression of air—the heating of a bicycle pump, for example. More about this in Chapter 17.

say it rises) by cooler air from above the water taking its place. The result is a sea breeze. At night the process reverses because the shore cools off more quickly than the water, and then the warmer air is over the sea (Figure 15-8). Build a fire on the beach and you'll notice that the smoke sweeps inward during the day and seaward at night.

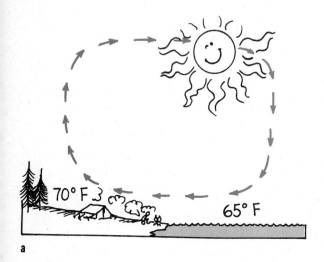
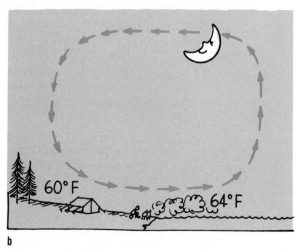

Figure 15-8
Convection currents produced by unequal heating. (*a*) Warmed air rises and cools as it expands. It then sinks and flows toward the warmed area to replace the air that has risen. The land is warmer than the water in the day (lower specific heat) and (*b*) cooler than the water at night, so the direction of air flow reverses.

Radiation

Heat from the sun somehow passes through space, then through the atmosphere and warms the earth's surface. This heat does not pass through the atmosphere by conduction, for air is a poor conductor. Nor does it pass through by convection, for convection begins only after the earth is warmed. We also know that neither convection nor conduction is possible in the empty space between our atmosphere and the sun. We can see that heat must be transmitted some other way—by **radiation**.* The energy so radiated is called *radiant energy.*

Although the sun is an obvious source of radiant energy, all objects continually radiate energy in a mixture of wavelengths. Objects at low temperatures emit long waves, just as long lazy waves are produced when you shake a rope with little energy (Figure 15-9). Higher-temperature

---

*Do not confuse radiation with radioactivity—reactions that involve the atomic nucleus and are characteristic of nuclear power plants and the like. Radiation here is electromagnetic radiation, "heat" waves of low-frequency light, which we will treat in detail in Parts 4, 5, and 6.

objects emit waves of shorter wavelengths. Objects of everyday temperatures emit waves mostly in the long-wavelength infrared region, between radio and light waves. It is the shorter-wavelength infrared radiation that our skin commonly experiences as heat. So it is common to refer to infrared radiation as *heat radiation*.

Figure 15-9
Shorter wavelengths are produced when the rope is shaken more vigorously.

Objects that are hot enough radiate some of their energy in the range of visible light. At a temperature of about 600°C an object begins to emit the longest waves we can see, red light. Higher temperatures produce a yellowish light. At 1200°C all the different waves to which the eye is sensitive are emitted and we see an object as "white hot." At the same time, however, the object continues to emit the long waves, thus radiating a wide variety of wavelengths. When any kind of matter absorbs these waves, the waves set the electrons in the matter into vibration, and temperature increases. If radiation in the wave range of infrared falls on our skin, it may excite the sensation of warmth. In this way, heat is transmitted from the radiating object to the absorbing object.

Figure 15-10
Types of radiant energy (electromagnetic waves).

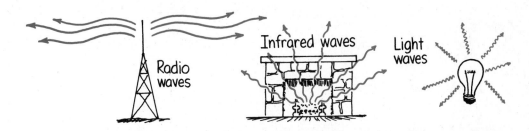

Question ▶ You can hold your fingers beside the candle without harm, but not above the flame. Why?

▶ **Answer**
Heat travels upward by air convection. Since air is a poor conductor, very little heat travels sideways.

Emission, Absorption, and Reflection of Radiation

All objects continually radiate energy. Why, then, doesn't the temperature of all objects continually decrease? The answer is that all objects also continually absorb radiant energy. If an object is radiating more energy than it is absorbing, its temperature does decrease; but if an object is absorbing more energy than it is emitting, its temperature increases. An object that is warmer than its surroundings emits more energy than it receives, and therefore it cools; an object colder than its surroundings is a net gainer of energy, and its temperature therefore increases. An object whose temperature is constant, then, emits as much radiant energy as it receives. If it receives none, it will radiate away all its available energy, and its temperature will approach absolute zero.

The rate at which an object radiates or absorbs radiant energy depends on the nature of the object and the difference between its temperature and the surrounding temperature. Emission and absorption occur at the surface of an object. A rough surface is a better absorber and emitter than a smooth surface because its irregularities give it more area, and its microscopic hills and valleys enable multiple scattering of radiation at the surface to occur. This is true both for long-wavelength heat radiation and shorter-wavelength light radiation. So we can see that a very good absorber of radiation reflects very little light and therefore appears black. Things that reflect little visible light appear black to us. A perfect absorber reflects no radiation and appears perfectly black.

A body that absorbs all the radiation incident upon it is called a *black-body*. Holes and cavities are practical examples of blackbodies. A keyhole in a closet door, for example, appears perfectly black. Small holes appear black because the radiation that enters is reflected from the inside walls many times and is partly absorbed at each reflection until none remains (Figure 15-12). For the same reason, the pupil of the eye is black (except when illuminated directly—with a flash camera, for example—in which case it appears pink).

Figure 15-11
The hole looks perfectly black and indicates a black interior, when in fact the interior has been painted a bright white.

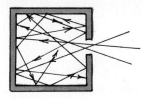

Figure 15-12
Radiation that enters the cavity has little chance of leaving before it is completely absorbed.

Figure 15-13
When the containers are filled with hot (or cold) water, the blackened one cools (or warms) faster.

If the walls of the hole or cavity are heated, the radiation given off in all directions is more than the incoming radiation, and there is a net escape of radiation from the hole. So we find that good absorbers are also good emitters. And poor absorbers are poor emitters. For example, a radio antenna constructed to be a good emitter of radio waves will also by its very design be a good receiver of radio waves. And a poorly designed transmitting antenna will also be a poor receiver. The same holds true for the atomic and molecular construction of matter. A blackbody is both an ideal absorber and an ideal radiator.

Whether a surface plays the role of net radiator or net absorber depends on whether its temperature is above or below the temperature of its surroundings. If the surface is hotter than the surrounding air, for example, it will be a net radiator and will cool. If the surface is colder than the surrounding air, it will be a net absorber and will become warmer.

The heat radiation that falls on a surface and is absorbed obviously cannot also be reflected; likewise, the heat radiation that falls on a surface and is reflected cannot be absorbed. Hence, a good absorber is a poor reflector of radiation, and a good reflector is a poor absorber.

Polished or mirrorlike surfaces are poor absorbers of both visible and heat radiation. This can be illustrated by placing thermometers in a pair of metal containers of the same size and shape, one having a brightly polished surface and the other a blackened surface (Figure 15-13). If they are filled with hot water, we will find that the container with the blackened surface cools faster. The blackened surface is a better radiator. Coffee will stay hot longer in a polished pot than in a blackened one. If we repeat the same experiment, but this time fill each container with ice water, we will find that the container with the blackened surface warms up faster. The blackened surface is also a better absorber of radiant energy.

Clean snow is a good reflector and therefore does not melt rapidly in sunlight. If the snow is dirty, it absorbs radiant energy from the sun and melts faster. Dropping black soot by aircraft on snowed-in mountainsides is a technique sometimes used in flood control. Controlled melting at favorable times rather than a sudden runoff of melted snow is thereby accomplished.

Light-colored buildings stay cooler in summer because they reflect much of the incoming radiant energy. In winter they stay warmer because they are also poor emitters and retain more of their internal energy. Paint your house a light color.

Question Why does a good absorber of radiant energy appear black?

▶ **Answer**

A good absorbing surface removes all visible colors when it converts all the radiant energy incident upon it to heat. It cannot both absorb and reflect the same wavelengths at the same time, so the absence of reflected light makes it black.

Figure 15-14
Patches of frost crystals
betray the hidden entrances
to mouse burrows. Each
cluster of crystals is frozen
mouse breath!

Cooling at Night by Radiation

Things that radiate more energy than they receive become cooler. This happens at night when solar radiation is absent. Objects out in the open radiate energy into the night and, because of the absence of warmer materials, may receive very little energy in return. They give out more energy than they receive and become cooler. If the object is a good conductor of heat—like metal, stone, or concrete—heat from the ground will be conducted to it, somewhat stabilizing its temperature. But materials such as wood, straw, and grass are poor conductors, and little heat is conducted into them from the ground. These insulating materials are net radiators and get *colder than the air*. It is common for frost to form on these kinds of materials even when the temperature of the air does not go down to freezing. Have you ever seen a frost-covered lawn or field on a chilly but above-freezing morning before the sun is up? The next time you see this, notice that the frost forms only on the grass, straw, or other poor conductors, while none forms on the cement, stone, or other good conductors.

Snow is a good example. During the day the snow gains very little heat energy from the sun because of its reflectivity, and at night it loses energy rapidly by radiating infrared radiation to space. Because of snow's poor conductivity, the ground conducts very little heat to it, and the surface of the snow becomes cooler than the surrounding air. Snow therefore has a significant cooling effect on the earth and atmosphere. The bottom of the snow remains at about ground temperature. That's why Midwestern wheat farmers like deep snows in winter—to protect their wheat fields from the harsh, cold weather.

Newton's Law of Cooling

An object at a different temperature from its surroundings will eventually come to a common temperature with its surroundings. A hot object will cool as it warms the surroundings, and an object cooler than its surroundings will warm up as the surroundings cool. The cooling *rate* of an object depends on how much hotter the object is than the surroundings. The temperature change per second of a hot apple pie will be more if the hot pie is put in a cold freezer than if put on the kitchen table.

When the pie cools in the freezer, the temperature difference is greater. A warm home will leak heat to the cold outside at a greater rate when there is a large difference between the temperature inside and outside. Keeping the inside of your home at a high temperature on a cold day is more costly than keeping it at a lower temperature; the smaller the temperature difference, the lower the cooling rate. At ordinary temperatures, the cooling rate—by conduction, convection, or radiation—is approximately proportional to the temperature difference, ΔT, between the object and its surroundings.

$$\text{Rate of cooling} \sim \Delta T$$

This is known as **Newton's law of cooling**. (Guess who was the first to establish this.)

A similar law holds for heating. If an object is cooler than its surroundings, its rate of warming up is also proportional to ΔT. Frozen food will warm up faster in a warm room than in a cold room.

The rate of cooling we experience on a cold day can be increased by the added convection of wind. We speak of this in terms of *wind chill*. For example, a wind chill of $-20°C$ means we are losing heat at the same rate as if the temperature were $-20°C$ without wind.

Question ▶	Since a hot cup of coffee loses heat more rapidly than a lukewarm cup of coffee, would it be correct to say that a hot cup of coffee will cool to room temperature before a not-so-hot cup of coffee?

The Greenhouse Effect

The earth and its atmosphere are warmed by radiant energy from the sun. The earth, in turn, emits what is called *terrestrial radiation*, much of which it loses to outer space. Changes in the average temperature of the earth are dictated by the difference in radiant energy entering and radiant energy leaving (Figure 15-15). During the last 500 000 years, the average temperature of the earth has fluctuated between 19°C and 27°C and is presently at the high point. The earth's temperature increases when either the radiant energy coming in increases or there is a decrease in the escape of terrestrial radiation.

Radiant energy from the sun, like that from any source as hot as the surface of the sun, is mainly in the range of ultraviolet, visible light, and short-wavelength infrared. Most of these short wavelengths pass rather freely through the atmosphere without being absorbed and are absorbed

▶ **Answer**

No! Although the rate of cooling is greater for the hotter cup, it has farther to cool to reach thermal equilibrium. The extra time is equal to the time it takes to cool to the initial temperature of the lukewarm cup of coffee. Cooling *rate* and cooling *time* are not the same thing.

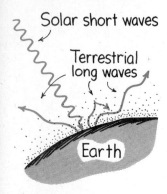

Figure 15-15
The hot sun emits short waves, and the cool earth re-emits long waves.

at the earth's surface. Radiant energy from the earth, like that from any cooler source, is mainly in the range of long-wavelength infrared. These longer wavelengths are not transmitted as freely through the atmosphere. Much of the terrestrial radiation is absorbed by water vapor and carbon dioxide in the earth's atmosphere. The atmosphere radiates much of this energy back to the earth and keeps the earth's temperature higher than it would be otherwise.

Because florists' greenhouses retain energy in a somewhat similar manner, the overall process is known as the **greenhouse effect** (Figure 15-16). Glass has the property of being transparent to waves of visible light and opaque to ultraviolet and infrared. Glass acts as sort of a one-way valve. It allows short waves of visible light to enter and prevents long waves from leaving. So when radiant energy from the sun (except for the ultraviolet) enters through the glass roof, it is absorbed by the interior of the greenhouse—mainly the soil and plants. The soil and plants, in turn, emit infrared radiation. This energy cannot get through the glass, and the greenhouse warms.

Figure 15-16
Glass is transparent to short-wavelength radiation but opaque to long-wavelength radiation due to the sun's high temperature. Reradiated energy from the plant is long wavelength because the plant has a relatively low temperature.

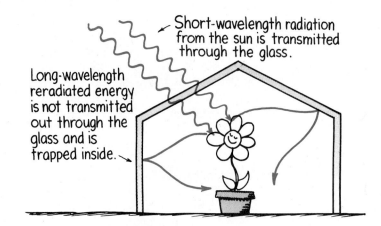

Short-wavelength radiation from the sun is transmitted through the glass.

Long-wavelength reradiated energy is not transmitted out through the glass and is trapped inside.

Question ▶ What does it mean to say that the greenhouse effect is like a one-way valve?

Interestingly enough, most of the warmth in a greenhouse is due to the role the enclosure plays in preventing convection. Outside the greenhouse, air warmed by contact with the ground rises and is continually being replaced via convection currents. But inside the greenhouse, the

▶ **Answer**
The transparent material, atmosphere for the earth and glass for the florist's greenhouse, passes only incoming short waves and blocks outgoing long waves. As a result, radiant energy is trapped within the "greenhouse."

soil warms only the air within the enclosure. This limited amount of air warms easily. So we find the earth is considerably warmed by the greenhouse effect, while ironically greenhouses are not.

Solar Power

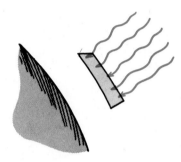

Figure 15-17
Over each square meter that is perpendicular to the sun's rays at the top of the atmosphere, the sun pours 1400 J of radiant energy each second. Hence the solar constant is 1.4 kJ/s/m², or 1.4 kW/m².

Step from the shade into the sunshine and we're noticeably warmed. The warmth we feel isn't so much because the sun is hot, for its surface temperature of nearly 6000°C is no hotter than the flames of some welding torches; we are warmed principally because the sun is so big. As a result, it emits enormous amounts of radiant energy, less than one part in a billion of which reaches the earth. The amount of radiant energy received each second over each square meter that is at right angles to the sun's rays at the top of the atmosphere is 1400 joules (1.4 kJ) (Figure 15-17). This amount of energy is called the **solar constant**. Expressed in terms of **solar power**, this is 1.4 kilowatts per square meter (1.4 kW/m²). The amount of solar power that reaches the ground is attenuated by the atmosphere and reduced by nonperpendicular elevation angles of the sun. As a result, the United States averages over the four seasons about 13 percent of the solar constant (0.18 kW/m²). This rate of heating corresponds to about twice the energy needed per day to heat or cool the average American house. This is part of the reason we see more and more homes using solar power for domestic heating and cooling.

Solar heating needs a distribution system to move solar energy from the collector to the storage or living space. When the distribution system requires external energy to operate fans or pumps, we have an active system. When the distribution is by natural means (conduction, convection, or radiation), we have a passive system. Even in the northern states, solar homes with either active or passive systems are essentially problem-free and economical.

On a larger scale, the problems of utilizing solar power are greater. First, there is the fact that no energy arrives at night. This calls for supplemental sources of energy or efficient solar-energy storage devices

Figure 15-18
Solar water heaters are covered with glass to provide a greenhouse effect, which further heats the water. Why are the collectors painted black?

Figure 15-19
A Boeing conception of a photovoltaic power satellite being constructed in low earth orbit. A space-shuttle orbiter (upper right) docks at the facility's assembly bay. To the left, an upper stage of a massive-lift launch vehicle approaches the facility to discharge its cargo of construction materials. The weightlessness in orbit allows the use of large weblike structures of a kind that would be crushed if used on earth. The satellite would be deployed in geosynchronous orbit after completion. The electricity produced would be converted to microwaves and beamed to earth.

Variations in weather, particularly cloud cover, produce a variable energy supply from day to day and from season to season. Even in clear daylight hours, the sun is high in the sky only part of the day. Solar-energy collecting and concentration systems, whether arrays of mirrors or photovoltaic cells, at this writing are not yet competitive with the costs of electrical power generation by conventional power sources. Projections indicate the story may be different by the turn of the century.

Of considerable interest is the idea of harvesting solar energy at the solar constant 24 hours per day using solar power plants in orbit, where weightlessness and the absence of wind and rain would permit the building of huge structures of low mass. These power plants could consist of giant banks of photovoltaic cells in geosynchronous orbits (similar to the orbits of communications satellites), which would render them always stationary with respect to any desired location on earth. These cells could collect solar energy continuously and convert it to electricity that would be fed to microwave generators aboard the satellites. The microwaves would be beamed to earth and picked up by antennae that could be located just about anywhere—on land, at sea, or in the desert.*

Whether large-scale solar plants are better located on earth or in space is debatable. The technological proficiency is now available for doing either, however, and is beyond debate. The essential considerations involve economic priorities rather than technology.

Figure 15-20
A climate-controlled city beneath a microwave-receiving antenna.

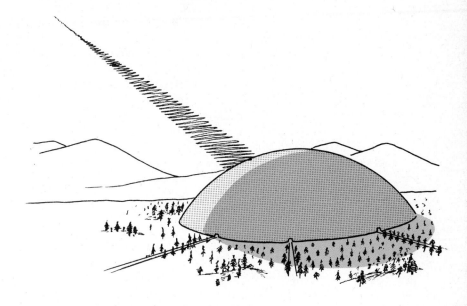

Excess Heat Problem

Figure 15-21
All the energy we consume ultimately becomes heat.

A radiative equilibrium exists that balances the amount of radiation the earth receives from the sun with the radiation the earth emits into space. This equilibrium results in an average temperature that supports life as we know it. The power generation in the present century is a new factor in the radiative equilibrium. We are all familiar with the problem of thermal pollution that is a by-product of power generation plants. The earth's environment is the heat sink for wasted energy. Power plants in orbit will not solve the problem. The amount of heat put into the earth's environment relates to not only power production but also energy consumption.

Suppose, for example, that enough power to supply all our needs were beamed to earth by orbiting solar power stations, and even suppose that all thermally polluting facilities—from steel mills to manufacturing plants—were also located in space. Then the heat sink for these facilities would be outer space rather than the planet Earth. This would significantly cut down the buildup of heat on earth, but as long as more and more energy is consumed on earth, more and more heat is the end product. When you make your toast in the morning, the extra heat your kitchen or breakfast

*The receiving antenna for such an arrangement might be a network of wires with about 80% open space. Sunlight would pass through it, but the microwaves would be absorbed by the antenna, making it safe for occupants beneath it. If we ever have cities covered by huge geodesic domes of the type proposed by Buckminster Fuller, these domes might be effective microwave-receiving antennae, since they would be transparent to sunlight and opaque to microwaves.

nook receives has little to do with the nature of the power plant that produced the electricity. All the energy we consume, whether from making toast, operating a TV, or running a power saw, ultimately becomes heat.

However you look at it, increased energy consumption on earth results in increased heating of the earth. This increase in heat must be carried away by terrestrial radiation, and to do this, the temperature of the radiator, the earth, increases—and climate changes. Warmer oceans result in increased evaporation and increased snowfalls in polar regions, which in turn result in increased glacial growth. The fraction of the earth that is presently beneath glaciers is about equal to the total area used for farmlands. Glacial growth and the corresponding larger areas of white snow reflect more solar radiation, which may well lead to a significant drop in global temperature. So overheating the earth might well trigger the next ice age! Or it might not. We don't know.

We can speculate about the long-term effects of worldwide climate changes. On one hand, it may turn out that overall changes will be tolerable and that civilizations will adapt and continue to function. On the other hand, the effects may be intolerable and be the final push to the colonization of space, where communities in orbit could easily control their temperatures by radiating excess energy into the enormous heat sink of space. Would space colonies thermally pollute the solar system in time? No, the energy used in a space facility is intercepted from a tiny fraction of the energy radiated from the sun, and whatever is discharged partially fills in what was taken away. It simply goes back to where it came from in the first place. Regulating the temperature of a colony in the vacuum of space should be an achievable task; doing the same on earth surrounded by its insulating atmosphere is a different story. On earth we must question the idea of continued growth. (Read Appendix IV, "Exponential Growth and Doubling Time"—important stuff.)

Thermos Bottle

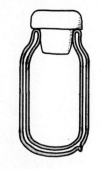

Figure 15-22
A Thermos bottle.

We can briefly summarize the ways in which heat is transferred by considering a device that inhibits these three methods—the common vacuum, or Thermos, bottle. The Thermos bottle is double-walled glass, with a vacuum between the walls. The two inner glass surfaces facing each other are silvered. When a hot liquid is poured into the bottle, it remains at nearly the same temperature for hours. This is because the transfer of heat by conduction, convection, and radiation is severely inhibited.

1. Heat transfer by *conduction* through the vacuum is impossible. Some heat escapes by conduction through the glass and stopper, but this is a slow process because glass and plastic or cork are poor conductors.
2. The vacuum also prevents heat loss through the walls of the bottle by *convection*.
3. Heat loss by *radiation* is prevented by the silvered surfaces of the walls, which reflect heat waves back into the bottle.

Summary of Terms

Conduction The transfer and distribution of heat energy that moves from molecule to molecule within a substance.

Convection The transfer of heat energy in a gas or liquid by means of currents in the heated fluid. The fluid moves, carrying energy with it.

Radiation The transfer of energy at the speed of light by means of electromagnetic waves.

Newton's law of cooling The rate of loss of heat with time from an object is proportional to the excess temperature of the substance over the temperature of its surroundings.

Greenhouse effect The heating effect of a medium such as glass or the earth's atmosphere that is transparent to the short-wavelength radiation of sunlight but opaque to long-wavelength terrestrial radiation. Energy of sunlight that enters the glass of a florist's greenhouse or the atmosphere of the earth is absorbed and reradiated at a longer wavelength that is consequently trapped, which produces heating.

Solar constant 1400 J/m^2 received from the sun each second at the top of the earth's atmosphere; expressed in terms of power, 1.4 kW/m^2.

Solar power Energy per unit time derived from the sun.

Review Questions

1. What are the three common ways in which heat is transferred?

Conduction

2. What is the role of "loose" electrons in heat conductors?

3. Distinguish between a conductor and an insulator.

4. Why does room-temperature tile feel cooler to the bare feet than the wooden floor?

5. Why are materials such as wood, fur, feathers, and even snow good insulators?

6. How does a blanket keep you warm on a cold night, even though it is not really a source of energy?

7. Why do we say that cold is not a tangible thing?

Convection

8. How is heat transferred from one place to another by convection?

Why Warm Air Rises

9. How does buoyancy relate to convection?

10. Why are fast-moving air molecules more likely to migrate upward than downward?

Why Expanding Air Cools

11. What happens to the pressure of air as it rises? What happens to its volume? What happens to its temperature?

12. How are the speeds of molecules of air affected when they are compressed by the action of a tire pump?

13. How are the speeds of molecules of air affected when a volume of air undergoes rapid expansion?

14. Does the rate at which molecules in a gas collide affect temperature, or is it the other way around? Explain.

15. Why is Millie's hand not burned when placing it above the escape valve of the pressure cooker (Figure 15-7)?

16. Why does the direction of coastal winds change from day to night?

Radiation

17. What exactly is radiant energy?

18. How do the wavelengths of radiant energy vary with the temperature of the radiating source?

Emission, Absorption, and Reflection of Radiation

19. Since all objects are absorbing energy from their surroundings, why doesn't the temperature of all objects continually increase?

20. Since all objects are emitting energy to their surroundings, why doesn't the temperature of all objects continually decrease?

21. What determines whether an object is a net absorber or net emitter?

22. Why does the pupil of the eye appear black?

23. Is a good absorber of radiation a good emitter or a poor emitter?

24. Which will normally cool faster, a black pot of hot water or a silvered pot of hot water? Explain.

25. Which will normally warm faster, a black pot of cold water or a silvered pot of cold water? Explain.

Cooling at Night by Radiation

26. What happens to the temperature of something that radiates energy without absorbing the same amount in return?

27. Will a good conductor that is in contact with the relatively warm earth become significantly colder than the earth when it radiates energy? Why or why not?

28. Will a good insulator that is in contact with the relatively warm earth become significantly colder than the earth when it radiates energy? Why or why not?

Newton's Law of Cooling

29. Why will a can of beverage cool faster in the freezer compartment than in the main part of a refrigerator?

30. Which will undergo the greater rate of cooling, a red-hot poker in a warm oven or a red-hot poker in a cold room (or do both cool at the same rate)?

31. Does Newton's law of cooling apply to warming as well as cooling?

The Greenhouse Effect

32. What is meant by *terrestrial radiation*?

33. Why is radiant energy from the sun composed of short waves and terrestrial radiation composed of relatively longer waves?

Solar Power

34. Distinguish between *active* and *passive* solar heating systems.

35. When will the technology exist for solar power stations in orbit?

36. How can solar energy from satellites be transferred to earth?

Excess Heat Problem

37. How does worldwide energy consumption relate to the average temperature of the world?

38. Will space colonies that get their energy from solar power not thermally pollute the solar system? Defend your answer.

The Thermos Bottle

39. What is the function of the silver surfaces of the Thermos bottle?

40. Heat cannot readily escape a Thermos bottle, so hot things inside stay hot. Will cold things inside a Thermos bottle likewise stay cold? Explain.

Home Projects

1. Hold the bottom end of a test tube full of cold water in your hand. Heat the top part in a flame until it boils. The fact that you can still hold the bottom shows that water is a poor conductor of heat. This is even more dramatic when you wedge chunks of ice at the bottom; then the water above can be brought to a boil without melting the ice. Try it and see.

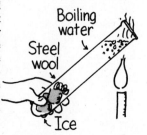

2. If you live where there is snow, do as Benjamin Franklin did nearly 2 centuries ago and lay samples of light and dark cloth on the snow. Note the difference in the rate of melting beneath the cloths.

3. Wrap a piece of paper around a thick metal bar and place it in a flame. Note that the paper will not catch fire. Can you figure out why? (Paper generally will not ignite until its temperature reaches 233°C.)

Exercises

1. Why is it difficult to estimate the temperature of things by touching them?

2. If 70°F air feels warm and comfortable to us, why does 70°F water feel cool when we swim in it?

3. At what common temperature will a block of wood and a block of metal both feel neither hot nor cold to the touch?

4. If you hold one end of a metal nail against a piece of ice, the end in your hand soon becomes cold. Does cold flow from the ice to your hand? Explain.

5. Silver is a very good conductor of heat. Is this quality favorable or unfavorable for silverware? Explain.

6. Why do restaurants serve baked potatoes wrapped in aluminum foil?

7. Many tongues have been injured by licking a piece of metal on a very cold day. Why would no harm result if a piece of wood were licked on the same day?

8. Wood is a better insulator than glass. Yet fiberglass is commonly used as an insulator in wooden buildings. Explain.

9. Visit a snow-covered cemetery and note the snow does not slope upward against the gravestones, but instead forms depressions as shown. Can you think of a reason for this?

10. If you were caught in freezing weather with only a candle for a heat source, would you be warmer in an Eskimo igloo or in a wooden shack?

11. When it is daytime on the moon, the moon-rock surface is hot enough to melt solder. Why is this not so on earth?

12. If you wish to cool something by placing it in contact with ice, should you put it on top of the ice block or put the ice block on top of it?

13. You can bring water in a paper cup to a boil by placing it in a hot flame. Why doesn't the paper cup burn?

14. Why can you comfortably hold your fingers close beside a candle flame, but not very close above the flame?

15. Why is it that you can safely hold your bare hand in a hot pizza oven for a few seconds, but if you momentarily touch the metal insides you'll burn yourself?

16. Wood conducts heat very poorly—it has a very low conductivity. Does wood still have a low conductivity if it is hot? Could you safely grab the wooden handle of a pan from a hot oven with your bare hand? Although the pan handle is hot, does much heat conduct from it to your hand if you do it quickly? Could you do the same with an iron handle? Explain.

17. Wood has a very low conductivity. Does it still have a low conductivity if it is very hot—that is, in the stage of smoldering red-hot coals? Could you safely walk across a bed of red-hot wooden coals with bare feet? Although the coals are hot, does much heat conduct from them to your feet if you step quickly? Could you do the same on red-hot iron coals? Explain.

18. In a still room, smoke from a cigarette will sometimes rise and then settle in the air before reaching the ceiling. Explain why.

19. Why would you expect a single helium atom to continually rise in an atmosphere of nitrogen and oxygen? Why doesn't it "settle off" like the smoke in the preceding question?

20. In a mixture of hydrogen and oxygen gases at the same temperature, which molecules move faster? Why?

21. One container is filled with argon gas and the other with krypton gas. If both gases have the same temperature, in which container are the atoms moving faster? Why?

22. If we warm a volume of air, it expands. Does it then follow that if we expand a volume of air, it warms? Explain.

23. How could you change the drawing in Figure 15-6 to make it illustrate the heating of air when it is compressed? Make a sketch of this case.

24. A snow-making machine used for ski areas consists of a mixture of compressed air and water blown through a nozzle. The temperature of the mixture may initially be well above the freezing temperature of water, yet crystals of snow are formed as the mixture is ejected from the nozzle. Explain how this happens.

25. What does the high specific heat of water have to do with convection currents in the air at the seashore?

26. What would be the most efficient color for steam radiators?

27. Why does a good *emitter* of heat radiation appear black at room temperature?

28. A number of bodies at different temperatures placed in a closed room will ultimately come to the same temperature. Would this thermal equilibrium be possible if good absorbers were poor emitters and if poor absorbers were good emitters? Explain.

29. From the rules that a good absorber of radiation is a good radiator and a good reflector is a poor absorber, state a rule relating the reflecting and radiating properties of a surface.

30. Suppose at a restaurant you are served coffee before you are ready to drink it. In order that it be hottest when you are ready for it, should you add cream to it right away or when you are ready to drink it?

31. Even though metal is a good conductor, frost can be seen on parked cars in the early morning even when the air temperature is above freezing. Can you explain this?

32. Why is whitewash sometimes applied to the glass of florists' greenhouses in the summer?

33. On a very cold sunny day you wear a black coat and a transparent plastic coat. Which should be worn on the outside for maximum warmth?

34. If the composition of the upper atmosphere were changed so that it permitted a greater amount of terrestrial radiation to escape, what effect would this have on the earth's climate? How about if the atmosphere reduced the escape of terrestrial radiation?

35. Is it important to convert temperatures to the Kelvin scale when we use Newton's law of cooling? Why or why not?

36. If you wish to save fuel and you're going to leave your warm house for a half hour or so on a very cold day, should you turn your thermostat down a few degrees, turn it off altogether, or let it remain at the room temperature you desire?

37. If you wish to save fuel and you're going to leave your cool house for a half hour or so on a very hot day, should you turn your air conditioning thermostat up a bit, turn it off altogether, or let it remain at the room temperature you desire?

38. As more energy is consumed on earth, the overall temperature of the earth tends to rise. Regardless of the increase in energy, however, the temperature does not rise indefinitely. By what process is an indefinite rise prevented? Explain your answer.

16 Change of State

The matter in our environment exists in four common states. Ice, for example, is the *solid* state of H_2O. Add energy, and you add motion to the rigid molecular structure, which breaks down to form H_2O in the *liquid* state, water. Add more energy, and the liquid changes to the *gaseous* state. Add still more energy, and the molecules break into ions and electrons, giving the *plasma* state. The state of matter depends on its temperature and the pressure that is exerted on it. Changes of state require changes in temperature or pressure or both—or, more generally, a transfer of energy.

Evaporation

Molecules in the liquid state bumble around haphazardly, repeatedly bumping into one another—sharing many different speeds and therefore many values of kinetic energy. Recall that the average kinetic energy per molecule determines the temperature of the liquid. At any time there are many molecules with less than the average kinetic energy and many with more than the average. Consider the very fastest molecules that happen to be moving upward near the surface. They may have enough energy to break free of the liquid. They can leave the surface and fly into the space above and join other molecules of gas. This is **evaporation**, a change of state from the surface of a liquid to a gas (or, more commonly, to a *vapor*).

Figure 16-1
The cloth covering on the sides of the canteen promotes cooling when it is wet. As the fastest-moving molecules evaporate from the wet cloth, the temperature of the cloth decreases and cools the metal, which in turn cools the water within. In this way the water in the canteen is cooled appreciably below air temperature.

288

Figure 16-2

Dogs have no sweat glands (except between the toes). They cool themselves by panting. In this way evaporation occurs in the mouth and within the bronchial tract.

It is the faster, more energetic molecules that evaporate, so the average kinetic energy of the molecules remaining in the liquid is lowered. Evaporation is therefore a cooling process. If you want to cool soup that is too hot, increase its rate of evaporation by putting it in a shallow wide bowl with a lot of surface area and then blow on it. The canteen shown in Figure 16-1 keeps cool because of evaporation. The cloth covering on the sides is kept wet so that water can evaporate from it and cool the liquid within.

The cooling effect of evaporation is strikingly evident when rubbing alcohol is poured on your back. The alcohol evaporates very rapidly, cooling the surface of the body quickly. The more rapid the evaporation, the faster the cooling.

When our bodies tend to overheat, our sweat glands produce perspiration. This is part of nature's thermostat, for the evaporation of perspiration cools us and helps us maintain a stable body temperature. Many animals do not have sweat glands and must cool themselves by other means (Figures 16-2 and 16-3).

Figure 16-3

Pigs have no sweat glands and therefore cannot cool by the evaporation of perspiration. Instead, they wallow in the mud to cool themselves.

Condensation

The process that is the converse of evaporation is **condensation**—the changing of a gas into a liquid. When gas molecules near the surface of a liquid are attracted to the liquid, they strike the surface with increased kinetic energy. This kinetic energy is absorbed by the liquid and increases the temperature of the liquid. We say that condensation is a warming process.

A dramatic example of the warming that results from condensation is the energy given up by steam when it condenses—a painful experience if it condenses on you. That's why a steam burn is much more damaging than a burn from boiling water of the same temperature; the steam gives up considerable energy when it condenses to a liquid and wets the skin. This energy release by condensation is utilized in steam-heating systems.

Figure 16-4
Heat is given up by steam when it condenses inside the radiator.

Water vapor need not be as hot as steam to give up energy when it condenses. You are warmed by condensation after you have taken a shower—even a cold shower—if you remain in the moist shower area. You quickly sense the difference if you step outside. Away from the moisture, net evaporation takes place quickly and you feel chilly. When you remain in the shower stall, even with the water off, the warming effect of condensation counteracts the cooling effect of evaporation. If as much moisture condenses as evaporates, you feel no change in body temperature. If condensation exceeds evaporation, you are warmed. If evaporation exceeds condensation, you are cooled. So now you know why you can dry yourself with a towel much more comfortably if you remain in the shower area. To dry yourself thoroughly, you can finish the job in a less moist area.

Spend a July afternoon in dry Tucson or Phoenix, where evaporation is appreciably greater than condensation. The result of this pronounced evaporation is a much cooler feeling than you would experience in a same-temperature July afternoon in New York City or New Orleans. In these humid locations, condensation noticeably counteracts evaporation, and you feel the warming effect as vapor in the air condenses on your skin. You are quite literally being "tattooed" by the impact of H_2O molecules in the air that slam into you. Put more mildly, you are warmed by the condensation of vapor in the air upon your skin.

Figure 16-5
If you're chilly outside the shower stall, step back inside and be warmed by the condensation of the excess water vapor there.

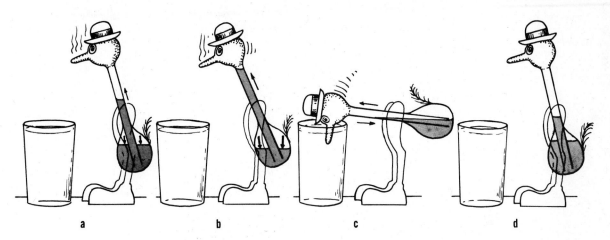

Figure 16-6
The toy drinking bird operates by the evaporation of ether inside its body and by the evaporation of water from the outer surface of its head. The lower body contains liquid ether, which evaporates rapidly at room temperature. As it (*a*) vaporizes, it (*b*) creates pressure (inside arrows), which pushes ether up the tube. Ether in the upper part does not vaporize because the head is cooled by the evaporation of water from the outer felt-covered beak and head. When the weight of ether in the head is sufficient, the bird (*c*) pivots forward, permitting the ether to run back to the body. Each pivot wets the felt surface of the beak and head, and the cycle is repeated.

Condensation in the Atmosphere

There is always some water vapor in the air. The amount of vapor depends on water supply and on the temperature of the air. More water molecules exist as vapor in hotter air than in colder air. We can understand this by thinking of a fly making a grazing contact with flypaper. At high speed it has enough momentum and energy to rebound from the flypaper into the air without sticking, but at low speed it more likely gets stuck. Similarly, when water vapor molecules collide at high speed, they are more likely to bounce from one another and remain in the vapor state. At low speeds they are more likely to stick together and become part of a liquid

▶ **Answer**
Not at all, for there is much activity taking place at the molecular level. Both evaporation and condensation occur continuously. The fact that the water level remains constant simply indicates equal rates of evaporation and condensation. As many molecules leave the surface by evaporation as return by condensation, so no *net* evaporation or condensation takes place. The two processes cancel each other.

(Figure 16-7). The faster that water molecules in the vapor state move, the less chance there is that they will condense to form droplets. So we see why more water molecules exist in the vapor state in warmer air than in cooler air.

Figure 16-7
Condensation of water vapor.

Fast-moving H₂O molecules rebound upon collision

Slow-moving H₂O molecules coalesce upon collision

There is a limit to the amount of humidity of air at a given temperature beyond which the vapor condenses to a liquid. Maximum humidity occurs when the maximum amount of vapor in air is reached—that is, when the air is *saturated*. When meteorologists report the *relative humidity*, they are comparing the amount of vapor in the air to the amount at saturation for the given air temperature. A relative humidity of 100 percent means that the air is saturated. A relative humidity of 50 percent, on the other hand, means there is half the amount of vapor in the air as when it is saturated at the same temperature. For the average person, weather is ideal when the temperature is about 20°C and the relative humidity is about 50 to 60 percent. When the relative humidity is high, condensation counteracts evaporation of perspiration, and we say the weather is muggy.

Fog and Clouds

Figure 16-8
Why is it common for clouds to form where there are updrafts of warm moist air?

Warm air rises. As it rises, it expands. As it expands, it chills. As it chills, water vapor molecules begin sticking together after colliding rather than bouncing off one another. If there are larger and slower-moving particles or ions present, water vapor condenses upon these particles, and we have a cloud. If these particles are not present, we can stimulate cloud formation by "seeding" the air with appropriate particles or ions.

Warm breezes blow over the ocean. When the moist air moves from warmer to cooler waters or from warm water to cool land, it chills. As it chills, water vapor molecules begin sticking rather than bouncing off one another when they have glancing collisions. Condensation takes place near ground level, and we have fog. The difference between fog and a cloud is basically altitude. Fog is a cloud that forms near the ground. Flying through a cloud is much like driving through fog.

Boiling

Under the right conditions, evaporation can take place beneath the surface of a liquid, forming bubbles of vapor that are buoyed to the surface where they escape. This change of state throughout a liquid rather than only at the surface is called **boiling**. Bubbles in the liquid can form only when the pressure of the vapor within the bubbles is great enough to resist the pressure of the surrounding water and atmospheric pressure. Unless the vapor pressure is great enough, the surrounding pressures will collapse any bubbles that may form. At temperatures below the boiling point, the vapor pressure in bubbles is not great enough, so bubbles do not form until the boiling point is reached. At this temperature, 100°C for water at atmospheric pressure, molecules are energetic enough to exert a vapor pressure as great as the pressure of the atmosphere.

Figure 16-9
The motion of molecules in the bubble of steam (much enlarged) creates a gas pressure that counteracts the atmospheric and water pressure against the bubble.

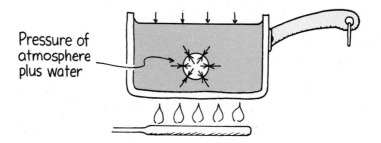

Pressure of atmosphere plus water

If atmospheric pressure is increased, the molecules in the vapor must move faster to exert a pressure within the bubble that is great enough to counteract the additional atmospheric pressure. This means that if we increase the pressure on the surface of a liquid, the boiling point of the liquid will be raised. The way a pressure cooker works is based on this fact. A pressure cooker has a tight-fitting lid that does not allow vapor to escape. As the evaporating vapor builds up inside the sealed pressure cooker, pressure on the surface of the liquid is increased, which prevents boiling. Bubbles that would normally form are crushed. Continued input of heat increases the temperature beyond 100°C. Boiling does not occur until the vapor pressure within the bubbles overcomes the increased pressure on the water. The boiling point is raised. Conversely, lowered pressure (as at high altitudes) decreases the boiling point of the liquid. So we see that boiling depends not only on temperature but on pressure as well.

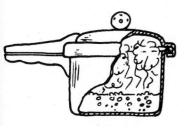

Figure 16-10
A pressure cooker.

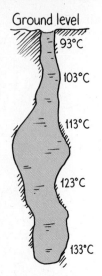

Ground level

93°C

103°C

113°C

123°C

133°C

Figure 16-11
An Old Faithful type of geyser.

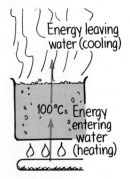

Energy leaving water (cooling)

100°C Energy entering water (heating)

Figure 16-12
Heating warms the water, and boiling cools it.

At high altitudes, water boils at a lower temperature. For example, in Denver, Colorado, the "mile-high city," water boils at 95°C instead of the 100°C boiling temperature characteristic of sea level. If you try to cook food in boiling water of a lower temperature, you must wait a longer time for proper cooking. A 3-minute boiled egg in Denver is yukky. If the temperature of the boiling water is too low, food would not cook at all. It is important to note that it is the high temperature of the water that cooks the food, not the boiling process itself.

Geysers

A geyser is a periodically erupting pressure cooker. It consists of a long, narrow, vertical hole into which underground streams seep (Figure 16-11). The column of water is heated by volcanic heat below to temperatures exceeding 100°C. This is because the vertical column of water exerts pressure on the deeper water, thereby increasing the boiling point. The narrowness of the shaft shuts off convection currents, which allows the deeper portions to become considerably hotter than the water surface. Water at the surface is less than 100°C, but the water temperature below, where it is being heated, is more than 100°C, high enough to permit boiling before water at the top reaches the boiling point. Boiling therefore begins near the bottom, where the rising bubbles push out the column of water above, and the eruption starts. As the water gushes out, the pressure on the remaining water is reduced. It then rapidly boils and erupts with great force.

Boiling Is a Cooling Process

Evaporation is a cooling process. So is boiling. At first thought, this may seem surprising—perhaps because we usually associate boiling with heating. But heating water is one thing; boiling is another. When 100°C water at atmospheric pressure is boiling, it is in thermal equilibrium. It is being cooled by boiling as fast as it is being heated by energy from the heat source (Figure 16-12). If cooling did not take place, continued application of heat to a pot of boiling water would result in a continued increase in temperature. The reason a pressure cooker reaches higher temperatures is because it prevents boiling, which in effect prevents cooling.

Question Since boiling is a cooling process, would it be a good idea to cool your hot and sticky hands by dipping them into boiling water?

▶ **Answer**

No, no, no! When we say boiling is a cooling process, we mean that the water (not your hands!) is being cooled relative to the higher temperature it would attain otherwise. Because of cooling, it remains at 100°C instead of getting hotter. A dip in 100°C water would be most uncomfortable for your hands!

Boiling and Freezing at the Same Time

We usually boil water by the application of heat. But we can boil water by the reduction of atmospheric pressure. We can dramatically show the cooling effect of evaporation and boiling when room-temperature water is placed in a vacuum jar (Figure 16-13). If the pressure in the jar is slowly reduced by a vacuum pump, the water will start to boil. The boiling process takes heat away from the water left in the dish, which cools to a lower temperature. As the pressure is further reduced, more and more of the slower-moving molecules boil away. Continued boiling results in a lowering of temperature until the freezing point of approximately 0°C is reached. Continued cooling by boiling causes ice to form over the surface of the bubbling water. Boiling and freezing are taking place at the same time! This must be witnessed to be appreciated. Frozen bubbles of boiling water are a remarkable sight.

Spray some drops of coffee into a vacuum chamber, and they, too, will boil until they freeze. Even after they are frozen, the water molecules will continue to evaporate into the vacuum until little crystals of coffee solids are left. This is how freeze-dried coffee is made. The low temperature of this process tends to keep the chemical structure of coffee solids from changing. When hot water is added, more of the original flavor of the coffee is retained. Boiling really is a cooling process!

Figure 16-13
Apparatus to demonstrate that water will freeze and boil at the same time in a vacuum. A gram or two of water is placed in a dish that is insulated from the base by a polystyrene cup.

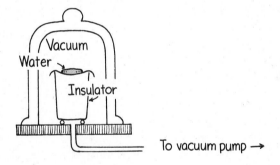

Melting and Freezing

Suppose you held hands with someone, and each of you started jumping around randomly. The more violently you jumped, the more difficult keeping hold would be. If you jumped violently enough, keeping hold would be impossible. Something like this happens to the molecules of a solid when it is heated. As heat is absorbed, the molecules vibrate more and more violently. If enough heat is absorbed, the attractive forces between the molecules will no longer be able to hold them together. The solid melts.

Freezing is the converse of this process. As energy is withdrawn from a liquid, molecular motion diminishes until finally the molecules, on the average, are moving slowly enough so that the attractive forces between them are able to cause cohesion. The molecules then vibrate about fixed positions and form a solid.

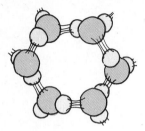

Figure 16-14
The open structure of pure ice crystals that normally fuse at 0°C. When other kinds of molecules or ions are introduced, crystal formation is interrupted, and the freezing temperature is lowered.

At atmospheric pressure, water freezes at 0°C—unless substances such as sugar or salt are dissolved in it. Then the freezing point is lower. In the case of salt, chlorine ions grab electrons from the hydrogen atoms in H_2O and impede crystal formation. The result of this impedance by "foreign" ions is that slower motion is required for the formation of the six-sided ice-crystal structures. As ice crystals do form, the impedance is intensified because the proportion of "foreign" molecules or ions among nonfused water molecules increases. Connections become more and more difficult. Only when the water molecules move slowly enough for attractive forces to play an unusually large part in the process can freezing be completed. The ice first formed is almost always pure H_2O. In general, adding anything to water lowers the freezing temperature. Antifreeze is a practical application of this process.

Regelation

Because H_2O molecules form open structures in the solid state (Figure 16-14), the application of pressure lowers the melting point of ice. The crystals are simply crushed to the liquid state. (The temperature of the melting point is only slightly lowered, 0.007°C for each additional atmosphere of added pressure.) When the pressure is removed, molecules crystallize and refreezing occurs. This phenomenon of melting under pressure and freezing again when the pressure is reduced is called **regelation**. It is one of the properties of water that make it different from other materials.

Regelation is neatly illustrated in Figure 16-15. A fine copper wire with weights attached to its ends is hung over a block of ice.* The wire will slowly cut its way through the ice, but its track will be left full of ice. So the wire and weights will fall to the floor, leaving the ice a solid block.

The making of snowballs is a good example of regelation. When we compress the snow with our hands, we cause a slight melting of the ice crystals; when pressure is removed, refreezing occurs and binds the snow together. Making snowballs is difficult in very cold weather because the pressure we can apply is not enough to melt the snow.

A skater skates on a thin film of water between the blade and the ice, which is produced by blade pressure and friction. As soon as the pressure is released, the water refreezes.

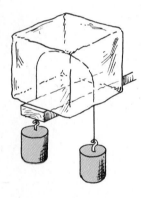

Figure 16-15
Regelation. The wire gradually passes through the ice without cutting it in half.

---

*Changes of state are occurring as ice melts and the water refreezes, and, as will be stressed in the next section and the rest of the chapter, energy is needed for these changes. When the water immediately above the wire refreezes, it gives up energy. How much? Enough to melt an equal amount of ice immediately under the wire. This energy must be conducted through the thickness of the wire. Hence this demonstration requires that the wire be an excellent conductor of heat. String, for example, won't do.

Energy and Changes of State

If we continually add heat to a solid or liquid, the solid or liquid will eventually change state. A solid will liquefy, and a liquid will vaporize. Energy is required for both liquefaction of a solid and vaporization of a liquid. Conversely, energy must be extracted from a substance to change its state in the direction from gas to liquid to solid (Figure 16-16).

Figure 16-16
Energy changes with change of state.

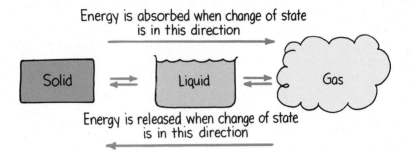

The cooling cycle of a refrigerator neatly employs the concepts shown in Figure 16-16. A liquid of low boiling point, usually Freon, is pumped into the cooling unit, where it turns into a gas and draws heat from the things stored there. The gas with its added energy is directed outside the cooling unit to coils located in the back, appropriately called condensation coils, where heat is given off to the air as the gas condenses to a liquid again. A motor pumps the fluid through the system, where it is made to undergo the cyclic process of vaporization and condensation. The next time you're near a refrigerator, place your hand near the condensation coils in the back and you'll feel the heat that has been extracted from inside.

An air conditioner employs the same principles and simply pumps heat energy from one part of the unit to another. When the roles of vaporization and condensation are reversed, the air conditioner becomes a heater.

So we see that a solid must absorb energy to melt, and a liquid must absorb energy to vaporize. Conversely, a gas must release energy to liquefy, and a liquid must release energy to solidify.

Question How does the formation of snow or rain affect the temperature of the air?

▶ **Answer**
The formation of snow or rain from water vapor is accompanied by an increase in atmospheric temperature. This is in accord with Figure 16-16, where gas, in this case the water vapor, releases energy in transforming to the liquid and/or solid state. So it is always a little warmer when it rains or snows than it would be otherwise!

The general behavior of many substances can be illustrated by a description of the changes of state of H_2O. To make the numbers simple, consider a 1-gram piece of ice at a temperature of $-50°C$ in a closed container, put on a stove to heat. A thermometer in the container reveals a slow increase in temperature up to 0°C. At 0°C, the temperature stops rising, even though heat is continually added. This heat melts the ice. In order for the whole gram of ice to melt, 80 calories are absorbed by the ice. The temperature will not begin rising until all the ice melts. Only when all the ice melts will each additional calorie absorbed by the water increase its temperature by 1°C until the boiling temperature, 100°C, is reached. Again, as heat is added, the temperature remains constant while more and more of the gram of water is boiled away and becomes steam. The water must absorb 540 calories of heat energy to vaporize the whole gram. Finally, when all the water has become steam at 100°C, the temperature begins to rise once more. It will continue to rise as long as heat is added. This process is graphed in Figure 16-17.

Figure 16-17

A graph showing the energy involved in the heating and the change of state of 1 g of H_2O.

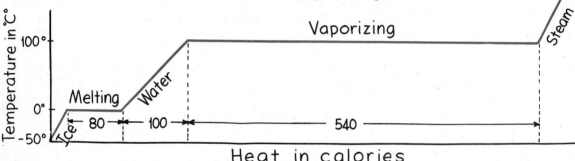

The 540 calories required to vaporize a gram of water is a relatively large amount of energy—much more than would be required to bring a gram of ice at absolute zero to boiling water at 100°C. Although the molecules in steam and boiling water at 100°C have the same average kinetic energy, steam has more potential energy because the molecules are relatively free of each other and are not bound together as in the liquid state. Steam contains a vast amount of energy that can be released during condensation.

The change of state sequence is reversible. If the molecules in a gram of steam condense to form boiling water, they transfer 540 calories of energy to the environment. When the water is cooled from 100°C to 0°C, 100 additional calories are transferred to the environment. When ice water fuses to become solid ice, 80 more calories of energy are transferred by the water to the environment. The vaporization of water requires 540 calories per gram, and the fusion (freezing) of water requires 80 calories.*

---

*These values are known as the *heat of vaporization* and the *heat of fusion*, respectively. In SI units, the heat of vaporization of water is 2.26 megajoules per kilogram (MJ/kg), and the heat of fusion of water is 0.335 MJ/kg.

Questions ▶

1. How much energy is transferred when a gram of steam at 100°C condenses to water at 100°C?

2. How much energy is transferred when a gram of boiling water at 100°C cools to ice water at 0°C?

3. How much energy is transferred when a gram of ice water at 0°C freezes to ice at 0°C?

4. How much energy is transferred when a gram of steam at 100°C turns to ice at 0°C?

Figure 16-18
On a cold day, hot water freezes faster than warm water because of the energy that leaves the hot water during rapid evaporation.

The large value of 540 calories per gram explains why under some conditions hot water will freeze faster than warm water.* This phenomenon occurs for water hotter than 80°C, and it is evident when the surface area that cools by rapid evaporation is large compared to the amount of water involved—like when you wash a car with hot water on a cold winter day or flood a skating rink with hot water, which melts, smooths out the rough spots, and refreezes quickly. The rate of cooling by rapid evaporation is very high because each evaporating gram draws at least 540 calories from the water left behind. This is an enormous amount of energy compared to the 1 calorie per Celsius degree that is drawn from each gram of water that cools by thermal conduction. Evaporation is truly a cooling process.

You dare not touch your dry finger to a hot skillet on a hot stove, but you can certainly do so without harm if you first wet your finger and touch the skillet briefly. You can even touch it a few times in succession as long as your finger remains wet. This is because energy that ordinarily would go into burning your finger goes instead into changing the state of the moisture on your finger. The energy converts the moisture to a vapor, which then provides an insulating layer between your finger and the skillet. Similarly, you are able to judge the hotness of a hot clothes iron.

Supervisor Paul Ryan of the Department of Public Works in Malden, Massachusetts, has for years used molten lead to seal pipes in certain plumbing operations. He startles onlookers by dragging his finger through

▶ **Answers**

1. One gram of steam at 100°C transfers 540 calories of energy when it condenses to become water at the same temperature.

2. One gram of boiling water transfers 100 calories when it cools 100°C to become ice water.

3. One gram of ice water at 0°C transfers 80 calories to become ice at 0°C.

4. One gram of steam at 100°C transfers to the surroundings a grand total of the above values, 720 calories, to become ice at 0°C.

---

*Hot water will not freeze before cold water, but will freeze before lukewarm water. Water at 100°C, for example, will freeze before water warmer than 60°C, but not before water cooler than 60°C.

Figure 16-19
Paul Ryan tests the hotness of molten lead by dragging his wetted finger through it.

Figure 16-20
Jearl Walker demonstrates change of state by walking barefoot on red-hot coals. His feet are protected only by natural perspiration, which in vaporizing absorbs otherwise harmful energy and creates an insulating layer of water vapor between the coals and the skin of his feet.

molten lead to judge its hotness (Figure 16-19). He is sure that the lead is very hot and his finger is thoroughly wet before he does this. (Do not try this on your own, for if the lead is not hot enough it will stick to your finger and you'll lose it!)

Similarly, one is able to walk barefoot on red-hot coals, as physics professor Jearl Walker demonstrates in front of his physics class (Figure 16-20). The coals are of wood that has a low thermal conductivity, so when surface heat passes to the feet, heat from the interior of the coal is slow to move out to the surface and into the feet. (Using ordinary coal would be a painful mistake!) Jearl's success is due in large part to the perspiration on the soles of his feet—perspiration facilitated by nervousness. On one occasion, the confidence he gained from his successful prior fire walk was accompanied by less nervousness and hence less perspiration. He burned his feet. Jearl calls this a "no-sweat situation."

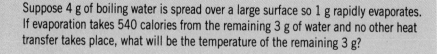

Question ▶ Suppose 4 g of boiling water is spread over a large surface so 1 g rapidly evaporates. If evaporation takes 540 calories from the remaining 3 g of water and no other heat transfer takes place, what will be the temperature of the remaining 3 g?

▶ **Answer**
The remaining 3 g will turn to 0°C ice under conditions where all 540 calories are taken from the remaining water (for example, when the surroundings are below freezing and don't contribute energy). 540 calories from 3 g means each gram gives up 180 calories. 100 calories from a gram of boiling water reduces its temperature to 0°C, and 80 more calories taken away turns it to ice. This is why hot water so quickly turns to ice in a freezing-cold environment.

Summary of Terms

Evaporation The change of state at the surface of a liquid as it passes to the gaseous state. This is caused by the random motion of molecules that occasionally escape from the liquid surface. Cooling of the liquid results.

Condensation The change of state from gas to liquid; the opposite of evaporation. Warming of the liquid results.

Boiling A rapid state of evaporation that takes place within the liquid as well as at its surface. As with evaporation, cooling of the liquid results.

Regelation The process of melting under pressure and the subsequent refreezing when the pressure is removed.

Review Questions

1. What are the four common states of matter?

Evaporation

2. Do all the molecules or atoms in a liquid have about the same speed or much different speeds?

3. What is evaporation, and why is it a cooling process? Exactly what is it that cools?

4. Why does a hot dog pant?

Condensation

5. What is condensation, and why is it a warming process? Exactly what is it that warms?

6. Why is a steam burn more damaging than a burn from boiling water of the same temperature?

7. Why do we feel uncomfortably warm on a hot and humid day?

Condensation in the Atmosphere

8. Why is there usually more water vapor in warm air than in cool air?

9. Why does water vapor in the air condense when it is chilled?

Fog and Clouds

10. Why does warm moist air form clouds when it rises?

11. What is the basic difference between a cloud and fog?

Boiling

12. Distinguish between evaporation and boiling.

13. Why does water not boil at 100°C when it is under higher pressure?

14. Why is the boiling point of water higher in a pressure cooker?

15. Is it the boiling process or the higher temperature that cooks food faster in a pressure cooker?

Geysers

16. How is a geyser similar to a pressure cooker?

17. Why will water at the bottom of a geyser not boil when it is 100°C?

18. Why does water at the bottom of a geyser get hotter and boil before water at the top?

19. What happens to the water pressure at the bottom of a geyser when some of the water above gushes out?

20. Why does the remaining water in a geyser erupt violently when some of the water above gushes out?

Boiling Is a Cooling Process

21. What is it that cools when boiling occurs—the water that leaves in the form of vapor or the water left behind?

22. We observe that the temperature of boiling water doesn't increase with continued heat input. Explain how this observation is evidence that boiling is a cooling process.

Boiling and Freezing at the Same Time

23. When will water boil at a temperature less than 100°C?

24. We can add heat to water and boil it to form steam. Isn't it a contradiction to say we can boil water to form ice? Explain.

Melting and Freezing

25. Why does increasing the temperature of a solid make it melt?

26. Why does decreasing the temperature of a liquid make it freeze?

27. Why does freezing of water not occur at 0°C when molecules other than H_2O are present?

28. Which freezes at the lower temperature, pure water or salt water?

Regelation

29. What happens to the hexagonal open structure of ice when sufficient pressure is applied to it?

30. What happens to the water that results from crushing crystals when the pressure is removed?

31. Why does a wire not simply cut a block of ice in half when it passes through the ice?

Energy and Changes of State

32. Does a liquid give off or absorb energy when it turns into a gas? When it turns into a solid?

33. Does a gas give off or absorb energy when it turns into a liquid?

34. Does a solid give off or absorb energy when it turns into a liquid?

35. Is Figure 16-16 consistent with or contrary to the statements that both evaporation and boiling are cooling processes and that condensation is a warming process? Defend your answer.

36. Is the food compartment in a refrigerator cooled by vaporization or by condensation of the refrigerating fluid?

37. How many calories are needed to change the temperature of 1 g of water by 1°C? To melt 1 g of ice at 0°C? To vaporize 1 g of boiling water at 100°C?

38. Why does the temperature of melting ice not rise when heat is applied?

39. Why does the temperature of boiling water not rise when heat is applied?

40. Why is the rate of cooling of hot water by evaporation greater than the rate of cooling of lukewarm water by thermal conduction?

41. Why is it important that your finger be wet if you touch it briefly to a hot clothes iron?

42. When we discussed in Chapter 4 the acceleration and high speeds you would encounter falling from great heights, no words of caution told you of the dangers of trying this on your own. Why are there words of caution in this chapter with respect to walking barefoot on hot coals and dipping fingers into molten lead?

Home Projects

1. Place a Pyrex funnel mouth down in a saucepan full of water so the straight tube of the funnel sticks above the water. Rest a part of the funnel on a nail or coin so water can get under it. Place the pan on a stove and watch the water as it begins to boil. Where do the bubbles form first? Why? As the bubbles rise, they expand rapidly and push water ahead of them. The funnel confines the water, which is forced up the tube and driven out at the top. Now do you know how a geyser and a coffee percolator work?

2. Watch the spout of a teakettle of boiling water. Notice that you cannot see the steam that issues from the spout. The cloud that you see farther away from the spout is not steam, but condensed water droplets. Now hold a candle in the cloud of condensed steam. Can you explain your observations?

3. You can make rain in your kitchen. Put a cup of water in a Pyrex saucepan or Silex coffeemaker and heat it slowly over a low flame. When the water is warm, place on top of it a saucer filled with ice cubes. As the water below is heated, droplets form at the bottom of the cold saucer and combine until they are large enough to fall, producing a steady "rainfall" as the water below is gently heated. How does this resemble and how does it differ from the way natural rain is formed?

4. Compare the temperature of boiling water with the temperature of a boiling solution of salt and water.

5. Suspend a heavy weight by copper wire over an ice cube. In a matter of minutes the wire will be pulled through the ice. The ice will melt beneath the wire and refreeze above it, leaving a visible path if the ice is clear.

6. Place a tray of nearly boiling water and a tray of water with a temperature between 60°C and boiling in a freezer. See which water will freeze first.

7. If you suspend a container of water in a large container of boiling water, the water in the inner container will reach 100°C but will not boil. Can you figure out why this is so?

Exercises

1. You can determine wind direction by wetting your finger and holding it up in the air. Explain.

2. Can you give two reasons why blowing over hot soup cools the soup?

3. Can you give two reasons why pouring a cup of hot coffee into the saucer results in faster cooling?

4. A covered glass of water sits for days with no drop in water level. Strictly speaking, can you say that nothing has happened, that no evaporation or condensation has taken place? Explain.

5. In the hypothetical case where all the molecules in a liquid had the same speed, would evaporation be a cooling process? Explain.

6. Where does the energy come from that keeps the dunking bird in Figure 16-6 operating?

7. An inventor claims to have developed a new perfume that lasts a long time because it doesn't evaporate. Comment on this claim.

8. Porous canvas bags filled with water are used by travelers in hot weather. When the bags are slung on the outside of a fast-moving car, the water inside is cooled considerably. Explain.

9. Why will wrapping a bottle in a wet cloth at a picnic often produce a cooler bottle than placing the bottle in a bucket of cold water?

10. The human body can maintain its customary temperature of 37°C on a day when the temperature is above 40°C. How is this done?

11. Why are icebergs often surrounded by fog?

12. How does Figure 16-7 help explain the moisture that forms on the inside of car windows when you're parking on a cool night?

13. You know that the windows in your warm house get wet on a cold day. But can moisture form on the windows if the inside of your house is cold on a hot day? How is this different?

14. Why do clouds often form above mountain peaks? (*Hint:* Consider the updrafts.)

15. Why will clouds tend to form above either a flat or a mountainous island in the middle of the ocean? (*Hint:* Compare the specific heat of the land with that of the water and the subsequent convection currents in the air.)

16. Why does the temperature of boiling water remain the same as long as the heating and boiling continue?

17. Water will boil spontaneously in a vacuum—on the moon, for example. Could you cook an egg in this boiling water? Explain.

18. Our inventor friend proposes a design of cookware that will allow boiling to take place at a temperature less than 100°C so that food can be cooked with less energy. Comment on this idea.

19. Your instructor hands you a closed flask of room-temperature water. When you hold it, the heat from your bare hands causes the water to boil. Quite impressive! How is this accomplished?

20. When you boil potatoes, will your cooking time be reduced with vigorously boiling water compared to gently boiling water? (Directions for cooking spaghetti call for vigorously boiling water—not to lessen cooking time, but to prevent something else. If you don't know what it is, ask a cook.)

21. Why does putting a lid over a pot of water on a stove shorten the time it takes for the water to come to a boil, whereas after the water is boiling, use of the lid only slightly shortens the cooking time?

22. In a nuclear submarine power plant, the temperature of the water in the reactor is above 100°C. How is this possible?

23. Why do many geysers erupt regularly?

24. Would regelation occur if ice crystals were not open-structured? Explain.

25. You'll slip on a smoothly polished floor much more readily than on an unpolished floor, but you'll slip more readily on bumpy ice than on smooth ice. Why is this so? (*Hint:* Why is ice slippery?)

26. People who live where snowfall is common will tell you that air temperatures are higher on snowy days than on clear days. Some misinterpret this by stating that snowfall can't occur on very cold days. Explain this misinterpretation.

27. How does melting ice change the temperature of surrounding air?

28. Air-conditioning units contain no water whatever, yet it is common to see water dripping from them when they're running on a hot day. Explain.

29. Some old-timers found that when they wrapped newspaper around the ice in their iceboxes, melting was inhibited. Discuss the advisability of this.

30. Why does dew form on a cold soft-drink can?

31. Why is it that in cold winters a tub of water placed in a farmer's canning cellar helps prevent canned food from freezing?

32. Why will spraying fruit trees with water before a frost help to protect the fruit from freezing?

33. There are several theories to explain how an ice age might begin. Consider this one: If the temperature of the world increases, evaporation of the oceans increases, precipitation increases, and snowfall increases. This results in more snow left in cooler places at the end of each summer, with successively increasing depths each winter that become hard-packed from the bottom up to become ice. All the while, the ice reflects more solar energy than would otherwise be absorbed. This cools the earth further and allows more freezing. Can you continue this sequence and see how the process reverses itself and how the ice age withdraws?

34. It so happens that drops of water placed in a hot skillet will quickly vaporize away, but drops in a very hot skillet will dance around for a longer time before vaporizing away. Can you think of a reason for this?

35. You have a friend who tells you that in exchange for a given sum of money, some people he met taught him techniques of mind over matter that allowed him to walk harmlessly with bare feet on red-hot coals. This is no jive, for after walking on wetted grass and then over smoldering wood embers his feet were completely unharmed. Furthermore, your friend urges that you raise money and learn these wonderful techniques yourself. What is your response?

17 Thermodynamics

The study of heat and its transformation to mechanical energy is called **thermodynamics** (which stems from Greek words meaning "movement of heat"). The science of thermodynamics was developed at the beginning of the last century, before the atomic and molecular theory of matter was understood. Whereas in previous chapters we described heat in terms of the microscopic behavior of jiggling atoms and molecules, in this chapter we invoke only macroscopic notions—such as mechanical work, pressure, and temperature—and their roles in energy transformations. Thermodynamics is a powerful theoretical science that bypasses the molecular details of the system altogether. Its foundation is the conservation of energy and the fact that heat flows from hot to cold and not the other way around. It provides the basic theory of heat engines, from steam turbines to fusion reactors, and the basic theory of refrigerators and heat pumps. We begin our study of thermodynamics with a look at one of its early concepts—a lowest limit of temperature.

Absolute Zero

In principle, there is no upper limit of temperature. As thermal motion increases, a solid object first melts and then vaporizes; as the temperature is further increased, molecules break up into atoms, and atoms lose some or all of their electrons, thereby forming a cloud of electrically charged

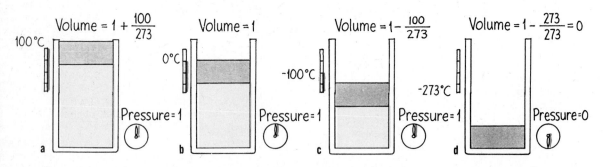

The volume of a gas changes by $\frac{1}{273}$ its volume at 0°C with each 1°C change in temperature when the pressure is held constant. (a) At 100°C the volume is $\frac{100}{273}$ greater than it is at (b) 0°C. (c) When the temperature is reduced to −100°C, the volume is reduced by $\frac{100}{273}$. (d) At −273°C the volume of the gas would be reduced by $\frac{273}{273}$ and so would be zero.

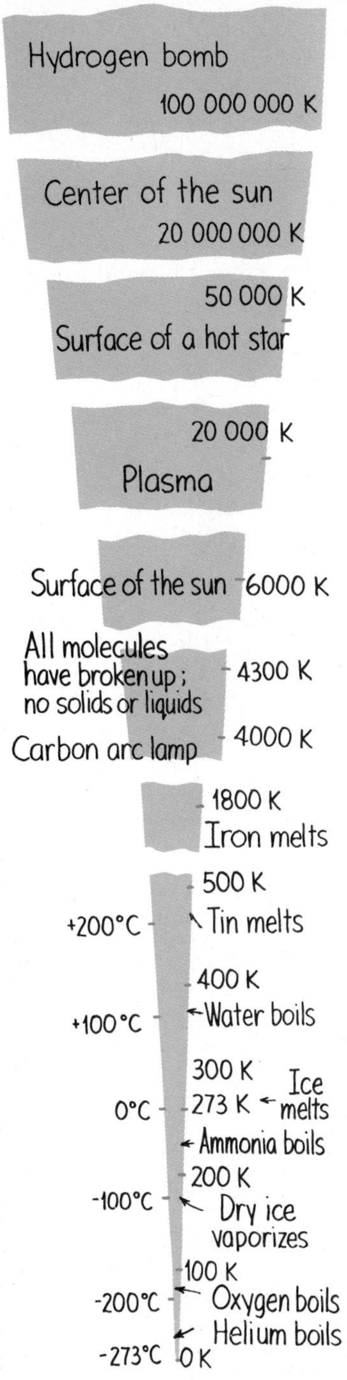

Hydrogen bomb
100 000 000 K

Center of the sun
20 000 000 K

50 000 K
Surface of a hot star

20 000 K
Plasma

Surface of the sun 6000 K

All molecules
have broken up; 4300 K
no solids or liquids

Carbon arc lamp 4000 K

1800 K
Iron melts

500 K
+200°C Tin melts

400 K
+100°C Water boils

300 K Ice
0°C 273 K melts
Ammonia boils

200 K
-100°C Dry ice
vaporizes

100 K
-200°C Oxygen boils
Helium boils
-273°C 0 K

Figure 17-2
Some absolute
temperatures.

particles—a plasma. This situation exists in stars, where the temperature is many millions of degrees Celsius.

In contrast, there is a definite limit at the other end of the temperature scale. Gases expand when heated and contract when cooled. Nineteenth-century experiments found that all gases, regardless of their initial pressures or volumes, change by $\frac{1}{273}$ of their volume at 0°C for each degree Celsius change in temperature, provided the pressure is held constant. So if a gas at 0°C were cooled down by 273°C, it would contract $\frac{273}{273}$ volumes and be reduced to zero volume. Clearly, we cannot have a substance with zero volume. It was also found that the pressure of any gas in any container of *fixed* volume would change by $\frac{1}{273}$ for each degree Celsius change in temperature. So a gas in a container of fixed volume cooled 273°C below zero would have no pressure whatsoever. In practice, every gas liquefies before it gets this cold. Nevertheless, these decreases by $\frac{1}{273}$ increments suggested the idea of a lowest temperature: −273°C. So there is a limit to coldness. When atoms and molecules lose all available kinetic energy, they reach the **absolute zero** of temperature.* At absolute zero, no more energy can be extracted from a substance and no further lowering of its temperature is possible. This limiting temperature is actually 273.15° below zero on the Celsius scale (and 459.69° below zero on the Fahrenheit scale).

On the absolute temperature scale, the Kelvin scale,† absolute zero is 0K (short for "0 Kelvin," rather than "0 degrees Kelvin"). There are no minus numbers in the Kelvin scale. Degrees on the Kelvin scale are calibrated with the same-sized divisions as on the Celsius scale. The melting point of ice is therefore 273.15K, and the boiling point of water is 373.15K.

Questions

1. Which is larger, a Celsius degree or a kelvin?
2. A piece of metal has a temperature of 0°C. If someone says a second, identical piece of metal is twice as hot, what do they mean? What is the temperature of the second piece of metal?

▶ **Answers**
1. Neither. They are equal.
2. To say a piece of metal is twice as hot means it has twice the internal energy. This means it has twice the absolute temperature, or two times 273K. This would be 546K, or 273°C.

---

*Even at absolute zero, molecules still have a small kinetic energy, called the *zero-point energy*. Helium, for example, has enough motion at absolute zero to keep it from freezing. The explanation for this involves quantum theory.

†Named after the famous British physicist Lord Scale, who coined the word *thermodynamics* and was the first to suggest this thermodynamic temperature scale.

Internal Energy

We all know that there is a vast amount of energy locked in all materials—this book, for example. The pages of this book are composed of molecules that are in constant motion. They have kinetic energy. Due to interactions with neighboring molecules, they also have potential energies. The pages can be easily burned, so we know they store chemical energy, which is really electric potential energy at the molecular level. We know there are vast amounts of energy associated with atomic nuclei. Then there is the "energy of being," described by the celebrated equation $E = mc^2$ (mass energy). Energy within a substance is found in these and other forms, which, when taken together, are called **internal energy**.* The internal energy in even the simplest substance can be quite complex. But in our study of heat changes and heat flow, we will be concerned only with the *changes* in the internal energy of a substance. Changes in temperature will indicate these changes in internal energy.

First Law of Thermodynamics

More than a hundred years ago heat was thought to be an invisible fluid called *caloric*, which flowed like water from hot objects to cold objects. Caloric was conserved in its interactions, and this discovery was the forerunner of the law of conservation of energy. In the mid-1880s, it became apparent that the flow of heat was nothing more than the flow of energy itself. The caloric theory of heat was gradually abandoned.† Today we view heat as a form of energy. Energy cannot be either created or destroyed.

When the law of energy conservation is applied to thermal systems, we call it the **first law of thermodynamics**:

> Whenever heat is added to a system, it transforms to an equal amount of some other form of energy.

By *system*, we mean a well-defined group of atoms, molecules, particles, or objects. The system may be the steam in a steam engine or it may be the whole earth's atmosphere. It can even be the body of a living creature. The important point is that we must be able to define what is contained within the system and what is outside it. If we add heat energy to the steam in a steam engine, to the earth's atmosphere, or to the body of a living creature, we will increase the internal energies of these things. And because of this added energy, these things may do work on external things. This added energy does one or both of two things: (1) it increases the internal energy of the system if it remains in the system or (2) it does

---

*If this book were poised at the edge of a table and ready to fall, it would possess gravitational potential energy; if it were tossed into the air, it would possess kinetic energy. But these are not examples of internal energy, because they involve more than just the elements of which the book is composed. They include gravitational interactions with the earth and motion with respect to the earth. If we wish to include these forms, we must do so in terms of a larger "system"—a system enlarged to include both the book and the earth. They are not part of the internal energy of the book itself.

†Popular ideas, when proved wrong, are seldom suddenly discarded. People tend to identify with the ideas that characterize their time; hence, it is often the young who are more prone to discover and accept new ideas and push the human adventure forward.

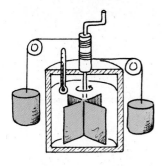

Figure 17-3
Paddle-wheel apparatus first used to compare heat energy with mechanical energy. As the weights fall, they give up potential energy and warm the water accordingly. This was first demonstrated by James Joule, for whom the unit of energy is named.

work on things external to the system if it leaves the system. So, more specifically, the first law states:

> **Heat added = increase in internal energy + external work done by the system.**

Note that the first law is an overall principle that is not concerned with the inner workings of the system itself. Whatever the details of molecular behavior in the system, the added heat energy serves only two functions: it increases the internal energy of the system or it enables the system to do external work (or both). Our ability to describe and predict the behavior of systems that may be too complicated to analyze in terms of atomic and molecular processes is one of the beauties of thermodynamics. Thermodynamics bridges the microscopic and macroscopic worlds.

Put an airtight can of air on a hot stove and heat it up. Since the can is of fixed size, no work is done on it, and all the heat that goes into the can increases the internal energy of the enclosed air. Its temperature rises. If the can is fitted with a movable piston, then the heated air can do work as it expands and push the piston outward. Can you see that the temperature of the enclosed air will be less than if no work were done on the piston? If heat is added to a system that does no external work, then the amount of heat added is equal to the increase in the internal energy of the system. If the system does external work, then the increase in internal energy is correspondingly less. The first law of thermodynamics is simply the thermal version of the law of conservation of energy.

Consider a given amount of heat supplied to a steam engine. The amount supplied will be evident in the increase in the internal energy of the steam and in the mechanical work done. The sum of the increase in internal energy and the work done will equal the heat input. In no way can energy output in any form exceed energy input.

Questions

1. If 100 J of energy is added to a system that does no external work, by how much is the internal energy of that system raised?
2. If 100 J of energy is added to a system that does 40 J of external work, by how much is the internal energy of that system raised?

Adding heat to a system so that it can do mechanical work is only one application of the first law of thermodynamics. If instead of adding heat to a system we do mechanical work on it, the first law tells us what we can expect: an increase in internal energy. A bicycle pump provides a good example. When we pump on the handle, the pump becomes hot.

▶ **Answers**
1. 100 J.
2. 60 J. We see from the first law that 100 J = 60 J + 40 J.

Why? Because we are primarily putting mechanical work into the system and raising its internal energy. If the process happens quickly enough so that very little heat is conducted from the system during compression, then most of the work input will go into increasing the internal energy, and the system becomes hotter.

Adiabatic Processes

A process wherein no heat enters or leaves a system is said to be *adiabatic* (from the Greek for "impassable"). **Adiabatic processes** can be achieved either by thermally insulating a system from its surroundings (with Styrofoam, for example) or by performing the process rapidly so that heat has little time to enter or leave. If we set the "heat added" part of the first law to zero, we see that changes in internal energy are equal to the work done on or by the system.* If work is done on a system—compressing it, for example—the internal energy increases. We raise its temperature. If work is done by the system—expanding against its surroundings, for example—the internal energy decreases. It cools.

A common example of an adiabatic process is the compression and expansion of the gases in the cylinders of an automobile engine (Figure 17-5). Compression and expansion occur in only hundredths of a second, which is too short a time for any appreciable amount of energy to leave the combustion chamber. For very high compressions, like those in diesel engines, the temperatures achieved are sufficient to ignite a fuel mixture without the use of spark plugs. Diesel engines have no spark plugs.

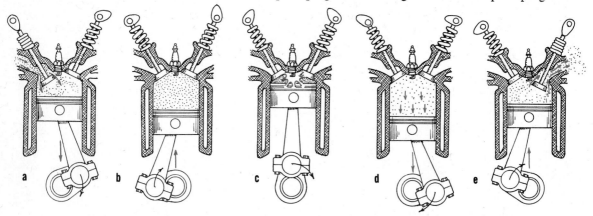

a b c d e

Figure 17-5

A four-cycle internal-combustion engine. (*a*) A fuel-air mixture from the carburetor fills the cylinder as the piston moves down. (*b*) The piston moves upward and compresses the mixture—adiabatically, since no heat transfer occurs. (*c*) The spark plug fires, ignites the mixture, and raises it to a high temperature. (*d*) Adiabatic expansion pushes the piston downward, the power stroke. (*e*) The burned gases are pushed out the exhaust pipe. Then the intake valve opens, and the cycle repeats.

---

*Δ Heat = Δ internal energy + work = 0.
Then we can say: − Work = Δ internal energy.

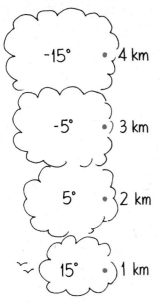

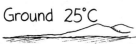

Figure 17-6
The temperature of a blob of dry air that expands adiabatically changes by about 10°C for each kilometer of elevation.

Figure 17-7
Chinooks, warm dry winds, occur when high-altitude air descends and is adiabatically warmed.

Meteorology and the First Law

Meteorologists use the first law of thermodynamics extensively in analyzing weather. Since changes in internal energy and the work done on or by a gas are proportional to changes in temperature and pressure in the gas, meteorologists express the first law in the following form:

Change in temperature ~ heat added (or subtracted) + pressure change

The temperature of the air may be changed by adding or subtracting heat, by changing the pressure of the air, or by both. Heat is added by solar radiation, by long-wave earth radiation, by moisture condensation, or by contact with the warm ground. An increase in air temperature results. The atmosphere may lose heat by radiation to space, by evaporation of rain falling through dry air, or by contact with cold surfaces. The result is a drop in air temperature.

There are many atmospheric processes, usually involving time scales of a day or less, in which the amount of heat added or subtracted is very small—small enough so that the process is nearly adiabatic. Then we have the adiabatic form of the first law:

Change in temperature ~ pressure change

Adiabatic processes in the atmosphere are characteristic of large masses of air, which we'll call *blobs*, which have dimensions on the order of kilometers. Because of this large size, mixing of different temperatures or pressures of air at the edges does not appreciably alter the overall composition of the blob, which behaves as if it were enclosed in a giant, tissue-light garment bag. As the blob of air flows up the side of a mountain, its pressure lessens, allowing it to expand and cool. The reduced pressure results in reduced temperature.* Measurements show that the

*Recall that in Chapter 15 we treated the cooling of rising air at the microscopic level by considering the behavior of colliding molecules. With thermodynamics, we consider only the macroscopic measurements of temperature and pressure to get the same results. It's nice to analyze things from more than one point of view.

Figure 17-8
A thunderhead is the result of the rapid adiabatic cooling of a rising mass of moist air.

temperature of a blob of dry air will decrease by 10°C for a decrease in pressure that corresponds to an increase in altitude of 1 kilometer. So dry air cools 10°C for each kilometer it rises (Figure 17-6). Air flowing over tall mountains or rising in thunderstorms or cyclones may change elevation by several kilometers. Thus, if a blob of dry air at ground level with a comfortable temperature of 25°C were lifted to 6 kilometers, the temperature would be a frigid −35°C. On the other hand, if air at a typical temperature of −20°C at an altitude of 6 kilometers descends to the ground, its temperature would be a whopping 40°C. A dramatic example of this adiabatic warming is the *chinook*—a wind that blows down from the Rocky Mountains across the Great Plains. Cold air moving down the slopes of the mountains is compressed into a smaller volume and is appreciably warmed. These changes in temperature are produced without any heat input or output whatsoever. The effect of expansion or compression on gases is quite impressive.*

A rising blob cools as it expands. But the surrounding air is cooler with increasing elevation also. The blob will continue to rise as long as it is warmer (less dense) than the surrounding air. If it gets cooler (denser) than its surroundings, it will sink. Under some conditions, large blobs of cold air sink and remain at a low level, with the result that the air above is warmer. When the upper regions of the atmosphere are warmer than the lower regions, we have a **temperature inversion**. Unless any rising warm air is less dense than this upper layer of warm air, it will rise no farther. It is common to see evidence of this over a cold lake where visible gases and particles, such as smoke, spread out in a flat layer above the lake rather than rise and dissipate higher in the atmosphere (Figure 17-9). Temperature inversions trap smog and other thermal pollutants. The smog of Los Angeles is trapped by such an inversion, caused by low-level cold air from the ocean over which is a layer of hot

Figure 17-9
The layer of campfire smoke over the lake indicates a temperature inversion. The air above the smoke is warmer than the smoke, and the air below is cooler.

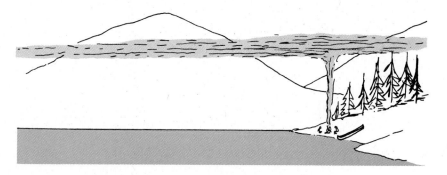

---

*Interestingly enough, when you're flying at high altitudes where outside air temperature is typically −35°C, you're quite comfortable in your warm cabin—but not because of heaters. The process of compressing outside air to a cabin pressure of nearly sea level would normally heat the air to a roasting 55°C (131°F). So air conditioners must be used to extract heat from the pressurized air.

air that moves over the mountains from the hotter Mojave Desert. The mountains aid in holding the trapped air (Figure 17-10). The mountains on the edge of Denver play a similar role in trapping smog beneath a temperature inversion.

Figure 17-10

Smog in Los Angeles is trapped by the mountains and a temperature inversion caused by warm air from the Mojave Desert overlying cool air from the Pacific Ocean.

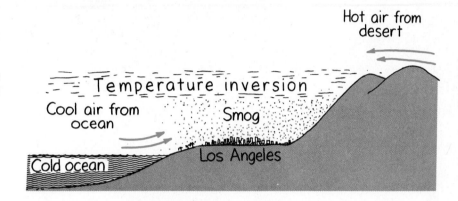

Questions ▶

1. If a blob of air initially at 0°C expands adiabatically while flowing upward alongside a mountain a vertical distance of 1 km, what will its temperature be? When it has risen 5 km?

2. What happens to the air temperature in a valley when cold air blowing across the mountain tops descends into the valley?

3. Imagine a giant dry-cleaner's garment bag full of air at a temperature of −10°C floating like a balloon with a string hanging from it 6 km above the ground. If you were able to yank it suddenly to the ground, what would its approximate temperature be?

Adiabatic blobs are not restricted to the atmosphere, and changes in these blobs do not necessarily happen quickly. Convection of some deep ocean currents takes thousands of years for circulation. The water masses are so huge and conductivities are so low that no appreciable quantities of heat are transferred to or from these blobs over these long periods of time. They are warmed or cooled adiabatically by changes in internal

---

▶ **Answers**

1. At 1-km elevation, its temperature will be −10°C; at 5 km, −50°C.

2. The air is adiabatically compressed and the temperature in the valley is increased. In this way, residents of some valley towns in the Rocky Mountains, such as Salida, Colorado, experience "banana-belt" weather in midwinter.

3. If it is pulled down so quickly that heat conduction is negligible, it would be adiabatically compressed by the atmosphere and its temperature would rise to a piping-hot 50°C (122°F), just as air is heated by compression in a bicycle pump.

Figure 17-11
Do convection currents in the earth's mantle drive the continents as they drift across the global surface? Do rising blobs of molten material cool faster or slower than the surrounding material? Do sinking blobs heat to temperatures above or below those of the surroundings? The answers to these questions are not known at this writing.

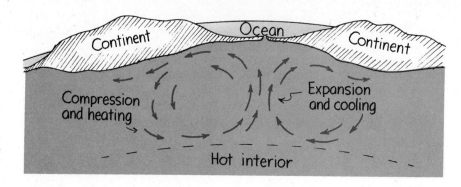

pressure. Changes in adiabatic ocean convection, as evidenced by the recurring El Niño events, have a great effect on the earth's climate. Ocean convection is influenced by the temperature of the ocean floor, which in turn is influenced by convection currents in the molten material that lies beneath the earth's crust (Figure 17-11). Knowledge of the behavior of molten material in the earth's mantle is harder to come by. Once a blob of hot liquid material deep within the mantle begins to rise, will it continue to rise to the earth's crust? Or will its rate of adiabatic cooling find it cooler and denser than its surroundings, at which point it will sink? Is convection self-perpetuating? Geophysical types are currently pondering these questions.

Second Law of Thermodynamics

Suppose we place a hot brick next to a cold brick in a thermally insulated region. We know that the hot brick will cool as it gives heat to the colder brick, which will warm. They will arrive at a common temperature: thermal equilibrium. No energy will be lost, in accordance with the first law of thermodynamics. But pretend the hot brick extracts heat from the cold brick and becomes hotter. Would this violate the first law of thermodynamics? Not if the cold brick becomes correspondingly colder so that the combined energy of both bricks remains the same. If this were to happen, it would not violate the first law. But it would violate the **second law of thermodynamics**. The second law distinguishes the direction of energy transformation in natural processes. The second law of thermodynamics can be stated in many ways, but most simply it is this:

Heat will never of itself flow from a cold object to a hot object.

Heat flows one way, downhill from hot to cold. Heat can be made to flow the other way, but only by imposing external effort. There is an enormous quantity of internal energy in the ocean, but without external effort all this energy cannot be used to light a single flashlight lamp. Gas expands adiabatically and cools. Once cooled, it will never spontaneously contract to its original volume and warm up to its original temperature. Outside work would be necessary. We could squeeze it back, but it would not do so of its own accord. To do so would violate the second law.

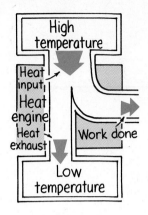

Figure 17-12
When heat flows in any heat engine from a high-temperature place to a low-temperature place, part of this heat can be turned into work. (If work is put into a heat engine, the flow of energy may go from a low-temperature to a high-temperature place, as in a refrigerator or air conditioner.)

The internal energy converted to useful work in heat engines flows from a hot region to a cold region. Every heat engine—whether it be a steam engine, gasoline or diesel engine, gas turbine, or jet engine—will (1) absorb internal energy from a reservoir of higher temperature, (2) convert some of this energy into mechanical work, and (3) discard the remaining energy to some lower-temperature reservoir, usually called a *sink* (Figure 17-12). In a gasoline engine, for example, the burning fuel in the combustion chamber is the high-temperature reservoir, mechanical work is done on the piston, and the discarded energy goes out as exhaust. The second law tells us that no heat engine can convert all the internal energy supplied into mechanical energy. Some heat is always exhausted in the process. Applied to heat engines, the second law may be stated:

When work is done by a heat engine running between two temperatures, T_{hot} and T_{cold}, only some of the input energy at T_{hot} can be converted to work; the rest is rejected as heat at T_{cold}.

There is always heat exhaust, which may be desirable or undesirable. Hot steam discharged in a laundry on a cold winter day may be quite desirable, while the same steam on a hot summer day is something else. When exhausted heat is undesirable, we call it *thermal pollution*.

Before the second law was understood, it was thought that for a very low friction heat engine, nearly all the input energy could be converted to useful work. But not so. It turns out that the fraction of internal energy that can be converted to useful work, even under ideal conditions, depends on the temperature difference between the hot reservoir and the cold reservoir (sink). The French physicist Sadi Carnot discovered this in 1824. His equation is

$$\text{Ideal efficiency} = \frac{T_{hot} - T_{cold}}{T_{hot}}$$

where T_{hot} is the temperature of the hot reservoir and T_{cold} the temperature of the cold one.* Whenever ratios of temperatures are involved, the absolute temperature scale must be used. So T_{hot} and T_{cold} are expressed in kelvins. As an example, suppose the hot reservoir in a steam turbine has a temperature of 400K (127°C) and the cold reservoir one of 300K (27°C). The ideal efficiency is

$$\frac{400 - 300}{400} = \frac{1}{4}$$

This means that even under *ideal* conditions, only 25 percent of the internal energy of the steam can be converted into work, while the remaining 75 percent is waste. Carnot's equation is by no means restricted to steam engines; it holds for heat engines of all kinds. In practice, friction is ever present in all engines, and efficiency is further reduced.

---

*For a derivation of this equation, see almost any introductory-level thermodynamics textbook. A good one is Keith Stowe, *Introduction to Statistical Mechanics and Thermodynamics* (New York: Wiley, 1984).

1. What would be the ideal efficiency of an engine if its hot reservoir and exhaust were the same temperature—say 400K?

2. What would be the ideal efficiency of a machine having a hot reservoir at 400K and a cold reservoir at absolute zero, 0K?

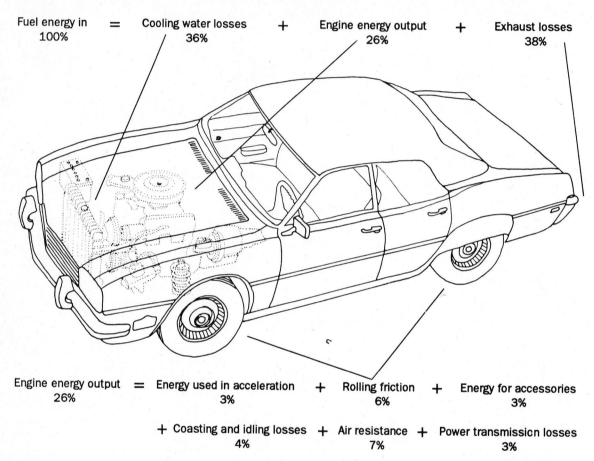

Fuel energy in = Cooling water losses + Engine energy output + Exhaust losses
100%   36%   26%   38%

Engine energy output = Energy used in acceleration + Rolling friction + Energy for accessories
26%   3%   6%   3%

+ Coasting and idling losses + Air resistance + Power transmission losses
4%   7%   3%

Figure 17-13

Only 26 percent of the heat energy produced in burning the gasoline becomes useful mechanical energy, and most of that is lost in friction and in overcoming air resistance. The losses shown are for a typical American car averaged over different driving situations.

The first law of thermodynamics states that energy can be neither created nor destroyed. The second law qualifies this by adding that the form energy takes in transformations "deteriorates" to less useful forms. Energy becomes more diffuse. The energy of gasoline burned in an automobile engine, for example, is not destroyed. It dissipates. Part of it goes into

▶ **Answers**

1. Zero efficiency; $\dfrac{400 - 400}{400} = 0$. So no work output is possible for any heat engine unless a temperature difference exists between the reservoir and the sink.

2. $\dfrac{400 - 0}{400} = 1$; only in this idealized case is an ideal efficiency of 100 percent possible.

moving the car; part heats the engine and surroundings; part may turn a generator and produce light energy for the headlights or supply energy for the radio; part goes out the exhaust. All these parts add up to the energy given up by the gasoline on combustion (Figure 17-13). All the energy exists, and none is really lost. But it is dissipated and is unavailable for driving another car. Like the Big Bang spreading through the universe, the energy from the gasoline will not exist in so concentrated a form again—at least not in this cycle of universal expansion.

Organized energy degenerates into disorganized energy. The organized energy in the form of electricity that goes to the electric lights in homes and office buildings degenerates into heat energy. This is a principal source of heating in many office buildings where the climate is moderate, such as the Transamerica Pyramid in San Francisco. Every bit of the electrical energy in the lamps, even the part that briefly exists in the form of light, is turned into heat energy, which is used to warm the buildings. This explains why the lights are on most of the time. The energy so generated has no further use. All the heat energy in the buildings cannot be reused to light a single lamp without some outside organizational effort. Heat is the graveyard of useful energy.

We see that the "quality" of energy is lowered with each transformation. Energy of an organized form tends to disorganized forms. In this broader regard, the second law can be stated thus:

Natural systems tend to proceed toward a state of greater disorder.

Figure 17-15
Push a heavy crate across a rough floor, and all your work will go into heating the floor and crate. Work against friction turns into disorganized energy.

You finally straighten up your room. Everything is dust-free and in its proper place. A week later it is cluttered again. You stack a pile of pennies on your table, all heads up. Somebody walks by and accidentally bumps into the table and the pennies topple to the floor below, certainly not all heads up. You're on the job at a large company and a co-worker puts up a joke sign that reads, "If you thing thinks are confused now,

Figure 17-16
Molecules of gas go from the bottle to the air and not vice versa.

just wait." These are examples of the tendency of the universe and all that is in it to become disordered. In the broadest sense, that is the message of the second law of thermodynamics. By *disordered*, we mean more random. For example, molecules of gas all moving in harmony make up an orderly state—and also an unlikely state. On the other hand, molecules of gas moving in haphazard directions and speeds make up a disorderly state—a more probable state. If you remove the lid of a bottle containing some gas, the gas molecules escape into the room and make up a more disorderly state (Figure 17-16). Relative order becomes disorder. You would not expect the reverse to happen; that is, you would not expect the molecules to spontaneously order themselves back into the bottle and thereby return to the more ordered containment. Such processes in which disorder goes to order are simply not observed to happen.

Disordered energy can be changed to ordered energy only at the expense of some organizational effort or work input. For example, water in a refrigerator freezes and becomes more ordered because work is put into the refrigeration cycle; gas can be ordered into a small region if work input is supplied to a compressor. But without some imposed work input, processes in which the net effect is an increase in order are not observed in nature.

Entropy

The idea of lowering the "quality" of energy is embodied in the idea of **entropy**.* Entropy is the measure of the *amount of disorder*. The second law states that, in the long run, entropy always increases. Gas molecules escaping from a bottle move from a relatively orderly state to a disorderly state. Disorder increases; entropy increases. Whenever a physical system is allowed to distribute its energy freely, it always does so in a manner such that entropy increases while the available energy of the system for doing work decreases.

Consider the old riddle, "How do you unscramble an egg?" The answer is simple: "Feed it to a chicken." But even then you won't get all your original egg back—egg making has its inefficiencies too. All living organisms, from bacteria to trees to human beings, extract energy from their surroundings and use it to increase their own organization. In living organisms, entropy decreases. But the order in life forms is maintained by increasing entropy elsewhere; life forms plus their waste products have a net increase in entropy.† Energy must be transformed into the living

---

*Entropy can be expressed as a mathematical equation, stating that the increase in entropy ΔS in an ideal thermodynamic system is equal to the amount of heat added to a system ΔQ divided by the temperature T of the system: $\Delta S = \Delta Q/T$.

†Interestingly enough, the American writer Ralph Waldo Emerson, who lived during the time the second law of thermodynamics was the new science topic of the day, philosophically speculated that not everything becomes more disordered with time and cited the example of human thought. Ideas about the nature of things grow increasingly refined and better organized as they pass through the minds of succeeding generations. Human thought is evolving toward more order.

Figure 17-17
Why is the motto of this contractor—"Increasing entropy is our business"—so appropriate?

system to support life. When it isn't, the organism soon dies and tends toward disorder.

The first law of thermodynamics is a universal law of nature for which no exceptions have been observed. The second law, however, is a probabilistic statement. Given enough time, even the most improbable states may occur; entropy may sometimes decrease. For example, the haphazard motions of air molecules could momentarily become harmonious in a corner of the room, just as a barrelful of pennies dumped on the floor could all come up heads. These situations are possible—but not probable. The second law tells us the most probable course of events, not the only possible one.

The laws of thermodynamics are often stated this way: You can't win (because you can't get any more energy out of a system than you put into it), you can't break even (because you can't even get as much energy out as you put in), and you can't get out of the game (entropy in the universe is always increasing).

Summary of Terms

Thermodynamics The study of heat and its transformation to other forms of energy.

Absolute zero The lowest possible temperature that a substance may have; the temperature at which molecules of a substance have their minimum kinetic energy.

Internal energy The total of all molecular energies—kinetic energy and potential energy—internal to a substance. *Changes* in internal energy are of principal concern in thermodynamics.

First law of thermodynamics A restatement of the law of energy conservation, usually as it applies to systems involving changes in temperature: The heat added to a system equals an increase in internal energy plus external work done by the system.

Adiabatic process A process, usually of expansion or compression, wherein no heat enters or leaves a system.

Temperature inversion The condition wherein the upper regions of the atmosphere are warmer than the lower regions.

Second law of thermodynamics Heat will never spontaneously flow from a cold object to a hot object. Also, no machine can be completely efficient in converting energy to work; some input energy is dissipated as heat. And finally, all systems tend to become more and more disordered as time goes by.

Entropy A measure of the disorder of a system. Whenever energy freely transforms from one form to another, the direction of transformation is toward a state of greater disorder and therefore toward one of greater entropy.

Suggested Reading

Stepp, R. D. *Making Theories to Explain the Weather.* 1983. This different and delightful little book is published by the author at the Physics Department, Humboldt State University, Arcata, Calif. 95521.

Review Questions

1. The word *thermodynamics* stems from which Greek words?

2. Is the study of thermodynamics concerned primarily with microscopic or macroscopic processes?

Absolute Zero

3. By how much does a volume of gas at 0°C reduce in size for each 1 Celsius degree decrease when the pressure is held constant?

4. By how much does the pressure of gas at 0°C reduce for each 1 Celsius degree decrease when the volume is held constant?

5. If we assume the gas does not condense to a liquid, what volume is approached for a gas at 0° cooled by 273 Celsius degrees?

6. What is the lowest possible temperature on the Celsius scale? On the Kelvin scale?

Internal Energy

7. Which is always greater, the molecular kinetic energy in a substance or the internal energy in a substance?

8. In the study of thermodynamics, is the principal concern the amount of internal energy in a substance or the changes in internal energy in a substance?

First Law of Thermodynamics

9. How does the law of the conservation of energy relate to the first law of thermodynamics?

10. What happens to the internal energy of a system when mechanical work is done on it? What happens to its temperature?

11. What is the relationship between heat added to a system and the internal energy and external work done by the system?

Adiabatic Processes

12. What condition is necessary for a process to be adiabatic?

13. If work is done *on* a system, does the internal energy of the system increase or decrease? If work is done *by* a system, does the internal energy of the system increase or decrease?

Meteorology and the First Law

14. Name at least six ways in which the temperature of the air can change.

15. What generally happens to the temperature of rising air?

16. What generally happens to the temperature of sinking air?

17. What is a chinook?

18. What is a temperature inversion?

19. Do adiabatic processes apply to water masses in the ocean? Defend your answer.

Second Law of Thermodynamics

20. How does the second law of thermodynamics relate to the direction of heat flow?

21. What three processes occur in every heat engine?

22. What exactly is thermal pollution?

23. If all friction could be removed from a heat engine, would it be 100 percent efficient? Explain.

24. What does it mean to say that energy becomes more diffuse when it transforms?

25. Give at least two examples to distinguish between organized energy and disorganized energy.

26. How much of the electrical energy transformed by a common light bulb becomes heat energy?

27. With respect to orderly and disorderly states, what do natural systems tend to do? Can a disorderly state ever transform to an orderly state? Explain.

Entropy

28. What is the physicist's term for *measure of messiness*?

29. What is the relationship between the second law of thermodynamics and entropy?

30. Distinguish between the first and second laws of thermodynamics in terms of whether or not exceptions occur.

Exercises

1. Consider a piece of metal that has a temperature of 10°C. If it is heated until it is twice as hot (has twice the internal energy), what would its temperature be?

2. On a chilly 10°C day, your friend who likes cold weather says she wishes it were twice as cold. Taking this literally, what temperature would this be?

3. If you shake a can of liquid back and forth, will the temperature of the liquid increase? (Try it and see.)

4. When air is compressed, why does its temperature increase?

5. When you pump a tire with a bicycle pump, the cylinder of the pump becomes hot. Give two reasons why this is so.

6. Is it possible to wholly convert a given amount of mechanical energy into heat? Is it possible to wholly convert a given amount of heat into mechanical energy? Cite examples to illustrate your answers.

7. Why do diesel engines need no spark plugs?

8. Everybody knows that warm air rises. So it might seem that the air temperature should be higher at the top of mountains than down below. But the opposite is most often the case. Why?

9. Imagine a giant dry-cleaner's bag full of air at a temperature of −35°C floating like a balloon with a string hanging from it 10 km above the ground. Estimate its temperature if you were able to yank it suddenly to the earth's surface.

10. We say that energy "deteriorates" in energy transformations. Does this contradict the first law of thermodynamics? Defend your answer.

11. The combined molecular kinetic energies of molecules in a very large container of cold water are greater than the combined molecular kinetic energies in a cup of hot tea. Pretend you partially immerse the teacup in the cold water and that the tea absorbs 10 units of energy from the water and becomes hotter, while the water that gives up 10 units of energy becomes cooler. Would this energy transfer violate the first law of thermodynamics? The second law of thermodynamics? Explain.

12. Why is *thermal pollution* a relative term?

13. Is it possible to construct a heat engine that produces no thermal pollution? Defend your answer.

14. Why is it advantageous to use steam as hot as possible in a steam-driven turbine?

15. How does the ideal efficiency of an automobile relate to the temperature of the engine and the temperature of the environment in which it runs? Be specific.

16. What happens to the efficiency of a heat engine when the temperature of the reservoir into which heat energy is rejected is lowered?

17. To increase the efficiency of a heat engine, would it be better to increase the temperature of the reservoir while holding the temperature of the sink constant or to decrease the temperature of the sink while holding the temperature of the reservoir constant? Explain.

18. What is the ideal efficiency of an automobile engine in which gasoline is heated to 2700K and the outdoor air is 300K?

19. Under what conditions would a heat engine be 100 percent efficient?

20. Suppose one wishes to cool a kitchen by leaving the refrigerator door open and closing the kitchen door and windows. What will happen to the room temperature? Why?

21. Why will a refrigerator with a fixed amount of food consume more energy in a warm room than in a cold room?

22. In buildings that are being electrically heated, is it at all wasteful to turn all the lights on? Is turning all the lights on wasteful if the building is being cooled by air conditioning?

23. Using both the first and the second law of thermodynamics, defend the statement that 100 percent of the electrical energy that goes into lighting a lamp is converted to thermal energy.

24. Is the total energy of the universe becoming more unavailable with time? Explain.

25. Water evaporates from a salt solution and leaves behind salt crystals that have a higher degree of molecular order than the more randomly moving molecules in the salt water. Has the entropy principle been violated? Why or why not?

26. Water put into a freezer compartment in your refrigerator goes to a state of less molecular disorder when it freezes. Is this an exception to the entropy principle? Explain.

PART 4

SOUND

18 Vibrations and Waves

Many things in nature wiggle and jiggle. We call a wiggle in time a *vibration*. A vibration cannot exist in an instant, but needs time to move to and fro. When we strike a bell the vibrations continue for some time before they die down. We call a wiggle in space and time a *wave*. A wave cannot exist in one place but must extend from one place to another. Light and sound are both vibrations that propagate through space as waves, but as two very different kinds. Sound is the propagation of vibrations through a material medium—a solid, liquid, or gas. If there is no medium to vibrate, then no sound is possible. On the other hand, light is a vibration of nonmaterial electric and magnetic fields—a vibration of pure energy. Light can pass through many materials, but it needs none; it can propagate through a vacuum—for example, between the sun and the earth. But the source of all waves—sound, light, or whatever—is something that is vibrating. For sound it can be the vibrating prongs of a tuning fork and for light the vibrations of electrons in an atom. Waves may be periodic vibrations, as in music, or nonperiodic wave pulses, as in the cracking sound of a firecracker or the snapping of one's fingers. We shall begin our study of vibrations and waves by considering the motion of a simple pendulum.

Vibration of a Pendulum

If we suspend a stone at the end of a piece of string, we have a simple pendulum. Pendulums swing to and fro with such regularity that they have long been used to control the motion of clocks. Galileo discovered that the time a pendulum takes to swing to and fro through small distances depends *only on the length of the pendulum and the acceleration of gravity.* * The time of a to-and-fro swing, called the *period*, does not depend on the mass or on the size of the arc through which it swings.

In addition to length, the period of a pendulum depends on the acceleration of gravity. Oil and mineral prospectors use very sensitive pendulums to detect slight differences in this acceleration, which is affected by the densities of underlying formations.

A long pendulum has a longer period than a shorter pendulum; that is, it swings to and fro less frequently than a short pendulum. When

---

*The exact relationship for the period of a simple pendulum is $T = 2\pi\sqrt{l/g}$, where T is the period, l is the length of the pendulum, and g is the acceleration of gravity.

Figure 18-1

Drop two masses, and they accelerate at *g*. Let them slide without friction down the same incline, and they slide together at the same fraction of *g*. Tie them to strings of the same length so they are pendulums, and they swing to and fro in unison. In all cases, the motions are independent of mass.

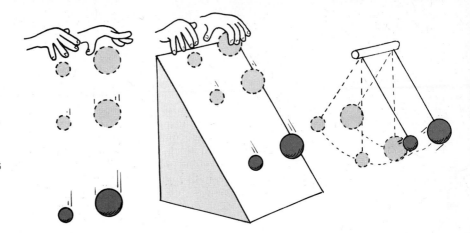

Figure 18-2

Frank Oppenheimer at San Francisco's Exploratorium demonstrates (*a*) a straight line traced by a swinging pendulum bob that leaks sand on the stationary conveyor belt. (*b*) When the conveyor belt is uniformly moving, a sine curve is traced.

walking, we allow our legs to swing with the help of gravity, like a pendulum. In the same way that a long pendulum has a greater period, a person with long legs tends to walk with a slower stride than a person with short legs. This is most noticeable in long-legged animals such as giraffes, horses, and ostriches, which run with a slower gait than do short-legged animals such as dachshunds, hamsters, and mice.

The to-and-fro motion of a pendulum bob along a single plane in a small arc is called *simple harmonic motion*. If a pendulum bob leaks sand while undergoing simple harmonic motion, it will trace out and retrace a short straight line (Figure 18-2*a*). Suppose, however, that we swing such a pendulum above a conveyor belt that moves in a perpendicular direction to the plane of the swinging pendulum (Figure 18-2*b*). The trace it now makes is called a **sine curve**. This curve is a pictorial representation of a wave.

a

b

Wave Description

The sine curve in Figure 18-3 represents a wave. The high points are called the *crests* and the low points are the *troughs*. The term **amplitude** refers to the distance from the midpoint (straight line) to the crest of the wave, or equivalently from the midpoint to the trough. So the amplitude is equal to the maximum displacement of the pendulum from its position of rest.

Figure 18-3
A sine curve.

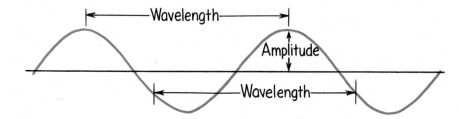

The distance from the top of one crest to the top of the next one is equal to the **wavelength**. Or, equivalently, the wavelength is the distance between successive identical parts of a wave. The wavelength of a wave at the beach is measured in meters, the wavelength of a ripple in a pond in centimeters, and the wavelength of light in millionths of a meter (micrometers).

Figure 18-4
Electrons in the transmitting antenna vibrate 940 000 times each second and produce 940-kHz radio waves.

How frequently a vibration occurs is described by its **frequency**. The frequency of a vibrating pendulum specifies the number of to-and-fro vibrations it makes in a given time (usually 1 second). A complete to-and-fro swing is one vibration. If it occurs in 1 second, we say the frequency is one vibration per second. If two vibrations occur in 1 second, the frequency is two vibrations per second. The unit of frequency is called the **hertz** (Hz), after Heinrich Hertz, who demonstrated the existence of radio waves in 1886. One vibration per second is 1 hertz, two vibrations per second is 2 hertz, and so on. Higher frequencies are measured in kilohertz (kHz), and still higher ones in megahertz (MHz). AM radio waves are broadcast in kilohertz; 940 kHz on your radio dial, for example, corresponds to radio waves that have a frequency of 940 000 vibrations per second. And 101 MHz on your FM dial corresponds to radio waves with a frequency of 101 000 000 hertz. These radio-wave frequencies are the frequencies at which electrons are forced to vibrate in the antenna of a radio station's transmitting tower. The source of all periodic waves is something that vibrates. The frequency of the vibrating source and the frequency of the wave it produces are the same.

If the frequency of an object is known, the time it takes to make a complete vibration, its **period**, can be calculated—and vice versa. If a pendulum makes two vibrations in 1 second, for example, its frequency is 2 hertz. The time needed to complete one vibration—that is, the period of vibration—is $\frac{1}{2}$ second. Or if the vibration period is 3 hertz, then the period is $\frac{1}{3}$ second. Frequency and period are reciprocals of each other:

$$\text{Frequency} = \frac{1}{\text{period}}$$

or vice versa:

$$\text{Period} = \frac{1}{\text{frequency}}$$

Questions

▶

1. What is the frequency in vibrations per second of a 60-Hz wave? What is its period?
2. Gusts of wind make the Sears Building in Chicago sway back and forth at a vibration frequency of about 0.1 Hz. What is its period of vibration?

Wave Motion

Most information about our surroundings comes to us in some form of waves. It is through wave motion that sounds come to our ears, light to our eyes, and electromagnetic signals to our radios and television sets. By *wave motion* we mean the transfer of energy from a source to a distant receiver without the transfer of matter between the two points.

Wave motion can be most easily understood by first considering its simplest case in a horizontally stretched string. If one end of such a string is shaken up and down, a rhythmic disturbance travels along the string. Each particle of the string moves up and down, while at the same time the disturbance moves along the length of the string. The medium returns to its initial condition after the disturbance has passed.

Perhaps a more familiar example of wave motion is a water wave. If a stone is dropped into a quiet pond, waves will travel outward in expanding circles, the centers of which are at the source of the disturbance. In this case we might think that water is being transported with the waves, since water is splashed onto previously dry ground when the waves meet the shore. We should realize, however, that barring obstacles the water

▶ **Answers**

1. A 60-Hz wave vibrates 60 times per second and has a period of $\frac{1}{60}$ of a second.
2. The period is 1/frequency = 1/(0.1 Hz) = 1/(0.1 vibration/s) = 10 s. Each vibration therefore takes 10 s.

will run back into the pond, and things will be much as they were in the beginning: the surface of the water will have been disturbed, but the water itself will have gone nowhere. Again, the medium returns to its initial condition after the disturbance has passed.

Let us consider another example of a wave to illustrate that what is transported from one place to another is a disturbance in a medium, not the medium itself. If you view a field of tall grass from an elevated position on a gusty day, you will see waves travel across the grass. The individual blades of grass do not leave their places; instead, they swing to and fro. Furthermore, if you stand in a narrow footpath, the grass that blows over the edge of the path, brushing against your legs, is very much like the water that doused the shore in our earlier example. While wave motion continues, the tall grass swings back and forth, vibrating between definite limits but going nowhere. When the wave motion stops, the grass returns to its initial position. It is characteristic of wave motion that the medium carrying the wave returns to its initial condition after the disturbance has passed.

Wave Speed

The speed of periodic wave motion is related to the frequency and wavelength of the waves. We can understand this by considering the simple case of water waves (Figures 18-5 and 18-6). Imagine that we fix our eyes at a stationary point on the surface of water and observe the waves passing by this point. If we count the number of crests of water passing this point each second (the frequency) and also observe the distance between crests (the wavelength), we can then calculate the horizontal distance that a particular crest travels each second. We will then know how frequently a distance equal to the wavelength is traveled—that is, the **wave speed**.

Figure 18-5
Water waves.

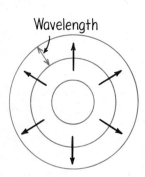

Figure 18-6
A top view of water waves.

For example, if two crests pass a stationary point each second and if the wavelength is 10 meters, then 2 × 10 meters of waves pass by in 1 second. The waves therefore travel at 20 meters per second. We can say

Wave speed = frequency × wavelength

This relationship holds true for all kinds of waves, whether they are water waves, sound waves, or light waves.

Figure 18-7
If the wavelength is 1 m, and one wavelength per second passes the pole, then the speed of the wave is 1 m/s.

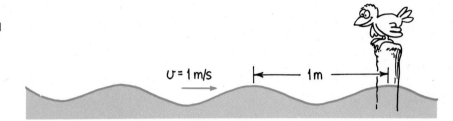

$\upsilon = 1$ m/s   1m

Questions ▶

1. If a train of freight cars, each 10 m long, rolls by you at the rate of three cars each second, what is the speed of the train?

2. If a water wave oscillates up and down three times each second and the distance between wave crests is 2 m, what is its frequency? Its wavelength? Its wave speed?

Transverse Waves

Fasten one end of a rope to a wall and hold the free end in your hand. If you suddenly jerk the free end up and then down, a pulse will travel along the rope and back (Figure 18-8). In this case the motion of the rope (up and down arrows) is at right angles to the direction of wave speed. The right-angled, or sideways, motion is called *transverse motion*. Now shake the rope in simple harmonic motion, and the series of pulses will produce a wave. Since the motion of the medium (the rope in this case) is transverse to the direction the wave travels, this type of wave is called a **transverse wave.**

Waves in the stretched strings of musical instruments and upon the surfaces of liquids are transverse. We will see later that electromagnetic waves, which make up radio waves and light, are also transverse.

▶ **Answers**

1. 30 m/s. We can see this two ways, the Chapter 2 way and the Chapter 18 way: from $\upsilon = d/t = 3 × 10$ m/1 s = 30 m/s, where d is the distance of train that passes you in t seconds. If we compare our train to wave motion, where wavelength corresponds to 10 m and frequency is 3 Hz, then Speed = frequency × wavelength = 3/s × 10 = 30 m/s.

2. The frequency of the wave is 3 Hz, its wavelength is 2 m, and its Wave speed = frequency × wavelength = 3 × 2 = 6 m/s. It is customary to express this as the equation $\upsilon = f\lambda$, where υ is wave speed, f is wave frequency, and λ (the Greek letter lambda) is wavelength.

Figure 18-8
A transverse wave.

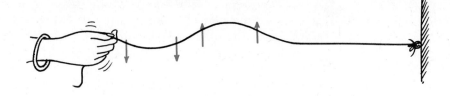

Longitudinal Waves

Not all waves are transverse. Sometimes the particles of the medium move to and fro in the same direction in which the wave travels. The particles move *along* the direction of the wave rather than at right angles to it. This produces a **longitudinal wave**.

Both a transverse and a longitudinal wave can be demonstrated with a spring or a Slinky, as shown in Figure 18-9. A transverse wave is demonstrated by shaking the end of a Slinky up and down. A longitudinal wave is demonstrated by shaking the end of the Slinky back and forth. In this case we see that the medium vibrates parallel to the direction of energy transfer. Part of the Slinky is compressed, and a wave of *compression* travels along the spring. In between successive compressions is a stretched region, called a *rarefaction*. Both compressions and rarefactions travel in the same direction along the Slinky.

Figure 18-9
Both waves transfer energy from left to right. When the end of the Slinky is shaken up and down, a transverse wave is produced. When it's shaken back and forth, a longitudinal wave is produced.

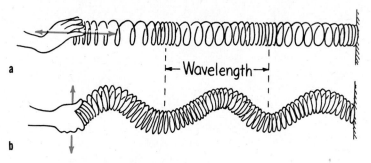

The wavelength of a longitudinal wave is the distance between successive compressions or, equivalently, the distance between successive rarefactions. The most common example of longitudinal waves is sound in air. Each molecule in the air vibrates to and fro about some equilibrium position as the waves move by. We will treat sound waves in detail in the next chapter.

Interference

Whereas a material object like a rock will not share its space with another rock, more than one vibration or wave can exist at the same time in the same space. If we drop two rocks in water, the waves produced by each can overlap and form an **interference pattern**. Within the pattern, wave effects may be increased, decreased, or neutralized.

When the crest of one wave overlaps the crest of another, their individual effects add together. The result is a wave of increased amplitude. This is called *constructive interference* (Figure 18-10). When the crest

Figure 18-10
Constructive and destructive interference in a transverse wave.

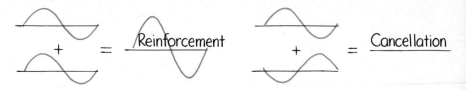

of one wave overlaps the trough of another, their individual effects are reduced. The high part of one wave simply fills in the low part of another. This is called *destructive interference*.

Wave interference is easiest to see in water. In Figure 18-11 we see the interference pattern made when two vibrating objects touch the surface of water. We can see the regions where a crest of one wave overlaps the trough of another to produce regions of zero amplitude. At points along these regions, the waves "arrive out of step." We say they are *out of phase* with each other.

Figure 18-11
Two sets of overlapping water waves produce an interference pattern.

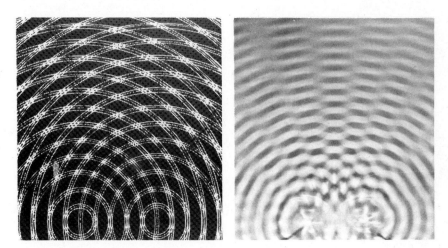

Interference is characteristic of all wave motion, whether the waves are water, sound, or light waves. We will treat the interference of sound in the next chapter and the interference of light in Chapter 28.

Standing Waves

If we tie a rope to a wall and shake the free end up and down, we produce a train of waves in the rope. The wall is too rigid to shake, so the waves are reflected back along the rope. By shaking the rope just right, we can cause the incident and reflected waves to form a **standing wave**, where parts of the rope, called the *nodes*, are stationary.

Standing waves are the result of interference. When two sets of waves of equal amplitude and wavelength pass through each other in opposite directions, the waves are steadily in and out of phase with each other. They produce stable regions of constructive and destructive interference (Figure 18-12).

Figure 18-12
The incident and reflected waves interfere to produce a standing wave.

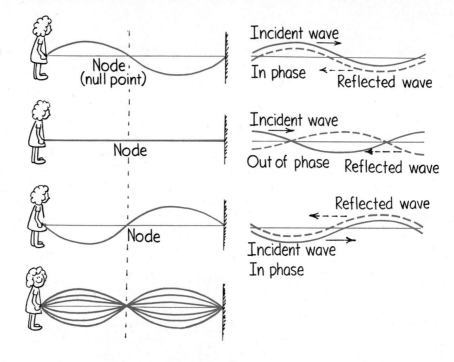

It is easy to make standing waves yourself. Tie a rope or, better, a rubber tube between two firm supports. Shake the tube from side to side with your hand near one of the supports. If you shake the tube with the right frequency, you will set up a standing wave as shown in Figure 18-13. Shake the tube with twice the frequency, and a standing wave of half the previous wavelength, two loops, will result. (The distance between successive nodes is a half wavelength; two loops make up a full wavelength.) Triple the frequency, and a standing wave with one-third the original wavelength, three loops, results—and so forth.

Figure 18-13
(a) Shake the rope until you set up a standing wave of one segment ($\frac{1}{2}$ wavelength). (b) Shake with twice the frequency and produce a wave with two segments (1 wavelength). (c) Shake with three times the frequency and produce three segments ($1\frac{1}{2}$ wavelengths).

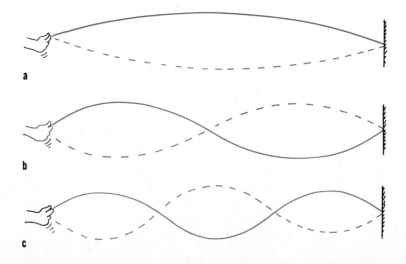

Standing waves are set up in the strings of musical instruments when plucked, bowed, or struck. They are set up in the air in an organ pipe and the air of a soda-pop bottle when air is blown over the top. Standing waves can be set up in a tub of water or a cup of coffee by sloshing it back and forth with the right frequency. Standing waves can be produced in either transverse or longitudinal waves.

Questions ▶

1. Is it possible for one wave to cancel another wave so that no amplitude remains?
2. Suppose you set up a standing wave of three segments, as shown in Figure 18-13c. If·you shake with twice as much frequency, how many wave segments will occur in your new standing wave? How many wavelengths?

Doppler Effect

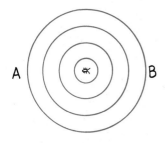

Figure 18-14
Top view of water waves made by a stationary bug jiggling in still water.

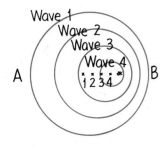

Figure 18-15
Water waves made by a bug swimming in still water.

A pattern of water waves produced by a bug jiggling its legs and bobbing up and down in the middle of a quiet puddle is shown in Figure 18-14. The bug is not going anywhere but is merely treading water in a fixed position. The waves it makes are concentric circles, because wave speed is the same in all directions. If the bug bobs in the water at a constant frequency, the distance between wave crests (the wavelength) is the same for all successive waves. Waves encounter point A as frequently as they encounter point B. This means that the frequency of wave motion is the same at points A and B or anywhere in the vicinity of the bug. This wave frequency is the same as the bobbing frequency of the bug.

Suppose the jiggling bug moves across the water at a speed less than the wave speed. In effect, the bug chases part of the waves it has produced. The wave pattern is distorted and is no longer concentric (Figure 18-15). The center of the outer wave was made when the bug was at the center of that circle. The center of the next smaller wave was made when the bug was at the center of that circle, and so forth. The centers of the circular waves move in the direction of the swimming bug. Although the bug maintains the same bobbing frequency as before, an observer at B would encounter the waves more often. The observer would encounter a higher frequency. This is because each successive wave has a shorter distance to travel and therefore arrives at B more frequently than if the bug weren't moving toward B. An observer at A, on the other hand, encounters a lower frequency because of the longer time between wave-crest arrivals. This occurs because each successive wave travels farther to A as a result of the bug's motion. This change in frequency due to the motion of the source (or receiver) is called the **Doppler effect**.

▶ **Answers**

1. Yes. This is called destructive interference. In a standing wave in a rope, for example, parts of the rope have no amplitude—the nodes.
2. If you impart twice the frequency to the rope, you'll produce a standing wave with twice as many segments. You'll have six segments. Since a full wavelength has two segments, you'll have three complete wavelengths in your standing wave.

Figure 18-16
The pitch of sound increases when the source moves toward you and decreases when the source moves away.

Water waves spread over the flat surface of the water. Sound and light waves, however, travel in three-dimensional space in all directions like an expanding balloon. Just as circular waves are closer together in front of the swimming bug, spherical sound or light waves ahead of a moving source are closer together and encounter a receiver more frequently.

The Doppler effect is evident when you hear the changing pitch of a car horn as the car drives by. When it approaches, the pitch is higher than normal (higher like a higher note on a musical scale). This is because the sound waves are encountering you more frequently. And when the car passes and moves away, you hear a drop in pitch because the waves are encountering you less frequently.

Question ▶ When a source moves toward you, is there an increase or a decrease in wave speed?

The Doppler effect also occurs for light. When a light source approaches, there is an increase in its measured frequency; when it recedes, there is a decrease in its frequency. An increase in frequency is called a *blue shift*, because the increase is toward the high-frequency, or blue, end of the color spectrum. A decrease in frequency is called a *red shift*, referring to the lower-frequency, or red, end of the color spectrum. The galaxies, for example, show a red shift in the light they emit. A measurement of this shift permits a calculation of their speeds of recession. A rapidly spinning star shows a red shift on the side turning away from us and a relative blue shift on the side turning toward us. This enables a calculation of the star's spin rate.

Wave Barriers

When the speed of a source is as great as the speed of the waves it produces, something interesting happens. A "wave barrier" is produced. Consider the bug in our previous example when it swims as fast as the wave speed. Can you see that the bug will keep up with the waves it produces? Instead of the waves getting ahead of the bug, they pile up (superimpose) on one another directly in front of the bug (Figure 18-17).

▶ **Answer**
Neither! It is the *frequency* of a wave that undergoes a change where there is motion of the source, not the wave speed. Be clear about the distinction between frequency and speed. How frequently a wave vibrates is altogether different from how fast it moves from one place to another.

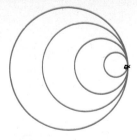

Figure 18-17
Wave pattern made by a bug swimming at wave speed.

The bug encounters a wave barrier. Much effort is required of the bug to swim over this barrier before it can swim faster than wave speed.

The same thing happens when an aircraft travels at the speed of sound. The waves overlap to produce a barrier of compressed air on the leading edges of the wings and other parts of the craft. Considerable thrust is required for the aircraft to push through this barrier. Once through, the craft can fly faster than the speed of sound without similar opposition. The craft is *supersonic*. It is like the bug, which once over its wave barrier finds the water ahead relatively smooth and undisturbed.

Bow Waves

When the bug swims faster than wave speed, ideally it produces a wave pattern as shown in Figure 18-18. It outruns the waves it produces. The waves overlap at the edges, and the pattern made by these overlapping waves is a V shape, called a **bow wave**, which appears to be dragging behind the bug. The familiar bow wave generated by a speedboat knifing through the water is a nonperiodic wave produced by the overlapping of many periodic circular waves.

Figure 18-18
Wave pattern made by a bug swimming faster than wave speed.

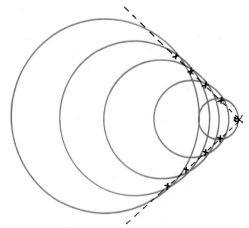

Some wave patterns made by sources moving at various speeds are shown in Figure 18-19. Note that after the speed of the source exceeds wave speed, increased speed produces a narrower V shape.*

Figure 18-19
Patterns made by a bug swimming at successively greater speeds. Overlapping at the edges occurs only when it travels faster than wave speed.

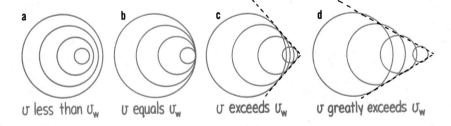

v less than v_w v equals v_w v exceeds v_w v greatly exceeds v_w

---

*Bow waves generated by boats in water are more complex than is indicated here. Our idealized treatment serves as an analogy for the production of the less complex shock waves in air.

Shock Waves

Figure 18-20
Shock waves of a bullet piercing a sheet of Plexiglas. Light is deflected as it passes through the compressed air that makes up the shock waves, which makes them visible.

Figure 18-21
A shock wave.

A speedboat knifing through the water generates a two-dimensional bow wave. A supersonic aircraft similarly generates a three-dimensional **shock wave**. Just as a bow wave is produced by overlapping circles that form a V, a shock wave is produced by overlapping spheres that form a cone. And just as the bow wave of a speedboat spreads until it reaches the shore of a lake, the conical wake generated by a supersonic craft spreads until it reaches the ground.

The bow wave of a speedboat that passes by can splash and douse you if you are at the water's edge. In a sense, you can say that you are hit by a "water boom." In the same way, when the conical shell of compressed air that sweeps behind a supersonic aircraft reaches listeners on the ground below, the sharp crack they hear is described as a **sonic boom**.

We don't hear a sonic boom from slower-than-sound, or subsonic, aircraft because the sound waves reach our ears one at a time and are perceived as one continuous tone. Only when the craft moves faster than sound do the waves overlap to encounter the listener in a single burst. The sudden increase in pressure is much the same in effect as the sudden expansion of air produced by an explosion. Both processes direct a burst of high-pressure air to the listener. The ear is hard pressed to distinguish between the high pressure from an explosion and the high pressure from many overlapping waves.

A water skier is familiar with the fact that next to the high hump of the V-shaped bow wave is a V-shaped depression. The same is true of a shock wave, which actually consists of two cones: a high-pressure cone generated at the bow of the supersonic aircraft and a low-pressure cone that follows at the tail of the craft. The edges of these cones are visible in the photograph of the the supersonic bullet in Figure 18-20. Between these two cones the air pressure rises sharply to above atmospheric pressure, then falls below atmospheric pressure before sharply returning to normal beyond the inner tail cone (Figure 18-22). This overpressure suddenly followed by underpressure intensifies the sonic boom.

A common misconception is that sonic booms are produced when an aircraft breaks through the sound barrier—that is, just as the aircraft

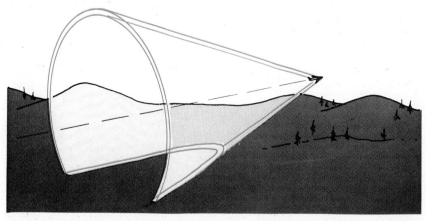

Figure 18-22
The shock wave is actually made up of two cones—a high-pressure cone with the apex at the bow and a low-pressure cone with the apex at the tail. A graph of the air pressure at ground level between the cones takes the shape of the letter N.

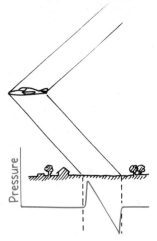

A *B* *C*

Figure 18-23
The shock wave has not yet reached listener A, but is now reaching listener B and has already reached listener C.

surpasses the speed of sound. This is the same as saying that a boat produces a bow wave when it overtakes its own waves. This is not so. The fact is that a shock wave and its resulting sonic boom are swept continuously behind an aircraft traveling faster than sound, just as a bow wave is swept continuously behind a speedboat. In Figure 18-23, listener B is in the process of hearing a sonic boom. Listener C has already heard it, and listener A will hear it shortly. The aircraft that generated this shock wave may have broken through the sound barrier hours ago!

It is not necessary that the moving source emit sound to produce a shock wave. Once an object is moving faster than the speed of sound, it will *make* sound. A supersonic bullet passing overhead produces a crack, which is a small sonic boom. If the bullet were larger and disturbed more air in its path, the crack would be more boomlike. When a lion tamer cracks a circus whip, the cracking sound is actually a sonic boom produced by the tip of the whip when it travels faster than the speed of sound. Both the bullet and the whip are not in themselves sound sources, but when traveling at supersonic speeds they produce their own sound as waves of air are generated to the sides of the moving objects.

Summary of Terms

Sine curve A wave form traced by simple harmonic motion that is uniformly moving in a perpendicular direction, like the wavelike path traced on a moving conveyor belt by a pendulum swinging at right angles above the moving belt.

Amplitude For a wave or vibration, the maximum displacement on either side of the equilibrium (midpoint) position.

Wavelength The distance between successive crests, troughs, or identical parts of a wave.

Frequency For a body undergoing simple harmonic motion, the number of vibrations it makes per unit time. For a series of waves, the number of waves that pass a particular point per unit time.

Hertz The SI unit of frequency. One hertz (symbol Hz) equals one vibration per second.

Period The time required for a vibration or a wave to make a complete cycle; equal to 1/frequency.

Wave speed The speed with which waves pass by a particular point:

$$\text{Wave speed} = \text{frequency} \times \text{wavelength}$$

Transverse wave A wave in which the individual particles of a medium vibrate from side to side in a direction perpendicular (transverse) to the direction in which the wave travels. Light consists of transverse waves.

Longitudinal wave A wave in which the individual particles of a medium vibrate back and forth in a direction parallel (longitudinal) to the direction in which the wave travels. Sound consists of longitudinal waves.

Interference pattern The pattern formed by superposition of different sets of waves that produces mutual reinforcement in some places and cancellation in others.

Standing wave A stationary wave pattern formed in a medium when two sets of identical waves pass through the medium in opposite directions.

Doppler effect The change in frequency of wave motion resulting from motion of the sender or receiver.

Bow wave The V-shaped wave made by an object moving across a liquid surface at a speed greater than the wave velocity.

Shock wave The cone-shaped wave made by an object moving at supersonic speed through a fluid.

Sonic boom The loud sound resulting from the incidence of a shock wave.

Review Questions

1. What is a *wiggle in time* called? A *wiggle in space and time*?
2. What is the source of all waves?
3. Distinguish between sound waves and light waves.

Vibration of a Pendulum
4. What feature about a pendulum makes it useful in clocks?
5. What is the period of a pendulum?
6. If a pendulum takes one second to make a complete to-and-fro swing, how great is its period?
7. The period of a certain pendulum is 2 s, and the period of another is 1 s. Which pendulum is longer?

Wave Description
8. How is a sine curve related to a wave?
9. Distinguish between these different parts of a wave: amplitude, wavelength, frequency, and period.
10. How many vibrations per second are represented in a radio wave of 101 MHz?
11. How do *frequency* and *period* relate to each other?

Wave Motion
12. Exactly what is it that moves from source to receiver in wave motion?
13. Does the medium in which a wave moves travel along with the wave itself? Give examples.

Wave Speed
14. What is the relationship among frequency, wavelength, and wave speed?
15. As the frequency of a wave of constant speed is increased, does the wavelength increase or decrease?

Transverse Waves
16. In what direction are the vibrations compared to the direction of wave travel in a transverse wave?

Longitudinal Waves
17. In what direction are the vibrations compared to the direction of wave travel in a longitudinal wave?
18. How do compressions and rarefactions of longitudinal waves compare to crests and troughs of transverse waves?

Interference
19. Distinguish between *constructive interference* and *destructive interference*.
20. What does it mean to say one wave is *out of phase* with another?
21. What kinds of waves are characterized by interference?

Standing Waves
22. What causes a standing wave?
23. What is a *node*?

Doppler Effect
24. Is it the frequency of a wave, the wave speed, or both that changes in the Doppler effect?
25. Is the Doppler effect characteristic of longitudinal waves, transverse waves, or both?
26. What do a *blue shift* and a *red shift* for light mean?

Wave Barriers
27. How do the speed of a source of waves and the speed of the waves themselves compare for the production of a wave barrier?

Bow Waves
28. How do the speed of a source of waves and the speed of the waves themselves compare for the production of a bow wave?
29. How does the V shape of a bow wave depend on the speed of the source?

Shock Waves

30. How is a bow wave similar to a shock wave?

31. How does the V shape of a shock wave depend on the speed of the source?

32. True or false: A sonic boom occurs only when an aircraft breaks through the sound barrier.

33. True or false: In order for an object to produce a sonic boom, it must be an emitter of sound.

Home Project

Tie a rubber tube, a spring, or a rope to a fixed support and produce standing waves. See how many nodes you can produce.

Exercises

1. Will a pendulum clock that is accurate at sea level gain time or lose time when located high in the mountains?

2. If a pendulum is shortened, does its frequency increase or decrease? What about its period?

3. You swing an empty suitcase to and fro at its natural frequency. If the case were filled with books, would the natural frequency be lower, greater, or the same as before?

4. Is the time required to swing to and fro (the period) on a playground swing longer or shorter when you stand rather than sit? Explain.

5. Why do short people tend to walk with quicker strides than tall people?

6. Why do you bend your arms when you run?

7. What kind of motion should you impart to the nozzle of a garden hose so that the resulting stream of water approximates a sine curve?

8. What kind of motion should you impart to a stretched coiled spring (or Slinky) to provide a transverse wave? A longitudinal wave?

9. If a gas tap is turned on for a few seconds, someone a couple of meters away will hear the gas escaping long before she smells it. What does this indicate about the way in which sound waves travel?

10. If we double the frequency of a vibrating object, what happens to its period?

11. What is the frequency of the second hand of a clock? The minute hand? The hour hand?

12. You dip your finger repeatedly into a puddle of water and make waves. What happens to the wavelength if you dip your finger more frequently?

13. How does the frequency of vibration of a small object floating in water compare to the number of waves passing it each second?

14. How far does a wave travel during one period?

15. A rock is dropped in water, and waves spread over the flat surface of the water. What becomes of the energy in these waves when they die out?

16. The wave patterns seen in Figure 18-5 are composed of circles. What does this tell you about the speed of waves moving in different directions?

17. Why is lightning seen before thunder is heard?

18. Would there be a Doppler effect if the source of sound were stationary and the listener in motion? Why or why not? In which direction should the listener move to hear a higher frequency? A lower frequency?

19. When you blow your horn while driving toward a stationary listener, an increase in frequency of the horn is heard by the listener. Would the listener hear an increase in horn frequency if he were also in a car traveling at the same speed in the same direction as you are? Explain.

20. Is there a Doppler effect when the motion of the source is at right angles to a listener? Explain.

21. How does the Doppler effect aid police in detecting speeding motorists?

22. How does the phenomenon of interference play a role in the production of bow or shock waves?

23. Does the conical angle of a shock wave open wider, narrow down, or remain constant as a supersonic aircraft increases its speed?

24. If the sound of an airplane does not come from the part of the sky where the plane is seen, does this imply that the airplane is traveling faster than the speed of sound? Explain.

25. Does a sonic boom occur at the moment when an aircraft exceeds the speed of sound? Explain.

26. Why is it that a subsonic aircraft, no matter how loud it may be, cannot produce a sonic boom?

27. Imagine a super-fast fish that is able to swim faster than the speed of sound in water. Would such a fish produce a "sonic boom"?

28. A skipper on a boat notices wave crests passing his anchor chain every 5 s. He estimates the distance between wave crests to be 15 m. He also correctly estimates the speed of the waves. What is this speed?

29. A weight suspended from a spring is seen to bob up and down over a distance of 20 cm twice each second. What is its frequency? Its period? Its amplitude?

30. Radio waves travel at the speed of light—300 000 km/s. What is the wavelength of radio waves received at 100 MHz on your radio dial?

19 Sound

If a tree fell in the middle of a deep forest hundreds of kilometers away from any living being, would there be a sound? Different people will answer this question in different ways. "No," some will say, "sound is subjective and requires a listener. If there is no listener, there will be no sound." "Yes," others will say, "a sound is not something in a listener's head. A sound is an objective thing." Discussions like this one often are beyond agreement because the participants fail to realize that they are arguing not about the nature of sound but about the definition of the word. Either side is right, depending on which definition is taken, but investigation can proceed only when a definition has been agreed on. The physicist usually takes the objective position and defines sound as a form of energy that exists whether or not it is heard and goes on from there to investigate its nature.

Origin of Sound

Most of the sounds we hear are waves produced by the vibrations of material objects. The sustained sounds from a piano, violin, or guitar are produced by the vibrating strings; the sound from a saxophone is produced by a vibrating reed; the sound from a flute is produced by a fluttering column of air blown by the mouthpiece. Your voice results from the vibration of your vocal cords.

In each of these cases a vibrating source sends a disturbance through the surrounding medium, usually air, in the form of longitudinal waves. Under ordinary conditions, the frequency of the vibrating source and the frequency of the sound waves produced are the same. We describe our subjective impression about the frequency of sound by the word *pitch*. A high-pitched sound like that from a piccolo has a high-vibration frequency, while a low-pitched sound like that from a foghorn has a low-vibration frequency. The human ear can normally hear pitches corresponding to the range of frequencies between a lower limit of about 20 hertz and an upper limit of about 20 000 hertz. (Many animals hear much higher frequencies.) As we grow older, the limits of this human hearing range shrink. Sound waves with frequencies below 20 hertz are called **infrasonic**, and those with frequencies above 20 000 are called **ultrasonic**. We cannot hear infrasonic and ultrasonic sound waves.

Nature of Sound in Air

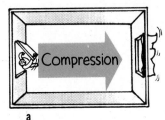

a

b

Figure 19-1
(*a*) When the door is opened, a compression travels across the room. (*b*) When the door is closed, a rarefaction travels across the room. (Adapted from *The New College Physics: A Spiral Approach* by A. V. Baez. Copyright © 1967 by W. H. Freeman and Company. Reprinted by permission.)

When we clap our hands, the sound produced is nonperiodic. It consists of a wave *pulse* that travels out in all directions. The pulse disturbs the air in the same way that a similar pulse would disturb a coiled spring or a Slinky. Each particle moves to and fro along the direction of the expanding wave.

For a clearer picture of this process, consider a long room as shown in Figure 19-1*a*. At one end is an open window with a curtain over it. At the other end is a door. When we open the door, we can imagine the door pushing the molecules next to it away from their initial positions and into their neighbors. The neighboring molecules, in turn, push into their neighbors, and so on, like a compression traveling along a spring, until the curtain flaps out the window. A pulse of compressed air has moved from the door to the curtain. This pulse of compressed air is called a **compression**.

When we close the door (Figure 19-1*b*), the door pushes neighboring air molecules out of the room. This produces an area of low pressure behind the door. Neighboring molecules then move into it, leaving a zone of lower pressure behind them. We say this zone of lower-pressure air is *rarefied*. Other molecules farther away from the door, in turn, move into these rarefied regions, and a disturbance again travels across the room. This is evidenced by the curtain, which flaps inward. This time the disturbance is a **rarefaction**.

Like all wave motion, it is not the medium itself that travels across the room, but the energy carried by the pulse. In both cases the pulse travels from the door to the curtain. We know this because in both cases the curtain moves after the door is opened or closed. If you continually swing the door open and closed in periodic fashion, you can set up a wave of periodic compressions and rarefactions that will make the curtain swing in and out of the window. On a much smaller but more rapid scale, this is what happens when a tuning fork is struck. The periodic vibrations of the tuning fork and the waves it produces are considerably higher in frequency and lower in amplitude than those caused by the swinging door. You don't notice the effect of sound waves on the curtain, but you are well aware of them when they meet your sensitive eardrums.

Consider sound waves in the tube shown in Figure 19-2. For simplicity, only the waves that travel in the tube are depicted. When the prong of the tuning fork next to the tube moves toward the tube, a compression enters the tube. When the prong swings away in the opposite direction,

Figure 19-2
Compressions and rarefactions traveling from the tuning fork through the tube.

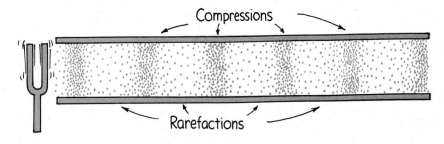

Figure 19-3

(*a*) The radio loudspeaker is a paper cone that vibrates in rhythm with an electric signal. The sound that is produced sets up similar vibrations in the microphone, which are displayed on an oscilloscope. (*b*) The shape of the waveform on the screen of the oscilloscope reveals information about the sound.

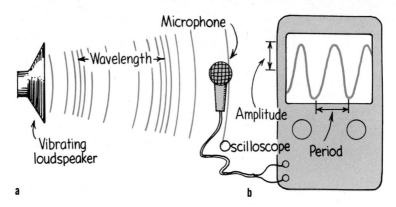

a   b

Figure 19-4

Waves of compressed and rarefied air, produced by the vibrating cone of the loudspeaker, make up the pleasing sound of music.

Media That Transmit Sound

a rarefaction follows the compression. It's like a Ping-Pong paddle moving to and fro in a room packed with Ping-Pong balls. As the source vibrates, a periodic series of compressions and rarefactions is produced. The frequency of the vibrating source and the frequency of the wave it produces are the same.

Pause to reflect on the physics of sound while you are quietly listening to your radio sometime. The radio loudspeaker is a paper cone that vibrates in rhythm with an electrical signal. Air molecules that constantly impinge on the vibrating cone of the speaker are themselves set into vibration. These in turn vibrate against neighboring molecules, which in turn do the same, and so on. As a result, rhythmic patterns of compressed and rarefied air emanate from the loudspeaker, showering the whole room with undulating motions. The resulting vibrating air sets your eardrum into vibration, which in turn sends cascades of rhythmic electrical impulses along the cochlear nerve canal and into the brain. And you listen to the sound of music.

Most sounds that we hear are transmitted through the air. However, any elastic substance—whether solid, liquid, gas, or plasma—can transmit sound.* Compared to solids and liquids, air is a relatively poor conductor of sound. You can hear the sound of a distant train more clearly if your ear is placed against the rail. Similarly, a watch placed on a table beyond hearing distance can be heard if you place your ear to the table. Native Americans often placed their ears to the ground to hear sounds that were inaudible in the air. The next time you go swimming, have a friend at some distance click two rocks together beneath the surface of the water while you are submerged. You will find that liquid is an excellent conductor of sound. If you and your friend ever happen to be in a vacuum, perform your last experiment and have her click two rocks together. You won't hear a thing! Sound will not travel in a vacuum. The transmission of sound requires a medium; if there is nothing to compress and expand, there can be no sound.

---

*Recall from Chapter 11 that an elastic substance like steel (in contrast to putty, which is an inelastic substance) has resilience and can transmit energy with little loss.

Speed of Sound

If we watch a person at a distance chopping wood or hammering, we can easily see that the blow takes place an appreciable time before its sound reaches our ears. Thunder is heard after a flash of lightning. These common experiences show that sound requires a recognizable time to travel from one place to another. The speed of sound depends on wind conditions, temperature, and humidity. It does not depend on the loudness or the frequency of the sound; all sounds travel at the same speed. The speed of sound in dry air at 0°C is about 330 meters per second, nearly 1200 kilometers per hour. Water vapor in the air increases this speed lightly. Sound travels faster through warm air than cold air. This is to be expected because the faster-moving molecules in warm air bump into each other more often and therefore can transmit a pulse in less time.* For each degree rise in temperature above 0°C, the speed of sound in air increases by 0.6 meter per second. So in air at a normal room temperature of about 20°C, sound travels at about 340 meters per second. In water, sound travels about four times as fast as it does in air, while in steel, the speed of sound is about fifteen times as great as in air.

Questions ▶

1. Do compressions and rarefactions in a sound wave travel in the same direction or in opposite directions from one another?
2. What is the approximate distance of a thunderstorm when you note a 3-s delay between the flash of lightning and the sound of thunder?

Reflection of Sound

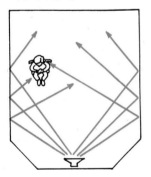

Figure 19-5
The angle of incident sound is equal to the angle of reflected sound.

We call the reflection of sound an *echo*. The fraction of sound energy that is reflected from a surface is large if the surface is rigid and smooth and less if the surface is soft and irregular. Sound energy that is not reflected is transmitted or absorbed.

Sound reflects from a smooth surface the same way that light does—the angle of incidence is equal to the angle of reflection (Figure 19-5). Sometimes when sound reflects from the walls, ceiling, and floor of a room, the surfaces are too reflective and the sound is garbled. This is due to multiple reflections called **reverberations**. However, if the reflective surfaces are too absorbent, the sound level is low and hall sounds are dull and lifeless. Reflection makes sound lively and full, as you know from singing in the shower. In the design of an auditorium or concert hall, a balance must be found between reverberation and absorption. The study of sound properties is called *acoustics*.

▶ **Answers**

1. They travel in the same direction.
2. Assuming the speed of sound in air is about 340 m/s, in 3 s it will travel (340 × 3) = 1020 m. We can assume no time delay for the light, so the storm is slightly more than 1 km away.

*The speed of sound in a gas is about $\frac{3}{4}$ the average speed of molecules.

Figure 19-6
The plastic plates above the orchestra reflect both light and sound. Adjusting them is quite simple: what you see is what you hear.

It is often advantageous to place highly reflective surfaces behind the stage to direct sound out to an audience. Above the stage in some concert halls are suspended reflecting surfaces. The ones in the symphony hall in San Francisco are large, shiny plastic surfaces that also reflect light (Figure 19-6). A listener can look up at these reflectors and see the reflected images of the members of the orchestra (the plastic reflectors are somewhat curved, which increases the field of view). Both sound and light obey the same law of reflection, so if a reflector is oriented so that you can see a particular musical instrument, rest assured that you will hear it also. Sound from the instrument will follow the line of sight to the reflector and then to you.

Refraction of Sound

Sound waves bend when parts of the wave fronts travel at different speeds. This happens in uneven winds or when sound is traveling through air of uneven temperatures. This bending of sound is called **refraction**. On a warm day, the air near the ground may be appreciably warmer than the rest of the air, so the speed of sound near the ground increases. Sound waves therefore tend to bend away from the ground, resulting in sound that does not seem to carry well.

We hear thunder when the lightning is reasonably close, but we often fail to hear thunder for distant lightning because of refraction. The sound travels slower at higher altitudes and bends away from the ground. The opposite often occurs on a cold day or at night when the layer of air near the ground is colder than the air above. Then the speed of sound near the ground is reduced. The higher speed of the wave fronts above causes a bending of the sound toward the earth, resulting in sound that can be heard over considerably longer distances (Figure 19-7).

Figure 19-7
The wave fronts of sound are bent in air of uneven temperatures.

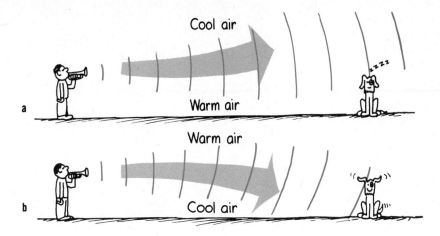

The refraction of sound occurs under water, where the speed of sound varies with temperature. This poses a problem for surface vessels that bounce ultrasonic waves off the bottom of the ocean to chart its features. This poses a blessing to submarines that wish to escape detection. Because of thermal gradients and layers of water at different temperatures, the refraction of sound leaves gaps or "blind spots" in the water. This is where submarines hide. If it weren't for refraction, submarines would be easy to detect.

The multiple reflections and refractions of ultrasonic waves are used by physicians in a technique for harmlessly "seeing" inside the body without the use of X rays. When high-frequency sound (ultrasound) enters the body, it is reflected more strongly from the outside of organs than from their interior, and a picture of the outline of the organs is obtained. When ultrasound is incident upon a moving object, the reflected sound has a slightly different frequency. Using this Doppler effect, a physician can "see" the beating heart of a fetus as early as 11 weeks (Figure 19-8).

The ultrasound echo technique may be relatively new to humans, but not to bats or dolphins. It is well known that bats emit ultrasonic squeaks and locate objects by their echos. Dolphins do this and more.* The ultrasonic waves emitted by a dolphin enable it to "see" through the bodies

Figure 19-8
A 5-month-old fetus displayed on a viewing screen by ultrasound.

---

*The primary sense of the dolphin is acoustic, for vision is not a very useful sense in the often murky and dark depths of the ocean. Whereas sound is a passive sense for us, it is an active sense for the dolphin who sends out sounds and then perceives its surroundings on the basis of the echoes that come back. What's more interesting, the dolphin can reproduce the sonic signals that paint the mental image of its surroundings; thus, the dolphin probably communicates its experience to other dolphins by communicating the full acoustic image of what is "seen," placing it directly in the minds of other dolphins. The dolphin needs no word or symbol for "fish," for example, but communicates an image of the real thing—perhaps with emphasis highlighted by selective filtering, as we similarly communicate a musical concert to others via various means of sound reproduction. Small wonder that the language of the dolphin is very unlike our own!

Figure 19-9
A dolphin emits ultrahigh-frequency sound to locate and identify objects in its environment. Distance is sensed by the time delay between sending sound and receiving its echo, and direction is sensed by differences in time for the echo to reach its two ears. A dolphin's main diet is fish and, since hearing in fish is limited to fairly low frequencies, they are not alerted to the fact they are being hunted.

of other animals and people. Skin, muscle, and fat are almost transparent to dolphins, so they "see" a thin outline of the body—but the bones, teeth, and gas-filled cavities are clearly apparent. Physical evidence of cancers, tumors, heart attacks, and even emotional state can all be "seen" by the dolphin—as humans have only recently been able to do with ultrasound.

Question   An oceanic depth-sounding vessel surveys the ocean bottom with ultrasonic sound that travels 1530 m/s in seawater. How deep is the water if the time delay of the echo from the ocean floor is 2 s?

Energy in Sound Waves

Wave motion of all kinds possesses energy of varying degrees. For electromagnetic waves we find a great amount of energy in X rays, somewhat less in ultraviolet radiation, and less still in visible sunlight. By comparison, the energy in sound is extremely small. That's because producing sound requires only a small amount of energy. For example, 10 000 000 people talking at the same time would produce a sound energy equal only to the energy needed to light a common flashlight. Hearing is possible only because of our remarkably sensitive ears. Very few microphones can detect sound that is softer than what we can hear.

Sound energy is dissipated in the form of heat in air. For waves of higher frequency, the sound energy is transformed into heat more rapidly than for waves of lower frequencies. As a result, sound of low frequencies will travel farther through air than sound of higher frequencies. That's why the foghorns of ships are of a low frequency.

▶ **Answer** 1530 m.

Forced Vibrations

If we strike an unmounted tuning fork, the sound from it may be rather faint. If we hold the same fork against a table and strike it with the same force, the sound is louder. This is because the table is forced to vibrate, and with its larger surface it will set more air in motion. The table will be forced into vibration by a fork of any frequency. This is a case of **forced vibration**. The vibration of a factory floor caused by the running of heavy machinery is an example of forced vibration. A more pleasing example is given by the sounding boards of stringed instruments.

Natural Frequency

Figure 19-10
The natural frequency of the smaller bell is higher than that of the larger bell, and it rings at a higher pitch.

When someone drops a wrench on a concrete floor, we are not likely to mistake its sound for that of a baseball bat. This is because the two objects vibrate differently when they strike the floor. Tap a wrench and the vibrations it makes are different from the vibrations of a baseball bat, or of anything else. When disturbed, any object composed of an elastic material will vibrate at its own special set of frequencies, which together form its special sound. We speak of an object's **natural frequency**, which depends on factors such as the elasticity and shape of the object. Bells and tuning forks, of course, vibrate at their own characteristic frequencies. And interestingly enough, most things from planets to atoms and almost everything else in between have a springiness to them and vibrate at one or more natural frequencies. A natural frequency is one at which the least amount of energy is required to produce forced vibrations.

Resonance

Figure 19-11
Pumping a swing in rhythm with its natural frequency produces a large amplitude.

When the frequency of forced vibrations on an object matches the object's natural frequency, a dramatic increase in amplitude occurs. This phenomenon is called **resonance**. Literally, *resonance* means "resounding," or "sounding again." Putty doesn't resonate because it isn't elastic, and a dropped handkerchief is too limp. In order for something to resonate, it needs a force to pull it back to its starting position and enough energy to keep it vibrating.

A common experience illustrating resonance occurs on a swing. When pumping a swing, we pump in rhythm with the natural frequency of the swing. More important than the force with which we pump is the timing. Even small pumps or small pushes from someone else, if delivered in rhythm with the frequency of the swinging motion, produce large amplitudes. A common classroom demonstration of resonance is illustrated with a pair of tuning forks adjusted to the same frequency and spaced a meter or so apart. When one of the forks is struck, it sets the other fork into vibration. This is a small-scale version of pushing a friend on a swing—it's the timing that's important. When a series of sound waves impinges on the fork, each compression gives the prong of the fork a tiny push. Since the frequency of these pushes corresponds to the natural frequency of the fork, the pushes will successively increase the amplitude of vibration. This is because the pushes occur at the right time and repeatedly occur in the same direction as the instantaneous motion of the fork.

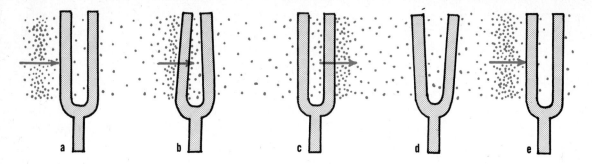

a   b   c   d   e

Figure 19-12

Stages of resonance. (a) The first compression meets the fork and gives it a tiny and momentary push; (b) the fork bends and then (c) returns to its initial position just at the time a rarefaction arrives and (d) overshoots in the opposite direction. Just when it returns to its initial position (e) the next compression arrives to repeat the cycle. Now it bends farther because it is moving.

If the forks are not adjusted for matched frequencies, the timing of pushes is off, and resonance will not occur. When you tune your radio set, you are similarly adjusting the natural frequency of the electronics in the set to match one of the many surrounding signals. The set then resonates to one station at a time instead of playing all stations at once.

Resonance is not restricted to wave motion. It occurs whenever successive impulses are applied to a vibrating object in rhythm with its natural frequency. In 1831, cavalry troops marching across a footbridge near Manchester, England, inadvertently caused the bridge to collapse when they marched in rhythm with the bridge's natural frequency. Since then, it has become customary to order troops to "break step" when crossing bridges. A more recent bridge disaster was caused by wind-generated resonance (Figure 19-13).

The effects of resonance are all about us. Resonance underscores not only the sound of music, but the color of autumn leaves, the height of ocean tides, the operation of lasers, and a vast multitude of phenomena that add to the beauty of the world about us.

Figure 19-13

In 1940, four months after being completed, the Tacoma Narrows Bridge in the state of Washington was destroyed by wind-generated resonance. The mild gale produced a fluctuating force in resonance with the natural frequency of the bridge, steadily increasing the amplitude until the bridge collapsed.

Interference

Figure 19-14

Wave interference for transverse and longitudinal waves.

Sound waves, like any waves, can be made to exhibit interference. Recall that wave interference was discussed in the last chapter. A comparison of interference for transverse waves and longitudinal waves is shown in Figure 19-14. In either case, when the crests of one wave overlap the crests of another wave, increased amplitude results. Or when the crest of one wave overlaps the trough of another wave, decreased amplitude results. In the case of sound, the crest of a wave corresponds to a compression, and the trough of a wave corresponds to a rarefaction. Interference occurs for both transverse and longitudinal waves.

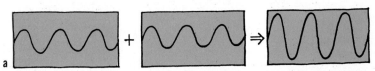

The superposition of two identical transverse waves in phase produces a wave of increased amplitude.

The superposition of two identical longitudinal waves in phase produces a wave of increased intensity.

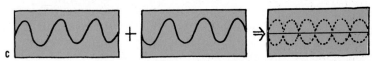

Two identical transverse waves that are out of phase destroy each other when they are superimposed.

Two identical longitudinal waves that are out of phase destroy each other when they are superimposed.

An interesting case of sound interference is illustrated in Figure 19-15. If you are an equal distance from two sound speakers that emit identical tones of constant frequency, the sound is louder because the effects of the two speakers add. The compressions and rarefactions of the tones arrive in step, or in phase. However, if you move to the side so that the paths from the speakers to you differ by a half wavelength, then the rarefactions from one speaker will be filled in by the compressions from the other speaker. This is destructive interference. It is just as if the crest of one water wave exactly filled in the trough of another water wave. If the region is devoid of any reflecting surfaces, little or no sound will be heard!

If the speakers emit a whole range of frequencies, not all wavelengths will destructively interfere for a given difference in path lengths. Interference of this type is usually not a problem, because there is usually enough reflection of sound to fill in canceled spots. Nevertheless, "dead spots" are sometimes evident in poorly designed theaters or music halls, where sound waves reflect off walls and interfere with nonreflected waves to produce zones of low amplitude. When you move your head a few centimeters in either direction, you may hear a noticeable difference.

Figure 19-15

Interference of sound waves. (*a*) Waves arrive in phase and interfere constructively when the path lengths from the speakers are the same. (*b*) Waves arrive out of phase and interfere destructively when the path lengths differ by half a wavelength (or $\frac{3}{2}$, $\frac{5}{2}$, etc.).

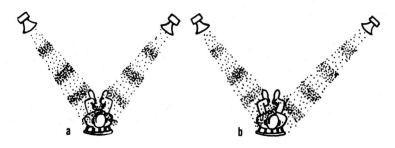

Beats

When two tones of slightly different frequencies are sounded together, a fluctuation in the loudness of the combined sounds is heard; the sound is loud, then faint, then loud, then faint, and so on. This periodic variation in the loudness of sound is called **beats** and is due to interference. Strike two slightly mismatched tuning forks, and because one fork vibrates at a different frequency than the other, the vibrations of the forks will be momentarily in step, then out of step, then in again, and so on. When the combined waves reach our ears in step—say, when a compression from one fork overlaps a compression from the other—the sound is a maximum. A moment later, when the forks are out of step, a compression from one fork is met with a rarefaction from the other, resulting in a minimum. The sound that reaches our ears throbs between maximum and minimum loudness and produces a tremolo effect.

Figure 19-16

The interference of two sound sources of slightly different frequencies produces beats.

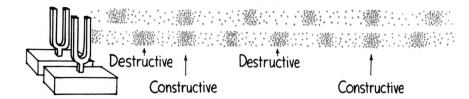

We can understand beats by considering the analogous case of two people walking side by side with different strides. At some moment they will be in step, a little later out of step, then in step again, and so on. Imagine that one person, perhaps with longer legs, takes exactly seventy steps in 1 minute, and the shorter person takes seventy-two steps in the same time. The shorter person gains two steps per minute on the taller person. A little thought will show that they will both be momentarily in step twice each minute. In general, if two people with different strides walk together, the number of times they are in step each minute is equal to the difference in the frequencies of the steps. This applies also to the pair of tuning forks. If one fork undergoes 264 vibrations each second and the other fork vibrates 262 times per second, they will be in step twice each second. A beat frequency of 2 hertz will be heard. The overall tone will correspond to the average frequency, 263 hertz.

Question ▶ What is the beat frequency when a 262-Hz and a 266-Hz tuning fork are sounded together? A 262-Hz and a 272-Hz fork?

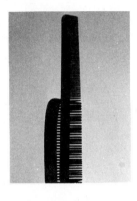

Figure 19-17
The unequal spacings of the combs produce a moiré pattern that is similar to beats.

If we overlap two combs of different teeth spacings, we'll see a moiré pattern that is related to beats (Figure 19-17). The number of beats per length will equal the difference in the number of teeth per length for the two combs.

Beats can occur with any kind of wave and provide a practical way to compare frequencies. To tune a piano, for example, a piano tuner listens for beats produced between a standard tuning fork and a particular string on the piano. When the frequencies are identical, the beats disappear. The members of an orchestra tune up by listening for beats between their instruments and a standard tone produced by a piano or some other instrument.

Beats are utilized by dolphins in surveying the motions of things around them. When a dolphin sends out sound signals, beats may be produced when the echoes it receives interfere with the sound it sends. When there is no relative motion between the dolphin and the object returning the sound, the sending and receiving frequencies are the same and no beats occur. But when there is relative motion, the echo has a different frequency due to the Doppler effect, and beats are produced when the echo and emitted sound combine. Dolphins rely more on sound than light, and it's nice that increases and decreases of beat frequencies are components of their acoustic images, just as similar differences in light frequencies are components of our visual images.

Radio Broadcasts

A radio receiver emits sound but, interestingly enough, it doesn't receive sound waves. A radio receiver, like a television set, receives *electromagnetic waves*—actually low-frequency light waves. These waves, which we will treat in detail in Part 6, are fundamentally different from sound waves—not only in their completely different nature, but in their extremely high frequencies, which are way beyond the range of human hearing.

▶ **Answer**
For the 262-Hz and 266-Hz forks, the ear will hear 264 Hz, which will beat at 4 Hz (266 − 262). For the 272-Hz and 262-Hz forks, 267 Hz will be heard, and some people will hear it throb ten times each second. Beat frequencies greater than 10 Hz are too rapid to be heard normally.

Amplitude modulation

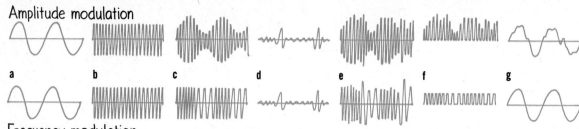

a   b   c   d   e   f   g

Frequency modulation

Figure 19-18

AM and FM radio signals.
(*a*) Sound waves enter a microphone. (*b*) Radio-frequency carrier wave produced by transmitter without sound signal. (*c*) Carrier wave modulated by signal. (*d*) Static interference. (*e*) Carrier wave and signal affected by static. (*f*) Radio receiver cuts out negative half of carrier wave. (*g*) Signal remaining is rough for AM because of static but is smooth for FM because the tips of the wave form are clipped without loss to the signal.

Every radio station has an assigned frequency at which it broadcasts. The electromagnetic wave transmitted at this frequency is the **carrier wave**. The relatively low frequency sound signal to be communicated is superimposed on the much higher frequency carrier wave in two principal ways: by slight variations in amplitude that match the audio frequency or by slight variations in frequency (Figure 19-18). This impression of the sound wave on the higher-frequency radio wave is **modulation**. When the amplitude of the carrier wave is modulated we call it AM, or **amplitude modulation**. AM stations broadcast in the range of 535 to 1605 kilohertz. When the frequency of the carrier wave is modulated, we call it FM, or **frequency modulation**. FM stations broadcast in the higher-frequency range of 88 to 108 megahertz. Amplitude modulation is like changing the brightness of a constant-color light bulb. Frequency modulation is like changing the color of a constant-intensity light bulb.

Turning the knob of a radio receiver to select a particular station is like adjusting the movable masses on the prongs of a tuning fork to make it resonate to the sound produced by another fork. In choosing a radio station you adjust the frequency of an electrical circuit inside the radio receiver to match and resonate to the frequency of the station you want. You sort out one carrier wave from many. Then the impressed sound signal is separated from the carrier wave, amplified, and fed to the loudspeaker. It's nice to hear only one station at a time!

Question   Is it correct to say that in every case, without exception, any radio wave travels faster than any sound wave?

▶ **Answer**

Yes, because any radio wave travels at the speed of light. A radio wave is an electromagnetic wave—in a very real sense, a low-frequency light wave. A sound wave, on the other hand, is a mechanical disturbance propagated through a material medium by material particles that vibrate against one another. In air, the speed of sound is about 340 m/s, about one-millionth the speed of a radio wave. Sound travels faster in other media, but in no case at the speed of light. No sound wave can travel as fast as light.

Summary of Terms

Infrasonic Describes a sound of a frequency too low to be heard by the normal human ear—below 20 hertz.

Ultrasonic Describes a sound of a frequency too high to be heard by the normal human ear—above 20 000 hertz.

Compression Condensed region of the medium through which a longitudinal wave travels.

Rarefaction Rarefied region, or region of lessened pressure, of the medium through which a longitudinal wave travels.

Reverberation Re-echoed sound.

Refraction The bending of a wave either through a nonuniform medium or from one medium to another, caused by differences in wave speed.

Forced vibration The setting up of vibrations in an object by a vibrating force.

Natural frequency A frequency at which an elastic object naturally tends to vibrate, so that minimum energy is required to produce a forced vibration or to continue vibration at that frequency.

Resonance The result of forced vibrations in an object when an applied frequency matches the natural frequency of the object.

Beats A series of alternate reinforcements and cancellations produced by the interference of two sets of superimposed waves of different frequencies, heard as a throbbing effect in sound waves.

Carrier wave The wave, usually of radio frequency, whose characteristics are modified in the process of modulation.

Modulation The process of impressing one wave system upon another of higher frequency.

Amplitude modulation (AM) A type of modulation in which the amplitude of the carrier wave is varied above and below its normal value by an amount proportional to the amplitude of the impressed wave.

Frequency modulation (FM) A type of modulation in which the frequency of the carrier wave is varied above and below its normal frequency by an amount that is proportional to the amplitude of the impressed signal. In this case, the amplitude of the modulated carrier wave remains constant.

Review Questions

1. How do physicists usually define sound?

Origin of Sound

2. What is the source of all sounds?

3. What is the relationship between *frequency* and *pitch*?

4. What is the average range of human hearing?

5. Distinguish between *infrasonic* and *ultrasonic* sound waves.

Nature of Sound in Air

6. Distinguish between a *compression* and a *rarefaction*.

7. Do compressions and rarefactions travel in the same or opposite directions from one another in a wave? Cite evidence to support your answer.

8. How does the paper cone of a radio loudspeaker emit sound?

Media That Transmit Sound

9. Compared to solids and liquids, how does air rank as a conductor of sound?

10. Why will sound not travel in a vacuum?

Speed of Sound

11. What common factors does the speed of sound depend upon? What common things does it *not* depend upon?

12. What is the speed of sound in dry air at 0°C?

13. Does sound travel faster in warm air than in cold air? Defend your answer.

14. How does the speed of sound in water and steel compare to the speed of sound in air?

Reflection of Sound

15. What is an *echo*?

16. What is the law of reflection for sound?

17. What exactly is a *reverberation*?

Refraction of Sound

18. What is the cause of refraction?

19. Does sound tend to bend upward or downward when its speed is less near the ground?

20. Why does sound sometimes refract under water?

21. There is a difference between the way we passively see our surroundings by daylight and the way we actively probe our surroundings with a searchlight in the darkness. Which of these ways of perceiving our surroundings is most like the way a dolphin perceives its environment?

Energy in Sound Waves

22. Which is normally greater, the energy in ordinary sound or the energy in ordinary light?

23. What ultimately becomes of the energy of sound in the air?

24. Why will sound of low frequencies travel farther in air than sound of high frequencies?

25. Why are foghorns low in frequency?

Forced Vibrations

26. Why will a struck tuning fork sound louder when it is held against a table?

27. Give at least three examples of forced vibration.

Natural Frequency

28. Give at least two factors that determine the natural frequency of an object.

29. How does the amount of energy required to produce forced vibrations in an object relate to the natural frequency of the object?

Resonance

30. Distinguish between *forced vibrations* and *resonance*.

31. What is required to make an object resonate?

32. How does a radio select one station at a time instead of playing all stations at once?

33. Why do troops "break step" when crossing a bridge?

Interference

34. Is it possible for one wave to cancel another? Defend your answer.

35. What kind of waves exhibit interference?

36. Distinguish between *constructive interference* and *destructive interference*.

37. What is responsible for "dead spots" in poorly designed theaters or concert halls?

Beats

38. What physical phenomenon underlies the production of beats?

39. What beat frequency will be heard when a 370-Hz and a 374-Hz tuning fork are sounded together?

40. How is the phenomenon of beats useful for tuning musical instruments?

Radio Broadcasts

41. How does a radio wave differ from a sound wave?

42. Distinguish between a carrier wave and a sound wave.

43. Distinguish between AM and FM.

44. How is tuning a radio like adjusting a pair of tuning forks for resonance?

Home Projects

1. Suspend the wire grille from a refrigerator or an oven from a string, the ends of which you hold to your ears. Let a friend gently stroke the grille with pieces of broom straw and other objects. The effect is best appreciated in a relaxed condition with your eyes closed. Be sure to try this!

2. In the bathtub, submerge your head and listen to the sound you make when clicking your fingernails together or tapping the tub beneath the water surface. Compare the sound with that you make when both the source and your ears are above the water. At the risk of getting the floor wet, slide back and forth in the tub at different frequencies and see how the amplitude of the sloshing waves quickly builds up when you slide in rhythm with the waves. (The latter of these projects is most effective when you are alone in the tub.)

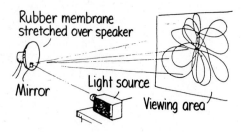

3. Stretch a piece of balloon rubber not too tightly over a radio loudspeaker. Glue a small, very lightweight piece of mirror, aluminum foil, or polished metal near one edge. Project a narrow beam of light on the mirror while your favorite music is playing and observe the beautiful patterns that are reflected on a screen or wall.

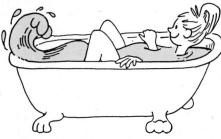

Rubber membrane stretched over speaker

Mirror

Light source

Viewing area

Exercises

1. Why do flying bees buzz?

2. A cat can hear sound frequencies up to 70 000 Hz. Bats send and receive ultrahigh-frequency squeaks up to 120 000 Hz. Which hears shorter wavelengths, cats or bats?

3. At the stands of a race track you notice smoke from the starter's gun before you hear it fire. Explain.

4. When a sound wave moves past a point in air, are there changes in the density of air at this point? Explain.

5. At the instant that a high-pressure region is created just outside the prongs of a vibrating tuning fork, what is being created inside between the prongs?

6. Why is it so quiet after a snowfall?

7. If a bell is ringing inside a bell jar, we can no longer hear it if the air is pumped out, but we can still see it. What differences in the properties of sound and light does this indicate?

8. Why is the moon described as a "silent planet"?

9. As you pour water into a glass, you repeatedly tap the glass with a spoon. As the tapped glass is being filled, does the pitch of the sound increase or decrease? (What should you do to answer this question?)

10. If the speed of sound depended on its frequency, how would distant music sound?

11. If the frequency of sound is doubled, what change will occur in its speed? In its wavelength?

12. Why does sound travel faster in warm air?

13. Why does sound travel faster in moist air? (*Hint:* At the same temperature, water vapor molecules have the same average kinetic energy as the heavier nitrogen and oxygen molecules in the air. How, then, do the average speeds of H_2O molecules compare with those of N_2 and O_2 molecules?)

14. Would the refraction of sound be possible if the speed of sound were unaffected by wind, temperature, and other conditions? Defend your answer.

15. Why can the tremor of the ground from a distant explosion be felt before the sound of the explosion can be heard?

16. What kinds of wind conditions would make sound more easily heard at long distances? Less easily heard at long distances?

17. What is the wavelength of a 340-Hz tone in air? What is the wavelength of a 34 000-Hz ultrasonic wave in air?

18. Ultrasonic waves have many applications in technology and medicine. One advantage is that large intensities can be used without danger to the ear. Cite another advantage of their short wavelength. (*Hint:* Why do microscopists use blue light rather than white light to see detail?)

19. An oceanic depth-sounding vessel surveys the ocean bottom with ultrasonic sound that travels 1530 m/s in seawater. How deep is the water if the time delay of the echo from the ocean floor is 6 s?

20. A bat flying in a cave emits a sound and receives its echo 1 s later. How far away is the cave wall?

21. Why is an echo weaker than the original sound?

22. How can you estimate the distance of a distant thunderstorm?

23. A rule of thumb for estimating the distance in kilometers between an observer and a lightning stroke is to divide the number of seconds in the interval between the flash and the sound by 3. Is this rule correct?

24. You watch a distant lady driving nails into her front porch at a regular rate of 1 stroke per second. You hear the sound of the blows exactly synchronized with the blows you see. And then you hear one more blow after you see her stop hammering. How far away is she?

25. If a single disturbance some unknown distance away sends out both transverse and longitudinal waves that travel with distinctly different speeds in the medium, such as in the ground during earthquakes, how could the origin of the disturbance be located?

26. Why will marchers at the end of a long parade following a band be out of step with marchers near the front?

27. Why do soldiers break step in marching over a bridge?

28. Why is the sound of a harp soft in comparison to other stringed instruments such as the piano, guitar, and bass fiddle?

29. Apartment dwellers will testify that bass notes are more distinctly heard from music played in nearby apartments. Why do you suppose lower-frequency sounds get through walls, floors, and ceilings more easily?

30. If the handle of a tuning fork is held solidly against a table, the sound from the tuning fork becomes louder. Why? How will this affect the length of time the fork keeps vibrating? Explain.

31. Why does a tuning fork eventually stop vibrating?

32. A special device can transmit sound out of phase from a noisy jackhammer to its operator using earphones. Over the noise of the jackhammer, the operator can easily hear your voice while you are unable to hear his. Explain.

33. What beat frequencies are possible with tuning forks of frequencies 256, 259, and 261 Hz, respectively?

34. Suppose a piano tuner hears 3 beats per second when listening to the combined sound from her tuning fork and the piano note being tuned. After slightly tightening the string, she hears 5 beats per second. Should she loosen or further tighten the string?

35. Contrast the means by which humans and dolphins perceive and communicate their environments.

20 Musical Sounds

Most of the sounds we hear are noises. The impact of a falling object, the slamming of a door, the roaring of a motorcycle, and most of the sounds from traffic in city streets are noises. Noise corresponds to an irregular vibration of the eardrum produced by some irregular vibration. If we make a diagram to indicate the pressure of the air on the eardrum as it varies with time, the graph corresponding to a noise might look like that shown in Figure 20-1a. The sound of music has a different character, having more or less periodic tones—or musical "notes." (Musical instruments can make noise as well!) The graph representing a musical sound has a shape that repeats itself over and over again (Figure 20-1b). Such graphs can be displayed on the screen of an oscilloscope when the electrical signal from a microphone is fed into the input terminal of this device.

Figure 20-1
Graphical representations of noise and music.

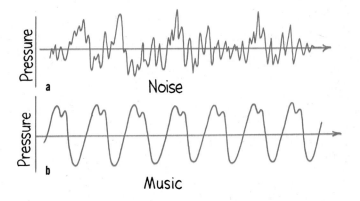

The line that separates music and noise is thin and subjective. To some contemporary composers, it is nonexistent. Some people consider contemporary music and music from other cultures to be noise. Differentiating these types of music from noise becomes a problem of aesthetics. However, differentiating traditional music—that is, Western classical music and most types of popular music—from noise presents no problem. A person with total hearing loss could distinguish between these by using an oscilloscope.

Musicians usually speak of musical tones in terms of three principal characteristics: pitch, loudness, and quality.

Pitch

The **pitch** of a sound corresponds to frequency. Rapid vibrations of the sound source produce a shrill high note, whereas slow vibrations produce a deep low note. We speak of the pitch of a sound in terms of its position in the musical scale. When concert A is struck on a piano, a hammer strikes two or three strings, each of which vibrates 440 times in 1 second. The pitch of concert A corresponds to 440 hertz.*

Different musical notes are obtained by changing the frequency of the vibrating sound source. This is usually done by altering the size, the tightness, or the mass of the vibrating object. A guitarist or violinist, for example, adjusts the tightness, or tension, of the strings of the instrument when tuning them. Then different notes can be played by altering the length of each string by "stopping" it with the fingers.

In wind instruments, the length of the vibrating air column can be altered (trombone and trumpet) or holes in the side of the tube can be opened and closed in various combinations (saxophone, clarinet, flute) to change the pitch of the note produced.

High-pitched sounds used in music are most often less than 4000 hertz, but the average human ear can hear sounds up to 20 000 hertz. Some people can hear tones of higher pitch than this, and so can most dogs. In general, the upper limit of hearing in people gets lower as they grow older. A high-pitched sound is often inaudible to an older person and yet may be clearly heard by a younger one. So by the time you can really afford that high-fidelity music system, you may not be able to appreciate the difference.

Loudness

Figure 20-2
James displays a sound signal on a cathode-ray oscilloscope.

The intensity of sound depends on pressure variations within the sound wave; it depends on the amplitude. (More specifically, as with any type of wave, intensity is proportional to the square of the amplitude.) Sound intensity is a purely objective and physical attribute of a sound wave and can be measured by various acoustical instruments (and the oscilloscope in Figure 20-2). **Loudness**, on the other hand, is a physiological sensation; it depends on intensity but in a complicated way. If you turned a radio up till it seemed about twice as loud as before, you would have to increase the power output and, therefore, the intensity by approximately eight times. Although the pitch of a sound can be judged very accurately, our ears are not very good at judging loudness. The loudest sounds we can tolerate have intensities a million million times greater than the faintest sounds.

The relative loudness of a sound the ear hears is called the sound level and is measured in decibels, a unit named after Alexander Graham Bell and abbreviated dB. Some common sounds and their sound levels are compared in Table 20-1.

---

*Interestingly enough, concert A varies from as low as 436 Hz to as high as 448 Hz in different symphony orchestras.

Table 20-1
Common sources and
sound levels

Source of sound	Sound level (dB)
Jet airplane 30 m away	140
Air-raid siren nearby	125
Amplified music	115
Riveter	95
Busy street traffic	70
Conversation in home	65
Quiet radio in home	40
Whisper	20
Rustle of leaves	10
Threshold of hearing	0

Decibel ratings are logarithmic: a sound of 10 decibels is 10 times as loud as 0 decibels, the threshold of hearing; however, 20 decibels is not twice as loud again, but 10 times again louder—or 100 times as loud as the threshold of hearing. Accordingly, 30 decibels is 1000 times the threshold of hearing and 40 decibels is 10 000 times. So 60 decibels represents sound intensity a million times greater than 0 decibels; 80 decibels represents sound 100 times as intense as 60 decibels. Physiological hearing damage begins at exposure to 85 decibels, the degree depending on the length of exposure and on frequency characteristics. Damage from loud sounds can be temporary or permanent, depending on whether the organs of Corti, the receptor organs in the inner ear, are impaired or destroyed. A single burst of sound can produce vibrations in the organs intense enough to tear them apart. Less intense, but severe, noise can interfere with cellular processes in the organs that cause their eventual breakdown. Unfortunately, the cells of these organs do not regenerate. You know that you'll ruin your sense of sight if you stare into a source of light as bright as the sun. Please don't ruin your sense of hearing by subjecting yourself to loud sounds.

Question ▶ Is hearing permanently impaired when attending concerts, clubs, or functions that feature very loud music?

▶ **Answer**

Yes, depending on how loud, how long, how near, and how often. Some music groups have emphasized loudness over quality. Tragically, as hearing becomes more and more impaired, members of the group (and their fans) require louder and louder sounds for the same perceived stimulation. Hearing loss caused by sounds is particularly common in the frequency range of 2000–5000 Hz. Human hearing is normally most sensitive around 3000 Hz.

Quality

We have no trouble distinguishing between the tone from a piano and a like-pitched tone from a clarinet. Each of these tones has a characteristic sound that differs in **quality**, or timbre. Most musical sounds are composed of a superposition of many frequencies called **partial tones**, or simply *partials*. The lowest frequency, called the **fundamental frequency**, determines the pitch of the note. Partial tones that are whole multiples of the fundamental frequency are called **harmonics**. A tone that has twice the frequency of the fundamental is the second harmonic, a tone with three times the fundamental frequency is the third harmonic, and so on (Figure 20-3).* It is the variety of partial tones that gives a musical note its characteristic quality.

Figure 20-3
Modes of vibration of a guitar string.

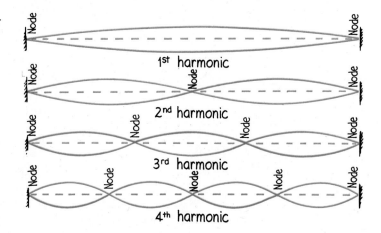

1ˢᵗ harmonic

2ⁿᵈ harmonic

3ʳᵈ harmonic

4ᵗʰ harmonic

Thus, if we strike middle C on the piano, we produce a fundamental tone with a pitch of about 262 hertz and also a blending of partial tones of two, three, four, five, and so on times the frequency of middle C. The number and relative loudness of the partial tones determine the quality of sound associated with the piano. Sound from practically every musical instrument consists of a fundamental and partials. Pure tones, those having only one frequency, can be produced electronically. The electronic synthesizer, for example, produces pure tones and mixtures of these to give a vast variety of musical sounds.

Figure 20-4
A composite vibration of the fundamental mode and the third harmonic.

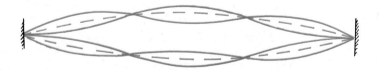

*Not all partial tones present in a complex tone are integer multiples of the fundamental. Unlike the harmonics of woodwinds and brasses, stringed instruments such as a piano produce "stretched" partial tones that are nearly, but not quite, harmonics. This is an important factor in tuning pianos and happens because the stiffness of the strings adds a little bit of restoring force to the tension.

The quality of a tone is determined by the presence and relative intensity of the various partials. The sound produced by a tone from the piano and the one produced by one of the same pitch from a clarinet have different qualities that the ear recognizes because their partials are different. A pair of tones of the same pitch with different qualities have either different partials or a difference in the relative intensity of the partials.

Musical Instruments

Figure 20-5
Sounds from the piano and clarinet differ in quality.

Conventional musical instruments can be grouped into one of three classes: those in which the sound is produced by vibrating strings, those in which the sound is produced by vibrating air columns, and those in which the sound is produced by *percussion* (the vibrating of a two-dimensional surface).

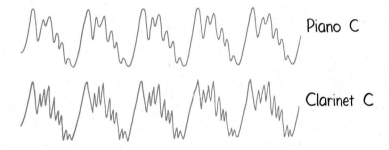

Piano C

Clarinet C

In a stringed instrument, the vibration of the strings is transferred to a sounding board and then to the air, with considerable dissipation of energy. Stringed instruments are low-efficiency producers of sound, so to compensate for this, we find relatively large string sections in orchestras. A smaller number of the high-efficiency wind instruments sufficiently balances a much larger number of violins.

In a wind instrument, the sound is a vibration of an air column in the instrument. There are various ways to set the air columns into vibration. In brass instruments such as trumpets, French horns, and trombones, vibrations of the player's lips interact with standing waves that are set up by acoustic energy reflected within the instrument by the flared bell. The lengths of the vibrating air columns are manipulated by valves that add or subtract extra segments. In woodwinds such as clarinets, oboes, and saxophones, a stream of air produced by the musician sets a reed vibrating, whereas in fifes, flutes, and piccolos, the musician blows air against the edge of a hole to produce a fluttering stream that sets the air columns into vibration.

In percussion instruments such as drums and cymbals, a two-dimensional membrane or elastic surface is struck to produce sound. The fundamental tone produced depends on the geometry, the elasticity, and, in some cases, the tension of the surface. Changes in pitch result from changing the tension in the vibrating surface; depressing the edge of a drum membrane with the hand is one way of accomplishing this. Different modes of vibration can be set up by striking the surface in different

places. In the kettledrum, the shape of the kettle changes the frequency of the drum. As in all musical sounds, the quality depends on the number and relative loudness of the partial tones.

Electronic musical instruments differ markedly from conventional musical instruments. Instead of strings that must be bowed, plucked, or struck, or reeds over which air must be blown, or diaphragms that must be tapped to produce sounds, some electronic instruments use electrons to generate the signals that make up musical sounds. Others start with sound from an acoustical instrument and then modify it. Electronic music demands of the composer and player an expertise beyond just knowing musicology. It brings a powerful new tool to the hands of the musician.

Musical Scales

As early as 530 BC Pythagoras found that notes played together on stringed instruments were pleasing to the ear when the ratios of the string lengths were the ratios of whole numbers, and he found them displeasing when the ratios were not (due in part to the beats produced). It was Galileo, who assisted his father, Vincenzo Galilei (a composer and musical theorist), in a set of experiments on the physical basis of harmony, who introduced the notion of frequency. A succession of notes of increasing frequency makes up a **musical scale**. Many scales exist that are the products of many cultures throughout various ages. The simplest scale in Western culture is the *just major scale*, the familiar "do-re-mi-fa-sol-la-ti-do," where each note is named by the letters A, B, C, D, E, F, and G. In this scale the pitch or frequency of successive notes goes up in the ratios given in Table 20-2. The ratio in the frequencies of two sounds is related to the musical interval of the two notes. The ratios in the table are those of adjacent frequencies. When two notes are in a 2:1 ratio, such as C′ above middle C, the interval is called an *octave*, the Latin word for the "eight" notes in this interval. Middle C is taken to be 264 hertz on this scale, and C′ is twice this—528 hertz. As we progress to notes higher than those in the table, each succeeding C has twice the frequency of the preceding C, each D has twice the frequency of the preceding D, and so on. The frequency of each note doubles with each octave (Figure 20-6).

The intervals between notes in the just scale, even the whole steps, are not all the same. The differences in pitch between the notes "mi" and "fa" and between "ti" and "do" are about half that of the other intervals. These intervals are called *half steps* (there are no "black notes" on the piano between these keys). To obtain these half steps when music is played in keys other than C, extra notes raised a half interval called

Figure 20-6
Piano keyboard. Low C (C‴) is 33 Hz, and successive harmonics double in frequency. The pitch that corresponds to 264 Hz (there are no "black notes" on the piano between these keys) is called *middle C*.

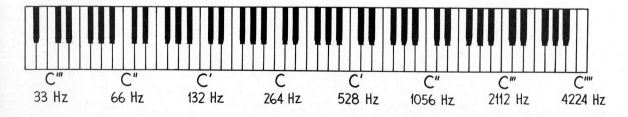

C‴	C″	C′	C	C′	C″	C‴	C⁗
33 Hz	66 Hz	132 Hz	264 Hz	528 Hz	1056 Hz	2112 Hz	4224 Hz

Table 20-2
Diatonic C Major scale

Note	Letter name	Frequency (Hz)	Frequency ratio	Interval
do	C	264		
			$\frac{9}{8}$	Whole
re	D	297		
			$\frac{10}{9}$	Whole
mi	E	330		
			$\frac{16}{15}$	Half
fa	F	352		
			$\frac{9}{8}$	Whole
sol	G	396		
			$\frac{10}{9}$	Whole
la	A	440		
			$\frac{9}{8}$	Whole
ti	B	495		
			$\frac{16}{15}$	Half
do	C'	528		

sharps (♯) or lowered a half interval called flats (♭) are required. For the key of C these are the black keys on a piano keyboard.

Because of the variation in musical intervals, when a piano has been tuned for a just scale in any one key, the frequencies are somewhat off for the other keys. For this reason, pianos are not usually tuned to the just scale, but to the *equal-tempered scale*. On this scale there are thirteen notes and twelve intervals in one octave, and the ratio between all successive notes is exactly the same (the twelfth root of 2, or 1.05946). This results in the same frequency for C♯ and D♭, D♯ and E♭, and so on, which is not generally the case for the just scale. The standard frequency for the equal-tempered scale is A = 440 hertz, the only note in the octave that agrees with its counterpart on the just scale. All other notes differ slightly between the two scales, as you can see by comparing Tables 20-2 and 20-3. The equal-tempered scale is now used for most music written in the Western world.

Table 20-3
Equal-tempered chromatic scale

Note	Frequency (Hz)	Frequency ratio	Interval
C	262	$\sqrt[12]{2}$	
			Half
C♯ or D♭	277	$\sqrt[12]{2}$	
			Half
D	294	$\sqrt[12]{2}$	
			Half
D♯ or E♭	311	$\sqrt[12]{2}$	
			Half
E	330	$\sqrt[12]{2}$	
			Half
F	349	$\sqrt[12]{2}$	
			Half
F♯ or G♭	370	$\sqrt[12]{2}$	
			Half
G	392	$\sqrt[12]{2}$	
			Half
G♯ or A♭	415	$\sqrt[12]{2}$	
			Half
A	440	$\sqrt[12]{2}$	
			Half
A♯ or B♭	466	$\sqrt[12]{2}$	
			Half
B	494	$\sqrt[12]{2}$	
			Half
C'	524	$\sqrt[12]{2}$	

Figure 20-7
A microscopic view of the grooves in a phonograph record.

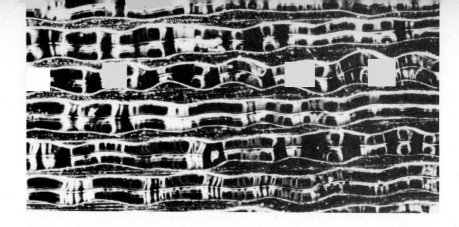

Fourier Analysis

Did you ever look closely at the grooves in a phonograph record? And did you notice the variations in the width of the grooves—variations that cause the phonograph needle that rides in the groove to vibrate? And did you ever wonder how all the distinct vibrations made by the various pieces of an orchestra are captured by the single-wave groove of the record? The sound of an oboe when captured by the groove of phonograph records and displayed on an oscilloscope screen looks like Figure 20-8a. This wave corresponds to the electronic signal produced by the vibrating needle. It also corresponds to the amplified signal that activates the loudspeaker of the sound system and to the amplitude of air vibrating against the eardrum. Figure 20-8b shows the wave form of a clarinet, and Figure 20-8c shows the wave form when oboe and clarinet are sounded together.

Figure 20-8
Wave forms of (a) an oboe, (b) a clarinet, and (c) the oboe and clarinet sounded together.

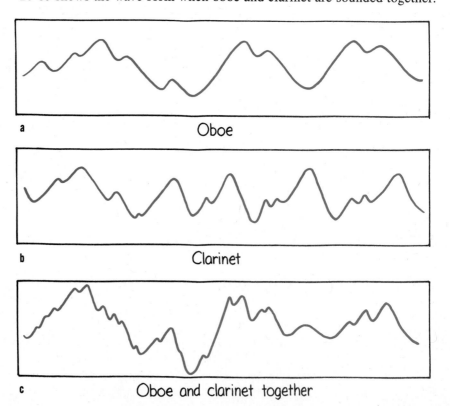

a   Oboe

b   Clarinet

c   Oboe and clarinet together

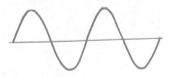

Figure 20-9
A sine wave.

The shape of the wave in Figure 20-8c is the net result of shapes *a* and *b* interfering with each other. If we know *a* and *b*, it is a simple thing to create *c*. But it is a far different problem to discern in *c* the shapes of *a* and *b* that make it up. Looking only at shape *c*, we cannot unscramble the oboe from the clarinet.

But play the record on the phonograph, and our ears will at once know what instruments are being played, what notes they are playing, and what their relative loudness is. Our ears break the overall signal into its component parts automatically.

In 1822 the French mathematician Joseph Fourier discovered a mathematical regularity to the component parts of periodic wave motion. He found that even the most complex periodic wave motion could be broken down into simple sine waves. A sine wave is the simplest of waves, having a single frequency (Figure 20-9). Fourier found that all periodic waves may be broken down into constituent sine waves of different amplitudes and frequencies. The mathematical operation for doing this is called **Fourier analysis**. We will not explain the mathematics here but simply point out that by such analysis one can find the pure sine tones that compose the tone of, say, a violin. When these pure tones are sounded together, as by striking a number of tuning forks or by selecting the proper keys on an electric organ, they combine to give the tone of the violin. The lowest-frequency sine wave is the fundamental and determines the pitch of the note. The higher-frequency sine waves are the partials that give the characteristic quality. Thus, the wave form of any musical sound is no more than a sum of simple sine waves.

Figure 20-10
The fundamental and its harmonics combine to produce a composite wave.

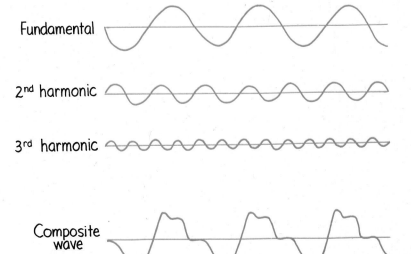

Since the wave form of music is a multitude of various sine waves, to duplicate sound accurately by radio, record player, or tape recorder we should have as large a range of frequencies as possible. The notes of a piano keyboard range from 27 hertz to 4200 hertz, but to duplicate the music of a piano composition accurately, the sound system must have a range of frequencies up to 20 000 hertz. The greater the range of the frequencies of an electrical sound system, the closer the musical output approximates the original sound, hence the wide range of frequencies in a high-fidelity sound system.

Our ear performs a sort of Fourier analysis automatically. It sorts out the complex jumble of air pulsations that reach it and transforms them into pure tones. And we recombine various groupings of these pure tones when we listen. What combinations of tones we have learned to focus our attention on determines what we hear when we listen to a concert. We can direct our attention to the sounds of the various instruments and discern the faintest tones from the loudest; we can delight in the intricate interplay of instruments and still detect the extraneous noises of others around us. This is a most incredible feat.

Figure 20-11
Does each hear the same music?

Laser Discs

You can experience the rich, full sounds of a string quartet or a symphony orchestra in your own room by means of a remarkable sound recording and reproduction technique called *digital audio*. Instead of an LP record and a conventional stylus, the digital player utilizes a laser beam, which is directed onto a plastic reflective disc. The player's pickup is a light sensor rather than a stylus.

Figure 20-12
The amplitude of the analog wave form over successive split seconds is recorded in binary code on the reflective surface of the laser disc.

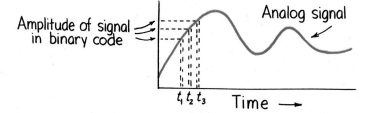

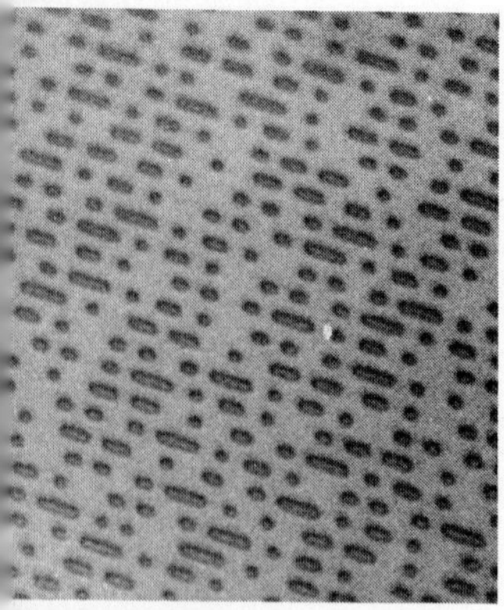

On a conventional phonograph record, the stylus is made to vibrate when it rides in the squiggly phonograph groove. The output is a signal like those shown in Figure 20-8. This type of continuous wave form is called an *analog* signal. The shape of the wave can be described by the numeric value of its amplitude during each split second (Figure 20-12). This numeric value can be expressed in a number system that is convenient for computers, called *binary*. In the binary code, any number can be expressed as a succession of ones and zeros; for example, the number 1 is 1, 2 is 10, 3 is 11, 4 is 100, 5 is 101, 17 is 10001, etc. So the shape of the analog wave form can be expressed as a series of "on" and "off" pulses that corresponds to a series of ones and zeros in binary code. That's where the laser disc comes in.

Instead of a squiggly phonograph groove, a laser audio disc has a series of microscopic pits about thirty times thinner than a human hair. When the laser beam falls on a flat portion of the reflective surface, it is reflected directly into the player's optical system; this gives an "on"

Figure 20-13
A microscopic view of the pits on a laser disc.

Figure 20-14
A tightly focused laser beam reads digital information represented by a series of pits on the laser disc.

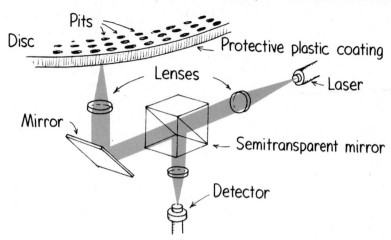

Pits

Disc

Protective plastic coating

Lenses

Laser

Mirror

Semitransparent mirror

Detector

pulse. When the beam is incident upon a passing pit, very little of the laser beam returns to the optical sensor; this gives an "off" pulse. A flickering of "on" and "off" pulses generates the "one" and "zero" digits of the binary code.

The rate at which these tiny pits on the disc are sampled is 44 100 times per second. A single audio disc is $\frac{1}{6}$ the size of a conventional LP and contains billions of bits of information. All this information is encoded on the reflective surface, which is covered with a protective layer of clear plastic. Since the laser beam is focused onto the signal surface below, it is immune to dust, scratches, and fingerprints. No more of the snap, crackle, and pop so characteristic of the familiar LP. And since the laser beam does not touch the disc, the disc never wears out—no matter how many times you play it.

The most extraordinary feature of the laser disc, however, is the quality of the sound. You can hear the difference.

Summary of Terms

Pitch The "highness" or "lowness" of a tone, as on a musical scale, which is principally governed by frequency. A high-frequency vibrating source produces a sound of high pitch; a low-frequency vibrating source produces a sound of low pitch.

Loudness The physiological sensation directly related to sound intensity or volume. Relative loudness, or sound level, is measured in decibels.

Quality The characteristic timbre of a musical sound, governed by the number and relative intensities of partial tones.

Partial tone One of the frequencies present in a complex tone. When a partial tone is an integer multiple of the lowest frequency, it is a harmonic.

Fundamental frequency The lowest frequency of vibration, or first harmonic. In a string the vibration makes a single segment.

Harmonic A partial tone that is an integer multiple of the fundamental. The vibration that begins with the fundamental vibrating frequency is the first harmonic, twice the fundamental is the second harmonic, and so on in sequence.

Musical scale A succession of notes of frequencies that are in simple ratios to one another. In Western culture the principal scales are the just major scale and the equal-tempered chromatic scale.

Fourier analysis A mathematical method that will resolve any periodic wave form into a series of simple sine waves.

Review Questions

1. Distinguish between noise and music.
2. What are the three principal characteristics of musical tones?

Pitch

3. How can the pitch of a guitar string be increased?
4. What is the range of normal human hearing? What happens to this range with age?

Loudness

5. How do the *intensities* of the loudest sounds we can tolerate compare to the lowest intensities we can hear?
6. Is the sound of 30 dB 30 times greater than the threshold of hearing, or 10^3 (a thousand) times greater?

Quality

7. What exactly determines the pitch of a note?
8. If the fundamental frequency of a note is 200 Hz, what is the frequency of the second harmonic? The third harmonic?
9. What exactly determines the musical quality of a note?
10. Why do the same notes plucked on a banjo and a guitar have distinctly different sounds?

Musical Instruments

11. What are the three principal classes of musical instruments?
12. Why do orchestras generally have a greater number of stringed instruments than wind instruments?

Musical Scales

13. What is the simplest musical scale in Western culture?
14. What is an *octave*?
15. Are the musical intervals the same for all adjacent notes on the just scale?
16. Are the musical intervals the same for all adjacent notes on the equal-tempered scale?

Fourier Analysis

17. What did Fourier discover about complex periodic wave patterns?
18. A high-fidelity sound system may have a frequency range that extends beyond the range of human hearing. Of what use is this extended range?

Laser Discs

19. How is the sound signal captured in a conventional phonograph record? How is the sound signal captured in a laser disc?

20. Why does a laser disc not wear out like a conventional phonograph record?

Home Projects

1. Test to see which ear has the better hearing by covering one ear and finding how far away your open ear can hear the ticking of a clock; repeat for the other ear. Notice also how the sensitivity of your hearing improves when you cup your ears with your hands.
2. With a strong magnifying glass, examine the grooves in phonograph records. If you have an old 78-RPM disk, compare the grooves with those of a $33\frac{1}{3}$-RPM disk.
3. Set a record into rotation on a record player. Place the edge of your fingernail in the groove and listen carefully. Notice that you hear only high-frequency sound. If you want to avoid damaging your record, try putting the needle on the record with the volume knob all the way down; put your ear just above the tone arm. Low fidelity!
4. Make the lowest-pitched sound you are capable of; then keep doubling the pitch to see how many octaves your voice can span.
5. On a sheet of graph paper, construct the composite wave of Figure 20-10 by superposing various vertical displacements of the fundamental and first two partial tones. Your instructor can show you how this is done. Then find the composite waves of partial tones of your own choosing.

Exercises

1. Explain how you can produce a low-pitched note on a guitar by altering (a) the length of the string, (b) the tension of the string, and (c) the thickness or the mass of the string.
2. Why is the thickness greater for the bass strings of a guitar?
3. Would a plucked guitar string vibrate for a longer or a shorter time if it had no sounding board? Why?
4. If you very lightly touch a guitar string at its midpoint, you can hear a tone that is one octave above the fundamental for that string. Explain.
5. If a guitar string vibrates in two segments, where can a tiny piece of folded paper be supported without flying off? How many pieces of folded paper could similarly be supported if the wave form were of three segments?

6. If the wavelength of a vibrating string is reduced, what effect does this have on the frequency of vibration and on the pitch?

7. The amplitude of a transverse wave in a stretched string is the maximum displacement of the string from its equilibrium position. What does the amplitude of a longitudinal sound wave in air correspond to?

8. Which of the two musical notes displayed one at a time on an oscilloscope screen has the higher pitch? Which is the louder?

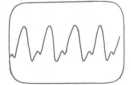 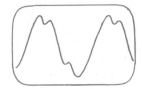

9. In a hi-fi speaker system, why is the woofer (low-frequency speaker) larger than the tweeter (high-frequency speaker)?

10. One person has a threshold of hearing of 5 dB and another of 10 dB. Which person has the more acute hearing?

11. How much more intense than the threshold of hearing is a sound of 10 dB? 30 dB? 60 dB?

12. How much more intense is a sound of 40 dB than a sound of 30 dB?

13. How is an organ able to imitate the sounds made by various musical instruments?

14. A person talking after inhaling helium gas has a high-pitched voice. One of the reasons for this is the higher speed of sound in helium than in air. Why does sound travel faster in helium?

15. Why does your voice sound fuller in the shower?

16. The frequency range for a telephone is between 500 and 4000 Hz. Why is a telephone inadequate for transmitting music?

17. A certain note has a frequency of 1000 Hz. What is the frequency of a note one octave above it? Two octaves above it? One octave below it? Two octaves below it?

18. How many octaves does normal human hearing span? How many octaves are on a common piano keyboard?

19. The author's range of hearing is from 20 Hz to 14 000 Hz (too many rock concerts in the mid-1960s to early 1970s). How many octaves can I hear?

20. If the fundamental frequency of a violin string is 440 Hz, what is the frequency of the second harmonic? The third?

21. At an outdoor concert, will the pitch of musical tones be affected on a windy day? Explain.

22. When water is poured into a glass, does the pitch of the sound of the filling glass increase or decrease? (Think about this; then try it and hear for yourself.)

23. As you pour water into a glass, repeatedly tap the glass with a spoon. Does the pitch of the tapped glass as it is being filled increase or decrease? (Think about this one also and then try it before confirming your thoughts and/or surprising yourself.)

24. Do all the people in a group hear the same music when they listen to it attentively? (Do all see the same sight when looking at a painting? Do all taste the same flavor when sipping the same wine? Do all perceive the same aroma when smelling the same perfume? Do all feel the same texture when touching the same fabric? Do all come to the same conclusion when listening to a logical presentation of ideas?)

PART 5

ELECTRICITY & MAGNETISM

21 Electrostatics

We find that electricity in one form or another underlies just about everything around us. It's in the lightning from the sky, it's in the spark beneath our feet when we scuff across a rug, and it's what holds atoms together to form molecules. The control of electricity is evident in technological devices of many kinds, from lamps to computers. In this technological age it is important to have an understanding of the basics of electricity and of how these basics can be manipulated to produce a prosperity unknown before recent times.

In this chapter we will investigate electricity at rest, or **electrostatics**, as it is called. This involves electric charges, the forces between them, the aura that surrounds them, and their behavior in materials. In the next chapter we will investigate the motion of electric charges, or *electric currents*, and the voltages that produce them and how they can be controlled. In Chapter 23 we will study the relationship of electric currents to magnetism and in Chapter 24 how magnetism and electricity can be controlled to operate motors and other electrical devices.

An understanding of electricity requires a step-by-step approach, for one concept is the building block for the next, and so on. So please put in extra care in the study of this material. It can be difficult, confusing, and frustrating if you're hasty. But with careful effort, it can be comprehensible and rewarding. Onward!

Electrical Forces

Consider a universal force like gravity, which varies inversely as the square of the distance but which is billions upon billions of times stronger than gravity. If there were such a force and if it were attractive like gravity, the universe would be pulled together into a tight sphere with all the matter in the universe pulled as close together as possible. But suppose this force were a repelling force, where every particle repelled every other particle. What then? The universe would be an ever-expanding gaseous cloud. Suppose, however, that the universe consisted of both attractive and repulsive particles, say, positives and negatives. Suppose that positives repelled positives but attracted negatives, and that negatives repelled negatives but attracted positives. Like kinds repel and unlike kinds attract (Figure 21-1). And suppose that there were equal numbers of each. What would the universe be like? It would be like the one we are living in. For there is such a force. We call it *electrical force*.

Figure 21-1
(*a*) Like charges repel.
(*b*) Unlike charges attract.

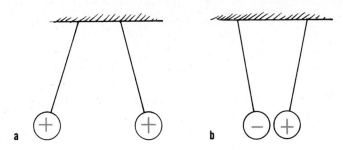

Clusters of positives and negatives have been pulled together by the enormous attraction of the electrical force. The result is that the huge forces have balanced out almost perfectly by forming compact and evenly mixed numbers of positives and negatives. Furthermore, between two bunches of such mixtures, there is almost no electrical attraction or repulsion. Any electrical forces between the earth and moon, for example, have been balanced out. In this way the much weaker gravitational force, which only attracts, is the predominant force between these bodies.

Electric Charges

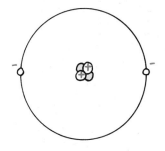

Figure 21-2
Model of a helium atom. The atomic nucleus is made up of two protons and two neutrons. The positively charged protons attract two negative electrons.

The terms *positive* and *negative* refer to electric *charge*, the fundamental quantity that underlies all electrical phenomena. Protons are positively charged and electrons are negatively charged. The attractive force between protons and electrons holds atoms together. Between neighboring atoms the negative electrons of one atom may at times be closer to the positive protons of another atom, so the attractive force between these charges is greater than the repulsive force, and the atoms combine to form a molecule. In fact, all the chemical bonding forces that hold atoms together to form molecules are electrical forces acting in small regions where the balance of attractive and repelling forces is not perfect. So anyone planning to study chemistry should first know something about electricity and before studying electricity should know something about atoms. Here are some important facts about atoms:

1. Every atom is composed of a positively charged nucleus, around which are distributed a number of negatively charged electrons.
2. The electrons of all atoms are identical. Each has the same quantity of negative charge and the same mass.
3. Protons and neutrons compose the nucleus. (The common form of hydrogen that has no neutrons is the only exception.) Protons are almost 2000 times more massive than electrons but carry an amount of positive charge equal to the negative charge of electrons. Neutrons are slightly more massive than protons and have no net charge.
4. All normal atoms have exactly as many electrons surrounding the nucleus as there are protons within the nucleus. Thus, a normal atom has no net charge.

Why don't protons pull the oppositely charged electrons into the nucleus? The answer offered about 70 years ago was that the electrons are not pulled into the nucleus by electrical force for the same reason the earth is not pulled into the sun by gravitational force: the electrons are held in orbit by the pull of the protons. This oversimplified explanation served as a starter for understanding the electrical nature of the atom; the terms *orbit* and *orbital* are still used, although the better word is *shell*, because the electrons do not stay in a plane of rotation the way planets do around the sun. The answer today has to do with the wave nature of electrons and quantum mechanics, which we will discuss in Chapters 30 and 31.

Why don't the protons in the nucleus mutually repel and fly apart? What holds the nucleus together? The answer is that in addition to electrical forces in the nucleus, there are even greater nonelectrical nuclear forces that are able to hold the protons together in spite of the electrical repulsion. We will treat this further in Chapter 32.

Conservation of Charge

Figure 21-3
Electrons are transferred from the fur to the rod. The rod is then negatively charged. Is the fur charged? How much compared to the rod? Positively or negatively?

In a neutral atom there are as many electrons as protons, so there is no net charge. The positive balances the negative exactly. If an electron is removed from an atom, then it is no longer neutral. The atom has one more positive charge (proton) than negative charge (electron) and is said to be positively charged. A charged atom is called an *ion*. A *positive ion* has a net positive charge. A *negative ion*, an atom with one or more extra electrons, is negatively charged.

Material objects are made of atoms, which means they are composed of electrons and protons (and neutrons as well). Objects ordinarily have equal numbers of electrons and protons and are therefore electrically neutral. But if there is a slight imbalance in the numbers, the object is electrically charged. An imbalance comes about when electrons are added or removed from an object. Although the innermost electrons in an atom are bound very tightly to the oppositely charged atomic nucleus, the outermost electrons of many atoms are bound very loosely and can be easily dislodged. How much work is required to tear an electron away from an atom varies for different substances. The electrons are held more firmly in rubber or plastic than in your hair, for example. Hence, when a comb is rubbed through your hair, electrons transfer from the hair to the comb. The comb then has an excess of electrons and is said to be *negatively charged*. Your hair, in turn, has a deficiency of electrons and is said to be *positively charged*. If you rub a glass or plastic rod with silk, you'll find that the rod becomes positively charged. The silk has a greater affinity for electrons than the glass or plastic rod. Electrons are rubbed off the rod and onto the silk.

So we see that an object having unequal numbers of electrons and protons is electrically charged. If it has more electrons than protons, it is negatively charged. If it has fewer electrons than protons, it is positively charged.

It is important to note that when we charge something, no electrons are created or destroyed. Electrons are simply transferred from one material to another. Charge is conserved. In every event, whether large-scale or at the atomic and nuclear level, the principle of *conservation of charge* has always been found to apply. No case of the creation or destruction of net electric charge has ever been found. The conservation of charge ranks with the conservation of energy and momentum as a significant fundamental principle in physics.

So any object that is electrically charged has an excess or deficiency of a given number of electrons. This means that the charge of the object is therefore a whole-number multiple of the charge of an electron. It cannot have a charge equal to the charge of $1\frac{1}{2}$ or $1000\frac{1}{2}$ electrons, for example. Charge is "grainy," or made up of elementary units called *quanta*. We say that charge is *quantized*, with the smallest quantum of charge being that of the electron (or proton). No smaller units of charge have ever been found.* All charged objects to date have a charge that is a whole-number multiple of the charge of a single electron.

Question ▶ If you walk across a rug and scuff electrons from your feet, are you negatively or positively charged?

Coulomb's Law

The electrical force, like gravitational force, decreases inversely as the square of the distance between charges. This relationship was discovered by Charles Coulomb in the eighteenth century and is called **Coulomb's law.** It states that for charged objects that are much smaller than the distance between them, the force between two charges varies directly as the product of the charges and inversely as the square of the separation distance. The force acts along a straight line from one charge to the other. Coulomb's law can be expressed as

$$F = k\frac{q_1 q_2}{d^2}$$

where d is the distance between the charged particles, q_1 represents the quantity of charge of one particle, q_2 represents the quantity of charge of the other particle, and k is the proportionality constant.

The unit of charge is called the **coulomb**, abbreviated C. It turns out that a charge of 1 C is the charge associated with 6.25 billion billion electrons. This might seem like a great number of electrons, but it only represents the amount of charge that passes through a common 100-watt light bulb in a little over a second.

▶ **Answer**

You have fewer electrons after you scuff your feet, so you are positively charged (and the rug is negatively charged).

---

*Within the protons and neutrons of the atomic nucleus, however, elemental charges equal to $\pm\frac{1}{3}$ or $\pm\frac{2}{3}$ the electron charge are associated with elementary particles—quarks. These particles apparently cannot be observed individually, but only in combinations that result in a net charge of 1, or a whole-number multiple of electrons, or 0 charge.

The proportionality constant k in Coulomb's law is similar to G in Newton's law of gravitation. Instead of being a very small number like G (6.67×10^{-11}), the electrical proportionality constant k is a very large number. It is approximately

$$k = 9\ 000\ 000\ 000\ \frac{\text{N·m}^2}{\text{C}^2}$$

or, in scientific notation, $k = 9 \times 10^9$ N·m^2/C^2. The units N·m^2/C^2 are not central to our interest here, but simply convert the right-hand side of the equation to the unit of force, the newton (N). What is important is the large magnitude of k. If, for example, a pair of like charges of 1 coulomb each were 1 meter apart, the force of repulsion between the two charges would be 9 billion newtons.* That would be about ten times the weight of a battleship! Obviously, such amounts of net charge do not usually exist in our everyday environment.

Questions ▶

1. The proton that makes up the nucleus of the hydrogen atom attracts the electron that orbits it. Relative to this force, does the electron attract the proton with less force, more force, or the same amount of force?

2. If a proton at a particular distance from a charged particle is repelled with a given force, by how much will the force decrease when the proton is three times as distant from the particle? Five times as distant?

3. What is the sign of charge of the particle in this case?

So Newton's law of gravitation for masses is similar to Coulomb's law for electric charges.† Whereas the gravitational force of attraction between particles such as an electron and a proton is extremely small, the electrical force between these particles is relatively enormous. The greatest difference between gravitation and electrical forces is that although gravitational forces are only attractive, electrical forces may be either attractive or repulsive.

---

*Contrast this to the gravitational force of attraction between two unit masses (kilograms) 1 m apart: 6.67×10^{-11} N. This is an extremely small force. For the force to be 1 N, the masses at 1 m apart would have to be nearly 123 000 kg each! Gravitational forces between ordinary objects are exceedingly small, and electrical forces (noncanceled) between ordinary objects are exceedingly huge. We don't sense them because the positives and negatives normally balance out, and even for highly charged objects, the imbalance of electrons to protons is normally less than one part in a trillion trillion.

†The similarities between these two forces have made some people think they may be different aspects of the same thing. Albert Einstein was one of these people, and he spent the latter part of his life searching with little success for a "unified field theory." More recently, the electrical force has been unified with one of the two nuclear forces, the weak force, which plays a role in radioactive decay. Physicists are still looking for a way to unify electrical and gravitational forces.

Conductors and Insulators

Metals conduct electricity well because one or more of the electrons in the outer shell of the atoms in a metal are not anchored to the nuclei of particular atoms, but are free to wander in the material. Such materials are called good **conductors**. Metals are good conductors of electricity for the same reason they are good conductors of heat. Electrons in their outer atomic shell are "loose."

The electrons in other materials, rubber and glass for example, are tightly bound and belong to particular atoms. They are not free to wander about other atoms in the material. These materials are poor conductors of electricity for the same reason they are generally poor conductors of heat. Such materials are called good **insulators**.

Figure 21-4
It is easier for electricity to flow through hundreds of kilometers of metal wire than through a few centimeters of insulating material.

All substances can be arranged in order of their ability to conduct electric charges. Those at the top of the list are the conductors and those at the bottom are the insulators. The ends of the list are very far apart. The conductivity of a metal, for example, can be more than a million trillion times greater than the conductivity of an insulator such as glass. In a common appliance cord, a charge flows through several meters of wire rather than flowing directly across from one wire to the other through a small fraction of a centimeter of rubber insulation.

▶ **Answers**

1. The same amount of force, in accord with Newton's third law—basic mechanics! Recall that a force is an interaction between two things, in this case between the proton and the electron. They pull on each other—equally.

2. It decreases to $\frac{1}{9}$ its original value; to $\frac{1}{25}$.

3. Positive.

Figure 21-5
Transistors.

Semiconductors

Whether a substance is classified as a conductor or an insulator depends on how tightly the atoms of the substance hold their electrons. Some materials, such as germanium and silicon, are neither good conductors nor good insulators, and fall in the middle of the range of electrical conductivity. These materials are fair insulators in their pure crystalline form but increase tremendously in conductivity when even one atom in 10 million is replaced with an impurity that adds or removes an electron from the crystal structure. These materials can be made to behave sometimes as insulators and sometimes as conductors, and are called **semiconductors**. Thin layers of semiconducting materials sandwiched together make up *transistors*, which are used to control the flow of currents in circuits, to detect and amplify radio signals, and to produce oscillations in transmitters; they also act as digital switches. These tiny solids were the first electrical components in which materials with different electrical characteristics were not interconnected by wires but physically joined in one structure. They require little power and in normal use last indefinitely.

A semiconductor will also conduct when light of the proper color shines on it. A pure selenium plate is normally a good insulator, and any electric charge built up on its surface will remain there for extended periods in the dark. If the plate is exposed to light, however, the charge leaks away almost immediately. If a charged selenium plate is exposed to a pattern of light, such as the pattern of light and dark that makes up this page, the charge would leak away only from the areas exposed to light. If a black plastic powder were brushed across its surface, the powder would stick only to the charged areas where the plate had not been exposed to light. Now if a piece of paper with an electric charge on the back were put over the plate, the black plastic powder would be drawn to the paper to form the same pattern as, say, the one on this page. If the paper were then heated to melt the plastic and fuse it to the paper, you might pay a dime for it and call it a Xerox copy.

Superconductors

At temperatures near absolute zero, certain metals acquire infinite conductivity (zero resistance to the flow of charge). Until very recently, it was generally thought that zero electrical resistance could be brought about only in certain metals at these very low temperatures. A big physics breakthrough occurred in 1987 with the discovery that various nonmetallic compounds exhibit zero resistance to the flow of charge at much higher temperatures—above 100K. These materials are called **superconductors**. Once electric current is established in a superconductor, the current will flow indefinitely. At this writing there is an enormous amount of interest in the physics community as to exactly why certain materials acquire superconducting properties. Explanations have generally to do with the wave nature of matter (quantum mechanics) and are being vigorously researched.

Charging

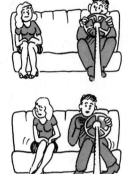

Figure 21-6
Charging by contact.

Figure 21-7
Charging by induction.

We charge things by transferring electrons from one place to another. We can do this by physical *contact*, as occurs when substances are rubbed together, or by placing charges in proximity to each other—by *induction*.

Charging by Contact

We are all familiar with the electrical effects produced when we rub substances together: we stroke a cat's fur and hear the crackle of sparks that are produced; we comb our hair, especially in front of a mirror in a dark room, so we can see as well as hear the sparks of electricity; we scuff our shoes across a rug and produce a spark when we reach for a door knob; or we do the same when sliding across plastic seat covers in an automobile (Figure 21-6). In all these cases electrons are being transferred by the close contact achieved when different substances are rubbed together, which is known as **charging by contact**.

Electrons can be transferred from one material to another by simply touching. A charged conducting rod placed in contact with a neutral object, for example, will transfer charge to the neutral object. If the object is a good conductor, the charge will spread quickly to all parts of its surface because the like charges repel one another. If it is a poor conductor, it may be necessary to touch the rod at several places on the object in order to get a more or less uniform distribution of charge. Charge is transferred from one material to another by contact.

Charging by Induction

If you bring a charged object *near* a conducting surface, you will induce electrons to move in the material even though there is no physical contact. Consider the two insulated metal spheres, A and B, in Figure 21-7. (*a*) They touch each other, so in effect they form a single non-charged conductor. (*b*) When a negatively charged rod is brought near A, positive charges in the metal are attracted toward the rod, and negative charges are repelled from the rod. This induces a charge redistribution. (*c*) If A and B are separated while the rod is still present, (*d*) they will each be equal and oppositely charged. This is **charging by induction**. The charged rod has never touched them, and it retains the same charge it had initially.

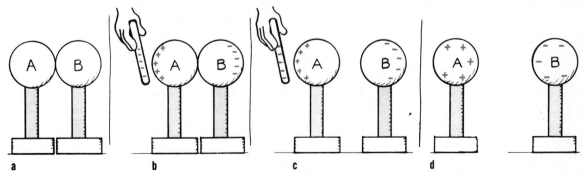

a   b   c   d

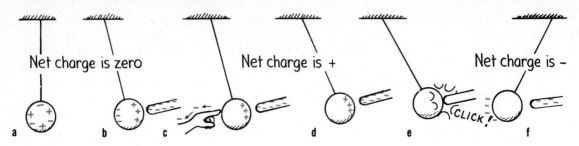

Net charge is zero   Net charge is +   Net charge is −

a   b   c   d   e   f

Figure 21-8
Charge induction by grounding. (*a*) Net charge on the metal ball is zero. (*b*) Charge redistribution
is induced by the presence of the charged rod. The net charge on the ball is still zero.
(*c*) Touching the negative side of the ball removes electrons by contact. (*d*) This leaves the ball
positively charged. (*e*) The ball is more strongly attracted to the negative rod, and when it touches,
charging by contact occurs. (*f*) The negative ball is repelled by the still somewhat negatively
charged rod.

We can similarly charge a single sphere if we touch it when the charges
are separated by induction. Consider the metal sphere that hangs from a
nonconducting string, as shown above in Figure 21-8. When we touch
the charged part of the metal surface with a finger, charges that repel
one another have a conducting path to a very large reservoir for electric
charge—the ground. Charge then flows from the metal sphere to the
ground. When we discharge a conductor by touching it, it is common to
say we are *grounding* it. We will return to this idea of grounding in the
next chapter when we discuss electric currents.

Questions ▶
1. Would the charges induced on the spheres A and B of Figure 21-7 on the previous
   page necessarily be exactly equal and opposite?
2. Why does the negative rod in Figure 21-7 have the same charge before and after the
   spheres are charged, but not when charging takes place as shown above in Figure
   21-8?

▶ **Answers**
1. The charges must be equal and opposite on both spheres, because each single
   positive charge on sphere A is the result of a single electron being taken from A
   and moved to B. This is like taking bricks from the surface of a brick road and
   putting them all on the sidewalk. The number of bricks on the sidewalk will be
   exactly matched by the number of holes in the road. Likewise, the number of extra
   electrons on B will exactly match the number of "holes" (positive charges) left
   in A. Remember that a positive charge is the equivalent of an absent electron.
2. In the charging process of Figure 21-7, no contact was made between the negative
   rod and either of the spheres. In Figure 21-8, however, the rod touched the positively charged sphere. A transfer of charge by contact reduced the negative charge
   on the rod.

Figure 21-9
The bottom of the negatively charged cloud induces a positive charge at the surface of the ground below.

Charging by induction occurs during thunderstorms. The negatively charged bottoms of clouds induce a positive charge on the surface of the earth below. Benjamin Franklin was the first to demonstrate this when his famous kite-flying experiment proved that lightning is an electrical phenomenon.* Lightning is an electrical discharge between the clouds and oppositely charged ground or between oppositely charged parts of clouds.

Franklin also found that charge leaks off sharp points, and thus he fashioned the first lightning rod. If a metal rod is placed above a building and connected to the ground, charge leaks off rather than building up. This continual leaking of charge prevents an inductive charge buildup, which would otherwise result in a sudden discharge between the cloud and the ground—a lightning strike. The primary purpose of the lightning rod, then, is to prevent a lightning strike from occurring. If sufficient cloud charge builds up to cause lightning in the vicinity of the building anyway, it will strike the rod and go down the connection to the ground instead of striking the building. The overall purpose of the lightning rod is to prevent a fire caused by lightning.

Charge Polarization

Charging by induction is not restricted to conductors. When a charged rod is brought near an insulator, there are no free electrons that can migrate throughout the insulating material. Instead, there is a rearrangement of the positions of charges within the atoms and molecules themselves (Figure 21-10). One side of the atom or molecule is induced into becoming more negative (or positive) than the opposite side. The atom or molecule is said to be **electrically polarized**. If the charged rod is negative, say, then the positive part of the atom or molecule is tugged in a direction

Figure 21-10
(*a*) The center of the negative electron cloud coincides with the center of the positive nucleus in an atom.
(*b*) When an external negative charge is brought nearby to the right, the electron cloud is distorted so the centers of negative and positive charge no longer coincide. The atom is electrically polarized.

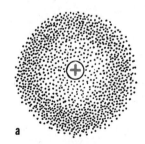 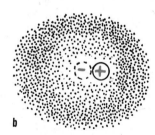

a   b

*Benjamin Franklin was careful to isolate himself from his apparatus and keep out of the rain when he conducted this experiment, so he wasn't electrocuted as were others who attempted to duplicate his experiment. In addition to being a great statesman, Franklin was a first-rate scientist. He introduced the terms *positive* and *negative* as they relate to electricity, established the one-fluid theory of electric currents, and contributed to our understanding of grounding and insulation. He also published a newspaper, formed the first fire insurance company, and invented a safer, more efficient stove—a very busy man! Only a task as important as helping to form the United States system of government kept him from spending even more of his time on his favorite activity—the scientific investigation of nature.

Figure 21-11

All the atoms or molecules near the surface become electrically polarized. Surface charges of equal magnitude and opposite sign are induced on opposite surfaces of the material.

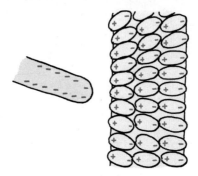

toward the rod, and the negative side of the atom or molecule is pushed in a direction away from the rod. The positive and negative parts of the atoms and molecules become aligned. They are electrically polarized (Figure 21-11).

We can see why electrically neutral bits of paper are attracted to a charged object. Molecules are polarized in the paper, with the sides of molecules opposite in sign to the charged object closest to it. Closeness wins, and the bits of paper experience a net attraction. Sometimes they will cling to the charged object and then suddenly fly off. This repulsion occurs because the paper bits acquire the same sign of charge as the charged object when they are in contact.

Rub an inflated balloon on your hair, and it becomes charged. Place the balloon against the wall, and it sticks. This is because the charge on

Figure 21-12

A charged comb attracts an uncharged piece of paper because the force of attraction for the closer charge is greater than the force of repulsion for the farther charge.

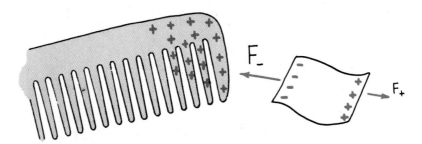

Question

A negatively charged rod is brought close to some small pieces of neutral paper. The positive sides of molecules in the paper are attracted to the rod and the negative sides of the molecules are repelled. Since negative and positive sides are equal in number, why don't the attractive and repulsive forces cancel out?

▶ **Answer**

The positive sides are simply closer to the rod. In accord with Coulomb's law, they therefore experience a greater electrical force than the farther-away negative sides. Hence we say that closeness wins. This greater force between positive and negative is attractive, so the neutral paper is attracted to the charged rod. Can you see that if the rod were positive, attraction would still occur?

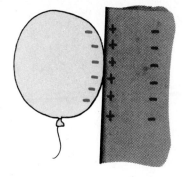

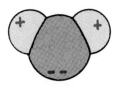

Figure 21-14

An H_2O molecule is an electric dipole.

the balloon induces an opposite surface charge on the wall. Closeness wins, for the charge on the balloon is slightly closer to the opposite induced charge than to the charge of the same sign (Figure 21-13). Many molecules, H_2O for example, are electrically polarized in their normal states. The distribution of electric charge is not perfectly even. There is a little more negative charge on one side of the molecule than the other (Figure 21-14). Such molecules are said to be *electric dipoles*.

Electric Field

Electrical forces, like gravitational forces, act between things that are not in contact with each other as well as between those that are. For both the electrical and gravitational cases, a *force field* exists that influences distance charges and masses respectively. The properties of space surrounding any mass can be considered to be so altered that another mass introduced to this region will experience a force. The "alteration in space" caused by a mass is called its *gravitational field*. We can think of any other mass as interacting with the field and not directly with the mass producing it. For example, when an apple falls from a tree, we say it is interacting with the mass of the earth, but we can also think of the apple as interacting with the gravitational field of the earth. The field plays an intermediate role in the force between bodies. It is common to think of distant rockets and the like as interacting with gravitational fields rather than with the masses of the earth and other bodies responsible for the fields. Just as the space around a planet and every other mass is filled with a gravitational field, the space around every electric charge is filled with an **electric field**—a kind of aura that extends through space.

Figure 21-15

A gravitational force holds the satellite in orbit about the planet (*a*), and an electrical force holds the electron in orbit about the proton (*b*). In both cases there is no contact between the bodies. We say that the orbiting bodies interact with the *force fields* of the planet and proton and are everywhere in contact with these fields. Thus, the force that one electric charge exerts on another can be described as the interaction between one charge and the field set up by the other.

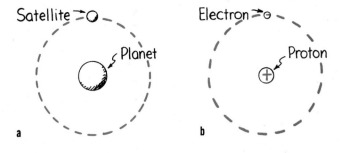

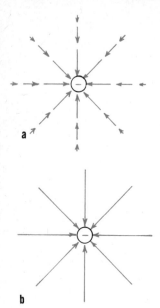

An electric field has both magnitude (strength) and direction. The magnitude of the field at any point is simply the force per unit of charge. If a charge q experiences a force F at some point in space, then the electric field E at that point is

$$E = \frac{F}{q}$$

The electric field is depicted with vector arrows in Figure 21-16a. The direction of the field is shown by the vectors and is defined to be the direction in which a small positive test charge at rest would be moved.* The direction of the force and that of the field at any point are the same. In the figure we see that all the vectors therefore point to the center of the negatively charged ball. If the ball were positively charged, the vectors would point away from the center because a positive test charge in the vicinity would be repelled.

A more useful way to describe an electric field is with electrical lines of force (Figure 21-16b). The lines of force shown in the figure represent a small number of the infinitely numerous possible lines that indicate the direction of the field. Where the lines are farther apart, the field is weaker. For an isolated charge, the lines extend to infinity; for two or more opposite charges, we represent the lines as emanating from a positive charge and terminating on a negative charge. Some electric field configurations are shown in Figure 21-17, and photographs of field patterns are shown in Figure 21-18. The photographs show bits of thread that are suspended in an oil bath surrounding charged conductors. The ends of the bits of threads are charged by induction and tend to line up end to end with the field lines, like iron filings in a magnetic field.

Figure 21-16
Electric field representations about a negative charge. (a) A vector representation. (b) A lines-of-force representation.

Figure 21-17
Some electric field configurations. (a) Lines of force about a single positive charge. (b) Lines of force for a pair of equal but opposite charges. Note that the lines emanate from the positive charge and terminate on the negative charge. (c) Uniform lines of force between two oppositely charged parallel plates.

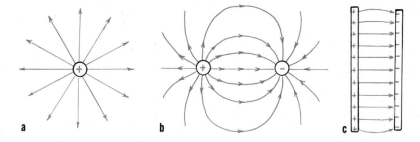

The electric field concept helps us understand not only the forces between isolated stationary charges, but also what happens when charges move. This motion is communicated to neighboring charges in the form of a field disturbance that emanates at the speed of light from the accelerating

---

*The test charge is small so that its electric field is small compared to the electric field already there. If the test charge were large, it might alter the charge distribution of the source of the field we are measuring. Recall in our study of heat the similar need for a thermometer of small mass when measuring the temperature of larger masses.

Figure 21-18
Bits of thread suspended in an oil bath surrounding charged conductors line up end-to-end along the direction of the field. (*a*) Equal and opposite charges. (*b*) Equal like charges. (*c*) Oppositely charged plates. (*d*) Oppositely charged cylinder and plate.

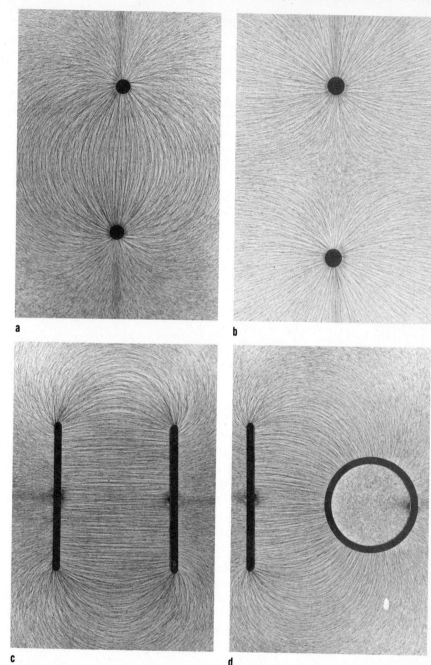

charge. We will learn that the electric field is a storehouse of energy and that energy can be transported over long distances in an electric field, which may be directed through and guided by metal wires or through empty space. We will return to this idea in the next chapter and later when we learn about electromagnetic radiation.

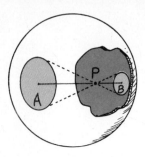

Figure 21-19
The test charge at P is attracted just as much to the greater number of farther charges at A as it is to the smaller number of closer charges at B. The net force on the test charge anywhere inside the conductor is zero. The electric field inside is therefore zero.

Electric Shielding

An important difference between electric and gravitational fields is that electric fields can be shielded by various materials, while gravitational fields cannot. The amount of shielding is characteristic of the medium within the field. For example, air diminishes the electric field between a pair of charges slightly more than a vacuum, while oil placed between the charges diminishes the field by about one-eightieth. Metal will completely shield an electric field.

Consider, for example, electric charge on a spherical metal shell. Because of mutual repulsion the charges will spread out uniformly over the outer surface of the shell. It is not difficult to see that the electrical force exerted on a sample test charge placed inside the shell at the exact center would be zero because opposing forces would balance in every direction. Interestingly enough, complete cancellation will occur anywhere inside the conducting sphere. Understanding why this is true requires more thought and involves the inverse-square law and a bit of geometry. Consider the test charge at point P in Figure 21-19, which is twice as far from the left side of the sphere as it is from the right side. If the electrical force between the test charge and the charges depended only on distance, then the test charge would be attracted only $\frac{1}{4}$ as much to the left side as to the right side (remember the inverse-square law: twice as far away means only $\frac{1}{4}$ the effect, three times as far away means only $\frac{1}{9}$ the effect, and so on). But the force depends also on the amount of charge. In the figure, note that the same solid angle subtends regions A and B, and that whatever the angle, A will have four times the area and therefore four times as much charge as region B. Since $\frac{1}{4}$ of 4 is equal to 1, a test charge at P is attracted equally to each side. Cancellation occurs. More thought will show that cancellation will occur anywhere inside the conductor. (Recall this argument back in Chapter 8 for the cancellation of gravity inside a hollow planet.)

If the conductor is not spherical, then the charge distribution will not be uniform. The charge distribution over conductors of various shapes is shown in Figure 21-20. Most of the charge on a conducting cube, for example, is mutually repelled to the corners. The neat thing is this: The exact charge distribution over the surface of a conductor is such that the electric field everywhere inside the conductor is zero. Look at it this way. If there were an electric field inside a conductor, then free electrons

Figure 21-20
The charges are distributed on the surface of all conductors in a way such that the electric field inside the conductor is zero.

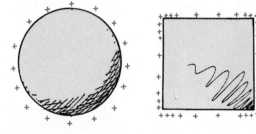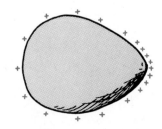

Figure 21-21
Electrons from the lightning bolt mutually repel to the outer metal surface. Although the electric field they set up may be great *outside* the car, the net electric field *inside* the car is zero.

inside the conductor would be set in motion. How far would they move? Until equilibrium is established—which is to say, when the positions of all the electrons produce a zero field inside the conductor.

We cannot shield ourselves from gravity, because gravity only attracts. There are no repelling parts of gravity to offset attracting parts. Shielding electric fields, however, is quite simple. Surround yourself or whatever you wish to shield with a conducting surface. Put this surface in an electric field of whatever field strength. The free charges in the conducting surface will arrange themselves on the surface of the conductor in such a way that all field contributions inside cancel one another. That's why certain electronic components are encased in metal boxes and why certain cables have a metal covering—to shield them from all outside electrical activity.

Question ▶ Small bits of aligned thread neatly show the electric fields in the four photos of Figure 21-18. But the threads are not aligned inside the cylinder in Figure 21-18*d*. Why?

▶ **Answer**
The electric field is shielded inside the cylinder, shown as a circle in the two-dimensional photograph. Hence the threads are without alignment. The electric field inside any conductor is zero—the only exception occurs when the conductor is carrying an electric current.

Electric Potential

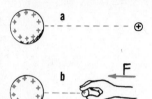

Figure 21-22

The small charge has more PE (potential energy) in (*b*) than in (*a*) because work was required to move it to the closer location.

When we studied energy in Chapter 6, we learned that an object may have gravitational potential energy because of its location in a gravitational field. Similarly, a charged object can have potential energy by virtue of its location in an electric field. Just as work is required to lift a massive object against the gravitational field of the earth, work is required to push a charged particle against the electric field of a charged body. This work increases the electric potential energy of the charged particle.* Consider the small positive charge located at some distance from a positively charged sphere in Figure 21-22a. If you push the small charge closer to the sphere (Figure 21-22b), you will expend energy to overcome electrical repulsion; that is, you will do work in pushing the charge against the electric field of the sphere. This work done in moving the small charge to its new location increases the energy of the small charge. We call the energy the charge now possesses by virtue of its location **electric potential energy**. If the charge is released, it accelerates in a direction away from the sphere, and its potential energy changes to kinetic energy.

If we push two charges instead, we do twice as much work, and the two charges in the same location have twice the electric potential energy as before; three charges have three times as much potential energy; ten charges ten times the potential energy; and so on. Rather than dealing with the total potential energy of a charged body, it is convenient when working with electricity to consider the electric potential energy per charge. We simply divide the amount of energy in any case by the amount of charge. For example, ten charges at a specific location will have ten times as much energy as one, but will also have ten times as much charge; so we see that the energy per charge remains the same at a particular field location whatever the amount of charge. The concept of potential energy per charge is called **electric potential**; that is,

$$\text{Electric potential} = \frac{\text{electric potential energy}}{\text{amount of charge}}$$

The unit of measurement for electric potential is the volt, so electric potential is often called *voltage*. A potential of 1 volt (V) equals 1 joule (J) of energy per 1 coulomb (C) of charge.

$$1 \text{ volt} = 1 \frac{\text{joule}}{\text{coulomb}}$$

Thus a 1.5-volt battery gives 1.5 joules of energy to every 1 coulomb of charge passing through the battery. Both the names *electric potential* and *voltage* are common, so either may be used. In this book the names will be used interchangeably.

The significance of voltage is that a definite value for it can be assigned to a location, whether or not a charge exists at that location. We can speak about the voltages at different locations in an electric field whether

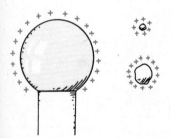

Figure 21-23

The larger test charge has more PE (potential energy) in the field of the charged dome, but the *electric potential* of any amount of charge at the same location is the same.

---

*This work is positive if it increases the electric potential energy of the charged particle and negative if it decreases it.

Figure 21-24
The sketch to the left shows the PE (potential energy) of a mass held in a gravitational field. The sketch to the right shows the PE of a charge held in an electric field. When released, how does the KE (kinetic energy) acquired by each compare to the decrease in PE?

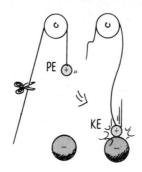

Figure 21-25
Although the voltage of the charged balloon is high, the electric potential energy is low because of the small amount of charge.

or not charges occupy those locations. Likewise with voltages at various locations in an electric circuit. In the next chapter you will see that the location of the positive terminal of a 12-volt battery is maintained at a voltage 12 volts higher than the location of the negative terminal. When a conducting medium connects this voltage difference, charges in the medium will move between these voltage-differing locations.

Rub a balloon on your hair, and the balloon becomes negatively charged—perhaps to several thousand volts! That would be several thousand joules of energy if the charge were 1 coulomb. However, 1 coulomb is a fairly respectable amount of charge. The charge on a balloon rubbed on hair is more typically much less than a millionth of a coulomb. Therefore, the amount of energy associated with the charged balloon is very, very small. A high voltage means a lot of energy only if a lot of charge is involved. Here we see there is a difference between electric potential energy and electric potential.

Questions ▶

1. If there were twice as many coulombs in the test charge near the charged sphere in Figure 21-24, would the electric potential energy of the test charge with respect to the charged sphere be the same or would it be twice as great? Would the electric potential of the test charge be the same or would it be twice as great?

2. What does it mean to say that an automobile battery "is 12 volts"?

▶ **Answers**

1. Twice as many coulombs would find the test charge with twice as much electric potential energy (this is because it would have taken twice as much work to put the charge at that location). But the electric potential would be the same. This is because the electric potential is total electric potential energy divided by total charge. For example, ten times the energy divided by ten times the charge will give the same value as two times the energy divided by two times the charge. Electric potential is not the same thing as electric potential energy. Be sure you understand this before you study further.

2. It means that one of the battery terminals is 12 V higher in potential than the other battery terminal. In the next chapter you will see that it also means that when a circuit is connected across these terminals, each coulomb of charge that makes up the resulting current will be given 12 J of energy.

Figure 21-26

A simple model of a
Van de Graaff generator.

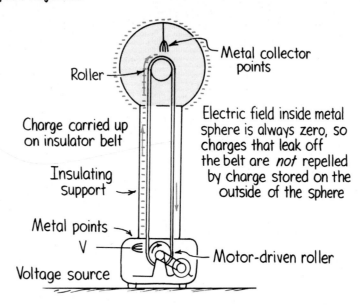

Roller

Metal collector
points

Charge carried up
on insulator belt

Electric field inside metal
sphere is always zero, so
charges that leak off
the belt are *not* repelled
by charge stored on the
outside of the sphere

Insulating
Support

Metal points

V

Motor-driven roller

Voltage source

Figure 21-27

Both the physics enthusiast
and the spherical dome of
the Van de Graaff generator
are charged to a high voltage. Why does her hair
stand out? (Courtesy W. B.
Saunders Company.)

Van de Graaff Generator

A common laboratory device for building up high voltages is the *Van de Graaff generator*. This is one of the lightning machines that mad scientists used in old science fiction movies. A simple model of the Van de Graaff generator is shown in Figure 21-26. A large, hollow metal sphere is supported by a cylindrical insulating stand. A motor-driven rubber belt inside the support stand passes a comb-like set of metal needles that are maintained at a high electric potential. Discharge by the points deposits a continuous supply of charge to the belt, which is carried up into the hollow conductor. Since the electric field inside the conductor is zero, the charge leaks onto metal points (tiny lightning rods) and is deposited on the inside of the sphere. The electrons repel each other to the outer surface of the conducting sphere. Static charge always lies on the outside surface of any conductor. This leaves the inside uncharged and able to receive more electrons as they are brought up the belt. The process is continuous and the charge builds up to a very high electric potential—on the order of millions of volts.

A sphere with a radius of 1 meter can be raised to a potential of 3 million volts before electrical discharge occurs through the air (because breakdown in air occurs at an electric field of about 3×10^6 volts per meter). The voltage can be further increased by increasing the radius of the sphere or by placing the entire system in a container filled with high-pressure gas. Van de Graaff generators can produce voltages as high as 20 million volts. These voltages are used to accelerate charged particles that can be used as projectiles for penetrating the nuclei of atoms. Touching one can be a hair-raising experience.

Summary of Terms

Electrostatics The study of electric charges at rest relative to one another (not *in motion*, as in electric currents).

Coulomb's law The relationship among electrical force, charge, and distance:

$$F = k \frac{q_1 q_2}{d^2}$$

If the charges are alike in sign, the force is repelling; if the charges are unlike, the force is attractive.

Coulomb The SI unit of electrical charge. One coulomb (symbol C) is equal to the total charge of 6.25×10^{18} electrons.

Conductor Any material through which charge easily flows when subject to an external electrical force.

Insulator Any material that resists charge flow through it when subject to an external electrical force.

Semiconductor A poorly conducting material, such as crystalline silicon or germanium, that can be made a better-conducting material by the addition of certain impurities or energy.

Superconductor A material in which the electrical resistance to the flow of electric current drops to near zero or zero under special circumstances that usually include low temperatures.

Charging by contact The transfer of charge from one substance to another by physical contact between substances.

Charging by induction The change in charge of a grounded object, caused by the electrical influence of electric charge close by but not in contact.

Electrically polarized Term applied to an atom or molecule in which the charges are aligned so that one side is slightly more positive or negative than the opposite side.

Electric field The energetic region of space surrounding a charged object. About a charged point, the field decreases with distance according to the inverse-square law, like a gravitational field. Between oppositely charged parallel plates, the electric field is uniform. A charged object placed in the region of an electric field experiences a force.

Electric potential energy The energy a charge possesses by virtue of its location in an electric field.

Electric potential The electric potential energy per amount of charge, measured in volts, and often called voltage:

$$\text{Voltage} = \frac{\text{electric energy}}{\text{amount of charge}}$$

Review Questions

Electrical Forces

1. How strong are electrical forces between an electron and a proton compared to gravitational forces between charged particles?

2. Why does gravitational force predominate over electrical force for astronomical bodies?

3. Why do electrical forces cancel out between the earth and moon, but not between atoms that are near one another?

Electric Charges

4. What part of an atom is *positively* charged and what part is *negatively* charged?

5. How does the charge of one electron compare to that of another electron?

6. How do the masses of electrons compare to the masses of protons? Neutrons?

7. How do the numbers of protons in the atomic nucleus normally compare to the number of electrons that orbit the nucleus?

Conservation of Charge

8. What is a positive ion? A negative ion?

9. What kind of charge does an object acquire when electrons are stripped from it?

10. What is meant by saying charge is *conserved*?

11. What is meant by saying charge is *quantized*?

12. What is the magnitude of the quantum of charge?

Coulomb's Law

13. How does a *coulomb* of charge compare to the charge of a single electron?

14. How is Coulomb's law similar to Newton's law of gravitation? How is it different?

15. What does the value of the proportionality constant, k, tell us about the relative strengths of electrical and gravitational forces?

16. How does the magnitude of electrical force between a pair of charges change when the charges are moved twice as far apart? Three times as far apart?

Conductors and Insulators

17. Why are metals good conductors of heat and electricity?

18. Why are materials such as glass and rubber good insulators?

Semiconductors

19. How does a *semiconductor* differ from a *conductor* or an *insulator*?

20. What is a transistor, and what are some of its functions?

Superconductors

21. How does a *superconductor* differ from an *ordinary conductor*?

Charging

22. What are the three common ways to charge an object?

Charging by Contact

23. Give an example of charging something by rubbing.

24. Give an example of charging something by touching.

Charging by Induction

25. Give an example of charging something by induction.

26. What do we do when we "ground" an object?

27. What exactly is lightning?

28. What is the primary purpose of the lightning rod? Secondary purpose?

Charge Polarization

29. How does an electrically *polarized* object differ from an electrically *charged* object?

30. What is normally the net charge of an electrically polarized object?

31. A piece of paper becomes polarized in the presence of, say, a negative charge. The positive side of the paper is attracted to the negative charge, and the negative side of the paper is repelled by the negative charge. So why don't these forces cancel out?

32. Will a balloon that is charged either positively or negatively stick to a neutral wall?

Electric Field

33. Give two examples of common force fields.

34. How is the direction of an electric field defined?

35. How is the magnitude of an electric field defined?

Electric Shielding

36. Why is there no electric field in the middle of a charged spherical conductor?

37. Is there an electric field at places other than the middle of a charged spherical conductor? Explain.

38. When charges mutually repel and distribute themselves on the surface of conductors, what is the effect inside the conductor?

39. What is the electric field inside a car struck by lightning?

Electric Potential

40. Distinguish between *electric potential energy* and *electric potential*.

41. What happens to the electric potential energy of a charge when positive work is done on it? What happens to the electric potential?

42. Why does the object with the most electric potential energy not necessarily have the most electric potential?

43. A balloon may easily be charged to several thousand volts. Does that mean it has several thousand joules of energy? Explain.

Van de Graaff Generator

44. What is the magnitude of the electric field inside the charged dome of the Van de Graaff generator?

45. Why does the girl's hair stand out in Figure 21-27?

Home Projects

1. Demonstrate charging by friction and discharging from points with a friend who stands at the far end of a carpeted room. With leather shoes, scuff your way across the rug until your noses are close together. This can be a delightfully tingling experience, depending on how dry the air is and how pointed your noses are.

2. Briskly rub a comb on your hair or a woolen garment and bring it near a small but smooth stream of running water. Is the stream of water charged?

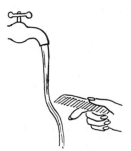

Exercises

1. We do not feel the gravitational forces between ourselves and the objects around us because these forces are extremely small. Electrical forces, in comparison, are extremely huge. Since we and the objects around us are composed of charged particles, why don't we usually feel electrical forces?

2. Why do clothes often cling together after tumbling in a clothes dryer?

3. When you remove your wool suit from the dry cleaner's garment bag, the bag becomes positively charged. Explain how this occurs.

4. When combing your hair, you scuff electrons from your hair onto the comb. Is your hair then positively or negatively charged? How about the comb?

5. At some automobile toll-collecting stations, a thin metal wire sticks up from the road and makes contact with cars before they reach the toll collector. What is the purpose of this wire?

6. Why are the tires for trucks carrying gasoline and other flammable fluids manufactured to conduct electricity?

7. An electroscope is a simple device consisting of a metal ball that is attached by a conductor to two fine gold leaves that are pro-tected from air distur-bances in a jar, as shown. When the ball is touched by a charged body, the leaves that normally hang straight down spread apart. Why? (Electroscopes are useful not only as charge detectors but also for meas-uring the quantity of charge: the more charge transferred to the ball, the more the leaves diverge.)

8. The leaves of a charged electroscope collapse in time. At higher altitudes they collapse more readily. Why is this true? (*Hint:* The existence of cosmic rays was first indicated by this observation.)

9. Would it be necessary for a charged body to actually touch the ball of the electroscope for the leaves to diverge? Defend your answer.

10. Strictly speaking, when an object acquires a posi-tive charge, what happens to its mass? When it acquires a negative charge? Think small.

11. How can you charge an object negatively with only the help of a positively charged object?

12. It is relatively easy to strip the outer electrons from a heavy atomic nucleus like that of uranium (which is then a uranium ion) but very difficult to strip the inner electrons. Why do you suppose this is so?

13. How does the magnitude of electric force compare between a pair of charged particles when they are brought to half their original distance of separation? To one-quarter their original distance? To four times their orig-inal distance? (What law guides your answers?)

14. If a large enough electric field is applied, even an insulator will conduct an electric current, as is evident in lightning discharges through the air. Explain how this happens, taking into account the opposite charges in an atom and how ionization occurs.

15. If you are caught outdoors in a thunderstorm, why should you not stand under a tree? Can you think of a reason why you should not stand with your legs far apart? Or why lying down can be dangerous? (*Hint:* Consider electric potential difference.)

16. Why is a good conductor of electricity also a good conductor of heat?

17. If you rub an inflated balloon against your hair and place it against a door, by what mechanism does it stick? Explain.

18. How are electrically neutral atoms and molecules able to electrically attract each other?

19. If you place a free electron and a free proton in an electric field, how will their accelerations compare? Their directions of travel?

20. A gravitational field vector points toward the earth; an electric field vector points toward an electron. Why do electric field vectors point away from protons?

21. By what specific means do the bits of fine threads line up in the electric fields shown in Figure 21-18?

22. If you put in 10 J of work to push a 1 C charge against an electric field, what will be its voltage with respect to its starting position? When released, what will be its kinetic energy if it flies past its starting position?

23. What is the voltage at the location of a 0.0001-C charge that has an electric potential energy of 0.5 J?

24. What safety is offered by staying inside an auto-mobile during a thunderstorm? Defend your answer.

25. Would you feel any electrical effects if you were inside the charged sphere of a Van de Graaff generator? Why or why not?

Electric Current

In the previous chapter we were introduced to the concept of electric potential, measured in volts. In this chapter we will see that this voltage acts like an "electrical pressure" that can produce a flow of charge, or *current* measured in amperes (or simply amps, abbreviated A), and that the *resistance* that restrains this flow is measured in ohms (Ω). We will see that when the flow is in only one direction we call it *direct current* (dc), and when it flows back and forth we call it *alternating current* (ac). The rate at which energy is transferred by electric current is *power*, which is measured in watts (W) or in thousands of watts, kilowatts (kW). We see here that there are many terms to be sorted out. This is easier to do when we have some understanding of the concepts these terms represent, and these concepts are better understood if we know how they relate to one another. We begin with the flow of electric charge.

Flow of Charge

Recall from our study of heat and temperature that when the ends of a conductor are at different temperatures, heat energy flows from the higher temperature to the lower temperature. The flow ceases when both ends reach the same temperature. Similarly, when the ends of an electrical conductor are at different electric potentials—when there is a **potential difference**—charge flows from the higher potential to the lower potential. The flow of charge persists until both ends reach the same potential. Without a potential difference, no flow of charge will occur. Connect one end of a wire to a charged Van de Graaff generator, for example, and the other end to the ground, and a surge of charge will flow through the wire. The flow will be brief, however, for the sphere will quickly reach a common potential with the ground.

Figure 22-1
Water flows from the reservoir of higher pressure to the reservoir of lower pressure. (*a*) The flow will cease when the difference in pressure ceases. (*b*) Water continues to flow because a difference in pressure is maintained with the pump.

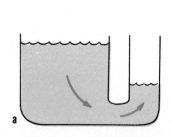
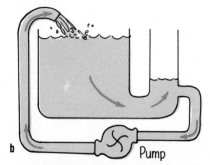

To attain a sustained flow of charge in a conductor, some arrangement must be provided to maintain a difference in potential while charge flows from one end to the other. The situation is analogous to the flow of water from a higher reservoir to a lower one (Figure 22-1*a*). Water will flow in a pipe that connects the reservoirs only as long as a difference in water level exists. The flow of water in the pipe, like the flow of charge in the wire that connects the Van de Graaff generator to the ground, will cease when the pressures at each end are equal (we imply this when we say that water seeks its own level). A continuous flow is possible if the difference in water levels—hence water pressures—is maintained with the use of a suitable pump (Figure 22-1*b*).

Electric Current

Just as water current is the flow of H_2O molecules, **electric current** is simply the flow of electric charge. In circuits of metal wires, electrons make up the flow of charge. This is because one or more electrons from each metal atom are free to move throughout the atomic lattice. These charge carriers are called *conduction electrons*. Protons, on the other hand, do not move because they are bound inside the nuclei of atoms that are more or less locked in fixed positions. In fluids, however, positive ions as well as electrons may compose the flow of electric charge.

The *rate* of electrical flow is measured in *amperes*. An ampere is the rate of flow of 1 coulomb of charge per second. (Recall that 1 coulomb, the standard unit of charge, is the electric charge of 6.25 billion billion electrons.) In a wire that carries 5 amperes, for example, 5 coulombs of charge pass any cross section in the wire each second. So that's a lot of electrons! In a wire that carries 10 amperes, twice as many electrons pass any cross section each second.

It is interesting to note that a current-carrying wire is not electrically charged. Under ordinary conditions, negative conduction electrons swarm through the atomic lattice made up of positively charged atomic nuclei. So there are as many electrons as protons in the wire. Whether a wire carries a current or not, the net charge of the wire is normally zero at every moment.

Figure 22-2
Each coulomb of charge that is made to flow in a circuit that connects the ends of this 1.5-V flashlight cell is energized with 1.5 J.

Figure 22-3
An unusual source of voltage. The electric potential between the head and tail of the electric eel (*Electrophorus electricus*) can be up to 600 V.

Voltage Sources

Charges do not flow of themselves. A sustained current requires a suitable pumping device to provide a difference in potential—to provide a voltage difference. An "electrical pump" is some sort of voltage source. If we charge one metal sphere positively and another negatively, we can develop a large voltage between the spheres. This voltage source is not a good electrical pump because when they are connected by a conductor, the potentials equalize in a single brief surge of moving charges. It is not practical. Chemical batteries or generators, on the other hand, are sources of energy in electric circuits and are capable of maintaining a steady flow.

Batteries and electric generators do work to pull negative charges away from positive ones. In chemical batteries, this work is done by the chemical disintegration of zinc or lead in acid, and the energy stored in the chemical bonds is converted to electric potential energy.* Generators separate charge by electromagnetic induction, a process we will describe in Chapter 24. The work done by whatever means in separating the opposite charges is available at the terminals of the battery or generator. This energy per charge provides the difference in potential (voltage) that provides the "electrical pressure" to move electrons through a circuit joined to these terminals.

The unit of electric potential (voltage) is the *volt*. A common automobile battery will provide an electrical pressure of 12 volts to a circuit connected across its terminals. Then 12 joules of energy are supplied to each coulomb of charge that is made to flow in the circuit.

Figure 22-4

Analogy between a simple hydraulic circuit (*a*) and an electrical circuit (*b*).

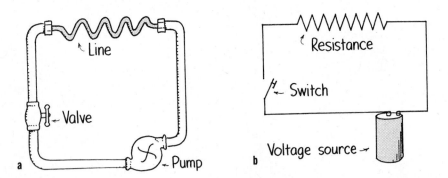

There is often some confusion about charge flowing *through* a circuit and voltage placed, or impressed, *across* a circuit. We can distinguish between these ideas by considering a long pipe filled with water. Water will flow *through* the pipe if there is a difference in pressure *across* or between its ends. Water flows from the high-pressure end to the low-pressure end. Only the water flows, not the pressure. Similarly, electric charge flows because of the differences in electrical pressure (voltage).

*You can find how this is done in almost any chemistry text.

You say that charges flow *through* a circuit because of an applied voltage *across* the circuit.* You don't say that voltage flows through a circuit. Voltage doesn't go anywhere, for it is the charges that move. Voltage produces current (if there is a complete circuit).

Electrical Resistance

A battery or generator of some kind is the prime mover and source of voltage in an electric circuit. How much current there is depends not only on the voltage but also on the **electrical resistance** the conductor offers to the flow of charge. This is similar to the rate of water flow in a pipe, which depends not only on the pressure behind the water but also on the resistance offered by the pipe itself. The resistance of a wire depends on the conductivity of the material and also on its thickness and length. Electrical resistance is less in thick wires. The longer the wire, of course, the greater the resistance. In addition, electrical resistance depends on temperature. The greater the jostling about of atoms within the conductor, the greater resistance the conductor offers to the flow of charge. For most conductors, increased temperature means increased resistance.† The resistance of some materials reaches zero at very low temperatures. These are the superconductors discussed briefly in the previous chapter.

Electrical resistance is measured in units called *ohms*. The Greek letter *omega*, Ω, is commonly used as the symbol for the ohm. This unit is named after Georg Simon Ohm, a German physicist who in 1826 discovered a simple and very important relationship among voltage, current, and resistance.

Ohm's Law

The relationship among voltage, current, and resistance is summarized by a statement called **Ohm's law**. Ohm discovered that the amount of current in a circuit is directly proportional to the voltage established across the circuit, and is inversely proportional to the resistance of the circuit. In short,

$$\text{Current} = \frac{\text{voltage}}{\text{resistance}}$$

Or, in units form,

$$\text{Amperes} = \frac{\text{volts}}{\text{ohms}}$$

---

*It is conceptually simpler to say that current flows through a circuit, but don't say this around somebody who is picky about grammar, for the expression "current flows" is redundant. More properly, charge flows—which *is* current.

†Carbon is an interesting exception. As temperature increases, more carbon atoms shake loose an electron. This increases the electric current. The resistance of carbon, in effect, lowers with increasing temperature. This and (primarily) its high melting point are why carbon is used in arc lamps.

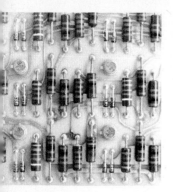

Figure 22-5
Resistors. The symbol of resistance in an electric circuit is —$\wedge\wedge\wedge$— .

So for a given circuit of constant resistance, current and voltage are proportional to each other.* This means we'll get twice the current for twice the voltage. The greater the voltage, the greater the current. But if the resistance is doubled for a circuit, the current will be half what it would be otherwise. The greater the resistance, the smaller the current. Ohm's law makes good sense.

Ohm's law tells us that a potential difference of 1 volt established across a circuit that has a resistance of 1 ohm will produce a current of 1 ampere. If 12 volts are impressed across the same circuit, the current will be 12 amperes. The resistance of a typical lamp cord is much less than 1 ohm, while a typical light bulb has a resistance of about 100 ohms. An iron or electric toaster has a resistance of 15–20 ohms. The low resistance permits a large current, which produces considerable heat. Inside electrical devices such as radio and television receivers, current is regulated by circuit elements called *resistors*, whose resistance may be a few ohms or millions of ohms.

Questions ▶

1. How much current will flow through a lamp that has a resistance of 60 Ω when 12 V are impressed across it?

2. What is the resistance of an electric frying pan that draws 12 A when connected to a 120-V circuit?

Ohm's Law and Electric Shock

What causes electric shock in the human body—current or voltage? The damaging effects of shock are the result of current passing through the body. From Ohm's law, we can see that this current depends on the voltage that is applied, and also on the electrical resistance of the human body. The resistance of one's body depends on its condition and ranges from about 100 ohms if soaked with salt water to about 500 000 ohms if the skin is very dry. If we touch the two electrodes of a battery with dry fingers, the resistance our body normally offers to the flow of charge is about 100 000 ohms. We usually cannot feel 12 volts, and 24 volts just barely tingles. If our skin is moist, 24 volts can be quite uncomfortable. Table 22-1 describes the effects of different amounts of current on the human body.

▶ **Answers**

1. 1/5 A. This is calculated from Ohm's law: 0.2 A = 12 V/60 Ω.
2. 10 Ω. Rearrange Ohm's law to read

$$\text{Resistance} = \frac{\text{voltage}}{\text{current}} = \frac{120 \text{ V}}{12 \text{ A}} = 10 \text{ }\Omega$$

*Many texts use V for voltage, I for current, and R for resistance, and express Ohm's law as $V = IR$. It then follows that $I = V/R$, or $R = V/I$, so if any two variables are known, the third can be found. Units are abbreviated V for volts, A for amperes, and Ω for ohms.

Table 22-1
Effect of electric currents
on the body

Current (A)	Effect
0.001	Can be felt
0.005	Is painful
0.010	Causes involuntary muscle contractions (spasms)
0.015	Causes loss of muscle control
0.070	Goes through the heart; serious disruption, probably fatal if current lasts for more than 1 s

Questions

1. At 100 000 Ω, how much current will flow through your body if you touch the terminals of a 12-V battery?

2. If your skin is very moist—so your resistance is only 1000 Ω—and you touch the terminals of a 12-V battery, how much current will you receive?

Figure 22-6
The bird can stand harmlessly on one wire of high potential, but it had better not reach over and grab a neighboring wire! Why not?

Many people are killed each year by current from common 120-volt electric circuits. If you touch a faulty 120-volt light fixture with your hand while your feet are on the ground, there is a 120-volt "electrical pressure" between your hand and the ground. Resistance to current flow is usually greatest between your feet and the ground, so the current is usually not enough to do serious harm. But if your feet and the ground are wet, there is a low-resistance electrical path between you and the ground. Your overall resistance is so lowered that the 120-volt potential difference across your body may produce a current greater than your body can withstand. Drops of water that collect around the on-off switch of devices such as a hair dryer can conduct current to the user. Although distilled water is a good insulator, the ions in ordinary water greatly reduce the electrical resistance. More dissolved materials, especially small amounts of salt, reduce the resistance even more. There is usually a layer of salt left from perspiration on your skin, which when wet lowers your skin resistance to a few hundred ohms or less. Handling electrical devices while taking a bath is a definite no-no.

To receive a shock, there must be a *difference* in electric potential between one part of your body and another part. Most of the current will pass along the path of least electrical resistance connecting these two points. Suppose you fell from a bridge and managed to grab onto a high-voltage power line, halting your fall. So long as you touch nothing else of different potential, you will receive no shock at all. Even if the wire is a few thousand volts above ground potential and even if you hang by it with two hands, no appreciable charge will flow from one hand to the

▶ **Answers**

1. The current through your body will be 12/100 000 A (0.00012 A).
2. You will receive 12/1000 A (0.012 A). Ouch!

Figure 22-7
The third prong connects the body of the appliance directly to ground. Any charge that builds up on an appliance is therefore conducted to the ground.

other. This is because there is no appreciable difference in electric potential between your hands. If, however, you reach over with one hand and grab onto a wire of different potential . . . zap! We have all seen birds perched on high-voltage wires. Every part of their bodies is at the same high potential as the wire, so they feel no ill effects.

Most electric plugs and sockets today are wired with three, instead of two, connections. The principal two flat prongs on a plug are for the current-carrying double wire, one part of which is "live" and the other neutral, while the third round prong is connected directly to the earth (Figure 22-7). Appliances such as irons, stoves, washing machines, and dryers are connected with these three wires. If the live wire accidentally comes in contact with the metal surface of the appliance, the current will be directed to ground and won't shock anyone who handles it.

Electric shock can overheat tissues in the body and disrupt normal nerve functions. It can upset the nerve center that controls breathing. In rescuing shock victims, the first thing to do is clear them from the electric supply with a dry wooden stick or some other nonconductor so that you don't get electrocuted yourself. Then apply artificial respiration.

Question ▶ What causes electric shock, current or voltage?

Direct Current and Alternating Current

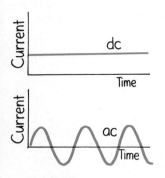

Figure 22-8
A time graph of dc and ac.

Electric current may be dc or ac. By *dc*, we mean **direct current**, which refers to the flowing of charges in *one* direction. A battery produces direct current in a circuit because the terminals of the battery always have the same sign. Electrons move from the repelling negative terminal toward the attracting positive terminal, always moving through the circuit in the same direction. Even if the current moves in unsteady pulses, so long as it moves in one direction only, it is dc.

Alternating current (*ac*) acts as the name implies. Electrons in the circuit are moved first in one direction and then in the opposite direction, alternating to and fro about relatively fixed positions. This is accomplished by alternating the direction of voltage at the energy source. Nearly all commercial ac circuits involve voltages and currents that alternate back and forth at a frequency of 60 cycles per second. This is 60-hertz current. In some countries, 25-hertz, 30-hertz, or 50-hertz current is used. Throughout the world, most residential and commercial circuits are ac because electric energy in the form of ac can be transmitted great distances with easy step-ups in voltage that result in lower heat losses in the wires. Why this is so will be discussed in Chapter 24. The primary use of electric current, whether dc or ac, is to transfer energy quietly, flexibly, and conveniently from one place to another.

▶ **Answer**
Electric shock *occurs* when current is produced in the body, which is *caused* by an impressed voltage.

Speed and Source of Electrons in a Circuit

Figure 22-9
The electric field lines between the terminals of a battery are directed through a conductor, which joins the terminals. A metal bar is shown here, but the conductor is usually an electric circuit. (If you do this, you won't be shocked, but the bar will heat quickly and may burn your hand!)

When we flip on the light switch on a wall and the circuit is completed, either ac or dc, the light bulb appears to glow immediately. When we make a telephone call, the electrical signal carrying our voice travels through the connecting wires at seemingly infinite speed. This signal is transmitted through the conductors at nearly the speed of light.

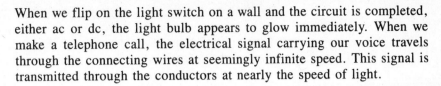
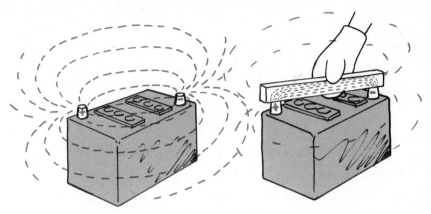

It is *not* the electrons that move at this speed.* Although electrons inside metal at room temperature have an average speed of a few million kilometers per hour, they make up no current because they are moving in all possible directions. There is no net flow in any preferred direction. But when a battery or generator is connected, an electric field is established inside the conductor. The electrons continue their random motions while simultaneously being nudged by this field. It is the electric field that travels through a circuit at nearly the speed of light. The conducting wire acts as a guide to the electric field lines that are established at the voltage source (Figure 22-9).

If the voltage source is dc, like the battery shown in Figure 22-9, the electric field lines are maintained in one direction in the conductor. Conduction electrons are accelerated by the field in a direction parallel to the field lines. Before they gain appreciable speed, they "bump into" the anchored metallic ions in their paths and lose some of their kinetic energy to them. This is why current-carrying wires become hot. These collisions interrupt the motion of the electrons, so the actual drift velocity of electrons is extremely low. In a typical dc circuit—the electrical system of an automobile, for example—electrons have a drift velocity that averages about a hundredth of a centimeter per second. At this rate it would take about 3 hours for an electron to travel through 1 meter of wire! So although an electric signal travels at nearly the speed of light in a wire, the electrons that move in response to this signal travel slower than a snail's pace.

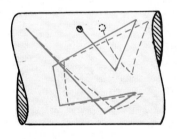

Figure 22-10
The solid lines depict a possible random path of an electron bounding about in an atomic lattice at a speed of about $\frac{1}{200}$ the speed of light. The dashed lines show an exaggerated view of how this path may be altered when an electric field is applied. The electron drifts toward the right with a speed much slower than a snail's pace. (This is quite idealized, for the electron wouldn't really encounter such similarly placed atoms and follow so similar a path when drifting.)

---

*Much effort and expense are spent in building particle accelerators that accelerate electrons and protons to speeds near that of light. If electrons in a common circuit traveled that fast, one would only have to bend a wire at a sharp angle and electrons traveling through the wire would possess so much momentum that they would fail to make the turn and fly off, providing a beam comparable to that produced by the accelerators!

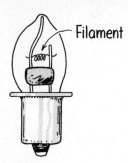
← Filament

Figure 22-11
The conduction electrons that surge to and fro in the filament of the lamp do not come from the voltage source. They are in the filament to begin with. The voltage source simply provides them with surges of energy.

In an ac circuit, the conduction electrons don't travel at all. They oscillate rhythmically to and fro about relatively fixed positions. When you talk to your friend on the telephone, it is the *pattern* of oscillating motion that is carried across town at nearly the speed of light. The electrons already in the wires vibrate to the rhythm of the traveling pattern.

A common misconception regarding electrical currents is that the current is propagated through the conducting wires by electrons bumping into one another—that an electrical pulse is transmitted in a manner similar to the way the pulse of a tipped domino is transferred along a row of closely spaced standing dominoes. This simply isn't true. The domino idea is a good model for the transmission of sound, but not for the transmission of electric energy. Electrons that are free to move in a conductor are accelerated by the electric field impressed upon them, but not because they bump into one another. True, they do bump into one other and other atoms, but this slows them down and offers resistance to their motion. Electrons throughout the entire closed path of a circuit all react simultaneously to the electric field.

This brings up another misconception regarding electricity. Many people think that the electrical outlets in the walls of their homes are a source of electrons—that electrons flow from the power utility through the power lines and into the wall outlets of their homes. This is not true. The outlets in homes are ac. Electrons do not travel through a wire in an ac circuit, but instead vibrate to and fro about relatively fixed positions. When you plug a lamp into an outlet, *energy* flows from the outlet into the lamp, not electrons. Energy is carried by the electric field and causes vibratory motion of the electrons that already exist in the lamp filament. Most of this electric energy is transformed into heat, while some of it takes the form of light. Power utilities do not sell electrons. They sell *energy*. You supply the electrons.

When you are jolted by an electric shock, the electrons making up the current in your body originate in your body. Electrons do not come out of the wire and through your body and into the ground. Energy does. The energy simply causes free electrons in your body to move in unison. Small movements tingle. Large movements can be fatal.

Electric Power

The moving charges in an electric current do work. This work, for example, can heat a circuit or turn a motor. The rate at which work is done—that is, the rate at which electric energy is converted into another form such as mechanical energy, heat, or light—is called **electric power**. Electric power is equal to the product of current and voltage.*

---

*Recall from Chapter 6 that Power = work/time; 1 W = 1 J/s. Note that the units for mechanical power and electrical power check (work and energy are both measured in joules):

$$\text{Power} = \frac{\text{charge}}{\text{time}} \times \frac{\text{energy}}{\text{charge}} = \frac{\text{energy}}{\text{time}}$$

Power = current × voltage

If the voltage is expressed in volts and the current in amperes, then the power is expressed in watts. So, in units form,

Watts = amperes × volts

If a lamp rated at 120 watts operates on a 120-volt line, you can see that it will draw a current of 1 ampere (120 watts = 1 ampere × 120 volts). A 60-watt lamp draws $\frac{1}{2}$ ampere on a 120-volt line. This relationship becomes a practical matter when you wish to know the cost of electrical energy, which is usually a few cents per kilowatt-hour depending on locality. A kilowatt is 1000 watts, and a kilowatt-hour represents the amount of energy consumed in 1 hour at the rate of 1 kilowatt.* Therefore, in a locality where electric energy costs 5 cents per kilowatt-hour, a 100-watt electric light bulb can be run for 10 hours at a cost of 5 cents, or a half cent for each hour. A toaster or iron, which draws much more current and therefore much more energy, costs several times as much to operate.

Figure 22-12
The power and voltage on the light bulb read "100 W 120 V." How many amperes will flow though the bulb?

Questions ▶

1. If a 120-V line to a socket is limited to 15 A by a safety fuse, will it operate a 1200-W hair dryer?
2. At 10¢/kWh, what does it cost to operate the 1200-W hair dryer for 1 h?

Types of Electric Circuits

Any path along which electrons can flow is a *circuit*. For a continuous flow of electrons, there must be a complete circuit with no gaps. A gap is usually provided by an electric switch that can be opened or closed to either cut off or allow energy flow. Most circuits have more than one device that receives electric energy. These devices are commonly connected in a circuit in one of two ways, *series* or *parallel*. When connected in series, they form a single pathway for electron flow between the terminals of the battery, generator, or wall socket (which is simply an extension of these terminals). When connected in parallel, they form branches, each of which is a separate path for the flow of electrons. Both series and parallel connections have their own distinctive characteristics. We shall briefly treat circuits using these two types of connections.

▶ **Answers**

1. From the expression Watts = amperes × volts, we see that A = 1200 W/120 V = 10 A, so the hair dryer will operate when connected to the circuit. But two hair dryers on the same socket will blow the fuse.
2. 12¢ (1200 W = 1.2 kW; 1.2 kW × 1 h × 10¢/1 kWh = 12¢).

---

*Since Power = energy/time, simple rearrangement gives Energy = power × time; thus, energy can be expressed in the unit *kilowatt-hours* (kWh).

Series Circuits

A simple **series circuit** is shown in Figure 22-13. Three lamps are connected in series with a battery. The same current exists almost immediately in all three lamps when the switch is closed. The current does not "pile up" in any lamp but flows *through* each lamp. Electrons that make up this current move in a direction from the negative terminal of the battery, through each of the resistive filaments in the lamps in turn, and then to the positive terminal of the battery (the same amount of current passes through the battery). This is the only path of the electrons through the circuit. A break anywhere in the path results in an open circuit, and the flow of electrons ceases. Burning out of one of the lamp filaments or simply opening the switch could cause such a break.

Figure 22-13

A simple series circuit. The 6-V battery provides 2 V across each lamp.

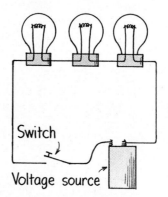

The circuit shown in Figure 22-13 illustrates the following important characteristics of series connections:

1. Electric current has but a single pathway through the circuit. This means that the current passing through the resistance of each electrical device is the same.
2. This current is resisted by the resistance of the first device, the resistance of the second, and that of the third also, so the total resistance to current is the sum of the individual resistances along the circuit path.
3. The current in the circuit is numerically equal to the voltage supplied by the source divided by the total resistance of the circuit. This is in accord with Ohm's law.
4. The total voltage impressed across a series circuit divides among the individual electrical devices in the circuit so that the sum of the "voltage drops" across the resistance of each individual device is equal to the total voltage supplied by the source. This follows from the fact that the amount of energy given to the total current is equal to the sum of energies given to each device.
5. The voltage drop across each device is proportional to its resistance. This follows from the fact that more energy is dissipated when a current passes through a large resistance than when the same current passes through a small resistance.

Questions ▶

1. What happens to current in other lamps if one lamp in a series circuit burns out?
2. What happens to the light intensity of each lamp in a series circuit when more lamps are added to the circuit?

It is easy to see the main disadvantage of a series circuit: if one device fails, current in the whole circuit ceases. Some cheap Christmas tree lights are connected in series. When one bulb burns out, it's fun and games (or frustration) trying to find which one to replace.

Most circuits are wired so that it is possible to operate several electrical devices, each independently of the other. In your home, for example, a lamp can be turned on or off without affecting the operation of other lamps or electrical devices. This is because these devices are connected not in series, but in parallel with one another.

Parallel Circuits

A simple **parallel circuit** is shown in Figure 22-14. Three lamps are connected to the same two points A and B. Electrical devices connected to the same two points of an electrical circuit are said to be *connected in parallel*. Electrons leaving the negative terminal of the battery need travel through only one lamp filament before returning to the positive terminal of the battery. In this case, current branches into three separate pathways from A to B. A break in any one path does not interrupt the flow of charge in the other paths. Each device operates independently.

Figure 22-14
A simple parallel circuit. A 6-V battery provides 6 V across each lamp.

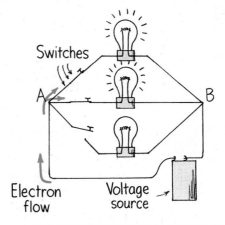

Switches

A   B

Electron flow   Voltage source

▶ **Answers**

1. If one of the lamp filaments burns out, the path connecting the terminals of the voltage source will break and current will cease. All lamps will go out.
2. The addition of more lamps in a series circuit results in a greater circuit resistance. This decreases the current in the circuit and therefore in each lamp, which causes dimming of the lamps. Energy is divided among more lamps, so the voltage drop across each lamp will be less.

The circuit shown in Figure 22-14 illustrates the following major characteristics of parallel connections:

1. Each device connects the same two points A and B of the circuit. The voltage is therefore the same across each device.
2. The total current in the circuit divides among the parallel branches. Since the voltage across each branch is the same, the amount of current in each branch is inversely proportional to the resistance of the branch.
3. The total current in the circuit equals the sum of the currents in its parallel branches.
4. As the number of parallel branches is increased, the overall resistance of the circuit is decreased. Overall resistance is lowered with each added path between any two points of the circuit. This means the overall resistance of the circuit is less than the resistance of any one of the branches.

Questions

1. What happens to the current in other lamps if one of the lamps in a parallel circuit burns out?
2. What happens to the light intensity of each lamp in a parallel circuit when more lamps are added in parallel to the circuit?

Parallel Circuits and Overloading

Electricity is usually fed into a home by way of two lead wires called *lines*. These lines are very low in resistance and are connected to wall outlets in each room. About 110–120 volts are impressed on these lines

---

▶ **Answers**

1. If one lamp burns out, the other lamps will be unaffected. The current in each branch, according to Ohm's law, is equal to voltage/resistance, and since neither voltage nor resistance is affected in the other branches, the current in those branches is unaffected. The total current in the overall circuit (the current through the battery), however, is decreased by an amount equal to the current drawn by the lamp in question before it burned out. But the current in any other single branch is unchanged.
2. The light intensity for each lamp is unchanged as other lamps are introduced (or removed). Only the total resistance and total current in the total circuit changes, which is to say the current in the battery changes. (There is resistance in a battery also, which we assume is negligible here.) As lamps are introduced, more paths are available between the battery terminals, which effectively decreases total circuit resistance. This decreased resistance is accompanied by an increased current, the same increase that feeds energy to the lamps as they are introduced. Although changes of resistance and current occur for the circuit as a whole, no changes occur in any individual branch in the circuit.

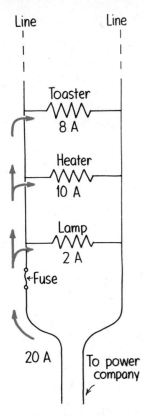

Figure 22-15
Circuit diagram for appliances connected to a household supply line.

Figure 22-16
A safety fuse.

by generators at the power utility. This voltage is applied to appliances and other devices that are connected in parallel by plugs to these lines. As more devices are connected to the lines, more pathways for current result in lowering of the combined resistance of the circuit. Therefore, a greater amount of current occurs in the lines. Lines that carry more than a safe amount of current are said to be *overloaded*.

We can see how overloading occurs by considering the circuit in Figure 22-15. The supply line is connected to an electric toaster that draws 8 amperes, to an electric heater that draws 10 amperes, and to an electric lamp that draws 2 amperes. When only the toaster is operating and drawing 8 amperes, the total line current is 8 amperes. When the heater is also operating, the total line current increases to 18 amperes (8 amperes to the toaster and 10 amperes to the heater). If you turn on the lamp, the line current increases to 20 amperes. Connecting any more devices increases the current still more. Connecting too many devices into the same line results in overheating that may cause a fire.

Safety Fuses

To prevent overloading in circuits, fuses are connected in series along the supply line. You can see the location of the fuse in Figure 22-15 and see that the entire line current must pass through the fuse. You can also see that if the fuse fails to pass current, then no current will reach the devices. The common fuse shown in Figure 22-16 is constructed with a wire ribbon that will heat up and melt at a given current. If the fuse is rated at 20 amperes, it will pass 20 amperes, but no more. A current above 20 amperes will melt the fuse, which "blows out" and breaks the circuit. Before a blown fuse is replaced, the cause of overloading should be determined and remedied. Often, insulation that separates the wires in a circuit wears away and allows the wires to touch. This effectively shortens the path of the circuit and is called a *short circuit*.

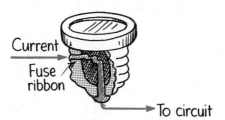

Circuits may also be protected by circuit breakers, which use magnets or bimetallic strips to open the switch. These are more and more replacing the type of fuse shown in Figure 22-16. Utility companies use circuit breakers to protect their lines all the way back to the generators.

Summary of Terms

Potential difference The difference in voltage between two points, measured in volts. It can be compared to the difference in water pressure between two containers: if two containers having different water pressures are connected by a pipe, water will flow from the one with the higher pressure to the one with the lower pressure until the two pressures are equalized. Similarly, if two points with a difference in potential are connected by a conductor, charge will flow from the one with the greater potential to the one with the smaller potential until the potentials are equalized.

Electric current The flow of electric charge that transports energy from one place to another. Measured in amperes, where 1 A is the flow of 6.25×10^{18} electrons per second.

Electrical resistance The property of a material that resists the flow of an electric current through it. Measured in ohms.

Ohm's law The statement that the current in a circuit varies directly with the potential difference or voltage and inversely with resistance:

$$\text{Current} = \frac{\text{voltage}}{\text{resistance}}$$

A potential difference of 1 V across a resistance of 1 Ω produces a current of 1 A.

Direct current (dc) An electric current flowing in one direction only.

Alternating current (ac) Electric current that repeatedly reverses its direction; the electric charges vibrate about relatively fixed points. In the United States the vibrational rate is 60 Hz.

Electric power The rate of energy transfer, or the rate of doing work; the amount of energy per unit time, which electrically can be measured by the product of current and voltage:

$$\text{Power} = \text{current} \times \text{voltage}$$

Measured in watts (or kilowatts), where $1 \text{ A} \times 1 \text{ V} = 1 \text{ W}$.

Series circuit An electric circuit with devices having resistances arranged in such a way that the same electric current flows through each of them.

Parallel circuit An electric circuit with two or more resistances arranged in branches in such a way that any single one completes the circuit independently of all the others.

Review Questions

Flow of Charge

1. What condition is necessary for the flow of heat energy to occur from one end of a metal bar to the other? For the flow of electric charge?

2. For electric circuits, what is analogous to the statement "Water seeks its own level"?

3. What condition is necessary for a sustained flow of electric charge through a conducting medium?

Electric Current

4. Why are *electrons*, rather than *protons*, the principal charge carriers in metal wires?

5. What exactly is an *ampere*?

6. Why is a current-carrying wire not electrically charged?

Voltage Sources

7. Name two kinds of "electric pumps."

8. How much energy is given to each coulomb of charge passing through a 6-V battery?

9. Does current flow *across* a circuit or *through* a circuit? Does voltage flow *across* a circuit or is it *established* across a circuit? Explain.

10. Does voltage produce current or does current produce voltage? Which is the cause and which is the effect?

Electrical Resistance

11. Will water flow more easily through a wide pipe or a narrow pipe? Will current flow more easily through a thick wire or a thin wire?

12. Is electrical resistance greater or lower in a copper wire that is heated?

Ohm's Law

13. What is the effect on current through a circuit of steady resistance when the voltage is doubled? What if both voltage and resistance are doubled?

14. How much current will flow through a radio speaker of 8 Ω resistance when 12 V are impressed across it?

15. What role do resistors normally play in an electric circuit?

Ohm's Law and Electric Shock

16. Which has the greater electrical resistance, wet skin or dry skin?

17. What happens to the resistance of your skin when you perspire?

18. Why is it a very poor idea to handle electrical devices while in the bathtub?

19. High voltage by itself does not produce electric shock. What does?

20. What is the function of the third prong on the plug of an electric appliance?

Direct Current and Alternating Current

21. Distinguish between *dc* and *ac*.

22. Which produces dc, a battery or a simple generator? Which produces ac?

23. What does it mean to say that a certain current is 60 Hz?

Speed and Source of Electrons in a Circuit

24. True or false: Electrons in a common battery-driven circuit travel at about the speed of light. Defend your answer.

25. Exactly what does flow through a circuit at about the speed of light?

26. Why does a wire that carries electric current become hot?

27. What is meant by *drift velocity*?

28. A tipped domino sends a pulse along a row of standing dominoes. Is this an analogy for the way electric current, sound, or both travel?

29. Exactly what is delivered through power lines into your home?

30. From where do the electrons originate that produce an electric shock when you touch a charged conductor?

Electric Power

31. What is the relationship among electric power, current, and voltage?

32. What is the unit of power?

33. Distinguish between a *kilowatt* and a *kilowatt-hour*.

Types of Electric Circuits

34. What is an *electric circuit*?

Series Circuits

35. In a circuit of two lamps in series, if the current through one lamp is 1 A, what is the current through the other lamp?

36. If 6 V are impressed across the above circuit and the voltage across the first lamp is 2 V, what is the voltage across the second lamp?

37. What is a main shortcoming of a series circuit?

Parallel Circuits

38. In a circuit of two lamps in parallel, if there are 6 V across one lamp, what is the voltage across the other lamp?

39. If the current through each of the two branches of a parallel circuit is the same, what does this tell you about the resistance of the two branches?

40. How does the total current through the branches of a simple parallel circuit compare to the current that flows through the voltage source?

41. As more doors are opened in a crowded room, the resistance to the motion of people trying to leave the room is reduced. How is this similar to what happens when more branches are added to a parallel circuit?

42. Are household circuits normally wired in *series* or in *parallel*?

Parallel Circuits and Overloading

43. How does the amount of line current in a home circuit differ from the amount of current that lights the lamp you are likely reading this by?

44. Why will too many electrical devices operating at one time often blow a fuse?

Safety Fuses

45. What is the function of a safety fuse, and why is it a bad idea to replace a blown fuse with a penny?

Home Projects

1. An electric cell is made by placing two plates of different materials that have different affinities for electrons in a conducting solution. You can make a simple 1.5-V cell by placing a strip of copper and a strip of zinc in a tumbler of salt water. The voltage of a cell depends on the materials used and the solution they are placed in, not the size of the plates. A battery is actually a series of cells.

An easy cell to construct is the citrus cell. Stick a paper clip and a piece of copper wire into a lemon. Hold the ends of the wire close together, but not touching, and place the ends on your tongue. The slight tingle you feel and the metallic taste you experience result from a slight current of electricity pushed by the citrus cell through the wires when your moist tongue closes the circuit.

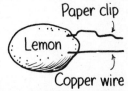

2. Examine the electric meter in your house. It is probably in the basement or on the outside of your house. You will see that in addition to the clocklike dials in the meter, there is a circular aluminum disk that spins between the poles of magnets when electric current goes into the house. The more electric current, the faster the disk turns. The speed of the disk is directly proportional to the number of watts used; for example, it will spin five times as fast for 500 W as for 100 W.

You can use the meter to determine how many watts an electrical device uses. First, see that all electrical devices in your home are disconnected (you may leave electric clocks connected, for the 2 watts they use will hardly be noticeable). The disk will be practically stationary. Then connect a 100-W bulb and note how many seconds it takes for the disk to make five complete revolutions. The black spot painted on the edge of the disk makes this easy. Disconnect the 100-W bulb and plug in a device of unknown wattage. Again, count the seconds for five revolutions. If it takes the same time, it's a 100-W device; if it takes twice the time, it's a 50-W device; half the time, a 200-W device; and so forth. In this way you can estimate the power consumption of devices fairly accurately.

Exercises

1. If a conductor is placed in contact with two separated objects charged to different electrical potential energies, can you say for certain which way charge will flow in the conductor? How about if the objects are charged to different electric potentials?

2. If an electric current flows from one object to another, what can we say about the relative magnitudes of the electric potentials of the two objects?

3. One example of a water system is a garden hose that waters a garden. Another is the cooling system of an automobile. Which of these exhibits behavior more analogous to that of an electric circuit? Why?

4. What happens to the brightness of light emitted by a lightbulb when the current that flows in it increases?

5. Your tutor tells you that an *ampere* and a *volt* really measure the same thing, and the different terms only serve to make a simple concept seem confusing. Why should you consider getting a different tutor?

6. In which of the circuits below does a current exist to light the bulb?

7. Does electric current flow out of a battery or through a battery? Does it flow into a light bulb or through a light bulb? Explain.

8. Energy is given to electric charge by pumping it from a low potential to a high potential. How do we get energy out of electricity?

9. Sometimes you hear someone say that a particular appliance "uses up" electricity. What is it that the appliance actually uses up and what becomes of it?

10. An electron moving in a wire collides again and again with atoms and travels an average distance between collisions that is called the *mean free path*. If the mean free path is less in some metals, what can you say about the resistance of these metals? For a given conductor, what can you do to lengthen the mean free path?

11. A simple lie detector consists of an electric circuit, one part of which is part of your body—like between your fingers. A sensitive meter shows the current that flows when a small voltage is applied. How does this technique indicate that a person is lying? (And when does this technique not tell when someone is lying?)

12. Only a small percentage of the electric energy fed into a common light bulb is transformed into light. What happens to the rest?

13. Why are thick wires rather than thin wires usually used to carry large currents?

14. Will a lamp with a thick or thin filament of the same length draw the most current?

15. Will the current in a light bulb connected to a 220-V source be greater or less than when the same bulb is connected to a 110-V source?

16. Which will do less damage, plugging a 110-V appliance into a 220-V circuit or plugging a 220-V appliance into a 110-V circuit? Explain.

17. Would the resistance of a 100-W bulb be greater or less than the resistance of a 60-W bulb? Assuming the filaments in each bulb are of the same length, which bulb has the thicker filament?

18. If a current of one- or two-tenths of an ampere flows into one of your hands and out the other, you will probably be electrocuted. But if the current flows into your hand and out the elbow above the same hand, you can survive even if the current is large enough to burn your flesh. Explain.

19. Would you expect to find dc or ac in the filament of a light bulb in your home? How about in an automobile?

20. The wattage marked on a light bulb is not an inherent property of the bulb but depends on the voltage to which it is connected, usually 110 or 120 V. How many amperes flow through a 60-W bulb connected in a 120-V circuit?

21. The damaging effects of electric shock result from the amount of current that flows in the body. Why, then, do we see signs that read "Danger—High Voltage" rather than "Danger—High Current"?

22. Comment on the warning sign shown in the sketch.

DANGER !
HIGH RESISTANCE
(1 000 000 000 Ω)

23. Why is the wingspan of birds a consideration in determining the spacing between parallel wires in a power line?

24. Estimate the number of electrons that a power company delivers annually to the homes of a typical city of 50 000 people.

25. If electrons flow very slowly through a circuit, why does it not take a noticeably long time for a lamp to glow when you turn on a distant switch?

26. Why is the speed of an electric signal so much greater than the speed of sound?

27. If a glowing light bulb is jarred and oxygen leaks inside, the bulb will momentarily brighten considerably before burning out. Putting excess current through a light bulb will also burn it out. What physical change occurs when a light bulb burns out?

28. Rearrange the equation Current = voltage/resistance to express *resistance* in terms of current and voltage. Then solve the following: A certain device in a 120-V circuit has a current rating of 20 A. What is the resistance of the device (how many ohms)?

29. Using the formula Power = current × voltage, find the current drawn by a 1200-W hair dryer connected to 120 V. Then, using the method you used in the previous exercise, find the resistance of the hair dryer.

30. Consider a pair of flashlight bulbs connected to a battery. Will they each glow brighter connected in series or in parallel? Will the battery run down faster if they are connected in series or in parallel?

31. If several bulbs are connected in series to a battery, they may feel warm to the touch but not visibly glow. What is your explanation?

32. In the circuit shown, how do the brightnesses of the identical light bulbs compare? Which light bulb draws the most current? What will happen if bulb A is unscrewed? If C is unscrewed?

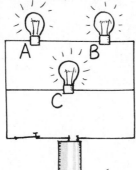

33. As more and more bulbs are connected in series to a flashlight battery, what happens to the brightness of each bulb? Assuming resistance inside the battery is negligible, what happens to the brightness of each bulb when more and more bulbs are connected in parallel?

34. What changes occur in the line current when more devices are introduced in a series circuit? In a parallel circuit? Why are your answers different?

35. It so happens that if too great a load is placed on a battery, the internal resistance of the battery is increased. This lowers the voltage that is supplied to the external circuit. If too many lamps are connected in parallel across a battery, will their brightness diminish? Explain.

36. Why are devices in household circuits almost never connected in series?

37. If a 60-W bulb and a 100-W bulb are connected in series in a circuit, through which bulb will there be the greater voltage drop? How about if they are connected in parallel?

38. The useful life an automobile battery has without being recharged is given in terms of ampere-hours. A typical 12-V battery has a rating of 60 ampere-hours, which means that a current of 60 A can be drawn for 1 h, 30 A can be drawn for 2 h, and so forth. Suppose you forget to turn off the headlights in your parked automobile. If each of the two headlights draws 3 A, how long will it be before your battery is "dead"?

39. How much does it cost to operate a 100-W lamp continuously for 1 month if the power utility rate is 8¢/kWh?

40. A 4-W night light is plugged into a 120-V circuit and operates continuously for 1 year. Find the following: (a) the current it draws, (b) the resistance of its filament, (c) the energy consumed in a year, and (d) the cost of its operation for a year at the utility rate of 8¢/kWh.

The levitation of the little magnet above the superconductive material indicates two forces of equal strength — an upward force I call a **magnetic force** and a downward force I call a **gravitational force**. But knowing their names is only the beginning of my education. I experiment and study further to cultivate a conceptual understanding of how surface currents induced in the superconductor provide the repelling magnetic force and how the weight of the little magnet is a gravitational interaction between its mass and the mass of the whole world. There's a big difference between **knowing the name** of a concept and **understanding** that concept.

23 Magnetism

The term *magnetism* comes from the region of Magnesia, an island in the Aegean Sea, where certain stones were found by the Greeks more than 2000 years ago. These stones, called *lodestones*, had the unusual property of attracting pieces of iron. Magnets were first fashioned into compasses and used for navigation by the Chinese in the twelfth century.

In the sixteenth century, William Gilbert, Queen Elizabeth's physician, made artificial magnets by rubbing pieces of iron against lodestone, and suggested that a compass always points north and south because the earth itself has magnetic properties. Later, in 1750, John Michell in England found that magnetic poles obey the inverse-square law, and his results were confirmed by Charles Coulomb. The subjects of magnetism and electricity developed independently of each other until 1820, when a Danish physicist named Hans Christian Oersted discovered that an electric current affects a magnetic compass.* He saw that magnetism was related to electricity. Shortly thereafter, the French physicist André Marie Ampère proposed that electric currents are the source of all magnetism.

Magnetic Forces

In Chapter 21 we discussed the forces that electrically charged particles exert on one another: the force between any two charged particles depends on the magnitude of the charge on each and their distance of separation, as specified in Coulomb's law. But Coulomb's law is not precisely true when the charges are moving with respect to each other. The force between electric charges depends also, in a complicated way, on the motion of the charges. We find that in addition to the force we call *electrical*, there is a force due to the motion of the charges that we call the **magnetic force**. Both electrical and magnetic forces are actually different aspects of the same phenomenon of electromagnetism.

Magnetic Poles

The forces that magnets exert on one another are similar to electrical forces, for they can both attract and repel without touching, depending on which ends of the magnets are held near one another. Like electrical

---

*We can only speculate about how often such relationships become evident when they "aren't supposed to" and are dismissed as "something wrong with the apparatus." Oersted, however, had the insight—characteristic of a good physicist—to see that nature was revealing another of its secrets.

Figure 23-1
A horseshoe magnet.

forces also, the strength of their interaction depends on the distance the two magnets are separated. Whereas electric charge is central to electrical forces, regions called *magnetic poles* give rise to magnetic forces.

A bar magnet suspended by a piece of string from its center will act as a compass. One end points northward, called the *north-seeking pole*, and the opposite end points southward, called the *south-seeking pole*. More simply, these are called the *north* and *south poles*. All magnets have both a north and a south pole. In a simple bar magnet these are located at the two ends. A common horseshoe magnet is simply a bar magnet that has been bent into a U shape. Its poles are also at its two ends.

When the north pole of one magnet is brought near the north pole of another magnet, they repel.* The same is true of a south pole near a south pole. If opposite poles are brought together, however, attraction occurs. We find that

Like poles repel; opposite poles attract.

This rule is similar to the rule for the forces between electric charges, where like charges repel one another and unlike charges attract. But there is a very important difference between magnetic poles and electric charges. Whereas electric charges can be isolated, magnetic poles cannot. Electrons and protons are entities by themselves. A cluster of electrons need not be accompanied by a cluster of protons and vice versa. But a north magnetic pole never exists without the presence of a south pole and vice versa. If you break a bar magnet in half, each half still behaves as a complete magnet. Break the pieces in half again, and you have four complete magnets. You can continue breaking the pieces in half and never isolate a single pole.† Even when your piece is one atom thick, there are two poles. This suggests that atoms themselves are magnets.

Question   Does every magnet necessarily have a north and south pole?

▶ **Answer**
Yes, just as every coin has two sides, a head and a tail. Some "trick" magnets may have more than one pair of poles, but nevertheless poles occur in pairs.

*The force of interaction between magnetic poles is given by $F \sim p_1 p_2 / d^2$, where p_1 and p_2 represent magnetic pole strengths and d represents the separation distance between the poles. Note the similarity of this relationship to Coulomb's law.
†Theoretical physicists have speculated for 60 years about the possible existence of discrete magnetic "charges," called *magnetic monopoles*. These tiny particles would carry either a single north or a single south magnetic pole and would be the counterparts to the positive and negative charges in electricity. Various attempts have been made to find monopoles, but none has proved successful. All known magnets always have at least one north and one south pole.

Magnetic Fields

Recall that the space that surrounds mass contains a gravitational field, and the space that surrounds electric charge contains an electric field. Interestingly enough, if the electric charge is moving, the region of space surrounding it is further altered. This alteration due to the motion of a charge is the **magnetic field**. We say a moving charge is surrounded by both an electric field and a magnetic field. Like the electric field, the magnetic field is a storehouse of energy. The greater the motion of the charge, the greater the magnitude of the magnetic field surrounding the charge. A magnetic field is produced by the motion of electric charge.*

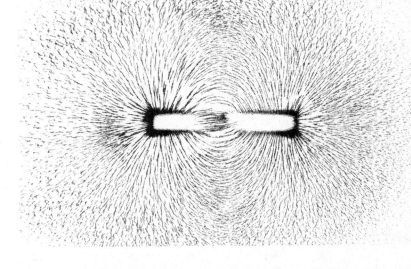

Figure 23-2
Top view of iron filings sprinkled on a sheet of paper on top of a magnet. The filings trace out a pattern of *magnetic field lines* in the surrounding space. Interestingly enough, the magnetic field lines continue inside the magnet (not revealed by the filings) and form closed loops. The source of the field is the motion of electrons in the iron atoms that compose the magnet.

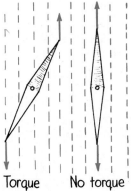

Torque   No torque

Figure 23-3
When the compass needle is not aligned with the magnetic field, the oppositely directed forces produce a pair of torques (called a *couple*) that twist the needle into alignment.

So if the motion of electric charges produces magnetism, where is this motion in a common bar magnet? The answer is: in the electrons in the atoms—although the magnet as a whole may be stationary, the magnet is composed of atoms whose electrons are in constant motion. Two kinds of electron motion contribute to magnetism: their spinning motion and their orbital motion. Electrons spin about their own axes like tops. A spinning electron is charge in motion. Electrons also revolve about the atomic nucleus—again, charge in motion. The field due to spinning is predominant in most materials.

So every atom is a tiny electromagnet. A pair of electrons spinning in the same direction makes up a stronger electromagnet. A pair of electrons spinning in opposite directions, however, has the opposite effect.

---

*Interestingly enough, since motion is relative, the magnetic field is relative. For example, when a charge moves by you, there is a definite magnetic field associated with the moving charge. But if you move along with the charge, so there is no motion relative to you, you will find no magnetic field associated with the charge. Magnetism is relativistic. In fact, it was Albert Einstein who first explained this when he published his first paper on special relativity, "On the Electrodynamics of Moving Charges." (More on relativity in Chapters 34 and 35.)

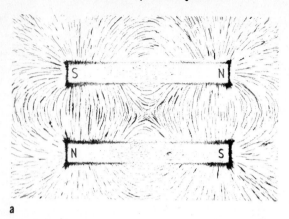

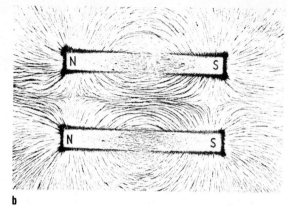

a

b

Figure 23-4
The magnetic field patterns for a pair of magnets.
(a) Opposite poles are parallel to each other, and (b) like poles are parallel to each other.

The magnetic fields of the two electrons cancel each other. This is why most substances are not magnets. In most atoms, the various fields cancel each other because the electrons spin in opposite directions. In materials such as iron, nickel, and cobalt, however, the fields do not cancel each other entirely. Each iron atom has four electrons whose spin magnetism is uncanceled. Each iron atom, then, is a tiny magnet. The same is true to a lesser degree for the atoms of nickel and cobalt.*

Magnetic Domains

Figure 23-5
The iron nails become induced magnets.

The magnetic field of individual iron atoms is so strong that interaction among adjacent atoms causes large clusters of them to line up with each other. These clusters of aligned atoms are called **magnetic domains**. Each domain is perfectly magnetized and is made up of billions of aligned atoms. The domains are microscopic (Figure 23-6), and there are many of them in a crystal of iron.

Not every piece of iron, however, is a magnet. This is because the domains in ordinary iron are not aligned. Consider a common iron nail: the domains in the nail are randomly oriented. They can be induced into alignment, however, when a magnet is brought nearby. (It is interesting to listen with an amplified stethoscope to the clickety-clack of domains undergoing alignment in a piece of iron when a strong magnet approaches.) The domains align themselves much as electrical charges in a piece of paper align themselves in the presence of a charged rod. When you remove the nail from the magnet, ordinary thermal motion causes most or all of the domains in the nail to return to a random arrangement. If the field of the permanent magnet is very strong, however, the nail may retain some permanent magnetism of its own after the two are separated.

Permanent magnets are made by simply placing pieces of iron or certain iron alloys in strong magnetic fields. Alloys of iron differ; soft iron

---

*Most common magnets are made from alloys containing iron, nickel, cobalt, and aluminum in various proportions. In these the electron spin contributes virtually all the magnetic properties. In the rare earth metals like gadolinium, the orbital motion is more significant.

Figure 23-6
A microscopic view of magnetic domains in a crystal of iron. Each domain consists of billions of aligned iron atoms.

is easier to magnetize than steel. It helps to tap the iron to nudge any stubborn domains into alignment. Another way of making a permanent magnet is to stroke a piece of iron with a magnet. The stroking motion aligns the domains in the iron. If a permanent magnet is dropped or heated, some of the domains are jostled out of alignment and the magnet becomes weaker.

Figure 23-7
Pieces of iron in successive stages of magnetism. The arrows represent domains; the head is a north pole and the tail a south pole. Poles of neighboring domains neutralize each other's effects, except at the ends.

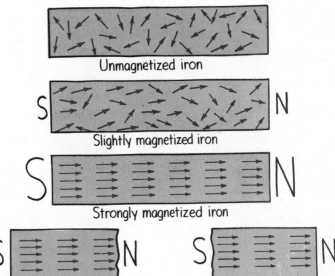

Unmagnetized iron

Slightly magnetized iron

Strongly magnetized iron

When a magnet is broken into two pieces, each piece is an equally strong magnet

Question   ▶   How can a magnet attract a piece of iron that is not magnetized?

▶ **Answer**
Domains in the unmagnetized piece of iron are induced into alignment by the magnetic field of the nearby magnet. See the similarity of this with Figure 21-12. Like the pieces of paper, pieces of iron will jump to a strong magnet when it is brought nearby. But unlike the paper, they are not then repelled. Can you think of the reason why?

Electric Currents and Magnetic Fields

A moving charge produces a magnetic field. A current of charges, then, also produces a magnetic field. The magnetic field that surrounds a current-carrying conductor can be demonstrated by arranging an assortment of compasses around a wire (Figure 23-8) and passing a current through it. The compasses line up with the magnetic field produced by current and show it to be a pattern of concentric circles about the wire. When the current reverses direction, the compass needles turn around, showing that the direction of the magnetic field changes also.

Figure 23-8
The compasses show the circular shape of the magnetic field surrounding the current-carrying wire.

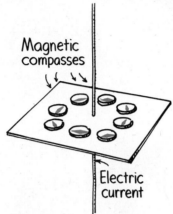

If the wire is bent into a loop, the magnetic field lines become bunched up inside the loop (Figure 23-9). If the wire is bent into another loop, overlapping the first, the concentration of magnetic field lines inside the double loop is twice as much as in the single loop. It follows that the magnetic field intensity in this region is increased as the number of loops is increased. The magnetic field intensity is appreciable for a current-carrying coil of wire with many loops.

Figure 23-9
Magnetic field lines about a current-carrying wire crowd up when the wire is bent into a loop.

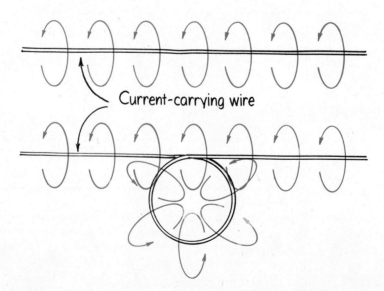

Current-carrying wire

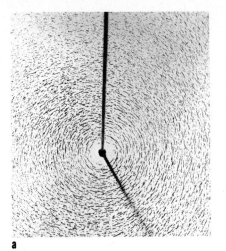
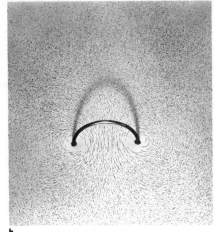
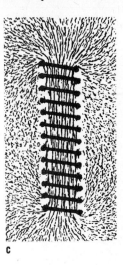

a   b   c

Figure 23-10
Iron filings sprinkled on paper reveal the magnetic field configurations about (a) a current-carrying wire, (b) a current-carrying loop, and (c) a coil of loops.

Electromagnets

If a piece of iron is placed in a current-carrying coil of wire, the magnetic domains in the iron are induced into alignment. This further increases the magnetic field intensity, and we have an **electromagnet**! Strong electromagnets are used to control charged particle beams in high-energy accelerators such as the supercollider. They also levitate and propel high-speed trains (Figure 23-11).

Electromagnets powerful enough to lift automobiles are a common sight in junkyards. The strength of these electromagnets is limited by over-heating of the current-carrying coils and saturation of magnetic domain alignment in the iron core. More powerful electromagnets omit the iron core altogether and use superconducting coils.

Figure 23-11
Conventional trains vibrate as they ride on rails at high speeds. This Japanese magnetically levitated train is capable of vibration-free high speeds, even in excess of 200 km/h.

Recall from Chapter 21 that in a superconductor there is no electrical resistance to limit the flow of electric charge and, therefore, no heating even for enormous currents. Electromagnets that utilize superconducting coils produce extremely strong magnetic fields—and do so very economically because there are no heat losses. At this writing the strength of superconducting electromagnets is limited by the breakdown of super-conductivity when the magnetic fields become too strong. Superconductors and superconducting magnets are presently generating a tremendous amount of interest and activity among physics types, for the stakes, both scientific and economic, are enormous.

Magnetic Force on Moving Charged Particles

A charged particle at rest will not interact with a static magnetic field. But if the charged particle moves in a magnetic field, the magnetic character of a charge in motion becomes evident. It experiences a deflecting force.* The force is greatest when the particle moves in a direction perpendicular to the magnetic field lines. At other angles, the force is less and becomes zero when the particle moves parallel to the field lines. In any case, the direction of the force is always perpendicular to the magnetic field lines and the velocity of the charged particle (Figure 23-12). So a moving charge is deflected when it crosses through a magnetic field, but when it travels parallel to the field no deflection occurs.

Figure 23-12
A beam of electrons is deflected by a magnetic field.

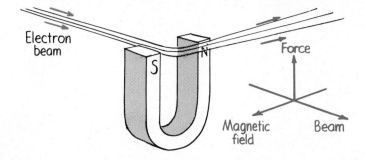

This sideways deflection is very different from the forces that occur in other interactions like the gravitation between masses, the electrostatic forces between charges, and the forces between magnetic poles. The force that acts on a moving charged particle does not act along the line that joins the sources of interaction, but instead acts perpendicularly to both the magnetic field and the charged particle beam.

We are fortunate that charged particles are deflected by magnetic fields. This fact is employed to spread electrons onto the inner surface of a TV tube and provide a picture. More interesting, charged particles from outer

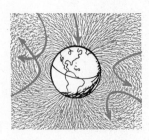

Figure 23-13
The magnetic field of the earth deflects many charged particles that make up cosmic radiation.

---

*When particles of electric charge q and velocity v move perpendicularly into a magnetic field of strength B, the force F on each particle is simply the product of the three variables: $F = qvB$. For nonperpendicular angles, v in this relationship must be the component of velocity perpendicular to B.

space are deflected by the earth's magnetic field. The intensity of harmful cosmic rays bombarding the earth's surface would be more intense otherwise.

Magnetic Force on Current-Carrying Wires

Simple logic tells you that if a charged particle moving through a magnetic field experiences a deflecting force, then a current of charged particles moving through a magnetic field experiences a deflecting force also. If the particles are trapped inside a wire when they respond to the deflecting force, the wire will also move (Figure 23-14).

Figure 23-14
A current-carrying wire experiences a force in a magnetic field. (Can you see this is a simple extension of Figure 23-12?)

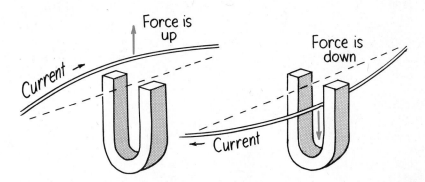

If we reverse the direction of current, the deflecting force acts in the opposite direction. The force is strongest when the current is perpendicular to the magnetic field lines. The direction of force is not along the magnetic field lines or along the direction of current. The force is perpendicular to both field lines and current. It is a sideways force.

We see that just as a current-carrying wire will deflect a magnetic compass, as discovered by Oersted in a high-school classroom in 1820, a magnet will deflect a current-carrying wire. Both cases show different effects of the same phenomenon. This discovery created much excitement, for almost immediately people began harnessing this force for useful purposes—with great sensitivity in electric meters and with great force in electric motors.

Electric Meters

The simplest meter to detect electric current is simply a magnet that is free to turn: compass. The next most simple meter is a compass in a coil of wires (Figure 23-15). When an electric current passes through the coil, each loop produces its own effect on the needle, so a very small current can be detected. A current-detecting instrument is called a *galvanometer*.

Figure 23-15
A very simple galvanometer.

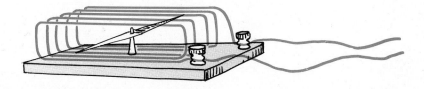

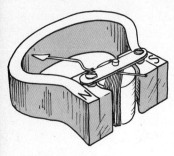

Figure 23-16
A common galvanometer design.

A more common design is shown in Figure 23-16. It employs more loops of wire and is therefore more sensitive. The coil is mounted for movement and the magnet is held stationary. The coil turns against a spring, so the greater the current in its windings, the greater its deflection. A galvanometer may be calibrated to measure current (amperes), in which case it is called an *ammeter*. Or it may be calibrated to measure electric potential (volts), in which case it is called a *voltmeter*.

Electric Motors

If we modify the design of the galvanometer slightly, we have an electric motor. The principal difference is that the current is made to change direction every time the coil makes a half rotation. After being forced to turn one half rotation, it overshoots just in time for the current to reverse, whereupon it is forced to continue another half rotation, and so on in cyclic fashion to produce continuous rotation.

Figure 23-17
Both the ammeter and the voltmeter are basically galvanometers. (The electrical resistance of the instrument is made to be very low for the ammeter and very high for the voltmeter.)

In Figure 23-18 we see the principle of the electromagnetic motor in bare outline. A permanent magnet produces a magnetic field in a region where a rectangular loop of wire is mounted to turn about the axis shown. When a current passes through the loop, it flows in opposite directions in the upper and lower sides of the loop (it has to do this because if charge flows into one end of the loop, it must flow out the other end).

Figure 23-18
A simplified motor.

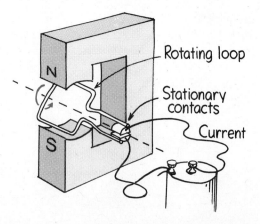

Rotating loop

Stationary contacts

Current

If the upper portion of the loop is forced to the left, then the lower portion is forced to the right, as if it were a galvanometer. But unlike a galvanometer, the current is reversed during each half revolution by means of stationary contacts on the shaft. The parts of the wire that brush against these contacts are called *brushes*. In this way, the current in the loop alternates so that the forces in the upper and lower regions do not change direction as the loop rotates. The rotation is continuous as long as current is supplied.

We have described here only a very simple dc motor. Larger motors, dc or ac, are usually made by replacing the permanent magnet by an electromagnet that is energized by the power source. Of course, more than a single loop is used. Many loops of wire are wound about an iron cylinder, called an *armature*, which then rotates when energized with electric current.

The advent of electric motors, needless to say, brought about the replacement of enormous human and animal toil, the world over. Electric motors have greatly changed the way people live.

Question ▶ What is the major similarity between a galvanometer and a simple electric motor? What is the major difference?

Earth's Magnetic Field

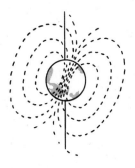

Figure 23-19
The earth is a magnet.

A suspended magnet or compass points northward because the earth itself is a huge magnet. The compass aligns with the magnetic field of the earth. The magnetic poles of the earth, however, do not coincide with the geographic poles, nor are they very close to the geographic poles. The magnetic pole in the northern hemisphere, for example, is located nearly 1800 kilometers from the geographic pole, somewhere in the Hudson Bay region of northern Canada. The other pole is located south of Australia (Figure 23-19). This means that compasses do not generally point to the true north. The discrepancy between the orientation of a compass and true north is known as the *magnetic declination*.

We do not know exactly why the earth itself is a magnet. The configuration of the earth's magnetic field is like that of a strong bar magnet placed near the center of the earth. But the earth is not a magnetized chunk of iron like a bar magnet. It is simply too hot for individual atoms to hold to a proper orientation. A better candidate is electric currents

▶ **Answer**

A galvanometer and a motor are similar in that they both employ coils positioned in a magnetic field. When a current passes through the coils, forces on the wires rotate the coils. The fundamental difference is that the maximum rotation of the coil in a galvanometer is one half turn, whereas in a motor the coil (armature) rotates through many complete turns. This is accomplished by alternating the current with each half turn of the armature.

within the earth's interior. About 2000 kilometers below the outer rocky mantle (which itself is almost 3000 kilometers thick) lies the molten part that surrounds the solid center. Most earth scientists think that moving charges looping around within the molten part of the earth create the magnetic field. Some earth scientists speculate that the electric currents are the result of convection currents—from heat rising from the central core (Figure 23-20), and that such convection currents combined with the rotational effects of the earth produce the earth's magnetic field. Because of its great size, the speed of moving charges need only be about a thousandth of a meter per second to account for the field. A firmer explanation awaits more study.

Whatever the cause, the magnetic field of the earth is not stable, but has wandered throughout geologic time. Evidence of this comes from analysis of the magnetic properties of rock strata. Iron atoms in a molten state are disoriented because of thermal motion, but a slight predominance of the iron atoms align with the magnetic field of the earth. When cooling and solidification occurs, this predominance records the direction of the earth's magnetic field in the resulting igneous rock. It's similar for sedimentary rocks, where magnetic domains in grains of iron that settle in sediments tend to align themselves with the earth's magnetic field and become locked into the rock that forms. The slight magnetism that results can be measured with sensitive instruments. As samples of rock are tested from different strata formed throughout geologic time, the magnetic field of the earth for different periods can be charted. This evidence shows that there have been times when the magnetic field of the earth has diminished to zero, and then the poles reversed themselves. More than twenty reversals have taken place in the past 5 million years. The most recent occurred 730 000 years ago. Prior reversals happened 870 000 and 950 000 years ago. Studies of deep sea sediments indicate the field was virtually switched off for 10 000 to 20 000 years just over 1 million years ago. We cannot predict when the next reversal will occur because the reversal sequence is not regular. But there is a clue in recent measurements that show a decrease of over 5 percent of the earth's magnetic field strength in the last 100 years. If this change is maintained, we may well have another reversal within 2000 years.

The reversal of magnetic poles is not unique to the earth. The sun's magnetic field reverses regularly, with a period of 22 years. This 22-year magnetic cycle has been linked, through evidence in tree rings, to periods of drought on earth. The long-known 11-year sunspot cycle is just half the time during which the sun gradually reverses its magnetic polarity.

Varying ion winds in the earth's atmosphere cause more rapid but much smaller fluctuations in the earth's magnetic field. Ions in this region are produced by the energetic interactions of solar ultraviolet rays and X rays with atmospheric atoms. The motion of these ions produces a small but important part of the earth's magnetic field. Like the lower layers of air, the ionosphere is churned by winds. The variations in these winds are responsible for nearly all fast fluctuations in the earth's magnetic field.

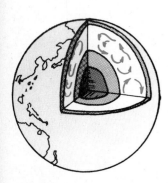

Figure 23-20

Convection currents in the molten parts of the earth's interior may drive electric currents to produce the magnetic field of the earth.

The universe is a shooting gallery of charged particles. They are called *cosmic rays* and are the nuclei of atoms stripped bare of their electrons. Their origin is uncertain; perhaps they are boiled off from stars or are nuclei that did not condense to form stars. In any event, they travel through space at fantastic speeds and make up the cosmic radiation that is hazardous to astronauts. Fortunately for those of us on the earth's surface, most of these charged particles are deflected away by the magnetic field of the earth. Some of them are trapped in the outer reaches of the earth's magnetic field and make up the Van Allen radiation belts (Figure 23-21).

Figure 23-21
The Van Allen radiation belts, shown here undistorted by the solar wind.

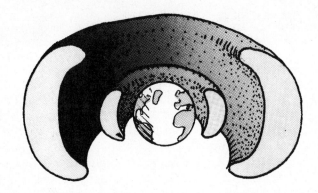

The Van Allen radiation belts consist of two doughnut-shaped rings, named after James A. Belts, who suggested their existence from data gathered by the U.S. satellite *Explorer I* in 1958.* The inner ring is about 3000 kilometers from the earth, and the outer ring, which is a larger and wider doughnut, is about 15 000 kilometers from the earth. Astronauts orbit at safe distances well below these belts of radiation. Most of the charged particles—protons and electrons—trapped in the outer belt probably come from the sun. Storms on the sun hurl charged particles out in great fountains, many of which pass near the earth and are trapped by its magnetic field. The trapped particles follow corkscrew paths around the magnetic field lines of the earth and bounce between the earth's magnetic poles high above the atmosphere. Disturbances in the earth's field often allow the ions to dip into the atmosphere, causing it to glow like a fluorescent lamp. This is the beautiful aurora borealis (northern lights).

The particles trapped in the inner belt probably originated from the earth's atmosphere. This belt is now masked by newer electrons that were produced by high-altitude hydrogen bomb explosions in 1962.

In spite of the earth's protective magnetic field, many cosmic rays reach the earth's surface. Some biological scientists speculate that the magnetic changes of the earth played a significant role in the evolution of life forms. One hypothesis is that in the early phases of primitive life, the earth's magnetic field was strong enough to hold off cosmic and solar

---

*Humor aside, the name is actually James A. Van Allen.

Figure 23-22
The aurora borealis (fluorescent lamp type) lighting of the sky caused by charged particles in the Van Allen belts striking atmospheric molecules.

radiations violent enough to destroy life. But during periods of zero strength, cosmic radiation and the spilling of the Van Allen belts increased mutation of the primitive life forms—not unlike the changing life forms produced by X rays in the famous heredity studies of fruit flies. Coincidences between the dates of increased life changes and the dates of the magnetic pole reversals lend support to this hypothesis.

Cosmic ray bombardment is greatest at the poles, because charged particles that hit the earth there do not travel *across* the magnetic field lines, but rather *along* the field lines and are not deflected. Cosmic ray bombardment decreases away from the poles and is smallest in equatorial regions. At sea level, about one to three particles strike each square centimeter each minute; this number increases rapidly with altitude. So cosmic rays are penetrating your body as you are reading this. (And even when you aren't reading this!)

Biomagnetism

Figure 23-23
The pigeon can sense direction because it has a built-in magnetic "compass" within its skull.

Certain bacteria biologically produce single-domain magnetite grains that they string together to form internal compasses. They then use these compasses to detect the dip of the earth's magnetic field. Equipped with a sense of direction, the organisms are able to locate food supplies. Amazingly, these bacteria south of the equator build the same single-domain magnets as their counterparts north of the equator, but then align them in opposite directions to coincide with the oppositely directed magnetic field in the southern hemisphere! Bacteria are not the only living organisms with built-in magnetic compasses: pigeons have recently been found to have multiple-domain magnetite magnets within their skulls that are connected with a large number of nerves to the pigeon brain. Pigeons have a magnetic sense, and not only can they discern longitudinal directions along the earth's magnetic field, but they can also detect latitude by the dip of the earth's field. Magnetic material has also been found in the abdomens of bees, whose behavior is affected by small magnetic fields. Searches of human beings to date show them to possess no such magnetic sense.

Summary of Terms

Magnetic force (1) Between magnets, it is the attraction of unlike magnetic poles for each other and the repulsion between like magnetic poles. (2) Between a magnetic field and a moving charge, it is a deflecting force due to the motion of the charge: the deflecting force is perpendicular to the motion of the charge and perpendicular to the magnetic field lines. This force is greatest when the charge moves perpendicular to the field lines and is smallest (zero) when moving parallel to the field lines.

Magnetic field The region of magnetic influence around a magnetic pole or a moving charged particle.

Magnetic domains Clustered regions of aligned magnetic atoms. When these regions themselves are aligned with one another, the substance containing them is a magnet.

Electromagnet A magnet whose field is produced by an electric current. Usually in the form of a wire coil with a piece of iron inside the coil.

Review Questions

1. The relationship between electricity and magnetism was discovered by whom and in what kind of room?

Magnetic Forces

2. The force between electric charges depends on the magnitude of charge, the distance of separation, and what else?

3. What is the origin of magnetic forces?

Magnetic Poles

4. Where are the magnetic poles located on a common bar magnet?

5. Where are the magnetic poles located on a common horseshoe magnet?

6. In what way is the rule for the interaction between magnetic poles similar to the rule for the interaction between electric charges?

7. In what way are *magnetic poles* very different from *electric charges*?

Magnetic Fields

8. An electric field surrounds an electric charge. What additional field surrounds a moving electric charge?

9. Why is *motion* a key word for magnetism?

10. What two kinds of motion are exhibited by electrons in an atom?

Magnetic Domains

11. What is a magnetic domain?

12. Why is iron magnetic and wood not?

13. Why will dropping an iron magnet on a hard floor make it a weaker magnet?

Electric Currents and Magnetic Fields

14. How do the directions of magnetic field lines about a current-carrying wire differ from the directions of electric field lines about a charge?

15. What happens to the direction of the magnetic field about an electric current when the direction of the current is reversed?

16. Why is the magnetic field strength greater inside a current-carrying loop of wire than about a straight section of wire?

Electromagnets

17. Why does a piece of iron in a current-carrying loop yield an even greater magnetic field strength?

18. Why are electromagnets stronger when constructed with superconducting coils?

Magnetic Force on Moving Charged Particles

19. In what direction relative to a magnetic field does a charged particle move in order to experience maximum deflecting force? Minimum deflecting force?

20. Both gravitational and electrical forces act in a direction along and parallel to the force fields. How is the direction of the magnetic force on a moving charge different?

21. What effect does the magnetic field about the earth have on cosmic ray bombardment?

Magnetic Force on Current-Carrying Wires

22. Since a magnetic force acts on a moving charged particle, does it make sense that a magnetic force also acts on a current-carrying wire? Defend your answer.

23. What relative direction between a magnetic field and a current-carrying wire results in greatest deflection? Smallest deflection?

24. What happens to the direction of deflection when the current in a wire is reversed?

Electric Meters

25. What is the function of a galvanometer?

26. What is a galvanometer called when calibrated to read current? Voltage?

Electric Motors

27. In what way is a galvanometer similar to an electric motor?

28. Why is it important that the direction of current change in the coil of an electric motor?

Earth's Magnetic Field

29. Why does a compass point northward? Will the needle point in the same direction when in the southern hemisphere?

30. What is meant by magnetic declination?

31. Why are magnetic domains in the core of the earth probably not permanently aligned?

32. How is the earth's magnetic field thought to be created?

33. What are *magnetic pole reversals*?

34. What are the *Van Allen radiation belts*?

35. What is the cause of the aurora borealis (northern lights)?

Biomagnetism

36. Name at least three creatures that are known to sense magnetic fields biologically.

Home Projects

1. Find the direction and dip of the earth's magnetic field lines in your locality. Magnetize a large steel needle or straight piece of steel wire by stroking it a couple of dozen times with a strong magnet. Run the needle through a cork and float it in a plastic or wooden container of water. The needle will point to the magnetic pole. Then remove both the needle and cork and press an unmagnetized common pin into each side of the cork. Rest the pins on the rims of a pair of drinking glasses so that the needle points to the magnetic pole. It should dip in line with the earth's magnetic field.

2. An iron bar can be easily magnetized by aligning it with the magnetic field lines of the earth and striking it lightly a few times with a hammer. The hammering jostles the domains so they can better fall into alignment with the earth's field. The bar can be demagnetized by striking it when it is in an east-west direction.

3. Bring a magnetic compass near the tops of iron or steel objects in your home (radiators, refrigerators, stoves, lamps, etc.). You will find that the north pole of the compass needle points to the tops of these objects, and the south pole of the compass needle points to the bottoms. This shows that the objects are magnets, having a south pole on top and a north pole on the bottom. What is the explanation for this? You will find that even cans of food that have been in a vertical position in the pantry are magnetized. Turn one over and test to see how many days it takes to lose and change its polarity.

Exercises

1. In what sense are all magnets electromagnets?

2. Since every iron atom is a tiny magnet, why aren't all iron materials themselves magnets?

3. "An electron always experiences a force in an electric field, but not necessarily in a magnetic field." Defend this statement.

4. The core of the earth is probably composed of iron and nickel, excellent metals for making permanent magnets. Why is it unlikely that the earth's core is a permanent magnet?

5. Why will a magnet attract an ordinary nail or paper clip, but not a wooden pencil?

6. Will either pole of a magnet attract a paper clip? Explain what is happening inside the attracted paper clip. (*Hint:* Consider Figure 21-12.)

7. One way to make a compass is to stick a magnetized needle into a piece of cork and float it in a glass bowl full of water. The needle will align itself with the magnetic field of the earth. Since the north pole of this compass is attracted northward, will the needle float toward the northward side of the bowl? Defend your answer.

8. What is the net magnetic force on a compass needle? By what mechanism does a compass needle line up with a magnetic field?

9. Since the iron filings that line up with the magnetic field of the bar magnet shown in Figure 23-2 are not themselves little magnets, by what mechanism do they align themselves with the field of the magnet?

10. The north pole of a compass is attracted to the north pole of the earth, yet like poles repel. Can you resolve this apparent dilemma?

11. Your friend says that when a compass is taken across the equator, it turns around and points in the opposite direction. Your other friend says this is not true, that southern-hemisphere types use the south pole of the compass to find direction. You're on; what do you say?

12. Why will a magnet placed in front of a television picture tube distort the picture? (*Note:* Do NOT try this with a color set. If you succeed in magnetizing the metal mask in back of the glass screen, you will have picture distortion even when the magnet is removed!)

13. Magnet A has twice the magnetic field strength of magnet B and at a certain distance pulls on magnet B with a force of 50 N. With how much force, then, does magnet B pull on magnet A?

14. A strong magnet attracts a paper clip to itself with a certain force. Does the paper clip exert a force on the strong magnet? If not, why not? If so, does it exert as much force on the magnet as the magnet exerts on it? Defend your answers.

15. To make a compass, point an ordinary iron nail along the direction of the earth's magnetic field and repeatedly beat on it for a few seconds with a hammer or a rock. Then suspend it at its center of gravity by a string. Why does the beating magnetize the nail?

16. When iron naval ships are built, the location of the shipyard and the orientation of the ship while in the shipyard are recorded on a brass plaque permanently fixed to the ship. Why?

17. Can an electron at rest with respect to a magnetic field be set into motion by the magnetic field? An electric field?

18. Magnetic fields can be used to trap plasmas in "magnetic bottles," but the plasma must be moving. Why?

19. A cyclotron is a device for accelerating charged particles in ever-increasing circular orbits to high speeds. The charged particles are subjected to both an electric field and a magnet field. One of these fields increases the speed of the charged particles, and the other field holds them in a circular path. Which field performs which function?

20. A magnetic field can deflect a beam of electrons, but it cannot do work on the electrons to speed them up. Why?

21. Two charged particles are projected into a magnetic field that is perpendicular to their velocities. If the charges are deflected in opposite directions, what does this tell you about them?

22. A beam of high-energy protons emerges from a cyclotron. Do you suppose there is a magnetic field associated with these particles? Why or why not?

23. Inside a laboratory room there is said to be either an electric field or a magnetic field, but not both. What experiments might be performed to establish what kind of field is in the room?

24. Why do astronauts keep to altitudes beneath the Van Allen radiation belts when doing space walks?

25. Residents of northern Canada are bombarded by more intense cosmic radiation than are residents of Mexico. Why is this so?

26. In a mass spectrometer, ions are directed into a magnetic field, where they are deflected and strike a detecting screen. If a variety of singly ionized atomic nuclei travel at the same speed through the magnetic field, would you expect all nuclei to be deflected by the same amount? Or would different nuclei be bent different amounts? What would you expect?

27. One way to shield a habitat in outer space from cosmic rays is with an absorbing blanket of some kind, like the atmosphere that protects the earth. Speculate on a second way for shielding that is also similar to earth shielding.

28. If you had two bars of iron—one magnetized and the other not—and no other equipment, how could you tell which bar was the magnet?

It's easy to spin the armature of my generator when the lamp is turned off and no current exists in the open circuit. But when the lamp is turned on and current is induced in the closed circuit, the armature is harder to spin. This is in accord not only with energy conservation, but with the laws of electromagnetic induction. Work must be done to move a current-carrying wire in a magnetic field— even though the current is induced by that same magnetic field. So the more current this generator induces, the more "electromagnetic repulsion" I encounter and the harder it is for me to spin its armature. How about that ⸮

24 Electromagnetic Induction

In the early 1800s, the only current-producing devices were voltaic cells, which produced small currents by dissolving metals in acids. These were the forerunners of our present-day batteries. In 1820 Oersted found that magnetism was produced by current-carrying wires. The question arose as to whether electricity could be produced from magnetism. The answer was provided in 1831 by two physicists, Michael Faraday in England and Joseph Henry in the United States—each working independently and without knowledge of the other. Their discovery changed the world by making electricity commonplace—powering industries by day and lighting up cities at night.

Electromagnetic Induction

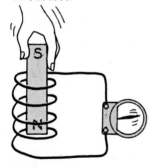

Figure 24-1
When the magnet is plunged into the coil, charges in the coil are set in motion; voltage is induced in the coil.

Figure 24-2
Voltage is induced in the wire loop whether the magnetic field moves past the wire or the wire moves through the magnetic field.

Faraday and Henry both discovered that electric current could be produced in a wire by simply moving a magnet in or out of a coil of wire (Figure 24-1). No battery or other voltage source was needed—only the motion of a magnet in a wire loop. They discovered that voltage was caused—or *induced*—by the relative motion between a wire and a magnetic field. Voltage is induced whether the magnetic field of a magnet moves near a stationary conductor or the conductor moves in a stationary magnetic field (Figure 24-2). The results are the same whether either or both move.

The greater the number of loops of wire that move in a magnetic field, the greater the induced voltage (Figure 24-3). Pushing a magnet into twice as many loops will induce twice as much voltage; pushing into ten times as many loops will induce ten times as much voltage; and so on. It may seem that we get something (energy) for nothing by simply increasing the number of loops in a coil of wire. But we don't: we find

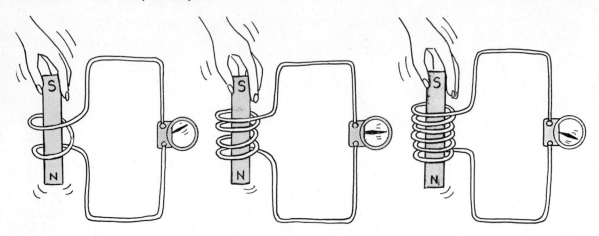

Figure 24-3

When a magnet is plunged into a coil of twice as many loops as another, twice as much voltage is induced. If the magnet is plunged into a coil with three times as many loops, then three times as much voltage is induced.

Figure 24-4

It is more difficult to push the magnet into a coil with many loops because the magnetic field of each current loop resists the motion of the magnet.

it is more difficult to push the magnet into a coil with more loops. This is because the induced voltage makes a current, which makes an electromagnet, which repels the magnet in our hand. So we do more work to induce more voltage (Figure 24-4). The amount of voltage induced depends on how fast the magnetic field lines are entering or leaving the coil. Very slow motion produces hardly any voltage at all. Quick motion induces a greater voltage. This phenomenon of inducing voltage by changing the magnetic field in a coil of wire is called **electromagnetic induction**.

Question   ▶   If you push a magnet into a coil, as shown in Figure 24-4, you'll feel a resistance to your push. Why is this resistance greater in a coil with more loops?

▶ **Answer**

Simply put, more work is required to induce the greater voltage in more loops. You can also look at it this way: when two magnets (electro- or permanent) are close to each other, the two magnets are either forced together or forced apart. In cases where one of the magnets is induced by motion of the other, the polarity of the fields is always such as to force the magnets apart. This is the resistive force you feel. Inducing more current in more coils simply increases the induced magnetic field strength and hence the resistive force.

Faraday's Law

Electromagnetic induction is summarized by **Faraday's law**, which states:

The induced voltage in a coil is proportional to the product of the number of loops and the rate at which the magnetic field changes within those loops.

The amount of *current* produced by electromagnetic induction depends not only on the induced voltage, but also on the resistance of the coil and the circuit that it connects.* For example, we can plunge a magnet in and out of a closed rubber loop and in and out of a closed loop of copper. The voltage induced in each is the same, providing each intercepts the same number of magnetic field lines. But the current in each is quite different. The electrons in the rubber sense the same voltage as those in the copper, but their bonding to the fixed atoms prevents the movement of charge that so freely occurs in the copper.

Voltage can be induced in a loop of wire in three apparently different ways: moving the loop near a magnet, moving a magnet near the loop, or changing a current in a nearby loop. In each case we have the important ingredient—a change in the amount of magnetic field in the loop.

Generators and Alternating Current

When a magnet is plunged into and back out of a coil of wire, the direction of the induced voltage alternates. As the magnetic field strength inside the coil is increased (magnet entering), the induced voltage in the coil is directed one way. When the magnetic field strength diminishes (magnet leaving), the voltage is induced in the opposite direction. The greater the frequency of field change, the greater the induced voltage. The frequency of the alternating voltage induced is equal to the frequency of the changing magnetic field within the loop.

It is more practical to induce voltage by moving the coil rather than moving the magnet. This can be done by rotating the coil in a stationary magnetic field (Figure 24-5). This arrangement is called a **generator**.

Figure 24-5
A simple generator. Voltage is induced in the loop when it is rotated in the magnetic field.

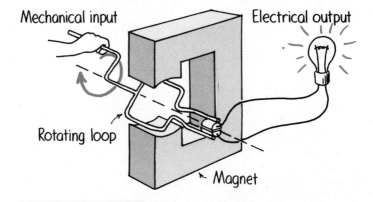

Mechanical input

Electrical output

Rotating loop

Magnet

---

*Current also depends on the "reactance" of the coil. Reactance is similar to resistance and is important in ac circuits; it depends on the number of loops in the coil and on the frequency of the ac source, among other things. We will not treat this complication.

Figure 24-6

(*a*) Motor effect: When a current moves to the right, there is a perpendicular upward force on the electrons. Since there is no conducting path upward, the wire is tugged upward along with the electrons. (*b*) Generator effect: When a wire with no initial current is moved downward, the electrons in the wire experience a deflecting force perpendicular to their motion. There is a conducting path in this direction that the electrons follow, thereby constituting a current.

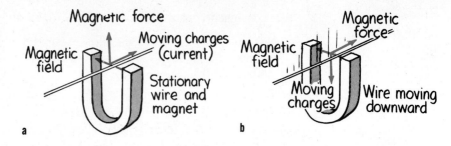

The construction of a generator is in principle identical to that of a motor. Only the roles of input and output are reversed. In a motor, electric energy is the input and mechanical energy the output; in a generator, mechanical energy is the input and electric energy the output. Both devices simply transform energy from one form to another.

It is interesting to compare the physics of a motor and a generator and to see that both operate under the same underlying principle: that moving charges experience a force that is perpendicular to both their motion and to the magnetic field they traverse (Figure 24-6). We will call the deflec-

Figure 24-7

As the loop rotates, there is a change in the number of magnetic field lines it encloses. It varies from a maximum at (*a*) to a minimum at (*c*) and back to a maximum again at (*e*). Induced voltage occurs at (*c*), where the greatest rate of change occurs.

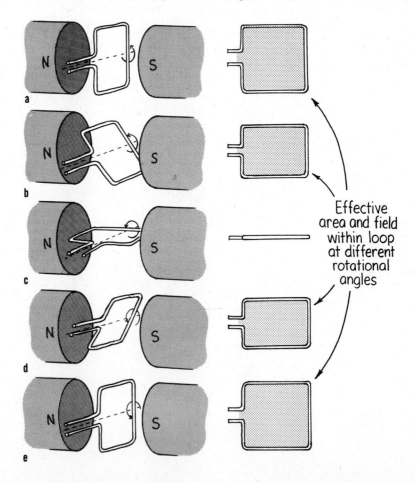

tion of the wire the *motor effect* and what happens as a result of the law of induction the *generator effect*. These effects are summarized in (*a*) and (*b*) of the figure. Study them. Can you see that the two effects are related?

We can see the details of electromagnetic induction in Figure 24-7. Note that when the loop of wire is rotated in the magnetic field, there is a change in the number of magnetic field lines within the loop. In (*a*) the loop has the largest number of lines inside it. As the loop rotates (*b*), it encircles fewer of the field lines until at (*c*) the loop lies along the field lines and encloses none at all. As rotation continues, it encloses more field lines (*d*) and reaches a maximum of lines when it has made a half turn (*e*). As rotation continues, the magnetic field inside the loop changes in cyclic fashion, with the greatest rate of change of field lines as it goes through the zero point. Hence the induced voltage is greatest at these points (Figure 24-8). Because the voltage induced by the generator alternates, the current produced is ac, an alternating current.* It changes magnitude and direction periodically. The alternating current in our homes is produced by generators standardized so that the current changes its magnitude and direction 60 cycles per second—60 hertz.

Figure 24-8
As the loop rotates, the magnitude and direction of the induced voltage (and current) changes. One complete rotation of the loop produces one complete cycle in voltage (and current).

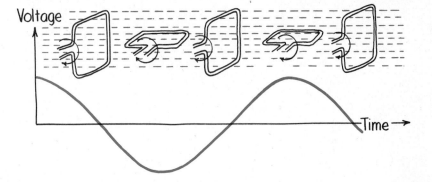

Power Production

Fifty years after Faraday and Henry discovered electromagnetic induction, Thomas Edison, Nikola Tesla, and George Westinghouse put their findings to practical use and showed the world that electricity could be generated reliably and in sufficient quantities to light entire cities.

Turbogenerator Power

Tesla built generators like those of today—but quite a bit more complicated than the simple model we have discussed. Tesla's generators had armatures consisting of bundles of copper wires that were made to spin within strong magnetic fields by means of a turbine, which in turn was spun by the energy of falling water or steam. The rotating loops of wire

*With appropriate brushes and by other means, the ac in the loop(s) can be taken off as dc to make a dc generator.

Figure 24-9
Steam drives the turbine, which is connected to the armature of the generator.

Steam

in the armature cut through the magnetic field of the surrounding electromagnets, thereby inducing alternating voltage and current.

We can look at this process from an atomic point of view. When the wires in the spinning armature cut through the magnetic field, oppositely directed electromagnetic forces act on the negative and positive charges. Electrons respond to this force by momentarily swarming relatively freely in one direction throughout the crystalline copper lattice; the copper atoms, which are actually positive ions, are forced in the opposite direction. But the ions are anchored in the lattice, so they hardly move at all. Only the electrons move, sloshing back and forth in alternating fashion with each rotation of the armature. The energy of this electronic sloshing is tapped at the electrode terminals of the generator.

MHD Power

An interesting device similar to the turbogenerator is the MHD (**mag**netohydrody**namic**) generator, which does away with a turbine and spinning armature altogether. Instead of making charges move in a magnetic field via a rotating armature, a plasma of electrons and positive ions expands through a nozzle and moves at supersonic speed through a magnetic field. Like the armature in a turbogenerator, the motion of charges through a magnetic field gives rise to a voltage and flow of current in accord with Faraday's law of induction. Whereas in a conventional generator "brushes" carry the current to the external load circuit, in the MHD generator the same function is performed by "electrodes" (Figure 24-10). Unlike the turbogenerator, the MHD generator can operate at any temperature to which the plasma can be heated, either by combustion or by nuclear processes. The high temperature results in a high thermodynamic efficiency, which means more power for the same amount of fuel and less waste heat. Efficiency is further boosted when the "waste" heat is used to turn water into steam and run a conventional steam-turbine generator.

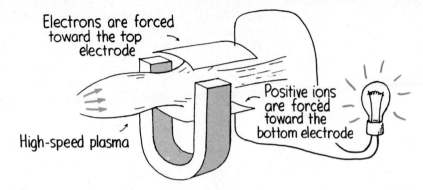

Electrons are forced toward the top electrode

Positive ions are forced toward the bottom electrode

High-speed plasma

Figure 24-10
A simplified MHD generator. The oppositely directed forces on positive and negative particles in the plasma beam effectively result in an induced direct electron current that flows from the bottom electrode, through the plasma, to the top electrode, then through the electric circuit (not shown), and back to the bottom electrode. There are no moving parts; only the plasma moves. In practice, superconducting electromagnets are used.

This substitution of a flowing plasma for rotating copper coils in a generator has become operational only recently because the technology to produce plasma of high-enough temperatures is new. Current plants use a high-temperature plasma formed by combustion of fossil fuels in air or oxygen.*

Generators of whatever kind, of course, don't produce energy—they simply convert energy from some other form to electric energy. Energy from the source, usually some type of fuel, is converted to mechanical energy either to drive the turbine or to produce the plasma, and the generator converts most of this to electrical energy. The electricity that is produced simply carries this energy to distant places. Some people think that electricity is a primary source of energy. It is not. It is a form of energy that must have a source.

Transformers

Primary

Secondary

Figure 24-11
Whenever the primary switch is opened or closed, voltage is induced in the secondary circuit.

It is interesting to see that electric energy can be carried across empty space from one device to another with the simple arrangement shown in Figure 24-11. Note that one coil is connected to a battery and the other is connected to a galvanometer. It is customary to refer to the coil connected to the power source as the *primary* (input) and to the other as the *secondary* (output). As soon as the switch is closed in the primary and current passes through its coil, a current occurs in the secondary also— even though there is no material connection between the two coils. Only a brief surge of current occurs in the secondary, however. Then, when the primary switch is opened, a surge of current again registers in the secondary but in the opposite direction.

This is the explanation. A magnetic field builds up around the primary when the current begins to flow through the coil. This means that the magnetic field is growing (that is, *changing*) about the primary. But since the coils are near each other, this changing field extends to the secondary coil, thereby inducing a voltage in the secondary. This induced voltage is only temporary, for when the current and the magnetic field of the

*The only large-scale MHD power plant that presently produces commercial electric power is located in the U.S.S.R. It is a joint collaboration of U.S. and Soviet scientists and engineers.

primary reach a steady state—that is, when the magnetic field is no longer changing—no further voltage is induced in the secondary. But when the switch is turned off, the current in the primary drops to zero. The magnetic field about the coil collapses, thereby inducing a voltage in the secondary coil, which senses the change. We see that voltage is induced whenever a magnetic field is *changing* through the coil, regardless of the reason.

Question ▶ When the switch of the primary in Figure 24-11 is opened or closed, the galvanometer in the secondary registers a current. But when the switch remains closed, no current is registered on the galvanometer of the secondary. Why?

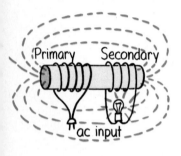

Figure 24-12
A simple transformer.

If you place an iron core inside the primary and secondary coils of the arrangement of Figure 24-11, the magnetic field about the primary is intensified by the alignment of magnetic domains. The field is also concentrated in the core and extends into the secondary, which intercepts more of the field change. The galvanometer will show greater surges of current when the switch of the primary is opened or closed. Instead of opening and closing a switch to produce the change of magnetic field, suppose that alternating current is used to power the primary. Then the rate at which the magnetic field changes in the primary (and hence in the secondary) is equal to the frequency of the alternating current. Now we have a **transformer** (Figure 24-12). A more efficient arrangement is shown in Figure 24-13.

If the primary and secondary have equal numbers of wire loops (usually called *turns*), then the input and output alternating voltages will be equal. There is no change. But if the secondary coil has more turns than the primary, the alternating voltage produced in the secondary coil will be greater than that produced in the primary. In this case, the voltage is said to be *stepped up*. If the secondary has twice as many turns as the primary, the voltage in the secondary will be double that of the primary.

We can see this with the arrangements in Figure 24-14. First consider the simple case of a single primary loop connected to a 1-volt alternating source and a single secondary loop connected to the voltmeter (*a*). The secondary intercepts the changing magnetic field of the primary, and a voltage of 1 volt is induced in the secondary. If another loop is wrapped around the core so the transformer has two secondaries (*b*), it intercepts

▶ **Answer**

When the switch remains in the closed position, there is a steady current in the primary and a steady magnetic field about the coil. This field extends to the secondary, but unless there is a *change* in the field, electromagnetic induction does not occur.

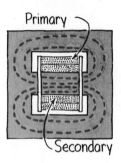

Primary

Secondary

Figure 24-13
A practical and more efficient transformer. The iron core guides the changing magnetic field lines.

Figure 24-14
(a) The voltage of 1 V induced in the secondary equals the voltage of the primary. (b) A voltage of 1 V is induced in the added secondary also because it intercepts the same magnetic field change from the primary. (c) The voltages of 1 V each induced in the two one-turn secondaries are equivalent to a voltage of 2 V induced in a single two-turn secondary.

the same magnetic field change. We see that 1 volt is induced in it also. There is no need to keep both secondaries separate, for we could join them (c) and still have a total induced voltage of 1 volt + 1 volt, or 2 volts. This is equivalent to saying that a voltage of 2 volts will be induced in a single secondary that has twice the number of loops as the primary. If the secondary is wound with three times as many loops, then three times as much voltage will be induced. Stepped-up voltage may light a neon sign or operate the picture tube in a television receiver.

If the secondary has fewer turns than the primary, the alternating voltage produced in the secondary will be *lower* than that produced in the primary. The voltage is said to be *stepped down*. This stepped-down voltage may safely operate a toy electric train. If the secondary has half as many turns as the primary, then only half as much voltage is induced in the secondary. So electric energy can be fed into the primary at a given alternating voltage and taken from the secondary at a greater or lower alternating voltage, depending on the relative number of turns in the primary and secondary coil windings.

The relationship between primary and secondary voltages with respect to the relative number of turns is given by

$$\frac{\text{Primary voltage}}{\text{Number of primary turns}} = \frac{\text{secondary voltage}}{\text{number of secondary turns}}$$

It might seem that we get something for nothing with a transformer that steps up the voltage. But we don't. When voltage is stepped up, more current is drawn by the primary. The transformer actually transfers energy from one coil to the other. The rate at which energy is transferred is called *power*. The power used in the secondary is supplied by the primary. The primary gives no more than the secondary uses, in accord

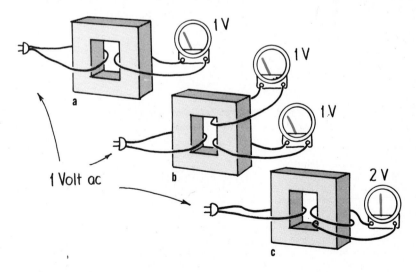

1 V

1 V

1 V

1 Volt ac

2 V

a

b

c

with the law of conservation of energy. If the slight power losses due to heating of the core are neglected, then

$$\text{Power into primary} = \text{power out of secondary}$$

Electric power is equal to the product of voltage and current, so we can say

$$(\text{Voltage} \times \text{current})_{\text{primary}} = (\text{voltage} \times \text{current})_{\text{secondary}}$$

The ease with which voltages can be stepped up or down with a transformer is the principal reason that most electric power is ac rather than dc.

Questions

1. If 100 V of ac are put across a 100-turn transformer primary, what will be the voltage output if the secondary has 200 turns?
2. Assuming the answer to the last question is 200 V and the secondary is connected to a floodlamp with a resistance of 50 Ω, what will be the ac current in the secondary circuit?
3. What is the power in the secondary coil?
4. What is the power in the primary coil?
5. What is the ac current drawn by the primary coil?
6. The voltage has been stepped up, and the current has been stepped down. Ohm's law says that increased voltage will produce increased current. Is there a contradiction here, or does Ohm's law not apply to circuits that have transformers?

Self-Induction

Current-carrying loops in a coil interact not only with loops of other coils but also with loops of the same coil. This is *self-induction*. A self-induced voltage is produced. This voltage is always in a direction opposing the changing voltage that produces it and is commonly called the "back electromotive force," or simply "back emf."* We won't treat self-induction

▶ **Answers**

1. From 100 V/100 primary turns = (?) V/200 secondary turns, you can see that the secondary puts out 200 V.
2. From Ohm's law, 200 V/50 Ω = 4 A.
3. Power = 200 V × 4 A = 800 W.
4. By the law of conservation of energy, the power in the primary is the same, 800 W.
5. 800 W = 100 V × (?) A, so you see the primary draws 8 A.
6. Ohm's law still holds, and there is no contradiction. The voltage induced across the secondary circuit, divided by the load (resistance) of the secondary circuit, equals the current in the secondary circuit. The current is stepped down in comparison to the larger current that is drawn in the primary circuit.

---

*This opposing direction of induction is called *Lenz's law* and is a consequence of the conservation of energy.

Figure 24-15
When the switch is opened, the magnetic field of the coil collapses. This sudden change in the field can induce a huge voltage.

and back emfs here, except to acknowledge a common and dangerous effect. A coil with a large number of turns has a large self-inductance. Suppose such a coil is used as an electromagnet and is powered with a dc source, perhaps a small battery. Current in the coil is then accompanied by a strong magnetic field. When we disconnect the battery by opening a switch, we had better be prepared for a surprise. When the switch is opened, the current in the circuit falls rapidly to zero and the magnetic field in the coil undergoes a sudden decrease (Figure 24-15). What happens when a magnetic field suddenly changes in a coil—even if it is the same coil that produced it? The answer is that a voltage is induced. The rapidly collapsing magnetic field with its store of energy may induce an enormous voltage, large enough to develop an arc across the switch—or you, if you are opening the switch! For this reason, electromagnets are connected to a circuit that absorbs excess charge and prevents the current from dropping too suddenly. This reduces the self-induced voltage. This is also, by the way, why you should not disconnect appliances by pulling out the plug instead of using the switch. The circuitry in the switch may provide a nonsudden change in current.

Power Transmission

Almost all electric energy sold today is in the form of ac, traditionally because of the ease with which it can be transformed from one voltage to another.* Large currents in wires produce heating of the wire and energy losses, so power is transmitted great distances at high voltages and correspondingly low currents (power = voltage × current). Power may be carried from the power plants to the cities at about 120 000 volts or more, stepped down to about 12 000 volts in the city, and finally stepped down again to provide the 120 volts used in household circuits.

Energy, then, is transferred from one system of conducting wires to another by electromagnetic induction. It is but a short step further to find that the same principles account for eliminating wires and sending energy from a radio-transmitter antenna to a radio receiver many kilometers away, and just a tiny step further to the transformation of energy of vibrating electrons in the sun to life energy on earth. The effects of electromagnetic induction are very far-reaching.

Figure 24-16
Power transmission.

---

*Nowadays, power utilities can transform dc voltages using semiconductor technology. Keep an eye on the present advances in superconductor technology, and watch for resulting changes in the way that power is presently transmitted.

Field Induction

Electromagnetic induction has thus far been discussed as the production of voltages and currents. Actually, the more fundamental *fields* underlie both voltages and currents. The modern view of electromagnetic induction holds that electric and magnetic fields are induced, which in turn gives rise to the voltages we have considered. Induction takes place whether or not a conducting wire or any material medium is present. In this more general sense, Faraday's law states:

> **An electric field is induced in any region of space in which a magnetic field is changing with time. The magnitude of the induced electric field is proportional to the rate at which the magnetic field changes. The direction of the induced electric field is at right angles to the changing magnetic field.**

There is a second effect, which is the counterpart to Faraday's law. It is the same as Faraday's law, except that the roles of electric and magnetic fields are interchanged. It is one of the many symmetries in nature. This effect was advanced by the British physicist James Clerk Maxwell in about 1860, and is known as **Maxwell's counterpart to Faraday's law**:

> **A magnetic field is induced in any region of space in which an electric field is changing with time. The magnitude of the induced magnetic field is proportional to the rate at which the electric field changes. The direction of the induced magnetic field is at right angles to the changing electric field.**

These statements are two of the most important statements in physics. They underlie an understanding of the nature of light and of electromagnetic waves in general.

In Perspective*

The ancient Greeks discovered that when a piece of amber (a natural plastic-like mineral) was rubbed, it picked up little pieces of papyrus. They found strange rocks on the island of Magnesia that attracted iron. Probably because the air in Greece was relatively humid, they never noticed or studied the static electric charge effects common in dry climates. Further development of our knowledge of electrical and magnetic phenomena did not take place until 400 years ago. The human world shrank as more was learned about electricity and magnetism. It became possible first to signal by telegraph over long distances, then to talk to another person many kilometers away through wires, then not only to talk but also to send pictures over many kilometers with no physical connections in between.

*Adapted from *The Feynman Lectures on Physics*, Vol. II, Chap. 1, pp. 1-10 and 1-11, by R. P. Feynman, R. B. Leighton, and M. Sands. Copyright © 1964 by Addison-Wesley, Reading, Mass. Reprinted by permission. Richard P. Feynman, a Nobel laureate in physics and professor of physics at Cal Tech, is considered by many physicists to be among the most brilliant and inspirational physicists of his time—certainly the most colorful. He died in 1988.

Energy, so vital to civilization, could be transmitted over hundreds of kilometers. The energy of elevated rivers was diverted into pipes that fed giant "waterwheels" connected to assemblages of twisted and interwoven copper wires that rotated about specially designed chunks of iron-revolving monsters called *generators*. Out of these, energy was pumped through copper rods as thick as your wrist and sent to huge coils wrapped around transformer cores, boosting it to high voltages for efficient long-distance transmission to cities. Then the transmission lines split into branches . . . then to more transformers . . . then more branching and spreading, until finally the energy of the river was spread throughout whole cities—turning motors, making heat, making light, working gadgetry. There was the miracle of hot lights from cold water hundreds of kilometers away—a miracle made possible by specially designed bits of copper and iron that turned because people had discovered the laws of electromagnetism.

These laws were discovered at about the time the American Civil War was being fought. From a long view of human history, there can be little doubt that events such as the American Civil War will pale into provincial insignificance in comparison with the more significant event of the nineteenth century: the discovery of the electromagnetic laws.

Summary of Terms

Electromagnetic induction The induction of voltage when a magnetic field changes with time. If the magnetic field within a closed loop changes in any way, a voltage is induced in the loop:

$$\text{Voltage induced} = -\text{no. of loops} \times \frac{\text{mag. field change}}{\text{time}}$$

This is a statement of Faraday's law. The induction of voltage is actually the result of a more fundamental phenomenon: the induction of an electric *field*, as defined for the more general case below.

Faraday's law An electric field is induced in any region of space in which a magnetic field is changing with time. The magnitude of the induced electric field is proportional to the rate at which the magnetic field changes. The direction of the induced field is at right angles to the changing magnetic field.

Generator A device that produces electric current by rotating a coil within a stationary magnetic field.

Transformer A device for transferring electric power from one coil of wire to another by means of electromagnetic induction.

Maxwell's counterpart to Faraday's law A magnetic field is induced in any region of space in which an electric field is changing with time. The magnitude of the induced magnetic field is proportional to the rate at which the electric field changes. The direction of the induced magnetic field is at right angles to the changing electric field.

Review Questions

Electromagnetic Induction

1. Exactly what was it that Michael Faraday and Joseph Henry discovered?

2. When a magnet is thrust into a coil of wire, voltage is induced in the wire, which in turn produces a current in the wire. Is each current-carrying loop of wire in the coil then an electromagnet?

3. What kind of interaction occurs between the *magnet* that is thrust into a coil and the *electromagnet* that the coil becomes? (Attraction or repulsion?)

4. Exactly what is it that must change for electromagnetic induction to occur?

Faraday's Law

5. Upon what two things does the current produced by electromagnetic induction depend?

6. What are the three ways that voltage can be induced in a wire?

Generators and Alternating Current

7. How does the frequency of induced voltage compare to how frequently a magnet is plunged in and out of a coil of wire?

8. What is the basic *difference* between a generator and an electric motor?

9. What is the basic *similarity* between a generator and an electric motor?

10. Where in the rotation cycle of a simple generator is the greatest rate of change of field lines? Where, then, is induced voltage at a maximum?

11. Why does the voltage induced in a generator alternate?

Power Production

12. Who discovered electromagnetic induction, and who put it to practical use?

Turbogenerator Power

13. What commonly supplies the energy input to a turbine?

MHD Power

14. What are the principal differences between an *MHD generator* and a *conventional generator*?

Transformers

15. What exactly does a transformer "transform" or, more correctly, transfer?

16. Why does a transformer require alternating current?

17. Can an efficient transformer step up energy? Explain.

18. If 10 V are impressed across the primary coil of a transformer, how many volts are induced in the secondary if it has five times the number of turns and is part of a complete circuit?

19. What name is given to the rate at which energy is transferred?

20. What is the principal advantage of *ac* over *dc*?

Self-Induction

21. When the magnetic field changes in a coil of wire, voltage in each loop of the coil is induced. Will voltage be induced in a loop if the source of the magnetic field is the coil itself?

Power Transmission

22. Why is power transmitted at high voltages over long distances?

23. Does the transmission of electric power require electrical conductors between the source and receiver? Defend your answer.

Field Induction

24. What is induced by the rapid alternation of a *magnetic field*?

25. What is induced by the rapid alternation of an *electric field*?

In Perspective

26. How can part of the energy of a cold river become the energy of a hot lamp hundreds of kilometers away?

Exercises

1. Why does an iron core increase the magnetic induction of a coil of wire?

2. Why are the armature and field windings of an electric motor usually wound on an iron core?

3. Why is a generator armature harder to rotate when it is connected to and supplying electric current to a circuit?

4. Will a cyclist coast farther if the lamp connected to his generator is turned off? Explain.

5. If your metal car moves over a wide, closed loop of wire embedded in a road surface, will the magnetic field of the earth within the loop be altered? Will this produce a current pulse? Can you think of a practical application for this at a traffic intersection?

6. At the security area of an airport, you walk through a weak magnetic field inside a coil of wire. What is the result of a small piece of nonmagnetized iron on your person that slightly alters the magnetic field in the coil?

7. A certain earthquake detector consists of a little box that contains a massive magnet suspended by sensitive springs. The magnet is surrounded by stationary coils of wire that are fastened to the box, which is firmly anchored to the earth. Explain how this device works, using two important principles of physics—one studied in Chapter 4 and the other in this chapter.

8. What is the primary difference between an electric *motor* and an electric *generator*?

9. How could the generator in Figure 24-5 be used as a motor?

10. Does the voltage output increase when a generator is made to spin faster? Explain.

11. Can a current-carrying loop be oriented in a uniform magnetic field in such a way that it does not tend to rotate? Explain.

12. If you place a metal ring in a region where a magnetic field is rapidly alternating, the ring may become hot to your touch. Why?

13. A length of wire is bent into a closed loop and a magnet is plunged into it, inducing a voltage and, consequently, a current in the wire. A second length of wire, twice as long, is bent into two loops of wire and a magnet is similarly plunged into it. Twice the voltage is induced, but the current is the same as that produced in the single loop. Why?

14. Two separate but similar coils of wire are mounted close to each other, as shown below. The first coil is connected to a battery and has a direct current flowing through it. The second coil is connected to a galvanometer. What do the galvanometer readings show when the current in the first coil is increasing? Decreasing? Remaining steady?

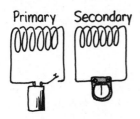

15. Why will more voltage be induced with the apparatus shown above if an iron core is inserted in the coils?

16. How does the current in the secondary of a transformer compare to the current in the primary when the secondary voltage is doubled?

17. In what sense can a transformer be thought of as an electrical lever?

18. Why can a hum usually be heard when a transformer is operating?

19. In the circuit shown, how many volts are impressed across and how many amps flow through the light bulb?

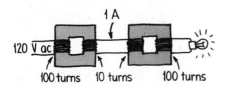

20. In the circuit shown, how many volts are impressed across and how many amps flow through the meter?

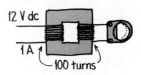

21. Will lamp A be lit in this circuit? Defend your answer.

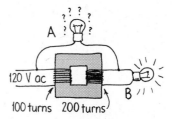

22. A model electric train requires 6 V to operate. If the primary coil of its transformer has 240 windings, how many windings should the secondary have if the primary is connected to a 120-V household circuit?

23. Neon signs require about 12 000 V for their operation. What should be the ratio of the number of loops in the secondary to the number of loops in the primary for a neon-sign transformer that operates off 120-V lines?

24. An induction coil in an automobile is actually a form of transformer that boosts 12 V to about 24 000 V. What is the ratio of secondary windings to primary windings? (The 12 V in a car is dc. In order to make the induction coil operate, the dc is switched on and off frequently.)

25. Your friend says that, according to Ohm's law, high voltage produces high current. Then your friend asks, So how can power be transmitted at high voltage and *low* current in a power line? What is your illuminating response?

26. When a bar magnet is dropped through a vertical length of copper pipe, it falls noticeably more slowly than when dropped through a vertical length of plastic pipe. Why?

27. What is wrong with this scheme? To generate electricity without fuel, arrange a motor to run a generator that will produce electricity that is stepped up with transformers so that the generator can run the motor and simultaneously furnish electricity for other uses.

28. Would electromagnetic waves exist if changing magnetic fields could produce electric fields, but changing electric fields could not in turn produce magnetic fields? Explain.

PART 6

LIGHT

25 Properties of Light

Light is the only thing we can really see. But what *is* light? We know that during the day the primary source of light is the sun, and the secondary source is the brightness of the sky. Other common sources are flames and, in modern times, white-hot filaments in light bulbs and glowing plasmas in glass tubes. In all cases we find that light originates from the accelerated motion of electrons. Light is an electromagnetic phenomenon, and it is only a tiny part of a larger whole—a wide range of electromagnetic waves called the *electromagnetic spectrum*. We begin our study of light by investigating its electromagnetic properties. In the next chapter we will discuss its appearance—color. In Chapter 27 we will learn how light behaves—how it reflects and refracts. Then we will learn about its wave nature in Chapter 28 and its quantum nature in Chapters 29 and 30.

Electromagnetic Waves

Figure 25-1
Shake an electrically charged object to and fro, and you produce an electromagnetic wave.

Shake the end of a stick back and forth in still water, and you'll produce waves on the water surface. Similarly shake an electrically charged rod to and fro in empty space, and you'll produce electromagnetic waves in space. This is because the moving charge is actually an electric current. What surrounds an electric current? The answer is a magnetic field. What surrounds a changing electric current? The answer is a changing magnetic field. Recall from the previous chapter that a changing magnetic field induces a changing electric field, in accordance with Faraday's law. And what does the changing electric field do? The changing electric field, in accordance with Maxwell's counterpart to Faraday's law, will induce a changing magnetic field. The vibrating electric and magnetic fields regenerate each other to make up an **electromagnetic wave**, which emanates (moves outward) from the vibrating charge. The strength of the induced fields very much depends on the speed of emanation. Let's see why this is so.

Electromagnetic Wave Velocity

The strength of each induced field depends not only on the vibrational rate, but also on the motion of the fields—how fast they emanate from the vibrating charge. The higher the speed, the greater the strength of the field that is induced—a high-speed changing magnetic field induces a stronger electric field than a low-speed changing magnetic field, and vice versa. If the speed of emanation is low, the strength of the induced field is weak. Too low a speed would mean the mutual induction would die out. But what of the energy of the fields in this case? The fields contain energy acquired from the vibrating charge; if the fields disappeared with no means of transferring energy to some other form, energy would be destroyed. So low-speed emanation of electric and magnetic fields is incompatible with the law of energy conservation.

Figure 25-2
The electric and magnetic fields of an electromagnetic wave are perpendicular to each other and to the direction of motion of the wave.

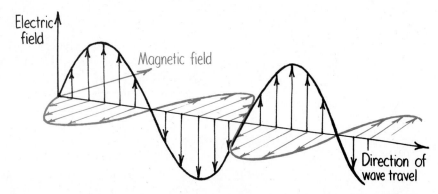

At emanation speeds too high, on the other hand, the fields would be induced to greater and greater magnitudes, with a crescendo of ever increasing energies—again, clearly a no-no with respect to energy conservation. At some critical speed, however, mutual induction would continue indefinitely, with neither a loss nor a gain in energy. From his equations of electromagnetic induction, Maxwell calculated the value of this critical speed and found it to be 300 000 kilometers per second. But this is the speed of light! Maxwell quickly realized that he had discovered the solution to one of the greatest mysteries of the universe—the nature of light. For if the electric charge is set into vibration within the incredible frequency range of 4.3×10^{14} to 7×10^{14} vibrations per second, the resulting electromagnetic wave will activate the "electrical antennae" in the retina of the eye. Light is simply electromagnetic waves in this range of frequencies! The lower frequency appears red, and the higher frequency appears violet.*

On the evening of Maxwell's discovery, he had a date with a young lady he was later to marry. While walking in a garden, his date remarked about the beauty and wonder of the stars. Maxwell asked how she would feel to know that she was walking with the only person in the world who knew what the starlight really was. For it was true. At that time, James Clerk Maxwell was the only person in the world to know that light of any kind is energy carried in waves of electric and magnetic fields that continually regenerate each other.

---

*It is common to describe sound and radio by *frequency* and light by *wavelength*. In this book, however, we stay with the single concept of frequency in describing light.

Question ▶ The magnitude of the constant speed of electromagnetic waves is a remarkable consequence of what central principle in physics?

Figure 25-3

The electromagnetic spectrum is a continuous range of waves extending from radio waves to gamma rays. The descriptive names of the sections are merely a historical classification, for all waves are the same in nature, differing principally in frequency and wavelength; all have the same speed.

The Electromagnetic Spectrum

In a vacuum, all electromagnetic waves move at the same speed and differ from one another in their frequency. The classification of electromagnetic waves according to frequency is the **electromagnetic spectrum** (Figure 25-3). Electromagnetic waves have been detected with a frequency as low as 0.01 hertz (Hz). Electromagnetic waves with frequencies on the order of several thousand hertz (kHz) are classified as radio waves. The very high frequency (VHF) television band of waves starts at about 50 million hertz (MHz). Still higher frequencies are called microwaves; these are followed by infrared waves, often called "heat waves." Further still is visible light, which makes up less than a millionth of 1 percent of the electromagnetic spectrum. The lowest frequency of light we can see with our eyes appears red. The highest visible frequencies are nearly twice the frequency of red and appear violet. Still higher frequencies are ultraviolet. These higher-frequency waves are more energetic and cause sunburns. Higher frequencies beyond ultraviolet extend into the X-ray and gamma-ray regions. There is no sharp boundary between these regions, which actually overlap each other. The spectrum is broken up into these arbitrary regions for classification.

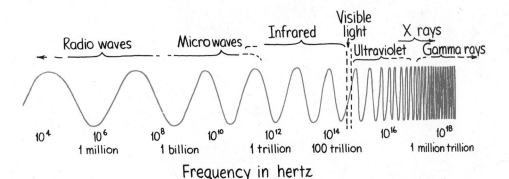

The concepts and relationships we treated earlier in our study of wave motion (Chapter 18) apply here. Recall that the frequency of a wave is the same as the frequency of the vibrating source. The same is true here: the frequency of the electromagnetic wave as it vibrates through space is identical to the frequency of the oscillating electric charge generating it.

▶ **Answer**

The underlying principle that dictates the speed of light and other electromagnetic waves is the conservation of energy.

Different frequencies result in different wavelengths—low frequencies produce waves of long wavelengths and high frequencies produce short wavelengths. For example, since the speed of the wave is 300 000 kilometers per second, an electric charge oscillating once per second (1 hertz) will produce a wave with a wavelength 300 000 kilometers long. This is because only one wavelength is generated in 1 second. If the frequency of oscillation were 10 hertz, then 10 wavelengths would be formed in 1 second, and the corresponding wavelength would be 30 000 kilometers long. A frequency of 10 000 hertz would produce wavelengths 30 kilometers long. So the higher the frequency of the vibrating charge, the shorter the wavelength of radiation.*

We tend to think of space as empty—but only because we cannot see the montages of electromagnetic waves that permeate every part of our surroundings. We see some of these waves, of course, as light. These waves constitute only a microportion of the electromagnetic spectrum. We are unconscious of radio waves, which engulf us every moment. Free electrons in every piece of metal on the earth's surface continually dance to the rhythms of these waves. They jiggle in unison with the electrons being driven up and down along radio- and television-transmitting antennae. A radio or television receiver is simply a device that sorts and amplifies these tiny currents. There is radiation everywhere. Our first impression of the universe is one of matter and void, but actually the universe is a dense sea of radiation in which occasional concentrates are suspended.

Question   Is it correct to say that a radio wave is a low-frequency light wave? Is a radio wave also a sound wave?

Transparent Materials

Light is energy carried in an electromagnetic wave that emanates from vibrating electrons in atoms. When light is incident upon matter, some of the electrons in the matter are forced into vibration. In this way, vibrations in the emitter are transmitted to vibrations in the receiver. This is similar to the way that sound is transmitted and received.

Thus the way a receiving material responds when light is incident upon it depends on the frequency of the light and the natural frequency of the

---

▶ **Answer**

Both a radio wave and light wave are electromagnetic waves, which originate in the vibrations of electrons. Radio waves have lower frequencies than light waves, so a radio wave may be considered to be a low-frequency light wave (and a light wave a high-frequency radio wave). But a sound wave is a mechanical vibration of matter and is not electromagnetic. A sound wave is fundamentally different from an electromagnetic wave. So a radio wave is definitely not a sound wave.

---

*The relationship is $c = f\lambda$, where c is the wave speed (constant), f is the frequency, and λ is the wavelength.

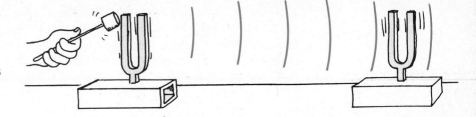

Figure 25-4
Just as a sound wave can force a sound receiver into vibration, a light wave can force electrons in materials into vibration.

electrons in the material. Visible light vibrates at a very high rate, some 100 trillion times per second (10^{14} hertz). If a charged object is to respond to these ultra-fast vibrations, it must have very, very little inertia. Electrons are light enough to vibrate at this rate.

Materials such as glass and water allow light to pass through in straight lines. We say they are **transparent** to light. To understand how light gets through a transparent material, visualize the electrons in an atom as if they were connected by springs (Figure 25-5).* When a light wave is incident upon them, they are set into vibration.

Materials that are springy (elastic) respond more to vibrations at some frequencies than others (Chapter 18). Bells ring at a particular frequency, tuning forks vibrate at a particular frequency, and so do the electrons of atoms and molecules. The natural vibration frequencies of an electron depend on how strongly it is attached to its atom or molecule. Different atoms and molecules have different "spring strengths." Electrons in glass have a natural vibration frequency in the ultraviolet range. When ultraviolet rays shine on glass, resonance occurs as the wave builds and maintains a large amplitude of vibration between the electron and the atomic nucleus, just as pushing someone at the resonant frequency on a swing builds a large amplitude. The energy the atom receives may be passed on to neighboring atoms by collisions, or it may be re-emitted. Resonating atoms in the glass can hold onto the energy of the ultraviolet light for quite a long time (about 100 millionths of a second). During this time the atom makes about 1 million vibrations, and it collides with neighboring atoms and gives up its energy as heat. Thus, glass is not transparent to ultraviolet.

At lower wave frequencies, like those of visible light, electrons in the glass are forced into vibration, but at less amplitude. The atom or molecule holds the energy for less time, with less chance of collision with neighboring atoms and molecules, and less energy transformed to heat.

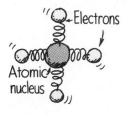

Figure 25-5
The electrons of atoms in glass are bound to the atomic nucleus as if they were connected by springs.

*Electrons, of course, are not really connected by springs. We are simply presenting a visual "spring model" of the atom to help us understand the interaction of light with matter. Physicists devise such conceptual models to understand nature, particularly at the submicroscopic level. The worth of a model lies not in whether it is "true," but whether it is useful. A good model not only is consistent with and explains observations, but also predicts what may happen. If predictions of the model are contrary to what happens, the model is usually either refined or abandoned. The simplified model that we present here—of an atom whose electrons vibrate as if on springs, with a time interval between absorbing energy and re-emitting energy—is quite useful for understanding how light passes through transparent material.

The energy of vibrating electrons is re-emitted as light. Glass is transparent to all the frequencies of visible light. The frequency of the re-emitted light that is passed from molecule to molecule is identical to the frequency of the light that produced the vibration in the first place. The principal difference is a slight time delay between absorption and re-emission.

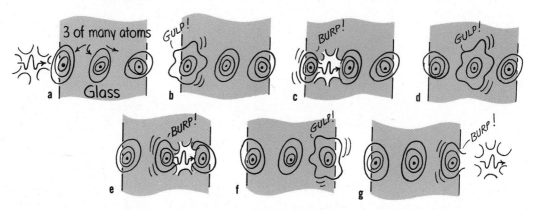

Figure 25-6

A light wave incident upon a pane of glass sets up vibrations in the molecules that produce a chain of absorptions and re-emissions, which in turn pass the light energy through the material and out the other side. Because of the time delay between absorptions and re-emissions, the light travels more slowly in the glass.

It is this time delay that results in a lower average speed of light through a transparent material (Figure 25-6). Light travels at different average speeds through different materials. We say *average speeds*, for the speed of light in a vacuum, whether in interstellar space or in the space between molecules in a piece of glass, is a constant 300 000 kilometers per second. We call this speed of light c.* Light travels a slight bit less than this in the atmosphere, which is usually rounded off as c. In water light travels at 75 percent of its speed in a vacuum, or $0.75\ c$. In glass light travels about $0.67\ c$, depending on the type of glass. In a diamond light travels at less than half its speed in a vacuum, only $0.41\ c$. When light emerges from these materials into air, it travels at its original speed, c.

Infrared waves, with frequencies lower than visible light, vibrate not only the electrons, but entire molecules in the structure of the glass. This vibration increases the internal energy and temperature of the structure, which is why infrared waves are often called *heat waves*. Glass is transparent to visible light, but not to ultraviolet and infrared light.

Figure 25-7

Glass blocks both infrared and ultraviolet, but is transparent to all the frequencies of visible light.

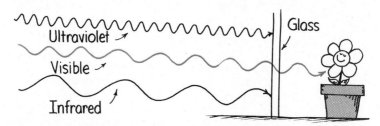

*The presently accepted value is 299 792 km/s, rounded to 300 000 km/s. (This corresponds to 186 000 mi/s.)

Questions ▶ 1. Why is glass transparent to visible light but opaque to ultraviolet and infrared?

2. Pretend that while you walk across a room, you make several momentary stops along the way to greet people who are "on your wavelength." How is this analogous to light traveling through glass?

3. In what way is it not analogous?

Opaque Materials

Most things around us are not transparent, but **opaque**—they absorb light without re-emission. Books, desks, chairs, and people are opaque. Vibrations given by light to their atoms and molecules are turned into random kinetic energy—into internal energy. They absorb light energy and become slightly warmer.

Metals are opaque. Interestingly enough, the outer electrons of atoms in metals are not bound to any particular atom. They are free to wander with very little restraint throughout the material (which is why metal conducts electricity and heat so well). The outer electrons have no connecting "springs," so when light shines on metal and sets these free electrons into vibration, their energy does not "spring" from atom to atom as it does in transparent materials. Reradiated energy is instead reflected. That's why metals are shiny.

The earth's atmosphere is transparent to visible light and some infrared, but, fortunately, quite opaque to high-frequency ultraviolet waves. The small amount of ultraviolet that does get through is responsible for sunburns. If it all got through we would be fried to a crisp. Clouds are semitransparent to ultraviolet, which is why you can get a sunburn on a cloudy day. Ultraviolet rays are not only harmful to the skin, but are also damaging to tar roofs. Now you know why tarred roofs are covered with gravel.

Figure 25-8

Metals are shiny because light that shines on them forces into vibration free electrons, which then emit their "own" light waves as reflection.

▶ **Answers**

1. The natural frequency of vibration for electrons in glass is the same as the frequency of ultraviolet light, so resonance in the glass occurs when ultraviolet waves shine on it. The energetic vibrations of electrons generate heat instead of wave re-emission, so the glass is opaque at higher frequencies. In the range of visible light, the forced vibrations of electrons in the glass are at smaller amplitudes—vibrations are more subtle, re-emission of light rather than the generation of heat occurs, and the glass is transparent. Lower-frequency infrared causes whole molecules, rather than electrons, to resonate, and again heat is generated and the glass is opaque.

2. Your average speed across the room would be less because of the time delays associated with your momentary stops. Likewise, the speed of light in glass is less because of the time delays in interactions with atoms along its path.

3. In the case of walking across the room, it is you who begin the walk and you who complete the walk. This is not analogous to the similar case of light, for according to our model for light passing through a transparent material, the light that is absorbed by an electron made to vibrate is not the same light that is re-emitted—even though the two, like identical twins, are indistinguishable.

Shadows

A thin beam of light is often called a *ray*. When we stand in the sunlight, some of the light is stopped while other rays pass on in a straight-line path. We cast a **shadow**—a region where light rays cannot reach.

If the shadow is close to its source, it is sharp-edged because the sun is so far away. Either a large faraway light source or a small nearby light source will produce a sharp shadow. A large nearby light source produces a somewhat blurry shadow (Figure 25-9). There is usually a dark part on the inside and a lighter part around the edges of a shadow. A total shadow is called an **umbra** and a partial shadow a **penumbra**. A pen-

Figure 25-9

A small light source produces a sharper shadow than a larger source.

a   b

Figure 25-10

An object held close to a wall casts a sharp shadow because no light can seep around to form penumbras. As the object is moved farther away, penumbras are formed and cut down on the umbra. When very far away, no shadow is evident because all the penumbras mix together into a big blur.

a   b

c

umbra appears where some of the light is blocked but where other light fills it in (Figure 25-10). This can happen where light from one source is blocked and light from another source fills in. A penumbra also occurs where light from a broad source is only partially blocked.

Both the earth and the moon cast shadows when sunlight is incident upon them. When the path of either of these bodies crosses into the shadow cast by the other, an eclipse occurs (Figure 25-11). A dramatic example of the umbra and penumbra occurs when the shadow of the moon falls on the earth—during a **solar eclipse**. Because of the large size of

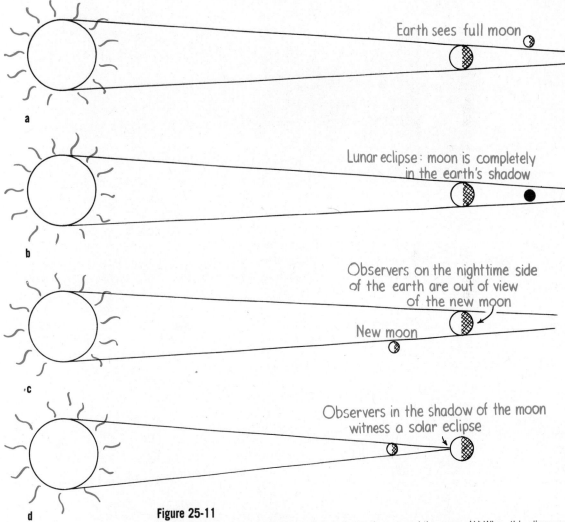

Figure 25-11

(a) A full moon is seen when the earth is between the sun and the moon. (b) When this alignment is perfect, the moon is in the earth's shadow, and a lunar eclipse is produced. (c) A new moon is seen when the moon is between the sun and the earth. (d) When this alignment is perfect, the moon's shadow falls on part of the earth to produce a solar eclipse.

Figure 25-12
Detail of the solar eclipse. A total eclipse is seen by observers in the umbra, and a partial eclipse is seen by observers in the penumbra. Most earth observers see no eclipse at all.

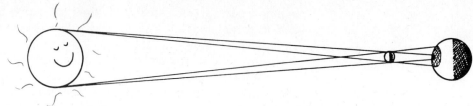

the sun, the rays taper to provide an umbra and a surrounding penumbra (Figure 25-12). If you stand in the umbra part of the shadow you experience darkness during the day—a total eclipse. If you stand in the penumbra, you experience a partial eclipse, for you see a crescent of the sun.* In a **lunar eclipse**, the moon passes into the shadow of the earth.

Questions ▶

1. Which type of eclipse—a solar eclipse, a lunar eclipse, or both—is dangerous to view with unprotected eyes?
2. Why are lunar eclipses more commonly seen than solar eclipses?

Seeing Light: The Eye

Light is the only thing we see with the most remarkable optical instrument known—the eye. A diagram of the human eye is shown in Figure 25-13.

Light enters the eye through the transparent cover called the *cornea*, which does about 70 percent of the necessary bending of the light before it passes through the pupil (which is an aperture in the iris). The light then passes through the lens, which is used only to provide the extra

▶ **Answers**

1. Only a solar eclipse is harmful when viewed directly. During a lunar eclipse, one views a very dark moon. It is not completely dark because the earth's atmosphere acts as a lens and bends some light into the shadow region. Interestingly enough, this is the light of red sunsets and sunrises all around the world, which is why the moon appears a faint deep red during a lunar eclipse.
2. The shadow of the relatively small moon on the large earth covers a very small part of the earth's surface. So only a relatively few people are in the shadow of the moon in a solar eclipse. But the shadow of the earth completely covers the moon during a total lunar eclipse, so everybody who views the nighttime sky can see the shadow of the earth on the moon.

---

*People are cautioned not to look at the sun at the time of a solar eclipse because the brightness and the ultraviolet light of direct sunlight are damaging to the eyes. This good advice is often misunderstood by those who then think that sunlight is more damaging at this special time. But staring at the sun when it is high in the sky is harmful whether or not an eclipse occurs. In fact, staring at the bare sun is more harmful than when part of the moon blocks it! The reason for special caution at the time of an eclipse is simply that more people are interested in looking at the sun during this time.

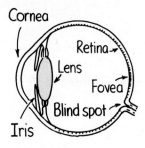

Figure 25-13
The human eye.

bending power needed to focus images of nearby objects on the layer at the back of the eye. This layer—the *retina*—is extremely sensitive, and until very recently it was more sensitive to light than any artificial detector made. Different parts of the retina receive light from different parts of the visual field outside. The retina is not uniform. There is a spot in the center of our field of view called the *fovea*, or region of most distinct vision. Much greater detail can be seen here than at the side parts of the eye. There is also a spot in the retina where the nerves carrying all the information exit; this is the *blind spot*. You can demonstrate that you have a blind spot in each eye if you hold this book at arm's length, close your left eye, and look at Figure 25-14 with your right eye only. You can see both the circle and the X at this distance. If you now move the book slowly toward your face, with your right eye fixed upon the circle, you'll reach a position about 20–25 centimeters from your eye where the X disappears. With both eyes open, you'll find no position where either the X or the circle disappears, because one eye "fills in" the part to which the other eye is blind. Be glad if you have two eyes.

Figure 25-14
The blind-spot experiment.

The retina is composed of tiny antennae that resonate to the incoming light. There are basically two kinds of antennae, the rods and the cones (Figure 25-15). As the names imply, some of the antennae are rod-shaped and some cone-shaped. There are three types of cones: those that are stimulated by low-frequency light, those stimulated by intermediate frequencies, and those stimulated by higher-frequency visible light. The rods predominate toward the periphery of the retina, while the three types of

Figure 25-15
Magnified view of the rods and cones in the human eye.

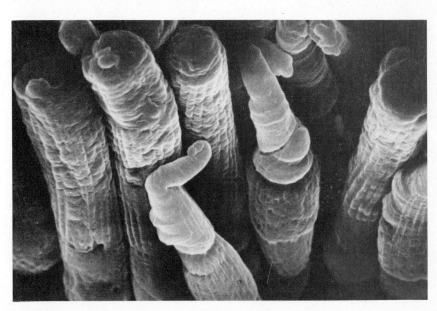

Figure 25-16
On the periphery of your vision you can see an object only if it is moving; you cannot see its color at all.

cones are denser toward the fovea. The cones are very dense in the fovea itself, and since they are packed so tightly, they are much finer or narrower than elsewhere in the retina. Color vision is possible because of the cones. Hence we see color most acutely by focusing an image on the fovea, where there are no rods. Primates and a species of ground squirrel are the only mammals that have the three types of cones and experience full color vision. The retinas of other mammals consist primarily of rods, which are sensitive only to lightness or darkness, like a black-and-white photograph or movie.

In the human eye, the number of cones decreases as we move away from the fovea. It's interesting that the color of an object disappears if viewed on the periphery of vision. This can be tested by having a friend enter your periphery of vision with some brightly colored objects. You will find that you can see the objects before you can see what color they are.

Another interesting fact is that the periphery of the retina is very sensitive to motion. Although our vision is poor from the corner of our eye, we are sensitive to anything moving there. We are "wired" to look for something jiggling to the side of our visual field, a feature that must have been important in our evolutionary development. So have your friend shake those brightly colored objects when she brings them into the periphery of your vision. If you can just barely see the objects when they shake, but not at all when they're held still, then you won't be able to tell what color they are (Figure 25-16). Try it and see!

Another distinguishing feature of the rods and cones is the intensity of light to which they respond. The cones require more energy before they will "fire" an impulse through the nervous system. If the intensity of light is very low, the things we see have no color. We see low intensities with our rods. Dark-adapted vision is almost entirely due to the rods, while vision in bright light is due to the cones. Stars, for example, look white to us. Yet most stars are actually brightly colored. A time exposure of the stars with a camera reveals reds and red-oranges for the "cooler" stars and blues and blue-violets for the "hotter" stars. The starlight is too weak, however, to fire the color-perceiving cones in the retina. So we see the stars with our rods and perceive them as white or, at best, as only faintly colored. Females have a slightly lower threshold of firing for the cones, however, and can see a bit more color than males. So if she says she sees colored stars and he says she doesn't, she is probably right!

The sensitivity of the rods decreases in bright light. We find that the rods "see" better toward the blue end of the color spectrum than the cones. The cones can see a deep red where the rods see no light at all. Red light may as well be black as far as the rods can tell. Thus, if you have two colored objects—say, blue and red—the blue will appear much brighter than the red in dim light, though the red might be much brighter than the blue in bright light. The effect is quite interesting. Try this: In a dark room, find a magazine or something that has colors, and, before you know for sure what the colors are, judge the lighter and darker areas. Then carry the magazine into the light. You should see a remarkable shift between the brightest and dimmest colors.*

The rods and cones in the retina are not connected directly to the optic nerve, but, interestingly enough, are connected to many other cells that are connected to each other. Many of these cells are interconnected, while only a few carry information to the optic nerve. Through these interconnections a certain amount of information is combined from several visual receptors and "digested" in the retina. In this way the light signal is "thought about" before it goes to the optic nerve and then to the main body of the brain. So some brain functioning occurs in the eye itself. The eye does some of our "thinking."

This thinking is betrayed by the iris, the colored part of the eye that expands and contracts and regulates the size of the pupil for admitting more or less light as the intensity of light changes. It so happens that the relative size of this enlargement or contraction is also related to our emotions. If we see, smell, taste, or hear something that is pleasing to us, our pupils automatically increase in size. If we see, smell, taste, or hear something repugnant to us, our pupils automatically contract. Many card players have betrayed the value of a hand by the size of their pupils! (The study of the size of the pupil as a function of attitudes is called *pupilometrics*.)

Figure 25-17
The size of your pupils depends on your mood.

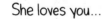
She loves you...   She loves you not?

The human eye can perceive degrees of brightness that range from about 500 million to 1. The difference in brightness between the sun and moon, for example, is about 1 million to 1. But, because of an effect called *lateral inhibition*, we don't perceive the actual differences in brightness. The brightest places in our visual field are prevented from outshining the rest, for whenever a receptor cell on our retina sends a strong brightness signal to our brain, it also signals neighboring cells to dim their responses. In this way, we even out our visual field, which allows us to discern detail in very bright areas and in dark areas as well. Lateral inhibition exaggerates the difference in brightness at the edges of places in our visual field. Edges, by definition, separate one thing from another. So we accentuate differences rather than similarities. The gray

*This phenomenon is called the *Purkinje effect* after Johannes Purkinje, the Czech physiologist who discovered it.

Figure 25-18
Both rectangles are equally bright. Cover the boundary between them with your pencil and see.

rectangle on the left in Figure 25-18 appears dimmer than the gray rectangle on the right when the edge that separates them is in our view. But cover the edge with your pencil or your finger, and they look equally bright. That's because both rectangles are equally bright; each rectangle is shaded lighter to darker, moving from left to right. Our eye concentrates on the boundary where the dark edge of the left rectangle joins the light edge of the right rectangle, and our eye-brain system assumes that the rest of the rectangle is the same. We pay attention to the boundary and ignore the rest.

Figure 25-19
Graph of brightness levels for the rectangles in Figure 25-18.

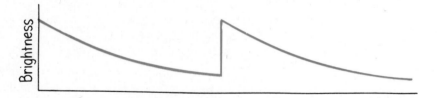

Questions to ponder: Is the way the eye picks out edges and makes assumptions about what lies beyond similar to the way we sometimes make judgments about other cultures and other people? Don't we in the same way tend to exaggerate the differences on the surface while ignoring the similarities and subtle differences within?

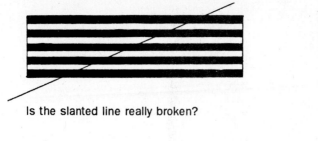

Is the slanted line really broken?

Are the dashes on the right really shorter?

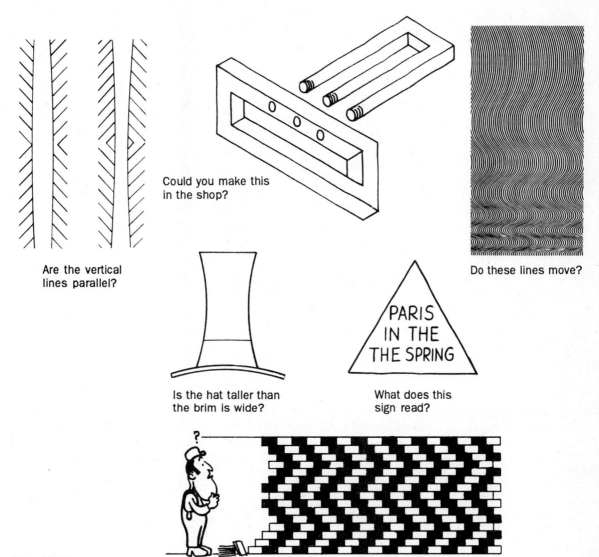

Are the vertical
lines parallel?

Could you make this
in the shop?

Do these lines move?

Is the hat taller than
the brim is wide?

PARIS
IN THE
THE SPRING

What does this
sign read?

Figure 25-20
Optical illusions.

Are the tiles really crooked?

Summary of Terms

Electromagnetic wave An energy-carrying wave emitted by vibrating electrons that is composed of oscillating electric and magnetic fields that regenerate one another.

Electromagnetic spectrum The range of electromagnetic waves extending in frequency from radio waves to gamma rays.

Transparent The term applied to materials through which light can pass in straight lines.

Opaque The term applied to materials that absorb light without re-emission and thus through which light cannot pass.

Shadow A shaded region that appears where light rays are blocked by an object.

Umbra The darker part of a shadow where all the light is blocked.

Penumbra A partial shadow that appears where some of the light is blocked and other light can fall.

Solar eclipse The event wherein the moon blocks light from the sun and casts its shadow on part of the earth.

Lunar eclipse The event wherein the moon passes into the shadow of the earth.

Review Questions

1. Does light make up a relatively large part or a relatively small part of the electromagnetic spectrum?

Electromagnetic Waves

2. What does a *changing magnetic field* induce?

3. What does a *changing electric field* induce?

4. What are the two components of an electromagnetic wave?

Electromagnetic Wave Velocity

5. Upon what two things does the strength of an induced electric or magnetic field depend?

6. What do electric and magnetic fields contain?

7. What is the relationship between energy conservation and the speed at which electromagnetic waves emanate from a vibrating charge?

8. What is light?

The Electromagnetic Spectrum

9. What is the principal difference between a *radio wave* and *light*?

10. What is the principal difference between *light* and an *X ray*?

11. How much of the measured electromagnetic spectrum does light occupy?

12. What color do the lowest visible frequencies appear? The highest?

13. How does the frequency of light compare to the frequency of the vibrating electron that produces it?

14. How does the wavelength of light compare to its frequency?

15. What is the wavelength of a wave that has a frequency of 1 Hz and travels at 300 000 km/s?

16. What is the wavelength of a wave that has a frequency of 2 Hz and travels at 300 000 km/s?

17. In what sense do we say that outer space is not really empty?

Transparent Materials

18. One tuning fork can force another to vibrate. How is this similar to light?

19. In what region of the electromagnetic spectrum is the resonant frequency of electrons in glass?

20. What is the fate of the energy in ultraviolet light that is incident upon glass?

21. What is the fate of the energy in visible light that is incident upon glass?

22. Why are your answers for Questions 20 and 21 different?

23. How does the frequency of re-emitted light compare to the frequency of the light that stimulates its re-emission?

24. How does the average speed of light in glass compare to its speed in a vacuum?

25. Why are infrared waves often called *heat waves*?

Opaque Materials

26. Why do opaque materials become warmer when light shines on them?

27. Why are metals shiny?

28. Why are there several different answers to the question: Is the earth's atmosphere transparent?

Shadows

29. Distinguish between an *umbra* and a *penumbra*.

30. Do the earth and the moon always cast shadows? What do we call the occurrence where one passes within the shadow of the other?

Seeing Light: The Eye

31. Distinguish between the *rods* and *cones* of the eye and between their functions.

32. What are the two unusual features for vision on the periphery of human vision?

33. Why do red-hot and blue-hot stars look white to the eye?

34. What is the evidence for saying that the eye does some of our thinking?

35. How are we able to discern detail in both very bright areas and very dim areas at the same time?

Home Projects

1. Compare the size of the moon on the horizon with its size higher in the sky. One way to do this is to hold at arm's length various objects that will just barely block out the moon. Experiment until you find something just right, perhaps a thick pencil or pen. You'll find the object will be less than a centimeter, depending on the length of your arms. Is the moon really bigger when near the horizon?

2. Monitor the change of size of your own eye pupil. Here's how. Punch a small hole in a piece of paper with a small-sized paper clip. Hold the paper a few centimeters from your eye and look at overhead lights or the sky through the hole. Don't focus on the pinhole, but relax and stare at the hole—and put it so close that you wouldn't be able to focus on it anyway (remove your glasses if you wear any). You see a nice circle. But that circle is not the hole. It's your pupil! To demonstrate this, cover the other eye with your free hand. Both pupils are locked, so if one changes, the other does as well. Covering the other eye and diminishing light to it make that pupil expand. You'll note they both expand as you see the circle of light expand while you look through the pinhole. Neat?

Exercises

1. Which waves have longer wavelengths: light waves, X rays, or radio waves?

2. In about 1675 the Danish astronomer Olaus Roemer found that light from eclipses of Jupiter's moon took an extra 1000 s to travel 300 000 000 km across the diameter of the earth's orbit around the sun. Show how this finding provided the first reasonably accurate measurement for the speed of light.

3. The sun is 1.50×10^{11} meters from the earth. How long does it take for the sun's light to reach the earth?

4. Beyond our nearest star, the sun, is Alpha Centauri 4.2×10^{16} meters away. If we received a radio message from this star today, how long ago would it have been sent?

5. What evidence can you cite to support the idea that light can travel in a vacuum?

6. Why would you expect the speed of light to be slightly less in the atmosphere than in a vacuum?

7. If you fire a bullet through a tree, it will slow down inside the tree and emerge at a speed less than the speed at which it entered. Does light, then, similarly slow down when it passes through glass and also emerge at a lower speed? Defend your answer.

8. Is glass transparent to frequencies of light that match its own natural frequencies? Explain.

9. What determines whether a material is transparent or opaque?

10. You can get a sunburn on a cloudy day, but you can't get a sunburn even on a sunny day if you are behind glass. Explain.

11. Suppose that sunlight falls on both a pair of reading glasses and a pair of dark sunglasses. Which pair of glasses would you expect to be warmer in sunlight? Defend your answer.

12. Why does a high-flying plane cast no shadow on the ground below, although a low-flying plane does?

13. Only some of the people on the daytime side of the earth can witness a solar eclipse when it occurs, whereas all the people on the nighttime side of the earth can witness a lunar eclipse when it occurs. Why is this so?

14. Would it be possible to have a lunar eclipse when the moon is in its crescent or half-moon phase? Or is the moon always at its fullest just before and after the earth's shadow passes over it? Explain.

15. Light from a location on which you concentrate your attention falls on your fovea, which contains only cones. If you wish to observe a weak source of light, like a faint star, why should you not look *directly* at the source?

16. Why do objects illuminated by moonlight lack color?

17. Why do we not see color at the periphery of our vision?

18. From your experimentation with Figure 25-14, is your blind spot located noseward from your fovea or to the outside of it?

19. When your hopeful sweetheart holds you and looks at you with contracted eye pupils and says, "I love you," should you believe it?

20. Can we infer that a person with large pupils is generally happier than a person with small pupils? Why not?

26 Color

Roses are red and violets are blue; colors intrigue artists and physics types too. To the physicist, the colors of objects are not in the substances of the objects themselves or even in the light they emit or reflect. Color is a physiological experience and is in the eye of the beholder. So when we say that light from a rose is red, in a stricter sense we mean that it appears red. Many organisms, including people with defective color vision, will not see the rose as red at all.

The colors we see depend on the frequency of the light we see. Different frequencies of light are perceived as different colors; the lowest frequency we detect appears to most people as the color red and the highest as violet. Between them range the infinite number of hues that make up the color spectrum of the rainbow. By convention these hues are grouped into the seven colors of red, orange, yellow, green, blue, indigo, and violet. These colors together appear white. The white light from the sun is a composite of all the visible frequencies.

Except for light sources such as lamps, lasers, and gas discharge tubes (which we will treat in Chapter 29), most of the objects around us reflect rather than emit light. They reflect only part of the light that is incident upon them, the part that gives them their color.

Selective Reflection

Figure 26-1
The colors of things depend on the colors of light that illuminate them.

A rose, for example, doesn't emit light; it reflects light (Figure 26-1). If we pass sunlight through a prism and then place a deep-red rose in various parts of the spectrum, the rose will appear brown or black in all parts of the spectrum except in the red. In the red part of the spectrum, the petals also will appear red, but the green stem and leaves will appear black. This shows that the red rose has the ability to reflect red light, but it cannot reflect other kinds of light; the green leaves have the ability to reflect green light and likewise cannot reflect other kinds of light. When the rose is held in white light, the petals appear red and the leaves appear green, because the petals reflect the red part of the white light and the leaves reflect the green part of the white light. To understand why objects reflect specific colors of light, we must turn our attention to the atom.

Light is reflected from objects in a manner similar to the way sound is "reflected" from a tuning fork when another that is nearby sets it into vibration. A tuning fork can be made to vibrate even when the frequen-

465

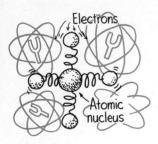

Figure 26-2
The outer electrons in an atom vibrate as if they were attached to the nucleus by springs. As a result, atoms and molecules behave somewhat like optical tuning forks.

cies are not matched, although at significantly reduced amplitudes. The same is true of atoms and molecules. We can think of atoms and molecules as three-dimensional tuning forks with electrons that behave as tiny oscillators that can vibrate as if attached by invisible springs—the spring model of the atom we discussed in the previous chapter. Electrons can be forced into vibration (oscillation) by the vibrating (oscillating) electric fields of electromagnetic waves.* Once vibrating, these electrons send out their own electromagnetic waves just as vibrating acoustical tuning forks send out sound waves.

Atoms (and molecules) have their own natural frequencies; electrons of one kind of atom can be set into vibration at frequencies that are different from the frequencies for other atoms. At the resonant frequencies where the amplitudes of oscillation are large, light is absorbed. But at frequencies below and above the resonant frequencies, light is re-emitted. If the material is transparent, the re-emitted light can travel through the material. For both opaque and transparent materials, the light re-emitted back into the medium from which it came is the reflected light.

Usually a material will absorb some frequencies of light and reflect the rest. If a material absorbs most visible frequencies and reflects red, for example, the material appears red. If it reflects all the visible frequencies, like the white part of this page, it will be the same color as the light that shines on it. If a material absorbs all the light that shines on it, it reflects none and is black.

Figure 26-3
The square on the left *reflects* all the colors illuminating it. In sunlight it is white. When illuminated with blue light, it is blue. The square on the right *absorbs* all the colors illuminating it. In sunlight it is warmer than the white square.

When white light falls on a flower, some of the frequencies are absorbed by the cells in the flower and some are reflected. Cells that contain chlorophyll absorb most of the frequencies and reflect the green part of the light that falls on it. The petals of a red rose, on the other hand, reflect primarily red light, with a small amount of blue. Interestingly enough, the petals of most yellow flowers, like daffodils, reflect red and green as well as yellow. Yellow daffodils reflect a broad band of frequencies. The reflected colors of most objects are not pure single-frequency colors, but are composed of a spread of frequencies.

---

*We use the words *oscillate* and *vibrate* interchangeably. Also, the words *oscillators* and *vibrators* have the same meaning.

An object can only reflect frequencies that are present in the illuminating light. The appearance of a colored object therefore depends on the kind of light used. A candle flame emits light that is deficient in blue; its light is yellowish. An incandescent lamp emits light that is richer toward the lower frequencies, enhancing the reds. In a fabric with a little bit of red in it, for example, the red will be more apparent under an incandescent lamp than when illuminated with a fluorescent lamp. Fluorescent lamps are richer in the higher frequencies, so blues are enhanced under them. With various kinds of illumination, it is difficult to tell the "true" color of objects. Colors appear different in daylight from how they appear when illuminated by either kind of lamp (Figure 26-5).

Figure 26-5
Color depends on the light source.

Selective Transmission

The color of a transparent object depends on the color of the light it transmits. A red piece of glass appears red because it absorbs all the colors that compose white light, except red, which it *transmits*. Similarly, a blue piece of glass appears blue because it transmits primarily blue and absorbs the other colors that illuminate it. The piece of glass contains dyes or *pigments*—fine particles that selectively absorb certain frequencies and selectively transmit others. From an atomic point of view, electrons in the pigment molecules are set into vibration by the illuminating light. Some of the frequencies are absorbed by the pigments, and others are re-emitted from molecule to molecule in the glass. The energy of the absorbed frequencies increases the kinetic energy of the molecules and the glass is warmed. Ordinary window glass is colorless because it transmits all visible frequencies equally well.

Figure 26-6

Only energy having the frequency of blue light is transmitted; energy of the other frequencies is absorbed and warms the glass.

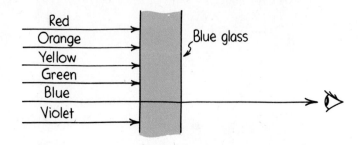

Red
Orange
Yellow
Green
Blue
Violet

Blue glass

Questions ▶

1. When red light shines on a red rose, why do the leaves become warmer than the petals?
2. When green light shines on a rose, why do the petals look black?
3. If you hold a match, a candle flame, or any small source of white light in between you and a piece of red glass, you'll see two reflections from the glass: one from the front surface and one from the back surface. What color reflections will you see?

Mixing Colored Light

The fact that white light from the sun is a composite of all the visible frequencies is easily demonstrated by passing sunlight through a prism and observing the rainbow-colored spectrum. The distribution of solar frequencies is uneven, being most intense in the yellow-green part of the spectrum. It is interesting to note that our eyes have evolved to have maximum sensitivity in this range. That's why it is more and more common for new fire engines to be painted yellow-green, particularly at airports where visibility is vital. This also explains why at night we see better under the illumination of yellow sodium-vapor lamps than under common tungsten-filament lamps of the same brightness.

Figure 26-7

The radiation curve of sunlight is a graph of brightness versus frequency. Sunlight is brightest in the yellow-green region, in the middle of the visible range.

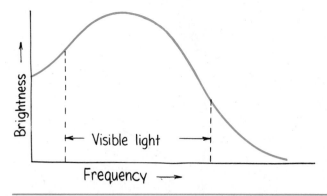

Brightness →

← Visible light →

Frequency ⟶

▶ **Answers**
1. The leaves absorb rather than reflect red light, so the leaves become warmer.
2. The petals absorb rather than reflect the green light.
3. Reflection from the top surface is white, because the light doesn't go far enough into the colored glass for absorption of nonred light. The back surface has only red to reach it, so red is what you see reflected.

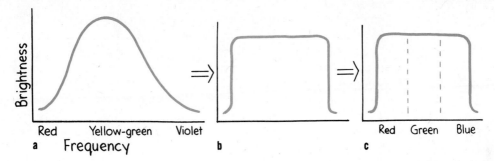

Figure 26-8

(a) Radiation curve of sunlight (b) simplified and (c) divided into three regions.

The graphical distribution of brightness versus frequency is called the *radiation curve* of sunlight (Figure 26-7). Most whites produced from reflected sunlight share this frequency distribution.

We have seen that all the visible frequencies mixed together produce white. Interestingly enough, the perception of white also results from the combination of only red, green, and blue light. We can understand this by greatly simplifying the solar radiation curve and dividing it into three regions as in Figure 26-8. Light in the lowest third of the spectral distribution stimulates the cones sensitive to low frequencies and appears red; light in the middle third stimulates the mid-frequency-sensitive cones and appears green; light in the high-frequency third stimulates the higher-frequency-sensitive cones and appears blue.

If we project these three colors on a screen, overlapping each other, their colors will add to produce white. When the beams of light reflect off a white screen, the light seen is an additive mixture because the lights are added together before the observer sees them. If two of the three colors are added, then another color will be produced (Figure 26-9). By

Figure 26-9

Color addition by the mixing of colored lights. When three projectors shine red, green, and blue light on a white screen, the overlapping parts produce different colors. See this in full color on the inside back cover.

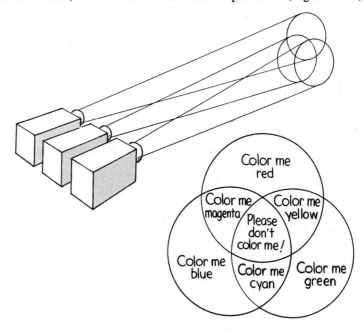

adding various amounts of red, green, and blue, the colors to which each of our three types of cones are sensitive, we can produce any color in the spectrum. For this reason, red, green, and blue are called the **additive primary colors**. A close examination of the picture on most color television tubes will reveal that the picture is an assemblage of tiny spots, each less than a millimeter across. When the screen is lit, some of the spots are red, some green, some blue; the mixtures of these primary colors at a distance provide a complete range of colors, plus white.*

Figure 26-10
The different regions of the simplified radiation curve for sunlight appear as the colors red, green, and blue. These are the *additive primary colors*.

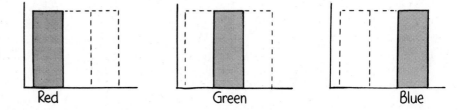

Red   Green   Blue

Mixing Colored Pigments

Every artist knows that if you mix red, green, and blue paint, the result will not be white, but a muddy dark brown. Red and green paint certainly do not combine to form yellow, as is the rule for additive mixtures. The mixing of paints is an entirely different process from the mixing of colored lights. Paint is composed of pigments—tiny solid particles that produce their characteristic colors by the processes of selective absorption or selective transmission of frequencies. When red and green pigments are combined, every color in the white light that illuminates them is absorbed, and the mixture is black. (In practice, absorption is less than 100 percent, so the mixture appears brown rather than black.)

Figure 26-11
Red + green = yellow. Red + blue = magenta (purple). Green + blue = cyan (turquoise). Yellow, magenta, and cyan are the *subtractive primary colors*.

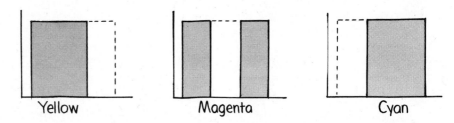

Yellow   Magenta   Cyan

We can understand color mixing by placing our special distribution "blocks" for red and green one atop the other (Figure 26-12). We have drawn them less tall and have indicated the regions of absorption by dark shading. We show white light incident from above. The red pigment shown on top (although order is not important) will absorb the green and blue parts of the white light. These parts are shaded to represent absorption.

---

*It's interesting to note that the "black" you see on the darkest scenes on a black-and-white TV tube is simply the color of the tube face itself, which is more a light gray than black. Because our eyes are sensitive to the contrast with the illuminated parts of the screen, we see this gray as black.

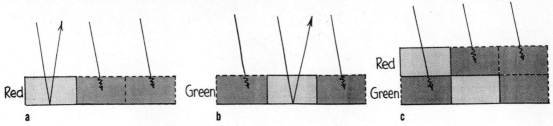

Figure 26-12
(a) Red pigment is represented by the three blocks. The darkness of the green and blue blocks indicate absorption. Only the red component of light gets through. (b) Green pigment is represented by three blocks, where red and blue components are absorbed. (c) A mixture of red and green pigments absorbs all the incident white light, and none is reflected.

The red pigment passes red, but this is absorbed by the green pigment (which absorbs red and blue and passes green). So incident white light is absorbed everywhere in the spectrum; no color is reflected.

Figure 26-13
When white light is incident on cyan pigment, the red component is absorbed. Yellow pigment absorbs the blue component. When cyan and yellow pigments are mixed, green is not absorbed and is the only component of white light reflected.

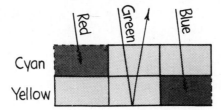

Consider, on the other hand, the mixing of cyan and yellow in Figure 26-13. This combination absorbs red and blue but reflects green. Cyan and yellow paints mixed together produce green. Similarly, the mixing of magenta and yellow paints produces red (Figure 26-14). By mixing magenta, cyan, and yellow pigments, we can produce the additive primaries—red, green, and blue. For this reason, we can summarize the results of color mixing by noting that any of the three additive primaries can be produced by a combination of any two of cyan, magenta, or yellow pigments. The mixture of absorbing pigments results in a subtraction of colors; the observer sees the light left over after absorption has taken place. For this reason, the latter three are called the **subtractive primary colors**. In painting or printing, the primaries are often said to be red, yellow, and blue. Here we are loosely speaking of magenta, yellow, and cyan. They can be combined to produce any color in the spectrum in painting or printing.

Figure 26-14
Magenta pigment absorbs green; yellow pigment absorbs blue. Only red is not absorbed by the mixture of magenta and yellow pigments. The mixture therefore appears red.

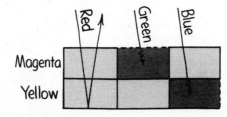

Exercise ▶ Complete the block diagram and predict the color that results from the mixing of cyan and magenta paints. Label and shade the appropriate areas as in Figures 26-13 and 26-14.

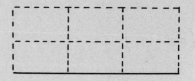

Color printing is an interesting application of color mixing. Three photographs (color separations) are taken of the illustration to be printed: one through a magenta filter, one through a yellow filter, and one through a cyan filter. This produces three negatives, each with a different pattern of exposed areas that corresponds to the filter used and the color distribution in the original illustration. Light is shone through these negatives onto metal plates specially treated to hold printer's ink only in areas that have been exposed to light. The ink deposits are regulated on different parts of the plate by tiny dots. Examine the colored pictures in a magazine with a magnifying glass and see how the overlapping dots of three colors give the appearance of many colors. Or look at a billboard up close.

Rules for Color Mixing

All the rules of color addition and subtraction can be deduced from the overlapping circles in Figure 26-9 and on the inside of the back cover of the book. The additive primaries are the red, green, and blue circles, and the subtractive primaries are located where the primary circles overlap. We see that a subtractive primary is the combination of two additive primaries.

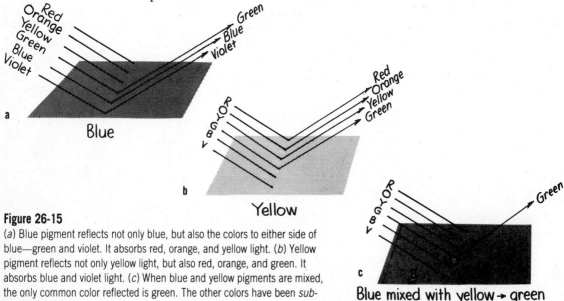

Figure 26-15
(a) Blue pigment reflects not only blue, but also the colors to either side of blue—green and violet. It absorbs red, orange, and yellow light. (b) Yellow pigment reflects not only yellow light, but also red, orange, and green. It absorbs blue and violet light. (c) When blue and yellow pigments are mixed, the only common color reflected is green. The other colors have been *subtracted* from the incident white light.

Notice that when an additive primary and its opposite subtractive primary combine additively—green with magenta, for example—they produce white. From the figure we can see that blue plus yellow produces white;* red plus cyan does also. Any two colors that add together to produce white are called **complementary colors**.

Questions ▶

1. From Figure 26-9, find the complements of cyan, of yellow, and of red.
2. Red + blue = _____ .
3. White − red = _____ .
4. White − blue = _____ .

Why the Sky Is Blue

If a beam of a particular frequency of sound is directed to a tuning fork of similar frequency, the tuning fork will be set into vibration and will effectively redirect the beam in many directions. The tuning fork *scatters* the sound. A similar process occurs with the scattering of light from atoms and particles that are far apart from one another, as in the atmosphere.†

We know that atoms behave like tiny optical tuning forks and re-emit light waves that shine on them. Very tiny particles do the same. The tinier the particle, the higher the frequency of light it will scatter. This is similar to the way small bells ring with higher notes than larger bells. The nitrogen and oxygen molecules and the tiny particles that make up

Figure 26-16
A beam of light falls on an atom and causes the electrons in the atom to vibrate. The vibrating electrons, in turn, re-emit light in various directions. Light is scattered.

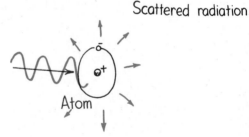

Scattered radiation

Incident beam

Atom

▶ **Answers**
1. Red, blue, cyan.
2. Magenta.
3. Cyan.
4. Yellow.

---

*This follows logically: since yellow is red + green, so blue + red + green = white.
†This type of scattering is called *Rayleigh scattering* (after Lord Scattering) and occurs whenever the scattering particles are much smaller than the wavelength of incident light and have resonances at frequencies higher than the scattered light. The shorter the wavelength of light, the more light is scattered.

Figure 26-17
In clean air the scattering of high-frequency light gives us a blue sky. When the air is full of particles larger than molecules, lower frequencies are also scattered. These add to give a whitish sky.

the atmosphere are like tiny bells that "ring" with high frequencies when energized by sunlight. Like sound from the bells, the re-emitted light is sent in all directions. It is scattered.

Most of the ultraviolet light from the sun is absorbed by a thin protective layer of ozone gas in the upper atmosphere. The remaining ultraviolet sunlight that passes through the atmosphere is scattered by atmospheric particles and molecules. Of the visible frequencies, violet is scattered the most, followed by blue, green, yellow, orange, and red, in that order. Red is scattered only a tenth as much as violet. Although violet light is scattered more than blue, our eyes are not very sensitive to violet light. The lesser amount of blue predominates in our vision, so we see a blue sky!

The blue of the sky varies in different places under different conditions. A principal factor is the water-vapor content of the atmosphere. On clear dry days the sky is a much deeper blue than on clear days with high humidity. Places where the upper air is exceptionally dry, such as Italy and Greece, have beautifully blue skies that have inspired painters for centuries. Where there are a lot of particles of dust and other particles larger than oxygen and nitrogen molecules, the lower frequencies of light are scattered more. This makes the sky less blue, and it takes on a whitish appearance. After a heavy rainstorm when the particles have been washed away, the sky becomes a deeper blue.

The grayish haze in the skies of large cities is the result of particles emitted by internal combustion engines (cars, trucks, and industrial plants). Even when idling, a typical automobile engine emits more than 100 billion particles per second. Most are invisible and provide a framework to which other particles adhere. These are the primary scatterers of lower frequency light. For the larger of these particles, absorption rather than scattering takes place and a brownish haze is produced. Yuk!

Why Sunsets Are Red

The lower frequencies of light are scattered the least by nitrogen and oxygen molecules. Therefore red, orange, and yellow light are transmitted through the atmosphere much more than violet and blue. Red, which is scattered the least, passes through more atmosphere than any other color. Therefore, when white light passes through a thick atmosphere, the higher frequencies are scattered most while the lower frequencies are transmitted with minimal scattering. Such a thicker atmosphere is presented to sunlight at sunset.

At noon the sunlight travels through the least amount of atmosphere to reach the earth's surface. Only a small amount of high-frequency light is scattered from sunlight, enough to make the sun somewhat yellow. As the day progresses and the sun is lower in the sky (Figure 26-18), the path through the atmosphere is longer, and more blue is scattered from the sunlight. The removal of blue leaves the transmitted light more reddish in appearance. The sun becomes progressively redder, going from yellow to orange and finally to a red-orange at sunset.*

Figure 26-18

A sunbeam must travel through more kilometers of atmosphere at sunset than at noon. As a result, more blue is scattered from the beam at sunset than at noon. By the time a beam of initially white light gets to the ground, only the lower frequencies survive to produce a red sunset.

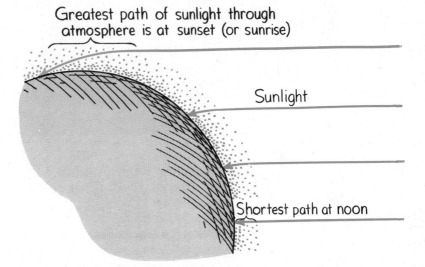

Greatest path of sunlight through atmosphere is at sunset (or sunrise)

Sunlight

Shortest path at noon

The colors of the sunset are consistent with our rules for color mixing. When blue is subtracted from white light, the complementary color that is left is yellow. When higher-frequency violet is subtracted, the resulting complementary color is orange. When medium-frequency green is subtracted, magenta is left. The combinations of resulting colors vary with atmospheric conditions, which change from day to day and give us a variety of sunsets.

---

*Sunsets and sunrises would be unusually colorful if particles larger than atmospheric molecules were more abundant in the air. This was the case all over world for three years following the eruption of the volcano Krakatoa in 1883, when micrometer-sized particles were spewed in abundance throughout the world's atmosphere.

Why Clouds Are White

Clusters of water molecules in a variety of sizes make up clouds. The different-sized clusters result in a variety of scattered frequencies: the tiniest, blue; slightly larger clusters, say, green; and still larger clusters, red. The overall result is a white cloud. Electrons close to one another in a cluster vibrate together and in step, which results in a greater intensity of scattered light than from the same number of electrons vibrating separately. Hence, clouds are bright!

Absorption occurs for larger droplets, and the scattered intensity is less. The clouds are darker. Further increase in the size of the drops causes them to fall to earth, and we have rain.

The next time you find yourself admiring a crisp blue sky, or delighting in the shapes of bright clouds, or watching a beautiful sunset, think about all those ultra-tiny optical tuning forks vibrating away—you'll appreciate these everyday wonders of nature even more!

Figure 26-19
A cloud is composed of a variety of particle sizes. The tiniest scatter blue, slightly larger scatter green, and still larger scatter reds. The overall result is a white cloud.

Why Water Is Greenish Blue

The color of water is not the beautiful deep blue that you often see on the surface of a lake or the ocean. That blue is the reflected color of the sky. The color of water itself, as you can see by looking at a piece of white material under water, is a pale greenish blue.

Although water is transparent to nearly all the visible frequencies of light, it strongly absorbs infrared waves. This is because water molecules resonate to the frequencies of infrared. The energy of the infrared waves is transformed into internal energy in the water, which is why sunlight warms water. Water molecules very weakly resonate in the visible red, which causes a slight absorption of red light in water. Red light is reduced to a quarter of its initial brightness by 15 meters of water. There is very little red light in the sunlight that penetrates below 30 meters of water. When red is taken away from white light, what color remains? This question can be asked in another way: what is the complementary color of red? The complementary color of red is cyan—a bluish green color. In seawater, the color of everything at these depths looks greenish.

Figure 26-20
Water is cyan because it absorbs red. The froth in the waves is white because, like clouds, it is composed of a variety of tiny clusters of water droplets that scatter all the visible frequencies.

Interestingly enough, many crabs and other sea creatures that appear black in deep water are found to be red when they are raised to the surface. At these depths, black and red look the same. Apparently the selection mechanism of evolution could not distinguish between black and red at such depths in the ocean.

So while the sky is blue because blue is strongly scattered by molecules in the atmosphere, water is bluish green because red is absorbed by molecules in the water. We see that the colors of things depend on which colors are scattered or reflected by molecules and also on which colors are absorbed by molecules.*

Figure 26-21
There are no blue pigments in the feathers of a blue jay. Instead there are tiny alveolar cells in the barbs of its feathers that scatter light—mainly high-frequency light. So a blue jay is blue for the same reason the sky is blue—scattering.

*Scattering by small, widely spaced particles in the irises of blue eyes, rather than any pigments, accounts for their color. Absorption by pigments accounts for brown eyes.

Questions

1. If molecules in the sky scattered low-frequency light more than high-frequency light, how would the colors of the sky and sunsets appear?

2. Distant dark mountains are bluish in color. What is the source of this blueness? (*Hint:* Exactly what is between us and the mountains we see?)

3. Distant snow-covered mountains reflect a lot of light and are bright. But they look yellowish, depending on how far away they are. Why do they appear yellowish? (*Hint:* What happens to the reflected white light in traveling from the mountain to us?)

Color Vision and Color Deficiency

Color is not in the world around you; color is in your head. The world around you is filled with a montage of vibrations—electromagnetic waves that stimulate the sensation of color when the vibrations interact with the cone-shaped receiving antennae in the retina of your eye. (The mechanics of color vision was discussed in Chapter 25.)

Some people have difficulties in seeing color, due mainly to cones that are either missing or malfunctioning. These people are *color-deficient*; when they match colors, the match-ups don't correspond to the choices most people make. Although the colors they see are different from what most of us see, they do see color; that's why the term *color blindness* is no longer used. Very few people are totally color-deficient, or color-blind, but about 10 percent of the population has difficulty of some kind in seeing colors. Most forms of partial color deficiency are inherited sex-linked anomalies: very few women are color-deficient, for both of a woman's parents must carry the defective gene—unlike men, who require the defective gene from only one parent.

▶ **Answers**

1. If low frequencies were scattered, the noontime sky would appear reddish orange. At sunset more reds would be scattered by the longer path of the sunlight, and the sunlight would be predominantly blue and violet. So sunsets would appear blue!

2. If we look at distant dark mountains, very little light from them reaches us, and the blueness of the atmosphere between us and the mountains predominates. The blueness is of the low-altitude "sky" between us and the mountains. That's why distant mountains look blue!

3. The reason that bright snow-covered mountains appear yellow is because the blue in the white light from the snowy mountains is scattered on its way to us. So by the time the light gets to us, it is weak in the high frequencies and strong in the low frequencies—hence it is yellowish. For greater distances, farther away than mountains are usually seen, they would appear orange for the same reason a sunset appears orange.

   Why do we see the scattered blue when the background is dark, but not when the background is bright? Because the scattered blue is faint. A faint color will show itself against a dark background, but not against a bright background. For example, when we look from the earth's surface at the atmosphere against the darkness of space, the atmosphere is sky blue. But astronauts above who look below through the same atmosphere to the bright surface of the earth do not see the same blueness.

The simplest way to describe color deficiency is by the number of colors that a person must mix together in order to match all the hues of the rainbow. People with normal vision need only the three additive primary colors to reproduce all the colors of the spectrum. All three types of cones are fully functional. People who are partially color-deficient are likely to have some kind of deficiency in one type of cone and need only two hues to reproduce all the colors they can see. The majority of partially color-deficient people can match all the hues they can see with combinations of only blue and yellow; they cannot distinguish between reds and greens, which take on a grayish appearance. (How many automobile accidents could have been avoided if traffic engineers had taken color deficiency into account when stoplights were first designed?) A few people are yellow-blue deficient, and all the colors they can see can be reproduced by combinations of red and green. The very few individuals who are totally color-blind need only one color (any color at all—they all seem the same) to reproduce all they can see. These few see no color at all; their world looks like a black-and-white movie.

Color deficiency is not usually an all-or-nothing thing. There are many people who actually can see all the colors and who need three hues to match the spectrum but who need abnormal amounts of one of the hues. A person who is green-weak sees the spring grass as being a grayish green, but at least is able to perceive green as a distinct color. A person who is *red-green blind*, however, cannot distinguish green at all, no matter how intense the source is.

Summary of Terms

Additive primary colors The three colors—red, blue, and green—that when added in certain proportions will produce any color in the spectrum.

Subtractive primary colors The three colors of absorbing pigments—magenta, yellow, and cyan—that when mixed in certain proportions will reflect any color in the spectrum.

Complementary colors Any two colors that when added produce white light.

Review Questions

1. What is the relationship between the frequency of light and its color?

Selective Reflection

2. What happens to light when it falls upon a material that has a natural frequency equal to the frequency of the light?

3. What happens to light when it falls upon a material that has a natural frequency above or below the frequency of the light?

4. Distinguish between the white of this page and the black of this ink in terms of what happens to the white light that falls on both.

5. How does the color of an object differ when illuminated by candle light and by the light from a fluorescent lamp?

Selective Transmission

6. What is a *pigment*?

7. What color light is transmitted through a piece of red glass?

8. Which will warm quicker in sunlight, a clear or a colored piece of glass? Why?

Mixing Colored Light

9. What is the evidence for the statement that white light is a composite of all the colors of the spectrum?

10. What is the color of the peak frequency of solar radiation?

11. What color light are our eyes most sensitive to?

12. What is a *radiation curve*?

13. What frequency ranges of the radiation curve do red, green, and blue light occupy?

14. Why are red, green, and blue called the *additive primary colors*?

Mixing Colored Pigments

15. Why does a mixture of red and green pigments appear blackish?

16. Why does a mixture of cyan and yellow pigments appear blue?

17. Why are magenta, yellow, and cyan called the *subtractive primary colors*?

18. If you look with a magnifying glass at pictures printed in full color in magazines, you'll notice three colors of ink plus black. What are these colors?

Rules for Color Mixing

19. What is the resulting color of equal intensities of red, blue, and green light combined?

20. What is the resulting color of equal intensities of blue light and green light combined?

21. What is the resulting color of equal intensities of red light and cyan light combined?

22. Why are red and cyan called *complementary colors*?

Why the Sky Is Blue

23. Which interacts more with high-pitched sounds, small bells or large bells?

24. Which interacts more with high-frequency light, small particles or large particles?

25. True or false: The sky is blue because oxygen and nitrogen molecules are blue in color.

26. Why does the sky sometimes appear whitish?

27. Why is the sky a deeper blue after a heavy rainstorm?

Why Sunsets Are Red

28. True or false: The sky at sunset is reddish because at sunset oxygen and nitrogen molecules are red in color.

29. Why does the sun look reddish at sunrise and sunset, but not at noon?

30. Why does the color of sunsets vary from day to day?

Why Clouds Are White

31. What is the evidence for a variety of particle sizes in a cloud?

32. What is the evidence for extra big particles in a rain cloud?

Why Water Is Greenish Blue

33. What part of the electromagnetic spectrum is most absorbed by water?

34. What part of the visible electromagnetic spectrum is most absorbed by water?

35. What color results when red is subtracted from white light?

36. Why does water appear cyan?

37. What color would a red crab appear very deep in the water?

Color Vision and Color Deficiency

38. What is the simplest way to describe color deficiency?

Home Projects

1. Stare at a piece of colored paper for 45 seconds or so. Then look at a plain white surface. The cones in your retina receptive to the color of the paper become fatigued, so you see an afterimage of the complementary color when you look at a white area. This is because the fatigued cones send a weaker signal to the brain. All the colors produce white, but all the colors minus one produce the complementary to the missing color. Try it and see!

2. Cut a disk a few centimeters or so in diameter from a piece of cardboard; punch two holes a bit off-center, big enough for a piece of string loop as shown in the sketch. Twirl the disk as shown, so the string winds up like a rubber band on a model airplane. Then tighten

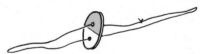

the string by pulling outward and the disk will spin. If half the disk is colored yellow and the other half blue, when it is spun the color will be mixed and appear nearly white (how close to white depends on the hues of the colors). Try this for other complementary colors.

3. Fashion a cardboard tube with closed ends of metal foil in which you punch holes with a pencil, one about 3 or so millimeters in diameter and the other twice as big. Look through the tube at the colors of things against the black background of the tube. You'll see colors that look very different from how they appear against ordinary backgrounds.

4. Simulate your own sunset: add a few drops of milk to a glass of water and look through it at a light bulb. The bulb appears to be red or pale orange, while light scattered to the side appears blue. Try it and see.

Exercises

1. Why will the leaves of a rose be heated more than the petals when illuminated with red light? What does this have to do with people in the hot desert wearing white clothes?

2. In a dress shop with only fluorescent lighting, a customer insists on taking dresses into the daylight at the doorway to check their color. Is she being reasonable? Explain.

3. If the sunlight were somehow green instead of white, what color garment would be most advisable on an uncomfortably hot day? On a very cold day?

4. Why do we not list black and white as colors?

5. Why are the interiors of optical instruments black?

6. What color would red cloth appear if it were illuminated by sunlight? By light from a neon sign? By cyan light?

7. Why does a white piece of paper appear white in white light, red in red light, blue in blue light, and so on for every color?

8. A spotlight is coated so that it won't transmit yellow light from its white-hot filament. What color is the emerging beam of light?

9. How could you use the spotlights at a play to make the yellow clothes of the performers suddenly change to black?

10. Suppose two flashlight beams are shined on a white screen, one beam through a pane of blue glass and the other through a pane of yellow glass. What color appears on the screen where the two beams overlap? Suppose, instead, that the two panes of glass are placed in the beam of a single flashlight. What then?

11. Does a color television work by color addition or by color subtraction? Which dots must be struck by electrons to create the color yellow? Magenta? White?

12. By reference to Figure 26-9, complete the following equations:

Yellow light + blue light = _____ light.
Green light + _____ light = white light.
Magenta + yellow + cyan = _____ light.

13. Check Figure 26-9 to see if the following three statements are accurate. Then fill in the last statement. (All colors are combined by the addition of light.)

Red + green + blue = white.
Red + green = yellow = white − blue.
Red + blue = magenta = white − green.
Green + blue = cyan = white − _____.

14. Why can't we see stars in the daytime?

15. Why is the sky a darker blue at high altitudes? (*Hint:* What color is the "sky" on the moon?)

16. Can stars be seen from the moon in the "daytime" when the sun is shining?

17. Will people in space habitats see the sun against a background of stars, or will they see stars only when they are in the shadow of their habitat? Explain.

18. What does the fact that one can get a sunburn in the shade have to do with the fact that the ultraviolet light not absorbed by the atmosphere is very much scattered by it?

19. Comment on the statement, "Oh, that beautiful red sunset is just the leftover colors that weren't scattered on their way through the atmosphere."

20. If the sky were composed of atoms that mainly scattered orange light rather than blue, what color would sunsets be?

21. Tiny particles, like tiny bells, scatter high-frequency waves more than low-frequency waves. Large particles, like large bells, mostly scatter low frequencies. Intermediate-sized particles and bells mostly scatter intermediate frequencies. What does this have to do with the whiteness of clouds?

22. Very big particles, like droplets of water, absorb more radiation than they scatter. What does this have to do with the darkness of rain clouds?

23. If the atmosphere of the earth were about fifty times thicker, would ordinary snowfall still seem white or would it be some other color? What color?

24. The atmosphere of Jupiter is more than 1000 km thick. From the surface of this planet, would you expect to see a white sun?

25. The refraction of all the sunsets and sunrises around the world causes the redness of the moon during a lunar eclipse. Explain how this is so.

27 Reflection and Refraction

Most of the things we see around us do not emit their own light. They are visible because they re-emit most of the light reaching their surface from a primary source, such as the sun or a lamp, or from a secondary source, such as the illuminated sky. When light falls on the surface of a material it is either re-emitted without change in frequency or is absorbed in the material and turned into heat.* Usually both of these processes occur in varying degrees. When the re-emitted light is returned into the medium from which it came, it is *reflected*, and the process is **reflection**. When the re-emitted light bends from its original course and proceeds in straight lines from molecule to molecule into a transparent material, it is *refracted*, and the process is **refraction**.

Reflection

Figure 27-1
Light interacts with atoms as sound interacts with tuning forks.

Principle of Least Time†

Law of Reflection

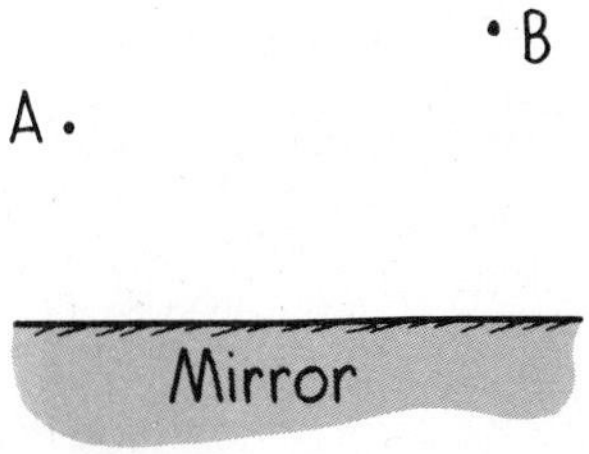

• B

A •

Mirror

Figure 27-2

When this page is illuminated with sunlight or lamplight, a general electron vibration is set up in the normal energy states of its atoms; whole electron clouds vibrate in response to the oscillating electric fields of the illuminating light. Even under bright sunlight, the amplitudes of these vibrations are less than 1 percent of the radius of the atomic nucleus. It is these tiny electron vibrations that re-emit the light by which we see the page. When the page is illuminated by white light, it appears white, which reveals the fact that the electrons are set into vibrations at all the visible frequencies. Very little absorption occurs. The ink on the page is a different story. Except for a bit of reflection, it absorbs all the visible frequencies and therefore appears black.

So on the surfaces of all the objects around us, the electron clouds of the atoms undergo slight vibrations under the influence of illuminating light. These tiny vibrations over a wide frequency range reflect the various colors of light by which we see these objects. Simply put, we say that we see these objects by the light they reflect.

We all know that light ordinarily travels in straight lines. In going from one place to another, light will take the most efficient path and travel in a straight line. This is true if there is nothing to obstruct the passage of light between the places under consideration. If light is reflected from a mirror, the kink in the otherwise straight-line path is described by a simple formula. If light is refracted, as when it goes from air into water, still another formula describes the deviation of light from the straight-line path. Before thinking of light with these formulas, we will first consider an idea that underlies all the formulas that describe light paths. This idea was discovered by the French scientist Pierre Fermat in about 1650, and it is called **Fermat's principle of least time**. His idea is this: out of all possible paths that light might take to get from one point to another, it takes the path that requires the *shortest time*.

We can understand reflection by the principle of least time. Consider the following situation. In Figure 27-2 we see two points, A and B, and an ordinary plane mirror beneath. How can we get from A to B in the shortest time? The answer is simple enough—go straight from A to B! But if we add the condition that the light must strike the mirror in going from A to B in the shortest time, the answer is not so easy. One way would be to go as quickly as possible to the mirror and then to B, as shown by the solid lines in Figure 27-3. This gives us a short path to the mirror but a very long path from the mirror to B. If we instead consider a point

---

*A less common fate is absorption followed by re-emission at lower frequencies—fluorescence (Chapter 29).

†This material and many of the examples of least time are adapted from *The Feynman Lectures on Physics*, Vol. I, Chap. 26, by R. P. Feynman, R. B. Leighton, and M. Sands. Copyright © 1963 by Addison-Wesley, Reading, Mass. Reprinted by permission.

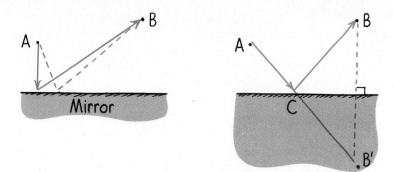

Figure 27-3   **Figure 27-4**

on the mirror a little to the right, we slightly increase the first distance, but we greatly decrease the second distance, and so the total path length shown by the dashed lines—and therefore the travel time—is less. How can we find the exact point on the mirror for which the time is shortest? We can find it very nicely by a geometric trick.

We construct on the opposite side of the mirror an artificial point, B', which is the same distance "through" and below the mirror as the point B is above the mirror (Figure 27-4). The shortest distance between A and this artificial point B' is simple enough to determine: it's a straight line. Now this straight line intersects the mirror at a point C, the precise point of reflection for the shortest path and hence the path of least time for the passage of light from A to B. Inspection will show that the distance from C to B equals the distance from C to B'. We see that the length of the path from A to B' through C is equal to the length of the path from A to B bouncing off point C along the way.

Further inspection and a little geometrical reasoning will show that the angle of incident light from A to C is equal to the angle of reflection from C to B. This is the **law of reflection**, and it holds for all angles (Figure 27-5):

   The angle of incidence equals the angle of reflection.

The law of reflection is illustrated with arrows representing light rays in Figure 27-6. Instead of measuring the angles of incident and reflected

Figure 27-5
Reflection.

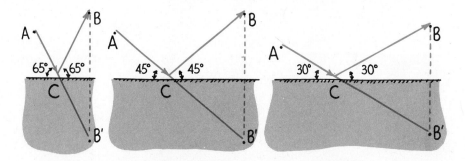

Figure 27-6
The law of reflection.

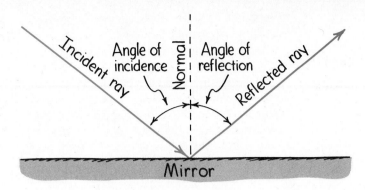

rays from the reflecting surface, it is customary to measure them from a line perpendicular to the plane of the reflecting surface. This imaginary line is called the *normal*. The incident ray, the normal, and the reflected ray all lie in the same plane.

Exercise ▶ The constructions of artificial points B′ in Figures 27-4 and 27-5 show how light encounters point C in reflecting from A to B. By similar construction, show that light that originates at B and reflects to A also encounters the same point C.

Plane Mirrors

Suppose a candle flame is placed in front of a plane mirror. Rays of light are sent from the flame in all directions. Figure 27-7 shows only four of the infinite number of rays leaving one of the infinite number of points on the candle. When these rays encounter the mirror, they are reflected at angles equal to their angles of incidence. The rays diverge from the flame and on reflection diverge from the mirror. These divergent rays appear to emanate from behind the mirror, from a point located where the rays seem to diverge (dashed lines). An observer sees an image of the candle at this point. The light rays do not actually come from this point, so the image is called a *virtual image*. The image is as far behind the mirror as the object is in front of the mirror, and image and object

▶ **Answer**

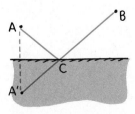

Construct an artificial point A′ as far below the mirror as A is above it; then draw a straight line from B to A′ to find C, as shown at the left. Both constructions superimposed, at right, show that C is common to both. We see that light will follow the same path if it goes in the opposite direction. Whenever you can see somebody else in a mirror, be assured that they can also see you.

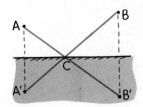

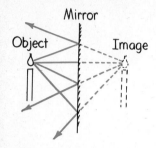

Mirror

Object   Image

Figure 27-7
A virtual image is formed
behind the mirror and is
located at the position where
the extended reflected rays
(dashed lines) converge.

Figure 27-8
(a) The virtual image formed
by a *convex* mirror (a mirror
that curves outward) is
smaller and closer to the
mirror than the object.
(b) When the object is close
to a *concave* mirror (a mirror
that curves inward like a
"cave"), the virtual image is
larger and farther away than
the object. In either case the
law of reflection holds for
each ray.

have the same size. When you view yourself in a mirror, for example, the size of your image is the same as the size your twin would appear if located as far behind the mirror as you are in front—as long as the mirror is flat.

When the mirror is curved, the sizes and distances of object and image are no longer equal. We will not get into curved mirrors in this text, except to say that the law of reflection still holds. A curved mirror behaves as a succession of flat mirrors, each at a slightly different angular orientation from the one next to it. At each point, the angle of incidence is equal to the angle of reflection (Figure 27-8). Note that in a curved mirror, unlike in a plane mirror, the normals (shown by the dashed black lines) at different points on the surface are not parallel to one another.

Whether the mirror is plane or curved, the eye-brain system cannot ordinarily tell the difference between an object and its reflected image. So the illusion that an object exists behind a mirror (or in some cases in front of a concave mirror) is merely due to the fact that the light from the object enters the eye in exactly the same manner, physically, as it would have entered if the object really were at the image location.

a   b

Questions ▶

1. What evidence can you cite to support the claim that the frequency of light does not change upon reflection?

2. If you wish to take a picture of your image while standing 5 m in front of a plane mirror, for what distance should you set your camera to provide sharpest focus?

▶ **Answers**

1. Simply stand in front of a mirror and compare the color of your shirt with the color of its image. The fact that the color is the same is evidence that the frequency of light doesn't change upon reflection.

2. You should set your camera for 10 m; the situation is equivalent to your standing 5 m in front of an open window and viewing your twin standing 5 m in back of the window.

Only part of the light that strikes a surface is reflected. On a surface of clear glass, for example, only about 4 percent is reflected from each surface, while on a clean and polished aluminum or silver surface, about 90 percent of incident light is reflected.

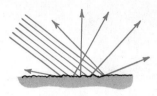

Figure 27-9
Diffuse reflection. Although the reflection of each single ray obeys the law of reflection, the many different surface angles that light rays encounter in striking a rough surface cause reflection in many directions.

Diffuse Reflection

When light is incident on a rough surface, it is reflected in many directions. This is called **diffuse reflection** (Figure 27-9). If the surface is so smooth that the distances between successive elevations on the surface are less than about one-eighth the wavelength of the light, there is very little diffuse reflection, and the surface is said to be *polished*. A surface therefore may be polished for radiation of long wavelength but not polished for light of short wavelength. The wire-mesh "dish" shown in Figure 27-10 is very rough for light waves and is hardly mirrorlike. But for long-wavelength radio waves it is polished and is an excellent reflector.

Light reflecting from this page is diffuse. The page may be smooth to a long radio wave, but to a fine light wave it is rough. Rays of light hitting on this page encounter millions of tiny flat surfaces facing in all directions. The incident light therefore is reflected in all directions. This is a desirable circumstance. It enables us to see objects from any direction or position. Most of our environment is seen by diffuse reflection.

Figure 27-10
The open-mesh parabolic dish is a diffuse reflector for short-wavelength light but polished for long-wavelength radio waves.

Figure 27-11
A magnified view of the surface of ordinary paper.

An undesirable circumstance related to diffuse reflection is the ghost image that occurs on a TV set when the TV signal bounces off buildings and other obstructions. For antenna reception, this difference in path lengths for the direct signal and the reflected signal produces a slight time delay. The ghost image is normally displaced to the right, the direction of scanning in the TV tube, because the reflected signal arrives at the receiving antenna later than the direct signal. Multiple reflections may produce multiple ghosts.

Refraction

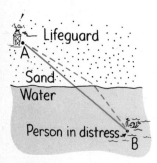

Figure 27-12
Refraction.

Recall from Chapter 25 that the average speed of light is lower in glass and other transparent materials than through empty space. Light travels at different speeds in different materials.* It travels at 300 000 kilometers per second in a vacuum, at a slightly lower speed in air, and at about three-fourths that speed in water. In a diamond, light travels at about 40 percent its speed in a vacuum. Light bends when it passes obliquely from one medium to another. This is *refraction*. It is common observation that a ray of light bends and takes a longer path when it encounters glass or water at a grazing angle. But the longer path taken is nonetheless the path requiring the least time. A straight-line path would take a longer time. We can illustrate this with the following situation.

Imagine that you are a lifeguard at a beach and you spot a person in distress in the water. We show the relative positions of you, the shoreline, and the person in distress in Figure 27-12. You are at point A, and the

---

*Just how much the speed of light differs from its speed in a vacuum is given by the index of refraction, n, of the material; n is the ratio of the speed of light in a vacuum to the speed of light in the material:

$$n = \frac{\text{speed of light in vacuum}}{\text{speed of light in material}}$$

For example, the speed of light in a diamond is $(1/2.4)c$, so $n = 2.4$. For a vacuum, $n = 1$.

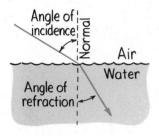

Figure 27-13
Refraction.

person is at point B. You can run faster than you can swim. Should you travel in a straight line to get to B? A little thought will show that a straight-line path would not be the best choice, because if you instead spent a little bit more time traveling farther on land, you would save a lot more time in swimming a lesser distance in the water. The path of shortest time is shown by the dashed-line path, which clearly is not the path of the shortest distance. The amount of bending at the shoreline of course depends on how much faster you can run than swim. The situation is similar for a ray of light incident upon a body of water, as shown in Figure 27-13. The angle of incidence is larger than the angle of refraction by an amount that depends on the relative speeds of light in air and in water.

Question ▶ Suppose our lifeguard in the preceding example were a seal instead of a human being. How would its path of least time from A to B differ?

Consider the pane of thick window glass in Figure 27-14. When light goes from point A through the glass to point B, it will go in a straight-line path. In this case, light encounters the glass perpendicularly, and we see that the shortest distance through both air and glass corresponds to the shortest time. But what about light that goes from point A to point

Figure 27-14
Refraction through glass. Although dashed line AC is the shortest path, light goes a slightly longer path through the air from A to a, then a shorter path through the glass to c, and then to C. The emerging light is displaced but parallel to the incident light.

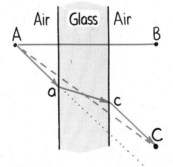

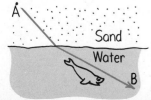

▶ **Answer**
The seal can swim faster than it can run and its path would bend as shown; likewise with light emerging from the bottom of a piece of glass into air.

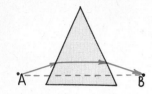

Figure 27-15
A prism.

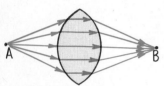

Figure 27-16
A curved prism.

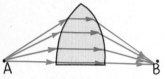

Figure 27-17
A converging lens.

C? Will it travel in the straight-line path shown by the dashed line? The answer is no, for if it did so it would be spending more time inside the glass, where light travels more slowly than in air. The light will instead take a less-inclined path through the glass. The time saved by taking the resulting shorter path through the glass more than compensates for the added time required to travel the slightly longer path through the air. The overall path is the path of least time. The result is a parallel displacement of the light beam, because the angles in and out are the same. You'll notice this displacement when you look through a thick pane of glass at an angle. The more acute the angle, the more pronounced the displacement.

Another example of interest is the prism, in which opposite faces of the glass are not parallel (Figure 27-15). Light that goes from point A to point B will not follow the straight-line path shown by the dashed line, because too much time would be spent in the glass. Instead, the light will follow the path shown by the solid line—a path that is a bit farther through the air—and pass through a thinner section of the glass to make its trip to point B. By this reasoning, one might think that the light should take a path closer to the upper vertex of the prism and seek the minimum thickness of glass. But if it did, the extra distance through the air would result in an overall longer time of travel. The angles of refraction provide the single path of least time.

Interestingly enough, a properly curved prism will provide many paths of equal time from a point A on one side to a point B on the opposite side (Figure 27-16). The curve decreases the thickness of the glass correctly to compensate for the extra distances light travels to points higher on the surface. For appropriate positions of A and B and for the appropriate curve on the surfaces of this modified prism, all light paths are of exactly equal time. In this case, all the light from A that is incident on the glass surface is focused on point B. We see that this shape is simply the upper half of a converging lens (treated in more detail later in this chapter).

Whenever we watch a sunset, we see the sun for several minutes after it has sunk below the horizon. The earth's atmosphere is thin at the top

Figure 27-18
Because of atmospheric refraction, when the sun is near the horizon it appears higher in the sky.

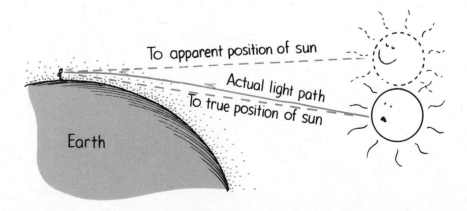

Figure 27-19
The sun is distorted by differential refraction.

and dense at the bottom. Since light travels faster in thin air than it does in dense air, light from the sun can get to us more quickly if, instead of just going in a straight line, it avoids the denser air by taking a higher and longer path to penetrate the atmosphere at a steeper tilt (Figure 27-18). Since the density of the atmosphere changes gradually, the light path bends gradually to produce a curved path. Interestingly enough, this path of least time provides us with a slightly longer period of daylight each day. Furthermore, when the sun (or moon) is near the horizon, the rays from the lower edge are bent more than the rays from the upper edge. This produces a shortening of the vertical diameter, causing the sun to appear elliptical (Figure 27-19).

We are all familiar with the mirage we see while driving on a hot road. The sky appears to be reflected from water on the distant road, but when we get there, the road is dry. Why is this so? The air is very hot just above the road surface and cooler above. Light travels faster through the less dense and thinner hot air than in the cool region. So light, instead of coming to us from the sky in straight lines, also has least-time paths by which it curves down into the hotter region near the road for a while before reaching our

Figure 27-20
A mirage is the result of refraction through the air.

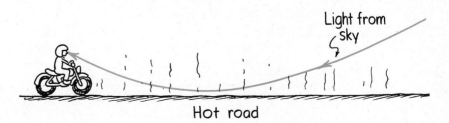

Hot road

Figure 27-21

A mirage. The apparent wetness of the road is not reflection of the sky by water but refraction of sky light through the warmer and less-dense air near the road surface.

eyes (Figure 27-20). A mirage is not, as many people mistakenly believe, a "trick of the mind." A mirage is formed by real light and can be photographed, as shown in Figure 27-21.

When we look at an object over a hot stove or over a hot pavement, we see a wavy, shimmering effect. This is due to the various least-time paths of light as it passes through varying temperatures and therefore varying densities of air. The twinkling of stars results from similar phenomena in the sky, where light passes through unstable layers in the atmosphere.

Question If the speed of light were the same in the various temperatures and densities of air, would there still be slightly longer daytimes, twinkling stars at night, mirages, and slightly squashed suns at sunset?

In the foregoing examples, how does light seemingly know what conditions exist and what compensations a least-time path requires? When approaching window glass at an angle, how does light know to travel a bit farther in air to save time in taking a less-inclined angle and therefore a shorter path through the pane of glass? In approaching a prism or a lens, how does light know to travel a greater distance in air to reach a thinner portion of the glass? How does light from the sun know to travel above the atmosphere an extra distance before taking a shortcut through the air to save time? How does sky light above know that it can reach

▶ **Answer**

No.

us in minimum time if it dips toward a hot road before tilting upward to our eyes? The principle of least time appears to be noncausal. It seems as if light has a mind of its own, that it can "smell" all the possible paths, calculate the times for each, and choose the one that requires the least time. Is this the case? These questions, as intriguing as they seem, are appropriate for some magical quality of light. They simply don't apply to a more straightforward physics interpretation of refraction.

Cause of Refraction

When light bends as it passes obliquely from one medium to another, we call the process *refraction*. The cause of refraction is the changing of the average speed of light in going from one transparent medium to another. We can understand this by considering the action of a pair of toy cart wheels connected to an axle as the wheels roll from a smooth sidewalk onto a grass lawn. If the wheels meet the grass at some angle (Figure 27-22), they will be deflected from their straight-line course. The direction of the rolling wheels is shown by the dashed line. Note that on meeting the lawn, where the wheels roll slower owing to interaction with the grass, the left wheel slows down first. This is because it meets the grass while the right wheel is still on the smooth sidewalk. The faster-moving right wheel tends to pivot about the slower-moving left wheel. It travels farther during the same time the left wheel travels a lesser distance in the grass. This action bends the direction of the rolling wheels toward the "normal," the lightly dashed line perpendicular to the grass-sidewalk border.

Figure 27-22
The direction of the rolling wheels changes when one part slows down before the other part.

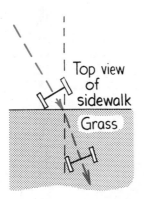

Figure 27-23
The direction of the light waves changes when one part of the wave slows down before the other part.

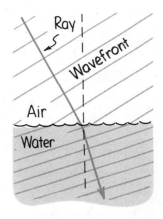

A light wave bends in a similar way (Figure 27-23). Note the direction of light shown by the solid arrow (the light ray) and wave fronts at right angles to the ray. (If the light source were close, the wave fronts would appear as segments of circles; but if we assume the distant sun is the source, the waves form practically straight lines.) In any case, we see that the wave fronts are everywhere perpendicular to the light rays. In the figure the wave meets the water surface at an angle, so the left portion of the wave slows down in the water while the part still in the air travels at speed *c*. The ray or beam of light remains perpendicular to the wave front and bends at the surface, just as the wheels bend to change direction when they roll from the sidewalk into the grass. In both cases the bending is caused by a change in speed.*

Sample wavefronts reflected from the top of a tree are shown in Figure 27-24. If the temperature of the air were uniform, the average speed of light would be the same in all parts of the air; light traveling toward the ground would meet the ground. But the air is warmer and less dense near the ground and the wavefronts pick up speed as they travel downward and bend upward. The observer sees a mirage.

Figure 27-24

A wave explanation of a mirage. Wave fronts of light travel faster in the hot air near the ground and bend upward.

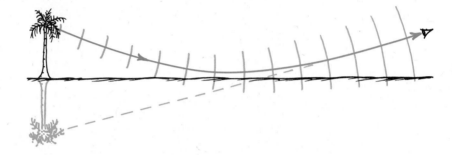

Figure 27-25

When light slows down in going from one medium to another, like going from air to water, it bends toward the normal. When it speeds up in traveling from one medium to another, like going from water to air, it bends away from the normal.

The refraction of light is responsible for many illusions; one of them is the apparent bending of a stick partly immersed in water. The submerged part seems closer to the surface than it really is. Likewise when you view a fish in the water. The fish appears nearer to the surface and closer (Figure 27-26). Because of refraction, submerged objects appear to be magnified. If we look straight down into water, an object submerged 4 meters beneath the surface will appear to be 3 meters deep.

We see that we can interpret the bending of light at the water surface in at least two ways. We can say that the light that reflects from the fish to reach the observer's eye does so in the least time by taking a shorter path upward toward the water surface and a correspondingly longer path through the air. In this view, least time dictates the path taken. Or we can say that the waves of light that happen to be directed upward at an angle toward the water surface are bent off-kilter as they speed up when emerging into the air and that these waves reach the observer's eye. In

---

*The quantitative law of refraction, called *Snell's law*, was first worked out in 1621 by W. Snell, a Dutch astronomer and mathematician: $n_1 \sin \theta_1 = n_2 \sin \theta_2$, where n_1 and n_2 are the indices of refraction of the media on either side of the surface and θ_1 and θ_2 are the respective angles of incidence and refraction. If three of these values are known, the fourth can be calculated from this relationship.

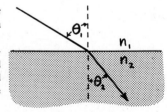

Figure 27-26
Because of refraction, a submerged object appears to be nearer to the surface than it actually is.

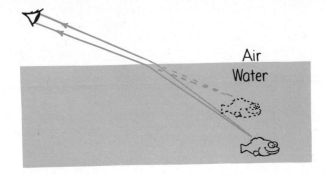

this view, the change in speed from water to air dictates the path taken, and this path turns out to be a least-time path. Whichever view we choose, the results are the same.

| **Question** | ▶ | If the speed of light were the same in all media, would refraction still occur when light passes from one medium to another? |

Dispersion

We know that the average speed of light is less than c in a transparent medium; how much less depends on the nature of the medium and on the frequency of light. The speed of light in a transparent medium depends on its frequency. Recall from Chapter 25 that frequencies of light that match the natural or resonant frequencies of the electron oscillators in the atoms and molecules of the transparent medium are absorbed, and frequencies near the resonant frequencies interact more often in the absorption/re-emission sequence and therefore travel more slowly. Since the natural or resonant frequency of most transparent materials is in the ultraviolet part of the spectrum, the higher frequencies of visible light travel more slowly than the lower frequencies. Violet travels about 1 percent more slowly in ordinary glass than does red light. The colors between red and violet travel at their own respective speeds.

Different frequencies of light travel at different speeds in transparent materials; because they travel at different speeds, they refract differently and bend by different amounts. When light is bent twice, as in a prism, the separation of the different colors of light is quite noticeable. This separation of light into colors arranged according to their frequency is called *dispersion* (Figure 27-27).

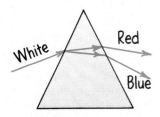

Figure 27-27
Dispersion through a prism makes the components of white light visible.

▶ **Answer**
No.

Figure 27-28
The rainbow is seen in a part of the sky opposite the sun and is centered on the "antisun."

Figure 27-29
Dispersion of sunlight by a single raindrop.

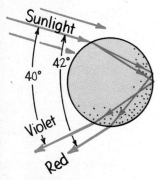

Rainbows

A most spectacular illustration of dispersion is the rainbow. The conditions for seeing a rainbow are the sun shining in one part of the sky and rain falling in the opposite part of the sky. When we turn our backs toward the sun, we see the spectrum of colors in a bow. Seen high enough from an airplane, the bow forms a complete circle. All rainbows would be completely round if the ground were not in the way.

The beautiful colors of rainbows are dispersed from the sunlight by thousands of tiny spherical drops that act like prisms. We can better understand this by considering an individual raindrop, as shown in Figure 27-29. Follow the ray of sunlight as it enters the drop near its top surface. Some of the light here is reflected (not shown), and the remainder is refracted into the water. At this first refraction, the light is dispersed into its spectrum colors, violet being deviated the most and red the least. Reaching the opposite side of the drop, each color is partly refracted out into the air (not shown) and partly reflected back into the water. Arriving at the lower surface of the drop, each color is again reflected (not shown)

Figure 27-30
Sunlight is incident on two sample raindrops as shown and emerges from them as dispersed light. The observer sees the red light from the upper drop and the violet light from the lower drop. Millions of drops produce the whole spectrum.

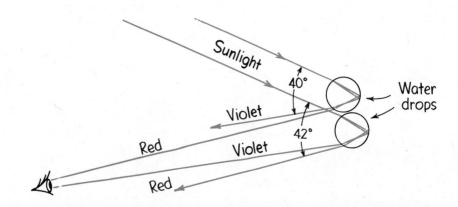

and refracted into the air. This second refraction is similar to that of a prism, where refraction at the second surface increases the dispersion already produced at the first surface.

Although each drop disperses a full spectrum of colors, an observer is in a position to see only a single color from any one drop (Figure 27-30). If violet light from a single drop reaches the eye of an observer, red light from the same drop is incident elsewhere toward the feet. To see red light, one must look to a drop higher in the sky. The color red will be seen when the angle between a beam of sunlight and the dispersed light is 42°. The color violet is seen when the angle between the sunbeams and dispersed light is 40°.

Figure 27-31
Only raindrops along the dashed line disperse red light to the observer at a 42° angle; hence, the light forms a bow.

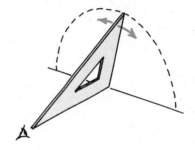

Figure 27-32
Double reflection in a drop produces a secondary bow.

Why does the light dispersed by the raindrops form a bow? The answer to this involves a little geometric reasoning. For simplicity, we will consider only the dispersion of red light. An observer can see red by looking upward so that there is a 42° angle between the sunbeams and the dispersed light that reaches the eyes. But one can also see red by looking sideways and in other directions at the same angle. Figure 27-31 shows that all the drops that disperse light at the same angle form a bow. Drops elsewhere may form somebody else's rainbow, but not the bow that our observer sees. Do you see why a rainbow is bow shaped?

Often a larger, secondary bow with colors reversed can be seen arching at a greater angle around the primary bow. We won't treat this secondary bow except to say that it is formed by similar circumstances and is a result of double reflection within the raindrops (Figure 27-32). Because of this extra reflection (and extra refraction loss), the secondary bow is much dimmer.

Question ▶ If light traveled at the same speed in raindrops as it does in air, would we still have rainbows?

▶ **Answer**
No.

Figure 27-33
Under the rainbow, the sky is bright where thousands of overlapping rainbows combine into a whitish glow.

Total Internal Reflection

Some Saturday night when you're taking your bath, fill the tub extra deep and bring a waterproof flashlight into the tub with you. Put the bathroom light out. Shine the submerged light straight up and then slowly tip it. Note how the intensity of the emerging beam diminishes and how more light is reflected from the water surface to the bottom of the tub. At a certain angle, called the **critical angle**, you'll notice that the beam no longer emerges into the air above the surface. The intensity of the emerging beam reduces to zero where it tends to graze the surface. When the flashlight is tipped beyond the critical angle (48° from the normal for water), you'll notice that all the light is reflected back into the tub. This is **total internal reflection**. The light striking the air-water surface obeys

Figure 27-34
Light emitted in the water is partly refracted and partly reflected at the surface. The length of the arrows indicates the proportions refracted and reflected. At and beyond the critical angle the beam is totally internally reflected.

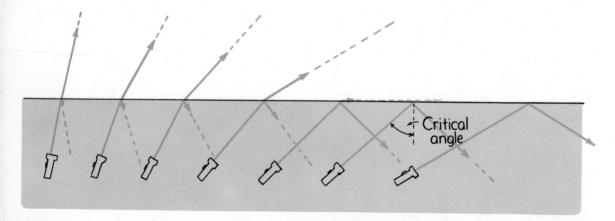

Critical angle

Figure 27-35

An observer underwater sees a circle of light at the still surface. Beyond a cone of 96° (twice the critical angle) an observer sees a reflection of the water interior or bottom.

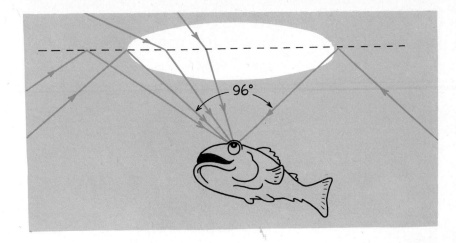

the law of reflection: the angle of incidence is equal to the angle of reflection. The only light emerging from the water surface is that which is diffusely reflected from the bottom of the bathtub. This procedure is shown in Figure 27-34. The proportion of light refracted and light internally reflected is indicated by the relative lengths of the arrows.

Your pet goldfish in the bathtub looks up to see a compressed view of the outside world (Figure 27-35). The 180° view from horizon to opposite horizon is seen through an angle of 96°—twice the critical angle. A lens that similarly compresses a wide view is called a *fisheye lens*.

The critical angle for glass is about 43°, depending on the type of glass. This means that within glass, light that is incident at angles greater than 43° will be totally internally reflected. No light will escape beyond this angle; instead, all of it will be reflected back into the glass. Whereas a silvered or aluminized mirror reflects only about 90 percent of incident light, glass prisms as shown in Figure 27-36 are more efficient. A little light is lost by reflection before it enters the prism, but once inside reflection on the 45°-slanted face is total—100 percent. Moreover, this light is not marred by any dirt or dust on the outside surface, which is the principal reason for the use of prisms instead of mirrors in many optical instruments.

Figure 27-36

Total internal reflection in a prism. In (*a*) the prism changes the direction of the light beam by 90°, in (*b*) by 180°, and in (*c*) it does not change the direction but instead turns the image upside down.

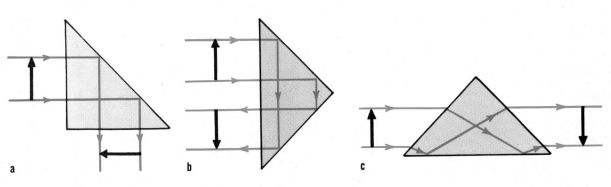

a

b

c

Figure 27-37

Total internal reflection in a pair of prisms.

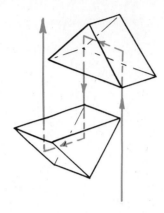

A pair of prisms each reflecting light through 180° is shown in Figure 27-37. Binoculars use pairs of prisms to lengthen the light path between lenses and thus eliminate the need for long barrels. So a compact set of binoculars is as effective as a longer telescope (Figure 27-38). Another advantage of prisms is that whereas the image of a straight telescope is upside down, reflection by the prisms in binoculars re-inverts the image, so things are seen right-side up.

Figure 27-38

Prism binoculars.

The critical angle for a diamond is about 24.5°, smaller than for any other known substance. The critical angle varies slightly for different colors, just as speed varies slightly for different colors. Once light enters a diamond gemstone, most is incident on the sloped backsides at angles greater than 24.5° and is totally internally reflected (Figure 27-39). Because of the great slowdown in speed as light enters a diamond, refraction is pronounced and there is great dispersion. Further dispersion occurs as the light exits through the many facets at its face. Hence we see unexpected flashes of a wide array of colors. Interestingly enough, when these flashes are narrow enough to be seen by only one eye at a time, the diamond sparkles.

Total internal reflection also underlies the operation of optical fibers, or light pipes (Figure 27-40). An optical fiber "pipes" light from one place to another by a series of total internal reflections, much as a bullet

Figure 27-39

Paths of light in a diamond. Rays that strike the inner surface of a diamond at angles greater than the critical angle, about 24.5° depending on color, are internally reflected and exit via refraction at the top surface.

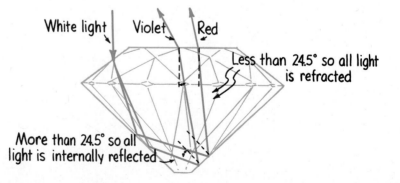

Figure 27-40
The light is "piped" from below by a succession of total internal reflections until it emerges at the top ends.

ricochets down a steel pipe. Light rays bounce along the inner walls, following the twists and turns of the fiber. Optical fibers are used in decorative table lamps and to illuminate instrument displays on automobile dashboards from a single bulb. Dentists use them with flashlights to get light where they want it. Bundles of thin flexible glass or plastic fibers are used to see what is going on in inaccessible places, such as the interior of a motor or a patient's stomach. They can be made small enough to snake through blood vessels or through tubes such as the urethra. Light shines down some of the fibers to illuminate the scene and is reflected back along others.

Figure 27-41
The hairs of the polar bear are transparent light pipes that direct ultraviolet light to its skin—which is guess what color? Black! So what's white and black and warm all over? A polar bear under the arctic sun.

Optical fibers are important in communications because they offer a practical alternative to copper wires and cables. In metropolitan areas, thin glass fibers now replace thick, bulky, and expensive copper cables to carry thousands of simultaneous telephone messages among the major switching centers. In many aircraft, control signals are fed from the pilot to the control surfaces by means of optical fibers. Signals are carried in the modulations of laser light. Unlike electricity, light is indifferent to temperature and fluctuations in surrounding magnetic fields, so the signal is clearer. Also, it is much less likely to be tapped by eavesdroppers.

Optical fibers in nature are found on the polar bear—the hairs of its fur! The hairs of a polar bear are actually transparent optical fibers. Its fur appears white because visible light is reflected from the rough inner surface of each hollow hair. However, the hairs trap ultraviolet light. Like light within an optical fiber, the radiant energy is conducted through the hairs to the bear's skin. The skin is very efficient at absorbing all the solar energy it can get, and is black.

Lenses

A very practical case of refraction occurs in lenses. We can understand a lens by analyzing equal-time paths as we did earlier, or we can assume that it consists of a set of several matched prisms and blocks of glass arranged in the order shown in Figure 27-42. The prisms and blocks refract incoming parallel light rays so they converge to (or diverge from) a point. The arrangement shown in Figure 27-42a converges the light, and we call such a lens a **converging lens**. Note that it is thicker in the middle.

Figure 27-42
A lens may be thought of as a set of prisms.

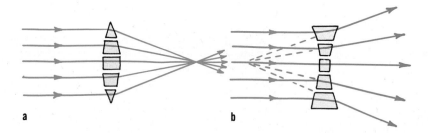

a   b

The arrangement in *b* is different. The middle is thinner than the edges, and it diverges the light; such a lens is called a **diverging lens**. Note that the prisms diverge the incident rays in a way that makes them appear to come from a single point in front of the lens. In both lenses the greatest deviation of rays occurs at the outermost prisms, for they have the greatest angle between the two refracting surfaces. No deviation occurs exactly in the middle, for in that region the glass faces are parallel to each other. Real lenses are not made of prisms, of course, as is indicated in Figure 27-42; they are made of a solid piece of glass with surfaces ground usually to a circular curve. In Figure 27-43 we see how smooth lenses refract waves.

Figure 27-43
Wave fronts travel more slowly in glass than in air. In (a) the waves are retarded more through the center of the lens, and convergence results. In (b) the waves are retarded more at the edges, and divergence results.

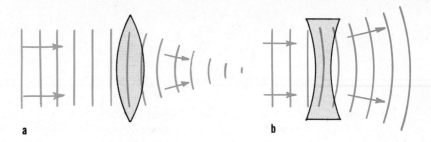

a   b

Some key features in lens description are shown for a converging lens in Figure 27-44. The *principal axis* of a lens is the line joining the centers of curvatures of its surfaces. The *focal point* is the point at which a beam of parallel light, parallel to the principal axis, converges. Incident parallel beams that are not parallel to the principal axis focus at points above or below the focal point. All such possible points make up a *focal plane*. Since a lens has two surfaces, it has two focal points and two focal planes. When the lens of a camera is set for distant objects, the film is in the focal plane behind the lens in the camera.

Figure 27-44
Key features of a converging lens.

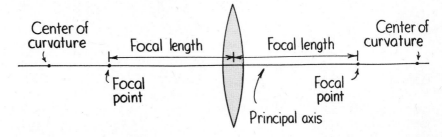

Center of curvature — Focal length — Focal length — Center of curvature — Focal point — Focal point — Principal axis

In the diverging lens, an incident beam of light parallel to the principal axis is not converged to a point, but is diverged—so the light appears to come from a point in front of the lens. The *focal length* of a lens, whether converging or diverging, is the distance between the center of the lens and its focal point. For a thin lens, the focal lengths on either side are equal, even when the curvatures on the two sides are different.

Figure 27-45
The moving patterns of bright and dark areas at the bottom of the pool result from the uneven surface of water, which behaves like a blanket of undulating lenses. A fish looking upward at the sun would see the sun shimmering in intensity. Because of similar irregularities in the atmosphere, we see the stars twinkle.

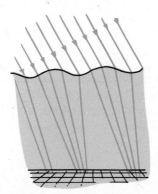

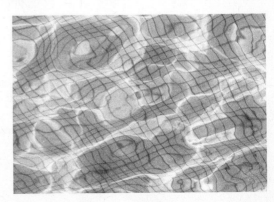

Image Formation by a Lens

At this moment, light is reflecting from your face onto the page of this book. Light that reflects from your forehead, for example, strikes every part of the page. Likewise for the light that reflects from your chin. Every part of the page is illuminated with reflected light from your forehead, your nose, your chin, and every part of your face. You don't see an image of your face on the page because there is too much overlapping of light. But put a barrier with a pinhole in it between your face and the page, and the light that reaches the page from your forehead will not overlap the light from your chin. Likewise for the rest of your face. Without this overlapping, there will be an image of your face on the page. It will be very dim, for very little light reflected from your face gets through the pinhole. To see it, you'd have to shield the page from other light sources.

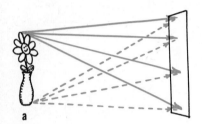 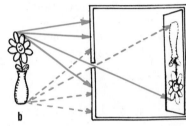 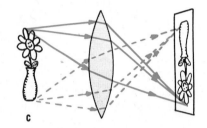

a   b   c

Figure 27-46

Image formation. (*a*) No image appears on the wall because rays from all parts of the object overlap all parts of the wall. (*b*) A single small opening in a barrier prevents overlapping rays from reaching the wall; a dim upside-down image is formed. (*c*) A larger opening admits more light and a lens converges the rays upon the wall without overlapping; more light makes a brighter image.

The first cameras had no lenses and admitted light through a small pinhole. You can see why the image is upside down by the sample rays in Figure 27-46*b*. Long exposure times were required because of the small amount of light admitted by the pinhole. A somewhat larger hole would admit more light, but overlapping rays would produce a blurry image. Too large a hole would allow too much overlapping and no image would be discernible. That's where a converging lens comes in (Figure 27-46*c*). The lens converges light onto the screen without the unwanted overlapping of rays. Whereas the first pinhole cameras were useful only for still objects because of the long exposure time required, moving objects can be taken with the lens camera because of the short exposure time. Now you know why photographs taken with lens cameras came to be called *snapshots*.

The simplest use of a converging lens is a magnifying glass. Magnification occurs when an image is observed through a wider angle with the use of a lens than without the lens. With unaided vision, an object far away is seen through a relatively small angle of view, while the same object when closer is seen through a larger angle of view. This wider angle permits the perception of greater detail. A magnifying glass is simply a converging lens that increases the angle of view.

Figure 27-47

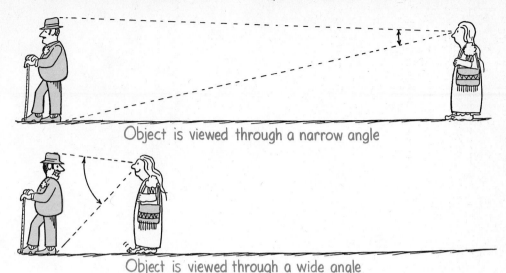

Object is viewed through a narrow angle

Object is viewed through a wide angle

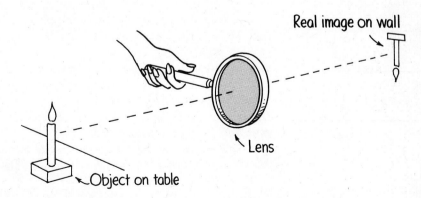

Figure 27-48
When an object is near a converging lens (inside its focal point *f*), the lens acts as a magnifying glass to produce a virtual image. The image appears larger and farther from the lens than the object.

Figure 27-49
When an object is far from a converging lens (beyond its focal point), a real upside-down image is formed.

When we use a magnifying glass, we hold it close to the object we wish to see magnified. This is because a converging lens will magnify only when the object is inside the focal point. The magnified image will be farther from the lens than the object, and it will be right-side-up. If a screen is placed at the image distance, no image will appear on it. This is because no light is actually directed to the image position. The rays that reach our eye, however, behave virtually as if they came from the image position, so we call this a **virtual image**.

When the object is far away enough to be outside the focal point of a converging lens, instead of a virtual image a **real image** is formed. Figure 27-49 shows a case in which a converging lens forms a real image on a screen. A real image is inverted. A similar arrangement is used for projecting slides and motion pictures on a screen and for projecting a real image on the film of a camera. Real images with a single lens will always be inverted.

Real image on wall

Lens

Object on table

Figure 27-50
A diverging lens forms a virtual, right-side-up image of Jamie and his cat.

A diverging lens, when used alone, produces only a diminished virtual image. It makes no difference how far or how near the object is. When a diverging lens is used alone the image is always virtual, erect, and smaller than the object. A diverging lens is often used as a "finder" on a camera. When you look at the object to be photographed through such a lens, you see a virtual image that approximates the same proportions as the photograph.

Question ▶ Why is the greater part of the photograph in Figure 27-50 out of focus?

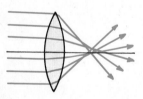

Figure 27-51
Spherical aberration.

Lens Defects

No lens provides a perfect image. A distortion in an image is called an **aberration**. By combining lenses in certain ways, aberrations can be minimized. For this reason, most optical instruments use compound lenses, each consisting of several simple lenses, instead of single lenses.

Spherical aberration results from light passing through the edges of a lens and focusing at a slightly different place from where light passing through the center of the lens focuses (Figure 27-51). This can be remedied by covering the edges of a lens, as with a diaphragm in a camera. Spherical aberration is corrected in good optical instruments by a combination of lenses.

▶ **Answer**

Both Jamie and his cat and the virtual image of Jamie and his cat are "objects" for the lens of the camera that took this photograph. Since the objects are at different distances from the camera lens, their respective images are at different distances with respect to the film in the camera. So only one can be brought into focus. The same is true of your eyes. You cannot focus on near and far objects at the same time.

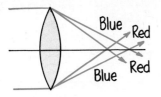

Figure 27-52
Chromatic aberration.

Chromatic aberration is the result of the different speeds of various colors and hence the different refractions they undergo (Figure 27-52). In a simple lens (as in a prism), red light and blue light do not come to focus in the same place. *Achromatic lenses*, which combine simple lenses of different kinds of glass, correct this defect.

The pupil of the eye regulates the amount of light that enters the eye by changing its size. Vision is sharpest when the pupil is smallest because light then passes through only the center of the eye's lens, where spherical and chromatic aberrations are minimal. Also, light bends the least through the center of a lens, so minimum focusing is required for a sharp image. An image that is formed by straight lines of light can appear in focus anywhere. You see better in bright light because your pupils are smaller.*

Astigmatism of the eye is a defect that results when the cornea is curved more in one direction than the other, somewhat like the side of a barrel. Because of this defect, the eye does not form sharp images. The remedy is eyeglasses with cylindrical lenses that have more curvature in one direction than in another.

Questions ▶

1. If light traveled at the same speed in glass as in air, would glass lenses still alter the direction of light rays?

2. Why is there chromatic aberration in light that passes through a lens, but no chromatic aberration in light that reflects from a mirror?

An option for those with poor sight in the last 500 years has been wearing spectacles or, in more recent times, contact lenses. It is interesting to note that at the present time there is an alternative to both spectacles and contacts for people with poor eyesight. Experimental and controversial techniques today allow eye surgeons to shave the cornea of the eye to a proper shape for normal vision. In tomorrow's world, the wearing of eyeglasses and contact lenses may be a thing of the past. We really do live in a rapidly changing world. And that can be nice.

▶ **Answers**

1. No.

2. Different frequencies travel at different speeds in a transparent medium and therefore refract at different angles, which produces chromatic aberration. But the angles at which light reflects has nothing to do with the frequency of light. One color reflects the same as any other reflected color. Mirrors are therefore preferable to lenses in telescopes because with reflection there is no chromatic aberration.

---

*If you wear glasses and ever misplace them, or if you find it difficult to read small print as in a telephone book, hold a pinhole (in a piece of paper or whatever) in front of your eye, close to the page. You'll see the print clearly, and because you're close, it will seem magnified. Try it and see!

Summary of Terms

Reflection   The return of light rays from a surface in such a way that the angle at which a given ray is returned is equal to the angle at which it strikes the surface. When the reflecting surface is irregular, light is returned in irregular directions; this is *diffuse reflection.*

Refraction   The bending of an oblique ray of light when it passes from one transparent medium to another. This is caused by a difference in the speed of light in the transparent media. When the change in medium is abrupt (say, from air to water), the bending is abrupt; when the change in medium is gradual (say, from cool air to warm air), the bending is gradual, which accounts for mirages.

Fermat's principle of least time   Light will take the path that requires the least time when it goes from one place to another.

Law of reflection   The angle of an incidence equals the angle of reflection. The incident and reflected rays lie in a plane that is normal to the reflecting surface.

Critical angle   The minimum angle of incidence at which a light ray is totally reflected within a medium.

Total internal reflection   The total reflection of light traveling in a medium when it strikes the surface of a less dense medium at an angle greater than the critical angle.

Converging lens   A lens that is thicker in the middle than at the edges and refracts parallel rays passing through it to a focus.

Diverging lens   A lens that is thinner in the middle than at the edges, causing parallel rays passing through it to diverge.

Virtual image   An image formed by light rays that do not converge at the location of the image. A virtual image is that reflected by a mirror; it cannot be displayed on a screen.

Real image   An image formed by light rays that converge at the location of the image. A real image can be displayed on a screen.

Aberration   A limitation on perfect image formation inherent, to some degree, in all optical systems.

Suggested Reading

Greenler, R. *Rainbows, Halos, and Glories.* New York: Cambridge University Press, 1980.

Review Questions

1. Distinguish between *reflection* and *refraction.*

Reflection

2. What does incident light do to the electron clouds that surround the atomic nucleus?

3. What do electron clouds do when they are made to oscillate?

Principle of Least Time

4. What is Fermat's principle of least time?

Law of Reflection

5. What is the law of reflection?

Plane Mirrors

6. Compared to the distance of an object in front of a plane mirror, how far behind the mirror is the image?

7. Does the law of reflection hold for curved mirrors? Explain.

8. How much light is reflected from the surface of clear glass?

Diffuse Reflection

9. Does the law of reflection hold for diffuse reflection? Explain.

10. How can a surface be polished for some waves and not others?

Refraction

11. Light bends when it passes obliquely from one medium to another. Does this bending increase or decrease the time of travel?

12. How does the angle at which a ray of light strikes a pane of window glass compare to the angle at which it passes out the other side?

13. How does the angle at which a ray of light strikes a prism compare to the angle at which it passes out the other side?

14. Does light travel faster in thin air or in dense air? What does this have to do with the duration of daylight?

15. What is a mirage?

16. Why do stars twinkle?

17. Does light "know" which path takes the least time from one place to another, or is there some other explanation?

Cause of Refraction

18. What is the cause of refraction?

19. When a cart wheel rolls from a smooth sidewalk onto a plot of grass, the interaction of the wheel with the blades of grass slows the wheel. What slows light when it passes from air into glass or water?

20. When light passes from one material into another, it bends toward the normal to the surface. Or it may bend away from the normal. When does it do which?

21. Does refraction make the bottom of a swimming pool seem deeper or shallower?

Dispersion

22. What happens to light of a certain frequency when it is incident upon a material that has the same natural frequency?

23. Which travels more slowly in glass, red light or violet light?

24. If light of different frequencies has different speeds in a material, does it also refract at different angles in the same material? Explain.

Rainbows

25. What is it that prevents all rainbows from being complete circles?

26. Does a single raindrop illuminated by sunlight disperse a single color or a spectrum of colors?

27. Does a viewer see a single color or a spectrum of colors dispersed from a single faraway drop?

28. Why is a secondary rainbow dimmer than a primary bow?

Total Internal Reflection

29. What is meant by *critical angle*?

30. When is light totally reflected in a glass prism?

31. When is light totally reflected in a diamond?

32. What are two advantages of using prisms in binoculars?

33. Light normally travels in straight lines, but it "bends" in an optical fiber. Explain.

Lenses

34. Distinguish between a *converging lens* and a *diverging lens*.

35. What is the *focal length* of a lens?

Image Formation by a Lens

36. Distinguish between a *virtual image* and a *real image*.

37. What kind of lens can be used to produce a real image? A virtual image?

Lens Defects

38. Distinguish between *spherical aberration* and *chromatic aberration*.

39. Why is vision sharpest when the pupils of the eye are very small?

40. What is astigmatism, and how can it be corrected?

Home Projects

1. Make a pinhole camera, as illustrated below. Cut out one end of a small cardboard box, and cover the end with tissue or wax paper. Make a clean-cut pinhole at the other end. (If the cardboard is thick, make it through a piece of tinfoil placed over an opening in the cardboard.) Aim the camera at a bright object in a darkened room, and you will see an upside-down image on the tissue paper. If in a dark room you replace the tissue paper with unexposed photographic film, cover the back so it is light tight, and cover the pinhole with a removable flap, you are ready to take a picture. Exposure times differ depending principally on the kind of film and amount of light. Try different exposure times, starting with about 3 seconds. Also try boxes of various lengths. You'll find everything in focus in your photographs, but the pictures will not have clear-cut, sharp outlines. The principal difference between your pinhole camera and a commercial one is the glass lens, which is larger than the pinhole and therefore admits more light in less time.

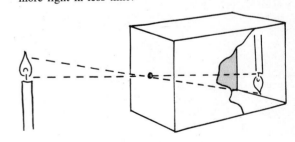

2. If you don't have a prism, you can produce a spectrum by placing a tray of water in bright sunlight. Lean a pocket mirror against the inside edge of the pan and adjust it until a spectrum appears on the wall or ceiling.

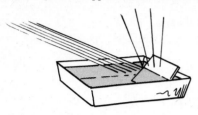

3. Stand a pair of mirrors on edge with the faces parallel to each other. Place an object such as a coin between the mirrors and look at the reflections in each mirror. Neat?

4. Set up two pocket mirrors at right angles and place a coin between them. You'll see four coins. Change the angle of the mirrors and see how many images of the coin you can see. With the mirrors at right angles, look at your face. Then wink. What do you see? You now see yourself as others see you. Hold a printed page up to the double mirrors and contrast its appearance with the reflection of a single mirror.

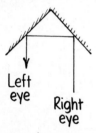

Left eye

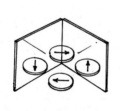

Right eye

5. Determine the magnification power of a lens by focusing on the lines of a ruled piece of paper. Count the spaces between the lines that fit into one magnified space, and you have the magnification power of the lens. You can do the same with binoculars and a distant brick wall. Hold the binoculars so that only one eye looks at the bricks through the eyepiece while the other eye looks directly at the bricks. The number of bricks seen with the unaided eye that will fit into one magnified brick gives the magnification of the instrument.

Magnified space

3 spaces fit into one magnified space

6. Look at the reflections of overhead lights from the two surfaces of eyeglasses, and you will see two fascinatingly different images. Why are they different?

Exercises

1. Fermat's principle is of least time rather than of least distance. Would least distance apply as well for reflection? For refraction? Why are your answers different?

2. Her eye at point P looks into the mirror. Which of the numbered cards can she see reflected in the mirror?

3. Cowboy Joe wishes to shoot his assailant by ricocheting a bullet off a mirrored metal plate. To do so, should he simply aim at the mirrored image of his assailant? Explain.

4. What must be the minimum length of a plane mirror in order for you to see a full view of yourself?

5. What effect does your distance from the plane mirror have in the above answer? (Try it and see!)

6. Hold a pocket mirror at almost arm's length from your face and note the amount of your face you can see. To see more of your face, should you hold the mirror closer or farther, or would you have to have a larger mirror? (Try it and see!)

7. The diagram shows a person and her twin at equal distances on opposite sides of a thin wall. Suppose a window is to be cut in the wall so each twin can see a complete view of the other. Show the size and location of the smallest window that can be cut in the wall to do the job. (*Hint:* Draw rays from the top of each twin's head to the other twin's eyes. Do the same from the feet of each to the eyes of the other.)

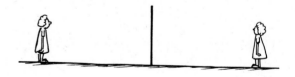

8. Suppose that you walk 1 m/s toward a mirror. How fast do you and your image approach each other?

9. What is wrong with the cartoon of the man looking at himself in the mirror? (Have a friend face a mirror as shown, and you'll see.)

10. Why is the lettering on the front of some vehicles "backward"?

11. A person in a dark room looking through a window can clearly see a person outside in the daylight, whereas the person outside cannot see the person inside. Explain.

12. Why is it difficult to see the roadway in front of you when driving on a rainy night?

13. Does the reflection of a scene in calm water look the same as the scene itself only upside down? (*Hint:* Can you ever see the reflection of the top of a stone that extends above water?)

14. Why does reflected light from the sun or moon appear as a column in the body of water as shown? How would it appear if the water surface were perfectly smooth?

15. Show with a simple diagram that when a mirror with a fixed beam incident on it is rotated through a certain angle, the reflected beam is rotated through an angle twice as large. (This doubling of displacement makes irregularities in ordinary window glass more evident.)

16. When you look at yourself in the mirror and wave your right hand, your beautiful image waves the left hand. Then why don't the feet of your image wiggle when you shake your head?

17. A pair of toy cart wheels are rolled obliquely from a smooth surface onto two plots of grass, a rectangular plot as shown in *a* and a triangular plot as shown in *b*.

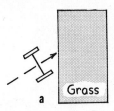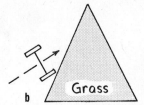

The ground is on a slight incline so that after slowing down in the grass, the wheels will speed up again when emerging on the smooth surface. Finish each sketch by showing some positions of the wheels inside the plots and on the other sides, thereby indicating the direction of travel.

18. If light of all frequencies traveled at the same speed in glass, how would white light appear after passing through a prism?

19. Which is more likely to sparkle, a nearby diamond or a diamond farther away? Explain.

20. A beam of light bends as shown in *a*, while the edges of the immersed square bend as shown in *b*. Do these pictures contradict each other? Explain.

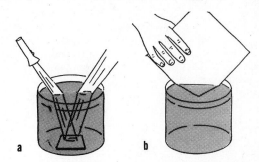

21. If you were spearing a fish, would you aim above, below, or directly at the observed fish to make a direct hit? If you instead used light from a laser as your "spear," would you aim above, below, or directly at the observed fish? Defend your answers.

22. If the fish in the previous exercise were small and blue and your laser light were red, what corrections should you make? Explain.

23. When a fish looks upward at an angle of 45°, does it see the sky or the reflection of the bottom beneath? Defend your answer.

24. If you were to send a beam of laser light to a space station above the atmosphere and just above the horizon, would you aim the laser above, below, or at the visible space station? Defend your answer.

25. What accounts for the large shadows cast by the ends of the thin legs of the water strider? What accounts for the ring of bright light around the shadows?

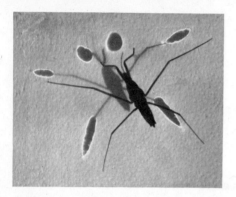

26. Transparent plastic swimming-pool covers called *solar heat sheets* have thousands of small lenses made up of air-filled bubbles. The lenses in these sheets are advertised to focus heat from the sun into the water and raise its temperature. Do you think the lenses of such sheets direct more solar energy into the water? Defend your answer.

27. Would the average intensity of sunlight measured by a light meter at the bottom of the pool in Figure 27-45 be different if the water were still?

28. Figure 27-28 shows the rainbow as an ellipse rather than as a circle to indicate it is being viewed from the side. Can one view a rainbow from the side so that it appears as the segment of an ellipse rather than the segment of a circle? Defend your answer.

29. Two observers standing apart from one another do not see the "same" rainbow. Explain.

30. A rainbow viewed from an airplane may form a complete circle. Where will the shadow of the airplane appear? Explain.

31. How is a rainbow similar to the halo sometimes seen around the moon on a night when ice crystals are in the upper atmosphere?

32. What is responsible for the rainbow-colored fringe commonly seen at the edges of a spot of white light from the beam of a lantern or slide projector?

33. What would be the effect of a pinhole camera (Figure 27-46*b* and Home Project 1) that has two pinholes instead of one? Multiple holes?

34. What condition must exist for a converging lens to produce a virtual image? Can a diverging lens ever produce a real image?

35. Can you take a photograph of your image in a plane mirror and focus the camera on both your image and the mirror frame? Explain.

36. In terms of focal length, how far behind the camera lens is the film located when very distant objects are being photographed?

37. Maps of the moon are actually upside down. Why is this so?

38. Why do older people who do not wear glasses read books farther away from their eyes than do younger people?

39. Rays of light parallel to the principal axis of a converging lens pass through the lens and come to a focus a certain distance from the lens—its focal point. When the lens is under water, will the focal point be longer, shorter, or the same distance as for air?

40. No glass is perfectly transparent. Because of reflections, about 92 percent of light energy is transmitted by an average sheet of clear, dust-free windowpane. The 8-percent loss is not noticeable through a single sheet, but through several sheets it is apparent. How much light is transmitted by two sheets, each of which transmits 92 percent?

28 Light Waves

Throw a rock into a quiet pool, and waves appear along the surface of the water. Strike a tuning fork, and waves of sound spread in all directions. Light a match, and waves of light similarly expand in all directions—at the enormously high speed of 300 000 kilometers per second. In this chapter we will study the wave nature of light. In the next chapter we will see that light has a particle nature as well, but for now we will investigate some of the wave properties of light: diffraction, interference, and polarization.

Huygens' Principle

Although Galileo is credited as being the first to design a pendulum device to operate a system of toothed wheels, it was a Dutchman, Christian Huygens, who made the first pendulum clock. Huygens is most remembered, however, for his idea about waves.* He proposed that light waves spreading out from a point source could be regarded as the overlapping of tiny secondary waves (Figure 28-2). In other words, wave fronts are made up of tinier wave fronts. This idea is called **Huygens' principle**.

Consider the spherical wave front in Figure 28-3. We can see that if all points along the wave front AA′ are sources of new wavelets, a short

Figure 28-1
Wave fronts of water waves.

Figure 28-2
These drawings are from Huygens' book *Treatise on Light*. Light from A (*left*) expands in wave fronts, every point of which (*right*) behaves as if it were a new source of waves. Secondary wavelets starting at *b,b,b,b* form a new wave front (*d,d,d,d*); secondary wavelets starting at *d,d,d,d* form still another new wave front (DCEF).

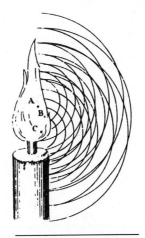

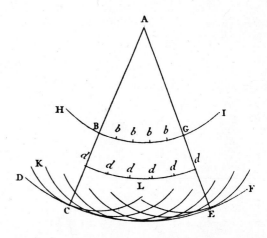

*In 1665, 20 years before Huygens made his hypothesis about wave fronts, the English physicist Robert Hooke first proposed a wave theory of light.

513

Figure 28-3
Huygens' principle applied to a spherical wave front.

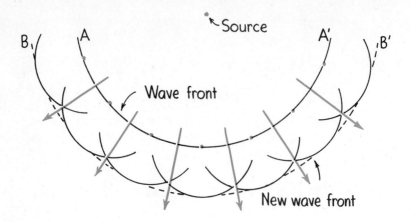

time later the new overlapping wavelets will form a new surface, BB′, which can be regarded as the envelope of all the wavelets. In the figure we show only a few of the infinite number of wavelets from a few secondary point sources that combine to produce the smooth envelope BB′.

Figure 28-4
Huygens' principle applied to a plane wave front.

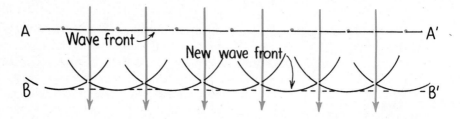

As the wave spreads, a segment appears less curved. Very far from the original source, the waves nearly form a plane—as do waves from the sun, for example. A Huygens wavelet construction for plane wave fronts is shown in Figure 28-4. We see the laws of reflection and refraction illustrated via Huygens' principle in Figure 28-5.

Figure 28-5
Huygens' principle applied to (a) reflection and (b) refraction.

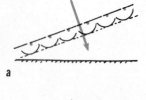

a

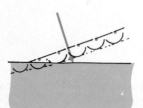

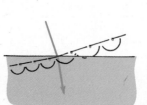

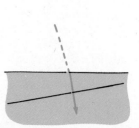

b

Figure 28-6

The oscillating meter stick makes plane waves in the tank of water. Water oscillating in the opening acts as a source of waves that fan out on the other side of the barrier.

Plane waves can be generated in water by successively dipping a horizontally held straightedge such as a meter stick into the surface (Figure 28-6). The photographs in Figure 28-7 are top views of a ripple tank, where plane waves are produced in water by an oscillating straightedge. The straightedge is not shown, but the plane waves produced are shown incident upon openings of various sizes. In *a*, where the opening is wide, we see the plane waves continue through the opening without change—except at the corners, where the waves are bent into the shadow region, as predicted by Huygens' principle. As the width of the opening is narrowed as in *b*, less and less of the incident wave is transmitted, and the spreading of waves into the shadow region becomes more pronounced. When the opening is small compared to the wavelength of the incident wave as in *c*, the truth of Huygens' idea that every part of a wave front can be regarded as a source of new wavelets becomes quite apparent. As the waves are incident upon the narrow opening, the water sloshing up and down in the opening is easily seen to act as a "point" source of the new waves that fan out on the other side of the barrier. We say the waves are *diffracted* as they spread into the shadow region.

Figure 28-7

Plane waves passing through openings of various sizes. The smaller the opening, the greater the bending of the waves at the edges.

a

b

c

Diffraction

In the previous chapter we saw that light can be bent from its ordinary straight-line path by reflection and refraction, and now we see another way light bends. Any bending of light by means other than reflection and refraction is called **diffraction**. The diffraction of plane water waves shown in Figure 28-7 occurs for all kinds of waves, including light waves.

When light passes through an opening that is large compared to the wavelength of light, it casts a shadow such as that shown in Figure 28-8a. We see a rather sharp boundary between the light and dark area

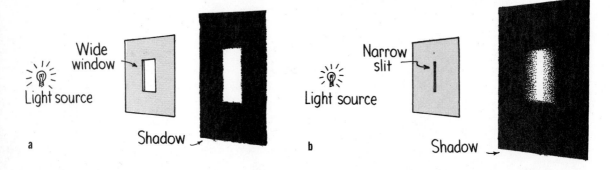

Figure 28-8

(a) Light casts a sharp shadow when the opening is large compared to the wavelength of the light. (b) When the opening is small, diffraction is apparent and the shadow is fuzzy.

of the shadow. But if we pass light through a thin razor slit in a piece of opaque cardboard, we see that the light diffracts (Figure 28-8b). The sharp boundary between the light and dark area disappears, and the light spreads out like a fan to produce a bright area that fades into darkness without sharp edges. The light is diffracted.

A graph of the intensity distribution for diffracted light through a single thin slit appears in Figure 28-9. Because of diffraction, there is a gradual change in light intensity rather than an abrupt change from dark to light. A photodetector sweeping across the screen would sense a gradual change from no light to maximum light. (Actually, there are slight fringes of intensity to either side of the main pattern; we will see shortly that these are evidence of interference that is more pronounced with a double slit or multiple slits.)

Diffraction is not confined by narrow slits or to openings in general but can be seen for all shadows. On close examination, even the sharpest shadow is blurred slightly at the edge (Figure 28-11).

The amount of diffraction depends on the wavelength of the wave compared to the size of the obstruction that casts the shadow. Long waves are better at filling in shadows, which is why foghorns emit low-frequency sound waves—to fill in any "blind spots." Likewise for radio waves of the standard AM broadcast band, which are very long compared to the size of most objects in their path. The wavelength of AM radio waves ranges from 186 to 560 meters, and the waves readily bend around buildings and other objects that might otherwise obstruct them. A long-wavelength radio wave doesn't "see" a relatively small building in its path—but a short-wavelength radio wave does. The radio waves of the FM band range from 2.8 to 3.4 meters and don't bend very well around

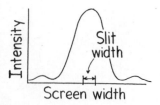

Figure 28-9

Graphic interpretation of diffracted light through a single thin slit.

Figure 28-10

(*a*) Waves tend to spread into the shadow region. (*b*) When the wavelength is about the size of the object, the shadow is soon filled in. (*c*) When the wavelength is short compared to the object, a sharp shadow is cast.

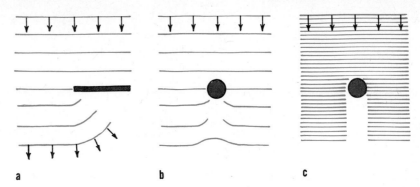

a   b   c

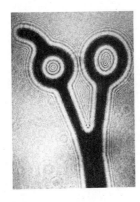

Figure 28-11

Diffraction fringes are evident in the shadows of laser light, which is of a single frequency. These fringes would be filled in by multitudes of other fringes if the source were white light.

buildings. This is one of the reasons that FM reception is often poor in localities where AM comes in loud and clear. In the case of radio reception, we don't wish to "see" objects in the path of radio waves, so diffraction is nice.

Diffraction is not so nice when we wish to see objects—like those under a microscope when the wavelengths of light are small compared to the objects we wish to see. Diffraction hinders our view. For objects appreciably larger than the wavelength of light, the effects of diffraction are relatively small. The details of small objects become less and less well defined as its size approaches the wavelength of light illuminating it. If the object is as small as or smaller than the illuminating wavelength, no structure, detail, or anything can be seen. No amount of magnification or perfection of microscope design can defeat this fundamental diffraction limit. To circumvent this problem, microscopists illuminate tiny objects with electron beams rather than light. They use electron microscopes, which take advantage of the fact that all matter has wave properties: a beam of electrons has a wavelength smaller than visible light. (We discussed this briefly back in Chapter 10, and will look at it in greater detail in Chapter 30.) So more detail can be seen with an electron microscope than with an optical microscope that is restricted to the longer waves of visible light. In an electron microscope, electric and magnetic fields, rather than optical lenses, are used to focus and magnify images.

Figure 28-12

(*a*) The shadow of a screw in laser light shows fringes produced by destructive interference of the diffracted light. (*b*) A longer exposure of the screw shows fringes within the shadow produced by constructive interference.

a   b

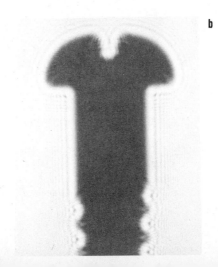

The fact that smaller details can be better seen with smaller wavelengths is neatly employed by the dolphin in scanning its environment with ultrasound. The echoes of long-wavelength sound give the dolphin an overall image of objects in its surroundings. To examine more detail, the dolphin emits sound of shorter wavelengths. Recall from Chapter 19 that the multiple reflections of very short waves provide the dolphin an acoustical image of the insides of the object scanned. The dolphin has always done naturally what the physician has only recently been able to do with an ultrasonic imaging device.

Question ▶ Why does a microscopist use blue light to illuminate the objects viewed?

Interference

Spectacular illustrations of diffraction are shown in Figure 28-12. Physicist Chuck Manka made these by placing pieces of photographic film in the shadow of a screw illuminated with spatially filtered laser light. The fringes appear in both figures. These fringes are produced by **interference**, which we first discussed in Chapter 18. Constructive and destructive interference is reviewed in Figure 28-13. We see that the adding, or superposition, of a pair of identical waves in phase with each other produces a wave of the same frequency but with twice the amplitude. If the waves are exactly one-half wavelength out of phase, their superposition results in complete cancellation. If they are out of phase by other amounts, partial cancellation occurs.

The interference of water waves is a common sight, as shown in Figure 28-14. In some places crests overlap crests, while in other places crests overlap troughs of other waves.

Under more carefully controlled conditions, interesting patterns are produced when two sources of waves are placed side by side (Figure 28-15).

Figure 28-13
Wave interference.

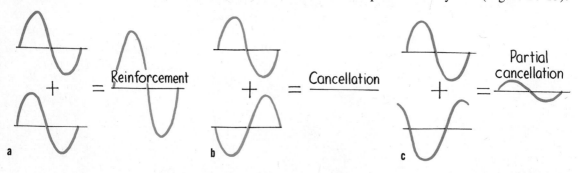

Answer
For the objects viewed, less diffraction results from the short waves of blue light compared to other, longer waves. So the microscopist sees more detail with short-wave blue light, just as a dolphin beautifully investigates fine detail in its environment by the echoes of ultra-short wavelengths of sound.

Figure 28-14
Interference of water waves.

Drops of water are allowed to fall at a controlled frequency into shallow tanks of water (ripple tanks) while their patterns are photographed from above. Note that areas of constructive and destructive interference are incident on the top edges of the ripple tanks and that the number of these regions and the distances between them depend on the distance between the wave sources and on the wavelength (or frequency) of the waves. Interference is not restricted to easily seen water waves, but is a property of all waves.

In 1801 the wave nature of light was convincingly demonstrated when the British physicist and physician Thomas Young performed his now-

Figure 28-15

Interference patterns of over-lapping waves from two vibrating sources.

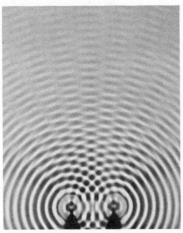
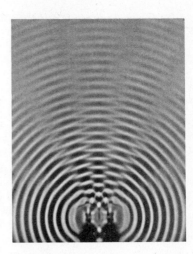

Figure 28-16

Thomas Young's original drawing of a two-source interference pattern. Letters C, D, E, and F mark regions of destructive interference.

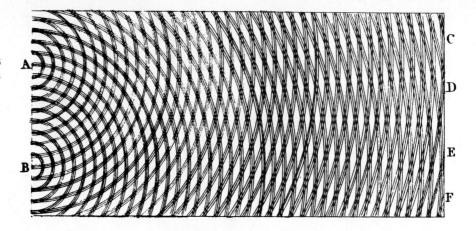

famous interference experiment.* Young found that light directed through two closely spaced pinholes recombined to produce fringes of brightness and darkness on a screen behind. The bright fringes of light resulted from light waves from the two holes arriving crest to crest, while the dark areas resulted from light waves arriving trough to crest. Figure 28-16 shows Young's drawing of the pattern of superimposed waves from the two sources. His experiment is now done with two closely spaced slits instead of pinholes, so the fringe patterns are straight lines (Figure 28-17).

Figure 28-17

When monochromatic light passes through two closely spaced slits, a striped interference pattern is produced.

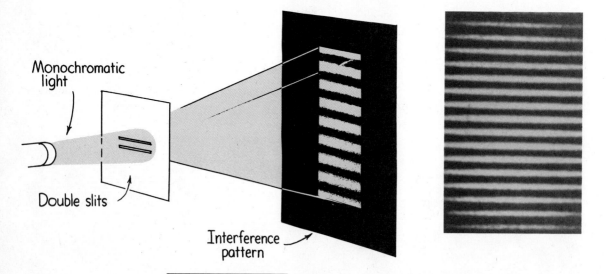

Monochromatic light

Double slits

Interference pattern

*Thomas Young read fluently at the age of 2; by 4, he had read the Bible twice; by 14, he knew eight languages. In his adult life he was a physician and scientist, contributing to an understanding of fluids, work and energy, and the elastic properties of materials. He was the first person to make progress in deciphering Egyptian hieroglyphics. No doubt about it—Thomas Young was a bright guy!

Bright area

Dark area

Figure 28-18
Bright fringes occur when waves from both slits arrive in phase; dark areas result from the overlapping of waves that are out of phase.

We see in Figure 28-19 how the series of bright and dark lines results from the different path lengths from the slits to the screen. For the central bright fringe, the paths from each slit are the same length, and the waves arrive in phase and reinforce each other. The dark fringes on either side of the central fringe result from one path being longer (or shorter) by one-half wavelength, where the waves arrive half a wavelength, or 180°, out of phase. The other sets of dark fringes occur where the paths differ by odd multiples of one-half wavelength: $\frac{3}{2}$, $\frac{5}{2}$, and so on.

Figure 28-19
Light from O passes through slits M and N and produces an interference pattern on the screen S.

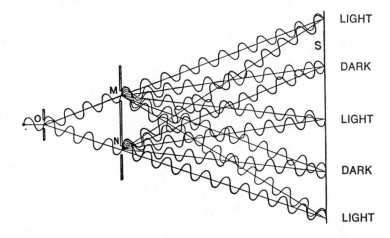

LIGHT

DARK

LIGHT

DARK

LIGHT

Questions ▶

1. If the double slits were illuminated with monochromatic red light, would the fringes be more widely or more closely spaced than if illuminated with monochromatic blue light?

2. Why is it important that monochromatic light is used?

▶ **Answers**

1. More widely spaced. Can you see in Figure 28-19 that a slightly longer—and therefore a slightly more displaced—path from the entrance slit to the screen would result for the longer waves of red light?

2. If light of various wavelengths were diffracted by the slits, dark fringes for one wavelength would be filled in with bright fringes for another, resulting in no distinct fringe pattern. If you haven't seen this, be sure to ask your instructor to demonstrate it.

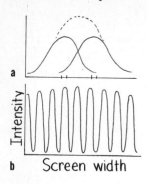

a

Intensity

b   Screen width

Figure 28-20
The light that diffracts through each of the double slits does not form a super-position of intensities as suggested in (a). The intensity pattern, because of interference, is as shown in (b).

Figure 28-21
A diffraction grating disperses light into colors by interference. It may be used in place of a prism in a spectrometer.

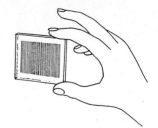

In performing this double-slit experiment, suppose we cover one of the slits so that light passes through only a single slit. Then light will fan out and illuminate the screen to form a simple diffraction pattern, as discussed earlier (Figure 28-9). If we cover the other slit and allow light to pass only through the slit just uncovered, we get the same illumination on the screen, only displaced somewhat because of the difference in slit location. If we didn't know better, we might expect that with both slits open, the pattern would simply be a superposition of the single-slit diffraction patterns, as suggested in Figure 28-20a. But this doesn't happen. Instead, the pattern formed is one of alternating light and dark bands, as shown in *b*. We have an interference pattern. Interference of light waves does not, by the way, create or destroy light energy; it merely redistributes it.

Interference patterns are not limited to single and double slits. A multitude of closely spaced slits makes up a *diffraction grating*. These devices, like prisms, disperse white light into colors. Whereas a prism separates the colors of light by refraction, a diffraction grating separates colors by interference. These are used in devices called *spectrometers*, which we will discuss in the next chapter, and more commonly in items such as costume jewelry and automobile bumper stickers. These materials are ruled with tiny grooves that diffract light into a brilliant spectrum of colors. This is also seen in the colors dispersed by the feathers of some birds and in the beautiful colors dispersed by the pits on the reflective surface of digital audio laser discs.

Single-Color Thin Film Interference

Another way interference fringes can be produced is by the reflection of light from the top and bottom surfaces of a thin film. A simple demonstration can be set up with a monochromatic light source and a couple of pieces of glass. A sodium-vapor lamp provides a good source of monochromatic light. The two pieces of glass are placed one atop the other, as shown in Figure 28-22. A very thin piece of paper is placed between the plates at one edge. This leaves a very thin wedge-shaped film of air between the plates. If the eye is in a position to see the reflected image of the lamp, the image will not be continuous but will be made up of dark and bright bands.

Sodium arc lamp

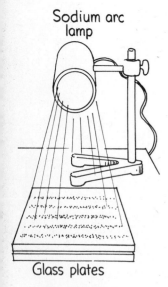

Glass plates

Figure 28-22
Interference fringes.

Figure 28-23
Reflection from the upper and lower surfaces of a "thin film of air."

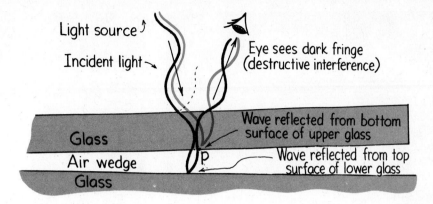

The cause of these bands is the interference between the waves reflected from the glass on the top and bottom surfaces of the air wedge, as shown in the exaggerated view in Figure 28-23. The light reflecting from point P comes to the eye by two different paths. In one of these paths the light is reflected from the top of the air wedge; in the other path it is reflected from the lower side. If the eye is focused on point P, both rays reach the same place on the retina of the eye. But these rays have traveled different distances and may meet in phase or out of phase, depending on the thickness of the air wedge—that is, on how much farther one ray has traveled than the other. When we look over the whole surface of the glass, we see alternate dark and bright regions—the dark portions where the air thickness is just right to produce destructive interference and the bright portions where the air wedge is just the proper amount thinner or thicker to result in the reinforcement of light. So the dark and bright bands are caused by the interference of light waves reflected from the two sides of the thin film.*

If the surfaces of the glass plates used are perfectly flat, the bands are uniform. But if the surfaces are not perfectly flat, the bands are distorted. The interference of light provides an extremely sensitive method for testing the flatness of surfaces. Surfaces that produce uniform fringes are said to be optically flat—that is, flat compared with the wavelength of visible light (Figure 28-24).

---

*Phase shifts at some reflecting surfaces also contribute to interference. For simplicity and brevity, our concern with this topic will be limited to this footnote: in short, when light in a medium is reflected at the surface of a second medium in which the speed of transmitted light is less (when there is a greater index of refraction), there is a 180° phase shift, but no phase shift occurs when the second medium is one that transmits light at a higher speed (and there is a lower index of refraction). In our air-wedge example, no phase shift occurs for reflection at the upper glass-air surface, and a 180° shift does occur at the lower air-glass surface. So at the apex of the air wedge where the thickness approaches zero, the phase shift produces cancellation, and the wedge is dark. Likewise with a soap film so thin that its thickness is appreciably smaller than the wavelength of light.

Figure 28-24
Optical flats used for testing the flatness of surfaces.

When a plano-convex lens is placed on an optically flat plate of glass and illuminated from above with monochromatic light, a series of light and dark rings is produced. This pattern is known as *Newton's rings* (Figure 28-25). These light and dark rings are the same kinds of fringes observed with plane surfaces. This is a useful testing technique in polishing precision lenses.

Figure 28-25
Newton's rings.

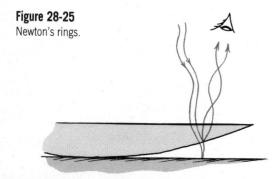

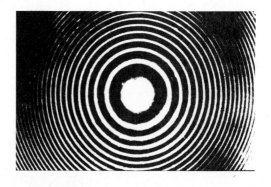

Question ▶ How would the spacings between Newton's rings differ when illuminated by red and by blue light?

▶ **Answer**
The rings would be more widely spaced for longer-wavelength red light than for the shorter waves of blue light. Do you see the geometrical reason for this?

Interference Colors by Reflection from Thin Films

We have all noticed the spectrum of colors reflected from a soap bubble or from gasoline on a wet street. We have seen the beautiful colors reflected from some bird feathers that seem to change in hue as the bird moves. These colors are produced by the *interference* of light waves. This phenomenon, often called *iridescence*, is observed in thin films.

A soap bubble appears iridescent in white light when the thickness of the soap film is about the same as the wavelength of light. Light waves reflected from the outer and inner surfaces of the film travel different distances. When illuminated by white light, the film may be just the right thickness at one place to cause the destructive interference of, say, yellow light. When yellow light is subtracted from white light, the mixture left will appear as the complementary color of yellow: blue. At another place where the film is thinner, a different color may be canceled by interference, and the light seen will be its complementary color. The same thing happens to gasoline on a wet street. Light reflects from both the upper gasoline surface and the lower gasoline-water surface. If the incident beam is monochromatic blue, as in Figure 28-26, the gasoline surface appears yellow. This is because the blue is subtracted from the white, leaving the complementary color: yellow. The different colors correspond to different thicknesses of the thin film, providing a vivid "contour map" of microscopic differences in surface "elevations."

Over a wider field of view, different colors can be seen even if the thickness of the gasoline film is uniform. This has to do with the apparent thickness of the film: light reaching the eye from different parts of the surface is reflected at different angles and traverses different thicknesses. If the light is incident at a grazing angle, for example, the ray transmitted to the gasoline's lower surface travels a longer distance. Longer waves are canceled in this case, and different colors appear.

Figure 28-26

The thin film of gasoline is just the right thickness that blue light reflected from the top surface of the gasoline is canceled by light of the same wavelength reflected from the water.

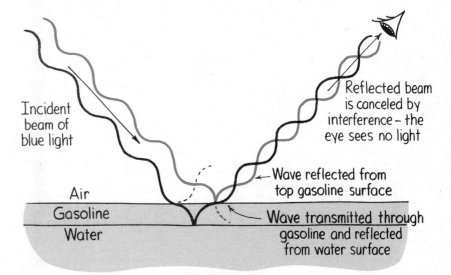

Dishes washed in soapy water and poorly rinsed have a thin film of soap on them. Hold such a dish up to a light source so that interference colors can be seen. Then turn the dish to a new position, keeping your eye on the same part of the dish, and the color will change. Light reflecting from the bottom surface of the transparent soap film is canceling light reflecting from the top surface of the soap film. Different wavelengths of light are canceled for different angles.

The wavelengths of light and other regions of the electromagnetic spectrum are measured with interference techniques. The principle of interference provides a means of measuring extremely small distances with great accuracy. Instruments called *interferometers*, which use the principle of interference, are the most accurate instruments known for measuring small distances.

Questions ▶

1. What color will appear to be reflected from a soap bubble in sunlight when its thickness is such that green light is canceled?

2. In the left column are the colors of certain objects. In the right column are various ways in which colors are produced. Match the right column to the left.

(a) yellow daffodil   (1) interference
(b) blue sky   (2) diffraction
(c) rainbow   (3) selective reflection
(d) peacock feathers   (4) refraction
(e) soap bubble   (5) scattering

Polarization

Interference and diffraction provide the best evidence that light is wavelike in nature. Wave motion in general can be either longitudinal or transverse. Sound in air travels in longitudinal waves, where the vibratory motion is *along* the direction of wave propagation. When we shake a taut rope, the vibratory motion that propagates along the rope is perpendicular, or *transverse*, to the rope. Both longitudinal and transverse waves exhibit interference and diffraction effects. Are light waves, then, longitudinal or transverse? The fact that light waves can be **polarized** demonstrates that they are transverse.

If we shake a taut rope up and down as in Figure 28-27, a transverse wave travels along the rope in a plane. We say that such a wave is *plane-polarized.** If the rope is shaken up and down vertically, the wave is

Figure 28-27
A vertically polarized plane wave and a horizontally polarized plane wave.

▶ **Answers**

1. The composite of all the visible wavelengths except green results in the complementary color, magenta. See Figure 26-9.
2. a—3; b—5; c—4; d—2; e—1.

*Light may also be circularly polarized and elliptically polarized, which are also transverse polarizations.

vertically plane-polarized; that is, the waves traveling along the rope are confined to a vertical plane. If we shake the rope from side to side, we produce a horizontally plane-polarized wave.

A single vibrating electron emits an electromagnetic wave that is also plane-polarized. The plane of polarization will be in the same vibrational direction of the electron. A vertically accelerating electron, then, emits light that is vertically polarized, while a horizontally accelerating electron emits light that is horizontally polarized (Figure 28-28).

Figure 28-28
(*a*) A vertically polarized plane wave from a charge vibrating vertically. (*b*) A horizontally polarized plane wave from a charge vibrating horizontally.

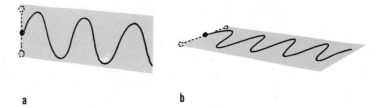

a   b

A common light source such as an incandescent lamp, a fluorescent lamp, a candle flame, or an arc lamp emits light that is nonpolarized. This is because there is no preferred vibrational direction for the accelerating charges producing the light. The number of planes of vibration might be as numerous as the accelerating charges producing them. A few planes are represented in Figure 28-29*a*. We can represent all these planes by radial lines (Figure 28-29*b*) or, more simply, by vectors in two mutually perpendicular directions (Figure 28-29*c*), as if we had resolved all the vectors of Figure 28-29*b* into horizontal and vertical components. This simpler schematic representation is a useful shorthand. In this case, the diagram represents nonpolarized light. Polarized light would be represented by a single vector.

Figure 28-29
Representations of planes of plane-polarized waves.

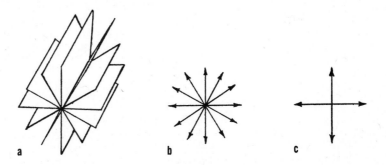

a   b   c

All transparent crystals of a noncubic natural shape have the property of transmitting light through two mutually perpendicular planes. Certain crystals* not only produce two internal beams polarized at right angles to each other but also strongly absorb one beam while transmitting the

*Called *dichroic*.

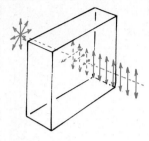

Figure 28-30
One component of the incident nonpolarized light is absorbed, resulting in emerging polarized light.

Figure 28-31
A rope analogy illustrates the effect of crossed polaroids.

other (Figure 28-30). Tourmaline is one such common crystal, but unfortunately the transmitted light is colored. Herapathite, however, does the job without discoloration. Microscopic crystals of herapathite are embedded between cellulose sheets in uniform alignment and are used in making Polaroid filters. Some Polaroid sheets consist of certain aligned molecules rather than tiny crystals.*

If you look at nonpolarized light through a Polaroid filter, you can rotate the filter in any direction, and the light will appear unchanged. But if this light is polarized, then as you rotate the filter, you can progressively cut off more and more of the light until it is blocked out. An ideal Polaroid will transmit 50 percent of incident nonpolarized light. That 50 percent is, of course, polarized. When two Polaroids are arranged so that their polarization axes are aligned, light will be transmitted through both (Figure 28-31). If their axes are at right angles to each other, no light will penetrate the pair. Actually, some of the shorter wavelengths do get through, but not to any significant degree. When Polaroids are used in pairs like this, the first one is called the *polarizer* and the second one the *analyzer*.

Much of the light reflected from nonmetallic surfaces is polarized. The glare from glass or water is a good example. Except for normal incidence, the reflected ray contains more vibrations parallel to the reflecting surface, while the transmitted beam contains more vibrations at right

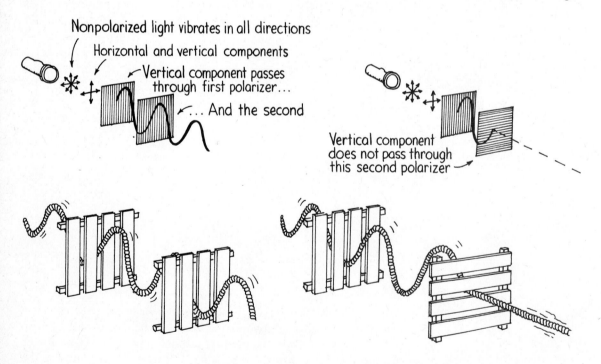

Nonpolarized light vibrates in all directions
Horizontal and vertical components
Vertical component passes through first polarizer...
...And the second

Vertical component does not pass through this second polarizer

---

*The molecules are polymeric iodine in a sheet of polyvinyl alcohol or polyvinylene.

a   b   c

Figure 28-32
Light is transmitted when the axes of the Polaroids are aligned (*a*), but absorbed when Dotty Jean rotates one so that the axes are at right angles to each other (*b*). When she inserts a third Polaroid at an angle between the crossed Polaroids, light is again transmitted (*c*). Why? (For the answer, after you have given this some thought see Appendix III, "More About Vectors.")

angles to the surface (Figure 28-34). Skipping flat rocks off the surface of a pond is analogous. When the rocks hit parallel to the surface, they easily reflect; but if they hit with their faces at right angles to the surface, they "refract" into the water. The glare from reflecting surfaces can be appreciably diminished with the use of Polaroid sunglasses. The polarization axes of the lenses are vertical, as most glare reflects from horizontal surfaces. Properly aligned Polaroid eyeglasses enable us to see the projection of stereoscopic movies or slides on a flat screen in three dimensions.

Figure 28-33
Polaroid sunglasses block out horizontally vibrating light. When the lenses overlap at right angles, no light gets through.

Figure 28-34
Most glare from nonmetallic surfaces is polarized. Here we see that the components of incident light that are parallel to the surface are reflected, while the components that are perpendicular to the surface are refracted into the medium. Since most of the glare we encounter is from horizontal surfaces, the polarization axes of Polaroid sunglasses are vertical.

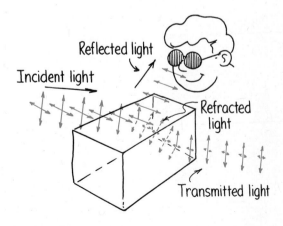

Question ▶ Which pair of glasses is best suited for automobile drivers? (The polarization axes are shown by the straight lines.)

a   b   c

Three-Dimensional Viewing

Figure 28-35
The crystal structure of ice in stereo. You'll see depth when your brain combines the views of your left eye looking at the left figure and your right eye looking at the right figure. To accomplish this, focus your eyes for distant viewing before looking at this page. Without changing your focus, look at the page, and each figure will appear double. Then adjust your focus so that the two inside images overlap to form a central composite image. Practice makes perfect. (If you instead *cross* your eyes to overlap the figures, near and far are reversed!)

Vision in three dimensions depends primarily on the fact that both eyes give their impressions simultaneously (or nearly so), each eye viewing the scene from a slightly different angle. To convince yourself that each eye sees a different view, hold an upright finger at arm's length and see how it appears to switch from left to right in front of the background as you alternately close each eye. The drawings of Figure 28-35 illustrate a stereo view of the crystal structure of ice.

The familiar hand-held stereoscopic viewer (Figure 28-38) simulates the effect of depth. In this device, there are two photographic transparencies (or slides) taken from slightly different positions. When they are viewed at the same time, the arrangement is such that the left eye sees

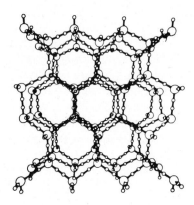
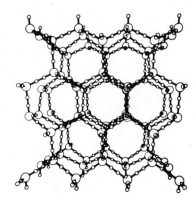

Figure 28-36
With your eyes focused for distant viewing, the second and fourth lines appear to be farther away; if you cross your eyes, the second and fourth lines appear closer.

The test of all knowledge is experiment.
Experiment is the *sole judge* of scientific "truth."
                    Richard P. Feynman

The test of all knowledge is experiment.
Experiment is the *sole judge* of scientific "truth."
                    Richard P. Feynman

▶ **Answer**
Glasses *a* are best suited because the vertical axis blocks horizontally polarized light, which composes much of the glare from horizontal surfaces. Glasses *c* are suited for viewing 3-D movies.

Figure 28-37
A stereo view of snowflakes. View these in the same way as Figure 28-35.

Figure 28-38
A stereoscopic viewer.

Figure 28-39
A 3-D slide show using Polaroids. The left eye sees only polarized light from the left projector, the right eye sees only polarized light from the right projector, and both views merge in the brain to produce depth.

the scene as photographed from the left, and the right eye sees it as photographed from the right. As a result, the objects in the scene sink into relief in correct perspective, giving apparent depth to the picture. The device is constructed so that each eye sees only the proper view. There is no chance for one eye to see both views. If you remove the slides from the hand viewer and project each view on a screen by slide projector (so that the views are superimposed), a blurry picture results. This is because each eye sees both views simultaneously. This is where Polaroid filters come in. If you place the Polaroids in front of the projectors so that one is horizontal and the other vertical, and you view the polarized image with polarized glasses of the same orientation, each eye will see the proper view as with the stereoscopic viewer (Figure 28-39). You then will see an image in three dimensions.

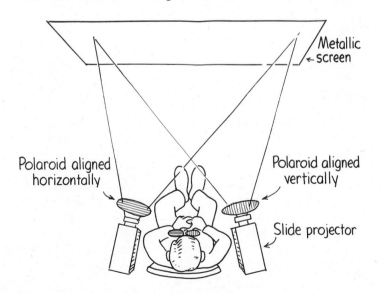

Metallic
← screen

Polaroid aligned
horizontally

Polaroid aligned
vertically

Slide projector

Colors by Transmission Through Polarizing Materials

Colors as vivid as those produced by interference from reflecting surfaces can also be produced under somewhat comparable conditions by transmission through certain transparent crystals. It so happens that all transparent crystals of a noncubic natural shape have the property of transmitting light in two mutually perpendicular planes and, most importantly, at two slightly different speeds. When light emerges from these crystals, the vibrations are confined to these planes. They are polarized.

When materials that are themselves composed of noncubic transparent crystals, like sugar, are placed between crossed Polaroids and illuminated with white light, colored light emerges. Various thicknesses of mica sheets between crossed Polaroids produce a wide assortment of colors. More spectacular is cellophane, which is composed of chain molecules that behave optically like microscopic noncubic crystals. When the cellophane is crumpled, a great variety of vivid colors is produced.

The explanation for this is not simple and requires a fairly good understanding of vectors. So unless you're into vectors, you may want to skip or skim the following paragraph and Figures 28-40 and 28-41.

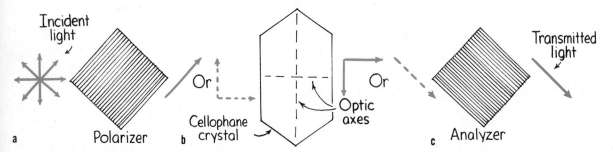

Figure 28-40

A single light wave traveling through cellophane between crossed Polaroids. The cellophane should be sandwiched between the Polaroids but is spread out for convenience. Follow the transformations of the light vector from incidence (a) to transmission (c). Figure 28-41 shows how the vertical component changes direction in the cellophane.

Consider a single wavelength of light that travels through cellophane sandwiched between two crossed Polaroids. We show this in a spread-out fashion in Figure 28-40. Illuminating light is incident upon the first Polaroid, the polarizer, which directs polarized light into the cellophane before finally emerging from the second Polaroid, the analyzer. We assume that the optic axes of the cellophane are at 45° with respect to either Polaroid axis. Incoming light passes through the polarizer and is polarized at 45° with respect to the horizontal and vertical axes of the cellophane. The 45° vector has a horizontal and vertical component, as shown by the dashed vectors, which will be transmitted along the axes of the cellophane. These components will travel through the cellophane at different speeds. As a result, one wave component gets ahead of the other. In the figure, we suppose that the thickness of the cellophane is such that the vertical component gains one-half wavelength over the horizontal component. The emerging vertical component in this case points down instead of up. The resultant vector is effectively rotated through 90° from the vector passed by the polarizer. It is then in a direction corresponding to the axis of the analyzer and passes through at maximum transmission.

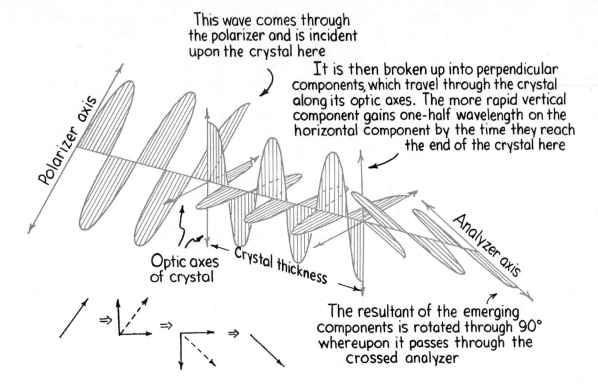

This wave comes through the polarizer and is incident upon the crystal here

It is then broken up into perpendicular components, which travel through the crystal along its optic axes. The more rapid vertical component gains one-half wavelength on the horizontal component by the time they reach the end of the crystal here

Polarizer axis

Analyzer axis

Optic axes of crystal

Crystal thickness

The resultant of the emerging components is rotated through 90° whereupon it passes through the crossed analyzer

Figure 28-41
A wave diagram of polarized light undergoing rotation through a crystal.

Only for this particular wavelength and this particular cellophane thickness will a 90° rotation of the incident vector occur. A wave diagram illustrating this case is shown in Figure 28-41. If the analyzer is rotated to a different position, some other rotated wavelength will correspond to the alignment with the axis of the analyzer. As the analyzer is turned, then, different colors are transmitted by the system.

The variety of colors that emerge from crumpled cellophane sandwiched between crossed Polaroids is spectacular—especially when projected on a large screen and one of the Polaroids is rotated in rhythm with your favorite music!

Question ▶ If all the crystals in a sample were perfectly aligned and the plane of incident polarized light were coincident with one of the transmission axes of the crystals, would colored light emerge through the second Polaroid?

▶ **Answer**
No, because the incident beam would not split into two components and travel at different speeds along both crystal axes to emerge in a shifted phase and rotated for color separation.

Holography

Perhaps the most exciting illustration of interference is the **hologram**, a two-dimensional photographic plate illuminated with laser light that allows you to see a faithful reproduction of a scene in three dimensions. The hologram was invented and named by Dennis Gabor in 1947, 10 years before lasers were invented. *Holo* in Greek means "whole," and *gram* in Greek means "message" or "information." A hologram contains the whole message or entire picture. When illuminated with laser light, the image is so realistic that you can actually look around the corners of objects in the image and see the sides.

In ordinary photography, a lens is used to form an image of an object on photographic film. Light reflected from each point on the object is directed by the lens only to a corresponding point on the film. And all the light that reaches the film comes from the object being photographed. In the case of holography, however, no image-forming lens is used. Instead, each point of the object being "photographed" reflects light to the *entire* photographic plate, so every part of the plate is exposed with light reflected from every part of the object. Most importantly, the light used to make a hologram must be of a single frequency and all parts exactly in phase: it must be *coherent*. If, for example, white light were used, the diffraction fringes for one frequency would be washed out by other frequencies. Only a laser can produce such light (we will treat lasers in detail in the next chapter). Holograms are made with laser light.

A conventional photograph is a recording of an image, but a hologram is a recording of the interference pattern resulting from the combination of two sets of wave fronts. One set of wave fronts is from light reflected by the object, and the other set is from a *reference beam*, which is diverted from the illuminating beam and sent directly to the photographic plate (Figure 28-42). The developed photograph has no recognizable image. The hologram is simply a hodgepodge of whirly lines, tiny fringed areas of varying density—dark where wave fronts from the object and reference beam arrived in phase and light where the wave fronts were out of phase. The hologram is a photographic pattern of microscopic interference fringes.

When a hologram is placed in a beam of coherent light, the light is diffracted through the microscopic fringes to produce wave fronts identical in form to the original wave fronts reflected by the object. When

Figure 28-42
A simplified arrangement for making a hologram. The laser light that exposes the photographic plate consists of two parts: the reference beam reflected from the mirror and light reflected from the object. The wave fronts of these two parts interfere to produce microscopic fringes on the photographic plate. The exposed and developed plate is then a hologram.

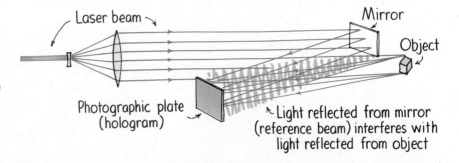

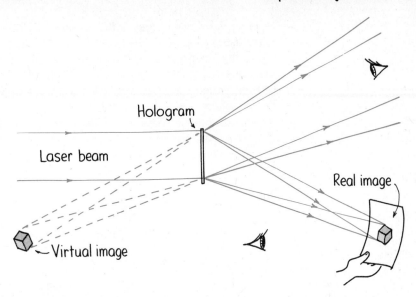

Figure 28-43
When laser light is transmitted through the hologram, the divergence of diffracted light produces a three-dimensional image that can be seen when looking *through* the hologram, like looking through a window. This is a *virtual image*, for it appears to be only in back of the hologram, like your virtual image in a mirror. By refocusing your eyes, you can see all parts of the virtual image, near and far, in sharp focus. Converging diffracted light produces a *real image* in front of the hologram that can be projected on a screen. Since the image is three-dimensional, you cannot see the entire image in sharp focus for any single position on a flat screen.

viewed with the eye or any other optical instrument, the diffracted wave fronts produce the same effect as the original reflected wave fronts. You look through the hologram and see a full, realistic three-dimensional image as though you were viewing the original object through a window. Parallax is evident when you move your head and see down the sides of the object or when you lower your head and look underneath the object. Holographic pictures are extremely realistic.

Interestingly enough, if the hologram is made on film, you can cut it in half and still see the entire image. And you can cut one of the pieces in half again, and again. This is because every part of the hologram has received and recorded light from the entire object. Similarly, light outside an open window fills the entire window, so you can see an outside view from every part of the open window. Because vast amounts of information are recorded in a tiny area, the film used for holograms must have a resolving power much greater than ordinary fine-grain photographic film. The optical storage of information via holograms is finding wide application in computers.

Even more interesting is holographic magnification. If holograms are made using a short-wavelength light and viewed with a longer-wavelength light, the resulting image is magnified in the same proportion as the wavelengths. Holograms made with X rays would be magnified thousands of times when viewed with visible light and appropriate geometric viewing arrangements. Because holograms require no lenses, the possibilities of an X-ray microscope are particularly attractive.

And as for television, today's two-dimensional screens may appear as quaint to your children as your grandparents' early radio receivers appear to you now.

Light is fascinating—especially when it is diffracted through the interference fringes of a hologram.

Summary of Terms

Huygens' principle The theory by which light waves spreading out from a point source can be regarded as the superposition of tiny secondary wavelets.

Diffraction The bending of light around an obstacle or through a narrow slit in such a way that fringes of light and dark or colored bands are produced.

Interference The superposition of waves producing regions of reinforcement and regions of cancellation. Constructive interference refers to regions of reinforcement; destructive interference refers to regions of cancellation. The interference of selected wavelengths of light produces colors known as *interference colors*.

Polarization The alignment of the electric vectors that make up electromagnetic radiation. Such waves of aligned vibrations are said to be *polarized*.

Hologram A two-dimensional microscopic diffraction pattern that shows three-dimensional optical images.

Suggested Reading

Falk, D. S., Brill, D. R., and Stork, D. *Seeing the Light: Optics in Nature.* New York: Harper & Row, 1985.

Spears, J. S., and Zollman, D. *The Fascination of Physics.* Menlo Park, Ca.: Benjamin/Cummings, 1985. A fascinating account of physics by a fascinating husband-and-wife team.

Review Questions

Huygens' Principle

1. According to Huygens, how does every point on a wave behave?

2. Will plane waves incident upon a small opening in a barrier fan out on the other side or continue as plane waves?

Diffraction

3. Is diffraction more pronounced through a small opening or through a large opening?

4. What is the relationship between the wavelength of a wave and the size of the opening through which diffraction occurs?

5. Which are more easily diffracted around buildings, AM or FM radio waves? Why?

6. What are some of the assets and liabilities of diffraction?

Interference

7. Is interference restricted to only some types of waves or does it occur for all types of waves?

8. What exactly did Thomas Young demonstrate in his famous experiment with light?

9. Does the energy of light disappear when light destructively interferes?

Single-Color Thin Film Interference

10. What is monochromatic light?

11. What accounts for the light and dark bands when monochromatic light reflects from a glass pane atop another glass pane?

12. What is the cause of Newton's rings?

13. Newton's rings show up nicely when viewed with monochromatic light—long wavelengths show wide-spaced rings and short wavelengths show close-spaced rings. Why are these rings not seen when viewed with white light?

Interference Colors by Reflection from Thin Films

14. What produces iridescence?

15. What causes the spectrum of colors seen in gasoline splotches on a wet street? Why are these not seen on a dry street?

16. What accounts for the different colors in either a soap bubble or a layer of gasoline on water?

17. If you look at a soap bubble from different angles so you're viewing different apparent thicknesses of soap film, will you see different colors? Explain.

Polarization

18. What phenomenon distinguishes between longitudinal and transverse waves?

19. Is polarization characteristic of light or all types of waves?

20. How does the plane of polarization of light compare to the plane of vibration of the electron that produced it?

21. Why will light pass through a pair of Polaroids when the axes are aligned but not when the axes are at right angles to each other?

22. How much ordinary light will an ideal Polaroid transmit?

23. When *ordinary* light is incident at a grazing angle upon water, what can you say about the *reflected* light?

24. What is the advantage of Polaroid sunglasses over regular sunglasses?

Three-Dimensional Viewing

25. Why would depth not be seen if you viewed duplicates of ordinary slides in a stereo viewer (Figure 28-38), rather than the pairs of slides taken with a stereo camera?

26. What role do polarization filters play in a 3-D slide show?

Colors by Transmission Through Polarizing Materials

27. What optical property do noncubic transparent crystals have?

28. The components of a light wave, which are in phase when they enter a noncubic crystal, are likely to be out of phase when they leave. What accounts for this?

29. When components of light are out of phase in exiting such a crystal, what happens to the direction of the resulting light vector?

30. If the direction of the exiting vector coincides with the polarization axis of the analyzer, does it get through? What happens if the vector is at right angles to the polarization axis?

Holography

31. How does a hologram differ from a conventional photograph?

32. How is *coherent* light different from ordinary light?

33. When coherent light is diffracted onto photographic film, interference fringes are produced. What would happen to these fringes if light of other frequencies were mixed in with the coherent light?

34. What would happen to the diffraction pattern of coherent light if light of the same frequency but of different phases were mixed in with the coherent light?

35. Why is coherent light necessary to make a hologram?

Home Projects

1. With a razor blade, cut a slit in a card and look at a light source through it. You can vary the size of the opening by bending the card slightly. See the interference fringes? Try it with two closely spaced slits.

2. Next time you're in the bathtub, froth up the soapsuds and notice the colors of highlights from the illuminating light overhead on each tiny bubble. Notice that different bubbles reflect different colors, due to the different thicknesses of soap film. If a friend is bathing with you, compare the different colors that you each see reflected from the same bubbles. You'll see that they're different—for what you see depends on your point of view!

3. Do this one at your kitchen sink. Dip a dark-colored coffee cup (dark colors make the best background for viewing interference colors) in dishwashing detergent, and then hold it sideways and look at the reflected light from the soap film that covers its mouth. Swirling colors appear as the soap runs down to form a wedge that grows thicker at the bottom with time. The top becomes thinner, so thin that it appears black. This tells us that its thickness is less than one-fourth the thickness of the shortest waves of visible light. Whatever its wavelength, light reflecting from the inner surface reverses phase, rejoins light reflecting from the outer surface, and cancels. The film soon becomes so thin it pops.

4. When you're wearing Polaroid sunglasses, view the glare from nonmetallic surfaces such as a road or body of water, tip your head from side to side, and see how the glare intensity changes as you vary the number of electric vector components aligned with the polarization axis of the glasses. Also notice the polarization of different parts of the sky when you hold the sunglasses in your hand and rotate them.

5. Place a bottle of corn syrup between two sheets of Polaroid. Place a white light source behind the syrup. Then look through the Polaroids and syrup and view spectacular colors as you rotate one of the Polaroids.

6. See spectacular interference colors with a polarized-light microscope. Any microscope, including an inexpensive toy microscope, can be converted into a polarized-light microscope by fitting a piece of Polaroid inside the eyepiece and taping another onto the stage of the microscope. Mix drops of naphthalene and benzene on a slide and watch the growth of crystals. Rotate the eyepiece and change the colors. Or simply observe the interference colors of pieces of cellophane.

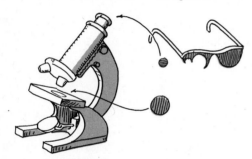

7. Make some slides for a slide projector by sticking some crumpled cellophane onto pieces of slide-sized Polaroid. (Also try strips of cellophane tape overlapped at different angles and experiment with different brands of transparent tape.) Project them onto a large screen or white wall and rotate a second, slightly larger piece of Polaroid in front of the projector lens in rhythm with your favorite music. You'll have your own light show.

Exercises

1. Diffraction for sound waves is much more evident in our everyday environment than for light waves. Why is this so?

2. How are interference fringes of light analogous to the sound of beats?

3. Why do *radio waves* diffract around buildings, while *light waves* do not?

4. Can you think of a reason why TV channels of lower numbers might give better pictures in regions of poor TV reception?

5. Two loudspeakers a meter or so apart emit pure tones of the same frequency and loudness. When a listener walks past in a path parallel to the line that joins the loudspeakers, the sound is heard to alternate from loud to soft. What is going on?

6. In the preceding exercise, suggest a path along which the listener could walk so as not to hear alternate loud and soft sounds.

7. Light illuminates two closely spaced thin slits and produces an interference pattern on a screen behind. How will the distance between the fringes of the pattern differ for red light compared to blue light?

8. What happens to the distance between interference fringes if the two slits are moved farther apart?

9. Why is Young's experiment more effective with slits than with the pinholes he first used?

10. For complete cancellation of light reflected out of phase from the two surfaces of a thin film, what else besides frequency and wavelength must be the same for both parts of the recombined wave?

11. A pattern of fringes is produced when monochromatic light passes through a pair of thin slits. Would such a pattern be produced by three parallel thin slits? By thousands of such slits? Give an example to support your answer.

12. The colors of peacocks and hummingbirds are the result not of pigments, but of ridges in the surface layers of their feathers. By what physical principle do these ridges produce colors?

13. The colored wings of many butterflies are due to pigmentation, but in others, such as the Morpho butterfly, the colors do not result from any pigmentation. When the wing is viewed from different angles, the colors change. How are these colors produced?

14. Why do the iridescent colors seen in some seashells (such as abalone shells) change as the shells are viewed from different positions?

15. When dishes are not properly rinsed after washing, different colors are reflected from their surfaces. Explain.

16. Why are interference colors more apparent for thin films than for thick films?

17. Will the light from two very close stars produce an interference pattern? Explain.

18. We explain the vivid colors seen when double reflection occurs by interference. Does interference also underlie the explanation for the colors produced by transmission with the aid of Polaroids? Explain.

19. Why do Polaroid sunglasses reduce glare, whereas nonpolarized sunglasses simply cut down on the total amount of light reaching the eyes?

20. How can you determine the polarization axis for a single sheet of Polaroid?

21. Most of the glare from nonmetallic surfaces is polarized, the axis of polarization being parallel to that of the reflecting surface. Would you expect the polarization axis of Polaroid sunglasses to be horizontal or vertical? Why?

22. How can a single sheet of Polaroid film be used to show that the sky is partially polarized? (Interestingly enough, unlike humans, bees and many insects can discern polarized light and use this ability for navigation.)

23. Why will an ideal Polaroid filter transmit 50 percent of incident nonpolarized light?

24. What percentage of light would be transmitted by two ideal Polaroids sandwiched with their polarization axes aligned? With their axes at right angles to each other?

25. Why does a painting look less flat when viewed with one eye? (If you're not on to this, look at paintings with one eye and see the difference.)

26. Why did practical holography have to await the advent of the laser?

27. How is magnification accomplished with holograms?

28. If you are viewing a hologram and you close one eye, will you still perceive depth? Explain.

I'm amazed how zillions of electrons boiling off the electrodes in this lamp jiggle back and forth in rhythm to the alternating voltage and crash into and excite zillions of atoms in the low-pressure mercury vapor. And as the mercury atoms de-excite, they emit ultraviolet photons that shower upon the thin, powdery coating of phosphors on the tube's inner surface. This rain of energy excites the phosphors, which in turn radiate a mishmash of lower frequencies that combine to produce white light—all of which enables me to read my physics book and discover more that is happening all around me!

29 Light Emission

Figure 29-1
Electrons occupy discrete
shells about the nucleus of
an atom.

If energy is pumped into a metallic antenna in such a way as to cause
free electrons to vibrate to and fro a few thousand times per second, a
radio wave is emitted. If free electrons could be made to vibrate to and
fro on the order of a million billion times per second, a visible light wave
would be emitted. But light is not produced from metallic antennae, nor
is it exclusively produced by atomic antennae via subtle oscillations of
electron shells, as discussed in previous chapters. We now distinguish
between light reflected, refracted, scattered, and diffracted by objects and
light emitted by objects. In this chapter we discuss the physics of light
sources—of light *emission*.

The details of light emission from atoms involve the transitions of
electrons from higher to lower energy states within the atom. This emission process can be understood in terms of the familiar planetary model
of the atom that we discussed in Chapter 10. Just as each element is
characterized by the number of electrons that occupy the shells surrounding the atomic nucleus, so also each element possesses its own characteristic pattern of electron shells, or energy states. These states are found
only at certain radii and energies. We say that these noncontinuous energy
states are *discrete*. We call these discrete states *quantum states*, and we'll
return to them in detail in the next two chapters. For now we'll concern
ourselves only with their role in light emission.

Excitation

An electron farther from the nucleus has a greater electric potential energy
with respect to the nucleus than an electron nearer the nucleus. We say
that the farthermost electron is at a higher energy state, or, equivalently,
a higher energy level. In a sense, this is similar to the energy of a spring
door or a pile driver. The wider the door is pulled open, the greater its

Figure 29-2
The different orbits in an
atom are like steps. When an
electron is raised to a higher
orbit, the atom is excited.
When the electron falls to a
lower step, it releases energy
in the form of light.

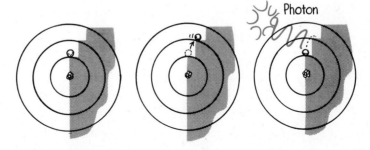

Photon

spring potential energy; the higher the ram of a pile driver is raised, the greater its gravitational potential energy.

When an electron is in any way raised to a higher energy level, the atom is said to be *excited*. The electron's higher level is only momentary, for like the pushed-open spring door, it soon returns to its stable state. The atom loses its temporarily acquired energy when the electron returns to a lower level and releases this energy as radiant energy. The atom has undergone the process of **excitation** and *de-excitation*.

As each element has its own number of electrons, each element also has its own characteristic set of energy levels. Electrons dropping from higher to lower energy levels in an excited atom emit with each jump a throbbing pulse of electromagnetic radiation called a *photon*, the frequency of which is related to the energy transition of the jump. We think of this photon as a localized corpuscle of pure energy—a "particle" of light—which is ejected from the atom. The frequency of the photon is directly proportional to its energy. In shorthand notation,

$$E \sim f$$

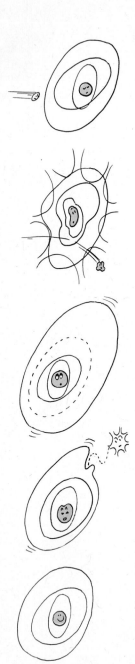

A photon in a beam of red light, for example, carries an amount of energy that corresponds to its frequency. A photon of twice the frequency has twice as much energy and is found in the ultraviolet part of the spectrum. If many atoms in a material are excited, many frequencies are emitted that correspond to the many different levels excited. These frequencies combine to give a characteristic color of light from each excited element.

The light emitted in the glass tubes in an advertising sign is a familiar example of excitation. The different colors in the sign correspond to the excitation of different gases, although it is common to refer to any of these as "neon." Only the red light is that of neon. At the ends of the glass tube that contains the neon gas are electrodes. Electrons are boiled off these electrodes and are jostled back and forth at high speeds by a high ac voltage. Millions of high-speed electrons vibrate back and forth inside the glass tube and smash into millions of target atoms; each smash boosts orbital electrons into higher energy levels by an amount of energy equal to the decrease in kinetic energy of the bombarding electron, and this energy is radiated as the characteristic red light of neon when the electrons fall back to their stable orbits. The process occurs and recurs many times, as neon atoms are continually undergoing a cycle of excitation and de-excitation. The overall result of this process is the transformation of electrical energy into radiant energy.

The colors of various flames are due to excitation. Different atoms in the flame emit colors characteristic of their energy-level spacings. Common table salt placed in a flame, for example, produces the characteristic yellow of sodium. Every element, excited in a flame or otherwise, emits its own characteristic color—a composite of the many colors that make up its characteristic spectrum.

Figure 29-3
Excitation and de-excitation.

Street lamps provide another example. City streets are no longer illuminated by incandescent lamps but are now illuminated with the light emitted by gases such as mercury vapor. Not only is the light brighter, it is less expensive. Whereas most of the energy in an incandescent lamp is converted to heat, most of the energy put into a mercury-vapor lamp is converted to light. The light from these lamps is rich in blues and violets and therefore is a different "white" from the light from an incandescent lamp. See if your instructor has a spare prism or diffraction grating you can borrow. Look through the prism or grating at the light from the street lamp and see the discreteness of the colors, which indicates the discreteness of the atomic levels. Also note that the colors from different mercury-vapor lamps are identical, showing that the atoms of mercury are identical.

The excitation/de-excitation process can be accurately described only by quantum mechanics. An attempt to view the process in terms of classical physics runs into contradictions. Classically, an accelerating electric charge emits electromagnetic radiation. Interestingly enough, an electron does accelerate in a transition from a higher to a lower energy level, for just as the innermost planets of the solar system have greater orbital speeds than those in the outermost orbits, the electrons in the innermost orbits of the atom have greater speeds. An electron gains speed in dropping to lower energy levels. Fine—the accelerating electron radiates a photon! But not so fine—the electron is continually undergoing acceleration (centripetal acceleration) in any orbit, whether or not it changes energy levels. According to classical physics, it should continually radiate energy. But it doesn't. All attempts to explain the emission of light by an excited atom in terms of a classical model have been unsuccessful. We have no simple model for conceptualizing this process. We shall simply say that light is emitted when electrons in an atom make a transition from a higher to a lower energy level and that the energy and frequency of the emitted photon are described by the relationship $E \sim f$.

Question ▶ Suppose a friend suggests that for a first-rate operation, the gaseous neon atoms in a neon tube should be periodically replaced with fresh atoms because the energy of the atoms tends to be used up with continued excitation, producing dimmer and dimmer light. What do you say to this?

▶ **Answer**

The neon atoms don't give out any energy that is not imparted to them by the electric current in the tube and therefore don't get "used up." Any single atom may be excited and re-excited without limit. If the light is, in fact, becoming dimmer and dimmer, it is probably because a leak exists. Otherwise there is no advantage whatsoever in changing the gas in the tube, for a "fresh" atom is indistinguishable from a "used" one. Both are ageless and older than the solar system.

Emission Spectra

Every element has its own characteristic pattern of electron energy levels and therefore emits its own characteristic pattern of light frequencies, its **emission spectrum**, when excited. This pattern can be seen when light is passed through a prism or, better, when it is first passed through a thin slit and then focused through a prism onto a viewing screen behind. Such an arrangement of slit, focusing lenses, and prism (or diffraction grating) is called a **spectroscope**, a useful instrument of modern science.

Figure 29-4
A simple spectroscope.

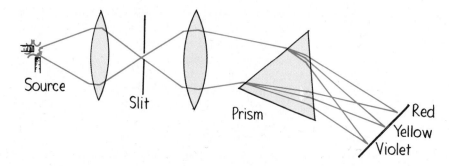

Each component color is focused at a definite position according to its frequency and forms an image of the slit on the screen, photographic film, or appropriate detector. The different-colored images of the slit are called *spectral lines*. A black-and-white picture of some typical spectral patterns labeled by wavelengths is shown in Figure 29-5, and more in their natural colors are shown on the inside of the back cover. It is customary to refer to colors in terms of their wavelengths rather than their frequencies. A given frequency corresponds to a definite wavelength.*

If the light given off by a sodium-vapor lamp is analyzed in a spectroscope, a single yellow line is produced—a single image of the slit. If we narrow the width of the slit, we find that this line is really composed of two very close lines. These lines correspond to the two predominant frequencies of light emitted by excited sodium atoms. The rest of the spectrum is dark. (Actually, there are many other lines, often too dim to be seen with the naked eye.)

The same happens with all glowing vapors. The light from a mercury-vapor lamp shows a pair of bright yellow lines close together (but in different positions from those of sodium), a very intense green line, and several blue and violet lines. A neon tube produces a more complicated pattern of lines. We find that the light emitted by each element in the vapor state produces its own characteristic pattern of lines. These lines correspond to the electron transitions between atomic energy levels and are as characteristic of each element as are fingerprints of people. The spectroscope therefore is widely used in chemical analysis.

---

*Recall from Chapter 18 that $v = f\lambda$, where v is the wave speed, f is the wave frequency, and λ (lambda) is the wavelength. For light, v is the constant c, so we see from $c = f\lambda$ the relationship between frequency and wavelength; namely, $f = c/\lambda$ and $\lambda = c/f$.

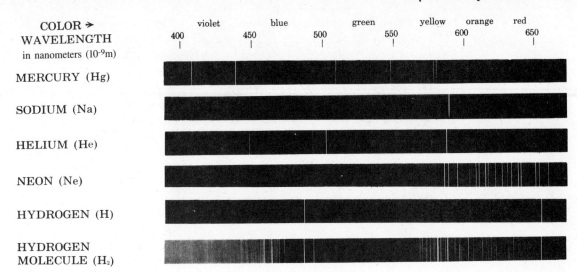

COLOR →
WAVELENGTH
in nanometers (10⁻⁹m)

MERCURY (Hg)

SODIUM (Na)

HELIUM (He)

NEON (Ne)

HYDROGEN (H)

HYDROGEN
MOLECULE (H₂)

Figure 29-5
Some typical spectral patterns.

The next time you see evidence of atomic excitation, perhaps the green flame produced when a piece of copper is placed in a fire, squint your eyes and see if you can imagine electrons jumping from one energy level to another in a pattern characteristic of the atom being excited—a pattern that gives off a color unique to that atom. That's what's happening!

| **Question** ▶ | Spectral patterns are not shapeless smears of light but, instead, consist of fine and distinct straight lines. Why is this so? |

Incandescence

Light that is produced as a result of high temperature has the property of **incandescence** (from a Latin word meaning "to grow hot"). Most incandescent light we are familiar with is white, like that from a common incandescent lamp. This seems to be quite different from the bright red light emitted from a cool neon gas tube. We know that a composite of all visible frequencies appears white, so does this mean that an infinite number of energy levels characterizes the tungsten atoms making up the filament of the incandescent lamp? The answer is no; if the filament were vaporized and then excited, the tungsten gas would emit a finite number of frequencies and produce an overall bluish color. Light emitted by atoms

▶ **Answer**

The spectral lines are simply images of the slit, which is itself a thin, straight opening through which light is admitted before being diffracted by the prism (or diffraction grating). When the slit is adjusted for its most narrow opening, closely spaced lines can be resolved (distinguished from one another). A wider slit admits more light, which permits easier detection of dimmer radiant energy but at the expense of resolution when closely spaced lines blur together.

Figure 29-6
The sound of an isolated bell rings with a clear and distinct frequency, whereas the sound emanating from a box of bells crowded together is discordant. Likewise with the difference between the light emitted from atoms in the gaseous state and that from atoms in the solid state.

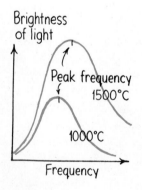

Figure 29-7
Radiation curves for an incandescent solid.

far from one another in the gaseous state is quite different from the light emitted by the same atoms closely packed in the solid state. This is analogous to the differences in sound from an isolated ringing bell and from a box crammed with ringing bells (Figure 29-6). In a gas the atoms are far apart. Electrons undergo transitions between energy levels within the atom quite unaffected by the presence of neighboring atoms. But when the atoms are closely packed, as in a solid, electrons of the outer orbits make transitions not only within the energy levels of their "parent" atoms but also between the levels of neighboring atoms. These energy-level transitions are no longer well defined but are altered by interactions between neighboring atoms, resulting in an infinite variety of transitions—hence the infinite number of radiant energy frequencies.

As might be expected, the interactions of electrons between neighboring atoms in a solid are proportional to temperature. The greater the molecular kinetic energy, the greater the number of interactions and resulting radiant energy. A plot of radiated energy over a wide range of frequencies for two different temperatures is shown in Figure 29-7 (recall that we treated the radiation curve for sunlight back in Chapter 26). As the solid is heated further, transitions between wider-spaced energy levels occur, and higher frequencies of radiation are emitted. The predominant frequency of emitted radiant energy, the *peak frequency*, is directly proportional to the absolute temperature of the emitter:

$$\bar{f} \sim T$$

We use the bar above the f to indicate the peak frequency, for many frequencies of radiant energy are emitted from the incandescent source. If the temperature of an object (in kelvins) is doubled, the peak frequency of emitted radiant energy is doubled. The electromagnetic waves of violet light have nearly twice the frequency of red light waves. A violet-hot star therefore has nearly twice the surface temperature of a red-hot star.* The temperature of incandescent bodies, whether they be stars or blast-furnace interiors, can be determined by measuring the peak frequency (or color) of radiant energy they emit.

Absorption Spectra

When we view white light from an incandescent source with a spectroscope, we see a continuous rainbow-colored spectrum. If a gas is placed between the source and the spectroscope, however, careful inspection will show that the spectrum is not quite continuous. This is an **absorption spectrum**, and there are dark lines distributed throughout it; these dark lines against a rainbow-colored background are like emission lines in reverse. These are *absorption lines*.

---

*If you study this topic further, you will find that the time rate at which an object radiates energy is proportional to the fourth power of its Kelvin temperature. So a doubling of temperature corresponds to a doubling of the frequency of radiant energy but a sixteen-fold increase in the rate of emission of radiant energy. So the object would emit sixteen times as many photons per second and therefore be sixteen times brighter.

Question ▶ From the radiation curves shown in Figure 29-7, which emits the higher average frequency of radiant energy—the 1000°C source or the 1500°C source? Which emits more radiant energy?

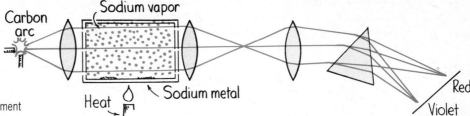

Figure 29-8
Experimental arrangement for demonstrating the absorption spectrum of a gas.

Atoms absorb light as well as emit light. An atom will most strongly absorb light having the frequencies to which it is tuned—the same frequencies it emits. When a beam of white light passes through a gas, the atoms of the gas absorb selected frequencies from the beam. This absorbed light is re-radiated, but in *all* directions instead of only in the direction of the incident beam. When the light remaining in the beam spreads out into a spectrum, the frequencies that were absorbed show up as dark lines in the otherwise continuous spectrum. The positions of these dark lines correspond exactly to the positions of lines in an emission spectrum of the same gas (Figure 29-9).

Figure 29-9
Emission and absorption spectra.

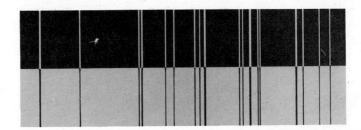

Although the sun is a source of incandescent light, the spectrum it produces, upon close examination, is not continuous. There are many absorption lines, called *Fraunhofer lines* in honor of the Bavarian optician J. D. Lines, who first observed and mapped them accurately. Similar lines are found in the spectra produced by the stars. These lines indicate that the sun and stars are each surrounded by an atmosphere of cooler gases that absorb some of the frequencies of light coming from the main body. Analysis of these lines reveals the chemical composition of the

▶ **Answer**

The 1500°C radiating source emits the higher average frequencies, as noted by the extension of the curve to the right. The 1500°C source is the brighter and also emits more radiant energy, as noted by its greater vertical displacement.

atmospheres of such sources. We find from these analyses that the stellar elements are the same elements that exist on earth. An interesting sidelight is that when spectroscopic measurements were made of the sun, spectral lines different from those measured in the laboratories were found. These lines identified a new element, which was named *helium*, after Helios, the sun. Helium was discovered in the sun before it was discovered on earth. How about that!

We can determine the speed of stars by studying the spectra they emit. Just as a moving sound source produces a Doppler shift in its pitch (Chapter 18), a moving light source produces a Doppler shift in its light frequency. The frequency (not the speed!) of light emitted by an approaching source increases, while the frequency of light from a receding source decreases. The corresponding spectral lines are displaced toward the red end of the spectrum for receding sources and toward the blue end of the spectrum for approaching sources. Since the universe is expanding, the galaxies show a red shift in their spectra.

You'll get into more applications of atomic spectra if you study astronomy, chemistry, or biology. We shall see in Chapter 31 how the spectra of elements enable us to determine atomic structure.

Fluorescence

So we see that thermal agitation or bombarding particles such as high-speed electrons is not the only means of imparting excitation energy to an atom. An atom may be excited by absorbing a photon of light. From the relationship $E \sim f$, we see that high-frequency light, such as ultraviolet, which lies beyond the visible spectrum, delivers much more energy per photon than lower-frequency light. Many substances undergo excitation when illuminated with ultraviolet light.

Materials that are excited by ultraviolet light emit visible light upon de-excitation. The action of these materials is called **fluorescence**. In these materials a photon of ultraviolet light collides with an atom of the material and gives up its energy in two parts. Part of the energy produces heat, increasing the kinetic energy of the entire atom. The other part of the energy goes into excitation, boosting an electron to a higher energy state. Upon de-excitation, this part of the energy is released as a photon of light. Since this photon has only part of the energy of the ultraviolet photon, its frequency is lower and falls into the visible part of the spectrum.

Figure 29-10

In fluorescence some of the absorbed energy of an ultraviolet photon goes into excitation and the remainder into heat. The photon then emitted is less energetic and therefore of a lower frequency than the ultraviolet photon.

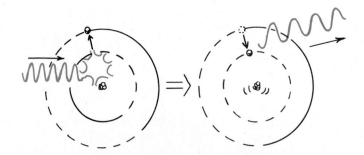

An electron boosted to a higher energy level by a photon of ultraviolet light may return to its stable orbit in several steps. Since the photon energy released at each step is less than the total energy originally in the ultraviolet photon, lower-frequency photons are emitted. Hence, ultraviolet light shining on the material causes it to glow an overall red, yellow, or whatever color is characteristic of the material. Fluorescent dyes are used in paints and fabrics to make them glow when bombarded with ultraviolet photons in sunlight. These are the Day-Glo colors, which are spectacular when illuminated with an ultraviolet lamp.

Detergents that make the claim of cleaning your clothes "whiter than white" use the principle of fluorescence. Such detergents contain a fluorescent dye that converts the ultraviolet light in sunlight into blue visible light, so clothes dyed in this way appear to reflect more blue light than they otherwise would. This makes the clothes appear whiter.*

The next time you visit a natural science museum, go to the geology section and take in the exhibit of minerals illuminated with ultraviolet light. You'll notice that different minerals radiate different colors. This is to be expected because different minerals are composed of different elements, which in turn have different sets of electron energy levels. Seeing the radiating minerals is a beautiful visual experience that is even more fascinating when integrated with your knowledge of nature's submicroscopic happenings. High-energy ultraviolet photons strike on the surface of the minerals, causing the excitation of atoms in the mineral structure; and the radiant energy of light frequencies corresponds exactly to the tiny energy-level spacings—and every excited atom emits its characteristic frequency, with no two different minerals giving off exactly the same color light. Beauty is in both the eye and the mind of the beholder.

Figure 29-11
An excited atom may deexcite in several combinations of jumps.

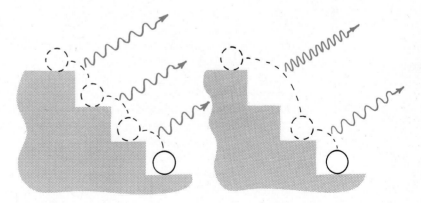

---

*Interestingly enough, the same detergents marketed in some countries are adjusted for a rosier, warmer effect.

Question ▶ Why would it be impossible for a fluorescent material to emit ultraviolet light when illuminated by infrared light?

Fluorescent Lamps

The fluorescent lamp consists of a cylindrical glass tube with electrodes at each end. Like in the neon sign tube, electrons are boiled from the electrodes and forced to vibrate at high speeds by the ac voltage. The tube is filled with very low pressure mercury vapor, which is excited by the impact of the high-speed electrons. As the energy levels in mercury are relatively far apart, the resulting emission of light is of high frequency, mainly in the ultraviolet region. This is the primary excitation process. The secondary process occurs when the ultraviolet light strikes *phosphors*—powdery materials on the inner surface of the tube. The phosphors are excited by the absorption of the ultraviolet photons and give off a multitude of lower frequencies that combine to produce white light. Different phosphors can be used to produce different colors of light.

Figure 29-12
A fluorescent tube. Ultraviolet (UV) light is emitted by gas in the tube excited by an AC electric current. The UV light, in turn, excites phosphors on the inner surface of the glass tube, which emit white light.

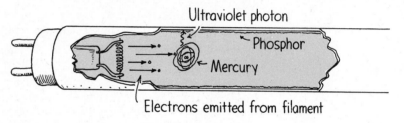

Phosphorescence

When excited, certain crystals as well as some large organic molecules remain in a state of excitement for a prolonged period of time. Their electrons are boosted into higher orbits and become "stuck." As a result, there is a time delay between the processes of excitation and de-excitation. Materials that exhibit this peculiar property are said to be *phosphorescent*. **Phosphorescence** occurs in the element phosphorus, which is used in luminous clock dials and in other objects that are made to glow in the dark. Atoms or molecules in these materials are excited by incident visible light. Rather than de-exciting immediately, as fluorescent materials do, many of the atoms remain in a state of excitement, sometimes for as long as several hours—although most undergo de-excitation rather quickly. If the source of excitation is removed—for example, if the lights are put out—an afterglow occurs while millions of atoms spontaneously undergo gradual de-excitation.

▶ **Answer**

Photon energy output would be greater than photon energy input, which would violate the law of conservation of energy.

A TV screen is slightly phosphorescent, the glow decaying rather quickly, but just slowly enough so that successive scans of the picture blend into one another. The afterglow of some phosphorescent light switches in the home may last more than an hour. Likewise for luminous clock dials, excited by visible light. Some older clock dials glow indefinitely in the dark, not because of a long time delay between excitation and de-excitation, but because they contain radium or some other radioactive material that continuously supplies energy to keep the excitation process going. Such dials are no longer common because of the potential harm of the radioactive material to the user, especially if in a wristwatch or pocket watch.

Many living creatures—from bacteria to fireflies and larger animals like jellyfish—chemically excite molecules in their bodies that give off light. We say that such living things are *bioluminescent*. Under some conditions certain fish become luminescent when they swim, but remain dark when still. Schools of these fish hang motionless and are not seen, but when alarmed streak the depths with sudden light, creating a sort of deep-sea fireworks. The mechanism of bioluminescence is not well understood and is currently being researched.

Lasers

The phenomena of excitation, fluorescence, and phosphorescence underlie the operation of a most intriguing instrument, the **laser** (light amplification by stimulated emission of radiation).* Stimulated emission is a relatively new technique, although Einstein predicted it in 1917. To understand how a laser operates, we must first discuss coherent light.

Figure 29-13
Incoherent white light contains waves of many frequencies (and wavelengths) that are out of phase with one another.

Light emitted by a common lamp is incoherent; that is, photons of many frequencies and many phases of vibration are emitted. The light is as incoherent as the footsteps on an auditorium floor when a mob of people are chaotically rushing about. Incoherent light is chaotic. A beam of incoherent light spreads out after a short distance, becoming wider and wider and less intense with increased distance.

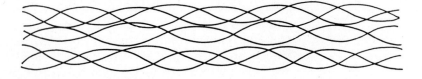

Figure 29-14
Light of a single frequency and wavelength is still out of phase.

*A word constructed from the initials of a phrase is called an *acronym*.

Even if the beam is filtered so that it is a single frequency (monochromatic), it is still incoherent, for the waves are out of phase with one another. The slightest differences in their directions result in a spreading with increased distance.

Figure 29-15
Coherent light: all the waves are identical and in phase.

A beam of photons having the same frequency, phase, and direction—that is, a beam of photons that travel exactly alike—is said to be *coherent*. Only a beam of coherent light will not noticeably spread and diffuse.

A laser is an instrument that produces a beam of coherent light. One model, the earliest, consists of a ruby crystal rod a few centimeters long that has been "doped" with impurities—atoms of chromium. The ruby rod is surrounded by a photoflash tube that sends high-intensity green light into the ruby (Figure 29-16). Excitation of the chromium atoms occurs as the outer electrons are boosted to higher energy levels. The electrons fall back to an intermediate level and then, more slowly, to their lower level. At this last stage they emit a photon of red light; the ruby fluoresces.

Figure 29-16
A ruby crystal laser.

Ruby crystal rod

Photoflash tube

Coherent laser light

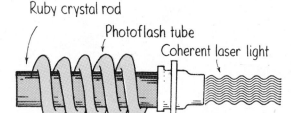

Photons of red light within the crystal trigger the de-excitation of neighboring atoms that are still in the excited state. The photons stimulated into radiation will be identical in frequency, phase, and direction to the incident photons. These photons then may further stimulate the radiation of other excited atoms, thereby producing a beam of coherent light. Most of this light initially escapes through the sides of the crystal in random directions. Light traveling along the crystal axis, however, is reflected from mirrors that have been coated to reflect selectively the desired wavelength of light. One mirror is totally reflecting, while the other is partially reflecting. The reflected waves reinforce each other after each round-trip reflection between the mirrors, thereby setting up a resonance condition in which the light builds up to an appreciable intensity.

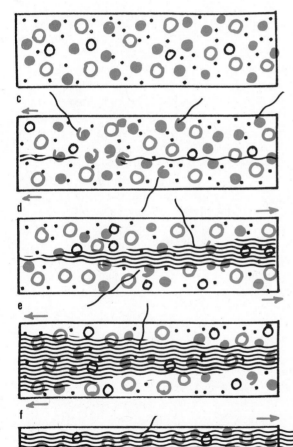

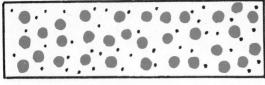

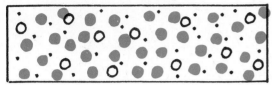

Figure 29-17
Laser action in a helium-neon laser.

(*a*) The laser consists of a narrow Pyrex tube that contains a low-pressure gas mixture consisting of 85% helium (small black dots) and 15% neon (large colored dots).

(*b*) When a high-voltage current zaps through the tube, it excites both helium and neon atoms to their usual higher states, which immediately undergo de-excitation, except for one state in the helium that is characterized by a prolonged delay before de-excitation. This is called a *metastable state*. Since this state is relatively stable, a sizable population of excited helium atoms (fine open circles) is built up. These wander about in the tube and act as an energy source for neon, which has an otherwise hard-to-come-by metastable state very close to the energy of the excited helium.

(*c*) When excited helium atoms collide with neon atoms in the ground state, the helium gives up its energy to the neon, which is boosted to its metastable state (bold open circles). The process continues, and the population of excited neon atoms soon outnumbers neon in the ground state. This inverted population in effect is waiting to radiate its energy.

(*d*) Some neon atoms eventually de-excite and radiate red photons in the tube. When this radiant energy passes other excited neon atoms, the latter are stimulated into emitting photons exactly in phase with the radiant energy that stimulated the emission. Photons pass out of the tube in irregular directions, giving it the familiar neon glow.

(*e*) Photons parallel to the tube reflect from specially coated parallel mirrors at the ends of the tube. The reflected photons stimulate the emission of other neon atoms, thereby producing a photon avalanche of the same frequency, phase, and direction.

(*f*) The photons flash to and fro between the mirrors, becoming amplified with each pass.

(*g*) Some "leak" out one of the mirrors, which is only partially reflecting. These make up the laser beam.

Figure 29-18
A helium-neon laser.

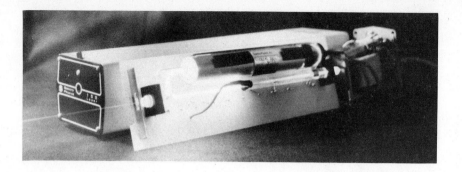

The light that escapes in pulses through the more transparent-mirrored end makes up the laser beam.

Shortly after the development of the ruby laser, physicists made a gas laser that produced a continuous beam. The first such laser used a mixture of helium and neon gases. When a high voltage is applied to electrodes at the ends of a tube of the mixed gases, the electrical discharge through the helium gas excites the atoms of the neon gas into a prolonged excited state (Figure 29-17). In addition to crystal lasers and gas lasers, other new types have joined the laser family: glass lasers, chemical and liquid lasers, and semiconductor lasers. Lasers now produce beams ranging from infrared through ultraviolet. Some models can be tuned to various frequency ranges. Most exciting is the prospect of an X-ray laser beam.

The laser is not a source of energy. It is simply a converter of energy that takes advantage of the process of stimulating emission to concentrate a certain fraction of its energy (commonly 1 percent) into radiant energy of a single frequency moving in a single direction. Like all devices, a laser can put out no more energy than is put in.

The list of practical applications of the laser is vast. In its simplest application, a laser beam is an unbendable chalk line and an unstretchable measuring stick, providing accuracy for the machinist, speed for construction workers, and economy for the surveyor. Lasers were used to ensure the straightness of the underwater tunnel in San Francisco Bay. Laser beams have been bounced off the moon and returned in order to measure the exact distance between the earth and the moon and to provide information on continental drift.

Some lasers are so intense (and concentrated) that eye surgeons use them to "weld" detached retinas back into place without making an incision. The light is simply brought to focus in the region where the welding is to take place.

Whereas radio wavelengths span hundreds of meters and television waves span many centimeters, laser wavelengths are measured in millionths of a centimeter. As a result, an enormous number of messages bunched into a very narrow band of frequencies can be carried on these ultrashort waves. Communications can be carried in a laser beam directed through space, through the atmosphere, or through optical fibers (light pipes) that can be bent like cables.

Figure 29-19
The Nova laser at Lawrence Livermore National Laboratory, used to study inertial confinement fusion. The world's largest and most powerful laser, the Nova, with its ten beams, is capable of producing up to 100 trillion W of power in a burst of light lasting for one billionth of a second.

Figure 29-20
A product's unique identification is coded in the UPC symbol. The price and name of the product are held in computer memory.

Lasers are used with computers. A laser gun burns microscopic holes, each of which represents a bit of information, into the surface of a computer tape. A second laser later "reads" them. In this way a trillion bits of data can be stored on a single standard-length reel of computer tape. A telephone directory for all the world's inhabitants could be stored on such a tape.

The laser is at work in supermarket checkout counters, where code-reading machines scan the universal product code (UPC) symbol printed on packages (Figure 29-20). Laser light is reflected from the bars and spaces and converted to an electric signal as the symbol is scanned. The signal rises to a high value when reflected from a bright space and falls to a low value when reflected from a dark bar. The high and low values are converted to 1's and 0's, respectively, and the information is in the binary code ready for computer processing.

Lasers measure and detect pollutants in exhaust gases. Different gases absorb light at characteristic wavelengths and leave their "fingerprints" on a reflected beam of laser light. The specific wavelength and amount of light absorbed are analyzed by a computer, which produces an immediate tabulation of the pollutants. The beam of a laser is used as a non-wearing "optical needle" for audio and video discs, as a knife to cut cloth rapidly and accurately in garment factories, in a much-improved technique for fingerprint detection, and for the production of holograms. A most important application of the laser is concentrating huge amounts of energy on hydrogen pellets and bringing them up to thermonuclear temperatures. The laser may someday allow us to control thermonuclear fusion.

Just after its invention in 1958, the laser was touted as being a solution looking for a problem. It didn't have to look very far or wait very long. It has ushered in a whole new technology—the promise of which we have only begun to tap. The future for laser applications seems unlimited.

Summary of Terms

Excitation The process of boosting one or more electrons in an atom or molecule from a lower to a higher energy level. An atom in an excited state will usually decay (de-excite) rapidly into a lower state by the emission of radiation. The energy of the radiation is proportional to its frequency: $E \sim f$.

Emission spectrum The distribution of wavelengths in the light from a luminous source.

Spectroscope An optical instrument that separates light into its constituent frequencies in the form of spectral lines.

Incandescence The state of glowing while at a high temperature, caused by electrons in vibrating atoms and molecules that are shaken in and out of their stable energy levels, emitting radiant energy in the process. The peak frequency of radiant energy is proportional to the absolute temperature of the heated substance:

$$\bar{f} \sim T$$

Absorption spectrum A continuous spectrum, like that of white light, interrupted by dark lines or bands that result from the absorption of certain frequencies of light by a substance through which the radiant energy passes.

Fluorescence The property of absorbing radiant energy of one frequency and re-emitting radiant energy of lower frequency. Part of the absorbed radiant energy goes into heat and the other part into excitation; hence, the emitted radiant energy has a lower energy, and therefore a lower frequency, than the absorbed radiant energy.

Phosphorescence A type of light emission that is the same as fluorescence except for a delay between excitation and de-excitation, which provides an afterglow. The delay is caused by atoms being excited to energy levels that do not decay rapidly. The afterglow may last from fractions of a second to hours, or even days, depending on the type of material, temperature, and other factors.

Laser (light amplification by stimulated emission of radiation) An optical instrument that produces a beam of coherent monochromatic light.

Review Questions

1. If electrons are made to vibrate to and fro at a few million hertz, radio waves are emitted. What class of waves would be emitted if electrons were made to vibrate to and fro at a few million billion hertz?

2. When light strikes a surface and is reflected, the behavior of electrons in the outer shells of molecules is one of subtle oscillations. What is the behavior of electrons when light is emitted from a glowing surface?

3. What does it mean to say an energy state is *discrete*?

Excitation

4. Which has the greatest energy with respect to the atomic nucleus, electrons in inner electron shells or electrons in outer electron shells?

5. What does it mean to say an atom is *excited*?

6. How is a ball rolling down a stairway analogous to an electron undergoing de-excitation?

7. What is the name given to a throbbing pulse of electromagnetic radiation?

8. What is the relationship between the *difference in energy* between energy levels and the *energy of the photon* that is ejected by a transition between those levels?

9. What is the relationship between the *energy* of a photon and its *vibrational frequency*?

10. Which has the higher *frequency*, red or blue light?

11. Which has the greater *energy* per photon, red or blue light?

12. An electron loses some of its kinetic energy when it bombards a neon atom in a glass tube. What becomes of this energy?

13. Can a neon atom in a glass tube be excited more than once?

14. What does the variety of colors displayed in the flame of a burning log represent?

15. Which converts more energy into heat, an incandescent lamp or a mercury-vapor lamp?

16. If you look through a prism or diffraction grating at the light emitted by a mercury-vapor lamp, you'll see a variety of colors. What do these colors represent?

Emission Spectra

17. What is it that every element has that makes it emit its own characteristic colors of light?

18. What is a *spectroscope*, and what does it do?

19. Why do spectral lines have a "line shape"?

Incandescence

20. What is the "color" of all the visible light frequencies taken together?

21. When a gas glows, discrete colors are emitted. When a solid glows, the colors are smudged. Why?

22. What is the relationship between the peak frequency of emitted light and the temperature of its incandescent source?

Absorption Spectra

23. Distinguish between *emission spectra* and *absorption spectra*.

24. How does the frequency of light emitted by a substance compare to the frequency of light absorbed by the same substance? (How does the frequency of radio waves that an antenna sends compare to the frequency of radio waves an antenna receives?)

25. Since an absorbing gas re-emits the light it absorbs, why are there dark lines in an absorption spectrum? That is, why doesn't the re-emitted light simply fill in the dark places?

26. Where was helium discovered?

27. How can astrophysicists tell whether a star is receding or approaching earth?

Fluorescence

28. Name three ways in which atoms can be excited.

29. Why is ultraviolet light rather than infrared light so effective in making certain materials fluoresce?

30. If an atom absorbs ultraviolet light and emits red light, what becomes of the "missing" energy?

Fluorescent Lamps

31. Distinguish between the primary and secondary excitation processes that occur in a fluorescent lamp.

32. What enables a fluorescent lamp to give off white light?

Phosphorescence

33. Distinguish between *fluorescence* and *phosphorescence*.

34. What is responsible for the afterglow of phosphorescent materials?

Lasers

35. Distinguish between *monochromatic light* and *coherent light*.

36. Does the red light emitted by a helium-neon laser come from helium or neon?

37. What is a *metastable state*?

38. In the operation of a helium-neon laser, why is it important that the metastable state of helium be relatively stable? What would be the effect of this state de-exciting too rapidly? (Refer to Figure 29-17.)

39. In the operation of a helium-neon laser, why is it important that the metastable state in the helium atom closely match the energy level of a more difficult-to-come-by metastable state in neon?

40. How is the inverted population of excited neon atoms in a helium-neon laser achieved?

41. What happens when a photon grazes a neon atom that is excited to a metastable state with the same energy as the photon? What happens when these two photons graze two other similarly excited neon atoms?

42. How does the avalanche of photons in a laser beam differ from the hordes of photons emitted by an incandescent lamp?

43. What is the typical efficiency of a common laser?

44. A friend speculates that scientists in a certain country have developed a laser that puts out far more energy than is put into it. Your friend asks for your response to this speculation. What is your response?

Home Project

Borrow a diffraction grating from your physics instructor. The common kind looks like a photographic slide, and light passing through it or reflecting from it is diffracted into its component colors by thousands of finely ruled lines. Look through the grating at the light from a sodium-vapor street lamp. If it's a low-pressure lamp, you'll see the nice yellow spectral "line" that dominates sodium light (actually, it's two closely spaced lines). If the street lamp is round, you'll see circles instead of lines; look through a slit cut in cardboard or whatever, and you'll see lines. What happens with the now-common high-pressure sodium lamps is more interesting. Because of the collisions of excited atoms, you'll see a smeared-out spectrum that is nearly continuous, almost like that of an incandescent lamp. Right at the yellow location where you'd expect to see the sodium line is a dark area. This is the sodium absorption band. It is due to the cooler sodium, which surrounds the high-pressure emission region. You should view this a block or so away so that the line, or circle, is small enough to allow the resolution to be maintained. Try this. It is very easy to see!

Exercises

1. Have you ever watched a fire and noticed that the burning of different materials often produces flames of different colors? Why is this so?

2. Green light is emitted when electrons in a substance make a particular energy-level transition. If blue light were instead emitted from the same substance, would it correspond to a greater or lesser energy-level transition?

3. Electrons lose kinetic energy when they bombard gases in glass tubes. Do they lose more kinetic energy or less kinetic energy when blue light rather than red light is emitted? Defend your answer.

4. Ultraviolet light causes sunburns, whereas visible light does not. Why is this so?

5. If we double the frequency of light, we double its energy. If we instead double the wavelength of light, how does its energy compare?

6. Why doesn't a neon sign finally "run out" of excited atoms and produce dimmer and dimmer light?

7. If light were passed through a round hole instead of a thin slit in a spectroscope, how would the spectral "lines" appear? Why would a hole be more disadvantageous than a slit?

8. If we investigate the light from a common neon tube and from a helium neon laser with a prism or diffraction grating, what striking difference do we see?

9. What is the evidence for the claim that iron exists in the cool gas surrounding the sun?

10. How might the Fraunhofer lines in the spectrum of sunlight that are due to absorption in the sun's atmosphere be distinguished from those due to absorption by gases in the earth's atmosphere?

11. Does atomic excitation occur in solids as well as in gases? How does the radiant energy from an incandescent solid differ from the radiant energy emitted by an excited gas?

12. A lamp filament is made of tungsten. Why do we get a continuous spectrum rather than a tungsten line spectrum when light from an incandescent lamp is viewed with a spectroscope?

13. Why are fluorescent colors so bright?

14. Why do different fluorescent minerals emit different colors when illuminated with ultraviolet light?

15. Your friend reasons that if ultraviolet light can activate the process called *fluorescence*, infrared light ought to also. Your friend looks to you for approval or disapproval of this idea. What is your position?

16. The forerunner to the laser involved microwaves rather than visible light. What does *maser* mean?

17. A ruby laser is activated by a photoflash tube that emits green light. Why would a laser composed of a green crystal and a photoflash tube that emits red light not work?

18. List these light sources in decreasing order of their efficiency for emitting light: laser, fluorescent lamp, incandescent lamp.

19. A laser cannot put out more energy than is put into it. A laser can, however, produce pulses of light with more power output than the power input required to run the laser. Explain.

20. We know that a lamp filament at 2500K emits white light. Does the lamp filament emit radiant energy at room temperature?

21. Since every object has some temperature, every object radiates energy. Why, then, can't we see most objects in the dark?

22. How do the relative temperatures compare for reddish, bluish, and whitish stars?

23. Most of the radiant energy emitted by a red-hot star is in the infrared region. Most of the radiant energy emitted by a violet-hot star is in the ultraviolet. Why is it there are no green-hot stars? (*Hint:* If there were a green-hot star, what would this tell you about the relative rates of individual molecular vibrations in the star? What would its radiation curve look like? Is this likely?)

24. Part *a* in the sketch below shows a radiation curve of an incandescent solid and its spectral pattern as produced with a spectroscope. Part *b* shows the "radiation curve" of an excited gas and its emission spectral pattern. Part *c* shows the curve produced when a cool gas is between an incandescent source and the viewer; the corresponding spectral pattern is left as an exercise for you to construct. Part *d* shows the spectral pattern of an incandescent source as seen through a piece of green glass; you are to sketch in the corresponding radiation curve.

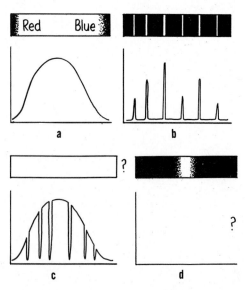

30 Light Quanta

The classical physics that we have so far studied deals with two categories of phenomena: particles and waves. According to our everyday experience, "particles" are tiny objects like bullets. They have mass and obey Newton's laws—they *travel* through space in straight lines unless a force acts upon them. Likewise, according to our everyday experience, "waves," like waves in the ocean, are phenomena that *extend* in space. When a wave travels through an opening or around a barrier, different parts of the wave interfere and diffraction results. Therefore, particles and waves are easy to distinguish from each other. In fact, they have properties that are mutually exclusive. Nonetheless, the question of how to classify light has been a mystery for centuries.

One of the earliest recorded theories about the nature of light is that of Plato, who thought that light consisted of streamers emitted by the eye. Euclid also held this view. The Pythagoreans, somewhat later, believed that light emanated from luminous bodies to the eye in the form of very fine particles, while Empedocles taught that light is composed of high-speed waves of some sort. In 1704 Isaac Newton described light as a stream of particles or corpuscles. He maintained this view despite his later experiment with light reflecting from glass plates, in which he noticed fringes of brightness and darkness (Newton's rings) and recognized that particles of light have some sort of wave nature. Christian Huygens, a contemporary of Newton, strongly advocated a wave theory of light.

With all this history as background, Thomas Young in 1801 performed "the double-slit experiment," which seemed to prove, finally, that light was a wave phenomenon. This view was reinforced in 1862 by Maxwell's finding that light was energy carried in oscillating electric and magnetic fields of electromagnetic waves. The wave view of light was confirmed experimentally by Hertz, 25 years later. In 1905, however, Albert Einstein published a Nobel Prize–winning paper that challenged the wave theory of light by demonstrating that light in its interactions with matter was confined not in continuous waves, as Maxwell envisioned, but in tiny packets of energy Einstein called *photons*. This strange behavior of light gave the initial glimpse of the quantum rules that govern the world of the atom. In this chapter we shall enter the world of the very small and look into some of the strange and exciting aspects of quantum reality.

Birth of the Quantum Theory

At the turn of the century, technology reached a level that enabled scientists to design experiments sensitive to the behavior of very small particles. With the discovery of the electron in 1897 and the investigation of radioactivity at about the same time, experimenters began to probe the atomic structure of matter. In 1900 the German theoretical physicist Max Planck announced his theoretical discovery that the energies of vibrating electrons in incandescent light sources were restricted to distinct values. As a result, radiation was emitted in discrete bundles of energy, which he called **quanta** (*quanta* is the plural form of *quantum*, just as *momenta* is the plural form of *momentum*). Planck's discovery began a revolution of ideas that has completely changed the way we think about the physical world. We will see that the rules we apply to the everyday macroworld, the Newtonian laws that work so well for large objects like baseballs and planets, simply don't apply to events in the microworld of the atom. In the macroworld the study of motion is called *mechanics*; in the microworld the study of the motion of quanta is called **quantum mechanics**. More broadly, the body of laws developed from 1900 to the late 1920s that describe all quantum phenomena of the microworld is known as *quantum physics*.

Quantization and Planck's Constant

Quantization, the idea that physical quantities come in discrete bundles, is certainly not a new idea to physics. Matter is quantized; the mass of a brick of gold, for example, is equal to some whole-number multiple of the mass of a single gold atom. Electricity is quantized, as electric charge is some whole-number multiple of the charge of a single electron.

Quantum physics states that in the microworld of the atom, the amount of energy in any system is quantized—that not all values of energy are possible. This is analogous to saying a campfire can only be so hot. It might burn at 450°C or it might burn at 451°C, but in no way can it burn at 450.5°C. Believe it? Well, you shouldn't, for as far as our macroscopic thermometers can measure, a campfire can burn at any temperature as long as it's above that required for combustion. But the energy of the campfire, interestingly enough, is the composite energy of a great number and great variety of elemental units of energy. A simpler example is the energy in a beam of laser light, which is a whole-number multiple of a single lowest value of energy—one quantum. The quanta of light, or electromagnetic radiation in general, are the photons.

Recall from the previous chapter that the frequency of a photon f is proportional to its energy: $E \sim f$. When the energy of a photon is divided by its frequency, the single number that results is the proportionality constant, called **Planck's constant**, h.* We shall see that Planck's constant is a fundamental constant of nature that serves to set a lower limit on the smallness of things. It ranks with the velocity of light as a basic

---

*Planck's constant, h, has the numerical value 6.6×10^{-34} J·s.

constant of nature and appears again and again in quantum physics. We can insert this constant in the above proportion and express it as an exact equation:

$$E = hf$$

This equation gives the smallest amount of energy that can be converted to light with frequency f. The radiation of light is not emitted continuously but is emitted as a stream of photons, each of which throbs at a frequency f and carries an energy hf.

Questions

1. What does the term *quantum* mean?
2. How much total energy is in a monochromatic beam composed of n photons of frequency f?

The new physics tells us that the physical world is a coarse, grainy place of discrete packets rather than the smooth, continuous place with which we are familiar. The "commonsense" world described by classical physics seems smooth and continuous because the quantum graininess is on a very small scale compared to the size of things in the familiar world. Planck's constant is small in terms of familiar units. For example, the smooth blends of black, white, and gray in the photographs in this book do not look smooth at all when viewed through a magnifying glass. You'll see that a photograph consists of many tiny dots. The image is quantized, something you do not notice from a distance. Using special tools, quantum physics provides a description of the grainy world of atoms and energy that blurs to make the world we know.

Physicists were reluctant to adopt Planck's revolutionary quantum notion. Before it could be taken seriously, the quantum idea would have to be verified by some means independent of light emission from an incandescent source. The verification was supplied less than 5 years later by Einstein, who extended Planck's ideas to explain the photoelectric effect in the Nobel Prize–winning paper we mentioned earlier.

Photoelectric Effect

In the latter part of the nineteenth century, several investigators noticed that light was capable of ejecting electrons from various metal surfaces. This is the **photoelectric effect**, now used in electric eyes, in the photographer's light meter, and in the sound track of motion pictures.

▶ **Answers**

1. A *quantum* is the smallest elemental unit of a quantity. Radiant energy, for example, is composed of many quanta, each of which is called a *photon*. So the more photons in a beam of light, the more energy in that beam.
2. The energy in a beam of light containing n quanta is $E = nhf$.

Figure 30-1
An apparatus used for measuring the photoelectric
effect.

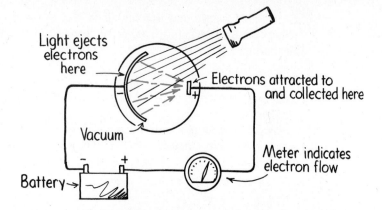

An arrangement for observing the photoelectric effect is shown in Figure 30-1. Light shining on the negatively charged photosensitive metal surface liberates electrons, which are attracted to the positive plate to produce a measurable current. If we instead charge this plate with just enough negative charge that it repels electrons, the current can be stopped. We can then calculate the energies of the ejected electrons from the easily measured potential difference between the plates.

The photoelectric effect was not particularly surprising to early investigators. The ejection of electrons could be accounted for by classical mechanics, which pictures the incident light waves building an electron's vibration up to greater and greater amplitudes until it finally breaks loose from the metal surface, just as water drops boil off the surface of hot water. It should take considerable time to heat the surface of a weak source sufficiently to boil off electrons. It was found instead that electrons are ejected as soon as the light is turned on—but not as many were ejected as with a strong source. Careful examination of the behavior of the photoelectric effect led to several observations that were quite contrary to the classical wave picture:

1. The time lag between turning on the light and the ejection of the first electrons was not affected by the brightness or frequency of the light.
2. The effect was easy to observe with violet or ultraviolet light but not with red light.
3. The rate at which electrons were ejected was proportional to the brightness of the light.
4. The maximum energy of the ejected electrons was not affected by the brightness of the light, but there were indications that the energy did depend on the frequency of the light.

The lack of any appreciable time lag was especially difficult to understand in terms of the wave picture. According to the wave theory, an electron in dim light should, after some delay, build up enough vibrational energy for ejection, while an electron in bright light should be ejected immediately. However, this didn't happen. It was not unusual to observe an electron being ejected immediately, even in the dimmest of

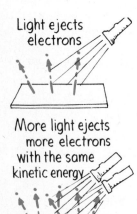

Light ejects
electrons

More light ejects
more electrons
with the same
kinetic energy

Figure 30-2
The photoelectric effect
depends on intensity.

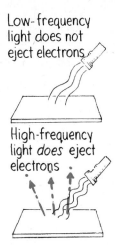

Low-frequency light does not eject electrons

High-frequency light *does* eject electrons

Figure 30-3
The photoelectric effect depends on frequency.

lights. The observation that the brightness of light in no way affected the energies of ejected electrons was perplexing. The stronger electric fields of brighter light did not cause electrons to be ejected at greater speeds. More electrons were ejected in brighter light, but not at greater speeds. A weak beam of ultraviolet light, on the other hand, produced a smaller number of ejected electrons but at much higher speeds. This was most puzzling.

Einstein produced the answer in 1905, the same year he explained Brownian motion and set forth his theory of special relativity. His clue was Planck's quantum theory of radiation. Planck had assumed that the emission of light in quanta was due to restrictions on the vibrating electrons that produced the light, which he considered to be discrete pulses of the electromagnetic waves described by Maxwell. Einstein, on the other hand, attributed the quantum properties to light itself and viewed radiation as a hail of particles. To emphasize this particle aspect, we speak of photons (by analogy with electrons, protons, and neutrons) whenever we are thinking of the particle nature of light. One photon is completely absorbed by each electron ejected from the metal. The absorption is an all-or-nothing process and is immediate, so there is no delay as "wave energies" build up.

A light wave has a broad front, and its energy is spread out along this front. For the light wave to eject a single electron from a metal surface, all its energy would somehow have to be concentrated on that one electron. But this is as improbable as an ocean wave hitting a beach and knocking only one single seashell far inland with an energy equal to the energy of the whole wave. Therefore, instead of thinking of light encountering a surface as a continuous train of waves, the photoelectric effect suggests we conceive of light encountering a surface or any detector as a succession of corpuscles, or photons. The energy of each photon is proportional to the frequency of light, while the number of photons has to do with its brightness.

Experimental verification of Einstein's explanation of the photoelectric effect was made 11 years later by the American physicist Robert Millikan. Every aspect of Einstein's interpretation was confirmed.

Questions

1. Will brighter light eject more electrons from a photosensitive surface than dimmer light of the same frequency?
2. Will high-frequency light eject a greater number of electrons than low-frequency light?

▶ **Answers**
1. Yes. The number of ejected electrons depends on the number of incident photons.
2. Not necessarily. The energy (not the number) of ejected electrons depends on the frequency of the illuminating photons. A bright source of blue light, for example, may eject more electrons than a dim violet source, but not more energetic electrons.

The photoelectric effect proves conclusively that light has particle properties. We cannot conceive of the photoelectric effect on the basis of waves. On the other hand, we have seen that the phenomenon of interference demonstrates convincingly that light has wave properties. We cannot conceive of interference in terms of particles. In classical physics, this appears to be and is contradictory. From the point of view of quantum physics, light has properties resembling both. It is "just like a wave" or "just like a particle," depending on the particular experiment. So we think of light as both, as a wave-packet. How about "wavicle"? Quantum physics calls for a new way of thinking.

Wave-Particle Duality

The wave and particle nature of light is evident in the formation of optical images. We understand the photographic image produced by a camera in terms of light waves, which spread from each point of the object, refract as they pass through the lens system, and converge to focus on the photographic film. The path of light from the object through the lens system and to the focal plane can be calculated using methods developed from the wave theory of light.

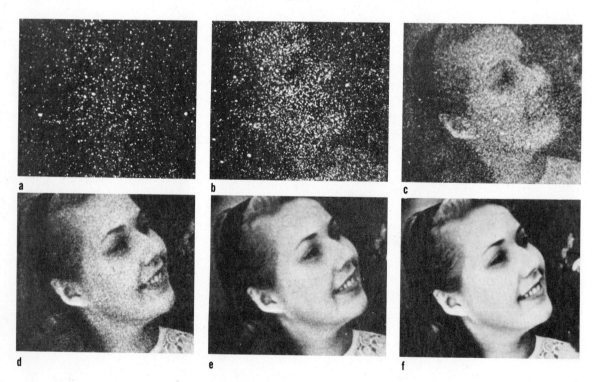

Figure 30-4
Stages of exposure revealing the photon-by-photon production of a photograph. The approximate numbers of photons at each stage were (a) 3×10^3, (b) 1.2×10^4, (c) 9.3×10^4, (d) 7.6×10^5, (e) 3.6×10^6, and (f) 2.8×10^7.

But now consider carefully the way in which the photographic image is formed. The photographic film consists of an emulsion that contains grains of silver halide crystal, each grain containing about 10^{10} silver atoms. Each photon that is absorbed gives up its energy hf to a single grain in the emulsion. This energy activates surrounding crystals in the entire grain and is used in development to complete the photochemical process. Many photons activating many grains produce the usual photographic exposure. When a photograph is taken with exceedingly feeble light, we find that the image is built up by individual photons that arrive independently and are seemingly random in their distribution. We see this strikingly illustrated in Figure 30-4, which shows how an exposure progresses photon by photon.

Double-Slit Experiment

Let's return to Young's double-slit experiment, which we discussed in terms of waves in Chapter 28. Recall that when we pass monochromatic light through a pair of closely spaced thin slits, we produce an interference pattern (Figure 30-5). Now let's consider the experiment in terms of photons. Suppose we dim our light source so that in effect only one

Figure 30-5
(a) Arrangement for double-slit experiment. (b) Photograph of interference pattern. (c) Graphic representation of pattern.

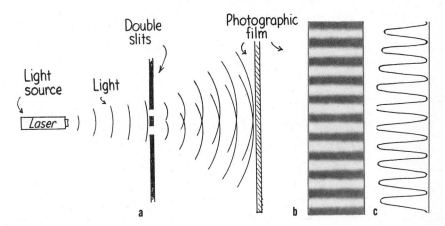

Figure 30-6
Stages of two-slit interference pattern. The pattern of individually exposed grains progresses from (a) 28 photons to (b) 1000 photons to (c) 10 000 photons. As more photons hit the screen, a pattern of interference fringes appears.

photon at a time reaches the barrier with the thin slits. If the film is exposed to the light for a very short time, the film becomes exposed as simulated in Figure 30-6a. Each spot represents the place where the film has been exposed to a photon. If light is allowed to expose the film for a longer time, a pattern of fringes emerges as in Figure 30-6b and c. This is quite amazing. Spots on the film are seen to progress photon by photon to form the same interference pattern characterized by waves!

a

b

c

Figure 30-7
Single-slit diffraction pattern.

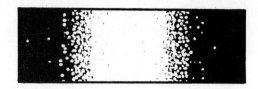

If we cover one slit so that photons hitting the photographic film can only pass through a single slit, the tiny spots on the film accumulate to form a single-slit diffraction pattern (Figure 30-7). We find that photons hit the film at places they would not hit if both slits were open! If we view this classically, we are perplexed and may ask how photons passing through the single slit "know" that the other slit is covered and therefore fan out to produce the wide single-slit diffraction pattern. Or, if both slits are open, how do photons traveling through one slit "know" that the other slit is open and avoid certain regions, proceeding only to areas that will ultimately fill to form the fringed double-slit interference pattern?* We have no satisfying answer to these questions, which presuppose that light obeys the rules of classical physics. What we do know is that *light behaves as a stream of photons when it interacts with the photographic film or other detectors and behaves as a wave in traveling from a source to the place where it is detected.* Light travels as a wave and hits as a stream of photons. Photons strike the film at places where we would expect to see constructive interference of waves. That light exhibits both wave and particle behavior is one of the interesting surprises that physicists have discovered in this century. Even more surprising has been the discovery that objects with mass also exhibit a dual wave-particle behavior.

Particles as Waves: Electron Diffraction

If light can have particle properties, why can't particles have wave properties? This question was posed by the French physicist Louis de Broglie while still a graduate student in 1924. His answer constituted his doctoral thesis in physics and later won him the Nobel Prize in physics. According to de Broglie, every particle of matter is somehow endowed with a wave to guide it as it travels. Under the proper conditions, then, every particle will produce an interference or diffraction pattern. All bodies—electrons, protons, atoms, mice, you, planets, suns—have a wavelength that is related to their momentum by

$$\text{Wavelength} = \frac{h}{\text{momentum}}$$

---

*From a Newtonian point of view, this wave-particle duality is indeed mysterious. This leads some people to believe that quanta have some sort of consciousness, with each photon or electron having a mind of its own. The mystery, however, is like beauty. It is in the mind of the beholder rather than in nature itself. We conjure models to understand nature, and when inconsistencies arise, we sharpen or change our models. The wave-particle duality of light doesn't fit the model devised by Newton. An alternate model is that quanta have minds of their own. Another model is quantum physics. In this book we subscribe to the latter.

Figure 30-8
Fringes produced by the diffraction (a) of light and (b) of an electron beam.

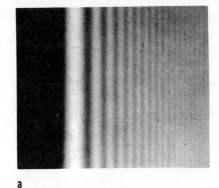
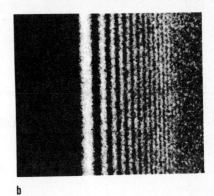

a   b

Figure 30-9
An electron microscope makes practical use of the wave nature of electrons. The wavelength of electron beams is typically thousands of times shorter than the wavelength of visible light, so the electron microscope is able to distinguish detail not visible with optical microscopes.

where h is Planck's constant. A body of large mass and ordinary speed has such a small wavelength that interference and diffraction are negligible; rifle bullets fly straight and do not pepper their targets far and wide with detectable interference patches.*

However, beams of electrons directed through double slits exhibit interference patterns, just as photons do. The double-slit experiment discussed in the last section can be performed with electrons as well as with photons. For electrons the apparatus is more complex, but the procedure

Figure 30-10
Detail of a male mosquito head as seen with a scanning electron microscope at a "low" magnification of 88 times.

---

*A bullet of mass 0.02 kg traveling at 330 m/s, for example, has a de Broglie wavelength of $\dfrac{6.6 \times 10^{-34}\ \text{J·s}}{(0.02\ \text{kg})(330\text{m/s})} = 10^{-34}$ m, an incredibly small size a million million million millionth the size of a hydrogen atom. An electron traveling at 2 percent the speed of light, on the other hand, has a wavelength 10^{-10} m, which is equal to the diameter of the hydrogen atom. Diffraction effects for electrons are measurable, whereas diffraction effects for bullets are not.

Figure 30-11
Electron interference patterns filmed from a TV monitor, showing the diffraction of a very low intensity electron-microscope beam through an electrostatic biprism.

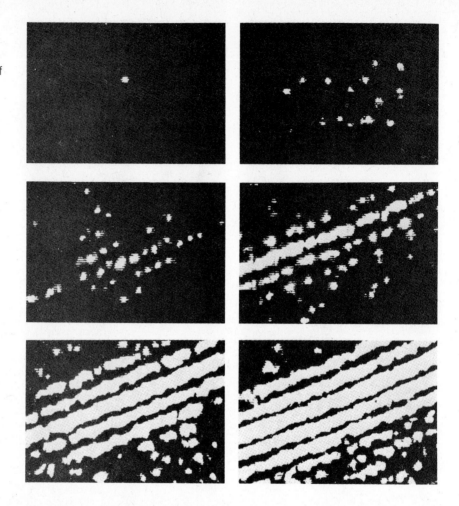

is essentially the same. The intensity of the source can be reduced to direct electrons one at a time through a double-slit arrangement, producing the same remarkable results as with photons. Like photons, electrons arrive at the screen as particles, but the *pattern* of arrival is wave-like. The angular deflection of electrons to form the interference pattern agrees perfectly with calculations using de Broglie's equation for the wavelength of an electron. This wave-particle duality is not restricted to photons and electrons. In Figure 30-11 we see the results of a similar procedure that uses a standard electron microscope and an electrostatic biprism. The electron beam of very low current density is directed at the prism; a pattern of fringes produced by individual electrons builds up step by step and is displayed on a TV monitor. The image is gradually filled by electrons to produce the interference pattern customarily associated with waves. Neutrons, protons, whole atoms, and to an immeasurable degree even high-speed rifle bullets exhibit a duality of particle and wave behavior.

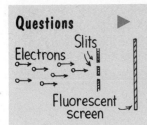

Slits

Electrons

Fluorescent
screen

1. If electrons behaved only like particles, what pattern would you expect on the screen after the electron passed through the double slits?

2. We don't notice the de Broglie wavelength for a pitched baseball. Is this because the wavelength is very large or because it is very small?

3. If an electron and a proton have the same de Broglie wavelength, which particle has the greater speed?

Uncertainty Principle

The wave-particle duality of quanta has inspired interesting discussions about the limits of our ability to accurately measure the properties of small objects. The discussions center on the idea that the act of measuring something in some way affects the quantity being measured.

We know, for example, that if we place a cool thermometer in a cup of hot coffee, the temperature of the coffee is altered as it gives heat to the thermometer. The measuring device alters the quantity being measured. But we can correct for these errors in measurement if we know the initial temperature of the thermometer, the masses and specific heats involved, and so forth. Such corrections fall well within the domain of classical physics; these are *not* the uncertainties of quantum physics. Quantum uncertainties stem from the wave-particle duality of matter and from an associated intrinsic limit in our ability to make accurate measurements at the quantum level. Innumerable experiments have shown that any measurement that in any way probes a system necessarily disturbs the system by at least one quantum of action, h—Planck's constant. So any measurement that involves interaction between the measurer and that being measured is subject to this minimum inaccuracy.

If we look at a cup of hot coffee across the room and judge that it is hot by observing the steam rising from it, this measurement involves no probing. This act of "measuring" neither adds nor subtracts energy from the coffee. Placing a thermometer in it is a different story. We physically interact with the coffee and thereby subject it to alteration. The quantum contribution to this alteration, however, is completely dwarfed by classical uncertainties and is negligible. Quantum uncertainties are significant only in the atomic and subatomic realm.

▶ **Answers**

1. If electrons behaved only like particles, they would form two bands, as indicated in *a*. Because of their wave nature, they actually produce the pattern shown in *b*.

a   b

2. We don't notice the wavelength of a pitched baseball because it is extremely small—on the order of 10^{-20} times smaller than the atomic nucleus.

3. The same wavelength means that the two particles have the same momentum. This means that the less massive electron must travel faster than the heavier proton.

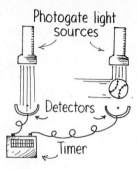

Photogate light
sources

Detectors

Timer

Figure 30-12
The ball's speed is mea-
sured by dividing the distance between the photo-
gates by the time difference
in crossing light paths. Photons hitting the ball alter its
motion much less than the
motion of an oil supertanker
is altered when a few fleas
bump into it.

Suppose we wish to measure the speed of a pitched baseball by passing it through a pair of photogates that are a known distance apart. The ball is timed as it interrupts beams of light in the gates. The accuracy of the ball's measured speed has to do with uncertainties in distance between the gates and the timing mechanisms. By comparison, interactions between the macroscopic ball and the photons it encounters are insignificant. But such encounters are not insignificant in the case of measuring submicroscopic things like electrons. Even a single photon bouncing off an electron appreciably alters the motion of the electron—and in an unpredictable way. If we wished to observe an electron and determine its whereabouts with light, the wavelength of the light would have to be very short. We fall into a dilemma. A short wavelength that can better "see" the tiny electron corresponds to a large quantum of energy, which greatly alters the electron's state of motion. If, on the other hand, we use a long wavelength that corresponds to a smaller quantum of energy, the change we induce to the electron's state of motion will be smaller, but the determination of its position by the coarser wave will be less accurate. The act of observing something as tiny as an electron produces a considerable uncertainty in either its position or its motion. Although this uncertainty is completely negligible for measurements of position and motion regarding everyday (macroscopic) objects, it is a predominant fact of life in the atomic domain.

The uncertainty of measurement in the atomic domain was first stated mathematically in 1927 by Werner Heisenberg of Germany and is called the **uncertainty principle**. The uncertainty principle is one of the fundamental principles in quantum mechanics. Heisenberg found that when the uncertainties in the measurement of momentum and position for a particle are multiplied together, the product must be equal to or greater than Planck's constant, h, divided by 2π, which is represented as $\hbar$ (called *h-bar*). We can state the uncertainty principle for momentum and position in a simple formula:

$$\Delta P \ \Delta X \geq \hbar$$

The Δ here means "uncertainty of": ΔP is the uncertainty of momentum (the symbol for momentum is conventionally P), and ΔX is the uncertainty of position. The product of these two uncertainties can be equal to or greater than ($\geq$) the size of $\hbar$. For minimum uncertainties, the product will equal $\hbar$; the product of larger uncertainties will be greater than $\hbar$. But in no case can the product of the uncertainties be less than $\hbar$. The significance of the uncertainty principle is that even in the best of conditions, the lower limit of uncertainty is $\hbar$. This means, for example, that if we wish to know the momentum of an electron with great accuracy, the corresponding uncertainty in position will be large. Or if we wish to know the position with great accuracy, the corresponding uncertainty in momentum will be large. The sharper one of these quantities is, the less sharp is the other. Only in the classical limit where $\hbar$ becomes zero could the uncertainties of both position and momentum be arbitrarily small. But Planck's constant is greater than zero, and we cannot in principle simultaneously know both quantities with absolute certainty.

The uncertainty principle operates similarly with energy and time. We cannot measure a particle's energy with complete precision in an infinitely short span of time. The uncertainty in our knowledge of energy, ΔE, and the duration taken to measure the energy, Δt, are related by the expression*

$$\Delta E \ \Delta t \geqslant \hbar$$

The greatest accuracy we can ever hope to attain is that case in which the product of the energy and time uncertainties equals $\hbar$. The more accurately we determine the energy of a photon, an electron, or a particle of whatever kind, the more uncertain we will be of the time during which it has that energy.

The uncertainty principle is relevant only to quantum phenomena. The inaccuracies in measuring the position and momentum of a baseball due to the interactions of observation, for example, are completely negligible. But the inaccuracies in measuring the position and momentum of an electron are far from negligible. This is because the uncertainties in the measurements of these subatomic quantities are comparable to the magnitudes of the quantities themselves.†

There is a danger in applying the uncertainty principle to areas outside of quantum mechanics. Some people conclude from statements about the interaction between the observer and the observed that the universe does not exist "out there," independent of all acts of observation, and that reality is created by the observer. Others interpret the uncertainty principle as nature's shield of forbidden secrets. Some critics of science use the uncertainty principle as evidence that science itself is uncertain. The reality of the universe whether observed or not, nature's secrets, and the uncertainties of science have very little to do with Heisenberg's uncertainty principle. The profundity of the uncertainty principle has to do with the unavoidable interaction between nature at the atomic level and the means by which we probe it.

---

*We can see that this is consistent with the uncertainty in momentum and position. Recall that Δ momentum = force $\times$ Δ time and that Δ energy = force $\times$ Δ distance. Then

$$\hbar = \Delta \text{ momentum } \Delta \text{ distance}$$

$$= (\text{force} \times \Delta \text{ time}) \ \Delta \text{ distance}$$

$$= (\text{force} \times \Delta \text{ distance}) \ \Delta \text{ time}$$

$$= \Delta \text{ energy } \Delta \text{ time}$$

†The uncertainties in measurements of momentum, position, energy, or time that are related to the uncertainty principle for a pitched baseball are only 1 part in about 10 million billion billion billion (10^{-34}). Quantum effects are negligible even for the swiftest bacterium, where the uncertainties are about 1 part in a billion (10^{-9}). Quantum effects become evident for atoms, where the uncertainties are about 1 part in a hundred (10^{-2}). For electrons moving in an atom, quantum uncertainties dominate, and we are in the full-scale realm of the quantum world.

Questions

1. Is Heisenberg's uncertainty principle applicable to the practical case of using a thermometer to measure the temperature of a glass of water?
2. A Geiger counter measures radioactive decay by registering the electrical pulses produced in a gas tube when radiation particles pass through it. The radiation particles emanate from a radioactive source—say, radium. Does the act of measuring the decay rate of radium alter the radium or its decay rate?
3. Can the quantum principle that we cannot observe something without changing it be reasonably extrapolated to support the claim that you can make a stranger turn around and look at you by staring intently at his back?

Complementarity

The realm of quantum physics is seemingly confusing. Light waves that interfere and diffract deliver their energy in particle packages of quanta. Electrons that move through space in straight lines and experience collisions as if they were particles distribute themselves spatially in interference patterns as if they were waves. In this confusion there is an underlying order. The behavior of light and electrons is confusing in the same way! Light and electrons both exhibit wave and particle characteristics.

The Danish physicist Niels Bohr, one of the founders of quantum physics, formulated an explicit expression of the wholeness inherent in this dualism. He called his expression of this wholeness the concept of

▶ **Answers**

1. No. Although we probably subject the temperature of water to a change by the act of probing it with a thermometer, especially one appreciably colder or hotter than the water, the uncertainties that relate to the precision of the thermometer are quite within the domain of classical physics. The role of uncertainties at the subatomic level is inapplicable here.

2. Not at all, because the interaction involved is between the Geiger counter and the radiation, not between the Geiger counter and the radium. It is the behavior of the radiation particles that is altered by measurement, not the radium from which they emanate. See how this ties into the next question.

3. No. Here we must be careful in defining what we mean by *observing*. If our observation involves probing (giving or extracting energy), we indeed change to some degree that which we observe. For example, if we shine a light source onto the person's back, our observation consists of probing, which, however slight, physically alters the configuration of atoms on his back. If he senses this, he may turn around. But simply staring intently at his back is observing in the passive sense. The light you receive or block by blinking, for example, has already left his back. So whether you stare, squint, or close your eyes completely, you in no physical way alter the atomic configuration on his back. Shining a light or otherwise probing something is not the same thing as passively looking at something. A failure to make the simple distinction between *probing* and *passive observation* is at the root of much nonsense that is said to be supported by quantum physics. Better support for the above claim would be positive results from a simple and practical test, rather than the assertion that it rides on the hard-earned reputation of quantum theory.

Figure 30-13
Opposites are seen to complement one another in the yin-yang symbol of Eastern cultures.

complementarity. According to this concept, quantum phenomena exhibit complementary (mutually exclusive) properties—appearing either as particles or as waves—depending on the type of experiment in which they manifest. Experiments designed to examine individual exchanges of energy and momentum bring out particlelike properties, while experiments designed to examine spatial distribution of energy bring out wavelike properties. The wavelike properties of light and particlelike properties of light complement one another, but both are necessary for the understanding of "light."

The idea that opposites are components of a wholeness is not new. Ancient Eastern cultures incorporated it as an integral part of their world view. This is demonstrated in the yin-yang diagram of T'ai Chi Tu (Figure 30-13). One side of the circle is called *yin*, and the other side is called *yang*. Where there is yin, there is yang. Only the union of yin and yang forms a whole. Where there is low, there is also high. Where there is night, there is also day. Where there is birth, there is also death. A whole person integrates yin (emotion, intuition, feminine, right brain) with yang (reason, logic, masculine, left brain). Each has aspects of the other. For Niels Bohr, the yin-yang diagram symbolized the principle of complementarity. In later life Bohr wrote on the implications of complementarity in many aspects of life. In 1947, when he was knighted for his contributions to physics, he chose for his coat of arms the yin-yang symbol.

Summary of Terms

Quantum theory The theory that energy is radiated in definite units called *quanta*, or *photons*. Just as matter is composed of atoms, radiant energy is composed of quanta. The theory further states that all material particles have wave properties.

Planck's constant A fundamental constant, h, which relates the energy of light quanta with their frequency:

$$h = 6.6 \times 10^{-34} \text{ joule-second}$$

Photoelectric effect The emission of electrons from a metal surface when light shines on it.

Uncertainty principle The principle formulated by Heisenberg that states that the ultimate accuracy of measurement is given by the magnitude of Planck's constant, h. Further, it is not possible to measure exactly both the position and the momentum of a particle at the same time, nor the energy and the time during which the particle has that energy simultaneously.

Complementarity The principle enunciated by Niels Bohr that states that the wave and particle models of either matter or radiation complement each other and when combined provide a fuller description of either one.

Suggested Reading

Cole, K. C. *Sympathetic Vibrations: Reflections of Physics as a Way of Life.* New York: Morrow, 1984. A delightful unraveling of the theories of Bohr, Einstein, and other developers of quantum physics, with emphasis on the human side of physics.

March, R. H. *Physics for Poets*, 2nd ed. New York: McGraw-Hill, 1978. This textbook presents an interesting account of contemporary physics and its historical roots.

Review Questions

1. Did the findings of Young, Maxwell, and Hertz support the wave theory or the particle theory of light?

2. Did Einstein's photon theory of light support the wave theory or the particle theory of light?

Birth of the Quantum Theory

3. What exactly did Max Planck consider quantized, the energy of vibrating electrons in matter or the energy of light itself?

4. Distinguish between the study of *mechanics* and the study of *quantum mechanics*.

Quantization and Planck's Constant

5. What does it mean to say that a certain quantity is *quantized*?

6. Why is the energy of a campfire not a whole-number multiple of a single quantum, although the energy of a laser beam is?

7. What is a quantum of light called?

8. In the previous chapter we learned the formula $E \sim f$. In this chapter we learned the formula $E = hf$. Explain the difference between these two formulas. What is h?

9. In the formula $E = hf$, does f stand for wave frequency, as defined in Chapter 18?

10. Which has the lower energy quanta, red light or blue light? Radio waves or X rays?

Photoelectric Effect

11. What evidence can you cite for the wave nature of light? For the particle nature of light?

12. Which are more successful in dislodging electrons from a metal surface, photons of violet light or photons of red light? Why?

13. Why won't a very bright beam of red light impart more energy to an electron than a feeble beam of violet light?

14. What exactly did Einstein consider quantized, the energy of vibrating electrons in matter or the energy of light itself?

15. Does the brightness of a beam of light primarily depend on the frequency of photons or on the number of photons?

16. Einstein proposed his explanation of the photoelectric effect in 1905. When were his views on this confirmed?

Wave-Particle Duality

17. Why do photographs look grainy when magnified?

18. Does light behave primarily as a wave or as a particle when it interacts with the crystals of matter in photographic film?

Double-Slit Experiment

19. Does light travel from one place to another in a wavelike way or a particlelike way?

20. Does light interact with a detector in a wavelike way or a particlelike way?

21. When does light behave as a wave? When does it behave as a particle?

Particles as Waves: Electron Diffraction

22. What evidence can you cite for the wave nature of particles?

23. When electrons are diffracted through a double slit, do they arrive at the screen in a wavelike way or a particlelike way? Is the pattern of hits wavelike or particlelike?

Uncertainty Principle

24. In which of the following are quantum uncertainties significant: measuring simultaneously the speed and location of a baseball; of a spitball; of an electron.

25. What is the uncertainty principle with respect to motion and position?

26. If measurements show a precise position for an electron, can those measurements show precise momentum also? Explain.

27. If measurements show a precise value of energy for an electron undergoing de-excitation, can those measurements show a precise time for this event as well? Explain.

28. Is there a distinction between passively observing something and actively probing something? In which case does one affect the behavior of that something?

Complementarity

29. What is the principle of complementarity?

30. How does the principle of complementarity regard the opposite wavelike and particlelike properties of light?

Exercises

1. Distinguish between *classical physics* and *quantum physics*.

2. How can a photon's energy be given by the formula $E = hf$ when f in the formula is a wave frequency?

3. Which photon has the greatest energy, an infrared, a visible, or an ultraviolet?

4. We speak of photons of red light and photons of green light. Can we speak of photons of white light? Why or why not?

5. A beam of red light and a beam of blue light have exactly the same energy. Which beam contains the greater number of photons?

6. As a solid is heated and begins to glow, why does it first appear red?

7. Silver bromide (AgBr) is a light-sensitive substance used in some types of photographic film. To cause exposure of the film, it must be illuminated with light having sufficient energy to break apart the molecules. Why do you suppose this film may be handled without exposure in a darkroom illuminated with red light? How about blue light? How about very bright red light as compared to very dim blue light?

8. Suntanning produces cell damage in the skin. Why is ultraviolet radiation capable of producing this damage, while visible radiation is not?

9. In the photoelectric effect, does brightness or frequency determine the number of electrons ejected per second?

10. Explain how the photoelectric effect is used to open automatic doors when someone approaches.

11. If you shine an ultraviolet light on the metal ball of a negatively charged electroscope (shown in Exercise 7 in Chapter 21), it will discharge. But if the electroscope is positively charged, it won't discharge. Can you venture an explanation?

12. When ultraviolet light falls on certain dyes, visible light is emitted. Why does this not happen when infrared light falls on these dyes?

13. Estimate the relative current readings in the ammeter in Figure 30-1 for illumination of the photosensitive plate by light of various colors, for various intensities.

14. Explain briefly how the photoelectric effect is used in the operation of at least two of the following: an electric eye, a photographer's light meter, the sound track of a motion picture.

15. Does the photoelectric effect *prove* that light is made of particles? Do interference experiments *prove* that light is composed of waves? (Is there a distinction between what something *is* and how it *behaves*?)

16. Does Einstein's explanation of the photoelectric effect invalidate Young's explanation of the double-slit experiment? Explain.

17. The camera that took the photograph of the woman's face (Figure 30-4) used ordinary lenses that are well known to refract waves. Yet the step-by-step formation of the image is evidence of photons. How can this be? What is your explanation?

18. If a proton and an electron have identical speeds, which has the longer wavelength?

19. One electron travels twice as fast as another. Which has the longer wavelength?

20. Does the de Broglie wavelength of a proton become longer or shorter as velocity increases?

21. If a cannonball and a BB have the same speed, which has the longer wavelength?

22. We don't notice the wavelength of moving matter in our ordinary experience. Is this because the wavelength is extraordinarily large or extraordinarily small?

23. What principal advantage does an electron microscope have over an optical microscope?

24. Would a beam of protons in a "proton microscope" exhibit greater or less diffraction than electrons of the same speed in an electron microscope? Defend your answer.

25. If Planck's constant, *h*, were equal to zero rather than a tiny number, how would the uncertainty principle differ?

26. Comment on the idea that the theory one believes determines the meaning of one's observations and not vice versa.

27. There is at least one electron on the tip of your nose. If somebody looks at it, will its motion be altered? How about if it is looked at with one eye closed? With two eyes, but crossed? Does the uncertainty principle apply here?

28. Do we alter that which we attempt to measure in a public opinion survey? Does the uncertainty principle apply here?

29. If a system has been determined by specific causes that are well understood, does it follow that the future of that system can be predicted? (Is there a distinction between a *determined* system and a *predictable* system?)

30. To measure the exact age of Old Methuselah, the oldest living tree in the world, a Nevada professor of dendrology, aided by an employee of the U.S. Bureau of Land Management, in 1965 cut the tree down and counted its rings. Is this an example of the uncertainty principle in its extreme or an example of arrogant and criminal stupidity?

PART 7

ATOMIC & NUCLEAR PHYSICS

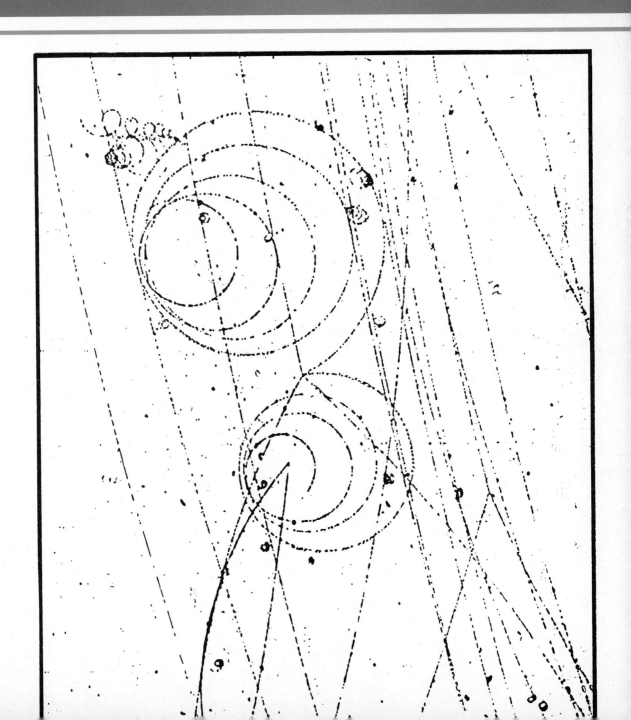

31 The Atom and the Quantum

We discussed the atom as a building block of matter in Chapter 10 and as an emitter of light in the preceding chapters. We know that the atom is composed of a central nucleus surrounded by a complex arrangement of electrons; the study of this atomic structure is called *atomic physics*. In this chapter we will outline some of the developments that led to our present understanding of atomic physics and trace these developments from classical to quantum physics. In the two following chapters, we will learn about the study of the structure of the atomic nucleus—*nuclear physics*. This knowledge and its implications are having a profound impact on human society.

We begin our study of atomic and nuclear physics with a brief account of some of the events at the beginning of the century that led to our current understanding of the atom.

Discovery of the Atomic Nucleus

A few years after Einstein announced the photoelectric effect, the British physicist Ernest Rutherford performed his now-famous gold-foil experiment. This 1909 experiment showed that the atom was mostly empty space, with most of its mass packed into the central region—the *nucleus*. Rutherford directed a beam of positively charged particles (alpha parti-

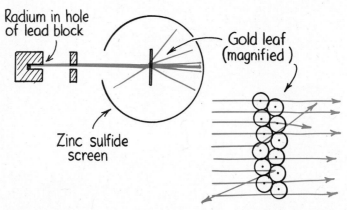

Figure 31-1
The occasional large-angle scattering of alpha particles from the gold atoms led Rutherford to the discovery of the small, very massive nuclei at their centers.

cles) from a radioactive source through a very thin gold leaf and measured the angles at which the particles were deflected from their straight-line path as they emerged. This was accomplished by noting spots of light on a zinc sulfide screen around the gold leaf (Figure 31-1). Most particles continued, as expected, in a more or less straight-line path through the thin foil. But, surprisingly, some particles were widely deflected. Some were even scattered back along their incident paths. It was like firing bullets at a piece of tissue paper and finding some bullets bouncing back. Rutherford reasoned that the particles that were undeflected traveled through empty regions of the gold foil, while the small number of particles that were deflected were repelled from the centers of gold atoms that must also be positively charged. Rutherford had discovered the atomic nucleus.

Atomic Spectra: Clues to Atomic Structure

During the period of Rutherford's experiments, chemists were using the spectroscope (discussed in Chapter 29) for chemical analysis, while physicists were busy trying to find order in the confusing arrays of spectral lines. It had long been noted that the lightest element, hydrogen, had a far more orderly spectrum than the other elements. An important sequence of lines in the hydrogen spectrum runs from a single line in the red region of the spectrum to one in the blue, then to several lines in the violet, followed by many in the ultraviolet. Spacing between successive lines becomes smaller and smaller from the first in the red to the last in the ultraviolet, until the lines become so close they seem to merge. A Swiss schoolteacher, J. J. Balmer, first expressed the positions of these lines in a single mathematical formula. Balmer, however, could give no reason why his formula worked so successfully. Using his formula, it was possible to locate lines that had not yet been measured. This success spurred the discovery of other similar formulas that fixed the line positions for other elements.

Figure 31-2
A portion of the hydrogen spectrum.

Another regularity in atomic spectra was found by J. Rydberg. He noticed that the sum of the frequencies of two lines in the spectrum of hydrogen sometimes equals the frequency of a third line. This relationship was later advanced as a general principle by W. Ritz and is called the **Ritz combination principle**. It states that the spectral lines of any element include frequencies that are either the sum or the difference of the frequencies of two other lines. Like Balmer, Ritz was unable to offer an explanation for this regularity. These regularities were the clues Danish physicist Niels Bohr used to understand the structure of the atom itself.

Bohr Model of the Atom

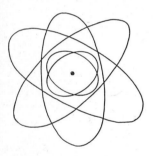

Figure 31-3
The Bohr model of the atom. Although this model is very oversimplified, it is still useful in understanding light emission.

In 1913 Bohr applied the quantum theory of Planck and Einstein to the nuclear atom of Rutherford and formulated the well-known planetary model of the atom.* Bohr reasoned that Planck's quantized electron energy states (discussed in the previous chapter) would correspond to electron orbits of different radii. He was able to calculate the energy states in the hydrogen atom and to show that they corresponded to quantized circular orbits. From the Balmer formula he was able to calculate the angular momentum of an electron in each orbit. Bohr asserted that the angular momentum is a whole-number multiple of $\hbar(h/2\pi)$, where h is Planck's constant, which links the energy and frequency of light.

According to Bohr, an electron orbiting farther from the nucleus is in a higher energy state than one in a closer orbit. Bohr reasoned that light is emitted when electrons make a transition from a higher to a lower orbit and that the frequency of emitted radiation is given by $E = hf$, where E is the difference in energy of the atom when the electron is in the different orbits.

Bohr's planetary model of the atom faced a major difficulty. Accelerating electrons radiate energy in the form of electromagnetic waves; an electron accelerating around a nucleus should radiate energy continuously. This radiating away of energy should cause the electron to lose orbital speed and spiral into the nucleus (Figure 31-4). Bohr boldly broke with classical physics by stating that the electron doesn't radiate light while it accelerates around the nucleus in a single orbit, but that radiation of light takes place only when the electron jumps orbit from a higher energy level to a lower energy level. The orbits farther from the nucleus have higher energy levels than closer orbits. The energy of the emitted photon is equal to the *difference* in energy between the two energy levels, $E = hf$. Color depends on the jump. So the quantization of light energy neatly corresponds to the quantization of electron energy.

Bohr's views, as outlandish as they seemed at the time, explained the regularities being found in atomic spectra. Bohr's explanation of the Ritz combination principle is shown in Figure 31-5. If an electron is raised to the third energy level, it can return to its initial level by a single jump from the third to the first level—or by a double jump, first to the second level and then to the first level. These two return paths will produce three spectral lines. Note that the energy jump along path A plus B is equal to the energy jump C. Since frequency is proportional to energy, the frequencies of light emitted along path A and path B when added equal the frequency of light emitted when the transition is along path C. Now we see why the sum of two frequencies in the spectrum is equal to a third frequency in the spectrum.

Bohr was able to account for X rays and show that they are emitted when electrons jump from outer to innermost orbits. He predicted X-ray

---

*The model has major defects, because the electrons do not revolve in a plane as planets do. Later the model was revised; *orbits* became *shells*. We use *orbit* because it is commonly used. Electrons are not just bodies, like planets, but are like waves arranged in a thin shell.

Figure 31-4
From a classical point of view, an electron accelerating around its orbit should emit radiation. This loss of energy should be accompanied by a decrease in orbital speed and a spiraling of the electron into the nucleus. But this does not happen.

Figure 31-5
Three of the many energy levels in an atom. The energy jumps A + B equal the energy jump C. The sum of the frequencies of light emitted from jump A and jump B therefore equal the frequency of light emitted from jump C.

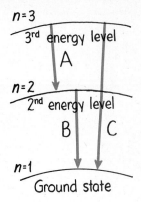

frequencies that were later experimentally confirmed. Bohr was also able to calculate the "ionization energy" of a hydrogen atom—the energy needed to knock the electron out of the atom completely. This also was verified by experiment.

Using measured X-ray and visible frequencies, it was possible to map energy levels of all the atomic elements. Bohr's model had electrons orbiting in neat circles (or ellipses) arranged in groups or shells. This model of the atom accounted for the general chemical properties of the elements and predicted properties of a missing element, which led to the discovery of hafnium.

Bohr solved the mystery of atomic spectra while providing an extremely useful model of the atom. He was quick to point out that his model was to be interpreted as a crude beginning, and the picture of electrons whirling like planets about the sun was not to be taken literally (to which popularizers of science paid no heed). His discrete orbits were conceptual representations of an atom whose later description involved quantum mechanics. Still, his planetary model of the atom, with electrons occupying discrete energy levels, underlies the more complex models of the atom today.

Questions

1. What is the maximum number of paths for de-excitation available to a hydrogen atom excited to the $n = 3$ level in changing to the ground state?

2. Two predominant spectral lines in the hydrogen spectrum, an infrared one and a red one, have frequencies 2.7×10^{14} Hz and 4.6×10^{14} Hz respectively. Can you predict a higher-frequency line in the hydrogen spectrum?

▶ **Answers**

1. Three, as shown in Figure 31-5.

2. The addition of frequencies $2.7 \times 10^{14} + 4.6 \times 10^{14} = 7.3 \times 10^{14}$ Hz, which happens to be the frequency of a violet line in the hydrogen spectrum. Can you see that if the infrared line is produced by a transition similar to path A in Figure 31-5 and the red line corresponds to path B, then the violet line corresponds to path C?

Relative Sizes of Atoms

The diameters of the electron orbits in the Bohr model of the atom are determined by the amount of electrical charge in the nucleus. For example, the positive proton in the hydrogen atom holds one electron in an orbit at a certain radius. If we double the positive charge in the nucleus, the orbiting electron will be pulled into a tighter orbit with half its former radius since the electrical attraction is doubled. This doesn't quite happen, however, because the double charge in the nucleus normally holds a second orbital electron, which diminishes the effect of the positive nucleus. This added electron makes the atom electrically neutral. The atom is no longer hydrogen. It is helium.

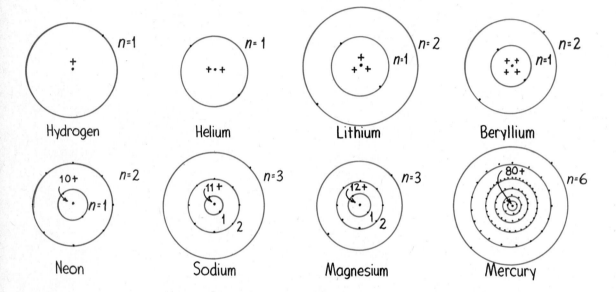

Figure 31-6
Orbital models for some light and heavy atoms drawn to approximate scale. Note that the heavier atoms are not appreciably larger than the lighter atoms. (Adapted from *Descriptive College Physics* by Harvey E. White. Copyright © 1971 by Litton Educational Publishing, Inc. Reprinted by permission of Van Nostrand Reinhold Company.)

The two orbital electrons assume an orbit characteristic of helium. An additional proton in the nucleus pulls the electrons into an even closer orbit and, furthermore, holds a third electron in a second orbit. This is the lithium atom, atomic number 3. We can continue with this process, increasing the positive charge of the nucleus and adding successively more electrons and more orbits all the way up to atomic numbers above 100—to the synthetic radioactive elements.*

We find that as the nuclear charge increases and additional electrons are added in outer orbits, the inner orbits shrink in size because of the stronger nuclear attraction. This means that the heavier elements are not much larger in diameter than the lighter elements. The diameter of the

---

*Each orbit will hold only so many electrons. A rule of quantum mechanics is that an orbit is filled when it contains a number of electrons given by $2n^2$, where n is 1 for the first orbit, 2 for the second orbit, 3 for the third orbit, and so on. For $n = 1$, there are 2 electrons; for $n = 2$, there are $2(2^2)$ or 8 electrons; for $n = 3$, there are a maximum of $2(3^2)$ or 18 electrons, etc. The number n is called the *principal quantum number*.

uranium atom, for example, is only about three hydrogen diameters, even though it is 238 times as massive. The schematic diagrams in Figure 31-6 are drawn approximately to the same scale.

We find that each element has an arrangement of electron orbits unique to that element. For example, the radii of the orbits for the sodium atom are the same for all sodium atoms, but different from the radii of the orbits for other kinds of atoms. When we consider the 92 naturally occurring elements, we find that there are 92 distinct patterns or orbits—a different pattern for each element.

Question ▶ What fundamental force dictates the size of an atom?

Explanation of Quantized Energy Levels: Electron Waves

So we see that photons are emitted when electrons make a transition from a higher to a lower energy level, and the frequency of the photon is equal to the energy-level difference divided by Planck's constant, h. If an electron jumps down a large energy-level difference, the emitted photon has a high frequency—perhaps ultraviolet. If an electron jumps through a lesser energy difference, the emitted photon is lower in frequency—perhaps a brief burst of red light. Each element has its own characteristic energy levels; thus, transitions of electrons between these levels result in each element emitting its own characteristic colors. Each of the elements emits its own unique pattern of spectral lines.

The idea that electrons may occupy only certain levels was very perplexing to early investigators and to Bohr himself. It was perplexing because the electron was considered to be a particle, a tiny BB whirling around the nucleus like a planet whirling around the sun. Just as a satellite can orbit at any distance from the sun, it would seem that an electron should be able to orbit around the nucleus at any radial distance—depending, of course, like the satellite, on its speed. Moving among all orbits would enable the electrons to emit all energies of light. But this doesn't happen. It can't. Why the electron occupies only discrete levels is understood by considering the electron to be not a particle, but a *wave*.

Louis de Broglie introduced the concept of matter waves in 1924. He hypothesized that a wave was associated with every particle and that the wavelength of a matter wave is inversely related to a particle's momentum. These **de Broglie matter waves** behave just like other waves; they can be reflected, refracted, and diffracted to demonstrate interference. Using the interference of matter waves, de Broglie showed that the values of angular momentum in Bohr's orbits were a natural consequence of standing electron waves. A Bohr orbit exists where an electron wave closes on itself (Figure 31-7). In this way it reinforces itself constructively

---

▶ **Answer**

The electrical force.

Figure 31-7

(a) Orbital electrons form standing waves only when the circumference of the orbit is equal to an integral number of wavelengths. In (b) the wave does not close in on itself in phase, and therefore it undergoes destructive interference.

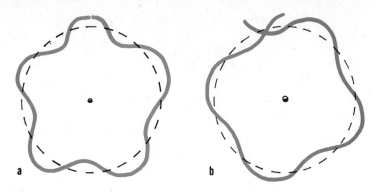

a   b

in each cycle, just as the wave on a violin string is constructively reinforced by its successive reflections. In this view the electron is thought of not as a particle located at some point in the atom but as though its mass and charge were spread out into a standing wave surrounding the atomic nucleus, the wavelength of which must fit evenly into the circumferences of the orbits. The circumference of the innermost orbit is equal to one wavelength of the electron wave. The second orbit has a circumference of two electron wavelengths, the third three, and so forth (Figure 31-8). This is similar to a chain necklace made of paper clips. No matter what size necklace is made, its circumference is equal to some multiple

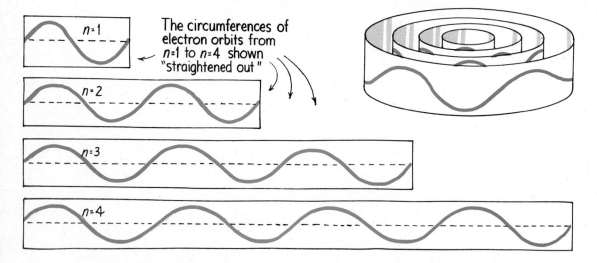

The circumferences of electron orbits from n=1 to n=4 shown "straightened out"

n=1

n=2

n=3

n=4

Figure 31-8

The electron orbits in an atom have discrete radii because the circumferences of the orbits are integral multiples of the electron wavelengths. The wavelengths differ for different elements (and also for different orbits within the elements), resulting in discrete orbital radii, or energy levels, which are characteristic of each element. The figure is greatly oversimplified, as the standing waves make up spherical and ellipsoidal shells rather than flat, circular ones.

of the length of a single paper clip.* Since the circumferences of electron orbits are discrete, it follows that the radii of these orbits, and hence the energy levels, are also discrete.

This view explains why electrons don't spiral closer and closer to the nucleus, causing atoms to shrink in on themselves. If each electron orbit is described by a standing wave, the circumference of the smallest orbit can be no smaller than one wavelength—no fraction of a wavelength is possible in a circular (or elliptical) standing wave.

Quantum Mechanics

The mid-1920s saw many changes in physics. Not only was light found to have particle properties, but particles were found to have wave properties. Starting with de Broglie's matter waves, the Austrian-German physicist Erwin Schrödinger formulated an equation that describes how matter waves change under the influence of external forces. Schrödinger's equation plays the same role in **quantum mechanics** that Newton's equation (acceleration = force/mass) plays in classical physics.† The matter waves in Schrödinger's equation are mathematical rather than physical entities, so the equation provides us with a purely mathematical rather than a visual model of the atom—which puts it beyond the scope of this book. So our discussion of it will be brief.‡

In **Schrödinger's wave equation**, the thing that "waves" is the non-material *matter wave amplitude*—a mathematical entity called a *wave function*, represented by the symbol ψ (the Greek letter psi). The wave function given by Schrödinger's equation represents the possibilities that can occur for a system. For example, the location of the electron in a hydrogen atom may be anywhere from the center of the nucleus to a radial distance approaching infinity. Its possible position and its probable position at a particular time are not the same. A physicist can calculate its probable position by multiplying the wave function by itself ($|\psi|^2$). This produces a second mathematical entity called a *probability density function*, which tells us at a given time the probability per unit volume for each of the possibilities represented by ψ.

Experimentally, there is a finite probability (chance) of finding an electron in a particular region at any instant—0 for never, 1 for always. The value of this probability lies between the limits 0 and 1. For example,

---

*For each orbit the electron has a unique speed, which determines its wavelength. Electron speeds are less and wavelengths longer for orbits of increasing radii; so to make our analogy accurate, we would have to use not only more paper clips to make increasingly longer necklaces but larger paper clips for each longer necklace as well.

†Schrödinger's wave equation, strictly for math types, is

$$\left(-\frac{\hbar^2}{2m}\nabla^2 + V \right)\psi = i\hbar\frac{\partial\psi}{\partial t}$$

‡Our short treatment of this complex subject is hardly conducive to any real understanding of quantum mechanics. At best it serves as a brief overview and possible introduction to further study. The reading suggested at the end of the chapter may be quite useful.

Figure 31-9

Probability distribution of an electron cloud.

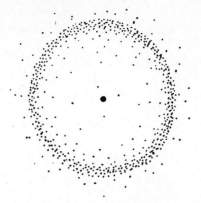

if the probability is 0.4 for finding the electron within a certain radius, this would signify a 40 percent chance of finding it there. So the Schrödinger equation cannot tell a physicist where an electron can be found in an atom at any moment, but only the *likelihood* of finding it there— or, for a large number of measurements, what fraction of measurements will find the electron in each region. For example, if the electron's position in the first Bohr energy state for hydrogen is repeatedly measured and the results plotted in the form of dots, the resulting pattern will resemble a sort of electron cloud (Figure 31-9). An individual electron may at various times be detected anywhere in this probability cloud; it even has an extremely small but finite probability of momentarily existing inside the nucleus. It is detected most of its time, however, close to an average distance from the nucleus, which fits the orbital radius described by Niels Bohr.

Figure 31-10

From the Bohr model of the atom to the modified model with de Broglie waves to the mathematical model of Schrödinger.

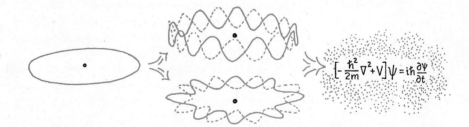

$$\left[-\frac{\hbar^2}{2m}\nabla^2 + V\right]\psi = i\hbar\frac{\partial\psi}{\partial t}$$

Most—but not all—physicists view quantum mechanics as fundamental to nature. Interestingly enough, Albert Einstein, one of the founders of quantum physics, never accepted it as fundamental; he considered the probabilistic nature of quantum phenomena as the outcome of a deeper physics. He stated, "Quantum mechanics is certainly imposing. But an inner voice tells me it is not yet the real thing. The theory says a lot, but does not really bring us closer to the secret of 'the Old One.'"

Questions

1. Consider 100 photons diffracting through a thin slit to form a diffraction pattern. If we detect 5 photons in a certain region in the pattern, what is the probability (between 0 and 1) of detecting a photon in this region?

2. Open a second identical slit and the diffraction pattern is one of bright and dark bands. Suppose the region where 5 photons hit before now has none; the probability of finding a photon there is now 0. A wave theory says waves that hit before are now canceled by waves from the other slit—that crests and troughs combine to 0. But our measurement is of photons that either make a hit or don't. How does a quantum mechanical view reconcile this?

Correspondence Principle

If a new theory is valid, it must account for the verified results of the old theory. This is the **correspondence principle**, first enunciated by Bohr. New theory and old must correspond; that is, they must overlap and agree in the region where the results of the old theory have been fully verified.*

When the techniques of quantum mechanics are applied to macroscopic systems rather than atomic systems, the results are essentially identical to those of classical mechanics. For a large system such as the motion of planets where classical physics is successful, the Schrödinger equation leads to results that differ from classical theory only by infinitesimal amounts. The two domains blend when the de Broglie wavelength is small compared to the dimensions of the system. Classical physics is less cumbersome and more useful when a system's dimensions are large compared with atomic dimensions. But at the atomic level, quantum physics reigns.

So we see that when quantum physics is applied to large orbits, it is smoothly linked with the classical physics of Newton. We will consider the very large in later chapters and see that Newtonian physics links with relativity theory in the domain of the very massive and/or the very fast.

▶ **Answers**

1. We have a 0.05 probability of detecting a photon at this location. In quantum mechanics we say $|\psi|^2 = 0.05$. Note that we are predicting the probability of an event, rather than predicting the event itself.

2. Light travels from the slits to the screen as waves, and strikes the screen as particles. In quantum mechanics we say the wave function ψ, which may be plus or minus, travels from the slits to the screen and determines where light will strike as a particle. The square of the sum of wave functions ψ_1 and ψ_2 for both slits will be zero in the region where no photons are detected. That is, $(|\psi_1 + \psi_2|)^2 = 0$.

---

*This is a general rule not only for good science, but for all good theory—even in areas as far removed from science as government, religion, and ethics.

Summary of Terms

Ritz combination principle The theory that the spectral lines of the elements have frequencies that are either the sums or the differences of the frequencies of two other lines.

de Broglie matter waves The associated wave properties of all particles of matter. The wavelength of a particle wave is related to its momentum and Planck's constant, h, by the relationship

$$\text{Wavelength} = \frac{h}{\text{momentum}}$$

Quantum mechanics The branch of quantum physics that deals with finding the probability amplitudes of matter waves, organized principally by findings of Werner Heisenberg (1925) and Erwin Schrödinger (1926).

Schrödinger's wave equation The fundamental equation of quantum mechanics, which interprets the wave nature of material particles in terms of probability wave amplitudes. It is as basic to quantum mechanics as Newton's laws of motion are to classical mechanics.

Correspondence principle The rule that a new theory is valid provided that, when it overlaps with the old, it agrees with the verified results of the old theory.

Suggested Reading

Cline, Barbara L. *The Questioners: Physicists and the Quantum Theory*. New York: Crowell, 1973. A fascinating account of the development of quantum physics with emphasis on the participating physicists.

Gamow, George. *Thirty Years That Shook Physics*. New York: Dover, 1985. A historical tracing of quantum theory by someone who was part of it.

Hey, A. J., and Walters, P. *The Quantum Universe*. New York: Cambridge University Press, 1987. A broad view of modern physics with many illustrations.

Pagels, H. R. *The Cosmic Code: Quantum Physics as the Language of Nature*. New York: Simon & Schuster, 1982. An elegant and highly recommended book for the general reader.

Trefil, J. *Atoms to Quarks*. New York: Scribner's, 1980. A nice development of quantum theory in the early chapters, which continue on to particle physics.

Review Questions

1. Distinguish between *atomic physics* and *nuclear physics*.

Discovery of the Atomic Nucleus

2. Why are alpha particles repelled by the atomic nucleus?

3. Why do most alpha particles fired through a piece of gold foil emerge undeflected?

4. Why do a few alpha particles fired through a piece of gold foil bounce backward?

Atomic Spectra: Clues to Atomic Structure

5. What did J. J. Balmer discover about atomic spectra?

6. What did J. Rydberg and W. Ritz discover about atomic spectra?

Bohr Model of the Atom

7. What relationship between electron orbits and light emission did Bohr postulate?

8. According to Bohr, can a single electron in one excited atom give off more than one photon when it returns to ground level?

9. What is the relationship between the difference in energy levels in an atom and the light it emits?

10. Did Bohr state that atoms are like tiny solar systems? Are they?

Relative Sizes of Atoms

11. Why is a helium atom smaller than a hydrogen atom?

12. Why are heavy atoms not appreciably larger than the hydrogen atom?

Explanation of Quantized Energy Levels: Electron Waves

13. How does treating the electron as a wave rather than as a particle solve the riddle of why electron orbits are discrete?

14. How many wavelengths are there in an electron wave in the first orbit? In the second orbit? In the nth orbit?

15. How can we explain why electrons don't spiral into the attracting nucleus?

Quantum Mechanics

16. What does a wave function ψ represent?

17. Distinguish between a *wave function* and a *probability density function*.

18. How does the probability cloud of the electron in a hydrogen atom compare with the orbital radius described by Niels Bohr?

Correspondence Principle

19. Why is the correspondence principle a test for the validity of an idea?

20. Would it be correct to say that an electron is a wave or a particle? Or would it be correct to say that an electron, whatever it is, *behaves* as if it were a wave or a particle?

Exercises

1. How does Rutherford's model of the atom account for the back-scattering of alpha particles directed at the gold leaf?

2. Why are spectral lines often referred to as "atomic fingerprints"?

3. Uranium is 238 times more massive than hydrogen. Why, then, isn't the diameter of the uranium atom 238 times that of the hydrogen atom?

4. If all atoms of the same kind were not of the same size, would crystal structure be affected? How?

5. How does the wave model of electrons orbiting the nucleus account for discrete energy values rather than arbitrary energy values?

6. How can a hydrogen atom, which has only one electron, have so many spectral lines?

7. Suppose that a certain atom possesses four distinct energy levels. Assuming that all transitions between levels are possible, how many spectral lines will this atom exhibit? Which transitions correspond to the highest-energy light emitted? To the lowest-energy light?

8. An electron de-excites from the fourth quantum level to the third and then directly to the ground state. Two photons are emitted. How do their combined energies compare to the energy of the single photon that would be emitted by de-excitation from the fourth level directly to the ground state?

9. A helium atom is more massive than a hydrogen atom. Why is it smaller than a hydrogen atom?

10. Why does no stable electron orbit exist in an atom for a circumference of 2.5 de Broglie wavelengths?

11. Distinguish between the wavelength and the amplitude of a matter wave.

12. If Planck's constant, h, were larger, would atoms be larger also? Defend your answer.

13. If the world of the atom is so uncertain and subject to the laws of probabilities, how can we accurately measure such things as light intensity, electric current, and temperature?

14. Why do we say that light has wave properties? Why do we say that light has particle properties?

15. Why do we say that electrons have particle properties? Why do we say that electrons have wave properties? Is there a contradiction here? Explain.

16. Comment on the statement, "If an electron is not a particle, then it must be a wave." (Do you hear statements like this often?)

17. What principle supports the fact that quantum physics gives the same results as classical physics for large numbers of quanta?

18. Is the correspondence principle relevant to the everyday macroworld?

19. Explain how Newton's first and second laws of motion satisfy the correspondence principle.

20. Richard Feynman in his book *The Character of Physical Law* states: "A philosopher once said, 'It is necessary for the very existence of science that the same conditions always produce the same results.' Well, they don't!" Who was speaking of classical physics, and who was speaking of quantum physics?

"Know nukes!" The natural heat of the earth that powers this geyser, like the heat that warms a natural hot spring or powers a volcano, comes from nuclear power—the radioactivity of minerals in the earth's core. Power from the atomic nucleus has been with us since the earth was formed and is not restricted to today's nuclear reactors, or "nukes," as they are called. How about that!

32 Atomic Nucleus and Radioactivity

In the last chapter we were concerned with the study of the clouds of electrons that make up the atom—*atomic physics*. In this chapter we will burrow beneath the electrons and go deeper into the atom—to the atomic nucleus. The study of the nucleus is connected to the chance discovery of radioactivity in 1896, which in turn is connected to the discovery of X rays 2 months earlier. So to begin our study of *nuclear physics* we'll consider X rays.

X Rays and Radioactivity

Before the turn of the century the German physicist Wilhelm Roentgen discovered a "new kind of ray" produced by an electron beam striking a glass surface. He named these **X rays**—rays of an unknown nature. Roentgen further discovered that a beam of electrons directed against a metal plate produced X rays that could pass through solid materials. Today we know that X rays are high-frequency electromagnetic waves, which usually are emitted by the excitation of the innermost orbital electrons of atoms. Whereas the electron current in a fluorescent lamp excites the outer electrons of atoms and produces ultraviolet and visible photons, a more energetic beam of electrons excites the innermost electrons and produces higher-frequency photons of X radiation.

X-ray photons have high energy and can pass through many layers of atoms before being absorbed or scattered. X rays do this when they pass through you to produce images of the inside of your body (Figure 32-1).

Two months after Roentgen announced his discovery of X rays, the French physicist Antoine Henri Becquerel tried to find out whether any elements spontaneously emitted X rays. To do this, he wrapped a photographic plate in black paper to keep out the light and then put pieces of various elements against the wrapped plate. He thought that if these materials emitted X rays, the rays would go through the paper and blacken the plate. He found that although most elements produced no effect, uranium did give out rays. It was soon discovered that similar rays are emitted by other elements, such as thorium, actinium, and two new elements discovered by Pierre and Marie Curie—polonium and radium. The emission of these rays was evidence of much more drastic changes in the atom than atomic excitation. These rays were the result not of changes in the

591

Figure 32-1

X rays emitted by excited metallic atoms in the electrode pass through the hand and expose the film.

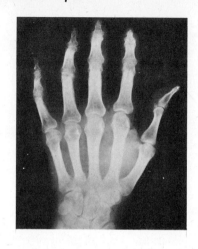

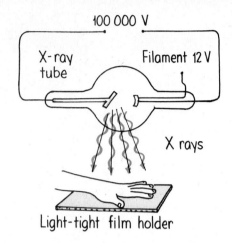

electron energy states of the atom, but of changes occurring within the central atomic core—the nucleus. These rays were the result of a spontaneous chipping apart of the atomic nucleus—radioactivity.

Alpha, Beta, and Gamma Rays

The radioactive elements emit three distinct types of rays, called by the first three letters of the Greek alphabet, α, β, γ—*alpha*, *beta*, and *gamma*. **Alpha rays** have a positive electrical charge, **beta rays** have a negative charge, and **gamma rays** have no charge at all. The three rays can be separated by putting a magnetic field across their paths (Figure 32-3).

Figure 32-2

A gamma ray is simply electromagnetic radiation, much higher in frequency and energy than light and X rays.

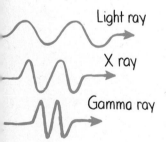

Figure 32-3

The bending of a narrow beam of rays in a magnetic field. The beam comes from a radioactive source placed at the bottom of a hole drilled in a lead block.

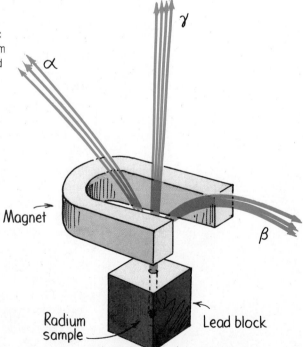

Further investigation has shown that an alpha ray is a stream of helium nuclei, and a beta ray is a stream of electrons. Hence, we often call these **alpha particles** and **beta particles**. A gamma ray is electromagnetic radiation (a stream of photons) whose frequency is higher than that of X rays. Whereas X rays originate in the electron cloud outside the atomic nucleus, gamma rays originate in the nucleus. Gamma photons provide information about nuclear structure, much as visible and X-ray photons give information about atomic electron structure.

Figure 32-4
Alpha particles are the least penetrating and can be stopped by a few sheets of paper. Beta particles will readily pass through paper, but not through a sheet of aluminum. Gamma rays penetrate several centimeters into solid lead.

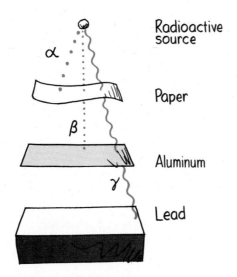

The Nucleus

As described earlier, the atomic nucleus occupies only a few quadrillionths the volume of the atom. Thus, the atom is mostly empty space. The nucleus is composed of **nucleons**, which when electrically charged are protons and when electrically neutral are neutrons. The neutron's mass is slightly greater than the proton's. Nucleons have nearly 2000 times the mass of electrons, so the mass of an atom is practically equal to the mass of its nucleus. The positive charge of the proton is the same in magnitude as the negative charge of the electron. When an electron is ejected from the nucleus (beta emission), a neutron becomes a proton.

Nuclear radii range from about 10^{-15} meter for hydrogen to about six times larger for uranium. The shapes of nuclei are generally spherical but deviate from that shape sometimes in the "football" way and sometimes in the "doorknob" way. The protons and neutrons tend to cluster into alpha particles, making a generally lumpy surface. A "skin" of neutrons is thought to make up the outer portion of the heavier nuclei. Just as there are energy levels for the orbital electrons of an atom, there are energy levels within the nucleus. Whereas electrons making transitions to lower orbits emit photons of light, similar changes of energy states within the nucleus result in the emission of gamma-ray photons. The emission of alpha particles can be understood from the viewpoint of quantum mechanics. Just as orbital electrons form a probability cloud about the nucleus, inside the radioactive nucleus there is a similar probability cloud for alpha particles (and other particles as well). Probability is extremely high that the alpha particle will be inside the nucleus, but if the probability wave extends outside, there is a finite chance that the alpha particle will be outside. Once outside, it is hurled violently away by electrical repulsion.

In addition to alpha, beta, and gamma rays, more than 200 various other particles have been detected coming from the nucleus when clobbered by energetic particles. We do not think of these so-called elementary particles as being buried within the nucleus and then popping out, just as we do not think of a spark as being buried in a match before it is struck. These particles come into being when the nucleus is disrupted. Regularities in the masses of these particles as well as the particular characteristics of their creation are explained by the existence of a fairly compact family of six subnuclear particles—the *quarks*.

Quarks are the fundamental building blocks of all nucleons. An unusual property of quarks is that they carry fractional electrical charges. One kind, the *up* quark, carries $+\frac{2}{3}$ the electron charge, and another kind, the down quark, has $-\frac{1}{3}$ the electron charge. (The name *quark*, inspired by a quotation from *Finnegans Wake* by James Joyce, was chosen in 1963 by Murray Gell-Mann, who first proposed their existence.) Each quark has an antiquark with opposite electric charge. The proton consists of the combination *up up down*, and the neutron of *up down down*. As with magnetic poles, no quarks have been isolated and experimentally observed; most investigators think quarks by nature cannot be isolated. We will not delve further into the nature of quarks.

Lighter particles like muons and electrons, and still lighter particles called *neutrinos*, make up a class of six particles called *leptons*, which are not composed of quarks. At the present time, the six quarks and six leptons (and their antiparticles) are thought to be the truly *elementary particles*, particles not composed of more elementary entities. They are the building blocks of which all matter is composed. Investigation of elementary particles is at the frontier of our present knowledge and the area of much current excitement and research.

Isotopes

The nucleus of a hydrogen atom consists of a single proton. Helium has two protons, lithium has three, and so forth. Every succeeding element in the list of elements has one more proton than the preceding element. In neutral atoms, there are as many protons in the nucleus as there are electrons outside the nucleus. The number of protons in the nucleus is the same as the atomic number.

The number of neutrons in the nucleus of a given element may vary somewhat. For example, every atom of chlorine has 17 protons and hence 17 orbital electrons that determine what chemical compounds it can make. But the number of accompanying neutrons varies. Atoms that have like numbers of protons but unlike numbers of neutrons are called **isotopes**. The two chief types of chlorine isotopes have 35 and 37 times the mass of a single nucleon. We denote these by $^{35}_{17}$Cl and $^{37}_{17}$Cl, where the numbers refer to the **atomic number** and the **atomic mass number** (approximately but not exactly equal to the *atomic mass*).

The mass number corresponds to the total number of nucleons in the nucleus. Since for both isotopes of chlorine 17 of these are protons, the lighter isotope has 35 − 17 = 18 neutrons, and the heavier isotope has 37 − 17 = 20 neutrons. In nature the isotopes of chlorine are found mixed, with three times as many $^{35}_{17}$Cl atoms as there are $^{37}_{17}$Cl atoms, so the average atomic mass of naturally occurring chlorine is about 35.5. Chlorine is only one of many elements found in nature that consist of two or more stable isotopes. Even hydrogen, the lightest element, has three known isotopes (Figure 32-5). The double-weight hydrogen isotope $^{2}_{1}$H is called *deuterium*. "Heavy water" is the name usually given to H_2O in which one or both of the H's are deuterium atoms. In all hydrogen compounds in nature, such as hydrogen gas and water, there is 1 atom of deuterium to about 6000 atoms of "ordinary" hydrogen. The triple-weight hydrogen isotope $^{3}_{1}$H, which is not stable but lives long enough to be a known constituent of atmospheric water, is called *tritium*. Tritium is present only in extremely minute amounts—less than 1 per 10^{17} atoms of ordinary hydrogen.

Figure 32-5
Three isotopes of hydrogen. Each nucleus has a single proton, which holds a single orbital electron, which in turn determines the chemical properties of the atom. The different number of neutrons changes the mass of the atom but not its chemical properties.

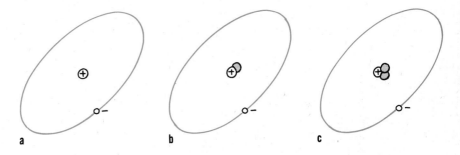

a   b   c

There are three isotopes of uranium that naturally occur in the earth's crust; the most common is $^{238}_{92}$U. In briefer notation we can drop the atomic number and simply say U-238. (We will see in the next chapter that it is the isotope U-235 that undergoes nuclear fission.) Of the 83 elements present on earth in significant amounts, 20 have a single stable isotope. The others have from 2 to 10 stable isotopes. Taking radioactive isotopes into account, over 2000 distinct isotopes are known.

Question

State the numbers of protons and neutrons in each of the following nuclei: $^{1}_{1}$H, $^{14}_{6}$C, $^{235}_{92}$U.

▶ **Answer**

1 proton and no neutron in $^{1}_{1}$H; 6 protons and 8 neutrons in $^{14}_{6}$C; 92 protons and 143 neutrons in $^{235}_{92}$U.

Why Atoms Are Radioactive

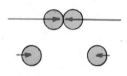

Figure 32-6
The nuclear strong interaction is a very short-range force. For nucleons very close or in contact, it is very strong. But a few nucleon diameters away it is nearly zero.

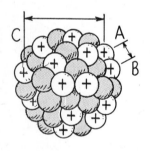

Figure 32-7
Proton A both attracts (nuclear force) and repels (electrical force) proton B, but mainly repels proton C (because the nuclear attraction is weaker at greater distances). The greater the distance between A and C, the more unstable the nucleus because of this repulsion. (The shaded particles are neutrons.)

The positively charged and closely spaced protons in a nucleus have huge electrical forces of repulsion between them. Why don't they fly apart because of this huge repulsive force? Because there is an even more formidable force within the nucleus—the nuclear force. Both neutrons and protons are bound to each other by this awesome attractive force. The nuclear force is much more complicated than the electrical force and is only now becoming understood. The principal part of the nuclear force, the part that holds the nucleus together, is called the *strong interaction.** The strong interaction is an attractive force that acts between protons, neutrons, and particles called *mesons*, all of which are called *hadrons*. This interaction acts over only a very short distance (Figure 32-6). It is very strong between nucleons less than 10^{-15} meter apart, but close to zero at greater separations. So the strong nuclear interaction is a short-range force. Electrical interaction, on the other hand, weakens as the inverse square of separation distance and is a relatively long-range force. So as long as protons are close together, as in small nuclei, the nuclear force easily overcomes the electrical force of repulsion. But for distant protons, like those on opposite edges of a large nucleus, the attractive nuclear force may be small in comparison to the repulsive electrical force. Hence, a larger nucleus is not as stable as a smaller nucleus.

The presence of the neutrons also plays a large role in nuclear stability. Although protons both attract and repel one another, neutrons only attract protons and other neutrons, and therefore may be considered a sort of nuclear cement. Many of the first 20 or so elements have equal numbers of neutrons and protons.

Even more nuclear cement is needed by the heavier elements. In fact, most of the mass of the heavier elements is composed of neutrons. For example, in U-238, which has 92 protons, there are 146 (238 − 92) neutrons. If the uranium nucleus were to have equal numbers of protons and neutrons, 92 protons and 92 neutrons, it would fly apart at once because of the electrical repulsion forces. Relative stability is found only if there are some 54 extra neutrons. This greater number of neutrons is required to compensate for the repulsion between distant protons. Even so, the U-238 nucleus is still unstable because of the electrical forces.

To put the matter in another way: there is an electrical repulsion between *every pair* of protons in the nucleus, but there is not a substantial nuclear attractive force between every pair (Figure 32-7). Every proton in the uranium nucleus exerts a repulsion on each of the other 91 protons—those near and those far. However, each proton (and neutron) exerts an appreciable nuclear attraction only on those nucleons that are near it.

We find that all nuclei having more than 82 protons are unstable. In this unstable environment, alpha and beta emissions take place. A nuclear force distinct from the strong interaction is responsible for beta emission.

---

*Fundamental to the strong interaction is the *color force* (which has nothing to do with visible color). This color force interacts between quarks and holds them together by the exchange of "gluons." Read more about this in H. R. Pagels, *The Cosmic Code: Quantum Physics as the Language of Nature* (New York: Simon & Schuster, 1982).

This weaker nuclear force is called the *weak interaction* and principally affects lighter particles like electrons and still lighter particles called *neutrinos*. These belong to the lepton class of particles. The mechanism for the emission of leptons goes beyond electrical repulsion and involves a quantum-mechanical explanation beyond the scope of this book.

Half-Life

The radioactive decay rate of an element is measured in terms of a characteristic time, the **half-life**. The half-life of a radioactive isotope is the time needed for half of any given quantity of that isotope to decay. Radium-226, for example, has a half-life of 1620 years. This means that half of any given specimen of radium-226 will be converted into some other element by the end of 1620 years. In the next 1620 years, half of the remaining radium will decay, leaving only one-fourth the original number of radium atoms. By a succession of disintegrations the radium ultimately becomes lead. After 20 half-lives, an initial quantity of radioactive atoms will be diminished 1 millionfold. The isotopes of some elements have a half-life of less than a millionth of a second, while uranium-238, for example, has a half-life of 4.5 billion years. The isotopes of each radioactive element have their own characteristic half-lives.

Figure 32-8
Every 1620 years the amount of radium decreases by half.

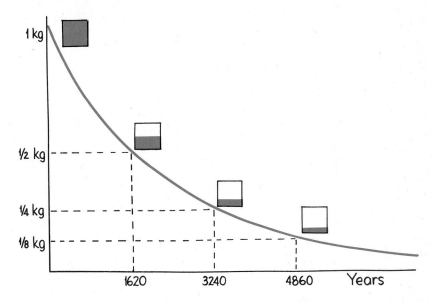

The neutron is stable inside the atomic nucleus, but by itself it is unstable and has a half-life of 12 minutes. Many elementary particles have even shorter half-lives. The muon, a close relative of the electron produced by the bombardment of cosmic rays against the upper atmosphere, has a half-life of 2 millionths of a second (2×10^{-6} s).* The

---

*The muon is currently of great interest to physicists because of its recently found ability to catalyze the fusion of hydrogen isotopes at ordinary temperatures. We will return to the muon when we study nuclear fusion in the next chapter.

shortest half-lives of elementary particles are on the order of 10^{-23} second, the time light takes to travel a distance equal to the diameter of a nucleus.

The half-lives of radioactive elements and elementary particles are remarkably constant, for rates of radioactive decay are not affected by any external conditions, however drastic. Wide temperature and pressure extremes, strong electric and magnetic fields, and even violent chemical reactions have no detectable effect on the rate of decay of a given element. Any of these stresses, although severe by ordinary standards, is far too mild to affect the nucleus in the deep interior of the atom.

How do we measure half-lives? Observing a specimen and waiting until half of it is decayed won't work if the half-life is long. The half-life of a radioactive substance is related to its rate of disintegration. The shorter the half-life of an element, the faster is the rate of disintegration and the more active is the element. The half-life can be determined from the rate of disintegration, which can be measured in the laboratory.

Half-lives can also be estimated with the use of quantum mechanics. The kinetic energy of an emerging alpha particle tells us its wavelength, which in turn gives its probability of leaking from the nucleus. This information enables the prediction of the half-life of the parent nucleus with some success.

Radiation Detectors

Ordinary thermal motions of atoms jostling against one another in a gas or liquid are not energetic enough to dislodge electrons, and the atoms remain neutral. But when an energetic particle such as an alpha or a beta particle shoots through matter, electrons one after another are knocked from the atoms in its path, leaving a trail of freed electrons and positively charged ions. This ionization process is responsible for the harmful effects of high-energy radiation in living cells. It also makes it relatively easy to trace the paths of high-energy particles. We will briefly discuss five radiation detection devices.

1. A *Geiger counter* consists of a central wire in a hollow metal cylinder filled with gas. An electrical voltage is applied across the cylinder and wire so that the wire is more positive than the cylinder. If radiation enters the tube and ionizes an atom in the gas, the freed electron is attracted to the positively charged central wire. As this electron is accelerated toward the wire, it collides with other atoms and knocks out more electrons, which in turn produce more and more electrons, and so on, resulting in a cascade of electrons moving toward the wire. This makes a short pulse of electric current, which activates a counting device connected to the tube. Amplified, this pulse of current makes the familiar clicking sound we associate with radiation detectors.

2. A *cloud chamber* shows a visible path of ionizing radiation in the form of fog trails. It consists of a cylindrical glass chamber closed at the upper end by a glass window and at the lower by a movable piston. Water vapor or alcohol vapor in the chamber can be saturated by adjusting the piston.

Figure 32-9

Some radiation detectors.
(*a*) A Geiger counter detects incoming radiation by its ionizing effect on enclosed gas in the tube. (*b*) A scintillation counter detects incoming radiation by flashes of light that are produced when charged particles or gamma rays pass through it.

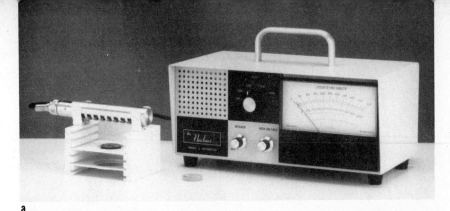

a

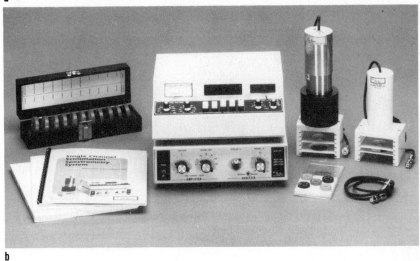

b

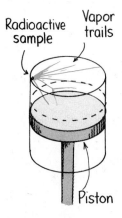

Figure 32-10
The cloud chamber.

The radioactive sample is placed inside the chamber, as shown in Figure 32-10, or outside the thin glass window. When radiation passes through the chamber, ions are produced along the paths. If the saturated air in the chamber is then suddenly cooled by motion of the piston, tiny droplets of moisture condense about these ions and form vapor trails showing the paths of the radiation. These are the atomic versions of the ice-crystal trails left in the sky by jet planes.

Even simpler is the continuous cloud chamber. This has a steady supersaturated vapor, because it sits on a slab of dry ice so there is a temperature gradient from near room temperature at the top to very low temperature at the bottom. In either version the fog tracks that form are illuminated with a lamp and may be seen or photographed through the glass top. The chamber may be placed in a strong electric or magnetic field, which will bend the paths in a manner that provides information about the charge, mass, and momentum of the radiation particles. Positively and negatively charged particles will bend in opposite directions.

3. The particle trails seen in a *bubble chamber* are minute bubbles of gas in liquid hydrogen (Figure 32-11 and the front cover of this book). The liquid hydrogen is heated under pressure in a glass and stainless steel

Figure 32-11
The 72-inch hydrogen bubble chamber at the Lawrence Berkeley Laboratory. Liquid hydrogen fills the bottom of the chamber. The rays to be studied enter the chamber through a window.

chamber to a point just short of boiling. If the pressure in the chamber is suddenly released at the moment an ion-producing particle enters, a thin trail of bubbles is left along the particle's path. All the liquid erupts to a boil, but in the few thousandths of a second before this happens, photographs are taken of the particle's brief-lived trail. As with the cloud chamber, a magnetic field in the bubble chamber reveals the charge and relative mass of the particles. Bubble chambers have been widely used by researchers in recent decades, but presently there is greater interest in spark chambers.

4. A *spark chamber* is a counting device that consists of an array of closely spaced parallel plates; every other plate is grounded, and plates in between are maintained at a high voltage (about 10 kV). Ions are produced in the gas between the plates as charged particles pass through the chamber. Discharge along the ionic path produces a visible spark between pairs of plates. A trail of many sparks reveals the path of the incident particle. A different design called a *streamer chamber* consists of only two widely spaced plates, between which an electric discharge, or "streamer," closely follows the path of the incident charged particle.

Figure 32-12
Tracks of elementary particles in a bubble chamber. Two particles have been destroyed at the points from which the spirals emanate, and four others are created in the collision.

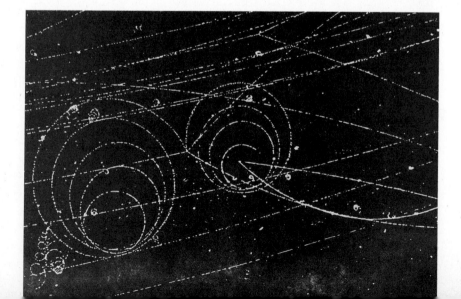

The principal advantage of spark and streamer chambers over the bubble chamber is that more events can be monitored in a given time.

5. A *scintillation counter* uses the fact that certain substances are easily excited and emit light when charged particles or gamma rays pass through them. Tiny flashes of light, or scintillations, are converted into electric signals by special photo-multiplier tubes. A scintillation counter is much more sensitive to gamma rays than a Geiger counter and, in addition, can measure the energy of charged particles or gamma rays absorbed in the detector.

Questions ▶

1. If a sample of radioactive isotopes has a half-life of 1 day, how much of the original sample will be left at the end of the second day? The third day?
2. Which will give a higher counting rate on a radiation detector, radioactive material that has a short half-life or radioactive material that has a long half-life?

Natural Transmutation of Elements

When a nucleus emits an alpha or a beta particle, a different element is formed. The changing of one chemical element to another is called **transmutation**. Consider uranium, for example. Uranium-238 has 92 protons and 146 neutrons. When an alpha particle is ejected, the nucleus is reduced by two protons and two neutrons (an alpha particle is a helium nucleus consisting of two protons and two neutrons). The 90 protons and 144 neutrons left behind are then the nucleus of a new element. This element is *thorium*. We can express this reaction as

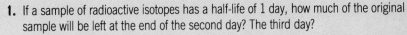

$$^{238}_{92}\text{U} \rightarrow {}^{234}_{90}\text{Th} + {}^{4}_{2}\text{He}$$

An arrow is used here to show that the $^{238}_{92}\text{U}$ changes into the other elements. When this happens, energy is released, partly in the form of gamma radiation, partly in the kinetic energy of the alpha particle ($^{4}_{2}\text{He}$), and partly in the kinetic energy of the thorium atom. In this and all such equations, the mass numbers at the top balance ($238 = 234 + 4$) and the atomic numbers at the bottom also balance ($92 = 90 + 2$).

Thorium-234, the product of this reaction, is also radioactive. When it decays, it emits a beta particle. Recall that a beta particle is an elec-

▶ **Answers**

1. One-fourth of the original sample will be left—the three-fourths that underwent decay is now a different element altogether. At the end of 3 days, $\frac{1}{8}$ of the original sample will remain.
2. The material with the shorter half-life is more active and will give a higher counting rate on a radiation detector.

tron—not an orbital electron, but one from the nucleus—from one of the neutrons. You may find it useful to think of a neutron as a combined proton and electron (although it's not really the case), because when the neutron emits an electron, it becomes a proton. A neutron is ordinarily stable when it is locked in the nucleus of an atom, but a free neutron is radioactive and has a half-life of 12 minutes. It decays into a proton by beta emission.* So in the case of thorium, which has 90 protons, beta emission leaves it with one less neutron and one more proton. The new nucleus then has 91 protons and is no longer thorium, but the element *protactinium*. Although the atomic number has increased by 1 in this process, the mass number (protons + neutrons) remains the same. The nuclear equation is

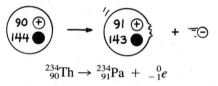

$$^{234}_{90}\text{Th} \rightarrow ^{234}_{91}\text{Pa} + ^{0}_{-1}e$$

We write an electron as $^{0}_{-1}e$. The 0 indicates that its mass is insignificant when compared to that of the protons and neutrons that alone contribute to the mass number. The -1 is the charge of the electron. Remember that this electron is a beta particle from the nucleus and not an electron from the surrounding electron cloud.

From the previous two examples, we can see that when an element ejects an alpha particle from its nucleus, the mass number of the resulting atom decreases by 4, and its atomic number *decreases* by 2. The resulting atom belongs to an element two spaces back in the periodic table (see the inside front cover). When an element ejects a beta particle (electron) from its nucleus, the mass of the atom is practically unaffected so there is no change in mass number, but its atomic number *increases* by 1. The resulting atom belongs to an element one place forward in the periodic table. Gamma emission results in no change in either the mass number or the atomic number. So we see that the emission of an alpha or beta particle by an atom produces a different atom in the periodic table. Alpha

---

*Beta emission is always accompanied by the emission of a neutrino, a neutral particle with about zero mass that travels at about the speed of light. The neutrino ("little neutral one") was predicted from theoretical calculations by Wolfgang Pauli in 1930 and detected in 1956. Neutrinos are hard to detect because they interact very weakly with matter. To capture a neutrino is extremely difficult. Whereas a piece of solid lead a few centimeters thick will stop most gamma rays from a radium source, it would take a piece of lead about 8 light-years thick to stop half the neutrinos produced in typical nuclear decays. Thousands of neutrinos are flying through you every second of every day, because the universe is filled with them. Only occasionally, one or two times a year or so, does a neutrino or two interact with the matter of your body.

At this writing, whether or not the neutrino has mass is not known. Present speculation is that if neutrinos do have any mass, they are so numerous that they may make up about 90% of the mass of the universe—enough to halt the present expansion and ultimately close the cycle from the Big Bang to the Big Crunch. Neutrinos may be the "glue" that holds the universe together.

emission lowers the atomic number and beta emission increases it. So we see that radioactive elements can decay backward or forward in the periodic table.*

The radioactive decay of $^{238}_{92}$U to $^{206}_{82}$Pb, an isotope of lead, is shown in Figure 32-13. The steps in the decay process are shown in the diagram, where each nucleus that plays a part in the series is shown by a burst. The vertical column containing the burst shows its atomic number, and the horizontal column shows its mass number. Each arrow that slants downward toward the left shows an alpha decay, and each arrow that points to the right shows a beta decay. Notice that some of the nuclei in the series can decay in both ways. This is one of several similar radioactive series that occur in nature.

Figure 32-13

U-238 decays to Pb-206 through a series of alpha and beta decays.

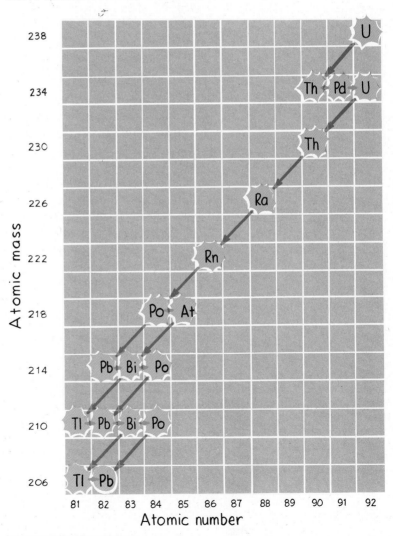

*Sometimes a nucleus emits a positron, which is the "antiparticle" of an electron. In this case, a proton becomes a neutron, and the atomic number is decreased.

Questions ▶

1. Complete the following nuclear reactions.

a. $^{226}_{88}\text{Ra} \rightarrow ^{?}_{?}? + ^{0}_{-1}e$

b. $^{209}_{84}\text{Po} \rightarrow ^{205}_{82}\text{Pb} + ^{?}_{?}?$

2. What finally becomes of all the uranium that undergoes radioactive decay?

Artificial Transmutation of Elements

The alchemists of old tried vainly for over 2000 years to cause the transmutation of one element to another. Enormous efforts were expended and elaborate rituals performed in the quest to change lead into gold. They never succeeded. Lead in fact can be changed to gold, but not by the chemical means employed by the alchemists. Chemical reactions involve alterations of the outermost shells of the electron clouds of atoms and molecules. To change an element from one kind to another, one must go deep within the electron clouds to the central nucleus, which is immune to the most violent chemical reactions. To change lead to gold, three positive charges must be extracted from the nucleus. Ironically enough, transmutations of atomic nuclei were constantly going on all around the alchemists, as they are around us today. Radioactive decay of minerals in rocks has been occurring since their formation. But this was unknown to the alchemists, and even if they had discovered these radiations, the atoms so transmuted would most likely have escaped their notice.

Ernest Rutherford, in 1919, was the first of many investigators to succeed in transmuting a chemical element. He bombarded nitrogen nuclei with alpha particles and succeeded in transmuting nitrogen into oxygen:

$$^{14}_{7}\text{N} + ^{4}_{2}\text{He} \rightarrow ^{17}_{8}\text{O} + ^{1}_{1}\text{H}$$

His source of alpha particles was a radioactive piece of ore. From a quarter of a million cloud-chamber tracks photographed on movie film, he showed seven examples of atomic transmutation. Analysis of tracks bent by a strong external magnetic field showed that when an alpha particle collided with a nitrogen atom, a proton bounced out and the heavy atom recoiled a short distance. The alpha particle disappeared. The alpha particle was absorbed in the process, transforming nitrogen to oxygen.

▶ **Answers**

1. **a.** $^{226}_{88}\text{Ra} \rightarrow ^{226}_{89}\text{Ac} + ^{0}_{-1}e$
   b. $^{209}_{84}\text{Po} \rightarrow ^{205}_{82}\text{Pb} + ^{4}_{2}\text{He}$

2. All uranium will ultimately become lead. On the way to becoming lead, it will exist as an element in a series, as indicated in Figure 32-13.

Since Rutherford's announcement in 1919, we have created many such nuclear reactions, first with natural bombarding projectiles from radioactive ores and then with still more energetic projectiles, protons and electrons hurled by giant particle accelerators. Artificial transmutation has produced the hitherto unknown synthetic elements from atomic number 93 to 109, the first eleven of which are named *neptunium, plutonium, americium, curium, berkelium, californium, einsteinium, fermium, mendelevium, nobelium,* and *lawrencium.* The purported discoverers of elements 104 and 105 have recommended the names *rutherfordium (Rf)* for element 104 and *hahnium (Ha)* for 105. Russian researchers, who also claim the discovery of these elements, have proposed different names. All these artificially made elements have short half-lives. If they ever existed naturally when the earth was formed, they have long since decayed.

Radioactive Isotopes

All elements have isotopes. Some of the isotopes of a given element may be radioactively stable, while most are quite radioactive. Radioactive isotopes of all the elements have been made by bombardment with neutrons and other particles. As such isotopes have become available and inexpensive, their use in scientific research and industry has expanded tremendously.

Figure 32-14
Tracking pipe leaks with radioactive isotopes.

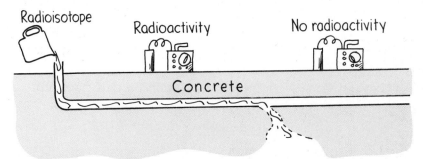

A small amount of radioactive isotopic material can be mixed with fertilizer for young plants. Once the plants are growing, we can easily measure how much fertilizer they have taken up by using a radiation detector. From such measurements, farmers know the proper amount of fertilizer to use. When used in this way, radioactive isotopes are called *tracers* (Figure 32-15).

Tracers are widely used in medicine to study the process of digestion and the way chemical substances move about in the body. Food containing a small amount of radioactive isotopic material is fed to a patient and traced through the body with a radiation detector. The same method can be used to study the circulation of the blood. Large amounts of radioactive iodine taken into the body are used to combat cancer of the thyroid gland. Since iodine tends to collect in the thyroid gland, the radioactive isotopes lodge where they can destroy the malignant cells. Radioactive isotopes are an increasingly powerful tool in medicine and biology.

Figure 32-15
Radioisotopes are used to check the action of fertilizers in plants and the progress of food in digestion.

Engineers can study how the parts of an automobile engine wear away during use by making the cylinder walls radioactive. While the engine is running, the piston rings rub against the cylinder walls. The tiny particles of radioactive metal that are worn away fall into the lubricating oil, where they can be measured with a radiation detector. By repeating this test with different oils, the engineer can determine which oil gives the least wear and longest life to the engine.

Tire manufacturers also employ radioactive isotopes. If a known fraction of the carbon atoms used in an automobile tire is radioactive, the amount of rubber left on the road when the car is braked can be estimated through a count of the radioactive atoms.

There are hundreds more examples of the use of radioactive isotopes. The important thing is that this technique provides a way to detect and count atoms in samples of materials too small to be seen with a microscope.

Carbon Dating

A constant bombardment of cosmic rays on the earth's atmosphere results in the transmutation of many atoms in the upper atmosphere. Protons and neutrons are scattered throughout the atmosphere. Most of the protons quickly capture stray electrons and become hydrogen atoms in the upper atmosphere, but the neutrons keep going for long distances because they have no charge and therefore do not interact electrically with matter. Sooner or later many of them collide with the nuclei of atoms in the lower atmosphere. If they interact with the nucleus of a nitrogen atom, the following reaction takes place:

$$^{14}_{7}N + ^{1}_{0}n \rightarrow ^{14}_{6}C + ^{1}_{1}H$$

So some of the nitrogen in the atmosphere is converted to carbon and hydrogen. Most carbon that exists is the stable C-12, $^{12}_{6}C$. Because of cosmic-ray bombardment, about 1 part in 10^{12} of the carbon in the atmosphere is C-14. Both of these isotopes join with oxygen and become carbon dioxide, which is taken in by plants. Hence, all plants have a tiny bit of radioactive C-14 in them. All animals eat plants (or eat plant-eating animals), and therefore have a little C-14 in them. All living things contain some C-14.

Carbon-14 is a beta emitter and decays back into nitrogen by the following reaction:

$$^{14}_{6}C \rightarrow ^{14}_{7}N + ^{0}_{-1}e$$

So the C-14 in living things changes into stable nitrogen. But because living things breathe, this decay is accompanied by a replenishment of

C-14, and a radioactive equilibrium is reached where there is a fixed ratio of C-14 to C-12. When a plant or an animal dies, however, replenishment stops. The percentage of C-14 steadily decreases—at a known rate. The longer an organism is dead, the less C-14 remains.*

The half-life of C-14 is about 5730 years, which means that half the C-14 atoms present in a body decay in that time. Half the remaining C-14 atoms decay in the following 5730 years, and so forth. The radioactivity of living things therefore gradually decreases at a steady rate after they die (Figure 32-16).

Figure 32-16
The radioactive carbon isotopes in the skeleton diminish by one-half every 5730 years.

We can measure the radioactivity of plants and animals today and compare this with the radioactivity of ancient organic matter. If we extract a small, but precise, quantity of carbon from an ancient wooden ax handle, for example, and find it has one-half as much radioactivity as an equal quantity of carbon extracted from a living tree, then the old wood must have come from a tree that was cut down or made from a log that died 5730 years ago. In this way, we can probe as much as 50 000 years into the past to find out such things as the age of ancient civilizations.

Carbon dating would be an extremely simple and accurate dating method if the amount of radioactive carbon in the atmosphere had been constant over the ages. But it hasn't been. Fluctuations in the sun's magnetic field as well as changes in the strength of the earth's magnetic field affect cosmic-ray intensities in the earth's atmosphere, which in turn produce fluctuations in the production of C-14. In addition, changes in the earth's climate affect the amount of carbon dioxide in the atmosphere. The oceans are great reservoirs of carbon dioxide. When the oceans are cold, they release less carbon dioxide into the atmosphere than when they are warm. These complications require complex corrective procedures for the accurate dating of ancient organic objects. Nevertheless, recalibration and refined techniques make carbon dating the most versatile chronometric method we have. Present laser-enrichment techniques using samples containing only a few milligrams of carbon are soon expected to double the current 50 000-year range. Similar results are now possible with a technique that bypasses a radioactive measure altogether. It involves an improved atomic acceleration-mass spectrometer device that effectively makes a C-14/C-12 count directly. There are also other techniques for assigning dates to relics of the past.

---

*A 1-g sample of contemporary carbon contains about 5×10^{22} atoms, 6.5×10^{10} of which are C-14 atoms, and has a beta disintegration rate of about 13.5 decays per minute.

Question ▶ Suppose an archeologist extracts a gram of carbon from an ancient ax handle and finds it one-fourth as radioactive as a gram of carbon extracted from a freshly cut tree branch. About how old is the ax handle?

Uranium Dating

The dating of older, but nonliving, things is accomplished with radioactive minerals, such as uranium. The naturally occurring isotopes U-238 and U-235 decay very slowly and ultimately become isotopes of lead—but not the common lead isotope Pb-208. For example, U-238 decays through several stages finally to become Pb-206, whereas U-235 finally becomes the isotope Pb-207. Some of the lead isotopes 206 and 207 that now exist were at one time uranium. The older the uranium-bearing rock, the higher the percentage of these remnant isotopes. We know the half-lives of the uranium isotopes and we can determine the percentage of each radiogenic lead isotope in a rock sample, so we can calculate the time of decay for uranium in the rock. In this way the ages of rocks have been found to be as much as 3.7 billion years. Samples from the moon, where there has been less obliteration of the early rocks than occurs on earth, have been dated at 4.2 billion years, which approaches the currently accepted 4.6-billion-year age of the earth and solar system.

Effects of Radiation on Humans

A common misconception is that radioactivity is something new. But radioactivity has been around far longer than the human race. It is as much a part of our environment as the sun and the rain. It is what warms the interior of the earth and makes it molten. In fact, radioactive decay inside the earth is what heats the water that spurts from a geyser or that wells up from a natural hot spring. The helium in a child's balloon is nothing more than the alpha particles that were produced by radioactive decay.

As Figure 32-17 shows, most of the radiation we encounter originates in the natural surroundings. It is in the ground we stand on and in the bricks and stones of surrounding buildings. This natural background radiation was present before humans emerged in the world. If our bodies couldn't tolerate it, we wouldn't be here. Even the cleanest air we breathe is somewhat radioactive due to cosmic ray bombardment. At sea level the protective blanket of the atmosphere reduces it, while at higher altitudes radiation is more intense. In Denver, the "mile-high city," a person receives more than twice as much radiation from cosmic rays as at sea level. A couple of round-trip flights between places as distant as New York and San Francisco exposes us to as much radiation as we receive in a normal chest X ray. The air time of airline personnel is limited because of this extra radiation.

▶ **Answer**

Assuming the ratio of C-14/C-12 was the same when the ax was made, the ax handle is two half-lives of C-14, about 11 460 years old.

Figure 32-17
Origins of radiation exposure for an average individual in the United States.

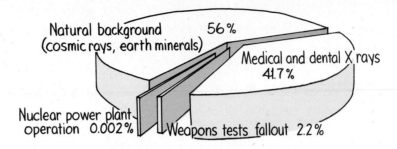

Natural background (cosmic rays, earth minerals) 56%

Medical and dental X rays 41.7%

Nuclear power plant operation 0.002%

Weapons tests fallout 2.2%

Exposure to radiation greater than normal background should be avoided because of the damage it can do. The cells of living tissue are composed of intricately structured molecules in a watery, ion-rich brine. When X radiation or nuclear radiation encounters this highly ordered soup, it produces chaos on the atomic scale. A beta particle, for example, passing through living matter collides with a small percentage of the molecules and leaves a randomly dotted trail of altered or broken molecules along with newly formed, chemically active ions and free radicals. Free radicals are unbonded, electrically neutral, very chemically active atoms or molecular fragments. The ions and free radicals may break even more molecular bonds or they may quickly form new strong bonds, creating molecules that may be useless or harmful to the cell. Gamma radiation produces a similar effect. As a high-energy gamma-ray photon moves through matter, it may rebound from an electron and give it a high kinetic energy. The electron then may career through the tissue, creating havoc in the ways described above. All types of high-energy radiation break or alter the structure of some molecules and create conditions in which other molecules will be formed that may be harmful to life processes.

The cells are able to repair most kinds of molecular damage if the radiation is not too intense. A cell could survive an otherwise lethal dose of radiation if the dose were spread over a long period of time to allow intervals for healing. When radiation is sufficient to kill a cell, the dead cells can be replaced by new ones. Important exceptions to this are most nerve cells, which are irreplaceable. Sometimes a cell will survive with a damaged DNA molecule. Defective genetic information will be transmitted to its daughter cell when it reproduces, and a cell mutation will occur. This will sometimes be insignificant. But if significant, it will probably result in cells that do not function as well as the original one. In rare cases it will be an improvement. A genetic change of this type could also be part of the cause of a cancer that will develop in the tissue at a much later time.

The concentration of disorder produced along the trajectory of a particle depends upon its energy, charge, and mass. Gamma-ray photons and very energetic beta particles cause the lowest intensity of damage. A particle of their type penetrates deeply with widely separated interactions, like a very fast BB fired through a hailstorm. Slow, massive, highly charged particles such as low-energy alpha particles are most disruptive.

They have collisions that are close together, more like a bull charging through a flock of sleepy sheep. They do not penetrate deeply because their energy is absorbed by many closely spaced collisions. Especially damaging particles of this type are the assorted nuclei (called *heavy primaries*) flung outward by the sun in solar flares. These include all the elements found on earth. Particles that approach the earth after leaving the sun are partly captured in the earth's magnetic field. The others are absorbed by collisions in the atmosphere, so practically none reach the earth. We are partly shielded from these dangerous particles by the very property that makes them a threat—their tendency to have many collisions close together. Astronauts do not have this protection, and they absorb large doses of radiation during the time they spend in space. Every few decades there is an exceptionally powerful solar flare that would almost certainly kill any conventionally protected astronaut far from the earth who is unprotected by its atmosphere and magnetic field.

Generally, radiation dosage is measured in *rads* (short for *radiation*), a unit of absorbed energy of ionizing radiation. The number of rads indicates the amount of radiation energy absorbed per gram of exposed material. However, when concerned with the potential ability of radiation to affect human beings, we measure the radiation in *rems* (roentgen *e*quivalent *m*an). In calculating the dosage in rems, we multiply the number of rads by a factor that allows for the different health effects of different types of radiation. For example, 1 rad of slow alpha particles has the same biological effect as 10 rads of fast electrons. We call both of these dosages 10 rems.

The average person in the United States is exposed to about 0.2 rem a year. This comes from within the body itself, from the ground, buildings, cosmic rays, diagnostic X rays, television, and so on. It varies widely from place to place on the planet, but it is strongest near the poles where the earth's magnetic field does not act as a shield. It is also stronger at higher altitudes where the atmosphere provides less protection.

The lethal dose of radiation is on the order of 500 rems; that is, a person has about a 50-percent chance of surviving a dose of this magnitude received over a short period of time. Under radiotherapy—the use of radiation to kill cancer cells—a patient may receive localized doses in excess of 200 rems each day for a period of weeks. A typical diagnostic chest X ray exposes a person to from 5 to 30 millirems, less than one ten-thousandth of the lethal dose. However, even small doses of radiation can produce long-term effects due to mutations within the body's tissues. And, because a small fraction of any X-ray dose reaches the gonads, some mutations occasionally occur that are passed on to the next generation. Medical X rays for diagnosis and therapy have a far larger effect on the human genetic heritage than any other artificial source of radiation. Perspective dictates we keep in mind that we normally receive significantly more radiation from natural minerals in the earth than from all artificial sources of radiation combined.

Figure 32-18
The internationally used symbol to indicate an area where radioactive material is being handled or produced.

Taking all causes into account, most of us will receive a lifetime exposure of less than 20 rems, distributed over several decades. This makes us a little more susceptible to cancer and other disorders. But more significant is the fact that all living beings have always absorbed natural radiation and that the radiation received in the reproductive cells has produced genetic changes in all species for generation after generation. Small mutations selected by nature for their contributions to survival over billions of years can gradually come up with some interesting organisms—*us*, for example!

Summary of Terms

X ray Electromagnetic radiation of higher frequencies than ultraviolet emitted by innermost electrons in excited atoms.

Alpha ray A stream of helium nuclei ejected by certain radioactive nuclei.

Beta ray A stream of beta particles ejected by certain radioactive nuclei.

Gamma ray High-frequency electromagnetic radiation emitted by the nuclei of radioactive atoms.

Alpha particle The nucleus of a helium atom, which consists of two neutrons and two protons, ejected by certain radioactive elements.

Beta particle An electron (or positron) emitted during the radioactive decay of certain nuclei.

Nucleon A nuclear proton or neutron; the collective name for either or both.

Quarks The elementary constituent particles of building blocks of nuclear matter.

Isotopes Atoms whose nuclei have the same number of protons but different numbers of neutrons.

Atomic number The number associated with an atom, which is equal to the number of protons in the nucleus or, equivalently, to the number of electrons in the electric cloud of a neutral atom.

Atomic mass number The number associated with an atom, which is equal to the number of nucleons in the nucleus.

Half-life The time required for half the atoms of a radioactive element to decay.

Transmutation The conversion of an atomic nucleus of one element into an atomic nucleus of another element through a loss or gain in the number of protons.

Review Questions

X Rays and Radioactivity

1. What is the principal difference between a beam of X rays and a beam of light?

2. What did the physicist Roentgen discover about a beam of electrons striking a glass surface?

3. What did the physicist Becquerel discover about uranium?

4. What two elements did Pierre and Marie Curie discover?

Alpha, Beta, and Gamma Rays

5. How do the electric charges of alpha, beta, and gamma rays differ?

6. How does the source differ for a beam of gamma rays and a beam of X rays?

The Nucleus

7. Give two examples of a nucleon.

8. How does the mass of a nucleon compare to the mass of an electron?

9. When beta emission occurs, what change takes place in an atomic nucleus?

10. In what way is the emission of gamma rays from a nucleus similar to the emission of light from an atom?

11. Why does an alpha particle leave at high speed once it gets outside an atomic nucleus?

12. What are *quarks*?

Isotopes

13. Distinguish between an *isotope* and an *ion*.

14. Distinguish between *atomic number* and *atomic mass number*.

15. Distinguish between *deuterium* and *tritium*.

Why Atoms Are Radioactive

16. Why does the repulsive electric force of protons in the atomic nucleus not cause the protons to fly apart?

17. Why do protons in a very large nucleus have a greater chance of flying apart by electrical repulsion?

18. Why is a smaller nucleus generally more stable than a larger nucleus?

19. Distinguish between a *hadron* and a *lepton*.

Half-Life

20. What is meant by *radioactive half-life*?

21. What is the half-life of R-226? Of a muon?

Radiation Detectors

22. What kind of trail is left when an energetic particle shoots through matter?

23. What radiation detectors operate primarily by sensing the trails left by energetic particles that shoot through matter?

Natural Transmutation of Elements

24. What exactly is a *transmutation*?

25. When thorium, atomic number 90, decays by emitting an alpha particle, what is the atomic number of the resulting nucleus?

26. When thorium decays by emitting a beta particle, what is the atomic number of the resulting nucleus?

27. How does the atomic mass change for each of the above two reactions?

28. What is the effect on the makeup of a nucleus when it emits an alpha particle? A beta particle? A gamma ray?

29. What is the long-range fate of all the uranium that exists in the world?

Artificial Transmutation of Elements

30. The alchemists of old believed that elements could be changed to other elements. Were they correct? Were they effective? Why or why not?

31. When did the first successful intentional transmutation of an element occur?

Radioactive Isotopes

32. How are radioactive isotopes produced?

33. What is a radioactive *tracer*?

Carbon Dating

34. Which is radioactive, C-12 or C-14?

35. Why is there more C-14 in new bones than in old bones of the same mass?

36. Why would the carbon dating method be useless for dating old coins but not old pieces of cloth?

Uranium Dating

37. Why is there lead mixed in with all deposits of uranium ores?

Effects of Radiation on Humans

38. From where does most of the radiation you encounter originate?

39. Which is worse, having cells in your body *damaged* by radiation or *killed* by radiation?

40. Is radioactivity in the world something relatively new? Defend your answer.

Home Project

Some watches and clocks have luminous hands that continuously glow. These have traces of radium bromide mixed with zinc sulfide (safer clock faces use light rather than radioactive disintegration as a means of excitation and become progressively dimmer in the dark). If you have a luminous watch or clock available, take it into a completely dark room and, after your eyes have become adjusted to the dark, examine the luminous hands with a very strong magnifying glass or the eyepiece of a microscope or telescope. You should be able to see individual tiny flashes, which together seem to be a steady source of light to the unaided eye. Each flash occurs when an alpha particle ejected by a radium nucleus strikes a molecule of zinc sulfide.

Exercises

1. X rays are most similar to which of the following: alpha, beta, or gamma rays?

2. Why is a sample of radioactive material always a little warmer than its surroundings?

3. Some people say that all things are possible. Is it at all possible for a hydrogen nucleus to emit an alpha particle? Defend your answer.

4. Why are alpha and beta rays deflected in opposite directions in a magnetic field? Why are gamma rays undeflected?

5. The alpha particle has twice the electric charge of the beta particle but deflects less than the beta in a magnetic field. Why is this so?

6. How would the paths of alpha, beta, and gamma radiations compare in an electric field?

7. Which type of radiation—alpha, beta, or gamma—results in the greatest change in mass number? Atomic number?

8. Which type of radiation—alpha, beta, or gamma—results in the least change in mass number? Atomic number?

9. In bombarding atomic nuclei with proton "bullets," why must the protons be accelerated to high energies to make contact with the target nuclei?

10. Why would you expect alpha particles to be less able to penetrate in materials than beta particles of the same energy?

11. Within the atomic nucleus, which interaction tends to hold it together and which interaction tends to push it apart?

12. What evidence supports the contention that the strong nuclear interaction is stronger than the electrical interaction at short internuclear distances?

13. If a sample of radioactive isotope has a half-life of 1 year, how much of the original sample will be left at the end of the second year? Third year? Fourth year?

14. A sample of a particular radioisotope is placed near a Geiger counter, which is observed to register 160 counts per minute. Eight hours later the detector counts at a rate of 10 counts per minute. What is the half-life of the material?

15. Radiation from a point source obeys the inverse-square law. If a Geiger counter 1 m from a small sample reads 360 counts per minute, what will be its counting rate 2 m from the source? 3 m?

16. Why do the charged particles in bubble chambers move in spiral paths rather than the circular or helical paths they would ideally follow?

17. From Figure 32-13, how many alpha and beta particles are emitted in the radioactive decay of a U-238 nucleus to Pb-206?

18. When $^{226}_{88}$Ra decays by emitting an alpha particle, what is the atomic number of the resulting nucleus? What is the resulting atomic mass?

19. When $^{218}_{84}$Po emits a beta particle, it transforms into a new element. What are the atomic number and atomic mass of this new element? What are they if the polonium instead emits an alpha particle?

20. State the number of neutrons and protons in each of the following nuclei: $^{2}_{1}$H, $^{12}_{6}$C, $^{56}_{26}$Fe, $^{197}_{79}$Au, $^{90}_{38}$Sr, and $^{238}_{92}$U.

21. How is it possible for an element to decay "forward in the periodic table"—that is, to decay to an element of higher atomic number?

22. When radioactive phosphorus (P) decays, it emits a positron. Will the resulting nucleus be another isotope of phosphorus? If not, what?

23. How could a physicist test the following statement? "Strontium-90 is a pure beta source."

24. Elements above uranium in the periodic table do not exist in any appreciable amounts in nature because they have short half-lives. Yet there are several elements below uranium in atomic number with equally short half-lives that do exist in appreciable amounts in nature. How can you account for this?

25. Your friend, fretful about living near a fission power plant, wishes to get away from radiation by traveling to the high mountains and sleeping out at night on granite outcroppings. What comment do you have about this?

26. Another friend has journeyed to the mountain foothills to escape the effects of radioactivity altogether. While bathing in the warmth of a natural hot spring she wonders aloud how the spring gets its heat. What do you tell her?

27. Coal contains minute quantities of radioactive materials, yet there is more environmental radiation surrounding a coal-fired power plant than a fission power plant. What does this indicate about the shielding that typically surrounds these power plants?

28. A friend produces a Geiger counter to check the local background radiation. It ticks. Another friend, who normally fears most that which is understood least, makes an effort to keep away from the region of the Geiger counter and looks to you for advice. What do you say?

29. Why is the carbon-dating technique currently not accurate for estimating the ages of materials older than 50 000 years?

30. The age of the Dead Sea Scrolls was found by carbon dating. Could this technique have worked if they were carved in stone tablets? Explain.

33 Nuclear Fission and Fusion

Nuclear Fission

The greater force
is nuclear

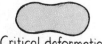

Critical deformation

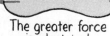

The greater force
is electrical

Figure 33-1
Nuclear deformation may result in repulsive electrical forces exceeding attractive nuclear forces, in which case fission occurs.

In 1939 two German scientists, Otto Hahn and Fritz Strassmann, made an accidental discovery that was to change the world. While bombarding a sample of uranium with neutrons in the hope of creating new and heavier elements, they were astonished to find chemical evidence for the production of barium, an element about half the mass of uranium. They were reluctant to believe their own results. News of this discovery reached Lise Meitner and Otto Frisch, refugees from Nazism working in Sweden, who proposed that the uranium nucleus, activated by neutron bombardment, had split in half. Meitner realized that the nuclear event was similar to the process of cell division in biology, so she named the nuclear process *fission*.

Nuclear fission involves the delicate balance within the nucleus between nuclear attraction and the electrical repulsion between charges. In nearly all nuclei the nuclear forces dominate. In uranium, however, this domination is tenuous. If the uranium nucleus is stretched into an elongated shape (Figure 33-1), the electrical forces may push it into an even more elongated shape. If the elongation passes a critical point, nuclear forces give way to electrical ones, and the nucleus separates. This is nuclear fission.

The absorption of a neutron by a uranium nucleus is apparently enough to cause such an elongation. The resultant fission process may produce any of several combinations of smaller nuclei. A typical example recorded in cloud chambers is

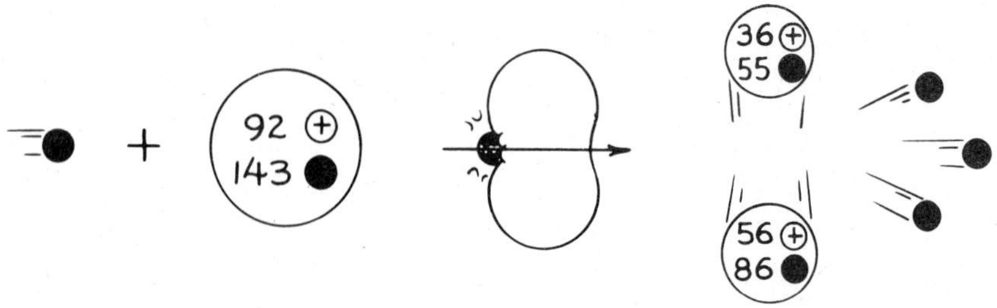

$$^{1}_{0}n + ^{235}_{92}U \rightarrow ^{91}_{36}Kr + ^{142}_{56}Ba + 3(^{1}_{0}n)$$

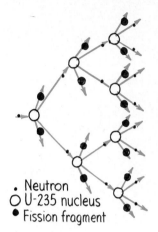

. Neutron
O U-235 nucleus
● Fission fragment

Figure 33-2
A chain reaction.

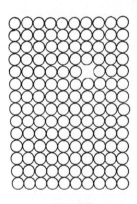

U-235
O U-238

Figure 33-3
Only 1 part in 140 of naturally occurring uranium is U-235.

This reaction releases energy of about 200 000 000 electron volts.* (By comparison, the explosion of the TNT molecule releases 30 electron volts.) The combined mass of the fission fragments and neutrons produced in fission is less than the mass of the original uranium atom. The tiny amount of missing mass converted to this awesome amount of energy is in accord with Einstein's relation $E = mc^2$. The energy of fission is mainly in the form of kinetic energy of the fission fragments that fly apart from one another, with some kinetic energy given to ejected neutrons and the rest to gamma radiation.

The scientific world was jolted by the news of nuclear fission—not only because of the enormous energy release but also because of the extra neutrons liberated in the process. A typical fission reaction releases an average of about two or three neutrons. These new neutrons can in turn cause the fissioning of two or three other atomic nuclei, releasing more energy and a total of from four to nine more neutrons. If each of these splits just one nucleus, the next step in the reaction will produce between 8 and 27 neutrons, and so on. Thus, a whole **chain reaction** can proceed at an ever-accelerating rate (Figure 33-2).

Why doesn't a chain reaction start in naturally occurring uranium deposits?† Chain reactions don't ordinarily happen because fission occurs mainly for the rare isotope U-235, which makes up only 0.7 percent of the uranium in pure uranium metal. The prevalent isotope U-238 absorbs neutrons but does not ordinarily undergo fission, so any chain reaction is quickly snuffed by the neutron-absorbing U-238 nuclei. With rare exceptions, naturally occurring uranium is too "impure" to undergo a chain reaction spontaneously.

If a chain reaction began in a tiny piece of pure U-235, the number of neutrons escaping from the surface without being captured might be so numerous that a growing chain reaction would be short-lived. Recall from Chapter 11 that small objects have a disproportionately large surface area compared to their small volumes—the greater the surface area, the greater the number of escaping neutrons. In a large piece of fissionable material, where the surface area is less in proportion to size, the results would be explosive. The minimum size of the material that will sustain a chain reaction is called the *critical size*, the mass of which is called the **critical mass**. If the mass of a fissionable material is greater than the critical mass, an explosion of enormous magnitude may take place.

Consider a large quantity of U-235 in two units, each smaller than the critical size and separated a short distance (Figure 33-4). Because of the relatively large surface area of each unit, neutrons readily escape and a

---

*The electron volt (eV) is defined as the amount of energy an electron acquires in accelerating through a potential difference of 1 V.

†There is evidence that to a small degree one *has* occurred—millions of years ago when isotopic abundances were different and occurred in unusually rich concentrations under very unusual circumstances. See *Scientific American*, July 1976.

chain reaction cannot develop. But if the pieces are suddenly driven together, the relative surface area is decreased. If the timing is right and the combined mass is greater than the critical mass, a violent explosion takes place. Such a device is a nuclear bomb.

Figure 33-4
Diagram of a simple atomic bomb.

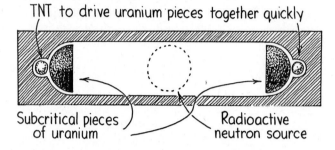

TNT to drive uranium pieces together quickly

Subcritical pieces of uranium

Radioactive neutron source

The uranium used in the Hiroshima blast was U-235 about the size of a baseball. Separating this much fissionable material from natural uranium was one of the principal and most difficult tasks of the secret Manhattan Project during World War II. Project scientists used two methods of isotope separation. One method depended on the fact that lighter U-235 moves at a slightly faster average speed than U-238 at the same temperature. In gaseous form, the faster isotope therefore has a higher rate of diffusion through a thin membrane or small opening, resulting in a slightly enriched U-235 gas on the other side (Figure 33-5). Diffusion through thousands of chambers ultimately produced a sufficiently enriched sample of U-235. The other, and less successful, method consisted of shooting uranium ions into a magnetic field. The smaller-mass U-235 ions were deflected more by the magnetic field than the U-238 ions and were collected atom by atom through a slit positioned to catch them. After a couple of years, both methods netted a few kilograms of U-235.

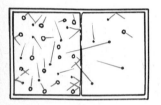

Figure 33-5
Lighter molecules move faster than heavier ones at the same temperature and diffuse more readily through a thin membrane.

Question ▶ Why will gaseous U-235 move slightly faster than U-238 at the same temperature?

Uranium-isotope separation today is more easily accomplished with a gas centrifuge. Uranium is mixed with fluorine and the uranium-hexafluorine gas is whirled in a drum at tremendously high rim speeds (on the order of 1500 kilometers per hour). Under centrifugal force, the heavier U-238 gravitates to the outside and gas rich in the lighter U-235 is extracted from the center. Engineering difficulties, only recently overcome, prevented the use of this method in the Manhattan Project.

▶ **Answer**

At the same temperature, both isotopes have the same kinetic energy ($\frac{1}{2}mv^2$). So the U-235 isotope with the lesser mass must have a correspondingly higher velocity.

Nuclear Reactors

A chain reaction cannot ordinarily take place in *pure* natural uranium, since it is mostly U-238. The neutrons released by fissioning U-235 atoms are fast neutrons, readily captured by U-238, which do not fission. A crucial experimental fact is that *slow* neutrons are far more likely to be captured by U-235 than by U-238.* If neutrons can be slowed down, there is an increased chance that a neutron released by fission will cause fission in another U-235 atom, even amid the more plentiful and otherwise neutron-absorbing U-238 atoms. This increase may be enough to allow a chain reaction to take place.

Figure 33-6

A detailed view of an atomic reactor. A controlled chain reaction occurs in the uranium slugs embedded in the graphite.

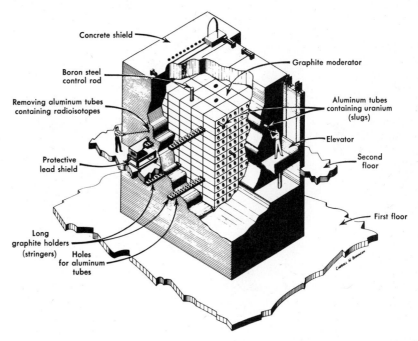

The Italian physicist Enrico Fermi reasoned that a chain reaction with ordinary uranium metal might be possible if the uranium were broken up into small lumps and separated by a material that would slow down neutrons. Fermi and his co-workers constructed the first nuclear reactor (or *atomic pile* as it was called) in the racquet courts beneath the stands of the University of Chicago's Stagg Field. They achieved the first self-sustaining controlled release of nuclear energy on December 2, 1942.

Three fates are possible for a neutron in ordinary uranium metal. It may (1) cause fission of a U-235 atom, (2) escape from the metal into nonfissionable surroundings, or (3) be absorbed by U-238 without causing fission. To make the first fate more probable, the uranium was divided into discrete parcels and buried at regular intervals in nearly 400 tons of

---

*This is similar to the selective absorption of different frequencies of light. Just as there are characteristic electron energy levels in an atom, there are similar energy levels within the nucleus.

Figure 33-7

The bronze plaque at Chicago's Stagg Field commemorates Enrico Fermi's historic fission chain reaction. Criticized in Italy for not giving the Fascist salute when presented with the 1938 Nobel Prize, Fermi never returned to Italy, and became an American citizen in 1945.

ON DECEMBER 2, 1942
MAN ACHIEVED HERE
THE FIRST SELF-SUSTAINING CHAIN REACTION
AND THEREBY INITIATED THE
CONTROLLED RELEASE OF NUCLEAR ENERGY

Figure 33-8

The now-demolished stands at the University of Chicago's Stagg Field as they looked in 1942 when in the racquet courts beneath them, a small group of scientists inaugurated the atomic age. They toasted in solemn silence by sipping wine from paper cups.

graphite, a familiar form of carbon. A simple analogy clarifies the function of the graphite: If a golf ball rebounds from a massive wall, it loses hardly any speed; but if it rebounds from a baseball, it loses considerable speed. The case of the neutron is similar. If a neutron rebounds from a heavy nucleus, it loses hardly any speed; but if it rebounds from a lighter carbon nucleus, it loses considerable speed. The graphite was said to "moderate" the neutrons.* The whole apparatus was called a *reactor*.

Now an uncontrolled reactor would not be a reactor, but a technical accident about to happen. A runaway chain reaction is prevented by control rods made of cadmium or boron metal, which easily absorb neutrons. The control rods are inserted into the reactor and absorb a selected fraction of the neutrons to maintain a steady rate of energy release. When fully inserted into the reactor, the control rods snuff the chain reaction. If the rods are fully removed, the reaction can build to a level capable of melting the reactor. Because "bomb-grade" isotopes are so highly diluted with U-238, an explosion like that of a nuclear bomb is not possible.

Heavy water, which is composed of the heavy hydrogen isotope deuterium, is an even more effective moderator. This is because ordinary hydrogen readily captures neutrons, while deuterium doesn't.

Plutonium

When a U-238 nucleus absorbs a neutron, no fission occurs. The nucleus instead emits a beta particle and becomes the first synthetic element beyond uranium—the transuranic element called *neptunium* (named after the first planet discovered via Newton's law of gravitation). Neptunium has a half-life of only 2.3 days, so it very soon emits a beta particle and becomes *plutonium* (named after Pluto, the second planet to be discovered via Newton's laws). The half-life of plutonium is about 24 000 years. The isotope Pu-239, like U-235, will undergo fission when it captures a neutron.

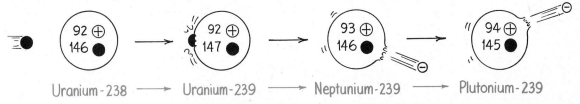

Uranium-238 ⟶ Uranium-239 ⟶ Neptunium-239 ⟶ Plutonium-239

Figure 33-9
After U-238 absorbs a neutron, it emits a beta particle, resulting in a neutron becoming a proton. The atom is no longer uranium, but neptunium. After the neptunium atom, in turn, emits a beta particle, it becomes plutonium.

Production of plutonium was the principal use of Fermi's atomic pile. Since plutonium is an element distinct from uranium, it can be separated from uranium in the pile by ordinary chemical methods. Consequently, the pile provides a process for making pure fissionable material far more easily than by separating the U-235 from natural uranium. The atomic bomb exploded over Nagasaki was a plutonium bomb.

Although the process of separating plutonium from uranium is simple in theory, it is very difficult in practice. This is because large quantities of radioactive fission products are formed in addition to plutonium. The radioactivity becomes so intense that the materials in the pile begin to fall apart physically. As a result, the pile has to be shut down after only a few grams of plutonium have been produced. The separation of plutonium from uranium must be done by remote control to protect the staff against radiation.

The element plutonium is chemically toxic in the same sense as are lead and arsenic. It attacks the nervous system and can cause paralysis; death can follow if the dose is sufficiently large. Fortunately, plutonium doesn't remain in its elemental form for long but rapidly combines with oxygen to form three compounds, PuO, PuO_2, and Pu_2O_3, all of which are chemically inert. They will not dissolve in water or in biological

---

▶ **Answers**

1. A moderator slows down neutrons that are normally too fast to be absorbed by fissionable isotopes.
2. Control rods absorb neutrons and thereby control the neutron flux and the rate of the chain reaction.

systems. These plutonium compounds do not attack the nervous system and have been found to be biologically harmless.

Plutonium in any form, however, is radioactively toxic. It is more toxic than uranium and less toxic than radium. Plutonium emits alpha particles that have a very short range and high energy transfer, which kill rather than mutate the cells they encounter (unlike the beta particles emitted by radium). Since damaged cells rather than dead cells contribute to cancer, plutonium ranks low as a cancer-producing substance. The greatest danger that plutonium presents to humans is its potential for use in nuclear fission bombs. Its usefulness is in breeder reactors.

Breeder Reactors

When U-238 is mixed with fissionable isotopes in a reactor, neutrons liberated by fission convert the relatively abundant U-238 into fissionable Pu-239. Or when Th-232 is mixed with fissionable isotopes, it similarly is converted to fissionable U-233. So in addition to the production of energy, fission fuel is bred from normally nonfissionable isotopes in the process. Such a reactor is a *breeder reactor.*

Figure 33-10
Pu-239 or U-233, like U-235, undergoes fission when it captures a neutron.

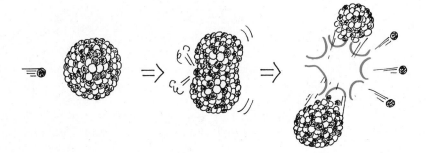

A breeder reactor effectively breeds more fuel than it consumes.* For about every two fissionable isotopes put into the reactor, three new fissionable isotopes are produced. This is like filling your gas tank with water, adding some gasoline, then driving your car and having more gasoline after your trip than you started with, at the expense of common water. After the initial cost of building a reactor, this is a very economical method of producing vast amounts of energy. With a breeder reactor, a power utility can, after a few years of operation, breed twice as much fuel as its original fuel.

The principal use of nuclear reactors in the world today, whether conventional or breeder, is the production of electric power. It is interesting to note that fission reactors are simply nuclear furnaces, which, like fossil-fuel furnaces, do nothing more elegant than boil water to produce

---

*Neutron capture is sufficient in a typical "nonbreeder" U.S. power reactor of the light-water type to convert almost 1 percent of the predominantly U-238 supply to plutonium over the 3-year period the fuel resides in the core. As the supply of U-238 decreases, a supply of Pu-239 increases and actually predominates over uranium in energy output near the end of the fuel cycle. To various extents, all reactors breed fissionable fuel.

steam for a turbine (Figure 33-11). The benefits of fission power are plentiful electricity; the conservation of the many billions of tons of coal, oil, and natural gas that every year are literally turned to heat and smoke and that in the long run may be far more precious as sources of organic molecules than as sources of heat; and the elimination of the megatons of sulfur oxides and other poisons that are put into the air each year by the burning of these fuels.

Figure 33-11

Diagram of a nuclear fission power plant.

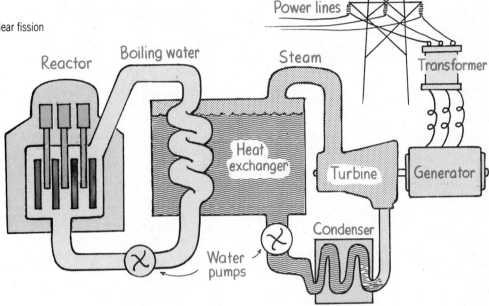

The drawbacks include the problems of storing radioactive wastes, the production of plutonium and the danger of nuclear weapons proliferation, the low-level release of radioactive materials into the air and ground-water, and, most importantly, the risk of an accidental release of large amounts of radioactivity.

Sound judgment requires us not only to compare the benefits and draw-backs of fission power but also to compare its benefits and drawbacks to those of alternate power sources. For a variety of reasons, public opinion in the United States and Europe is now against fission power plants. Nukes are on the decline and fossil-fuel power plants are on the upswing.

Question   In a breeder reactor, what is bred from what?

▶ **Answer**

Fissionable Pu-239 is bred from nonfissionable U-238, or fissionable U-233 is bred from nonfissionable Th-232.

Mass-Energy Equivalence

Figure 33-12
Work is required to pull a
nucleon from an atomic
nucleus.

Figure 33-13
The mass spectrometer. Ions
are directed into the semicircular "drum," where they
are swept into semicircular
paths by a strong magnetic
field. Because of inertia,
heavier ions are swept into
curves of large radii and
lighter ions are swept into
curves of smaller radii. The
radius of the curve is directly
proportional to the mass of
the ion. Using C-12 as a
standard, the masses of all
the elements and their isotopes are easily determined.

From Einstein's mass-energy equivalence, $E = mc^2$, we can think of
mass as congealed energy. The more energy associated with a particle,
the greater its mass. Would the mass of a nucleon be the same whether
it was located inside or outside an atomic nucleus? Wouldn't it have more
mass energy outside a nucleus than inside? We can answer this by considering the work that would be required to separate the nucleons from
a nucleus. Recall that work, which is expended energy, is equal to the
product of force and distance. Then think of the magnitude of force required
to pull nucleons apart through a sufficient distance to overcome the
attractive nuclear force. Tremendous work would be required. The work
we would have to put into such an endeavor would be evident in the
energy of the protons and neutrons. They would have more energy outside the nucleus. This extra energy would be equal to the energy or work
required to separate them. This energy, in turn, would be evident in mass.
Hence, the mass of nucleons outside a nucleus is greater than the mass
of the same nucleons when locked inside a nucleus.

The experimental verification of this conclusion is one of the triumphs
of modern physics. The masses of nucleons and the isotopes of the various elements can be measured with an accuracy of 1 part per million or
better. One means of doing this is with the *mass spectrometer* (Figure
33-13).

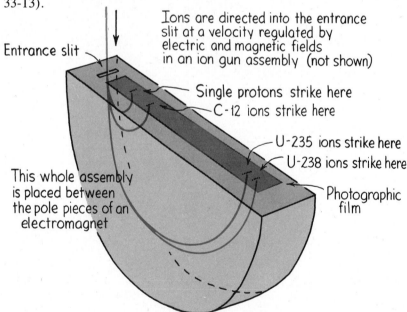

Ions are directed into the entrance
slit at a velocity regulated by
electric and magnetic fields
in an ion gun assembly (not shown)

Entrance slit

Single protons strike here
C-12 ions strike here

U-235 ions strike here
U-238 ions strike here

This whole assembly
is placed between
the pole pieces of an
electromagnet

Photographic
film

In the mass spectrometer, charged ions are directed into a magnetic
field and deflect into circular arcs. The greater the inertia of the ion, the
more it resists deflection and the greater the radius of its curved path.
The magnetic force sweeps heavier ions into larger arcs and lighter ions
into shorter arcs. The ions pass through exit slits where they may be
collected, or they strike photographic film where they may be simply
detected. An isotope is chosen as a standard, and its position on the film

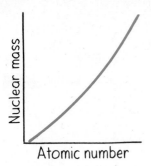

Figure 33-14
The plot shows how nuclear mass increases with increasing atomic number.

Figure 33-15
The plot shows that the average mass of a particular nucleon depends on the atomic number of the nucleus it is a part of. Individual nucleons are most massive in the lightest nuclei, least massive in the iron nucleus, and intermediate in the heaviest nuclei. When light nuclei fuse, the product nucleus is less massive than the sum of its parts; when the heaviest nuclei fission, the parts have less mass than the original nucleus. The mass defect, in both cases, is converted to energy.

of the mass spectrometer is used as a reference point. The standard is the common isotope of carbon, $^{12}_{6}C$. The mass of the carbon-12 nucleus is assigned the value of 12 atomic mass units. The atomic mass unit (amu) is defined to be precisely one-twelfth the mass of the common carbon-12 atom. With this reference, the amu's of the other atomic nuclei are measured. The masses of the proton and neutron are greater alone than when in a nucleus. They are 1.00728 and 1.00866 amu, respectively.

A plot of nuclear masses for the elements from hydrogen through uranium is shown in Figure 33-14. The curve slopes upward with increasing atomic number as expected: elements are more massive as atomic number increases. A more interesting curve results if we plot the nuclear mass per nucleon from hydrogen through uranium (Figure 33-15).

To obtain the nuclear mass *per nucleon*, we divide the nuclear mass by the number of nucleons in the particular nucleus. We find that the masses of protons and neutrons are different when combined in different nuclei. The mass of a proton is greatest as the nucleus of the hydrogen atom and becomes successively less and less massive as it occurs in atoms of increasing atomic number. The proton is least massive when in the iron atom. Beyond iron, the process reverses itself. Protons (and neutrons) become successively more and more massive with increasing atomic number, all the way to uranium and the transuranic elements.

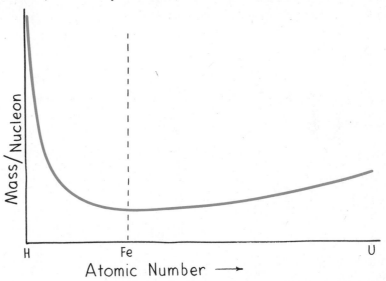

From the graph we can see why energy is released when a uranium nucleus is split into nuclei of lower atomic number. The masses of the fission fragments lie about halfway between uranium and hydrogen on the horizontal scale of the graph. Most important, the masses of these nucleons in the fission fragments are *less than* the masses of the same nucleons when combined in the uranium nucleus. The protons and neutrons decrease in mass. This mass difference is evident in the kinetic energy of the fission fragments—the 200 000 000 electron volts yielded by each uranium nucleus undergoing fission.

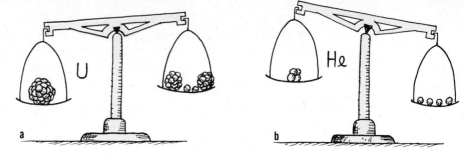

Figure 33-16

The mass of a nucleus is not equal to the sum of the masses of its parts. (*a*) The fission fragments of a heavy nucleus like uranium are less massive than the uranium nucleus. (*b*) Two protons and two neutrons are more massive in their free states than when combined to form a helium nucleus.

We can think of the mass-per-nucleon curve as an energy hill that starts at the highest point (hydrogen), slopes steeply down to the lowest point (iron), and then slopes more gradually up to uranium. Iron is at the bottom of the energy hill and is the most stable nucleus. It is also the strongest nucleus; more energy is required to pull each proton or neutron from the iron nucleus than from any other.

How about the left side of the hill? We can also gain energy by going from hydrogen toward iron. But to go in this direction is to *combine* or *fuse* nuclei. And this is fusion. Whereas energy is released in fissioning the heavy elements, energy is released in fusing the light elements. In *both* cases there is a decrease in the mass per nucleon, which causes a huge energy release. From the curve we see that any atom to the left or right of iron is a potential source of energy. The higher we are on the hill to either the right or the left and the farther we roll down the hill, the greater the amount of energy available. No energy can be released by the iron nucleus. If we split it, the total mass of the fragments is greater than the mass of the iron before splitting. If we fuse a pair of iron nuclei, the mass of the resulting nucleus is greater than the two masses before fusing. In this special case, energy would have to be put into either the fissioning or the fusing of iron; no energy would be liberated.

We can see by the curve that elements above iron cannot give off energy if fused. The mass of the fused nuclei would be greater than the sum of the masses before fusion. This would *require* energy, not *liberate* it. Likewise, elements below iron would yield no energy if fissioned. The sum of the masses of the fission fragments would be greater than that of the original nucleus, again requiring energy rather than liberating it. Energy is released only when there is a decrease in mass.

In both fission and fusion reactions, the amount of matter that is converted to energy is less than 0.1 percent. This is true whether the process takes place in bombs or in the stars.

Figure 33-17

Fictitious example: the "hydrogen magnets" weigh more when apart than when together. (Adapted from *The New College Physics: A Spiral Approach* by Albert V. Baez. Copyright © 1967 by W. H. Freeman and Company. Reprinted by permission.)

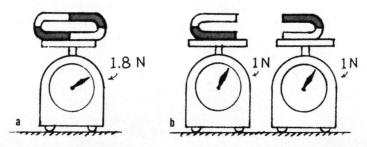

Nuclear Fusion

In the process of fusing hydrogen isotopes to form helium nuclei, **nuclear fusion**, the "shrinking" mass of the nucleons involved can produce as much as 26 700 000 electron volts of energy. The mass lost in the fusion of hydrogen isotopes appears as kinetic energy, most of which is carried away by neutrons expelled at nearly one-fifth the speed of light from the reaction. When the neutrons are stopped and captured, the energy of fusion is turned into heat, electricity, or other forms of energy.

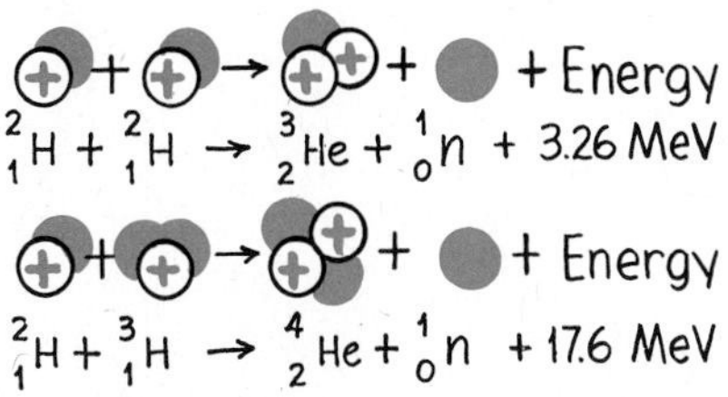

Figure 33-18
Fusion reactions.

$$\bigoplus + \bigoplus \rightarrow \bigoplus\bigoplus + \bullet + \text{Energy}$$

$$^2_1\text{H} + {}^2_1\text{H} \rightarrow {}^3_2\text{He} + {}^1_0\text{n} + 3.26\,\text{MeV}$$

$$\bigoplus + \bigodot \rightarrow \bigoplus\bigodot + \bullet + \text{Energy}$$

$$^2_1\text{H} + {}^3_1\text{H} \rightarrow {}^4_2\text{He} + {}^1_0\text{n} + 17.6\,\text{MeV}$$

Although the fusion energy per reaction of individual hydrogen atoms is less than the energy given up by the fissioning of individual uranium atoms, gram for gram fusion is several times more energy-producing than the fission process. This is because there are more hydrogen atoms in a gram of hydrogen than there are heavier uranium atoms in a gram of uranium.

Elements heavier than hydrogen and lighter than iron give off energy when fused, but much less per reaction than hydrogen when fused. The fusion of these heavier elements occurs in the advanced stages of a star. The energy released per gram during the various fusion stages from helium to iron amounts only to about one-fifth the energy released in the fusion of hydrogen to helium. Hydrogen, most notably in the form of deuterium, is the choicest fuel for fusion.

For a fusion reaction to take place, the nuclei must collide at very high speeds in order to overcome their mutual electrical repulsion. The required speeds correspond to the extremely high temperatures found in the sun and stars. We call fusion reactions in the sun and other stars **thermonuclear** reactions—that is, the welding together of atomic nuclei by high temperature. In the high temperatures of the sun, approximately 657 million tons of hydrogen are converted into 653 million tons of a mixture of helium and neutrons each second. The missing 4 million tons of mass become the kinetic energy of helium and neutrons. This kinetic energy heats the sun. Such reactions are, quite literally, nuclear burning. Thermonuclear reactions are analogous to ordinary chemical combustion. In both chemical and nuclear burning, high temperature starts the reaction, and the release of energy by the reaction maintains a high enough temperature to spread the fire. The net result of the chemical reaction is a combination of atoms into more tightly bound molecules. In nuclear burning, the high temperature starts a reaction or series of reactions with the net result of producing more tightly bound nuclei. The difference between chemical and nuclear burning is essentially one of scale.

Prior to the development of the atomic bomb, the temperatures required to initiate nuclear fusion on earth were unattainable. When it was found

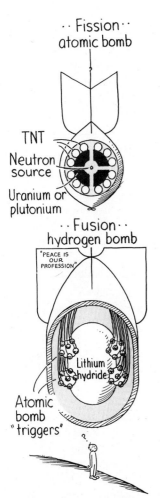

Fission
atomic bomb

TNT
Neutron
source
Uranium or
plutonium

Fusion
hydrogen bomb

"PEACE IS
OUR
PROFESSION"

Lithium
hydride

Atomic
bomb
"triggers"

Figure 33-19
Fission and fusion bombs.

that the temperatures inside an exploding atomic bomb are four to five times the temperature at the center of the sun, the thermonuclear bomb was but a step away. This first hydrogen bomb was detonated in 1952. Whereas the critical mass of fissionable material limits the size of a fission bomb (atomic bomb), no such limit is imposed on a fusion bomb (thermonuclear, or hydrogen, bomb). Just as there is no limit to the size of an oil-storage depot, there is no theoretical limit to the size of a fusion bomb. Like the oil in a storage depot, any amount of nuclear fuel can be stored with safety until it is ignited. Although a mere match can ignite an oil depot, nothing less energetic than a fission bomb can ignite a thermonuclear bomb. We can see that there is no such thing as a "baby" hydrogen bomb. It cannot be less energetic than its fuse, which is an atomic bomb.

The hydrogen bomb is another example of a new discovery applied to destructive rather than constructive purposes. The constructive side of the picture is the controlled release of vast amounts of clean energy.

Controlling Fusion

Carrying out fusion reactions under controlled conditions ordinarily requires temperatures of millions of degrees. Producing and sustaining such high temperatures are the goals of much current research. High temperatures are obtained in electric arcs, where injected gases are ionized and become plasmas. Further heating is accomplished by electromagnetic resonance techniques and magnetic compression.

There is the problem of finding a material that can be used to contain the plasma. All known materials melt and vaporize below 4000°C, while fusion generally requires temperatures exceeding 100 million degrees. The situation is not hopeless, however. A magnetic field is nonmaterial, can exist at any temperature, and can exert powerful forces on charged particles in motion. "Magnetic walls" of sufficient strength have been designed to contain plasmas in a kind of magnetic straitjacket. The magnetic fields can be regulated to compress the plasma and produce fusion temperatures. The plasma will not melt the walls of the electromagnets even if its temperature is a billion degrees, because its total heat content is very small due to its very low density. To have a controlled reactor rather than a bomb, plasma densities are kept low, about 10 000 times lower than atmospheric density. The plasma must also be kept off the walls of the electromagnets, not because it will melt them but because contact will slow the ions and therefore cool the plasma.

Some ions at 1 million degrees are moving fast enough to overcome electrical repulsion and slam together, but the energy output of these fusion reactions is small compared to the energy used to heat the plasma. Even at 100 million degrees, more energy must be put into the plasma than will be given off by fusion. At about 350 million degrees, the fusion reactions will produce enough energy to be self-sustaining. At this *ignition temperature*, nuclear burning yields a sustained power output without

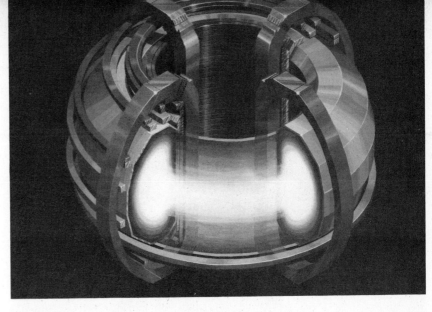

Figure 33-20
Plasma is constrained by magnetic fields in the vacuum vessel of a tokamak fusion device.

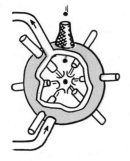

Figure 33-21
Fusion with multiple laser beams. Frozen pellets of deuterium are rhythmically dropped into synchronized laser cross fire. The resulting heat is carried off by molten lithium to produce steam.

further input of energy. A steady feeding of hydrogen isotopes is all that is needed to produce continuous power.

Fusion has already been achieved in several devices, but instabilities in the plasma current have thus far prevented a sustained reaction. One of the most difficult problems has been to devise a field system that will hold the plasma in a stable position long enough for an ample number of ions to fuse. A variety of magnetic-confinement devices are the subject of much present-day research.

Another promising approach uses high-energy lasers to bypass magnetic confinement altogether. One technique is to align an array of laser beams at a common point and drop solid pellets of hydrogen isotopes through the synchronous cross fire (Figure 33-21). The energy of the multiple beams crushes the deuterium-tritium fuel to a density more than 20 times that of lead. A fusion "burn" consumes most of the hydrogen to produce helium nuclei and neutrons that fly apart with kinetic energies equivalent to the mass loss of the fusion reaction. This outward propagation occurs in about a tenth of a billionth of a second and produces several hundred times more energy than is delivered by the laser beams that compress and ignite the pellets. Like the succession of small fuel/air explosions in an automobile engine's cylinders that convert into a smooth flow of mechanical power, the successive ignition of dropping pellets in a fusion power plant may similarly produce a steady stream of electrical power.* The success of this technique requires precise timing,

---

*The rate of pellet fusion is 5 per second on the projected Cascade power plant, now on the drawing boards at Lawrence Livermore Laboratory (for comparison, approximately 20 explosions per second occur in each automobile engine cylinder in a car that travels at highway speed). Such a plant could produce 1000 million W of electric power, enough to supply a city of about 600 000 people. Five fusion burns per second will provide about the same power as 60 L of fuel oil or 70 kg of coal per second from conventional power plants.

Figure 33-22
Pellet chamber at Lawrence Livermore Laboratory. The laser source is Nova, the most powerful laser in the world, which directs 10 beams into the target region.

for the necessary compression must occur before a shock wave causes the pellet to disperse. Success also requires the development of more efficient lasers, so the electricity generated will be greater than the amount required to operate the lasers.

Still other approaches involve the bombardment of fuel pellets not by laser light but by beams of electrons, light ions, and beams of heavy ions. The most novel of the new approaches involves the introduction of muons to take the place of electrons in hydrogen and effectively neutralize the proton's charge (see next section). Whatever the method, we are still looking forward to the great day—break-even day—when one of the fusion schemes will sustain a yield of at least as much energy as is required to initiate it. The aftermath will probably be the solution to power production the world over. This would be ideal, for although the inner chamber of the fusion device will be radioactive due to high-energy neutrons, that radioactivity is low-level—with a half-life measured in days and years compared to tens of thousands of years for fission reactions. Fusion produces clean, nonradioactive helium (good for children's balloons). Fusion reactors are also inherently incapable of a "runaway" accident because fusion requires no critical mass. Furthermore, there is no air pollution because there is no combustion. The problem of thermal pollution, characteristic of conventional steam-turbine plants, can be avoided by direct generation of electricity with MHD generators or by similar techniques using charge-particle fuel cycles that employ a direct energy conversion.

The fuel for nuclear fusion is hydrogen, the most plentiful element in the universe. The simplest reaction is the fusion of the hydrogen isotopes deuterium ($^2_1$H) and tritium ($^3_1$H), both found in ordinary water. For example, 30 liters of seawater contain 1 gram of deuterium, which when fused releases as much energy as 10 000 liters of gasoline or 80 tons of TNT. Natural tritium is much more scarce, but given enough to get started, a controlled thermonuclear reactor will breed it from deuterium in ample quantities. Because of the abundance of fusion fuel, the amount of energy that can be released in a controlled manner is virtually unlimited.

Questions

1. Fission and fusion are opposite processes, yet each releases energy. Isn't this contradictory?
2. Would you expect the temperature of the core of a star to increase or decrease as a result of the fusion of intermediate elements to manufacture heavy elements?

Cold Nuclear Fusion

The fusion of hydrogen isotopes may circumvent the need for high temperatures altogether. Nonthermonuclear fusion can be accomplished by effectively neutralizing the positive charge of the proton by a very nearby negative particle—but not by an electron. In a hydrogen atom, the electron orbits the nucleus at a relatively large distance that is dictated by the relatively long wavelength matter wave of the tiny electron. When the electron is replaced by a *muon*, an elementary particle that has the same negative charge as the electron but a mass more than 200 times greater, the relatively short wavelength matter wave puts it 200 times closer to the proton. The tight orbit of the muon about the nucleus results in a smaller version of the hydrogen atom, a "muonic atom." In the muonic atom the muon orbits so close to the nucleus that the muon and nucleus appear as a single neutral particle to distant protons. Thus, there is no electrical repulsion between a muonic atom and a hydrogen nucleus or any charged particle. Voilà! Ordinary thermal motion is all that is

▶ **Answers**

1. No, no, no! This is contradictory only if the same element is said to release energy by both the processes of fission and fusion. Only the fusion of light elements and the fission of heavy elements result in a decrease in nucleon mass and a release of energy.
2. Energy is absorbed and the star core tends to cool at this late stage of its evolution. This, however, allows the star to collapse, which produces an even greater temperature.

needed for these muonic atoms and hydrogen nuclei to bump into one another, whereupon they form a short-lived molecule, then fuse. This is *cold nuclear fusion*, known as muon-catalyzed fusion.*

Most of the energy of muon-catalyzed fusion is in the form of kinetic energy—mainly of the neutrons that are ejected in the reaction. The neatest thing about this type of fusion is that most reactions eject the muons as well, which go on to catalyze other fusion reactions. Surrounding atoms are essentially unaffected, except for the increased temperature from the high kinetic energies of the neutrons. There are several ways in which the kinetic energy of the neutrons can be harnessed, the simplest of which is the conversion to heat to drive turbines for generating electricity.

Figure 33-23

(*a*) In an ordinary hydrogen atom, a proton that penetrates the electron cloud is electrically repelled by the positive nucleus. High temperatures are required to slam the protons together. (*b*) When a muon takes the place of the electron in a hydrogen atom, the result is a tighter version of hydrogen—a muonic atom. If a proton penetrates the muon cloud, its close proximity to the positive nucleus finds it in the clutches of the strong nuclear force. Fusion rather than electrical repulsion occurs. (In practice the muonic atom penetrates the clouds of deuterium and tritium molecules.)

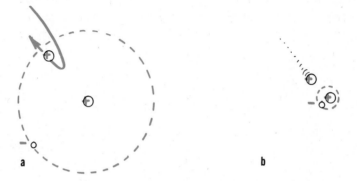

a   b

The not-so-neat thing about this is that muons are short-lived and are hard to come by. Muons are found in nature in secondary cosmic rays produced when primary cosmic rays collide with the upper atmosphere. Muons can be created by colliding high-energy ions from a particle accelerator with ordinary matter such as carbon. The collisions produce particles called *pions*, which quickly decay to make both positive and negative muons in a process much like the one that occurs when cosmic rays bombard the atmosphere. Muons themselves are unstable and have a half-life of about 2 microseconds. If muon-catalyzed fusion is to produce commercial power, the muon must catalyze enough reactions in its short lifetime to do more than just power the accelerator that generates the muons to begin with.

Recent experiments are encouraging. Negatively charged muons are fired into a gaseous mixture of deuterium and tritium molecules; they

---

*Muon-catalyzed fusion was suggested on theoretical grounds in the late 1940s by F. C. Frank and Andrei D. Sakharov. A decade later Luis W. Alvarez and his colleagues found evidence of muon-catalyzed fusion in bubble chamber tracks at the University of California at Berkeley. This discovery generated much excitement; enthusiasm waned, though, when calculations showed that most muons catalyzed only a single fusion before decay, producing too little energy and too few muons to catalyze later reactions. The Alvarez group had studied reactions involving only ordinary hydrogen and deuterium. More current findings involving a mixture of deuterium and tritium show much more promising results. See the article "Cold Nuclear Fusion" by Johann Rafelski and Steven E. Jones in the July 1987 issue of *Scientific American*.

collide with and replace orbital electrons about the nuclei of these molecules. When the muon enters its tight orbit about one of these isotopes,
the molecule breaks apart leaving a slow-moving muonic atom that easily
invades the nucleus of any deuterium or tritium molecules it encounters.
Fusion occurs, and the muon is released to initiate the process again. To
increase the number of reactions, the temperature of the gas is adjusted
to produce resonance between the energy absorbed by the molecules and
their vibrational states. Some investigators report the occurrence of this
resonance at about 900°C, which produces well over 100 fusions per
muon. Higher temperatures dampen the reactions, which means a muon-
catalyzed fusion reactor would not be susceptible to runaway reactions
or meltdowns. In any case, the muon-catalyzed fusion reaction is not a
chain reaction such as takes place in a fission reactor, since the reaction
does not produce more muons than come in. Once the muon is gone the
reaction stops. The low temperature and low power density of muon-
catalyzed fusion also makes it a poor contender as a weapon of warfare.
The most optimistic schemes for muon production fall far short of enough
reactions at one time to produce bomblike amounts of energy.

Cold nuclear fusion, whether caused by muons or particles other than
muons, is presently the focus of much current interest and research. If
it does not turn out to be a serious contender for an economically viable
method of generating energy, it at least demonstrates that fusion can be
initiated by means simpler than high-temperature plasmas or giant lasers.

Fusion Torch and Recycling

A fascinating application for the abundant energy that fusion of whatever
kind may provide is the *fusion torch*, a star-hot flame or high-temperature
plasma into which all waste materials—whether liquid sewage or solid
industrial refuse—could be dumped. In the high-temperature region the
materials would be reduced to their constituent atoms and separated by
a mass-spectrometer-type device into various bins ranging from hydrogen
to uranium. In this way, a single fusion plant could not only dispose of
thousands of tons of solid wastes per day but also provide a continuous
supply of fresh raw material—thereby closing the cycle from use to reuse.

This would be a major turning point in materials economy (Figure
33-24). Our present concern for recycling materials will reach a grand
fruition with this or a comparable achievement, for it would be recycling
with a capital *R*! Rather than gut our planet further for raw materials,
we'd be able to recycle our existing stock over and over again, adding
new material only to replace the relatively small amounts that are lost.
Fusion power can produce abundant electrical power, can desalinate water,
can help to cleanse our world of pollution and wastes, can recycle our
materials, and in so doing can provide the setting for a better world—
not in the far-off future, but perhaps in our own time. If and when fusion
power plants become a reality, they are likely to have an even more profound impact upon almost every aspect of human society than did the
harnessing of electromagnetic energy in the last century.

Figure 33-24

A closed materials economy could be achieved with the aid of the fusion torch. In contrast to present systems (a), which are based on inherently wasteful linear material economies, a stationary-state system (b) would be able to recycle the limited supply of material resources, thus alleviating most of the environmental pollution associated with present methods of energy utilization. (Adapted from "The Prospects of Fusion Power" by William C. Gough and Bernard J. Eastlund. Copyright © 1971 by Scientific American, Inc. Reprinted by permission.)

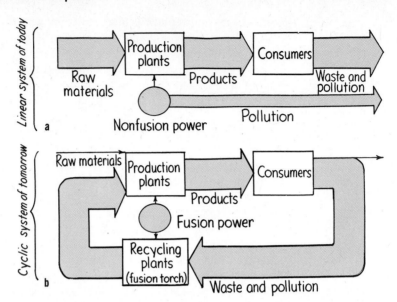

When we think of our continuing evolution, we can see that the universe is well suited to those who will live in the future. If people are one day to dart about the universe in the same way we are able to jet about the world today, their supply of fuel is assured. The fuel for fusion is found in every part of the universe, not only in the stars but also in the space between them. About 91 percent of the atoms in the universe are estimated to be hydrogen. For people of the future, the supply of raw materials is also assured; all the elements result from fusing more and more hydrogen nuclei together. Simply put, if you fuse 8 deuterium nuclei, you have oxygen; 26, you have iron; and so forth. Future humans might synthesize their own elements and produce energy in the process, just as the stars do. Humans may one day travel to the stars in ships fueled by the same energy that makes the stars shine.

Summary of Terms

Nuclear fission The splitting of the nucleus of a heavy atom, such as uranium-235, into two main parts, accompanied by the release of much energy.

Chain reaction A self-sustaining reaction that, once started, steadily provides the energy and matter necessary to continue the reaction.

Critical mass The minimum mass of fissionable material in a reactor or nuclear bomb that will sustain a chain reaction.

Nuclear fusion The combination of the nuclei of light atoms to form heavier nuclei, with the release of much energy.

Thermonuclear fusion Nuclear fusion produced by high temperature.

Review Questions

Nuclear Fission

1. What is the role of electrical forces in nuclear fission?

2. How does the energy release of a single uranium fission compare to that of a molecule of TNT when it explodes?

3. Which has the greater mass, a uranium nucleus before it undergoes fission or the fission fragments after fission?

4. When a nucleus undergoes fission, what is the role of the neutrons that are ejected?

5. Why does a chain reaction not occur in uranium mines?

6. Which has more total surface area, a whole apple or an apple cut in two pieces?

7. Which has more surface area, two separate pieces of uranium or the same pieces stuck together?

8. Which will leak more neutrons, two separate pieces of uranium or the same pieces stuck together?

9. Is an energetic chain reaction more likely to occur in two separate pieces of U-235 or in the same pieces stuck together?

10. What were the two methods used to separate U-235 from U-238 in the Manhattan Project during World War II?

Nuclear Reactors

11. What was the function of graphite in the first atomic reactor?

12. What are the two means of controlling the chain reaction in a nuclear reactor?

Plutonium

13. What is the outcome of U-238 absorbing a neutron?

14. What is the outcome of U-239 emitting a beta particle?

15. What is the outcome of Np-239 emitting a beta particle?

16. What do U-235 and Pu-239 have in common?

17. When is plutonium chemically toxic and when is it not?

18. Is plutonium radioactively more toxic or less toxic than uranium?

Breeder Reactors

19. What is the effect of putting small amounts of Pu-239 with large amounts of U-238?

20. Name three isotopes that undergo nuclear fission.

21. How does a breeder reactor breed nuclear fuel?

Mass-Energy Equivalence

22. Is work required to pull a nucleon out of an atomic nucleus? Does the nucleon, once outside, then have more potential energy? Does a nucleon with more energy have more mass?

23. Does a nucleon have more mass or less mass outside an atomic nucleus?

24. Is the amount of difference in mass for a nucleon outside an atomic nucleus the same or different for different nuclei?

25. What is the basic difference between the graphs of Figure 33-14 and Figure 33-15?

26. In what atomic nucleus is a proton most massive? Least massive?

27. How do the masses of nucleons in uranium compare to the masses of nucleons in the fission fragments of uranium?

28. What becomes of the missing mass?

29. Which atomic nucleus is the most tightly bound?

30. If an iron nucleus were fissioned, would the fission fragments have more mass per nucleon or less mass per nucleon?

31. If a pair of iron nuclei were fused, would the product nucleus have more mass per nucleon or less mass per nucleon?

Nuclear Fusion

32. Which releases more energy, the fissioning of a uranium atom or the fusing of a pair of hydrogen atoms? Which releases more energy, the fissioning of atoms in a gram of uranium or the fusing of atoms in a gram of hydrogen? Why are your answers different?

33. When a pair of hydrogen isotopes are fused, do the nucleons in the product nucleus have more mass or less mass?

34. To yield energy from helium, should it be fissioned or fused?

35. What exactly is *thermonuclear* fusion?

36. How is nuclear burning similar to chemical burning?

Controlling Fusion

37. What kind of containers are used to contain million-degree plasmas?

38. How is fusion accomplished with lasers?

39. How do the product particles of fusion reactions differ from the product particles of fission reactions?

Cold Nuclear Fusion

40. Which is smaller in size, an atom consisting of a proton and electron or an atom consisting of a proton and a muon?

41. What is the net charge of an atom consisting of a proton nucleus and a muon?

42. To which can a stray proton get closer, the nucleus of a conventional hydrogen atom or the nucleus of a muonic hydrogen atom? Why?

43. What happens when a stray proton gets very close to the nucleus of a muonic atom?

44. How are muons produced?

Fusion Torch and Recycling

45. If a star-hot flame is positioned between a pair of large electrically charged plates, one positive and the other negative, and materials dumped into the flame are dissociated into bare nuclei and electrons, in which direction will the nuclei move? In which direction will the electrons move?

46. Suppose the negative plate has a hole in it so that atomic nuclei that move toward and pass through it end up as a beam. Further suppose the beam is then directed between the pole pieces of a powerful electromagnet. Will the beam of charged nuclei continue in a straight line or will the beam be deflected?

47. Supposing the beam is deflected, will all nuclei, both light and heavy, deflect by the same amount? (Can you see that this device is a version of the mass spectrometer and that it effectively separates atomic nuclei?)

48. In a bucketful of seawater are minute amounts of gold. You can't separate them from the water with an ordinary magnet, but if the bucket is dumped into the fusion torch we are describing, a magnet can separate them. If hydrogen atoms are collected in bin #1 and uranium atoms are collected in bin #92, what will be the bin number for gold?

49. What would be the outcome of dumping sewage and refuse in a fusion torch instead of into the ocean or underground?

50. What is the fuel for nuclear fusion? Which atom makes up the primary building block for all other atoms? What is the most abundant element in the universe? Why will there be no shortage of energy and material in the "hydrogen age"?

Exercises

1. Why is it that uranium ore doesn't spontaneously explode?

2. Why will nuclear fission probably not be used for powering automobiles?

3. How is chemical burning similar to a nuclear chain reaction?

4. Why does a neutron make a better nuclear bullet than a proton or an electron?

5. Why will the escape of neutrons be proportionally less in a large piece of fissionable material than in a smaller piece?

6. Which shape is likely to produce a larger critical mass, a cube or a sphere? Explain.

7. Does the surface area as a whole increase or decrease when pieces of fissionable material are assembled into one piece? Does this assembly increase or decrease the probability of an explosion?

8. U-235 releases an average of 2.5 neutrons per fission, while Pu-239 releases an average of 2.7 neutrons per fission. Which of these elements might you therefore expect to have the smaller critical mass?

9. Which will provide the faster chain reaction, U-235 or Pu-239?

10. Why does plutonium not occur in appreciable amounts in natural ore deposits?

11. Discuss and make a comparison of pollution by conventional fossil-fuel power plants and nuclear-fission power plants.

12. The water that passes through a reactor core does not pass into the turbine. Instead, heat is transferred to a separate water cycle that is entirely outside the reactor. Why it this done?

13. Is the mass of an atomic nucleus greater or less than the sum of the masses of the nucleons composing it?

14. The energy release of nuclear fission is tied to the fact that the heaviest nuclei weigh about 0.1 percent more per nucleon than nuclei near the middle of the periodic table of elements. What would be the effect on energy release if the 0.1 percent figure were 1 percent?

15. To predict the approximate energy release of either a fission or a fusion reaction, explain how a physicist makes use of the curve of Figure 33-15 or a table of nuclear masses and the equation $E = mc^2$.

16. Which process would release energy from gold, fission or fusion? From carbon? From iron?

17. If uranium were to split into three segments of equal size instead of two, would more energy or less energy be released? Defend your answer in terms of Figure 33-15.

18. Suppose the curve of Figure 33-15 for mass per nucleon versus atomic number took the shape of the curve shown in Figure 33-14. Then would nuclear fission reactions produce energy? Would nuclear fusion reactions produce energy? Defend your answers.

19. The "hydrogen magnets" in Figure 33-17 weigh more when apart than when combined. What would be the basic difference if the fictitious example instead consisted of "uranium magnets"?

20. Which produces more energy, the fissioning of a single uranium atom or the fusing of a pair of deuterium atoms? The fissioning of a gram of uranium or the fusing of a gram of deuterium? (Why are your answers different?)

21. Why is there, unlike fission fuel, no limit to the amount of fusion fuel that can be safely stored in one locality?

22. If a fusion reaction produces no appreciable radioactive isotopes, why does a hydrogen bomb produce significant radioactive fallout?

23. List at least two major advantages to power production by fusion rather than by fission.

24. Nuclear fusion is a present hope for abundant energy in the near future. Yet the energy that has always sustained us has been the energy of nuclear fusion. Explain.

25. What effect on the mining industry can you foresee in the disposal of urban waste by a fusion torch coupled with a mass spectrometer?

26. The world has never been the same since the discovery of electromagnetic induction and its applications to electric motors and generators. Speculate and list some of the worldwide changes that are likely to follow the advent of successful fusion reactors.

PART 8

RELATIVITY

Albert Einstein (1879–1955)

Albert Einstein was born in Ulm, Germany, on March 14, 1879. According to popular legend, he was a slow child and learned to speak at a much later age than average; his parents feared for a while that he might be mentally retarded. Records from the elementary school he attended in Munich show, however, that Einstein was remarkably gifted in mathematics, physics, and playing the violin. He rebelled at the prevailing spirit of militarism in Germany and at the practice of education by regimentation and rote, and was expelled just as he was preparing to drop out of high school at the age of 15. Largely for business reasons, his family moved to Italy, and young Einstein renounced his German citizenship and went to live with family friends in Switzerland. There he was allowed to take the entrance examination for the renowned Swiss Federal Institute of Technology in Zurich, at 2 years younger than the normal age. He did not pass the examination, mainly because of his difficulties with the French language. He spent a year at a Swiss preparatory school in Aarau, where he was "promoted with protest in French." He tried the entrance exams again at Zurich and passed. As a student, he cut many lectures, preferring to study on his own, and in 1900 he succeeded in passing his examinations by cramming with the help of a friend's meticulous notes. He said later of this, "after I had passed the final examination, I found the consideration of any scientific problem distasteful to me for an entire year." During this year he became a citizen of Switzerland; he accepted a temporary summer teaching position and tutored two young high school students. He advised their father, a high school teacher himself, to remove the boys from school, where, he maintained, their natural curiosity was being destroyed. Einstein's job as a tutor was short-lived.

It was not until 2 years after graduation that he got a steady job as a patent examiner at the Swiss Patent Office in Berne. Einstein held this position for over 7 years. He found the work rather interesting, sometimes stimulating his scientific imagination, but mainly freeing him of financial worries while providing time to ponder the problems in physics that puzzled him.

With no academic connections whatsoever, and with essentially no contact with other physicists, he laid out the main lines along which twentieth-century theoretical physics has developed. In 1905, at the age of 26, he earned his Ph.D. in physics and published three major papers. The first was on the quantum theory of light, including an explanation of the photoelectric effect, for which he won the 1921 Nobel Prize in physics. The second paper was on the statistical aspects of molecular theory and Brownian motion, a proof for the existence of atoms. His third and most famous paper was on special relativity. In 1915 Einstein published a paper on the theory of general relativity in which he presented a new theory of gravitation that included Newton's theory as a special case. These trail-blazing papers have had a profound effect on the course of modern physics and on our understanding of the physical universe.

Einstein's concerns were not limited to physics. He lived in Berlin during World War I and denounced the German militarism of this time, expressing a deeply felt conviction that warfare should be abolished and an international organization founded to govern disputes between nations. In 1933, while Einstein was visiting the United States, Hitler came to power. Einstein spoke out against Hitler's racial and political policies and resigned his position at the University of Berlin. Hitler responded by putting a price on his head. Einstein then accepted a research position at the Institute for Advanced Study in Princeton, New Jersey. In 1940 he became an American citizen. Einstein was first a German citizen, then a Swiss citizen, then a German citizen again, and finally an American citizen. But more than that, he was a citizen of the world.

Einstein believed that the universe is indifferent to the human condition, asserting that moral questions were of utmost importance for human existence and that in order for humanity to continue, it must create a moral order. This, he felt, was particularly urgent in the nuclear age. Nuclear bombs, Einstein remarked, had changed everything but our way of thinking. He was a man of unpretentious disposition with a deep concern for the welfare of his fellow beings. C. P. Snow, who was acquainted with Einstein, gives us this assessment:

> Einstein was the most powerful mind of the twentieth century, and one of the most powerful that ever lived. He was more than that. He was a man of enormous weight of personality, and perhaps most of all, of moral stature. . . . I have met a number of people whom the world calls great; of these, he was by far, by an order of magnitude, the most impressive. He was—despite the warmth, the humanity, the touch of the comedian—the most different from other men.*

Albert Einstein was a great scientist; but he was much more than that—he was a great human being.

---

*From a review of *The Born-Einstein Letters, 1916–1955*; translated by Irene Born (New York: Walker, 1971).

34 Special Theory of Relativity

Albert Einstein's 1905 paper on relativity, titled "On the Electrodynamics of Moving Bodies," showed that the relative motion of electrical charges gave rise to magnetic forces. This interrelationship between electricity and magnetism, Einstein stated, was a consequence of the more profound interrelationship between space and time. Just as the forces between electric charges are affected by motion, the very measurements of space and time are also affected by motion. All measurements of space and time depend on relative motion. For example, the length of a rocket ship poised on its launching platform and the ticks of clocks within are found to be different when the same ship is moving at high speeds. It has always been common sense that we change our position in space when we move, but Einstein flouted common sense and stated that in moving we also change our rate of proceeding into the future—time itself is altered. Einstein went on to show that a consequence of the interrelationship between space and time is an interrelationship between mass and energy, given by the famous equation $E = mc^2$.

Figure 34-1
Front row, left to right: A. A. Michelson, Albert Einstein, and Robert Millikan.

These are the ideas that make up this chapter—the ideas of special relativity—ideas so remote from your everyday experience that understanding them is quite difficult. It will be enough to become acquainted with these ideas, so be patient with yourself if you don't understand them right away. Perhaps your grandchildren will find them very much a part of their everyday experience, in which case they should find an understanding of relativity considerably less difficult.

Motion Is Relative

Recall from Chapter 3 that whenever we talk about motion, we must always specify the position from which motion is being observed and measured. For example, a person who walks along the aisle of a moving train may be walking at a speed of 1 kilometer per hour relative to his seat, but 60 kilometers per hour relative to the railroad station. We call the place from which motion is observed and measured a **frame of reference**. An object may have different velocities relative to different frames of reference.

To measure the speed of an object, we first choose a frame of reference and pretend that the frame of reference is standing still. Then we measure the speed with which the object moves relative to the frame of reference. In the foregoing example, if we pretend the train is standing still, then the speed of the walking person is 1 kilometer per hour. If we pretend that the ground is standing still, then the speed of the walking person is 60 kilometers per hour. But the ground is not really still, for the earth spins like a top about its polar axis. Depending on how near the train is to the equator, the speed of the walking person may be as much as 1600 kilometers per hour relative to a frame of reference at the center of the earth. And the center of the earth is moving relative to the sun. If we place our frame of reference on the sun, the speed of the person walking in the train, which is on the orbiting earth, is on the order of 110 000 kilometers per hour. And the sun is not at rest, for it orbits the center of our galaxy, which moves with respect to other galaxies.

Michelson-Morley Experiment

Isn't there some reference frame that is still? Isn't space itself still, and can't measurements be made relative to still space? In 1887 the American physicists A. A. Michelson and E. W. Morley attempted to answer these questions by performing an experiment that was designed to measure the motion of the earth through space. Because light travels in waves, it was then assumed that something in space vibrates—a mysterious something called *ether*, thought to fill all space and serve as a frame of reference attached to space itself. The physicists used a very sensitive apparatus called an *interferometer* to make their observations (Figure 34-2). In this instrument, a beam of light from a monochromatic source was separated into two beams with paths at right angles to each other; these were then recombined to show whether there was any difference in speed over the two paths. The interferometer was set with one path parallel to the motion of the earth in its orbit; then either Michelson or Morley carefully watched for any changes in speed as the apparatus was rotated to put the other path parallel to the motion of the earth. The interferometer was sensitive enough to measure the changes in the speed of light expected from either adding or subtracting the earth's orbital speed of 30 kilometers per second to or from the 300 000-kilometer-per-second speed of light. But no changes were observed. None. Something was wrong with the sensible idea that the speed of light measured by a moving receiver should be its usual speed in a vacuum, c, plus or minus the contribution from the motion of the source or receiver. Many repetitions and variations of the Michelson and Morley experiment by many investigators showed the same null result. This was a most puzzling fact at the turn of the century.

Figure 34-2
The Michelson-Morley interferometer, which splits a light beam into two parts and then recombines them to form an interference pattern after they have traveled different paths. Rotation was accomplished in their experiment by floating the massive sandstone slab in mercury. This schematic diagram shows how the half-silvered mirror splits the beam into two rays. The clear glass assured that both rays traverse the same amount of glass. Four mirrors at each end were used to lengthen the paths.

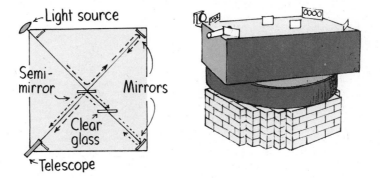

One interpretation of the bewildering result was suggested by the Irish physicist G. F. FitzGerald, who proposed that the length of the experimental apparatus shrank in the direction in which it was moving by just the amount required to counteract the presumed variation in the speed of light. FitzGerald's hypothesis accounted for the discrepancy, but since he had no suitable theory for why this was so, it wasn't taken seriously.

Einstein's interpretation was very different. if the speed of light does not conform to traditional or classical ideas, and if speed is simply a ratio of distance through space to a corresponding interval of time, Einstein recognized that the classical ideas of space and time were suspect. He rejected the classical idea that space and time were independent of each other and developed, with simple postulates, a profound relationship between space and time.

Postulates of the Special Theory of Relativity

Einstein saw no need for the ether. Gone with the stationary ether was the notion of an absolute frame of reference. All motion is relative, not to any stationary hitching post in the universe, but to arbitrary frames of reference. A rocket ship cannot measure its speed with respect to empty space, but only with respect to other objects. If, for example, rocket ship A drifts past rocket ship B in empty space, spaceman A and spacewoman B will each observe the relative motion, and from this observation each will be unable to determine who is moving and who is at rest, if either.

Figure 34-3
The speed of light is measured to be the same in all frames of reference.

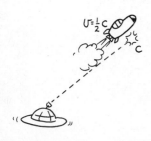

Figure 34-4
The speed of a light flash emitted by the space station is measured to be c by observers on both the space station and the rocket ship.

This is a familiar experience to a passenger on a train who looks out his window and sees the train on the next track moving by his window. He is aware only of the relative motion between his train and the other train and cannot tell which train is moving. He may be at rest relative to the ground and the other train may be moving, or he may be moving relative to the ground and the other train may be at rest, or they both may be moving relative to the ground. The important point here is that if you were in a train with no windows, there would be no way to determine whether the train was moving with uniform velocity or was at rest. This is the first of Einstein's postulates of the special theory of relativity:

All laws of nature are the same in all uniformly moving frames of reference.

On a jet airplane going 700 kilometers per hour, for example, coffee pours as it does when the plane is at rest; we swing a pendulum and it swings as it would if the plane were at rest on the runway. There is no physical experiment we can perform to determine our state of uniform motion. The laws of physics within the uniformly moving cabin are the same as those in a stationary laboratory.

Any number of experiments can be devised to detect accelerated motion, but none can be devised, according to Einstein, to detect that state of uniform motion. We can measure uniform velocity only with respect to some frame of reference.

It would be very peculiar if the laws of physics varied for observers moving at different speeds. It would mean, for example, that a pool player on a smooth-moving ocean liner would have to adjust her style of play to the speed of the ship, or even to the season as the earth varies in its orbital speed about the sun. It is our common experience that no such adjustment is necessary, and no experiment, however precise, has shown any difference at all. That is what the first postulate of relativity means.

One of the questions that Einstein as a youth asked his schoolteacher was, "What would a light beam look like if you traveled along beside it?" According to classical physics, the beam would be at rest to such an observer. The more Einstein thought about this, the more convinced he became of its impossibility. He finally came to the conclusion that even if an observer could travel at the speed of light, he would still measure the light leaving him at 300 000 kilometers per second. This was the second postulate in his special theory of relativity:

The speed of light in free space will be found to have the same value regardless of the motion of the source or the motion of the observer; that is, the speed of light is invariant.

To illustrate this statement, consider a rocket ship departing from the space station shown in Figure 34-4. A flash of light is emitted from the station at 300 000 kilometers per second, or c. Regardless of the velocity of the rocket, an observer in the rocket sees the flash of light pass her at the same speed c. If a flash is sent to the station from the moving rocket, observers on the station will measure the speed of the flash to be c. The speed of light is measured to be the same regardless of the speed of the source or receiver. *All* observers who measure the speed of light will find it has the same value c. We will see that the explanation for this has to do with the relationship between space and time.

Simultaneity

An interesting consequence of Einstein's second postulate occurs with the concept of **simultaneity**. We say that two events are simultaneous if they occur at the same time. Consider, for example, a light source in the exact center of the compartment of a rocket ship (Figure 34-5). When the light source is switched on, light spreads out in all directions at speed c. Because the light source is equidistant from the front and back ends of the compartment, an observer in the middle of the compartment finds that light reaches the front end at the same instant it reaches the back end. This occurs whether the ship is rest or moving at constant velocity. The events of hitting the back end and hitting the front end occur *simultaneously*.

Figure 34-5
From the point of view of the observer who travels with the compartment, light from the source travels equal distances to both ends of the compartment and therefore strikes both ends simultaneously.

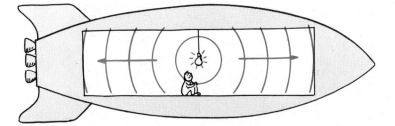

But time in one frame of reference may be different from time in another frame. To an outside observer who views the same two events in another frame of reference, say from a planet not moving with the ship, these same two events are *not* simultaneous. As light travels out from the source, this observer sees the ship move forward, so the back of the compartment moves toward the beam while the front moves away from it. The beam going to the back of the compartment therefore has a shorter distance to travel than the beam going forward (Figure 34-6). Since the speed of light is the same in both directions, this observer sees the event of light hitting the back of the compartment *before* seeing the event of light hitting the front of the compartment. (Of course, we are making the assumption that the observer can discern these slight differences.) A little thought will show that an observer in another rocket ship that passes the ship in the opposite direction would report that the light reaches the front of the compartment first.

Two events that are simultaneous in one frame of reference are not simultaneous in a frame moving relative to the first frame.

Figure 34-6

The same events of light striking the front and back of the compartment are not simultaneous from the point of view of an observer in a different frame of reference. Because of the ship's motion, light that strikes the back of the compartment doesn't have as far to go and strikes sooner than light that strikes the front of the compartment.

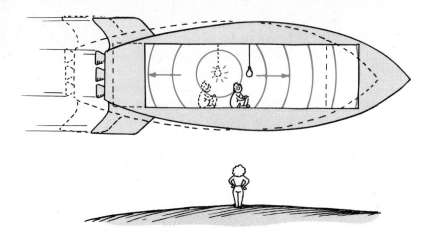

(This is true for all motion except one in which the observer moves perpendicular to the events in such a way as to be always equidistant from both events.) This nonsimultaneity of events in one frame that are simultaneous in another is a purely relativistic result—a consequence of light always having the same speed for all observers.

Questions ▶

1. How is the nonsimultaneity of hearing thunder *after* seeing lightning similar to relativistic simultaneity?

2. Suppose that the planet-bound observer in Figure 34-6 sees a pair of lightning bolts simultaneously strike the front and rear ends of the high-speed rocket ship compartment. Will the lightning strikes be simultaneous to an observer in the middle of the rocket ship compartment? (We assume here that an observer can detect any slight differences in time for light to travel from the ends to the middle of the compartment.)

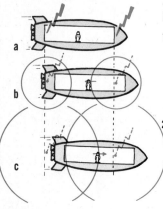

▶ **Answers**

1. It isn't! The duration between hearing thunder and seeing lightning has nothing to do with moving observers or relativity. In such a case you simply correct for the time the signal (sound, light, carrier pigeon, or whatever) takes to reach you. The relativity of simultaneity is a genuine discrepancy between observations made by observers in relative motion, and not a simple mistake that arises from a failure to account for the time of travel of sound or light signals.

2. No; an observer in the middle of the compartment will see the lightning that hits the front end of the compartment before seeing the lightning that hits the rear end. This is shown in positions (a), (b), and (c) to the left. In (a), we see both lightning bolts striking the ends of the compartment simultaneously. In position (b), light from the front lightning bolt reaches the observer. Slightly later, in (c), light from the rear lightning bolt reaches the observer.

Space-Time

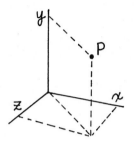

Figure 34-7
Point P can be specified with three numbers: the distance along the x axis, the y axis, and the z axis.

$$\frac{SPACE}{TIME} = \frac{SPACE}{TIME} = c$$

Figure 34-8
All space and time measurements of light are unified by c.

When we look up at the stars, we realize that we are actually looking backward in time. The stars we see farthest away are the stars we are seeing longest ago. The more we think about this, the more apparent it becomes that space and time must be intimately tied together.

The space we live in is three-dimensional; that is, we can specify the position of any location in space with three dimensions. Loosely speaking, these dimensions would be how far over, how far across, and how far up or down. For example, if we are at the corner of a rectangular room and wish to specify the position of any point in the room, we can do this with three numbers. The first would be the number of meters the point is along a line joining the left wall and the floor; the second would be the number of meters the point is along a line joining the adjacent right wall and the floor; and the third would be the number of meters the point lies above the floor or along the vertical line joining the walls at the corner. Physicists speak of these three lines as the coordinate axes of a reference frame (Figure 34-7). Three numbers—the distances along the x axis, the y axis, and the z axis—will specify the position of a point in space.

We specify the size of objects with three dimensions. A box, for example, is described by its length, width, and height. But the three dimensions do not give a complete picture. There is a fourth dimension—time. The box was not always a box of given length, width, and height. It began as a box only at a certain point in time, on the day it was made. Nor will it always be a box. At any moment it may be crushed, burned, or destroyed. So the three dimensions of space are a valid description of the box only during a certain specified period of time. We cannot speak meaningfully about space without implying time. Things exist in **space-time**. Each object, each person, each planet, each star, each galaxy exists in what physicists call "the space-time continuum."*

Two side-by-side observers at rest relative to one another share the same reference frame. Both would agree on measurements of space and time between given events, so we say they share the same realm of space-time. If there is relative motion between them, however, the observers will not agree on these measurements of space and time. At ordinary speeds, differences in their measurements are imperceptible, but at speeds near the speed of light—so-called relativistic speeds—the differences are appreciable. Each observer is in a different realm of space-time, and her measurements of space and time differ from the measurements of an observer in some other realm of space-time. The measurements differ not haphazardly but in such a way that each observer will always measure the same ratio of space and time for light in the reference frame of the

---

*Space and time may be quantized points in a four-dimensional space-time lattice. From the sizes of elementary particles and minimum separation between colliding particles, we find that there seems to be an elemental unit of distance of 6.6×10^{-16} m, and we find the lifetimes of all known elementary particles are consistent with being an integral number of "chronons" of about 2×10^{-23} s.

other (or in any reference frame); the greater the measured distance in space, the greater the measured interval of time. This constant ratio of space and time for light, *c*, is the unifying factor between different realms of space-time and is the essence of Einstein's second postulate.

Time Dilation

Time is not an absolute but is relative to the motion between the observer and the event being observed. Imagine, for example, that we are somehow able to observe a flash of light bouncing back and forth between a pair of parallel mirrors. If the distance between the mirrors is fixed, then the arrangement constitutes a sort of "light clock," because the back and forth trips of the light flash take equal intervals of time (Figure 34-9).

Figure 34-9
A light clock. Light will bounce up and down between parallel mirrors and "tick off" equal intervals of time.

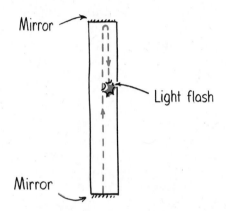

Mirror

Mirror

Light flash

Figure 34-10
(*a*) An observer moving with the spaceship observes the light flash moving vertically between the mirrors of the light clock. (*b*) An observer who is passed by the moving ship observes the flash moving along a diagonal path.

Suppose our light clock is inside a transparent high-speed spaceship. If we travel along with the ship and watch the light clock (Figure 34-10*a*), we will see the flash of light reflecting straight up and down between the two mirrors, just as it would if the spaceship were at rest. Our observations will show no relativistic effects because there is no relative motion between us and our light clock; we share the same reference frame in space-time.

If, instead, we stand at some relative rest position and observe the spaceship whizzing by us at an appreciable speed—say, half the speed

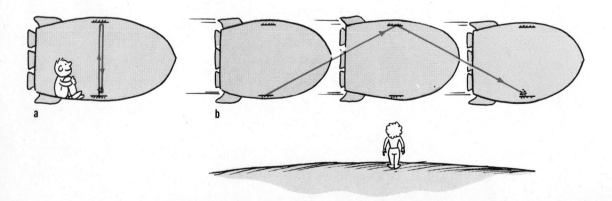

a

b

of light—things are quite different. We no longer share the same reference frame, for in this case there is relative motion between the observer and the observed. We will not see the path of the light in simple up-and-down motion as before. Because the light flash keeps up with the horizontally moving light clock, we will see the flash following a diagonal path (Figure 34-10*b*). Notice that the flash travels a *longer distance* as it moves between the mirrors in our position of space-time than it does in the reference frame of an observer riding with the ship. Since the speed of light is the same in all reference frames (Einstein's second postulate), the flash must travel for a longer time between the mirrors in our frame than in the reference frame of an observer on board. This follows from the definition of speed, simply stated, as a ratio of distance to time.

Figure 34-11
The longer distance taken by the light flash in following the diagonal path must be divided by a correspondingly longer time interval to yield an unvarying value for the speed of light.

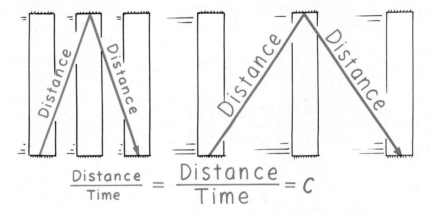

$$\frac{\text{Distance}}{\text{Time}} = \frac{\text{Distance}}{\text{Time}} = C$$

The longer diagonal distance must be divided by a correspondingly longer time interval to yield an unvarying value for the speed of light. Thus, from our relative position of rest, we measure a longer time interval between ticks when a clock is in motion than when it is at rest. We have considered a light clock in our example, but the same is true for any kind of clock. Moving clocks appear to run slow. This stretching out of time is called **time dilation**, which has nothing to do with the mechanics of clocks, but instead arises from the nature of time itself. Time dilation applies to *all* properly functioning timepieces.

Figure 34-12
When we see the rocket at rest, we see it traveling at the maximum rate in time: 24 hours per day. If we see the rocket traveling at the maximum rate through space (the speed of light), we see its time standing still.

The exact relationship of time dilation for different frames of reference in space-time can be derived from Figure 34-10 with simple geometry and algebra.* The relationship between the time t_0 in the observer's own frame of reference and the relative time t measured in another frame of reference is

$$t = \frac{t_0}{\sqrt{1 - \dfrac{v^2}{c^2}}}$$

where v represents the relative velocity between the observer and the observed and c is the speed of light. Because no material object can travel at or beyond the speed of light, the ratio v/c is always less than 1; likewise for v^2/c^2. For $v = 0$, this ratio is zero, and for everyday speeds where v is negligibly small compared to c, it's practically zero. Then we see $1 - (v^2/c^2)$ has a value of 1, as has $\sqrt{1 - (v^2/c^2)}$, and we find $t = t_0$ and time intervals appear the same in both systems. For higher speeds, v/c is between zero and 1, and $1 - (v^2/c^2)$ is less than 1; likewise, $\sqrt{1 - (v^2/c^2)}$. So t_0 divided by a value less than 1 produces a value greater than t_0, an elongation, a dilation of time.

---

*The light clock is shown in three successive positions in the figure below. The diagonal lines represent the path of the light flash as it starts from the lower mirror at position 1, moves to the upper mirror at position 2, and then back to the lower mirror at position 3. Distances on the diagram are marked ct, vt, and ct_0, which follows from the fact that the distance traveled by a uniformly moving object is equal to its speed multiplied by the time.

The symbol t_0 represents the time it takes for the flash to move between the mirrors as measured from a frame of reference fixed to the light clock. This is the time for straight up or down motion. The speed of light is c, and the path of light is seen to move a vertical distance ct_0. This distance between mirrors is at right angles to the motion of the light clock and is the same in both reference frames.

The symbol t represents the time it takes the flash to move from one mirror to the other as measured from a frame of reference in which the light clock moves with speed v. Since the speed of the flash is c and the time to go from position 1 to position 2 is t, the diagonal distance traveled is ct. During this time t, the clock (which travels horizontally at speed v) moves a horizontal distance vt from position 1 to position 2.

These three distances make up a right triangle in the figure, in which ct is the hypotenuse and ct_0 and vt are legs. A well-known theorem of geometry (the Pythagorean theorem) states that the square of the hypotenuse is equal to the sum of the squares of the other two sides. If we apply this to the figure, we obtain:

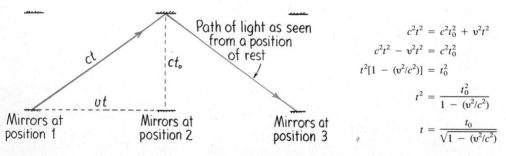

Path of light as seen from a position of rest

Mirrors at position 1   Mirrors at position 2   Mirrors at position 3

$$c^2t^2 = c^2t_0^2 + v^2t^2$$

$$c^2t^2 - v^2t^2 = c^2t_0^2$$

$$t^2[1 - (v^2/c^2)] = t_0^2$$

$$t^2 = \frac{t_0^2}{1 - (v^2/c^2)}$$

$$t = \frac{t_0}{\sqrt{1 - (v^2/c^2)}}$$

To consider some numerical values, assume that v is equal to 50 percent the speed of light. Then we substitute $0.5c$ for v in the time-dilation equation and after some arithmetic find that $t = 1.15\, t_0$. This means that if we viewed a clock on a spaceship traveling at half the speed of light, we would see the second hand take 1.15 minutes to make a revolution, whereas if it were at rest, we would see it take 1 minute. If the spaceship passes us at 87 percent the speed of light, $t = 2t_0$, we would measure time events on the spaceship taking twice the usual intervals. The hands of a clock on the ship would turn only half as fast as those on our own clock. Events on the ship would seem to take place in slow motion. At 99.5 percent the speed of light, $t = 10t_0$, we would see the second hand of the spaceship's clock take 10 minutes to sweep through a revolution requiring 1 minute on our clock.

To put these figures another way, at 99.5 percent c, the moving clock would appear to run a tenth of our rate; its hands would tick only 6 seconds while our clock's second hand ticks 60 seconds. At 87 percent c, the moving clock ticks at half rate and shows 30 seconds to our 60 seconds; at 50 percent c, the moving clock ticks $1/1.15$ as fast and ticks 52 seconds to our 60 seconds. We see that moving clocks run slow.

Nothing is unusual about a moving clock itself; it is simply ticking to the rhythm of a different time. The faster a clock moves, the slower it appears to run as viewed by an observer not moving with the clock. If it were possible for an observer to watch a clock pass by at the speed of light, the clock would not appear to be running at all. This observer would measure the interval between ticks to be infinite. The clock would be ageless! If our observer were moving with the clock, however, the clock would not show any slowing down of time. To him the clock would be operating normally. This is because there would be no motion between the observer and the observed. The v in the time-dilation equation would then be zero, and $t = t_0$; they share the same frame in space-time. If the person who whizzes past us checked a clock in our reference frame, however, he would find our clock to be running as slowly as we find his to be. We each see each other's clock running slow. There is really no contradiction here, for it is physically impossible for two observers moving at different velocities to refer to one and the same realm of space-time. The measurements made in one realm of space-time need not agree with the measurements made in another realm of space-time. The measurement they will always agree on, however, is the speed of light.

Time dilation has been confirmed in the laboratory innumerable times with atomic particle accelerators. The lifetimes of fast-moving radioactive particles increase as the speed goes up, and the amount of increase is just what Einstein's equation predicts.

Time dilation has been confirmed also for not-so-fast motion. In 1971 four cesium-beam atomic clocks were twice flown on regularly scheduled commercial jet flights around the world, once eastward and once westward, to test Einstein's theory of relativity with macroscopic clocks. The clocks indicated different times after their round trips. Relative to the atomic time scale of the U.S. Naval Observatory, the observed time differences, in billionths of a second, were in accord with relativistic prediction.

This all seems very strange to us only because it is not our common experience to deal with measurements made at relativistic speeds or atomic-clock-type measurements at ordinary speeds. The theory of relativity does not make common sense. But common sense, according to Einstein, is that layer of prejudices laid down in the mind prior to the age of 18. If we spent our youth zapping through the universe in high-speed spaceships, we would probably be quite comfortable with the results of relativity.

Questions

1. If you are moving in a spaceship at a high speed relative to the earth, would you notice a difference in your pulse rate? In the pulse rate of the people back on earth?

2. Will observers A and B agree on measurements of time if A moves at half the speed of light relative to B? If both A and B move together at half the speed of light relative to the earth?

3. Does time dilation mean that time really passes more slowly in moving systems or that it only seems to pass more slowly?

The Twin Trip

A dramatic illustration of time dilation is the classic case of the identical twins, one an astronaut who takes a high-speed round-trip journey while the other stays home on earth. When the traveling twin returns, he is younger than the stay-at-home twin. How much younger depends on the relative speeds involved. If the traveling twin maintains a speed of 50 percent the speed of light for 1 year (according to clocks aboard the spaceship), 1.15 years will have elapsed on earth. If the traveling twin maintains a speed of 87 percent the speed of light for a year, then 2 years will have elapsed on earth. At 99.5 percent the speed of light, 10 earth

▶ Answers

1. There would be no relative speed between you and your own pulse, which share the same frame of reference, so you would notice no relativistic effects in your own pulse. There would, however, be a relativistic effect between you and people back on earth. You would find their pulse rate slower than normal (and they would find your pulse rate slower than normal). Relativity effects are always attributed to the other guy.

2. When A and B have different motions relative to each other, each will observe a slowing of time in the frame of reference of the other. So they will not agree on measurements of time. When they are moving in unison, however, they share the same frame of reference and will agree on measurements of time. They will see each other's time as passing normally, and they will each see events on earth in the same slow motion.

3. The slowing of time in moving systems is not merely an illusion resulting from motion. Time really does pass more slowly in a moving system compared to one at relative rest, as we shall see in the next section. Read on!

years would pass in one spaceship year. At this speed the traveling twin would age a single year while the stay-at-home twin ages 10 years.

One question often arises: Since motion is relative, why isn't it just as well the other way around? Why wouldn't the traveling twin return to find his stay-at-home twin younger than himself? We will show that from the frames of reference of both the earthbound twin and traveling twin, it is the earthbound twin who ages more.

Figure 34-13
The traveling twin does not age as fast as the stay-at-home twin.

First, consider a spaceship hovering at rest relative to a distant planet. Suppose the spaceship sends regularly spaced brief flashes of light to the planet (Figure 34-14). Some time will elapse before the flashes get to the planet, just as 8 minutes elapse before sunlight gets to the earth. The light flashes will encounter the receiver on the planet at speed c. Since there is no relative motion between the sender and receiver, successive flashes will be received as frequently as they are sent. For example, if a flash is sent from the ship every 6 minutes, then after some initial delay, the receiver will receive a flash every 6 minutes. With no motion involved, there is nothing unusual about this.

Figure 34-14
When no motion is involved, the light flashes are received as frequently as the spaceship sends them.

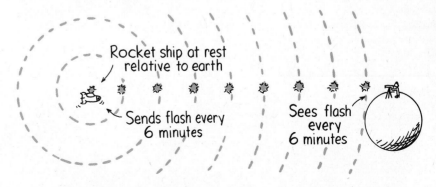

Rocket ship at rest relative to earth

Sends flash every 6 minutes

Sees flash every 6 minutes

When motion is involved, the situation is quite different. It is important to note that the speed of the flashes will still be c, no matter how the ship or receiver may move. How frequently the flashes are seen, however, very much depends on the relative motion involved. When the ship travels toward the receiver, the receiver sees the flashes more frequently. This happens not only because time is altered due to motion, but mainly because each succeeding flash has less distance to travel as the ship gets closer to the receiver. If the spaceship emits a flash every 6 minutes, the flashes will be seen at intervals of less than 6 minutes. Suppose the ship is traveling fast enough for the flashes to be seen twice as frequently. Then they are seen at intervals of 3 minutes (Figure 34-15).

Figure 34-15
When the sender moves toward the receiver, the flashes are seen more frequently.

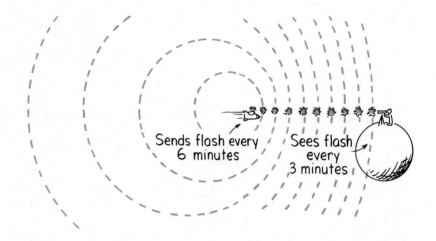

If the ship recedes from the receiver at the same speed and still emits flashes at 6-minute intervals, these flashes will be seen half as frequently by the receiver, that is, at 12-minute intervals (Figure 34-16). This is mainly because each succeeding flash has a longer distance to travel as the ship gets farther away from the receiver.

Figure 34-16
When the sender moves away from the receiver, the flashes are spaced farther apart and are seen less frequently.

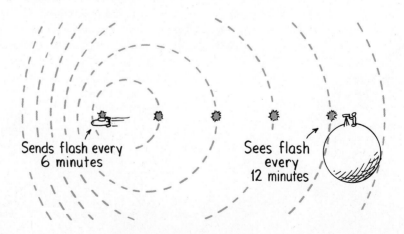

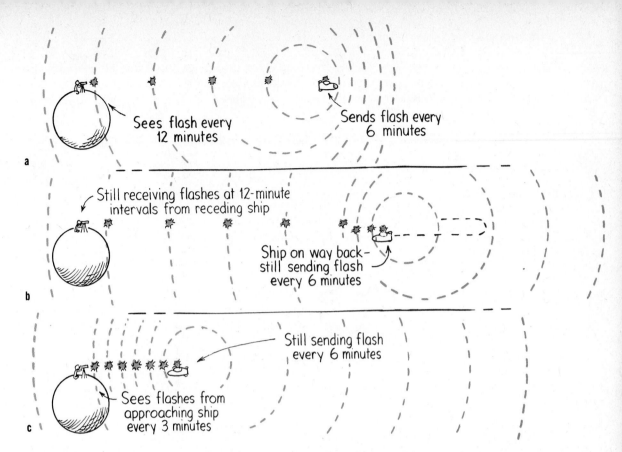

Sees flash every
12 minutes

Sends flash every
6 minutes

a

Still receiving flashes at 12-minute
intervals from receding ship

Ship on way back—
still sending flash
every 6 minutes

b

Still sending flash
every 6 minutes

Sees flashes from
approaching ship
every 3 minutes

c

Figure 34-17
The spaceship emits flashes
each 6 min during a 2-h trip.
During the first hour, it
recedes from the earth. During the second hour, it
approaches the earth.

The effect of moving away is just the opposite of moving closer to the receiver. So if the flashes are received twice as frequently when the spaceship is approaching (6-minute flash intervals are seen every 3 minutes), they are received half as frequently when it is receding (6-minute flash intervals are seen every 12 minutes).*

---

*This reciprocal relationship (halving and doubling of frequencies) is a consequence of the constancy of the speed of light and can be illustrated with the following example: Suppose a sender on earth emits flashes 3 min apart to a distant observer on a planet that is at rest relative to the earth. The observer, then, sees a flash every 3 min. Now suppose a second observer travels in a spaceship between the earth and the planet at a speed great enough to allow him to see the flashes half as frequently—6 min apart. This halving of frequency occurs for a speed of recession of $0.6c$. We can see that the frequency will double for a $0.6c$ speed of approach by supposing that the spaceship emits its own flash every time it sees an earth flash, that is, every 6 min. How does the observer on the distant planet see these flashes? Since the earth flashes and the spaceship flashes travel together at the same speed c, the observer will see not only the earth flashes every 3 min but the spaceship flashes every 3 min as well. So although a person on the spaceship emits flashes every 6 min, the observer sees them every 3 min at twice the emitting frequency. So for a speed of recession where frequency appears halved, frequency appears doubled for the same speed of approach. If the ship were traveling faster so that the frequency of recession were $\frac{1}{3}$ or $\frac{1}{4}$ as much, then the frequency of approach would be three- or fourfold, respectively. This reciprocal relationship does not hold for waves that require a medium. In the case of sound waves, for example, a speed that results in a doubling of emitting frequency for approach produces $\frac{2}{3}$ (not $\frac{1}{2}$) the emitting frequency for recession.

The light flashes make up a light clock. In the frame of reference of the receiver, events that take 6 minutes in the spaceship are seen to take 12 minutes when the spaceship recedes and only 3 minutes when the ship is approaching.

Questions

1. If the spaceship travels for 1 h and emits a flash every 6 min, how many flashes will be emitted?

2. The ship sends equally spaced 6-min flashes while approaching the receiver at constant speed. Will these flashes be equally spaced when they encounter the receiver?

3. If the receiver sees these flashes at 3-min intervals, how much time will occur between the first and the last flash (in the frame of reference of the receiver)?

Let's apply this doubling and halving of flash intervals to the twins. Suppose the traveling twin recedes from the earthbound twin at the same high speed for 1 hour and then quickly turns around and returns in 1 hour. The traveling twin takes a round trip of 2 hours, according to all clocks aboard the spaceship. This trip will not be seen to take 2 hours from the earth frame of reference, however. We can see this with the help of the flashes from the ship's light clock.

As the ship recedes from the earth, it emits a flash of light every 6 minutes. These flashes are received on earth every 12 minutes. During the hour of going away from the earth, a total of ten flashes are emitted. If the ship departs from the earth at noon, clocks aboard the ship read 1 PM when the tenth flash is emitted. What time will it be on earth when this tenth flash reaches the earth? The answer is 2 PM. Why? Because the time it takes the earth to receive 10 flashes at 12-minute intervals is 10 × (12 minutes), or 120 minutes (= 2 hours).

Suppose the spaceship is somehow able to turn around suddenly in a negligibly short time and return at the same high speed. During the hour of return it emits ten more flashes at 6-minute intervals. These flashes are received every 3 minutes on earth, so all ten flashes come in 30 minutes. A clock on earth will read 2:30 PM when the spaceship completes its 2-hour trip. We see that the earthbound twin has aged half an hour more than the twin aboard the spaceship!

The result is the same from either frame of reference. Consider the same trip again, only this time with flashes emitted from the earth at

---

▶ **Answers**

1. The ship will emit a total of ten flashes in 1 h, since (60 min)/(6 min) = 10.

2. Yes; as long as the ship moves at constant speed, the equally spaced flashes will be seen equally spaced but more frequently. (If the ship accelerated while sending flashes, then they would not be seen at equally spaced intervals.)

3. All ten flashes will be seen in 30 min, since 10 × (3 min) = 30 min.

Figure 34-18
The trip that takes 2 h in the frame of reference of the spaceship takes $2\frac{1}{2}$ h in the earth's frame of reference.

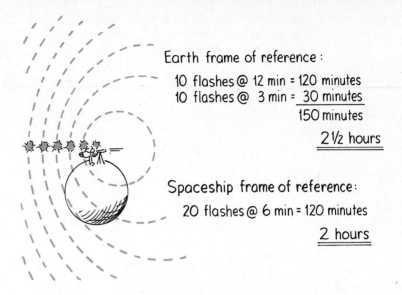

Earth frame of reference :

10 flashes @ 12 min = 120 minutes
10 flashes @ 3 min = 30 minutes
150 minutes

2 1/2 hours

Spaceship frame of reference:

20 flashes @ 6 min = 120 minutes

2 hours

regularly spaced 6-minute intervals in earth time. From the frame of reference of the receding spaceship, these flashes are received at 12-minute intervals (Figure 34-19a). This means that five flashes are seen

Figure 34-19
Flashes sent from earth at 6-min intervals are seen at 12-min intervals by the ship when it recedes and at 3-min intervals when it approaches.

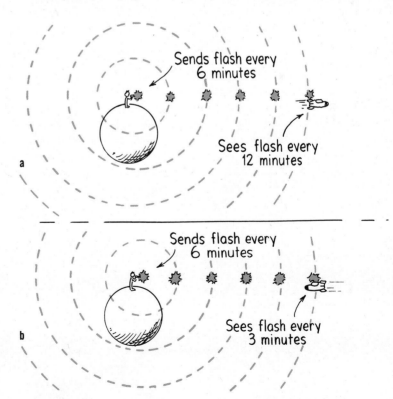

Sends flash every
6 minutes

Sees flash every
12 minutes

a

Sends flash every
6 minutes

Sees flash every
3 minutes

b

by the spaceship during the hour of receding from earth. During the spaceship's hour of approaching, the light flashes are seen at 3-minute intervals (Figure 34-19b), so twenty flashes will be seen.

So we see that the spaceship receives a total of twenty-five flashes during its 2-hour trip. According to clocks on the earth, however, the time it took to emit the twenty-five flashes at 6-minute intervals was 25 × (6 minutes), or 150 minutes (= 2.5 hours). This is shown in Figure 34-20.

Figure 34-20
A time interval of $2\frac{1}{2}$ h on earth is seen to take 2 h in the spaceship's frame of reference.

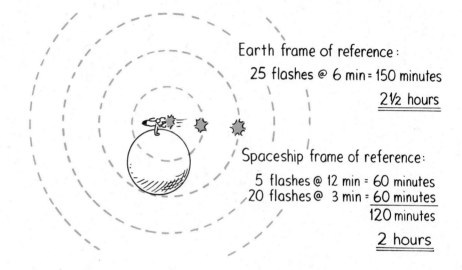

Earth frame of reference:

25 flashes @ 6 min = 150 minutes

2½ hours

Spaceship frame of reference:

5 flashes @ 12 min = 60 minutes

20 flashes @ 3 min = 60 minutes

120 minutes

2 hours

So both twins agree on the same results, with no dispute as to who ages more. While the stay-at-home twin remains in a single reference frame, the traveling twin has experienced two different frames of reference, separated by the acceleration of the spaceship in turning around. The spaceship has in effect experienced two different realms of time, while the earth has experienced a still different but single realm of time. The twins can meet again at the same place in space only at the expense of time.

Question   Since motion is relative, can't we say as well that the rocket ship is at rest and the earth moves, in which case the twin on the rocket ship ages more?

▶ **Answer**

No, not unless the earth then undergoes the turnaround and returns, as our rocket ship did in the twin-trip example. The situation is not symmetrical, for one twin remains in a single reference frame in space-time during the trip while the other makes a distinct change of reference frame, as evidenced by the acceleration in turning around.

Space Travel

One of the old arguments advanced against the possibility of human interstellar travel is that our life span is too short—at least for the distant stars. It was argued, for example, that the nearest star (after the sun), Alpha Centauri, is 4 light-years away, and a round trip even at the speed of light would require 8 years.* The center of our galaxy is some 30 000 light-years away, so it has been reasoned that a person traveling even at the speed of light would require a lifetime of 30 000 years to make such a voyage. But these arguments fail to take into account time dilation. Time for a person on earth and time for a person in a high-speed rocket ship are not the same.

Figure 34-21
From the earth frame of reference, light takes 30 000 years to travel from the center of the galaxy to our solar system. From the frame of reference of a high-speed spaceship, the trip takes less time. From the frame of reference of light itself, the trip takes no time. There is no time in a speed-of-light frame of reference.

A person's heart beats to the rhythm of the realm of time it is in; and one realm of time seems the same as any other realm of time to the person, but not to an observer who stands outside the person's frame of reference—for she can see the difference. For example, astronauts traveling at 99 percent the speed of light could go to the star Procyon (11.4 light-years distant) and back in 23.3 years. It would take light itself 22.8 years to make the same round trip. Because of time dilation, it would seem that only 3 years had gone by to the astronauts. This is what all their clocks would tell them—and biologically they would be only 3 years older. It would be the space officials greeting them on their return who would be 23.3 years older!

At higher speeds the results are even more impressive. At a rocket speed of 99.90 percent the speed of light, travelers could travel slightly more than 70 light-years in a single year of their own time; at 99.99 percent the speed of light, this distance would be pushed appreciably farther than 200 light-years. A 5-year trip for them would take them farther than light travels in 1000 earth-time years!

Present technology does not permit such journeys. Radiation is the greatest problem. A ship traveling at speeds close to the speed of light would encounter interstellar particles just as if it were on the launching

---

*A light-year is the distance light travels in 1 year, about 9.45×10^{12} km.

pad and a steady stream of particles shot by an atomic accelerator were incident on it. No way of shielding such intense particle bombardment for prolonged periods of time is presently known. And if somehow a way were devised to solve this problem, there would be the problem of energy and fuel. Spaceships traveling at relativistic speeds would require billions of times the energy used to put a space shuttle into orbit. Even some kind of interstellar ramjet that scooped up interstellar hydrogen gas for burning in a fusion reactor would have to overcome the enormous retarding effect of scooping up the hydrogen at high speeds. The practicalities of such space journeys are so far prohibitive.

If and when these problems are overcome and space travel becomes a routine experience, people will have the option of taking a trip and returning to any future century of their choosing. For example, one might depart from earth in a high-speed ship in the year 2050, travel for 5 years or so, and return in the year 2500. One could live among the earthlings of that period for a while and depart again to try out the year 3000 for style. People could keep jumping into the future with some expense of their own time, but they could not trip into the past. They could never return to the same era on earth that they bid farewell to. Time, as we know it, travels one way—forward. Here on earth we move constantly into the future at the steady rate of 24 hours per day. A deep-space astronaut leaving on a deep-space voyage must live with the fact that, upon her return, much more time will have elapsed on earth than she has subjectively experienced on her voyage. The credo of all star travelers, whatever their physiological condition, will be permanent farewell.

We can see into the past, but we cannot go into the past. For example, we experience the past when we look at the night skies. The starlight impinging on our eyes left those stars dozens, hundreds, even millions of years ago. What we see is the stars as they were long ago. We are thus eyewitnesses to ancient history—and can only speculate about what may have happened to the stars in the interim.

If we are looking at light that left a star, say, 100 years ago, then it follows that any sighted beings in that solar system are seeing us by light that left *here* 100 years ago and that, further, if they possessed super-telescopes, they might very well be able to eyewitness earthly events of a century ago—the aftermath of the American Civil War, for instance. They would see our past, but they would still see events in a forward direction; they would see our clocks running clockwise.

We can speculate about the possibility that time might just as well move counterclockwise into the past as clockwise into the future. In fact, the whole question of "time reversal" is discussed in great mathematical detail and is being studied by physicists in a number of exquisite and elaborate experiments with subatomic particles. It is hypothesized that particles that move faster than light and backward in time, called *tachyons*, may in fact exist. Experiments to detect them have proved unpromising to date. Nearly all physicists conclude that the only direction of time is forward.

This conclusion is blithely ignored in a limerick that is a favorite with scientist types:

> There was a young lady named Bright
> Who traveled much faster than light.
> She departed one day
> In an Einsteinian way
> And returned on the previous night.

Even with our heads fairly well into relativity, we may still unconsciously cling to the idea that there is an absolute time and compare all these relativistic effects to it—recognizing that time changes this way and that way for this speed and that speed, yet feeling that there still is some basic or absolute time. We may tend to think that the time we experience on earth is fundamental and that other times differ from it. This is understandable; we're earthlings. But the idea is confining. From the point of view of observers elsewhere in the universe, we may be moving at relativistic speeds; they see us moving in slow motion. They may see us living lifetimes a hundred times as long as theirs, just as with supertelescopes we would see them living lifetimes a hundredfold ours. There is no universally standard time. None.

We think of time and then we think of the universe. We think of the universe and we wonder about what went on before the universe began. We wonder about what will happen if the universe ceases to exist in time. But the concept of time applies to events and entities within the universe—not to the universe as a whole. Time is "in" the universe; the universe is not "in" time. Without the universe, there is no time; no before, no after. Likewise, space is "in" the universe; the universe is not "in" a region of space. There is no space "outside" the universe. Space-time exists within the universe. Think about that!

Length Contraction

As objects move through space-time, space as well as time undergoes changes in measurement. The lengths of objects appear to be contracted when they move by us at relativistic speeds. This **length contraction** was first proposed by FitzGerald and mathematically expressed by Hendrick A. Lorentz, a Dutch physicist. It is referred to as the Lorentz-FitzGerald contraction. We express it mathematically as

$$L = L_0 \sqrt{1 - \frac{v^2}{c^2}}$$

where v is the relative velocity between the observed object and the observer, c is the speed of light, L is the measured length of the moving object, and L_0 is the measured length of the object at rest.* Suppose that

---

*Note that we do not explain how this equation comes about (nor those that follow). We simply state them as "guides to thinking" about the ideas of special relativity.

an object is at rest and $v = 0$. Upon substitution of $v = 0$ in the equation, we find $L = L_0$, as we would expect. At 87 percent the speed of light, an object would appear to be contracted to half its original length. At 99.5 percent the speed of light, it would seem to contract to one-tenth its original length. If the object moved at c, its length would be zero. This is one of the reasons we say that the speed of light is the upper limit for the speed of any moving object. Another ditty popular with the science heads is this one:

> There was a young fencer named Fisk,
> Whose thrust was exceedingly brisk.
> So fast was his action
> The Lorentz-FitzGerald contraction
> Reduced his rapier to a disk.

As Figure 34-22 indicates, contraction takes place only in the direction of motion. If an object is moving horizontally, no contraction takes place vertically.

Figure 34-22

The Lorentz-FitzGerald contraction. As speed increases, length in the direction of motion decreases. Lengths in the perpendicular direction do not change.

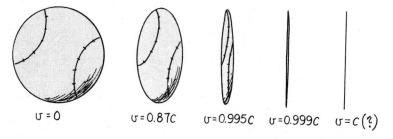

$v = 0$ $v = 0.87c$ $v = 0.995c$ $v = 0.999c$ $v = c\,(?)$

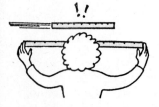

Figure 34-23

The meter stick is measured to be half as long when traveling at 87 percent the speed of light relative to the observer.

Do objects really contract at relativistic speeds? Well, if we attempt to check this out (in principle) and travel alongside the moving object with a meter stick, we note nothing at all unusual about the length of the object. An observer at rest watching this experiment would report that the reason we don't measure the contraction, which is obvious to him, is that our meter stick is shortened as well—a shortened meter stick can't detect a similarly shortened distance. But from our frame of reference, moving with the object, there is no contraction. This is because our relative velocity with respect to the object being measured is zero. The v in the Lorentz-FitzGerald equation refers to the relative velocity between the observed and the observer. And to us that is zero, so $L = L_0$. It is important to stress this point. Reality in one frame of reference is not the same reality in another frame of reference. The object doesn't contract in a reference frame where $v = 0$, but does contract in a reference frame where v is greater than zero.

We actually measure the distortion of space when we measure such a contraction, just as we measure the distortion of time itself when we find clocks running slow. The distortions are of space and time between different realms of space-time.

Question

  A rectangular billboard in space has the dimensions 10 m × 20 m. How fast and in what direction with respect to the billboard would a space traveler have to pass for the billboard to appear square?

Increase of Mass with Speed

If we push on an object, it will accelerate. If we maintain a steady push, it will accelerate to higher and higher speeds. If we push with a greater and greater force, the acceleration in turn will increase and the speeds attained should seemingly increase without limit. But there is a speed limit in the universe—c. In fact, we cannot accelerate any material object enough to reach the speed of light, let alone surpass it.

Why this is so can be understood from Newton's second law. Recall that the acceleration of a object depends not only on the impressed force, but on the mass as well: $a = F/m$. Einstein stated that when work is done to increase the velocity of an object, its mass increases as well. So an impressed force produces less and less acceleration as speed increases. The relationship between mass and speed is given by

$$m = \frac{m_0}{\sqrt{1 - \dfrac{v^2}{c^2}}}$$

Here m represents the measured mass or relativistic mass of an object pushed to any speed v. The symbol m_0 is the *rest mass*, the mass the object would have at rest. Again, v represents the relative velocity between the observer and the observed.

The faster a particle is pushed, the more its mass increases, thereby resulting in less and less response to the accelerating force. An investigation of the relativistic mass equation shows that as v approaches c, m approaches infinity! A particle pushed to the speed of light would have infinite mass and would require an infinite force, which is clearly impossible. Therefore, we say that no material particle can be accelerated to the speed of light.

Atomic particles have been accelerated to speeds in excess of 99 percent the speed of light, however. Their masses increase thousandsfold, as evidenced when a beam of particles, usually electrons or protons, is directed into a deflecting magnetic field. The more massive particles do not bend as readily as they would if they did not undergo the relativistic mass increase (Figure 34-24). They strike their targets at positions predicted by the relativistic equation. This mass increase must be compen-

▶ **Answer**

The space traveler would have to travel at $0.87c$ in a direction parallel to the longer side of the board.

sated for in circular accelerators like cyclotrons and bevatrons, where mass dictates the radius of curvature. One of the advantages of the linear accelerator is that the particle beam travels in a straight-line path, and mass changes do not produce deviations from a straight-line path. Mass increase with velocity is an everyday fact of life to physicists working with high-energy particles.

Figure 34-24
If the mass of the electrons did not increase with speed, the beam would follow the dashed line. But because of the increased inertia, the high-speed electrons in the beam are not deflected as much.

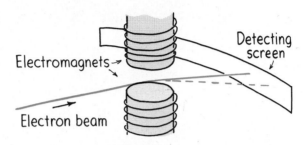

Mass-Energy Equivalence

The most remarkable aspect of special relativity is Einstein's law of the mass-energy equivalence. We have seen that the energy pumped into nuclear particles in an accelerator increases their mass. This is because energy and mass are equivalent to each other.

We all know that the energy that goes into accelerating the particles comes from a power plant someplace. Energy generated at the power plant probably results from either the chemical combustion or the nuclear reaction of certain fuels. If the process is chemical combustion, the masses of hydrocarbon molecules produced by combustion are reduced by about one part in 10^9. If the process is nuclear fission, the masses of the fission fragments after reaction are reduced by about one part in 1000. That is, the masses of the fission fragments are about a thousandth less than the mass of the initial uranium atoms before the reaction. In either the chemical or the nuclear case, mass is given off when energy is given off. Now comes the interesting question: how much mass is gained by the accelerator particles? If we neglect the inefficiencies of power transmission, the fuel fragments lose just as much mass as the particles at the accelerator gain! So the power company delivers mass every bit as much as it delivers energy. To say it delivers one is to say it delivers the other, because mass and energy in a practical sense are one and the same. When something gains energy, it gains mass. When something loses energy, it loses mass.

Einstein realized that anything with mass—even if it is not moving—has energy. Conversely, anything with energy—even if it is not matter (such as light or microwaves, for example)—has mass! The amount of energy E is related to the amount of mass m by the most celebrated equation of the twentieth century:

$$E = mc^2$$

Figure 34-25
When a power plant is said to deliver 90 million MJ of energy to its consumers, it is equivalent to saying that it delivers 1 g of energy to its consumers. This is because mass and energy are one and the same.

The c^2 is the conversion factor for energy units and mass units.* Because of the large magnitude of c, the speed of light, a small mass corresponds to an enormous quantity of energy. For example, the energy equivalent of a single gram of matter is greater than the energy used daily by the populations of our largest cities.

The change in mass when energy changes is so slight that it has not been detected until recent times. When we strike a match, for example, a chemical reaction occurs. Phosphorus atoms in the match head rearrange themselves and combine with oxygen in the air to form new molecules. The resulting molecules have very slightly less mass than the separate phosphorus and oxygen molecules. From a mass standpoint, the whole is slightly less than the sum of its parts, but not by very much—by only about one part in 10^9.

According to Einstein, the missing mass has not been destroyed. It has been carried off in the guise of radiant energy and kinetic energy. If E is the amount of energy given off, the "missing" mass is just E/c^2. For all chemical reactions that give off energy, there is a corresponding decrease in mass. It is important to realize that the total amount of mass

---

*When c is in meters per second and m is in kilograms, then E will be in joules. If the equivalence of mass and energy had been understood long ago when physics concepts were first being formulated, there would probably be no separate units for mass and energy. Furthermore, with a redefinition of space and time units, c could equal 1, and $E = mc^2$ would simply be $E = m$.

Figure 34-26
In 1 second, 4.5 billion tons of rest mass are converted to radiant energy in the sun. The sun is so massive, however, that in 1 million years only one ten-millionth of the sun's rest mass will have been converted to radiant energy.

is the same before and after the reaction *if you remember to include the mass equivalent of the energy.* Similarly, the total energy is the same before and after the reaction if you include the energy equivalent of the rest mass (the mass of matter at rest) involved. Mass is conserved and energy is conserved. Some people say that mass *is* converted to energy (or vice versa). It is really more accurate to say that rest mass is converted to "pure" energy (energy that is not due to rest mass). The total amount of mass does not change, nor does the total amount of energy.

In nuclear reactions, the decrease in rest mass is considerably more than in chemical reactions—about one part in 10^3. Rest mass is converted to heat and radiant energy in the sun and other stars by the process of thermonuclear fusion. It is this decrease in rest mass that bathes the solar system with radiant energy and nourishes life. The present stage of thermonuclear fusion in the sun has been going on for the past 5 billion years, and there is sufficient hydrogen fuel for fusion to last another 5 billion years. It is nice to have such a big sun!

The equation $E = mc^2$ is not restricted to chemical and nuclear reactions. Any change in energy corresponds to a change in mass. The increased kinetic energy of a baseball is accompanied by a slight increase in mass. The filament of a light bulb energized with electricity has more mass than when it is turned off. A hot cup of tea has more mass than the same cup of tea when cold. A wound-up spring clock has more mass than the same clock when unwound. But these examples involve incredibly small changes in mass—too small to be measured by conventional methods. For large energy changes, the effect is more noticeable. Electric energy provides a good example. Work is required to bring together electrically charged particles of the same electrical sign. A fantastically huge amount of work would be required to confine a single gram of bare electrons in a small region because of electrical repulsion. One gram of electrons confined to a sphere 10 centimeters in diameter would have a mass of 10 trillion kilograms!

The mass of a single electron is merely the energy equivalence of the work required to compress its charge. If its charge were infinitely spread out, the amount of work needed to squeeze it down to the classical radius of the electron, divided by c^2, would equal its mass.*

The equation $E = mc^2$ is more than a formula for the conversion of rest mass into pure energy, or vice versa. It states that energy and mass are the same. Mass is simply congealed energy. If you want to know how much energy is in a system, measure its mass. For an object at rest, its energy is its mass. It is energy itself that is hard to shake.

---

*San Francisco Sidewalk Astronomer John Dobson speculates that just as a clock becomes more massive when we do work on it by winding it against the resistance of its spring, the mass of the entire universe is nothing more than the energy that has gone into winding it up against mutual gravitation. In this view the mass of the universe is equivalent to the work done in spacing it out. So electrons have mass because their charge is confined, and the atoms that make up the universe have mass because they are dispersed.

The first direct proof that radiant energy could be converted to its mass form was found in 1932 in a photograph emulsion used in high-altitude balloons to catch cosmic rays. C. D. Anderson, an American physicist, found that a photon of gamma radiation that had entered the emulsion had changed into a pair of particles. One of the particles was an electron. The other particle, which had the same mass as an electron but had a *positive* charge rather than a negative charge, was given the name *positron*. A positron is the *antiparticle* of an electron. When a gamma ray changes from energy to mass, a pair of particles is produced. One of the particles is the antiparticle of the other, with the same mass and spin but with opposite charge. The antiparticle doesn't last long. It soon encounters a particle that is its antiparticle, and the pair is annihilated, sending out a gamma ray in the process.

Question ▶ Can we look at the equation $E = mc^2$ another way and say that matter transforms into pure energy when it is traveling at the speed of light squared?

Correspondence Principle

Einstein's mass-energy relationship is valid only insofar as the transformation equations for mass, length, and time are valid. Before any new theory can be accepted, it must satisfy the correspondence principle. Recall from Chapter 31 that this principle states that any new theory or any new description of nature must agree with the old where the old gives correct results. If the equations of special relativity are valid, they must correspond to those of classical mechanics when speeds much less than the speed of light are considered.

The relativity equations for time, length, and mass are:

$$t = \frac{t_0}{\sqrt{1 - \dfrac{v^2}{c^2}}}$$

$$L = L_0 \sqrt{1 - \frac{v^2}{c^2}}$$

$$m = \frac{m_0}{\sqrt{1 - \dfrac{v^2}{c^2}}}$$

▶ **Answer**

No, no, no! There are several things wrong with that statement. As matter is propelled faster, its mass increases rather than decreases. In fact, its mass approaches infinity. At the same time, its energy approaches infinity. It has more mass and more energy. Matter cannot be made to move at the speed of light, let alone the speed of light squared (which is not a speed!). $E = mc^2$ simply means that energy and mass are two sides of the same coin.

Note that these equations each reduce to Newtonian values for speeds that are very small compared to c. Then the ratio $(v/c)^2$ is very small, and for everyday speeds may be taken to be zero. The relativity equations become

$$t = \frac{t_0}{\sqrt{1 - 0}} = t_0$$

$$L = L_0 \sqrt{1 - 0} = L_0$$

$$m = \frac{m_0}{\sqrt{1 - 0}} = m_0$$

So for everyday speeds, the length, mass, and time of moving objects are essentially unchanged. The equations of special relativity hold for all speeds, although they are significant only for speeds near the speeds of light.

Einstein's theory of relativity has raised many philosophical questions. What, exactly, is time? Can we say that it is nature's way of seeing to it that everything does not all happen at once? And why does time seem to move in one direction? Has it always moved *forward*? Are there other parts of the universe where time moves *backward*? Is it likely that our three-dimensional perception of a four-dimensional world is only a beginning? Could there be a fifth dimension? A sixth dimension? A seventh dimension? And if so, what would the nature of these dimensions be? Perhaps these unanswered questions will be answered by the physicists of tomorrow. How exciting!

Summary of Terms

Frame of reference A vantage point (usually a set of coordinate axes) with respect to which position and motion may be described.

Postulates of the special theory of relativity (1) All laws of nature are the same in all uniformly moving frames of reference. (2) The speed of light in free space will be found to have the same value regardless of the motion of the source or the motion of the observer; that is, the speed of light is invariant.

Simultaneity The condition wherein events occur or operate at the same time. Two events that are simultaneous in one frame of reference are not simultaneous in a frame moving relative to the first frame.

Space-time The four-dimensional continuum in which all things exist: three dimensions are the coordinates of space and the fourth is of time.

Time dilation The apparent slowing down of time for an object moving at relativistic speeds.

Length contraction The apparent shrinking of an object moving at relativistic speeds.

Mass-energy equivalence The relationship between mass and energy as given by the equation $E = mc^2$.

Suggested Reading

Einstein, Albert. *The Meaning of Relativity*. Princeton, N.J.: Princeton University Press, 1950. Written for the average person by Einstein himself.

Epstein, Lewis C. *Relativity Visualized*. San Francisco: Insight Press, 1983.

Gamow, George. *Mr. Tompkins in Wonderland*. New York: Macmillan, 1940. An excellent and very interesting little book.

Gardner, Martin. *The Relativity Explosion*. New York: Vintage Books, 1976.

Taylor, Edwin F., and John A. Wheeler. *Spacetime Physics*. San Francisco: W. H. Freeman, 1966.

Review Questions

Motion Is Relative

1. If you walk at 1 km/h down the aisle of a bus that speeds along the road at 60 km/h, what is your speed relative to the ground?

2. In the previous question, what is your approximate speed relative to the sun as you walk down the aisle of the bus?

Michelson-Morley Experiment

3. What did Michelson and Morley discover about the speed of light?

4. What hypothesis did G. F. FitzGerald make to explain the findings of Michelson and Morley?

5. What idea did Einstein reject to explain the findings of Michelson and Morley?

Postulates of the Special Theory of Relativity

6. Cite at least three examples of Einstein's first postulate.

7. Cite at least three examples of Einstein's second postulate.

Simultaneity

8. Inside the moving compartment of Figure 34-5, light travels a certain distance to the front end and a certain distance to the back end of the compartment. How do these distances compare as seen in the frame of reference of the moving rocket?

9. How do these distances in Question 8 compare as seen in the frame of reference of an observer on a stationary planet?

10. Why are events that are simultaneous in one frame of reference not simultaneous in a frame moving relative to the first frame?

Space-Time

11. How many coordinate axes are usually used to describe three-dimensional space?

12. What does the fourth dimension measure?

13. Under what condition will you and a friend share the same realm of space-time?

14. Under what condition will you and a friend not share the same realm of space-time?

15. What is special about the ratio of space traveled and time taken for light?

Time Dilation

16. Suppose the parallel mirrors of a light clock were 150 000 km apart. In the frame of reference of the light clock, how much time would be required for a pulse of light to make a round trip between the mirrors?

17. Suppose the parallel mirrors of a light clock were 150 km apart. In the frame of reference of the light clock, how much time would be required for a pulse of light to make a round trip between the mirrors?

18. Would your answers to the previous two questions be different if your measurements were made from a frame of reference that is moving relative to the light clock? Explain.

19. Time is required for light to travel along a path from one point to another. If this path is seen to be longer because of motion, what happens to the time it takes for light to travel this longer path?

20. What do we call the "stretching out of time"?

21. How do measurements of time differ for events in a frame of reference that moves at 50 percent the speed of light relative to us?

22. How do measurements of time differ for events in a frame of reference that moves at 99.5 percent the speed of light relative to us?

23. Suppose a particular clock accurately shows time passing half as fast in a particular frame of reference as in our own. What is the velocity of this frame of reference relative to us?

24. If we see somebody's clock running slow due to relative motion, will they see our clocks running slow also? Or will they see our clocks running fast? Explain.

25. What is the evidence for time dilation?

26. What does Einstein say about common sense?

The Twin Trip

27. When a flashing light approaches you, each flash that reaches you has a shorter distance to travel. What effect does this have on how frequently you receive the flashes?

28. When a flashing light source approaches you, does the speed of light or the frequency of light—or both—increase?

29. If a flashing light source moves toward you fast enough so the duration between flashes seems half as long, how will the duration between flashes seem if the source is moving away from you at the same speed?

30. How many frames of reference does the stay-at-home twin experience in the twin trip? How many frames of reference does the traveling twin experience?

Space Travel

31. What two main obstacles prevent us from traveling today throughout the universe at relativistic speeds?

32. In this century, humans can enjoy jet hopping, traveling from one country to another by way of jet aircraft. In a future century, humans may enjoy "century hopping." How would this be accomplished?

Length Contraction

33. How long would a meter stick appear if it were traveling like a properly thrown spear at 99.5 percent the speed of light?

34. How long would the meter stick in the previous question appear if it were traveling with its length perpendicular to its direction of motion? (Why is your answer different from the previous question?)

35. If you were traveling in a high-speed rocket ship, would meter sticks on board appear to you to be contracted? Defend your answer.

Increase of Mass with Speed

36. What happens to the mass of an object that is pushed to higher speeds?

37. What is meant by "rest mass"? If an object has a rest mass of 1 kg, will its mass be greater, less, or the same if it is accelerated to a high speed?

38. What would be the mass of an object if pushed to the speed of light?

Mass-Energy Equivalence

39. The masses of particles in research accelerators gain appreciable mass when they are accelerated to speeds near the speed of light. Does this mass increase consume more energy from the power utility that services the accelerator? Explain.

40. In the preceding question, the power utility gets its energy from the mass of fuel, whether coal, oil, or atomic nuclei. If we neglect all inefficiencies at the power plant, in the transmission lines, and at the accelerator, how does the mass increase of the accelerated particles compare with the decrease in mass of the fuel at the power plant? Explain.

41. Does the equation $E = mc^2$ apply only to reactions that involve the atomic nucleus?

42. What is the origin of solar power?

43. What evidence is there for the equivalence of mass and energy?

Correspondence Principle

44. How does the correspondence principle relate to special relativity?

45. Do the relativity equations for time, length, and mass hold true for everyday speeds? Explain.

Exercises

1. A person riding on the roof of a freight-train car fires a gun pointed forward. Compare the velocity of the bullet along the ground when the train is at rest to that when the train is moving. Discuss the velocity of the bullet along the freight car if the gun is fired when the train is at rest and when it's in motion.

2. Suppose instead that the person riding on top of the freight car shines a searchlight beam in the direction in which the train is traveling. Does the light beam travel faster along the ground than it would if he had shined it while standing at rest on the ground? Explain.

3. Why was the Michelson and Morley experiment considered a failure? (Have you ever encountered other examples where failure has to do not with the lack of ability but with the impossibility of the task?)

4. When you drive down the highway you are moving through space. What else are you moving through?

5. In Chapter 25 we learned that light travels more slowly in glass than in air. Does this contradict the theory of relativity?

6. Two events that occur at the same point in space are simultaneous in a certain frame of reference. Are they simultaneous in all other reference frames? Explain.

7. Event A occurs before event B in a certain frame of reference. How could event B occur before event A in some other frame of reference?

8. We might think of the speed of light as a kind of speed limit in the universe—at least for the four-dimensional universe we comprehend. No material particle can attain or surpass this limit even when a continuous, unyielding force is exerted on it. Why is this so?

9. Since there is an upper limit on the speed of a particle, does it follow that there is also an upper limit on its momentum? On its kinetic energy? Explain.

10. Light travels a certain distance in, say, 20 000 years. How is it possible that an astronaut could travel slower than the speed of light and travel as far in a 20-year trip?

11. Could a human being who has a life expectancy of 70 years possibly make a round-trip journey to a part of the universe thousands of light-years distant? Explain.

12. A twin who makes a long trip at relativistic speeds returns younger than his stay-at-home twin sister. Could he return before his twin sister was born? Defend your answer.

13. Is it possible for a son or daughter to be biologically older than his or her parents? Explain.

14. If you were in a rocket ship traveling away from the earth at a speed close to the speed of light, what changes would you note in your pulse? In your mass? In your volume? Explain.

15. If you were on earth monitoring a person in a rocket ship traveling away from the earth at a speed close to the speed of light, what changes would you note in his pulse? In his mass? In his volume? Explain.

16. Consider a high-speed rocket equipped with a flashing light source. If the frequency of flashes when approaching is increased by a factor of 2, by how much is the period (time interval between flashes) changed? Is this period constant for a constant relative speed? For accelerated motion? Defend your answer.

17. You observe a spaceship moving away from you at speed v_1, half the speed of light. A rocket is fired from the spaceship, straight ahead so that it also moves away from you. Suppose that from the spaceship it is fired at half the speed of light, v_2, relative to the spaceship. It so happens the speed of the rocket relative to you is not the speed of light. The relativistic addition of velocities is given by

$$V = \frac{v_1 + v_2}{1 + \frac{v_1 v_2}{c^2}}$$

Substitute $0.5c$ for both v_1 and v_2 and show that the velocity V of the rocket relative to you is $0.8c$.

18. Pretend that the spaceship of the previous question is somehow traveling at c with respect to you, and it fires a rocket at speed c with respect to itself. Use the equation to show that the speed of the rocket with respect to you is still c!

19. Substitute small values of v_1 and v_2 in the preceding equation and show that for everyday velocities V is practically equal to $v_1 + v_2$.

20. How does the measured density of a body compare when at rest to when it is moving?

21. As a meter stick that has a rest mass of 1 kg moves past you, your measurements show it to have a mass of 2 kg. If your measurements show it to have a length of 1 m, in what direction is the stick pointing?

22. In the preceding exercise, if the stick is moving in a direction along its length (like a properly thrown spear), how long will it appear to you?

23. If a high-speed spaceship appears shrunken to half size, by how much will measurements of its mass differ?

24. The "2-mile" linear accelerator at Stanford University in California "appears" to be less than a meter long to the electrons that travel in it. Explain.

25. Electrons that are accelerated in the Stanford accelerator gain thousands of times more mass by the time they reach the end of their trip as they had when they started. In theory, if you could travel with them, would you notice an increase in their mass? In the mass of the target they are about to hit? Explain.

26. The electrons that illuminate the screen in a typical television picture tube travel at nearly one-fourth the speed of light and have an increased mass of nearly 3 percent. Does this relativistic effect tend to increase or decrease your electric bill?

27. How might the principle of correspondence be used to establish the validity of theories outside the domain of physical science?

28. What does the equation $E = mc^2$ mean?

29. Muons are elementary particles that are formed high in the atmosphere by the interactions of cosmic rays with gases in the upper atmosphere. Muons are radioactive and have average lifetimes of about two-millionths of a second. Even though they travel at almost the speed of light, they are so high that very few should be detected at sea level—at least according to classical physics. Laboratory measurements, however, show that muons in great proportions do reach the earth's surface. What is the explanation?

30. When we look out into the universe, we see into the past. John Dobson, founder of the San Francisco Sidewalk Astronomers, says that we cannot even see the backs of our own hands now—in fact, we can't see anything now. Do you agree? Explain.

31. One of the fads of the future might be "century hopping," where occupants of high-speed spaceships would depart from the earth for several years and return centuries later. What are the present-day obstacles to such a practice?

32. Is the statement by the philosopher Kierkegaard that "Life can only be understood backwards; but it must be lived forwards" consistent with the theory of special relativity?

35 General Theory of Relativity

The special theory of relativity is "special" in the sense that it treats the invariance of the laws of nature principally for uniformly moving reference frames—that is, reference frames that are not accelerated. Recall that Einstein postulated in 1905 that no observation made inside an enclosed chamber could determine whether the chamber was at rest or moving with constant velocity; that is, no mechanical, electrical, optical, or any other physical measurement that one could perform inside a closed compartment in a smoothly riding train traveling along a straight track (or in an airplane flying through still air with the window curtains drawn) could possibly give any information as to whether the train was moving or at rest (or the plane airborne or at rest on the runway). But if the track were not smooth and straight (or if the air were turbulent), the situation would be entirely different: uniform motion would give way to accelerated motion, which would be easily noticed. Einstein's conviction that the laws of nature should be expressed in the same form in every frame of reference, accelerated as well as nonaccelerated, was the primary motivation that led him to the general theory of relativity—a new theory of gravitation. Underlying it is the idea that the effects of gravitation and acceleration cannot be distinguished from one another.

Figure 35-1
Everything is weightless on the inside of a nonaccelerating spaceship far away from gravitational influences.

Principle of Equivalence

Einstein imagined himself in a spaceship far away from gravitational influences. In such a spaceship at rest or in uniform motion relative to the distant stars, he and everything within the ship would float freely; there would be no "up" and no "down." But when the rocket motors were activated and the ship was accelerating, things would be different; phenomena similar to gravity would be observed. The wall adjacent to the rocket motors would push up against any occupants and become the floor, while the opposite wall would become the ceiling. Occupants in the ship would be able to stand on the floor and even jump up and down. If the acceleration of the spaceship were equal to g, the occupants could well be convinced the ship was not accelerating, but was at rest on the surface of the earth.

To examine this new "gravity" in the accelerating spaceship, Einstein considered the consequence of dropping two balls, say one of wood and the other of lead. When the balls were released, they would continue to move upward side by side with the velocity of the ship at the moment of release. If the ship were moving at *constant velocity* (zero acceleration), the balls would remain suspended in the same place since both the ship and the balls would move the same amount. But since the spaceship was accelerating, the floor would move upward faster than the balls, which would soon be intercepted by the floor (Figure 35-3). Both balls, regardless of their masses, would meet the floor at the same time. Remembering Galileo's demonstration at the Leaning Tower of Pisa, occupants of the ship might be prone to attribute their observations to the force of gravitation.

Figure 35-2
When the spaceship accelerates, an occupant inside feels "gravity."

Figure 35-3
To an observer inside the accelerating ship, a lead ball and a wood ball appear to fall together when released.

The two interpretations of the falling balls are equally valid, and Einstein incorporated this equivalence, or impossibility of distinguishing between gravitation and acceleration, in the foundation of his general theory of relativity. The **principle of equivalence** states that observations made in an accelerated reference frame are indistinguishable from observations made in a Newtonian gravitational field. This equivalence would be relatively unimportant if it applied only to mechanical phenomena, but Einstein went further and stated that the principle holds for all natural phenomena; it holds for optical and all electromagnetic phenomena as well.

Bending of Light by Gravity

A ball thrown sideways in a stationary spaceship in a gravity-free region will follow a straight-line path relative both to an observer inside the ship and to a stationary observer outside the spaceship. But if the ship is accelerating, the floor overtakes the ball just as in our previous example. An observer outside the ship still sees a straight-line path, but to an observer in the accelerating ship, the path is curved; it is a parabola. The same holds true for a beam of light (Figure 35-4).

a
b

Figure 35-4

(a) An outside observer sees a horizontally thrown ball travel in a straight line, and since the ship is moving upward while the ball travels horizontally, the ball strikes the wall below a point opposite the window. (b) To an inside observer, the ball bends as if in a gravitational field.

Imagine that a light ray enters the spaceship horizontally through a side window and reaches the opposite wall after a very short time. The outside observer sees that the light ray enters the window and moves horizontally along a straight line with constant velocity toward the opposite wall. But the spaceship is accelerating upward, and during the time the light takes to reach the wall, the ship changes its position so that the ray will not meet a point exactly opposite the window, but will hit the wall slightly below. To an inside observer, the light ray is deflected downward toward the floor just as the thrown ball curves toward the floor (Figure 35-5). The curvature of the slow-moving ball is very pronounced;

but if the ball were somehow thrown horizontally across the spaceship cabin at a velocity equal to that of light, both curvatures would be the same.

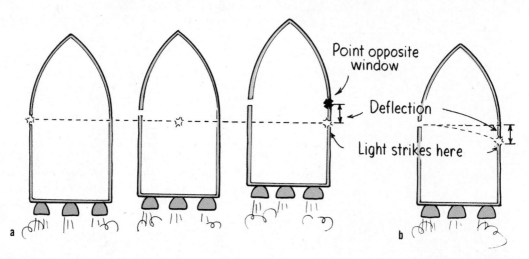

Point opposite window

Deflection

Light strikes here

a   b

Figure 35-5

(a) An outside observer sees light travel horizontally in a straight line, but like the ball in Figure 35-4, it strikes the wall slightly below a point opposite the window. (b) To an inside observer, the light bends as if responding to a gravitational field.

How can an observer inside the ship attribute this bending of light to gravitation? According to Newton's physics, gravitation is an interaction between the masses; a moving ball curves because of the interaction between its mass and the mass of the earth. But what of light, which is pure energy and is massless? To account for the bending of light in a gravitational field, a Newtonian observer might attribute a mass to light or think of its energy in terms of its "mass-equivalent." Even Einstein held that light is a stream of particlelike photons that bend in a gravitational field exactly as any material object would if it were moving with a speed equal to that of light—but not because of any masslike properties of light. Light bends if it travels in a space-time geometry that is bent. We shall see later in this chapter that the presence of mass results in the bending or warping of space-time; and, by the same token, a bending or warping of space-time reveals itself as a mass. The mass of the earth is too small to appreciably warp the surrounding space-time, which is practically flat, so any such bending of light in our immediate environment is not ordinarily detected. Close to bodies of mass much greater than the earth's, however, the bending of light is large enough to detect.

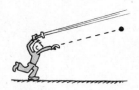

Figure 35-6

The trajectory of a flashlight beam is identical to the trajectory of a baseball "thrown" at the speed of light. Both paths curve equally in a uniform gravitational field.

Einstein predicted that measurements of starlight passing close to the sun would be deflected by an angle of 1.75 seconds of arc—large enough to be measured. Although stars are not visible when the sun is in the sky, the deflection of starlight can be observed during an eclipse of the sun. (Measuring this deflection has become a standard practice at every total eclipse since the first measurements were made during the total eclipse of 1919.) A photograph taken of the darkened sky around the eclipsed sun reveals the presence of the nearby bright stars. The positions of the

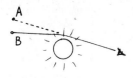

Figure 35-7
Starlight bends as it grazes the sun. Point A shows the apparent position; point B shows the true position.

stars are compared with those in other photographs of the same area taken at other times in the night with the same telescope. In every instance, the deflection of starlight has supported Einstein's prediction (Figure 35-7).

Light bends in the earth's gravitational field also—but not as much. We don't notice it only because the amount of deflection is tiny compared to the correspondingly vast distance that light travels due to its high speed. For example, in a constant gravitational field of 1 *g*, a beam of horizontally directed light will "fall" a vertical distance of 4.9 meters in 1 second (just as a baseball would), but will travel a horizontal distance of 300 000 kilometers. Its curve would hardly be noticeable when you're this far from the beginning point. But if the light traveled 300 000 kilometers in multiple reflections between idealized parallel mirrors, the effect would be quite noticeable (Figure 35-8). (Doing this would make a dandy home project for extra credit—like earning credit for a Ph.D.)

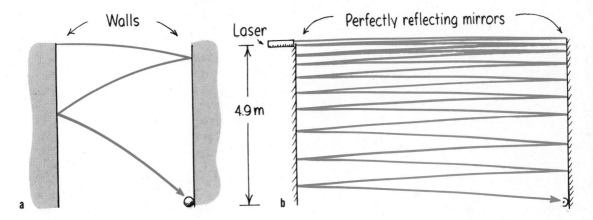

Figure 35-8
(*a*) If a ball is horizontally projected between a vertical pair of parallel walls, it will bounce back and forth and fall a vertical distance of 4.9 m in 1 s. (*b*) If a horizontal beam of light is directed between a vertical pair of perfectly parallel ideal mirrors, it will reflect back-and-forth and fall a vertical distance of 4.9 m in 1 s. The number of back-and-forth reflections is overly simplified in the diagram; if the mirrors were 300 km apart, for example, 1000 reflections would occur in 1 s.

Question ▶ Why do we not notice the bending of light in our everyday environment?

▶ **Answer**
Only because it travels so fast; just as over a short distance we do not notice the curved path of a high-speed bullet, we do not notice the curving of a light beam.

Gravity and Time: Gravitational Red Shift

According to Einstein's general theory of relativity, gravitation causes time to slow down. The stronger the gravitational field, the greater the slowing down of time. We can understand this by applying the principle of equivalence and time dilation to an accelerating frame of reference.

Imagine our accelerating reference frame to be a large rotating disk. Suppose we measure time from three identical clocks, one placed on the disk at its center, a second placed on the rim of the disk, and the third at rest on the nearby ground (Figure 35-9). From the laws of special relativity, we know that the clock attached to the center, since it is not moving with respect to the ground, should run at the same rate as the clock on the ground—but not at the same rate with respect to the clock attached to the rim of the disk. The clock at the rim is in motion with respect to the ground and should therefore be observed to be running more slowly than the ground clock—and therefore more slowly than the clock at the center of the disk. Although the clocks on the disk are attached to the same frame of reference, they do not run synchronously; the outer clock runs slower than the inner clock.

Figure 35-9
Clocks 1 and 2 are on an accelerating disk, and clock 3 is at rest. Clocks 1 and 3 run at the same rate, while clock 2 runs slower. From the point of view of an observer at clock 3, clock 2 runs slow because it is moving. From the point of view of an observer at clock 1, clock 2 runs slow because it is in a stronger centrifugal force field.

This difference in time would be the same for observers on the rotating disk and for observers at rest on the ground. Interpretations of the time difference for each observer are not the same, however. To the observer on the ground, the slower rate of the clock on the rim is due to its motion. But to an observer on the rotating disk, the disk clocks are not in motion with respect to each other; instead, a centrifugal force acts on the clock at the rim, while no such force acts on the clock at the center. The observer on the disk would say that the clock placed in the stronger force field will run slower. By interpreting centrifugal force as a force of gravitation and applying the principle of equivalence, we must conclude that clocks in strong gravitational fields of force run slower than clocks in weak fields of force. This slowing down will apply to all "clocks," whether physical, chemical, or biological. An executive working on the ground floor of a tall city skyscraper will age more slowly than her twin sister working on the top floor. The difference is very small, only a few millionths of a second per decade, because the difference in the earth's gravitational field at the bottom and top of the skyscraper is very small. For larger differences in gravitational field intensity, like those at the surface of the sun compared to the surface of the earth, the differences in time should be more pronounced. A clock at the surface of the sun should run measurably slower than a clock at the surface of the earth. Years before

Figure 35-10
The stronger a gravitational field, the slower a clock runs. A clock at the surface of the earth runs slower than a clock farther away.

he completed his general relativity theory, Einstein suggested a way to measure this when he formulated the principle of equivalence in 1907.

All atoms emit light at specific frequencies characteristic of the vibrational rate of electrons within the atom. Every atom is therefore a "clock," and a slowing down of atomic vibration indicates the slowing down of such clocks. An atom on the sun, where gravitation is strong, should emit light of a lower frequency (slower vibration) than light emitted by the same kind of atom on the earth. Since red light is at the low-frequency end of the visible spectrum, a lowering of frequency shifts the color toward the red. This effect is called the **gravitational red shift**. The gravitational red shift is observed in light from the sun, but various disturbing influences prevent accurate measurements of this tiny effect. It wasn't until 1960 that an entirely new technique, using high-frequency gamma rays from radioactive atoms, permitted incredibly precise and confirming measurements of the gravitational slowing of time between the top and bottom floors of a laboratory building at Harvard University.*

So measurements of time depend not only on relative motion, as we learned in the last chapter, but also on the relative gravitational field strengths of the regions in which the events are taking place and being measured. Just as time dilation in special relativity is relative to the *differences* in motion between the observed frame of reference and the frame of reference from which observations are being made, the gravitational red shift of general relativity is relative to the *differences* in gravitational field strengths at the location of the event and the location of the observer of the event. Where gravitation is seen to be stronger, events are seen to proceed more slowly into the future. As viewed from earth, a clock will be measured to tick more slowly on the surface of a star than on earth. If the star shrinks, the resulting increase in gravitation at its surface will be seen to be accompanied by a corresponding slowing of time, and we would measure longer intervals between the ticks of the star clock. But if we made our measurements of the star clock from the star itself, we would notice nothing unusual about the clock's ticking.

Suppose, for example, that an indestructible volunteer stands on the surface of a giant star that begins collapsing. We, as outside observers, will note a progressive slowing of time on the clock of our volunteer as the gravitational field increases. The volunteer himself, however, does not notice any differences in his own time. He is viewing events within his own frame of reference, and he notices nothing unusual about his own time. As the collapsing star proceeds toward becoming a black hole and time proceeds normally from the viewpoint of the volunteer, we on the outside perceive time for the volunteer as approaching a complete

---

*In the late 1950s, shortly after Einstein's death, the German physicist Rudolph Mössbauer discovered an important effect in nuclear physics that provides an extremely accurate method of using atomic nuclei as atomic clocks. The *Mössbauer effect*, for which its discoverer was awarded the Nobel Prize, has many practical applications. In late 1959 Drs. Pound and Rebka at Harvard University realized still another application that was a test for general relativity and performed the confirming experiment.

stop; we see him frozen in time with an infinite duration between the ticks of his clock or the beats of his heart. From our view, his time stops completely. The gravitational red shift, instead of being a tiny effect, is dominating.

It is important to note the relativistic nature of time in both special relativity and general relativity. In both theories, there is no way that you can extend the duration of your own experience. Others moving at different speeds or in different gravitational fields may attribute a great longevity to you, but your longevity is seen from their frame of reference—never your own. Changes in time are always attributed to "the other guy."

Questions ▶

1. Will a person at the top of a skyscraper age more than or less than a person at ground level?
2. Will a person at ground level age more than or less than a person at the bottom of a very deep well?

Gravity and Space: Motion of Mercury

Figure 35-11
A precessing elliptical orbit.

From the special theory of relativity, we know that measurements of space as well as of time undergo transformations when motion is involved. Likewise with the general theory: measurements of space differ in different gravitational fields—for example, close to and far away from the sun.

Planets orbit the sun and stars in elliptical orbits and move periodically into regions of weaker and stronger gravitational fields. Einstein directed his attention to the varying gravitational fields experienced by the planets orbiting the sun and found that the elliptical orbits of the planets should *precess* (Figure 35-11)—independently of the Newtonian influence of other planets. Near the sun, where the gravitational field is stronger, the rate of precession should be the greatest; and far from the sun, where the field is weak, any deviations from Newtonian mechanics should be virtually unnoticeable.

Mercury is the planet nearest the sun. The gravitational force between the sun and Mercury is greater than that between the sun and any of the more distant planets. If the orbit of any planet exhibits a measurable precession, it should be Mercury, and the fact that the orbit of Mercury does precess had been a mystery to astronomers since the early 1800s! Careful measurements showed that Mercury precesses about 573 seconds of arc per century. Perturbations by the other planets were found to account

▶ **Answers**

1. More; a person at the top of a skyscraper is in a slightly weaker gravitational field than a person at ground level.
2. Less; a person at the surface of the earth is in a stronger gravitational field than a person in a very deep well. (Recall from Chapter 8 that the gravitational field deep inside the earth is less than at the surface and is zero at the earth's center.)

for the precession—except for 43 seconds of arc per century more than the calculated value. Even after all known corrections due to possible perturbations by other planets had been applied, the calculations of physicists and astronomers failed to account for the extra 43 seconds of arc. Either Venus was extra massive or a never-discovered other planet (called Vulcan) was pulling on Mercury. And then came the explanation of Einstein, whose general relativity field equations applied to Mercury's orbit predict the extra 43 seconds of arc per century!

The mystery of Mercury was solved, and a new theory of gravity was recognized. Newton's law of gravitation, which had stood as an unshakable pillar of science for more than 2 centuries, was found to be a special limiting case of Einstein's more general theory. If the gravitational fields are comparatively weak, Newton's law turns out to be a good approximation of the new law—enough so that Newton's law, which is easier to work with mathematically, is the law that today's scientists use most of the time, except for cases involving enormous gravitational fields.

Gravity, Space, and a New Geometry

We can begin to understand that measurements of space are altered in a gravitational field by again considering the accelerated frame of reference of our rotating disk. Suppose we measure the circumference of the outer rim with a measuring stick. Recall the Lorentz-FitzGerald contraction from special relativity: the measuring stick will appear contracted to any observer not moving along with the stick, while an identical measuring stick moving much more slowly near the center will be nearly unaffected (Figure 35-12). All distance measurements along a *radius* of the rotating disk should be completely unaffected by motion, because motion is perpendicular to the radius. Since only distance measurements parallel to and around the circumference are affected, the ratio of circumference to diameter when the disk is rotating is no longer the fixed constant π (3.14159 . . .), but is a variable depending on angular speed and the diameter of the disk. According to the principle of equivalence, the rotating disk is equivalent to a stationary disk with a strong gravitational field near its edge and a progressively weaker gravitational field toward its center. Measurements of distance, then, will depend on the strength of gravitational field, even if no relative motion is involved. Gravity causes space to be non-Euclidean; the laws of Euclidean geometry taught in high school are no longer valid when applied to objects in the presence of strong gravitational fields.

Figure 35-12
A measuring stick along the edge of the rotating disk appears contracted, while a measuring stick farther in and moving more slowly is not contracted as much. A measuring stick along a radius is not contracted at all. When the disk is not rotating, C/D = π; but when the disk is rotating, C/D ≠ π and Euclidean geometry is no longer valid. Likewise in a gravitational field.

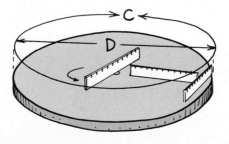

The familiar rules of Euclidean geometry pertain to various figures you can draw on a flat surface. The ratio of the circumference of a circle to its diameter is equal to π; all the angles in a triangle add up to 180°; the shortest distance between two points is a straight line. The rules of Euclidean geometry are valid in flat space, but if you draw these figures on a curved surface like a sphere or a saddle-shaped object, the Euclidean rules no longer hold (Figure 35-13). If you measure the sum of the angles for a triangle in space, you call the space flat if the sum is equal to 180°, spherelike or positively curved if the sum is larger than 180°, and saddlelike or negatively curved if it is less than 180°.

Figure 35-13
The sum of the angles for a triangle drawn (a) on a plane surface = 180°, (b) on a spherical surface is greater than 180°, and (c) on a saddle-shaped surface is less than 180°.

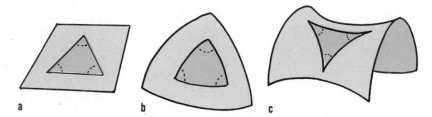

a   b   c

Of course the lines forming the triangles in Figure 35-13 are not "straight" from the three-dimensional view, but they are the "straightest" or *shortest* distances between two points if we are confined to the curved surface. These lines of shortest distance are called *geodesic lines* or simply **geodesics**.

The path of a light beam follows a geodesic. Suppose three experimenters on Earth, Venus, and Mars measure the angles of a triangle formed by light beams traveling between these three planets. The light beams bend when passing the sun, resulting in the sum of the three angles being larger than 180° (Figure 35-14). So the space around the sun is positively curved. The planets that orbit the sun travel along four-dimensional geodesics in this positively curved space-time. Freely falling objects, satellites, and light rays all travel along geodesics in four-dimensional space-time.

Figure 35-14
The light rays joining the three planets form a triangle. Since light passing near the sun bends, the sum of the angles of the resulting triangle is greater than 180°.

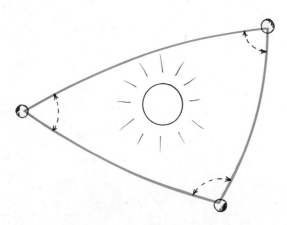

The whole universe may have an overall curvature. If it is negatively curved, it is open-ended and extends without limit; if it is positively curved, it closes in on itself. The surface of the earth, for example, forms a closed curvature, so if you travel along a geodesic, you come back to your starting point. Similarly, if the universe were positively curved, it would be closed, so if you could look infinitely into space through an ideal telescope, you would see the back of your own head! (This is assuming that you waited a long enough time or that light traveled infinitely fast.)

Figure 35-15

The geometry of the curved surface of the earth differs from the Euclidean geometry of flat space. Note that the sum of the angles for an equilateral triangle where the sides equal $\frac{1}{4}$ the earth's circumference is clearly greater than 180°, and the circumference is only twice its diameter instead of 3.14 times its diameter. Euclidean geometry is also invalid in curved space.

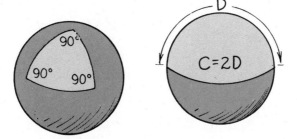

General relativity, then, calls for a new geometry: a geometry not only of curved space but of curved time as well—a geometry of curved four-dimensional space-time. The mathematics of this geometry is too formidable to present here. The essence, however, is that gravity is a manifestation of space-time geometry; a gravitational field is a geometrical warping of space-time. The presence of mass results in the curvature or warping of space-time; by the same token, a curvature of space-time reveals itself as mass. Instead of visualizing gravitational forces between masses, we abandon altogether the notion of force and instead think of masses responding in their motion to the curvature or warping of the space-time they inhabit. It is the bumps, depressions, and warpings of geometrical space-time that *are* the phenomena of gravity.

Figure 35-16

A two-dimensional analogy of four-dimensional warped space-time. Space-time near a star is curved in a way similar to the surface of a waterbed when a heavy ball rests on it.

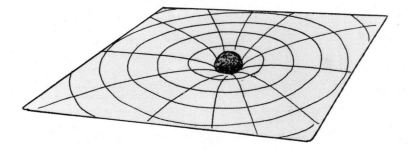

We cannot visualize the four-dimensional bumps and depressions in space-time because we are three-dimensional beings. We can get a glimpse of this warping by considering a simplified analogy in two dimensions: a heavy ball resting on the middle of a waterbed. The more massive the ball, the greater it dents or warps the two-dimensional surface. A marble rolled across the bed, but away from the ball, will roll in a relatively straight-line path, whereas a marble rolled near the ball will curve as it rolls across the indented surface. If the curve closes upon itself, its shape is an ellipse. The planets that orbit the sun similarly travel along four-dimensional geodesics in the warped space-time about the sun.

Gravitational Waves

Every object has mass and therefore makes a bump or depression in the surrounding space-time. When an object moves, the surrounding warp of space and time moves to readjust to the new position. These readjustments produce ripples in the overall geometry of space-time. This is similar to moving the ball that rests on the surface of the waterbed. A disturbance ripples across the waterbed surface in waves; if we move a more massive ball, then we get a greater disturbance and the production of even stronger waves. The ripples travel outward from the gravitational sources at the speed of light and are called **gravitational waves**.

Any moving object produces a gravitational wave. In general, the more massive the moving object and the more violent its motion, the stronger the resulting gravitational wave. But even the strongest waves produced by ordinary astronomical events are extremely weak—the weakest known in nature. For example, the gravitational waves emitted by a vibrating electric charge are a trillion trillion trillion times weaker than the electromagnetic waves emitted by the same charge.

Shake your hand back and forth: you have just produced a gravitational wave. It is not very strong, but it exists.

Newtonian and Einsteinian Gravitation

When Einstein formulated his new theory of gravitation, he realized that if his theory was valid, his field equations must reduce to Newtonian equations for gravitation in the weak-field limit. He showed that Newton's law of gravitation is a special case of the broader theory of relativity. Newton's law of gravitation is still an accurate description of most of the interactions between bodies in the solar system and beyond. From Newton's law, one can calculate the orbits of comets and asteroids and even predict the existence of undiscovered planets. Even today, when computing the trajectories of space probes to the moon and planets, only ordinary Newtonian theory is used. This is because the gravitational field of these bodies is very weak, and from the viewpoint of general relativity, the surrounding space-time is essentially flat. But for regions of more intense gravitation, where space-time is more appreciably curved, Newtonian theory cannot adequately account for various phenomena—like the precession of Mercury's orbit close to the sun and, in the case of stronger fields, the gravitational red shift and other apparent distortions in measurements of space and time. These distortions reach their limit in the case of a star that collapses to a black hole, where space-time completely folds over on itself. Only Einsteinian gravitation reaches into this domain.

We saw in Chapter 31 that Newtonian physics is linked at one end with quantum theory, whose domain is the very light and very small— tiny particles and atoms. And now we have seen that Newtonian physics is linked at the other end with relativity theory, whose domain is the very massive and very large.

We do not see the world the way the ancient Egyptians, Greeks, or Chinese did. It is unlikely that people in the future will see the universe as we do. Our view of the universe may be quite limited, and perhaps filled with misconceptions, but it is most likely clearer than the views of others before us. Our view today stems from the findings of Copernicus, Galileo, Newton, and, more recently, Einstein—findings that were often opposed on the grounds that they diminished the importance of humans in the universe. In the past the idea of importance was seen as rising above nature—being apart from nature. We have expanded our vision since then by enormous effort, painstaking observation, and an unrelenting desire to comprehend our surroundings. Seen from today's understanding of the universe, we find our importance in being very much a part of nature, not apart from it. We are the part of nature that is becoming more and more conscious of itself.

Summary of Terms

Principle of equivalence Local observations made in an accelerated frame of reference are indistinguishable from observations made in a Newtonian gravitational field.

Gravitational red shift The shift in wavelength toward the red end of the spectrum experienced by light leaving the surface of a massive object, as predicted by the general theory of relativity.

Geodesic The shortest path between points on any surface.

Gravitational wave A gravitational disturbance made by a moving mass that propagates through space-time.

Suggested Reading

Einstein, Albert. *Relativity: The Special and General Theory.* New York: Crown, 1961. (Orig. pub. 1916.)

Kaufmann, William J. *The Cosmic Frontiers of General Relativity.* Boston: Little, Brown, 1977.

Review Questions

1. What is the principal difference between *special relativity* and *general relativity*?

Principle of Equivalence

2. Compared with the number of pushups she could do on earth, how many could an occupant do in a spaceship that accelerates at *g* far from earth gravity?

3. Exactly what is *equivalent* in the principle of equivalence?

Bending of Light by Gravity

4. Compare the bending of the paths of baseballs and photons by a gravitational field.

5. Why must the sun be eclipsed when measuring the deflection of nearby starlight?

Gravity and Time: Gravitational Red Shift

6. Which runs faster, a clock at the top of the Sears Tower in Chicago or a clock on the shore of Lake Michigan?

7. What is the effect of a strong gravitational field on the frequency of light?

8. What is the effect of a strong gravitational field on measurements of time?

Gravity and Space: Motion of Mercury

9. Of all the planets, why is Mercury the best candidate for our finding evidence of the relationship of gravitation to space?

Gravity, Space, and a New Geometry

10. A measuring stick placed along the circumference of a rotating disk will appear contracted, but if it is oriented along a radius, it will not. Explain.

11. The ratio of circumference to diameter for measured circles on a disk equals π when the disk is at rest, but not when the disk is rotating. Explain.

12. Is the two-dimensional surface of the earth positively or negatively curved? Why?

Gravitational Waves

13. A star 10 light-years away explodes and produces gravitational waves. How long will it take these waves to reach the earth?

14. Why are gravitational waves so difficult to detect?

Newtonian and Einsteinian Gravitation

15. Does Einstein's theory of gravitation invalidate Newton's theory of gravitation? Explain.

Exercises

1. An astronaut awakes in her closed capsule, which actually sits on the moon. Can she tell whether her weight is the result of gravitation or of accelerated motion? Explain.

2. An astronaut is provided a "gravity" when the ship's engines are activated to accelerate the ship. This requires the use of fuel. Is there a way to accelerate and provide "gravity" without the sustained use of fuel? Explain.

3. In his famous novel *Journey to the Moon*, Jules Verne stated that occupants in a spaceship would shift their orientation from up to down when the ship crossed the point where the moon's gravitation became greater than the earth's. Is this correct? Defend your answer.

4. What happens to the separation distance between two people if they both walk north at the same rate from two different places on the earth's equator? And just for fun, where on the world is a step in every direction a step south?

5. We readily note the bending of light by reflection and refraction, but why is it we do not ordinarily notice the bending of light by gravity?

6. Why do we say that light travels in straight lines? Is it strictly accurate to say that a laser beam provides a perfectly straight line for purposes of surveying? Explain.

7. At the end of 1 s, a horizontally fired bullet has dropped a vertical distance of 4.9 m from its otherwise straight-line path in a gravitational field of 1 g. By what distance would a beam of light drop from its otherwise straight-line path if it traveled in a uniform field of 1 g for 1 s? For 2 s?

8. Light changes its energy when it "falls" in a gravitational field. This change in energy is not evidenced by a change in speed, however. What is the evidence for this change in energy?

9. Would we notice a slowing down or speeding up of a clock at the bottom of a very deep well?

10. If we witness events taking place on the moon, where gravitation is weaker than on earth, would we expect to see a gravitational red shift or a gravitational blue shift? Explain.

11. Why will the gravitational field intensity increase on the surface of a shrinking star?

12. Will a clock at the equator run slightly faster than an identical clock at one of the earth's poles?

13. Splitting hairs, should a person who worries about growing old live at the top or at the bottom of a tall apartment building?

14. Splitting hairs, if you shine a beam of colored light to a friend above in a high tower, will the color of light your friend receives be the same color you emit? Explain.

15. Is the color of light emitted from the surface of a massive star red-shifted or blue-shifted?

16. From our frame of reference on earth, objects slow to a stop before falling into black holes in space. With ideal telescopes, could we "see" these objects as they hover about the black holes? (Would they emit or reflect waves in the visible part of the spectrum?)

17. Would an astronaut falling into a black hole see the surrounding universe red-shifted or blue-shifted?

18. Why does the gravitational attraction between the sun and Mercury vary? Would it vary if the orbit of Mercury were perfectly circular?

19. Do binary stars (double-star systems that orbit about a common center of mass) radiate gravitational waves? Why or why not?

20. With respect to Newton's theory of gravitation, how does Einstein's theory of gravitation obey the correspondence principle?

Epilogue

Nature bestows on us an enormously high prize—being alive. An even higher prize is our capacity to comprehend and understand this nature. This book is an attempt to kindle that capacity, open new insights, and lead to a better understanding of many of the things you wondered about as a child. To add to your knowledge of why the sky is blue, how the planets orbit, and how the sun shines, I hope to have communicated the excitement, beauty, and flavor of the physics that surrounds you daily.

You may forget most of the facts, formulas, and diagrams that are a part of your understanding of physics and forget most of the insights experienced. But the flavor you'll retain—your discovery and understanding that the diverse phenomena about us are beautifully tied together by surprisingly few relationships: the laws of nature. We are as subject to these laws as is the blueness of the sky, the falling of the moon, and the shining of the stars. For we are a part of nature—not apart from it. We are the conscious part of nature that is investigating itself!

Your education is ultimately the flavor left over after the facts, formulas, and diagrams have been forgotten.

Paul G. Hewitt

Two major systems of measurement prevail in the world today: the *United States Customary System* (USCS, formerly called the British system of units), used in the United States of America and in Burma, and the *Système International* (SI) (known also as the international system and as the metric system), used everywhere else. Each system has its own standards of length, mass, and time. The units of length, mass, and time are sometimes called the *fundamental units* because, once they are selected, other quantities can be measured in terms of them.

United States Customary System

Based on the British Imperial System, the USCS is familiar to everyone in the United States. It uses the foot as the unit of length, the pound as the unit of weight or force, and the second as the unit of time. The USCS is presently being replaced by the international system—rapidly in science and technology (all 1988 Department of Defense contracts) and some sports (track and swimming), but so slowly in other areas and in some specialties it seems the change may never come. For example, we will continue to buy seats on the 50-yard line. Camera film is in millimeters but computer disks are in inches.

For measuring time there is no difference between the two systems except that in pure SI the only unit is the *second* (s, not sec) with prefixes; but, in general, minute, hour, day, year, and so on, with two or more lettered abbreviations (hr, not h), are accepted in the USCS.

Système International

During the 1960 International Conference on Weights and Measures held in Paris, the SI units were defined and given status. Table I-1 shows SI units and their symbols. SI is based on the *metric system*, originated by French scientists after the French revolution in 1791. The orderliness of this system makes it useful for scientific work, and it is used by scientists all over the world. The metric system branches into two systems of units. In one of these the unit of length is the meter, the unit of mass is the kilogram, and the unit of time is the second. This is called the *meter-kilogram-second* (mks) system and is preferred in physics. The other branch is the *centimeter-gram-second* (cgs) system, which because of its smaller values is favored in chemistry. The cgs and mks units are related to each other as follows: 100 centimeters equal 1 meter; 1000 grams equal 1 kilogram. Table I-2 shows several units of length related to each other.

Table I-1 Table conversions between different units of length

Unit of length	Kilometer	Meter	Centimeter	Inch	Foot	Mile
1 kilometer	= 1	1000	100 000	39 370	3280.84	0.62140
1 meter	= 0.00100	1	100	39.370	3.28084	6.21×10^{-4}
1 centimeter	= 1.0×10^{-5}	0.0100	1	0.39370	0.032808	6.21×10^{-6}
1 inch	= 2.54×10^{-5}	0.02540	2.5400	1	0.08333	1.58×10^{-5}
1 foot	= 3.05×10^{-4}	0.30480	30.480	12	1	1.89×10^{-4}
1 mile	= 1.60934	1609.34	160 934	63 360	5280	1

Table I-2
Some prefixes

Prefix	Definition
micro-	One-millionth: a microsecond is one-millionth of a second
milli-	One-thousandth: a milligram is one-thousandth of a gram
centi-	One-hundredth: a centimeter is one-hundredth of a meter
kilo-	One thousand: a kilogram is 1000 grams
mega-	One million: a megahertz is 1 million hertz

One major advantage of a metric system is that it uses the decimal system, where all units are related to smaller or larger units by dividing or multiplying by 10. The prefixes shown in Table I-3 are commonly used to show the relationships among units.

Table I-3
SI units

Quantity	Unit	Symbol
Length	meter	m
Mass	kilogram	kg
Time	second	s
Force	newton	N
Energy	joule	J
Current	ampere	A
Temperature	kelvin	K

Figure I-1
The standard kilogram.

Meter

The standard of length for the metric system originally was defined in terms of the distance from the north pole to the equator. This distance was thought at the time to be close to 10 000 kilometers. One ten-millionth of this, the meter, was carefully determined and marked off by means of scratches on a bar of platinum-iridium alloy. This bar is kept at the International Bureau of Weights and Measures in France. The standard meter in France has since been calibrated in terms of the wavelength of light—it is 1 650 763.73 times the wavelength of orange light emitted by the atoms of the gas krypton-86. The meter is now defined as being the length of the path traveled by light in a vacuum during a time interval of 1/299 792 458 of a second.

Kilogram

The standard unit of mass, the kilogram, is a block of platinum, also preserved at the International Bureau of Weights and Measures located in France. The kilogram equals 1000 grams. A gram is the mass of 1 cubic centimeter (cc) of water at a temperature of 4° Celsius. (The standard pound is defined in terms of the standard kilogram; the mass of an object that weighs 1 pound is equal to 0.4536 kilogram.)

Second

The official unit of time for both the USCS and the SI is the second. Until 1956 it was defined in terms of the mean solar day, which was divided into 24 hours. Each hour was divided into 60 minutes and each minute into 60 seconds. Thus, there were 86 400 seconds per day, and the second was defined as 1/86 400 of the mean solar day. This proved unsatisfactory because the rate of rotation of the earth is gradually becoming slower. In 1956 the mean solar day of the year 1900 was chosen as the standard on which to base the second. In 1964, the second was officially defined as the time taken by a cesium-133 atom to make 9 192 631 770 vibrations.

Newton

One newton is the force required to accelerate 1 kilogram at 1 meter per second per second. This unit is named after Sir Isaac Newton.

Joule

One joule is equal to the amount of work done by a force of 1 newton acting over a distance of 1 meter. In 1948 the joule was adopted as the unit of energy by the International Conference on Weights and Measures. Therefore, the specific heat of water at 15°C is now given as 4185.5 joules per kilogram Celsius degree. This figure is always associated with the mechanical equivalent of heat—4.1855 joules per calorie.

Ampere

The ampere is defined as the intensity of the constant electric current that, when maintained in two parallel conductors of infinite length and negligible cross section and placed 1 meter apart in a vacuum, would produce between them a force equal to 2×10^{-7} newton per meter length. In our treatment of electric current in this text, we have used the not-so-official but easier-to-comprehend definition of the ampere as being the rate of flow of 1 coulomb of charge per second, where 1 coulomb is the charge of 6.25×10^{18} electrons.

Kelvin

The fundamental unit of temperature is named after the scientist William Thomson, Lord Kelvin. The kelvin is defined to be 1/273.15 the thermodynamic temperature of the triple point of water (the fixed point at which ice, liquid water, and water vapor coexist in equilibrium). This definition was adopted in 1968 when it was decided to change the name *degree Kelvin* (°K) to *kelvin* (K). The temperature of melting ice at atmospheric pressure is 273.15K. The temperature at which the vapor pressure of pure water is equal to standard atmospheric pressure is 373.15K (the temperature of boiling water at standard atmospheric pressure).

Measurements of Area and Volume

Figure I-2
Unit square.

Figure I-3
Unit volume.

Area

The unit of area is a square that has a standard unit of length as a side. In the USCS it is a square whose sides are each 1 foot in length, called 1 square foot and written 1 ft^2. In the international system it is a square whose sides are 1 meter in length, which makes a unit of area of 1 m^2. In the cgs system it is 1 cm^2. The area of a given surface is specified by the number of square feet, square meters, or square centimeters that would fit into it. The area of a rectangle equals the base times the height. The area of a circle is equal to πr^2, where $\pi = 3.14$ and r is the radius of the circle. Formulas for the surface areas of other objects can be found in geometry textbooks.

Volume

The volume of an object refers to the space it occupies. The unit of volume is the space taken up by a cube that has a standard unit of length for its edge. In the USCS one unit of volume is the space occupied by a cube 1 foot on an edge and is called 1 cubic foot, written 1 ft^3. In the metric system it is the space occupied by a cube whose sides are 1 meter (SI) or 1 centimeter (cgs). It is written 1 m^3 or 1 cm^3 (or cc). The volume of a given space is specified by the number of cubic feet, cubic meters, or cubic centimeters that will fill it.

In the USCS volumes can also be measured in quarts, gallons, and cubic inches as well as in cubic feet. There are 1728 ($12 \times 12 \times 12$) cubic inches in 1 ft^3. A U.S. gallon is a volume of 231 in^3. Four quarts equal 1 gallon. In the SI volumes are also measured in liters. A liter is equal to 1000 cm^3.

Scientific Notation

It is convenient to use a mathematical abbreviation for large and small numbers. The number 50 000 000 can be obtained by multiplying 5 by 10, and again by 10, and again by 10, and so on until 10 has been used as a multiplier seven times. The short way of showing this is to write the number 5×10^7. The number 0.0005 can be obtained from 5 by using 10 as a divisor four times. The short way of showing this is to write 5×10^{-4} for 0.0005. Thus, 3×10^5 means $3 \times 10 \times 10 \times 10 \times 10 \times 10$, or 300 000; and 6×10^{-3} means $6/(10 \times 10 \times 10)$, or 0.006. Numbers expressed in this shorthand manner are said to be in scientific notation.

$$
\begin{aligned}
1\,000\,000 &= 10 \times 10 \times 10 \times 10 \times 10 \times 10 &&= 10^6 \\
100\,000 &= 10 \times 10 \times 10 \times 10 \times 10 &&= 10^5 \\
10\,000 &= 10 \times 10 \times 10 \times 10 &&= 10^4 \\
1000 &= 10 \times 10 \times 10 &&= 10^3 \\
100 &= 10 \times 10 &&= 10^2 \\
10 &= 10 &&= 10^1 \\
1 &= 1 &&= 10^0 \\
0.1 &= 1/10 &&= 10^{-1} \\
0.01 &= 1/100 = 1/10^2 &&= 10^{-2} \\
0.001 &= 1/1000 = 1/10^3 &&= 10^{-3} \\
0.0001 &= 1/10\,000 = 1/10^4 &&= 10^{-4} \\
0.00001 &= 1/100\,000 = 1/10^5 &&= 10^{-5} \\
0.000001 &= 1/1\,000\,000 = 1/10^6 &&= 10^{-6}
\end{aligned}
$$

We can use scientific notation to express some of the physical data often used in physics.

$$
\begin{aligned}
\text{Speed of light in a vacuum} &= 2.9979 \times 10^8 \text{ m/s} \\
\text{1 astronomical unit (A.U.),} & \\
\text{(average earth-sun distance)} &= 1.50 \times 10^{11} \text{ m} \\
\text{Average earth-moon distance} &= 3.84 \times 10^8 \text{ m} \\
\text{Equatorial radius of the sun} &= 6.96 \times 10^8 \text{ m} \\
\text{Equatorial radius of Jupiter} &= 7.14 \times 10^7 \text{ m} \\
\text{Equatorial radius of the earth} &= 6.37 \times 10^6 \text{ m} \\
\text{Equatorial radius of the moon} &= 1.74 \times 10^6 \text{ m} \\
\text{Average radius of hydrogen atom} &= 5 \times 10^{-11} \text{ m} \\
\text{Mass of the sun} &= 1.99 \times 10^{30} \text{ kg} \\
\text{Mass of Jupiter} &= 1.90 \times 10^{27} \text{ kg} \\
\text{Mass of the earth} &= 5.98 \times 10^{24} \text{ kg} \\
\text{Mass of the moon} &= 7.36 \times 10^{22} \text{ kg} \\
\text{Proton mass} &= 1.6726 \times 10^{-27} \text{ kg} \\
\text{Neutron mass} &= 1.6749 \times 10^{-27} \text{ kg} \\
\text{Electron mass} &= 9.1 \times 10^{-31} \text{ kg} \\
\text{Electron charge} &= 1.602 \times 10^{-19} \text{ C}
\end{aligned}
$$

Appendix II More About Motion

When we describe the motion of something, we say how it moves relative to something else (Chapter 3). In other words, motion requires a reference frame (an observer, origin, and axes). We are free to choose this frame's location and to have it moving relative to another frame. When our frame of motion has zero acceleration, it is called an *inertial frame*. In an inertial frame, force causes an object to accelerate in accord with Newton's laws. When our frame of reference is accelerated, we observe fictitious forces and motions (Chapter 7). Observations from a carousel, for example, are different when it is rotating and when it is at rest. Our description of motion and force depends on our "point of view."

We distinguish between *speed* and *velocity* (Chapters 2 and 3). Speed is how fast something moves, or the time rate of change of position (excluding direction): a *scalar* quantity. Velocity includes direction of motion: a *vector* quantity whose magnitude is speed. Objects moving at constant velocity move the same distance in the same time in the same direction.

Another distinction between speed and velocity has to do with the difference between distance and net distance, or *displacement*. Speed is *distance per duration* while velocity is *displacement per duration*. Displacement differs from distance. For example, a commuter who travels 10 kilometers to work and back travels 20 kilometers, but has "gone" nowhere. The distance traveled is 20 kilometers and the displacement is zero. Although the instantaneous speed and instantaneous velocity have the same value at the same instant, the average speed and average velocity can be very different. The average speed of this commuter's round trip is 20 kilometers divided by the total commute time—a value greater than zero. But the average velocity is zero. In science, displacement is often more important than distance. (To avoid information overload, we have not treated this distinction in the text.)

Acceleration is the rate at which velocity changes. This can be a change in speed only, a change in direction only, or both. Negative acceleration is often called *deceleration*.

In Newtonian space and time, space has three dimensions—length, width, and height—each with two directions. We can go, stop, and return in any of them. Time has one dimension, with two directions—past and future. We cannot stop or return, only go. In Einsteinian space-time, these four dimensions merge (Chapter 34).

Computing Velocity and Distance Traveled on an Inclined Plane

Recall from Chapter 2 Galileo's experiments with inclined planes. On page 22 we considered a plane tilted such that the speed of a rolling ball increases at the rate of 2 meters per second each second—an acceleration of 2 m/s^2. So at the instant it starts moving its velocity is zero, and 1 second later it is rolling at 2 m/s, at the end of the next second 4 m/s, the end of the next second 6 m/s, and so on. The velocity of the ball at any instant is simply

$$\text{Velocity} = \text{acceleration} \times \text{time}$$

Or, in shorthand notation,

$$v = at$$

(It is customary to omit the multiplication sign, $\times$, when expressing relationships in mathematical form. When two symbols are written together, such as the at in this case, it is understood that they are multiplied.)

How fast the ball rolls is one thing; how *far* it rolls is another. To understand the relationship between acceleration and distance traveled, we must first investigate the relationship between instantaneous velocity and *average velocity*. If the ball shown in Figure II-1 starts from rest, it will roll a distance of 1 meter in the first second. Question: what will be its average speed? The answer is 1 m/s (because it covered 1 meter in the interval of 1 second). But we have seen that the *instantaneous velocity* at the end of the first second is 2 m/s. Since the acceleration is uniform, the average velocity in any time interval is found the same way we usually find the average of any two numbers: add them and divide by 2. (Be careful not to do this when acceleration is not uniform!) So if we add the initial speed (zero in this case) and the final speed of 2 m/s and then divide by 2, we get 1 m/s for the average velocity.

In each succeeding second we see the ball roll a longer distance down the same slope in Figure II-2. Note the distance covered in the second time interval is 3 meters. This is because the average speed of the ball in this interval is 3 m/s. In the next 1-second interval the average speed is 5 m/s, so the distance covered is 5 meters. It is interesting to see that successive increments of distance increase as a *sequence of odd numbers*. Nature clearly follows mathematical rules!

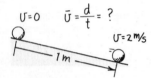

Figure II-1
The ball rolls 1 m down the incline in 1 s and reaches a speed of 2 m/s. Its average speed, however, is 1 m/s. Do you see why?

Question ▶ During the span of the second time interval, the ball begins at 2 m/s and ends at 4 m/s. What is the *average speed* of the ball during this 1-s interval? What is its *acceleration*?

▶ **Answer**

$$\text{Average speed} = \frac{\text{beginning} + \text{final speed}}{2} = \frac{2 \text{ m/s} + 4 \text{ m/s}}{2} = \frac{6 \text{ m/s}}{2} = 3 \text{ m/s}$$

$$\text{Acceleration} = \frac{\text{change in velocity}}{\text{time interval}} = \frac{4 \text{ m/s} - 2 \text{ m/s}}{1 \text{ s}} = \frac{2 \text{ m/s}}{1 \text{ s}} = 2 \text{ m/s}^2$$

Figure II-2
If the ball covers 1 m during its first second, then in each successive second it will cover the odd-numbered sequence of 3, 5, 7, 9 m, and so on. Note that the total distance covered increases as the square of the total time.

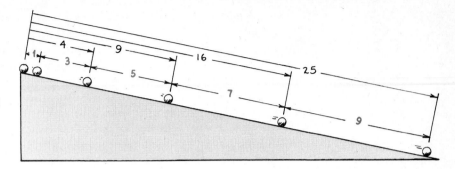

Investigate Figure II-2 carefully and note the *total* distances covered as the ball accelerates down the plane. The distances go from zero to 1 meter in 1 second, zero to 4 meters in 2 seconds, zero to 9 meters in 3 seconds, zero to 16 meters in 4 seconds, and so on in succeeding seconds. The sequence for *total distances* covered is of the *squares of the time*. We'll investigate the relationship between distance traveled and the square of the time for constant acceleration more closely in the case of free fall.

Computing Distance When Acceleration Is Constant

How far will an object released from rest fall in a given time? To answer this question, let us consider the case in which it falls freely for 3 seconds, starting at rest. Neglecting air resistance, the object will have a constant acceleration of about 10 meters per second each second (actually more like 9.8 m/s^2 but we want to make the numbers easier to follow).

$$\text{Velocity at the } \textit{beginning} = 0 \text{ m/s}$$
$$\text{Velocity at the } \textit{end} \text{ of 3 seconds} = (10 \times 3) \text{ m/s}$$
$$\textit{Average} \text{ velocity} = \tfrac{1}{2} \text{ the sum of these two speeds}$$
$$= \tfrac{1}{2} \times (0 + 10 \times 3) \text{ m/s}$$
$$= \tfrac{1}{2} \times 10 \times 3 = 15 \text{ m/s}$$
$$\text{Distance traveled} = \text{average velocity} \times \text{time}$$
$$= (\tfrac{1}{2} \times 10 \times 3) \times 3$$
$$= \tfrac{1}{2} \times 10 \times 3^2 = 45 \text{ m}$$

We can see from the meanings of these numbers that

$$\text{Distance traveled} = \tfrac{1}{2} \times \text{acceleration} \times \text{square of time}$$

This equation is true for an object falling not only for 3 seconds but for any length of time, as long as the acceleration is constant. If we let *d* stand for the distance traveled, *a* for the acceleration, and *t* for the time, the rule may be written, in shorthand notation,

$$d = \tfrac{1}{2} at^2$$

This relationship was first deduced by Galileo. He reasoned that if an object falls for, say, twice the time, it will fall with *twice the average speed.* Since it falls for *twice* the time at *twice* the average speed, it will fall *four* times as far. Similarly, if an object falls for *three* times the time, it will have an average speed *three* times as great and will fall *nine* times as far. Galileo reasoned that the total distance fallen should be proportional to the *square* of the time.

In the case of objects in free fall, it is customary to use the letter g to represent the acceleration instead of the letter a (g because acceleration is due to *gravity*). While the value of g varies slightly in different parts of the world, it is approximately equal to 9.8 m/s^2 (32 ft/s^2). If we use g for the acceleration of a freely falling object (negligible air resistance), the equations for falling objects starting from a rest position become

$$v = gt$$
$$d = \tfrac{1}{2} gt^2$$

Questions ▶

1. An auto starting from rest has a constant acceleration of 4 m/s^2. How far will it go in 5 s?

2. How far will an object released from rest fall in 1 s? In this case the acceleration is $g = 9.8$ m/s^2.

3. If it takes 4 s for an object to freely fall to the water when released from the Golden Gate Bridge, how high is the bridge?

Much of the difficulty in learning physics, like learning any discipline, has to do with learning the language—the many terms and definitions. Speed is somewhat different from velocity, and acceleration is vastly different from speed or velocity. Mass and weight are related but are different from each other. Similarly for work, heat, and temperature. Please be patient with yourself as you find learning the similarities and the differences among physics concepts is not an easy task.

▶ **Answers**

1. Distance $= \tfrac{1}{2} \times 4 \times 5^2 = 50$ m
2. Distance $= \tfrac{1}{2} \times 9.8 \times 1^2 = 4.9$ m
3. Distance $= \tfrac{1}{2} \times 9.8 \times 4^2 = 78.4$ m

Notice that the units of measurement when multiplied give the proper units of meters for distance:

$$d = \tfrac{1}{2} \times 9.8 \text{ m/s}^2 \times 16 \text{ s}^2 = 78.4 \text{ m}$$

Appendix III More About Vectors

Vectors and Scalars

DIRECTION

SIZE

Figure III-1

A *vector* quantity is a directed quantity—one that must be specified not only by magnitude (size) but by direction as well. Recall from Chapter 3 that velocity is a vector quantity. Other examples are force, acceleration, and momentum. In contrast, a *scalar* quantity can be specified by magnitude alone. Some examples of scalar quantities are speed, time, temperature, and energy.

Vector quantities may be represented by arrows. The length of the arrow tells you the magnitude of the vector quantity, and the arrowhead tells you the direction of the vector quantity. Such an arrow drawn to scale and pointing appropriately is called a *vector.*

Adding Vectors

Vectors that add together are called *component* vectors. The sum of component vectors is called a *resultant*.

To add two vectors, make a parallelogram with two component vectors acting as two of the adjacent sides (Figure III-2). (Here our parallelogram is a rectangle.) Then draw a diagonal from the origin of the vector pair; this is the resultant (Figure III-3).

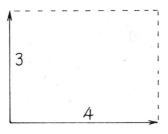

Figure III-2

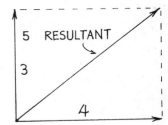

Figure III-3

Caution: Do not try to mix vectors! We cannot add apples and oranges, so velocity vector combines only with velocity vector, force vector combines only with force vector, and acceleration vector combines only with acceleration vector—each on its own vector diagram. If you ever show different kind of vectors on the same diagram, use different colors or some other method of distinguishing the different kinds of vectors.

Finding Components of Vectors

Recall from Chapter 3 that to find a pair of perpendicular components for a vector, first draw a dotted line through the tail end of the vector (in the direction of one of the desired components). Second, draw another dotted line through the tail end of the vector at right angles to the first dotted line. Third, make a rectangle whose diagonal is the given vector. Draw in the two components. Here we let **F** stand for "total force," **U** stand for "upward force," and **S** stand for "sideways force."

Figure III-4

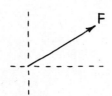

Figure III-5

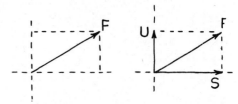

Figure III-6

Examples

1. A man pushing a lawnmower applies a force that pushes the machine forward and also against the ground. In Figure III-7, **F** represents the force applied by the man. We can separate this force into two components. The vector **D** represents the downward component, and **S** is the sideways component, the force that moves the lawnmower forward. If we know the magnitude and direction of the vector **F**, we can estimate the magnitude of the components from the vector diagram.

Figure III-7

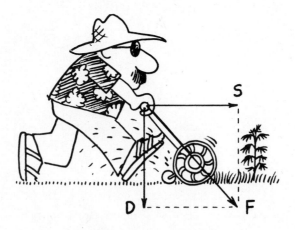

2. Would it be easier to push or pull a wheelbarrow over a step? Figure III-8 shows a vector diagram for each case. When you push a wheelbarrow, part of the force is directed downward, which makes it harder to get over the step. When you pull, however, part of the pulling force

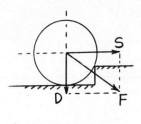

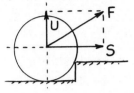

Figure III-8

Figure III-9

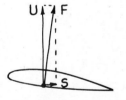

is directed upward, which helps to lift the wheel over the step. Note that the vector diagram suggests that pushing the wheelbarrow may not get it over the step at all. Do you see that the height of the step, the radius of the wheel, and the angle of the applied force determine whether the wheelbarrow can be pushed over the step? We see how vectors help us analyze a situation so that we can see just what the problem is!

3. If we consider the components of the weight of an object rolling down an incline, we can see why its speed depends on the angle. Note that the steeper the incline, the greater the component **S** becomes and the faster the object rolls. When the incline is vertical, **S** becomes equal to the weight, and the object attains maximum acceleration, 9.8 meters per second squared.

   There are two more force vectors that are not shown: the normal force **N**, which is equal and oppositely directed to **D**, and the friction force **f**, acting at the barrel-plane contact.

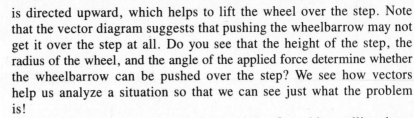

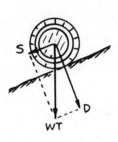

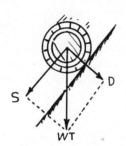

Figure III-10

4. When moving air strikes the underside of an airplane wing, the force of air impact against the wing may be represented by a single vector perpendicular to the plane of the wing (Figure III-10). We represent the force vector as acting midway along the lower wing surface, where the dot is, and pointing above the wing to show the direction of the resulting wind impact force. This force can be broken up into two components, one sideways and the other up. The upward component, **U**, is called *lift*. The sideways component, **S**, is called *drag*. If the aircraft is to fly at constant velocity at constant altitude, then lift must equal the weight of the aircraft and the thrust of the plane's engines must equal drag. The magnitude of lift (and drag) can be altered by changing the speed of the airplane or by changing the angle (called *angle of attack*) between the wing and the horizontal.

5. Consider the satellite moving clockwise in Figure III-11. Everywhere in its orbital path, gravitational force *F* pulls it toward the center of the host planet. At position A we see *F* separated into two components: *f*, which is tangent to the path of the projectile, and *f'*, which is perpendicular to the path. The relative magnitudes of these components in comparison to the magnitude of *F* can be seen in the imaginary

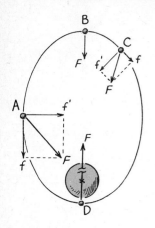

Figure III-11

rectangle they compose; *f* and *f'* are the sides, and *F* is the diagonal. We see that component *f* is along the orbital path but against the direction of motion of the satellite. This force component reduces the speed of the satellite. The other component, *f'*, changes the direction of the satellite's motion and pulls it away from its tendency to go in a straight line. So the path of the satellite curves. The satellite loses speed until it reaches position B. At this farthest point from the planet (apogee), the gravitational force is somewhat weaker but perpendicular to the satellite's motion, and component *f* has reduced to zero. Component *f'*, on the other hand, has increased and is now fully merged to become *F*. Speed at this point is not enough for circular orbit, and the satellite begins to fall toward the planet. It picks up speed because the component *f* reappears and is in the direction of motion as shown in position C. The satellite picks up speed until it whips around to position D (perigee), where once again the direction of motion is perpendicular to the gravitational force, *f'* blends to full *F*, and *f* is nonexistent. The speed is in excess of that needed for circular orbit at this distance, and it overshoots to repeat the cycle. Its loss in speed in going from D to B equals its gain in speed from B to D. Kepler discovered that planetary paths are elliptical, but never knew why. Do you?

6. Refer to the Polaroids held by Dotty Jean back in Chapter 28, in Figure 28-34. In the first picture (*a*), we see that light is transmitted through the pair of Polaroids because their axes are aligned. The emerging light can be represented as a vector aligned with the polarization axes of the Polaroids. When the Polaroids are crossed (*b*), no light emerges because light passing through the first Polaroid is perpendicular to the polarization axes of the second Polaroid, with no components along its axis. In the third picture (*c*), we see that light is transmitted when a third Polaroid is sandwiched at an angle between the crossed Polaroids. The explanation for this is shown in Figure III-12.

Figure III-12

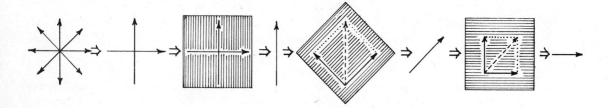

Sailboats

Sailors have always known that a sailboat can sail downwind, in the direction of the wind. Sailors have not always known, however, that a sailboat can sail upwind, against the wind. One reason for this has to do with a feature that is common only to recent sailboats—a finlike keel that extends deep beneath the bottom of the boat to ensure that the boat will knife through the water only in a forward (or backward) direction. Without a keel, a sailboat could be blown sideways.

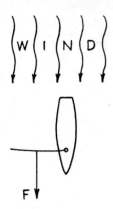

Figure III-13

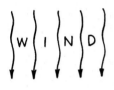

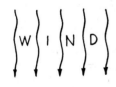

Figure III-14

Figure III-15

Figure III-13 shows a sailboat sailing directly downwind. The force of wind impact against the sail accelerates the boat. Even if the drag of the water and all other resistance forces are negligible, the maximum speed of the boat is the wind speed. This is because the wind will not make impact against the sail if the boat is moving as fast as the wind. The sail would simply sag. If there is no force, then there is no acceleration. The force vector in Figure III-13 *decreases* as the boat travels faster. The force vector is maximum when the boat is at rest and the full impact of the wind fills the sail, and is minimum when the boat travels as fast as the wind. If the boat is somehow propelled to a speed faster than the wind (by way of a motor, for example), then air resistance against the front side of the sail will produce an oppositely directed force vector. This will slow the boat down. Hence, the boat when driven only by the wind cannot exceed wind speed.

If the sail is oriented at an angle, as shown in Figure III-14, the boat will move forward, but with less acceleration. There are two reasons for this:

1. The force on the sail is less because the sail does not intercept as much wind in this angular position.
2. The direction of the wind impact force on the sail is not in the direction of the boat's motion, but is perpendicular to the surface of the sail. Generally speaking, whenever any fluid (liquid or gas) interacts with a smooth surface, the force of interaction is perpendicular to the smooth surface.* The boat does not move in the same direction as the perpendicular force on the sail, but is constrained to move in a forward (or backward) direction by its keel.

We can better understand the motion of the boat by resolving the force of wind impact, *F*, into perpendicular components. The important component is that which is parallel to the keel, which we label *K*, and the other component is perpendicular to the keel, which we label *T*. It is the component *K*, as shown in Figure III-15, that is responsible for the forward motion of the boat. Component *T* is a useless force that tends to tip the boat over and move it sideways. This component force is offset by the deep keel. Again, maximum speed of the boat can be no greater than wind speed.

---

*You can do a simple exercise to see that this is so. Try bouncing one coin off another on a smooth surface, as shown. Note that the struck coin moves at right angles (perpendicular) to the contact edge. Note also that it makes no difference whether the projected coin moves along path A or path B. See your instructor for a more rigorous explanation, which involves momentum conservation.

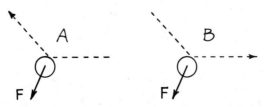

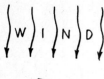

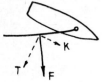

Figure III-16

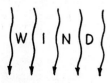

Figure III-17

Many sailboats sailing in directions other than exactly downwind (Figure III-16) with their sails properly oriented can exceed wind speed. In the case of a sailboat cutting across the wind, the wind may continue to make impact with the sail even after the boat exceeds wind speed. A surfer, in a similar way, exceeds the velocity of the propelling wave by angling his surfboard across the wave. Greater angles to the propelling medium (wind for the boat, water wave for the surfboard) result in greater speeds. A sailcraft can sail faster cutting across the wind than it can sailing downwind.

As strange as it may seem, maximum speed for most sailcraft is attained by cutting into (against) the wind, that is, by angling the sailcraft in a direction upwind! Although a sailboat cannot sail directly upwind, it can reach a destination upwind by angling back and forth in a zigzag fashion. This is called *tacking*. Suppose the boat and sail are as shown in Figure III-17. Component K will push the boat along in a forward direction, angling into the wind. In the position shown, the boat can sail faster than the speed of the wind. This is because as the boat travels faster, the impact of wind is increased. This is similar to running in a rain that comes down at an angle. When you run into the direction of the downpour, the drops strike you harder and more frequently; but when you run away from the direction of the downpour, the drops don't strike you as hard or as frequently. In the same way, a boat sailing upwind experiences greater wind impact force, while a boat sailing downwind experiences a decreased wind impact force. In any case the boat reaches its terminal speed when opposing forces cancel the force of wind impact. The opposing forces consist mainly of water resistance against the hull of the boat. The hulls of racing boats are shaped to minimize this resistive force, which is the principal deterrent to high speeds.

Iceboats (sailcraft equipped with runners for traveling on ice) encounter no water resistance and can travel at several times the speed of the wind when they tack upwind. Although ice friction is nearly absent, an iceboat does not accelerate without limits. The terminal velocity of a sailcraft is determined not only by opposing friction forces but also by the change in relative wind direction. When the boat's orientation and speed are such that the wind seems to shift in direction, so the wind moves parallel to the sail rather than into it, forward acceleration ceases—at least in the case of a flat sail. In practice, sails are curved and produce an airfoil that is as important to sailcraft as it is to aircraft. The effects are discussed in Chapter 13.

Appendix IV Exponential Growth and Doubling Time*

One of the most important things we seem not to perceive is the process of exponential growth. We think we understand how compound interest works, but we can't get it through our heads that a fine piece of tissue paper folded upon itself fifty times (if that were possible) would be more than 20 million kilometers thick. If we could, we could "see" why our income buys only half what it did 4 years ago, why the price of everything has doubled in the same time, why populations and pollution and nuclear bombs seem to (in fact, really do) proliferate out of control.†

When a quantity such as money in the bank, population, or the rate of consumption of a resource steadily grows at a fixed percent per year, we say the growth is exponential. Money in the bank may grow at 8 percent per year; electric power generating capacity in the United States grew at about 7 percent per year for the first three-quarters of the century. The important thing about exponential growth is that the time required for the growing quantity to increase its size by a fixed fraction is constant. So the time required for the quantity to double in size (increase by 100 percent) is also constant. For example, if the population of a growing city takes 12 years to double from 10 000 to 20 000 inhabitants and its growth remains steady, in the next 12 years the population will double to 40 000, in the next 12 years to 80 000, and so on.

There is an important relationship between the percent growth rate and its *doubling time*, the time it takes to double a quantity:‡

$$\text{Doubling time} = \frac{69.3}{\text{percent growth per unit time}} \approx \frac{70}{\%}$$

So to estimate the doubling time for a steadily growing quantity, we simply divide the number 70 by the percentage growth rate. For example, the 7 percent growth rate of electric power generating capacity in the

---

*This appendix is drawn from material by University of Colorado physics professor Albert A. Bartlett, who strongly asserts, "The greatest shortcoming of the human race is man's inability to understand the exponential function." See Professor Bartlett's still timely article, "Forgotten Fundamentals in the Energy Crisis" (*American Journal of Physics*, September 1978) or his revised version (*Journal of Geological Education*, January 1980).

†K. C. Cole, *Sympathetic Vibrations* (New York: Morrow, 1984).

‡For exponential decay we speak about half-life, the time required for a quantity to reduce to half its value. This case is treated in Chapter 32.

United States means that in the past the capacity has doubled every 10 years (70%/[7%/year] = 10 years). A 2 percent growth rate for world population means the population of the world doubles every 35 years (70%/[2%/year] = 35 years). A city planning commission that accepts what seems like a modest 3.5 percent growth rate may not realize that this means that doubling will occur in 70/3.5 or 20 years; that's double capacity for such things as water supply, sewage treatment plants, and other municipal services every 20 years.

Figure IV-1

An exponential curve. Notice that each of the successive equal time intervals noted on the horizontal scale corresponds to a doubling of the quantity indicated on the vertical scale. Such an interval is called the doubling time.

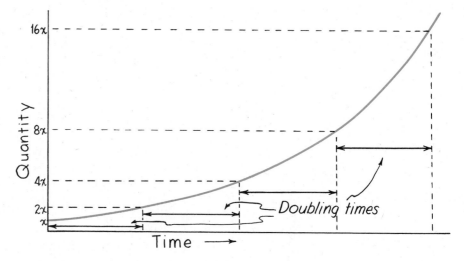

What happens when you put steady growth in a finite environment? Consider the growth of bacteria that grow by division, so that one bacterium becomes two, the two divide to become four, the four divide to become eight, and so on. Suppose the division time for a certain strain of bacteria is 1 minute. This is then steady growth—the number of bacteria grows exponentially with a doubling time of 1 minute. Further, suppose that one bacterium is put in a bottle at 11:00 AM and that growth continues steadily until the bottle becomes full of bacteria at 12:00 noon. Consider seriously the following question.

Question ▶ When was the bottle half-full?

It is startling to note that at 2 minutes before noon the bottle was only $\frac{1}{4}$ full. Table IV-1 summarizes the amount of space left in the bottle in the last few minutes before noon. If you were an average bacterium in the bottle, at which time would you first realize that you were running out of space? For example, would you sense there was a serious problem

▶ **Answer** 11:59 AM; the bacteria will double in number every minute!

at 11:55 AM when the bottle was only 3 percent filled ($\frac{1}{32}$) and had 97 percent of open space (just yearning for development)? The point here is that there isn't much time between the moment that the effects of growth become noticeable and the time when they become overwhelming.

Suppose that at 11:58 AM some farsighted bacteria see that they are running out of space and launch a full-scale search for new bottles. Luckily, at 11:59 AM they discover three new empty bottles, three times as much space as they had ever known. This quadruples the total resource space ever known to the bacteria, for they now have a total of four bottles, whereas before the discovery they had only one. Further suppose that thanks to their technological proficiency, they are able to migrate to their new habitats without difficulty. Surely, it seems to most of the bacteria that their problem is solved—and just in time.

Figure IV-2

Table IV-1
The last minutes
in the bottle

Time	Part full (%)	Part empty
11:54 AM	$\frac{1}{64}$ (1.5%)	$\frac{63}{64}$
11:55 AM	$\frac{1}{32}$ (3%)	$\frac{31}{32}$
11:56 AM	$\frac{1}{16}$ (6%)	$\frac{15}{16}$
11:57 AM	$\frac{1}{8}$ (12%)	$\frac{7}{8}$
11:58 AM	$\frac{1}{4}$ (25%)	$\frac{3}{4}$
11:59 AM	$\frac{1}{2}$ (50%)	$\frac{1}{2}$
12:00 noon	Full (100%)	None

Question ▶ If the bacteria growth continues at the unchanged rate, what time will it be when the three new bottles are filled to capacity?

We see from Table IV-2 that quadrupling the resource extends the life of the resource by only two doubling times. In our example the resource is space—but it could as well be coal, oil, uranium, or any nonrenewable resource.

Table IV-2
Effects of the discovery
of three new bottles

Time	Effect
11:58 AM	Bottle 1 is $\frac{1}{4}$ full
11:59 AM	Bottle 1 is $\frac{1}{2}$ full
12:00 noon	Bottle 1 is full
12:01 PM	Bottles 1 and 2 are both full
12:02 PM	Bottles 1, 2, 3, and 4 are all full

▶ **Answer** 12:02 PM!

Continued growth and continued doubling lead to enormous numbers. In two doubling times, a quantity will double twice ($2^2 = 4$; quadruple) in size; in three doubling times, its size will increase eightfold ($2^3 = 8$); in four doubling times, it will increase sixteenfold ($2^4 = 16$); and so on.

This is best illustrated by the story of the court mathematician in India who years ago invented the game of chess for his king. The king was so pleased with the game that he offered to repay the mathematician, whose request seemed modest enough. The mathematician requested a single grain of wheat on the first square of the chessboard, two grains on the second square, four on the third square, and so on, doubling the number of grains on each succeeding square until all squares had been used. At this rate there would be 2^{63} grains of wheat on the sixty-fourth square. The king soon saw that he could not fill this "modest" request, which amounted to more wheat than had been harvested in the entire history of the earth!

Figure IV-3

A single grain of wheat placed on the first square of the chessboard is doubled on the second square, this number is doubled on the third square, and so on, presumably for all 64 squares. Note that each square contains one more grain than all the preceding squares combined. Does enough wheat exist in the world to fill all 64 squares in this manner?

It is interesting and important to note that the number of grains on any square is one grain more than the total of all grains on the preceding squares. This is true anywhere on the board. Note from Table IV-3 that when eight grains are placed on the fourth square, the eight is one more than the total of seven grains that were already on the board. Or the thirty-two grains placed on the sixth square is one more than the total of thirty-one grains that were already on the board. We see that in one doubling time we use more than all that had been used in all the preceding growth!

So if we speak of doubling energy consumption in the next however many years, bear in mind that this means in these years we will consume more energy than has heretofore been consumed during the entire preceding period of steady growth. And if power generation continues to use predominantly fossil fuels, then except for some improvements in efficiency, we would burn up in the next doubling time a greater amount of coal, oil, and natural gas than has already been consumed by previous power generation; and except for improvements in pollution control, we can expect to discharge even more toxic wastes into the environment than the millions upon millions of tons already discharged over all the previous years of industrial civilization; and we would expect more human-made calories of heat to be absorbed by the earth's ecosystem than have been absorbed in the entire past! At the previous 7 percent annual growth rate in energy production, all this would occur in one doubling time of a single decade. If over the coming years the annual growth rate remains at half this value, 3.5 percent, then all this would take place in a doubling time of 2 decades. Clearly this cannot continue!

Table IV-3
Filling the squares on the chessboard

Square number	Grains on square	Total grains thus far
1	1	1
2	2	3
3	4	7
4	8	15
5	16	31
6	32	63
7	64	127
.	.	.
.	.	.
.	.	.
64	2^{63}	$2^{64} - 1$

The consumption of a nonrenewable resource cannot grow exponentially for an indefinite period, because the resource is finite and its supply finally expires. The most drastic way this could happen is shown in the graph in Figure IV-4a, where the rate of consumption, such as barrels of oil per year, is plotted against time, say in years. In such a graph the colored area under the curve represents the supply of the resource. We see that when the supply is exhausted, the consumption ceases altogether. This sudden change is rarely the case, for the rate of extracting the supply falls as it becomes more scarce. This is shown in Figure IV-4b. Note that the area under the curve is equal to the area under the curve in a. Why? Because the total supply is the same in both cases. The principal difference is in the time taken to finally extinguish the supply. History shows that the rate of production of a nonrenewable resource rises and falls in a nearly symmetric manner, as shown by the curve in c. The time during which production rates rise is approximately equal to the time during which these rates fall to zero or near zero. If we fit the data for U.S. oil production in the lower forty-eight states to such a curve, we find that we are just to the right of the peak. This suggests that one-half of the recoverable petroleum that was ever in the ground in the U.S. has

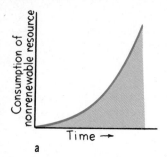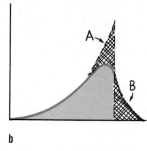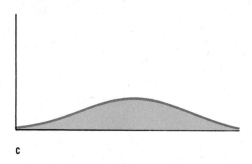

a   b   c

Figure IV-4

(*a*) If the exponential rate of consumption for a nonrenewable resource continues until it is depleted, consumption falls abruptly to zero. The shaded area under this curve represents the total supply of the resource. (*b*) In practice the rate of consumption levels off and then falls less abruptly to zero. Note that the crosshatched area A is equal to the crosshatched area B. Why? (*c*) At lower consumption rates, the same resource lasts a longer time.

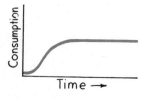

Figure IV-5

A curve showing the rate of consumption of a renewable resource such as agricultural or forest products, where a steady rate of production and consumption can be maintained for a long period, providing this production is not dependent upon the use of a nonrenewable resource that is waning in supply.

already been used and that in the future the domestic petroleum rate of production can only decrease. The U.S. production curve peaked in 1970, and by 1979 nearly half the U.S. consumption was imported. Each year we consume more oil than during the previous year.

Production rates for all nonrenewable resources decrease sooner or later. Only production rates for renewable resources, such as agriculture or forest products, can be maintained at steady levels for long periods of time (Figure IV-5), provided such production does not depend on waning nonrenewable resources such as petroleum. Much of today's agriculture is so petroleum-dependent that it can be said that modern agriculture is simply the process whereby land is used to convert petroleum into food. The implications of petroleum scarcity go far beyond rationing of gasoline for cars or fuel oils for home heating.

Power production from whatever sources will not meet the present increasing demands in this growing world. Fusion power may characterize the next century, but only the most optimistic forecasters see it as providing even a tiny fraction of our power needs during the first half of the twenty-first century. Even though the production of nuclear fission power had spectacular growth after its harnessing in 1942, it took 30 years of development to equal the annual energy consumption of firewood in the United States. Nuclear fusion will be vastly more complicated than fission, so even if nuclear fusion were successfully harnessed today, how long would it take before it could play a significant role in major power production?

However, the important questions are more basic: Is growth really good? Is bigger really better? Is it true that if we don't grow, we will stagnate? In answering these questions, bear in mind that human growth is an early phase of life that continues normally through adolescence. Physical growth stops when physical maturity is reached. What do we say of growth that continues in the period of physical maturity? We say that such growth is obesity—or, worse, cancer.

Questions to Ponder

1. According to a French riddle, a lily pond starts with a single leaf. Each day the number of leaves doubles, until the pond is completely covered by leaves on the 30th day. On what day was the pond half covered? One-quarter covered?

2. In an economy that has a steady inflation rate of 7 percent per year, in how many years does a dollar lose half its value?

3. At a steady inflation rate of 7 percent, what will be the price every 10 years for the next 50 years for a theater ticket that now costs $10? For a suit of clothes that now costs $100? For a car that now costs $10 000? For a home that now costs $100 000?

4. If the sewage treatment plant of a city is just adequate for the city's current population, how many sewage treatment plants will be necessary 42 years later if the city grows steadily at 5 percent annually?

5. In 1986 the population growth rate for the United States was 0.6 percent, for Mexico 2.6 percent, and for Kenya (the highest growth rate in the world) 4.1 percent (taking into account births, deaths, and immigration). At these rates, how long would it take for the population in each of these countries to double?

6. If world population doubles in 40 years and world food production also doubles in 40 years, how many people then will be starving each year compared to now?

7. A continued world population growth rate of 1.9 percent per year would produce a density of one person per square meter in 550 years. True or false: World population growth rate will sooner or later be zero.

8. Suppose you get a prospective employer to agree to hire your services for wages of a single penny for the first day, 2 pennies the second day, and double each day thereafter providing the employer keep to the agreement for a month. What will be your total wages for the month?

9. In the preceding exercise, how will your wages for only the 30th day compare to your total wages for the previous 29 days?

10. We hear often that reserves of nonrenewable resources such as coal, oil, and natural gas are "scarce," "abundant," or "superabundant." Why are these terms meaningless without referring also to their consumption rates?

11. Oil has been produced in the United States for about 125 years. If there remains undiscovered in the country as much oil as all that has been used, what is wrong with the argument that the remaining oil would be sufficient for another 125 years?

12. Present estimates are that one-eighth of the oil in the world has already been consumed. With respect to the example of the multiplying bacteria discussed in this appendix, how many "minutes are there until noon"?

13. How would your answer to the preceding exercise be different if new oil deposits were discovered that were equal in size to all those ever known?

14. Coal is relatively "abundant" in the United States today only because the growth in annual production of coal was zero from 1910 to the mid-1970s. In the previous half-century prior to 1910, coal production grew at a steady rate of almost 7 percent per year. Had this rate continued, the expiration of United States coal reserves would occur between the years 1965 and 1990, depending on low and high estimates of reserve sizes. Why does it violate good sense to talk of "abundant reserves" and continued growth at the same time?

15. When dealing with steady growth, is it necessary to have an accurate estimate of the size of a resource in order to make a reliable estimate of how long the resource will last? (Use Figure IV-4 to explain your answer.)

16. If fusion power were harnessed today, the abundant energy resulting would probably sustain and even further encourage our present appetite for continued growth and in a relatively few doubling times produce an appreciable fraction of the solar power input to the earth. Make an argument that the current delay in harnessing fusion is a blessing for the human race.

Glossary

Aberration The distortion in an image produced by a lens or system of lenses.

Absolute temperature scale A temperature scale that has its zero point at $-273.16°C$. Temperatures in the absolute scale are designated in kelvins (K).

Absolute zero The lowest possible temperature that any substance may have; the temperature at which the molecules of a substance have their minimum kinetic energy. Absolute zero is 0K; 0°C is 273.16K.

Absorption spectrum A continuous spectrum, like that of white light, interrupted by dark lines or bands that result from the absorption of certain frequencies of light by a substance through which the light passes.

Acceleration The rate at which an object's velocity changes with time; the change in velocity may be in magnitude (speed) or direction or both.

Acceleration due to gravity (g) The acceleration of a freely falling object. Its value near the earth's surface is about 9.8 meters per second each second.

Additive primary colors The three colors—red, blue, and green—that when added together in various proportions will produce any color in the spectrum.

Adhesion Molecular attraction between two surfaces making contact.

Adiabatic process A change in gas volume with no heat entering or leaving the system.

Air resistance The friction that acts on something moving through air.

Alloy A substance composed of two or more metals or of a metal and a nonmetal.

Alpha particle Nucleus of a helium atom, composed of two protons and two neutrons, ejected by radioactivity.

Alpha ray A stream of alpha particles ejected by certain radioactive nuclei.

Alternating current (ac) Electric current that rapidly reverses in direction. The electric charges vibrate about relatively fixed positions, usually at the rate of 60 hertz.

Ampere The unit of electric current, the flow of 1 coulomb of charge per second.

Amplitude For a wave or vibration, the maximum displacement on either side of the equilibrium (midpoint) position.

Amplitude modulation (AM) A type of *modulation* in which the amplitude of the carrier wave is varied above and below its normal value by an amount proportional to the amplitude of the impressed signal.

Aneroid barometer An instrument used to measure atmospheric pressure; based on the movement of the lid of a metal box, rather than on the movement of a liquid.

Angstrom An outdated unit of length equal to 10^{-10} meter. Atoms have a radius of 1 to 2 angstroms.

Angular momentum The product of an object's rotational inertia and angular velocity about a particular axis. For an object small compared to the radial distance, it is the product of its mass, linear speed, and radius of curvature.

Antimatter Matter composed of atoms with negative nuclei and positive electrons.

Antiparticle Particle having the same mass as a normal particle but a charge of the opposite sign. The antiparticle of the electron is the positron.

Apogee The point in an elliptical orbit farthest from the focus around which orbiting takes place.

Archimedes' principle An immersed object, submerged or floating, is buoyed up by a force equal to the weight of fluid displaced.

Archimedes' principle for air An object surrounded by the atmosphere is buoyed upward by a force equal to the weight of displaced air.

Astigmatism A defect of the eye caused when the cornea is curved more in one direction than another.

Atmospheric pressure   At the surface of the earth, the weight of air above that presses on surfaces; 1.01×10^5 newtons per square meter, or 101 kPa (14.7 pounds per square inch); used as a gas pressure unit.

Atom   The smallest particle of an element that has all the element's chemical properties, composed of a nucleus and a number of surrounding electrons.

Atomic mass number   The number associated with an atom that is the same as the number of nucleons in its nucleus.

Atomic mass unit (amu)   The standard unit of atomic mass, which is equal to one-twelfth the mass of an atom of carbon-12, arbitrarily given the value of exactly 12.

Atomic nucleus   The core of an atom, consisting of two basic nucleons, protons and neutrons. The protons have positive electric charge, giving the nucleus a positive electric charge; the neutrons have no electric charge.

Atomic number   A number designating an atom, which is the same as the number of protons in the nucleus or the same as the number of extranuclear electrons in a neutral atom.

Avogadro's principle   Equal volumes of all gases at the same temperature and pressure contain the same number of molecules, 6.02×10^{23} in one mole (a mass in grams equal to the molecular mass of the substance in atomic mass units).

Axis   The straight line about which rotation takes place.

Barometer   A device that measures atmospheric pressure.

Beats   A sequence of alternating reinforcement and cancellation of two sets of superimposed waves differing in frequency, heard as a throbbing sound.

Bernoulli's principle   Pressure of a fluid on a surface decreases as the fluid's velocity relative to the surface increases.

Beta particle   An electron (or a positron) emitted during the radioactive decay of an atomic nucleus.

Beta ray   A stream of beta particles.

Big Bang   The primordial explosion thought to have resulted in an expanding universe.

Bimetallic strip   Two strips of different metals welded or riveted together; used in thermostats.

Black body radiation   Radiation emitted by a perfect emitter of radiation (a black body); an ideal black body absorbs all the radiation it receives, and hence appears black at low temperatures.

Black hole   The configuration of a massive star that has undergone gravitational collapse, in which gravitation is so strong that even the star's own light cannot escape.

Boiling   Change from liquid to gas occurring beneath the surface of a liquid; rapid vaporization. The liquid loses energy, the gas gains it.

Bow wave   The V-shaped wave made by an object moving across a liquid surface at a speed greater than the wave speed.

Boyle's law   The product of pressure and volume is a constant for a given mass of confined gas regardless of changes in either pressure or volume individually, so long as temperature remains unchanged:

$$P_1 V_1 = P_2 V_2$$

Breeder reactor   Nuclear fission reactor producing power and more fuel than it consumes by transmuting nonfissionable isotopes into fissionable isotopes.

Brownian motion   Random movement of very small particles suspended in a fluid that results from collisions with molecules.

Buoyancy   The apparent loss of weight of an object due to buoyant force when it is immersed in a fluid.

Buoyant force   The net upward force that a fluid exerts on an immersed object, floating or submerged, due to the weight of displaced fluid, independent of the object's weight.

Calorie   Energy required to raise the temperature of one gram of water one degree Celsius; 4.186 joules. 1000 calories, c, equal one Calorie, C.

Capillarity   The rising or sinking of a liquid into a small vertical space due to adhesion between the liquid and the material and to surface tension of the liquid.

Carbon dating   The process of determining the time that has elapsed since death by measuring the radioactivity of remaining carbon-14 isotopes.

Carrier wave   A high-frequency radio wave modified by a lower-frequency wave.

Celsius temperature   Temperature on a scale that assigns the value 0°C to the freezing point of water and the value of 100°C to the boiling point of water at standard pressure.

Center of gravity   The average position of weight or the point associated with an object where gravity force is considered to act.

Center of mass   The average position of mass or the point associated with an object where its mass can be considered to be concentrated.

Centrifugal force   An outward force caused by rotation.

Centripetal force   A center-directed force that causes an object to follow a curved or circular path.

Chain reaction A self-sustaining reaction that, once started, steadily provides the energy and matter necessary to continue the reaction.

Charge polarization The spatial separation of positive and negative charges by the electrical alignment of molecules.

Charging by contact Transfer of charge between objects by rubbing or by simple touching.

Charging by induction Redistribution of charges in and on objects caused by the electrical influence of a charged object close by but not in contact.

Chemical reaction A process in which the rearrangement and exchange of atoms from one molecule to another occur.

Complementarity The principle that the wave and particle models of matter and radiation complement each other.

Complementary colors Any two colors that when added produce white light.

Components The parts into which a vector can be separated and that act in different directions from the vector.

Compression The region of increased pressure in a longitudinal wave.

Condensation The change of state from vapor to liquid; the opposite of evaporation. Warming of the liquid results.

Conduction, electrical The easy flow of electric charge through a material subjected to an impressed electrical force.

Conduction, heat The transfer and distribution of energy from molecule to molecule in an object by means of electron and molecular collisions.

Conductor Material that conducts heat or electricity easily.

Conservation Principle that the amount of a quantity such as energy, mass, momentum, or electrical charge within a system remains constant when the system otherwise undergoes changes.

Convection Transfer of energy in a fluid by means of currents in the heated fluid.

Converging lens A lens that is thicker in the middle than at the edges and that refracts parallel rays passing through it to a focus.

Cornea The transparent covering over the eye.

Correspondence principle A new theory is valid provided that when it overlaps with the old theory, it conforms to the verified results of the old theory.

Cosmic ray One of various high-speed particles that travel throughout the universe and originate in violent events in stars.

Cosmology Study of the origin and development of the entire universe.

Coulomb The SI unit of electrical charge; the charge of 6.25×10^{18} electrons.

Coulomb's law The relationship among electric force, charge, and distance:

$$F = k\, q_1 q_2 / d^2$$

If the charges are alike in sign, the force is repulsive; if the charges are unlike, the force is attractive.

Critical angle The angle at which light no longer is refracted when it meets a surface, but is instead totally internally reflected.

Critical mass Minimum mass of fissionable material in a reactor or nuclear bomb that will sustain a chain reaction.

Current Fluid flow (gas or liquid); also electron flow (electric). Alternating current (ac) rapidly and repeatedly reverses direction; direct current (dc) flows in one direction.

de Broglie matter waves All particles have wave properties; in de Broglie's equation, the product of momentum and wavelength equals Planck's constant.

Density Mass density is mass per volume; weight density is weight per volume; in general, any item per space element (e.g., number of dots per area).

Deuterium An isotope of hydrogen in which the nucleus consists of a single neutron in addition to the single proton.

Diffraction The bending of waves around a barrier, such as an obstacle or the edges of an opening.

Diffraction grating A series of closely spaced parallel slits used to separate colors of light by interference.

Dispersion The separation of light into colors arranged according to their frequency—for example, by interaction with a prism or diffraction grating.

Diverging lens A lens that is thinner in the middle than at the edges, causing parallel rays passing through it to diverge.

Doppler effect Change in frequency of sound or light due to relative motion of source and receiver.

Efficiency Ratio of result to effort, output to input, involving energy.

Elastic collision A collision in which no energy is transformed to heat.

Elasticity Property of a solid of returning to the original size and shape after deformation.

Elastic limit The distance of stretching or compressing beyond which an elastic material will not return to its original state.

Electric current The flow of electric charge that transports energy from one place to another. Measured in *amperes*, where 1 ampere is the flow of 6.25×10^{18} electrons (or protons) per second.

Electric field The energetic region of space surrounding a charged object. About a charged point, the field decreases with distance according to the inverse-square law. Between oppositely charged parallel plates the electric field is uniform. A charged object placed in the region of an electric field experiences a force.

Electric polarization The separation of charge in an object so that one part bears a positive charge and another part bears an equal negative charge.

Electric potential The electric potential energy per amount of charge, measured in *volts*, and often called *voltage*:

$$\text{Voltage} = \text{electrical energy/charge}$$

Electric power The rate of electrical energy transfer or the rate of doing work, which can be measured by the product of current and voltage:

$$\text{Power} = \text{current} \times \text{voltage}$$

Measured in *watts* (or *kilowatts*), where 1 ampere × 1 volt = 1 watt.

Electric resistance The property of a material that causes it to resist the flow of an electric current through it. Measured in *ohms*.

Electrodynamics The study of moving electric charge, as opposed to **electrostatics**.

Electromagnetic induction The induction of voltage when a magnetic field changes with time. If the magnetic field within a closed loop changes in any way, a voltage is induced in the loop:

$$\text{Voltage induced} \sim \frac{-\text{no. of loops} \times \text{mag. field change}}{\text{change in time}}$$

For the more general case of field induction, see **Faraday's law**.

Electromagnetic radiation The transfer of energy by the rapid oscillations of electromagnetic fields, which travel in the form of waves called *electromagnetic waves*.

Electromagnetic spectrum The range of frequencies over which electromagnetic radiation can be propagated. The lowest frequencies are associated with radio waves; microwaves have a higher frequency, and then infrared waves, light, ultraviolet radiation, X rays, and gamma rays in sequence.

Electromotive force (emf) Any force that gives rise to an electric current. A battery or a generator is a source of emf.

Electron The negative particles in the shell of an atom.

Electrostatics The study of electric charge at relative rest, as opposed to **electrodynamics**.

Element A substance composed of atoms that all have the same atomic number and therefore the same chemical properties.

Ellipse A closed curve of oval shape wherein the sum of the distances from any point on the curve to two internal focal points is a constant.

Emission spectrum A continuous or partial spectrum of wavelengths resulting from the characteristic frequencies of light from a luminous source.

Energy Anything that can change the condition of matter. Commonly defined circularly as the ability to do work; actually only describable (like pornography) by examples.

Entropy A measure of the degree of disorder in a substance or system.

Equilibrium The state of an object when not acted upon by a net force or net torque. An object in equilibrium may be at rest or moving at uniform velocity—that is, not accelerating.

Escape velocity The velocity that a projectile, space probe, etc., must reach to escape the gravitational influence of the earth or celestial body to which it is attracted.

Ether A hypothetical medium that was formerly thought to be required for the propagation of electromagnetic waves.

Evaporation The change of state at the surface of a liquid as it becomes vapor, resulting from the random motion of molecules that occasionally escape from the liquid surface; the opposite of condensation. Cooling of the liquid results.

Excitation The process of boosting one or more electrons in an atom or a molecule from a lower to a higher energy level. An atom in an excited state will usually decay (de-excite) rapidly into a lower state by the emission of radiation. The frequency and energy of emitted radiation are related by

$$E = hf$$

where h is Planck's constant.

Fact A close agreement by competent observers of a series of observations of the same phenomenon.

Fahrenheit temperature scale A temperature scale with the freezing point of water assigned the value 32°F and the boiling point of water 212°F.

Faraday's law   An electric field is induced in any region of space in which a magnetic field is changing with time. The magnitude of the induced electric field is proportional to the rate at which the magnetic field changes; the direction of the induced field is at right angles to the changing magnetic field.

Fermat's principle of least time   Light will take the path that requires the least time when it goes from one place to another.

Fluid   Anything that flows; in particular, a liquid or a gas.

Fluorescence   The property of absorbing radiation at one frequency followed by its re-emission at a lower frequency.

Focal length   Distance between the center of a mirror or lens and its focal point.

Focal point   Point at which light rays parallel to the axis of a mirror or lens come to a focus.

Focus   Point at which straight lines intersect.

Force   Any influence that can cause an object to be accelerated, measured in *newtons*.

Forced vibration   The setting up of vibrations in an object by a vibrating force.

Fourier analysis   A mathematical method that will resolve any periodic wave form into a sum of simple sine waves.

Frame of reference   A vantage point (usually a set of coordinate axes) with respect to which the position and motion of an object may be described.

Free fall   Motion of an object under the influence of gravitational pull only.

Freezing   The change of state from the liquid to the solid form; the opposite of *melting*. Energy is released by the substance undergoing freezing.

Frequency   For a vibrating object, the number of vibrations it makes per unit time. For a wave, the number of waves that pass a particular point per unit time.

Frequency modulation (FM)   A type of *modulation* in which the frequency of the carrier wave is varied above and below its normal frequency by an amount proportional to the amplitude of the impressed signal and in which the amplitude of the modulated carrier wave remains constant.

Friction   Forces resisting motion between one set of molecules and another due to electrical attraction and repulsion, usually between two solid surfaces; *static* before motion starts and *kinetic* during motion.

Fulcrum   The pivot point of a lever.

Galvanometer   An instrument used to measure electric current. With the proper combination of resistors, it can be converted to an *ammeter* or a *voltmeter*.

Gamma ray   High-frequency electromagnetic radiation emitted by the nuclei of radioactive atoms.

Gas   The state of matter beyond the liquid state, wherein molecules fill whatever space is available to them, taking no definite shape.

General theory of relativity   Einstein's generalization of special relativity, which features a geometric theory of gravitation.

Generator   A device that produces electric current by rotating a coil within a stationary magnetic field.

Geodesic   Shortest path between points on a surface.

Gravitation   Attraction between objects due to mass.

Gravitational field   The space surrounding a massive body in which another massive body experiences a force of attraction.

Gravitational potential energy   The stored energy that an object possesses by virtue of its elevated position in a gravitational field.

Gravitational red shift   The shift toward longer wavelength experienced by light leaving the surface of a massive object, as predicted by the general theory of relativity.

Gravitational wave   A gravitational disturbance that propagates through space-time, produced by a moving mass.

Greenhouse effect   The result of sunlight (high-frequency) energy that passes through the atmosphere, is absorbed, and then is radiated at a lower frequency that cannot escape through the atmosphere, thus increasing the temperature of the atmosphere.

Half-life   Time required for half the atoms in a radioactive element to decay; can be applied to any exponentially decreasing process.

Harmonics   Modes of vibrations that begin with a lowest or fundamental vibrating frequency and continue as a sequence of tones that are integral multiples of the fundamental frequency.

Heat   The energy that flows from one object to another by virtue of a difference in temperature. Measured in *calories* or *joules*.

Hertz   Unit of frequency; one cycle (complete vibration) per second.

Holography   The process in which three-dimensional optical images are produced by illuminating a microscopic interference pattern (hologram) with monochromatic light.

Hooke's law The extension x of an elastic object is directly proportional to the stretching force F that is applied:

$$F \sim x, \text{ or } F = -kx$$

Humidity Absolute humidity is the mass of water per volume of air. Relative humidity is absolute humidity at that temperature divided by the maximum possible, usually given as a percent.

Huygens' principle Light waves spreading out from a light source can be regarded as a superposition of tiny secondary wavelets.

Hypothesis An educated guess; a reasonable explanation of an observation or experimental result that is not fully accepted as factual.

Impulse The product of the force acting on an object and the time during which it acts, equal to the change in momentum that results.

Incandescence The state of an object glowing while at a high temperature, caused by electrons in vibrating atoms and molecules that are shaken in and out of their stable energy levels, emitting radiation in the process. The peak frequency of radiation is proportional to the absolute temperature of a heated substance:

$$\bar{f} \sim T$$

Inertia The fundamental property of inert material tending to resist changes in its state of motion.

Inertial frame of reference An unaccelerated vantage point in which Newton's laws hold exactly.

Insulator Any material through which charge strongly resists flow when subject to an impressed electrical force; a nonconductor.

Interference Superposition of waves, producing regions of reinforcement (constructive) and regions of cancellation (destructive).

Internal energy The total of all molecular energies, both kinetic energy and potential energy, internal to a substance. Changes in internal energy are of principal concern in thermodynamics.

Inverse-square law The intensity of an effect is related to the inverse square of the distance from the cause:

$$\text{Intensity} \sim \frac{1}{\text{distance}^2}$$

Gravity follows an inverse-square law, as do electric, magnetic, light, sound, and radiation phenomena.

Ion An electrically charged atom with either an excess or a deficiency of electrons compared to the number of protons in the nucleus.

Ionization The process of removing or adding electrons to or from the atomic nucleus.

Iris The colored part of the eye that surrounds the pupil.

Isotopes Atoms whose nuclei have the same number of protons but various numbers of neutrons.

Joule The SI unit of work or energy in any form.

Kelvin The SI unit of temperature.

Kepler's laws of planetary motion

Law 1: Each planet moves in an elliptical orbit with the sun at one focus.

Law 2: The line from the sun to any planet sweeps out equal areas of space in equal time intervals.

Law 3: The squares of the times of revolution (or years) (T) of the planets are proportional to the cubes of their average distances (R) from the sun. ($T^2 \sim R^3$ for all planets.)

Kilogram The standard unit of mass; 1 kilogram is the amount of mass that a force of 1 newton will accelerate 1 meter per second squared.

Kinetic energy Energy of motion, described by the relationship

$$\text{Kinetic energy} = \tfrac{1}{2}mv^2$$

Laser (**l**ight **a**mplification by **s**timulated **e**mission of **r**adiation) An optical instrument that produces a beam of coherent light.

Law A general hypothesis or statement about the relationship of natural quantities that has been tested over and over again and has not been contradicted. Also known as a *principle*.

Law of reflection The angle of incident radiation equals the angle of reflected radiation.

Law of universal gravitation Every object in the universe attracts every other object with a force that for two objects is proportional to the masses of the objects and inversely proportional to the square of the distance separating them:

$$F \sim m_1 m_2/d^2, \text{ or } F = Gm_1 m_2/d^2$$

Length contraction The apparent shrinking of an object moving at a speed that is a significant fraction of the speed of light.

Lens A piece of glass or other transparent material that can bring light to a focus.

Lever A simple machine, made of a bar that pivots around a fixed point.

Light The visible part of the electromagnetic spectrum.

Linear motion Motion along a straight-line path.

Liquid The state of matter between the solid and gaseous states in which the matter possesses a definite volume but no definite shape: it takes on the shape of its container.

Longitudinal wave A wave in which the individual particles of a medium vibrate back and forth in the direction in which the wave travels—for example, sound.

Lorentz contraction Contraction of an object in its direction of motion due to that motion; part of Einstein's special theory of relativity (see also **Length contraction**).

Loudness The physiological sensation directly related to sound intensity or volume. Relative loudness, or noise level, is measured in *decibels*.

Mach number The ratio of the speed of an object to the speed of sound. For example, an aircraft traveling at the speed of sound is rated Mach 1.0; traveling at twice the speed of sound, Mach 2.0.

Magnetic domain A clustered region of aligned magnetic atoms. When numerous regions themselves are aligned with each other, the substance containing them is a magnet.

Magnetic field The region of "altered space" that will interact with the magnetic properties of a magnet. It is located mainly between the opposite poles of a magnet or in the energetic space about an electric charge in motion.

Magnetic force Between magnets, it is the attraction of unlike magnetic poles for each other and the repulsion of like magnetic poles. Between a magnetic field and a moving charge, the moving charge is deflected from its path in the region of a magnetic field; the deflecting force is perpendicular to the magnetic field lines. This force is maximum when the charge moves perpendicularly to the field lines and is minimum (zero) when the charge moves parallel to the field lines.

Magnetic monopole A hypothetical particle having a single north or south magnetic pole, analogous to a positive or negative electric charge.

Magnetic pole One of the regions on a magnet that produce magnetic force.

Mass The quantity of matter in an object; the measurement of the inertia or sluggishness that an object exhibits in response to any effort made to start it, stop it, or change in any way its state of motion; a form of energy.

Mass-energy equivalence The relationship between mass and energy as given by the equation

$$E = mc^2$$

Maxwell's counterpart to Faraday's law A magnetic field is induced in any region of space in which an electric field is changing with time. The magnitude of the induced magnetic field is proportional to the rate at which the electric field changes. The direction of the induced magnetic field is at right angles to the changing electric field.

Melting The change of state from the solid to the liquid form; the opposite of *freezing*. Energy is absorbed by the substance that is melting.

MHD (magnetohydrodynamic) power The generation of electric power by interaction of a plasma and a magnetic field.

Mirage A floating image that appears in the distance and is caused by refraction of light in the atmosphere.

Model A representation of an idea created to make the idea more understandable.

Modulation Impressing a signal wave system on a higher frequency carrier wave, AM for amplitude signals and FM for frequency signals.

Molecule The smallest unit of a particular substance; a specific cluster of atoms with different properties from the separate atoms themselves.

Momentum The product of the mass of an object and its velocity.

Net force The resultant of all the forces that act on an object.

Neutrino Near-massless, uncharged particle emitted with an electron during beta decay.

Neutron star A star that has undergone a gravitational collapse in which electrons are compressed into protons to form neutrons.

Newton SI unit of force; a force of 1 newton accelerates a mass of 1 kilogram 1 meter per second each second.

Newton's law of cooling The rate of heat loss from an object is proportional to the excess temperature of the object over the temperature of its surroundings.

Newton's laws of motion

Law 1: Every object continues in its state of rest or of uniform motion in a straight line unless it is compelled to change that state by forces impressed upon it.

Law 2: The acceleration of an object is directly proportional to the net force acting on the object and inversely proportional to the mass of the object.

Law 3: To every action force, there is an equal and opposite reaction force.

Node Point of zero amplitude in a standing wave.

Nonlinear motion Any motion not along a straight-line path.

Normal A line that is perpendicular to a surface.

Normal force The component of support force perpendicular to a supporting surface. For an object resting on a horizontal surface, it is the upward force that balances the weight of the object.

Nuclear fission Splitting of the nucleus of a heavy atom, such as uranium-235, into parts, releasing much energy.

Nuclear fusion The combining of the nuclei of light atoms to form heavier nuclei, releasing much energy.

Nucleon A nuclear proton or neutron; the collective name for either or both.

Nucleus The positively charged core of an atom.

Ohm The SI unit of electrical resistance.

Ohm's law The current in a circuit varies in direct proportion to the potential difference or emf and in inverse proportion to resistance:

$$\text{Current} = \frac{\text{voltage}}{\text{resistance}}$$

A potential difference of 1 volt across a resistance of 1 ohm produces a current of 1 ampere.

Overtones Tones produced by vibrations that usually are multiples of the lowest, or fundamental, vibrating frequency.

Parabola The curved path followed by a projectile acting only under the influence of gravity.

Parallel circuit An electric circuit with two or more resistances arranged in branches in such a way that any single one completes the circuit independently of all others, providing two or more paths.

Pascal The SI unit of pressure.

Pascal's principle The pressure applied to a fluid confined in a container is transmitted undiminished throughout the fluid and acts in all directions.

Perigee The point in an elliptical orbit closest to the focus about which orbiting takes place.

Period The time required for a vibration or a wave to make a complete cycle; equal to 1/frequency.

Periodic table A scheme of ordering the elements to show the periodicity of similar chemical properties, as shown on the inside front cover of this book.

Perturbation The deviation of an orbiting object from its normal path, caused by an additional gravitational interaction.

Phosphorescence A type of light emission that is the same as fluorescence except for a delay between excitation and de-excitation. The delay is caused by atoms being excited to energy levels that do not decay rapidly. The afterglow thus provided may last from fractions of a second to hours or even days, depending on the type of material, temperature, and other factors.

Photoelectric effect The emission of electrons from a metal surface when light shines on it.

Photon A light corpuscle, or the basic packet of electromagnetic radiation. Just as matter is composed of atoms, light is composed of photons (quanta).

Pigment A material that selectively absorbs colored light.

Pitch The "highness" or "lowness" of a tone, as on a musical scale, which is governed by frequency. A high-frequency vibrating source produces a sound of high pitch; a low-frequency vibrating source produces a sound of low pitch.

Planck's constant A fundamental constant, h, that relates the energy and frequency of light quanta:

$$h = 6.6 \times 10^{-34} \text{ J·s}$$

Plasma Hot matter (beyond the gaseous state) that is composed of electrically charged particles. Most of the matter in the universe is in the plasma state.

Polarization The alignment of the electric vectors that make up electromagnetic radiation. Such waves of aligned vibrations are said to be *polarized*.

Positron The antiparticle of an electron; a positively charged electron.

Postulate A fundamental assumption.

Postulates of the special theory of relativity (1) All laws of nature are the same in all uniformly moving frames of reference. (2) The velocity of light in free space will be found to have the same value regardless of the motion of the source or the motion of the observer; that is, the velocity of light is invariant.

Potential difference The difference in electric potential, or voltage, between two points, which can be compared to a difference in water level between two containers. Measured in *volts*.

Potential energy The stored energy that an object possesses because of its position with respect to other objects.

Power The time rate at which work is performed:

$$\text{Power} = \text{work/time}$$

Pressure The ratio of the amount of force to the area over which the force is distributed:

$$\text{Pressure} = \text{force/area}$$

$$\text{Liquid pressure} = \text{weight density} \times \text{depth}$$

Primary colors The three colors—red, blue, and green—that, when added in certain proportions, will produce any color in the spectrum. (The effect is due to three types of retinal cells in the eye.)

Principle of equivalence Observations in a gravitational reference frame are equivalent to those made in an accelerating reference frame.

Prism A triangular piece of material such as glass that separates incident light by refraction into its component colors.

Projectile Any object that is projected by some force and continues in motion by virtue of its own inertia.

Pupil The part of the eye that admits the passage of light.

Quality The characteristic timbre of a musical sound, governed by the number and relative intensities of the overtones.

Quantum An elemental unit of a quantity.

Quantum mechanics The branch of quantum physics that deals with finding the probability amplitudes of matter waves; basic departure from classical mechanics.

Quantum theory The physical theory based on the idea that energy is radiated in definite units called *quanta* or *photons*. Just as matter is composed of atoms, radiant energy is composed of quanta. Further, all material particles have wave properties.

Quarks The elementary constituent particles or building blocks of nuclear matter.

Radiation The transfer of energy by means of electromagnetic waves or high-speed particles.

Rarefaction The region of reduced pressure in a longitudinal wave.

Rate How fast something happens or how much something changes per unit of time.

Real image An image formed by the actual convergence of light rays, which can be displayed on a screen.

Reflection The bouncing of light rays from a surface in such a way that the angle at which a given ray is returned is equal to the angle at which it strikes the surface. When the reflecting surface is irregular, light is returned in irregular directions and is called *diffuse reflection*.

Refraction The bending of an oblique ray of light when it passes from one transparent medium to another, caused by a difference in the speed of light in the transparent media. When the change in medium is abrupt (say, from air to water), the bending is abrupt; when the change in medium is gradual (say, from cool air to warm air), the bending is gradual, accounting for *mirages*.

Regelation The process of melting under pressure and the subsequent refreezing when the pressure is removed.

Relative Regarded in relation to something else.

Relative humidity The ratio of the amount of water vapor in a sample of air to the amount of water vapor the sample of air is capable of supporting at a given temperature.

Resolution A method of separating a vector into its component parts.

Resonance The setting up of vibrations in an object at its natural vibration frequency by a vibrating force or wave having the same (or submultiple) frequency.

Resultant The geometric sum of two or more vectors.

Retina The layer of light-sensitive tissue at the back of the eye.

Reverberation Re-echoed sound.

Ritz combination principle The spectral lines of elements have frequencies that are either the sums or the differences of frequencies of two other lines.

Rotational inertia The property of an object that resists any change in its state of rotation. If at rest, it tends to remain at rest; if rotating, it tends to remain rotating and will continue to do so unless interrupted.

Satellite A projectile or small celestial body that orbits a larger celestial body.

Scalar quantity A quantity that may be specified by magnitude, without regard to direction. Examples are mass, volume, speed, and temperature.

Scaling The study of how size affects the relationship of weight, strength, and surface area.

Scattering The emission in random directions of light that encounters particles small compared to the wavelength of light; more often at short wavelengths (blue) than at long wavelengths (red).

Schrödinger wave equation The fundamental equation of quantum mechanics, which interprets the wave nature of material particles in terms of probability wave amplitudes. It is as basic to quantum mechanics as Newton's laws of motion are to classical mechanics.

Scientific method An orderly method for gaining, organizing, and applying new knowledge.

Semiconductor A device made of material not only with properties that fall between a conductor and an insulator but with resistance that changes abruptly when other conditions change, such as temperature, voltage, and electric or magnetic field.

Series circuit An electric circuit with devices that have resistance arranged so that the same electric current flows through all of them.

Shock wave The cone-shaped wave made by an object moving at supersonic speed through a fluid.

SI (Système International) Modern system of definitions and metric notation, now spreading throughout the academic, industrial, and commercial community—the United States of America last.

Simple harmonic motion A vibratory or periodic motion, like that of a pendulum, in which the force acting on the vibrating body is proportional to its displacement from its central equilibrium position and acts toward that position.

Sine curve A wave form traced by simple harmonic motion that is uniformly moving in a direction perpendicular to the vibration direction, like the wavelike path traced on a moving conveyor belt by a pendulum swinging at right angles above the moving belt.

Sliding friction The contact force produced by the rubbing together of the surface of a moving object with the material over which it slides.

Solar constant The 1400 joules per square meter received from the sun each second at the top of the earth's atmosphere. Expressed in terms of power, it is 1.4 kilowatts per square meter.

Solar power Energy per unit time derived from the sun.

Solid The state of matter characterized by definite volume and shape.

Sonic boom The loud sound resulting from the incidence of a shock wave.

Sound A longitudinal wave phenomenon that consists of successive compressions and rarefactions of the medium through which the wave travels.

Space-time The four-dimensional continuum in which all things exist; three dimensions are the coordinates of space, and the fourth dimension is time.

Special theory of relativity A formulation of the consequences of the absence of a universal frame of reference. It has two postulates: (1) All laws of nature are the same in all uniformly moving frames of reference. (2) The velocity of light in free space will be found to have the same value regardless of the motion of the observer; that is, the velocity of light is invariant.

Specific heat The quantity of heat per unit mass required to raise the temperature of a substance by 1K or, equivalently, 1°C. Measured in joules per kilogram kelvin or calories per gram Celsius degree.

Spectroscope An optical instrument that separates light into its frequencies in the form of spectral lines; a *spectrometer* is an instrument that can also measure the frequencies.

Speed The time rate at which distance is covered by a moving object.

Standing wave A stationary wave pattern formed in a medium when two sets of identical waves pass through the medium in opposite directions.

Static friction The force between two objects at relative rest by virtue of contact that tends to oppose sliding.

Streamline The smooth path of a small region of fluid in steady flow.

Sublimation The direct conversion of a substance from the solid to the vapor state, or vice versa, without passing through the liquid state.

Subtractive primary colors The three colors of absorbing pigments—magenta, yellow, and cyan—that, when mixed in certain proportions, will reflect any color in the spectrum.

Superconductor A material that is a perfect conductor with zero resistance to the flow of electric charge.

Surface tension The tendency of the surface of a liquid to contract in area and thus behave like a stretched rubber membrane.

Technology A method and means of solving practical problems by implementing the findings of science.

Temperature A measure of the average kinetic energy per molecule in an object given in degrees Celsius or Fahrenheit or in kelvins.

Temperature inversion The condition wherein the upper regions of the atmosphere are warmer than the lower regions.

Terminal velocity The velocity attained by an object wherein the resistive forces counterbalance the driving forces, so motion is without acceleration.

Terrestrial radiation Radiant energy emitted from the earth after having been absorbed from the sun.

Theory A synthesis of a large body of information that encompasses well-tested and verified hypotheses about certain aspects of nature.

Thermodynamics The physics of the interrelationships between heat and other forms of energy, characterized by two principal laws:

First law: A restatement of the law of conservation of energy as it applies to systems involving changes in temperature. Whenever heat is added to a system, it transforms to an equal amount of some other form of energy.

Second law: Heat cannot be transferred from a colder body to a hotter body without work being done by an outside agent.

Thermonuclear Pertaining to nuclear fusion caused by high temperatures.

Time dilation   The apparent slowing down of time for an object moving at relativistic speeds.

Torque   The product of force and lever-arm distance, which tends to produce rotation.

Total internal reflection   The total reflection of light traveling in a medium when it is incident on the surface of a less dense medium at an angle greater than the critical angle.

Transformer   A device for transforming electrical power from one coil of wire to another by means of electromagnetic induction.

Transmutation   Changing an element into another by changing the number of protons in the atom nucleus.

Transverse wave   A wave in which the individual particles of a medium vibrate from side to side perpendicularly (transversely) to the direction in which the wave travels. The vibrations along a stretched string are transverse waves. The term applies also to nonmaterial waves where the periodically changing quantity (electric field) has a direction at right angles to the direction of the wave propagation.

Uncertainty principle   The ultimate accuracy of measurement is given by the magnitude of Planck's constant, h. Further, it is not possible to measure exactly both the position and the momentum of a particle at the same time, nor the energy and the time associated with a particle simultaneously.

Vector   An arrow drawn to scale, used to represent a vector quantity.

Vector quantity   A quantity that has both magnitude and direction. Examples are force, velocity, acceleration, torque, and electric and magnetic fields.

Velocity   The specification of the speed of an object and its direction of motion, a vector quantity.

Vibration   A "wiggle in time"; an oscillation.

Virtual image   An illusionary image that is seen by an observer through a lens but cannot be projected on a screen.

Volt   The SI unit of electric potential:

$$1 \text{ V} = 1 \text{ J/C}$$

Voltage   Electrical "pressure" or a measure of electrical potential difference.

Volume   The quantity of space an object occupies.

Watt   The SI unit of power:

$$1 \text{ W} = 1 \text{ J/s}$$

Wave   A "wiggle in space and time"; a disturbance propagated from one place to another with no actual transport of matter.

Wavelength   The distance between successive crests, troughs, or identical parts of a wave.

Wave velocity   The speed with which waves pass by a particular point:

$$\text{Wave velocity} = \text{frequency} \times \text{wavelength}$$

Weight   The force of the earth's gravitational attraction for any object on, below, or above the earth's surface.

Weightlessness   A condition wherein apparent gravitational pull is lacking.

Work   The product of the force exerted and the distance through which the force moves:

$$W = Fd$$

X ray   Electromagnetic radiation emitted by atoms when the innermost orbital electrons undergo excitation.

Index of Names

Index of Topics

The Conceptual Physics Photo Album

Conceptual Physics is a very personalized book, and this is reflected in the many photographs that are mainly of friends and family. The full-page chapter openers with the cartoon-style ballooned statements are as follows: Chapter 1 opens with little Jenny Jones, daughter of Bill Jones of the BJ Restaurant in Pattaya, Thailand. Uncle Ben is Benoit Cyr, also of Pattaya, Thailand. The three children who open Chapter 4 include little Debbie and Natalie Limogan, the children of my dear friends, Hideko and Herman Limogan. The little boy in the middle is Debbie's number one boyfriend, Genichiro Nakada, all of Daly City, CA. The little boy who opens Chapter 7 is my youngest son, James, at the age of 8. James was tragically killed in a car accident in Salida, Colorado, on February 29, 1988. He was 24 years old. Other photos of my dear son James are on pages 128, 356, and 506. Fortunately, he left a grandson, Manuel, who opens Chapter 10. Manuel, who looks very much like James looked at the same age, is also shown on pages 345 and 661. My wife Millie is on page 273. The lovely girl on page 179 is my daughter Leslie, now a geology type. My oldest son Paul, now in the U.S. Navy, is on pages 109 and 308. The little girl with the balloons who opens Chapter 13 is Manuel's cousin, Allie Hernandez, who lives in Pocatello, Idaho. The little chef who opens Chapter 15 is Joshua Laddin, the son of my dear friends Larry and Bessie Laddin, of San Francisco. The little boy with the laser disc who opens the Musical Sounds chapter is Brian Robinson, son of my dear friend Paul Robinson (Figure 6-17). Brian's brother and sister, David and Kristen Robinson, are shown on page 135. The little girl with the levitating magnet that opens Chapter 23 is Cathy Whitlatch, daughter of my CCSF colleague Norman Whitlatch. Dear friend Jane Perlas with

son Earl open Chapter 24. Veronica Roses with her magnifying glass, son James' childhood playmate, has been a chapter opener since the 1st edition. She now opens Chapter 26. The little boy commenting on the fluorescent lamp who opens Chapter 29 is Brandon Cadelinia, son of friends Sam and Myrna Cadelinia. The son of my dear friends Eric and Aghila Bergmark, little Jeremy at the geyser, opens Chapter 32. Jeremy's dad teaches science at Sellwood Middle School, in Portland, OR.

That's lovely K.K. Lee in the touching photo on page 68, dear friend who was my gal Friday on the 3rd and 4th editions. K.K. and husband Victor Ng have just given birth to a little girl, who they've promised for the next edition. Dotty Jean Hewitt with the polaroids on page 529 is my niece, who has grown up to be a teacher at Johnson and Wales, in Providence, RI. Her dad, my brother Dave, took her photos and also the photo of the water wheel to illustrate the solid, liquid and gaseous states of matter for the Part 2 opener, Properties of Matter. The group of people listening to a passing band on page 364 include my Mom (passed away in 1983), her second husband (my father died suddenly in 1960 while I was in college), my sister (a theologian at Wesley Theological Seminary in Washington, DC), and her husband and children: left to right, niece Joan, niece Cathy, my mother's husband, John Downs, my brother-in-law John Suchocki, my Mom, nephew John, sister Marjorie, and little Angie Harmon. (Nephew John, now a chemist, will be helping his Uncle Paul write a forthcoming physical science textbook in the near future.)

Other friends whose photos appear in the book are identified in the Acknowledgements page in the frontmatter of the text book.

VISIBLE-REGION SPECTRA FOR SELECTED ELEMENTS

Courtesy of Bausch & Lomb

Tungsten Lamp

Iron Arc

Molecular Hydrogen

Atomic Hydrogen

Neon

Barium

Fraunhofer Lines